Statistics

Fourth Edition

DAVID FREEDMAN
ROBERT PISANI
ROGER PURVES

Statistics

Fourth Edition

W • W • NORTON & COMPANY

NEW YORK • LONDON

Printed in the United States of America.

Cartoons by Dana Fradon and Leo Cullum

The text of this book is composed in Times Roman.
Composition by Integre Technical Publishing Company, Inc.
Manufacturing by R. R. Donnelley.

Library of Congress Cataloging-in-Publication Data

Freedman, David, 1938–
Statistics. — 4th ed. / David Freedman, Robert Pisani, Roger Purves.
p. cm.
Rev. ed. of: Statistics / David Freedman . . . [et al.], 3rd ed. ©1998.
Includes bibliographical references and index.
ISBN 0-393-92972-8
ISBN 13-978-0-393-92972-0
1. Mathematical statistics. I. Pisani, Robert. II. Purves, Roger. III. Statistics. IV. Title.
QA276.F683
519.5—dc21

W.W. Norton & Company, Inc., 500 Fifth Avenue, New York, N.Y. 10110
http://www.wwnorton.com

W.W. Norton & Company Ltd., Castle House, 75/76 Wells Street, London W1T 3QT

5 6 7 8 9 0

To Jerzy Neyman (1894–1981)

Born in Russia, Neyman worked in Poland and England before coming to the United States in 1938. He was one of the great statisticians of our time.

Contents

PART V. CHANCE VARIABILITY

PART VI. SAMPLING

PART VIII. TESTS OF SIGNIFICANCE

Preface

What song the Sirens sang, or what name Achilles assumed when he hid among women, though puzzling questions, are not beyond all conjecture.
—SIR THOMAS BROWNE (ENGLAND, 1605–1682)

TO THE READER

We are going to tell you about some interesting problems which have been studied with the help of statistical methods, and show you how to use these methods yourself. We will try to explain why the methods work, and what to watch out for when others use them. Mathematical notation only seems to confuse things for many people, so this book relies on words, charts, and tables; there are hardly any x's or y's. As a matter of fact, even when professional mathematicians read technical books, their eyes tend to skip over the equations. What they really want is a sympathetic friend who will explain the ideas and draw the pictures behind the equations. We will try to be that friend, for those who read our book.

WHAT IS STATISTICS?

Statistics is the art of making numerical conjectures about puzzling questions.

- What are the effects of new medical treatments?
- What causes the resemblance between parents and children, and how strong is that force?
- Why does the casino make a profit at roulette?
- Who is going to win the next election? by how much?
- How many people are employed? unemployed?

These are difficult issues, and statistical methods help a lot if you want to think about them. The methods were developed over several hundred years by people who were looking for answers to their questions. Some of these people will be introduced later.

AN OUTLINE

Part I is about designing experiments. With a good design, reliable conclusions can be drawn from the data. Some badly-designed studies are discussed too—so you can see the pitfalls, and learn what questions to ask when reading about a study. Study design is perhaps our most important topic; that is why we start there. The ideas look simple, but appearances may be deceptive: part I has a lot of depth.

Studies typically produce so many numbers that summaries are needed. Descriptive statistics—the art of summarizing data—is introduced in part II. Histograms, the average, the standard deviation, and the normal curve are all considered. The discussion continues in part III, where the focus is on analyzing relationships, for instance, the dependence of income on education. Here, correlation and regression are the main topics.

Much statistical reasoning depends on the theory of probability, discussed in part IV; the connection is through chance models, which are developed in part V. Coins, dice, and roulette wheels are the main examples in parts IV and V. The expected value and standard error are introduced; probability histograms are developed, and convergence to the normal curve is discussed.

Statistical inference—making valid generalizations from samples—is the topic of parts VI–VIII. Part VI is about estimation. For instance, how does the Gallup Poll predict the vote? Why are some methods for drawing samples better than others? Part VII uses chance models to analyze measurement error, and to develop genetic theory. Part VIII introduces tests of significance, to judge whether samples are consistent with hypotheses about the population. As parts VI–VIII show, statistical inferences depend on chance models. If the model is wrong, the resulting inference may be quite shaky.

Nowadays, inference is the branch of statistics most interesting to professionals. However, non-statisticians often find descriptive statistics a more useful branch, and the one that is easier to understand. That is why we take up descriptive statistics before inference. The bare bones of our subject are presented in chapters 1 to 6, 13, 16 to 21, 23, and 26. After that, the reader can browse anywhere. The next chapters to read might be 8, 10, 27, and 29.

EXERCISES

The sections in each chapter usually have a set of exercises, with answers at the back of the book. If you work these exercises as they come along and check the answers, you will get practice in your new skills—and find out the extent to which you have mastered them. Every chapter (except 1 and 7) ends with a set of review exercises. The book does not give answers for those exercises. Chapters 6, 15, 23, and 29 also have "special review exercises," covering all previous material. Such exercises must be answered without the clues provided by context.

When working exercises, you might be tempted to flip backward through the pages until the relevant formula materializes. However, reading the book backward will prove very frustrating. Review exercises demand much more than formulas. They call for rough guesses and qualitative judgments. In other words, they require a good intuitive understanding of what is going on. The way to develop that understanding is to read the book forward.

Why does the book include so many exercises that cannot be solved by plugging into a formula? The reason is that few real-life statistical problems can be solved that way. Blindly plugging into statistical formulas has caused a lot of confusion. So this book teaches a different approach: thinking.

GRAPHICS

As in previous editions, extensive use is made of computer graphics to display the data. Working drawings, however, are done freehand; the reader is encouraged to make similar sketches, rather than being intimidated by too much precision. The book still features cartoons by Dana Fradon of *The New Yorker*.

What's New in the Fourth Edition?

Of the making of books, there is no end.
—Ecclesiastes

The principal change is to the data. Statistics, like people, show wear and tear from aging. Fortunately or unfortunately, data are easier to rejuvenate. We started the first edition in 1971, and completed the fourth in 2006. These past 35 years were years of rapid change, as commentators have doubtless observed since prehistoric times.

There was explosive growth in computer use. Other technical developments include email (+), the world wide web (+), Windows (±), cell phones (±), and call centers with voice-activated menus (−). SAT scores bottomed out around 1990, and have since been slowly going up (chapter 5). Educational levels have been steadily increasing (chapter 4), but reading skills may—or may not—be in decline (chapter 27).

The population of the United States increased from 200 million to 300 million (chapter 24). There was corresponding growth in higher education. Over the period 1976 to 1999, the number of colleges and universities increased from about 3,000 to 4,000 (chapter 23). Student enrollments increased by about 40%, while the professoriate grew by 60%. The number of male faculty increased from 450,000 to 600,000; for women, the increase was 175,000 to 425,000. Student enrollments shifted from 53% male to 43% male.

There were remarkable changes in student attitudes (chapters 27, 29). In 1970, 60% of first-year students thought that capital punishment should be abolished; by 2000, only 30% favored abolition. In 1970, 36% of them thought that "being very well off financially" was "very important or essential"; by 2000, the figure was 73%.

The American public gained a fraction of an inch in height, and 20 pounds in weight (chapter 4). Despite the huge increase in obesity, there were steady gains in life expectancy—about 7 years over the 35-year period. Gain in life expectancy is a process ("the demographic transition") that started in Europe around 1800. The trend toward longer lives has major societal implications, as well as ripple effects on our exercises.

Family incomes went up by a factor of four, although much of the change represents a loss of purchasing power in the dollar (chapter 3). Crime rates peaked somewhere around 1990, and have fallen precipitously since (chapters 2, 29). Jury awards in civil cases once seemed out of control, but have declined since the 1990s along with crime rates. (See chapter 29; is this correlation or causation?) Our last topic is a perennial favorite: the weather. We have no significant changes to report (chapters 9, 24).*

ACKNOWLEDGMENTS FOR THE FOURTH EDITION

Technical drawings are by Dale Johnson and Laura Southworth. Type was set in TeX by Integre. Nick Cox (Durham), Russ Lyons (Indiana), and Sam Rose (Berkeley) gave us detailed and useful feedback. Máire Ní Bhrolcháin (Southampton), David Card (Berkeley), Rob Hollister (Swarthmore), Josh Palmer (Berkeley), Diana Petitti (Kaiser Permanente), and Philip Stark (Berkeley) helped us navigate the treacherous currents of the scholarly literature, and the even more treacherous currents of the world wide web.

ACKNOWLEDGMENTS FOR PREVIOUS EDITIONS

Helpful comments came from many sources. For the third edition, we thank Mike Anderson (Berkeley), Dick Berk (Pennsylvania), Jeff Fehmi (Arizona), David Kaye (Arizona), Steve Klein (Los Angeles), Russ Lyons (Indiana), Mike Ostland (Berkeley), Erol Pekoz (Boston), Diana Petitti (Kaiser Permanente), Juliet Shaffer (Berkeley), Bill Simpson (Winnipeg), Terry Speed (Berkeley), Philip Stark (Berkeley), and Allan Stewart-Oaten (Santa Barbara). Ani Adhikari (Berkeley) participated in the second edition, and had many good comments on the third edition.

The writing of the first edition was supported by the Ford Foundation (1973–1974) and by the Regents of the University of California (1974–75). Earl Cheit and Sanford Elberg (Berkeley) provided help and encouragement at critical times. Special thanks go to our editor, Donald Lamm, who somehow turned a permanently evolving manuscript into a book. Finally, we record our gratitude to our students, and other readers of our several editions and innumerable drafts.

---

*Most of the data cited here come from the *Statistical Abstract of the United States*, various editions. See chapter notes for details. On trends in life expectancy, see Dudley Kirk, "Demographic transition theory," *Population Studies* vol. 50 (1996) pp. 361–87.

PART I

Design of Experiments

1

Controlled Experiments

Always do right. This will gratify some people, and astonish the rest.
—MARK TWAIN (UNITED STATES, 1835–1910)

1. THE SALK VACCINE FIELD TRIAL

A new drug is introduced. How should an experiment be designed to test its effectiveness? The basic method is *comparison*.[1] The drug is given to subjects in a *treatment group*, but other subjects are used as *controls*—they aren't treated. Then the responses of the two groups are compared. Subjects should be assigned to treatment or control *at random*, and the experiment should be run *double-blind*: neither the subjects nor the doctors who measure the responses should know who was in the treatment group and who was in the control group. These ideas will be developed in the context of an actual field trial.[2]

The first polio epidemic hit the United States in 1916, and during the next forty years polio claimed many hundreds of thousands of victims, especially children. By the 1950s, several vaccines against this disease had been discovered. The one developed by Jonas Salk seemed the most promising. In laboratory trials, it had proved safe and had caused the production of antibodies against polio. By 1954, the Public Health Service and the National Foundation for Infantile Paralysis (NFIP) were ready to try the vaccine in the real world—outside the laboratory.

Suppose the NFIP had just given the vaccine to large numbers of children. If the incidence of polio in 1954 dropped sharply from 1953, that would seem to prove the effectiveness of the vaccine. However, polio was an epidemic disease whose incidence varied from year to year. In 1952, there were about 60,000 cases; in 1953, there were only half as many. Low incidence in 1954 could have meant that the vaccine was effective—or that 1954 was not an epidemic year.

The only way to find out whether the vaccine worked was to deliberately leave some children unvaccinated, and use them as controls. This raises a troublesome question of medical ethics, because withholding treatment seems cruel. However, even after extensive laboratory testing, it is often unclear whether the benefits of a new drug outweigh the risks.[3] Only a well-controlled experiment can settle this question.

In fact, the NFIP ran a controlled experiment to show the vaccine was effective. The subjects were children in the age groups most vulnerable to polio—grades 1, 2, and 3. The field trial was carried out in selected school districts throughout the country, where the risk of polio was high. Two million children were involved, and half a million were vaccinated. A million were deliberately left unvaccinated, as controls; half a million refused vaccination.

This illustrates the method of comparison. Only the subjects in the treatment group were vaccinated: the controls did not get the vaccine. The responses of the two groups could then be compared to see if the treatment made any difference. In the Salk vaccine field trial, the treatment and control groups were of different sizes, but that did not matter. The investigators compared the rates at which children got polio in the two groups—cases per hundred thousand. Looking at rates instead of absolute numbers adjusts for the difference in the sizes of the groups.

Children could be vaccinated only with their parents' permission. So one possible design—which also seems to solve the ethical problem—was this. The children whose parents consented would go into the treatment group and get the vaccine; the other children would be the controls. However, it was known that higher-income parents would more likely consent to treatment than lower-income parents. This design is biased against the vaccine, because children of higher-income parents are more vulnerable to polio.

That may seem paradoxical at first, because most diseases fall more heavily on the poor. But polio is a disease of hygiene. A child who lives in less hygienic surroundings is more likely to contract a mild case of polio early in childhood, while still protected by antibodies from its mother. After being infected, these children generate their own antibodies, which protect them against more severe infection later. Children who live in more hygienic surroundings do not develop such antibodies.

Comparing volunteers to non-volunteers biases the experiment. The statistical lesson: the treatment and control groups should be as similar as possible, except for the treatment. Then, any difference in response between the two groups is due to the treatment rather than something else. If the two groups differ with respect to some factor other than the treatment, the effect of this other factor might be *confounded* (mixed up) with the effect of treatment. Separating these effects can be difficult, and confounding is a major source of bias.

For the Salk vaccine field trial, several designs were proposed. The NFIP had originally wanted to vaccinate all grade 2 children whose parents would consent, leaving the children in grades 1 and 3 as controls. And this design was used in many school districts. However, polio is a contagious disease, spreading through contact. So the incidence could have been higher in grade 2 than in grades 1 or 3. This would have biased the study against the vaccine. Or the incidence could have been lower in grade 2, biasing the study in favor of the vaccine. Moreover, children in the treatment group, where parental consent was needed, were likely to have different family backgrounds from those in the control group, where parental consent was not required. With the NFIP design, the treatment group would include too many children from higher-income families. The treatment group would be more vulnerable to polio than the control group. Here was a definite bias against the vaccine.

Many public health experts saw these flaws in the NFIP design, and suggested a different design. The control group had to be chosen from the same population as the treatment group—children whose parents consented to vaccination. Otherwise, the effect of family background would be confounded with the effect of the vaccine. The next problem was assigning the children to treatment or control. Human judgment seems necessary, to make the control group like the treatment group on the relevant variables—family income as well as the children's general health, personality, and social habits.

Experience shows, however, that human judgment often results in substantial bias: it is better to rely on impersonal chance. The Salk vaccine field trial used a chance procedure that was equivalent to tossing a coin for each child, with a 50–50 chance of assignment to the treatment group or the control group. Such a procedure is objective and impartial. The laws of chance guarantee that with enough subjects, the treatment group and the control group will resemble each other very closely with respect to all the important variables, whether or not these have been identified. When an impartial chance procedure is used to assign the subjects to treatment or control, the experiment is said to be *randomized controlled*.[4]

Another basic precaution was the use of a *placebo*: children in the control group were given an injection of salt dissolved in water. During the experiment the subjects did not know whether they were in treatment or in control, so their response was to the vaccine, not the idea of treatment. It may seem unlikely that subjects could be protected from polio just by the strength of an idea. However, hospital patients suffering from severe post-operative pain have been given a "pain killer" which was made of a completely neutral substance: about one-third of the patients experienced prompt relief.[5]

Still another precaution: diagnosticians had to decide whether the children contracted polio during the experiment. Many forms of polio are hard to diagnose, and in borderline cases the diagnosticians could have been affected by knowing whether the child was vaccinated. So the doctors were not told which group the child belonged to. This was *double blinding*: the subjects did not know whether they got the treatment or the placebo, and neither did those who evaluated the responses. This randomized controlled double-blind experiment—which is about the best design there is—was done in many school districts.

How did it all turn out? Table 1 shows the rate of polio cases (per hundred thousand subjects) in the randomized controlled experiment, for the treatment group and the control group. The rate is much lower for the treatment group, decisive proof of the effectiveness of the Salk vaccine.

Table 1. The results of the Salk vaccine trial of 1954. Size of groups and rate of polio cases per 100,000 in each group. The numbers are rounded.

The randomized controlled double-blind experiment		
	Size	*Rate*
Treatment	200,000	28
Control	200,000	71
No consent	350,000	46

The NFIP study		
	Size	*Rate*
Grade 2 (vaccine)	225,000	25
Grades 1 and 3 (control)	725,000	54
Grade 2 (no consent)	125,000	44

Source: Thomas Francis, Jr., "An evaluation of the 1954 poliomyelitis vaccine trials—summary report," *American Journal of Public Health* vol. 45 (1955) pp. 1–63.

Table 1 also shows how the NFIP study was biased against the vaccine. In the randomized controlled experiment, the vaccine cut the polio rate from 71 to 28 per hundred thousand. The reduction in the NFIP study, from 54 to 25 per hundred thousand, is quite a bit less. The main source of the bias was confounding. The NFIP treatment group included only children whose parents consented to vaccination. However, the control group also included children whose parents would not have consented. The control group was not comparable to the treatment group.

The randomized controlled double-blind design reduces bias to a minimum—the main reason for using it whenever possible. But this design also has an important technical advantage. To see why, let us play devil's advocate and assume that the Salk vaccine had no effect. Then the difference between the polio rates for the treatment and control groups is just due to chance. How likely is that?

With the NFIP design, the results are affected by many factors that seem random: which families volunteer, which children are in grade 2, and so on. However, the investigators do not have enough information to figure the chances for the outcomes. They cannot figure the odds against a big difference in polio rates being due to accidental factors. With a randomized controlled experiment, on the other hand, chance enters in a planned and simple way—when the assignment is made to treatment or control.

The devil's-advocate hypothesis says that the vaccine has no effect. On this hypothesis, a few children are fated to contract polio. Assignment to treatment or control has nothing to do with it. Each child has a 50–50 chance to be in treatment or control, just depending on the toss of a coin. Each polio case has a 50–50 chance to turn up in the treatment group or the control group.

Therefore, the number of polio cases in the two groups must be about the same. Any difference is due to the chance variability in coin tossing. Statisticians understand this kind of variability. They can figure the odds against a difference as large as the observed one. The calculation will be done in chapter 27, and the odds are astronomical—a billion to one against.

2. THE PORTACAVAL SHUNT

In some cases of cirrhosis of the liver, the patient may start to hemorrhage and bleed to death. One treatment involves surgery to redirect the flow of blood through a *portacaval shunt*. The operation to create the shunt is long and hazardous. Do the benefits outweigh the risks? Over 50 studies have been done to assess the effect of this surgery.[6] Results are summarized in table 2 below.

Table 2. A study of 51 studies on the portacaval shunt. The well-designed studies show the surgery to have little or no value. The poorly-designed studies exaggerate the value of the surgery.

	Degree of enthusiasm		
Design	*Marked*	*Moderate*	*None*
No controls	24	7	1
Controls, but not randomized	10	3	2
Randomized controlled	0	1	3

Source: N. D. Grace, H. Muench, and T. C. Chalmers, "The present status of shunts for portal hypertension in cirrhosis," *Gastroenterology* vol. 50 (1966) pp. 684–91.

There were 32 studies without controls (first line in the table): 24/32 of these studies, or 75%, were markedly enthusiastic about the shunt, concluding that the benefits definitely outweighed the risks. In 15 studies there were controls, but assignment to treatment or control was not randomized. Only 10/15, or 67%, were markedly enthusiastic about the shunt. But the 4 studies that were randomized controlled showed the surgery to be of little or no value. The badly designed studies exaggerated the value of this risky surgery.

A randomized controlled experiment begins with a well-defined patient population. Some are eligible for the trial. Others are ineligible: they may be too sick

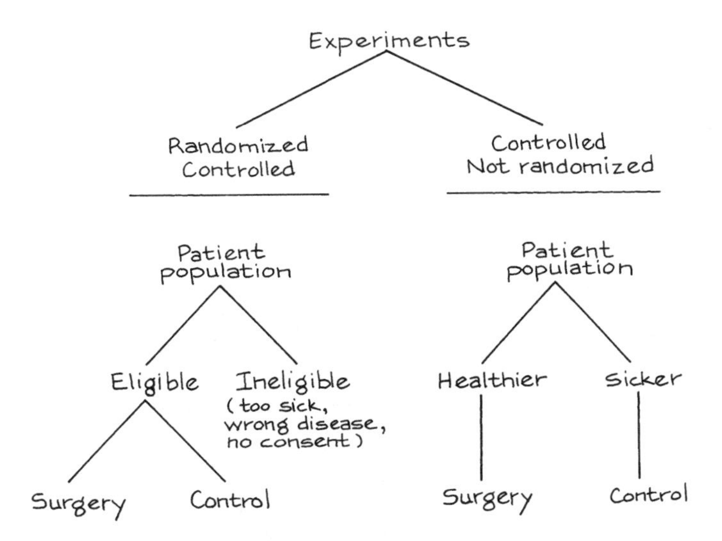

to undergo the treatment, or they may have the wrong kind of disease, or they may not consent to participate (see the flow chart at the bottom of the previous page). Eligibility is determined first; then the eligible patients are randomized to treatment or control. That way, the comparison is made only among patients who could have received the therapy. The bottom line: the control group is like the treatment group. By contrast, with poorly-controlled studies, ineligible patients may be used as controls. Moreover, even if controls are selected among those eligible for surgery, the surgeon may choose to operate only on the healthier patients while sicker patients are put in the control group.

This sort of bias seems to have been at work in the poorly-controlled studies of the portacaval shunt. In both the well-controlled and the poorly-controlled studies, about 60% of the surgery patients were still alive 3 years after the operation (table 3). In the randomized controlled experiments, the percentage of controls who survived the experiment by 3 years was also about 60%. But only 45% of the controls in the nonrandomized experiments survived for 3 years.

In both types of studies, the surgeons seem to have used similar criteria to select patients eligible for surgery. Indeed, the survival rates for the surgery group are about the same in both kinds of studies. So, what was the crucial difference? With the randomized controlled experiments, the controls were similar in general health to the surgery patients. With the poorly controlled studies, there was a tendency to exclude sicker patients from the surgery group and use them as controls. That explains the bias in favor of surgery.

Table 3. Randomized controlled experiments vs. controlled experiments that are not randomized. Three-year survival rates in studies of the portacaval shunt. (Percentages are rounded.)

	Randomized	*Not randomized*
Surgery	60%	60%
Controls	60%	45%

3. HISTORICAL CONTROLS

Randomized controlled experiments are hard to do. As a result, doctors often use other designs which are not as good. For example, a new treatment can be tried out on one group of patients, who are compared to "historical controls:" patients treated the old way in the past. The problem is that the treatment group and the historical control group may differ in important ways besides the treatment. In a controlled experiment, there is a group of patients eligible for treatment at the beginning of the study. Some of these are assigned to the treatment group, the others are used as controls: assignment to treatment or control is done "contemporaneously," that is, in the same time period. Good studies use contemporaneous controls.

The poorly-controlled trials on the portacaval shunt (section 2) included some with historical controls. Others had contemporaneous controls, but assignment to the control group was not randomized. Section 2 showed that the design of a study matters. This section continues the story. Coronary bypass surgery is a widely used—and very expensive—operation for coronary artery disease. Chalmers and associates identified 29 trials of this surgery (first line of table 4). There were 8 randomized controlled trials, and 7 were quite negative about the value of the operation. By comparison, there were 21 trials with historical controls, and 16 were positive. The badly-designed studies were more enthusiastic about the value of the surgery. (The other lines in the table can be read the same way, and lead to similar conclusions about other therapies.)

Table 4. A study of studies. Four therapies were evaluated both by randomized controlled trials and by trials using historical controls. Conclusions of trials were summarized as positive (+) about the value of the therapy, or negative (−).

Therapy	*Randomized controlled*		*Historically controlled*	
	+	−	+	−
Coronary bypass surgery	1	7	16	5
5-FU	0	5	2	0
BCG	2	2	4	0
DES	0	3	5	0

Note: 5-FU is used in chemotherapy for colon cancer; BCG is used to treat melanoma; DES, to prevent miscarriage.
Source: H. Sacks, T. C. Chalmers, and H. Smith, "Randomized versus historical controls for clinical trials," *American Journal of Medicine* vol. 72 (1982) pp. 233–40.[7]

Why are well-designed studies less enthusiastic than poorly-designed studies? In 6 of the randomized controlled experiments on coronary bypass surgery and 9 of the studies with historical controls, 3-year survival rates for the surgery group and the control group were reported (table 5). In the randomized controlled experiments, survival was quite similar in the surgery group and the control group. That is why the investigators were not enthusiastic about the operation—it did not save lives.

Table 5. Randomized controlled experiments vs. studies with historical controls. Three-year survival rates for surgery patients and controls in trials of coronary bypass surgery. Randomized controlled experiments differ from trials with historical controls.

	Randomized	*Historical*
Surgery	87.6%	90.9%
Controls	83.2%	71.1%

Note: There were 6 randomized controlled experiments enrolling 9,290 patients; and 9 studies with historical controls, enrolling 18,861 patients.
Source: See table 4.

Now look at the studies with historical controls. Survival in the surgery group is about the same as before. However, the controls have much poorer survival

rates. They were not as healthy to start with as the patients chosen for surgery. Trials with historical controls are biased in favor of surgery. Randomized trials avoid that kind of bias. That explains why the design of the study matters. Tables 2 and 3 made the point for the portacaval shunt; tables 4 and 5 make the same point for other therapies.

The last line in table 4 is worth more discussion. DES (diethylstibestrol) is an artificial hormone, used to prevent spontaneous abortion. Chalmers and associates found 8 trials evaluating DES. Three were randomized controlled, and all were negative: the drug did not help. There were 5 studies with historical controls, and all were positive. These poorly-designed studies were biased in favor of the therapy.

Doctors paid little attention to the randomized controlled experiments. Even in the late 1960s, they were giving the drug to 50,000 women each year. This was a medical tragedy, as later studies showed. If administered to the mother during pregnancy, DES can have a disastrous side-effect 20 years later, causing her daughter to develop an otherwise extremely rare form of cancer (clear-cell adenocarcinoma of the vagina). DES was banned for use on pregnant women in 1971.[8]

4. SUMMARY

1. Statisticians use the *method of comparison*. They want to know the effect of a *treatment* (like the Salk vaccine) on a *response* (like getting polio). To find

out, they compare the responses of a *treatment group* with a *control group*. Usually, it is hard to judge the effect of a treatment without comparing it to something else.

2. If the control group is comparable to the treatment group, apart from the treatment, then a difference in the responses of the two groups is likely to be due to the effect of the treatment.

3. However, if the treatment group is different from the control group with respect to other factors, the effects of these other factors are likely to be *confounded* with the effect of the treatment.

4. To make sure that the treatment group is like the control group, investigators put subjects into treatment or control at random. This is done in *randomized controlled experiments*.

5. Whenever possible, the control group is given a *placebo*, which is neutral but resembles the treatment. The response should be to the treatment itself rather than to the idea of treatment.

6. In a *double-blind* experiment, the subjects do not know whether they are in treatment or in control; neither do those who evaluate the responses. This guards against bias, either in the responses or in the evaluations.

2

Observational Studies

That's not an experiment you have there, that's an experience.
—SIR R. A. FISHER (ENGLAND, 1890–1962)

1. INTRODUCTION

Controlled experiments are different from *observational studies*. In a controlled experiment, the investigators decide who will be in the treatment group and who will be in the control group. By contrast, in an observational study it is the subjects who assign themselves to the different groups: the investigators just watch what happens.

The jargon is a little confusing, because the word *control* has two senses.

- A *control* is a subject who did not get the treatment.
- A *controlled experiment* is a study where the investigators decide who will be in the treatment group and who will not.

Studies on the effects of smoking, for instance, are necessarily observational: nobody is going to smoke for ten years just to please a statistician. However, the treatment-control idea is still used. The investigators compare smokers (the treatment or "exposed" group) with non-smokers (the control group) to determine the effect of smoking.

The smokers come off badly in this comparison. Heart attacks, lung cancer, and many other diseases are more common among smokers than non-smokers. So there is a strong *association* between smoking and disease. If cigarettes cause disease, that explains the association: death rates are higher for smokers because cigarettes kill. Thus, association is circumstantial evidence for causation. However, the proof is incomplete. There may be some hidden confounding factor which makes people smoke and also makes them get sick. If so, there is no point in quitting; that will not change the hidden factor. Association is not the same as causation.

Statisticians like Joseph Berkson and Sir R. A. Fisher did not believe the evidence against cigarettes, and suggested possible confounding variables. Epidemiologists (including Sir Richard Doll in England, and E. C. Hammond, D. Horn, H. A. Kahn in the United States) ran careful observational studies to show these alternative explanations were not plausible. Taken together, the studies make a powerful case that smoking causes heart attacks, lung cancer, and other diseases. If you give up smoking, you will live longer.[1]

Observational studies are a powerful tool, as the smoking example shows. But they can also be quite misleading. To see if confounding is a problem, it may help to find out how the controls were selected. The main issue: was the control group really similar to the treatment group—apart from the exposure of interest? If there is confounding, something has to be done about it, although perfection cannot be expected. Statisticians talk about *controlling for* confounding factors in an observational study. This is a third use of the word *control*.

One technique is to make comparisons separately for smaller and more homogeneous groups. For example, a crude comparison of death rates among smokers and non-smokers could be misleading, because smokers are disproportionately male and men are more likely than women to have heart disease anyway. The difference between smokers and non-smokers might be due to the sex difference. To eliminate that possibility, epidemiologists compare male smokers to male non-smokers, and females to females.

Age is another confounding variable. Older people have different smoking habits, and are more at risk for lung cancer. So the comparison between smokers and non-smokers is done separately by age as well as by sex. For example, male smokers age 55–59 are compared to male non-smokers age 55–59. This controls for age and sex. Good observational studies control for confounding variables. In the end, however, most observational studies are less successful than the ones on smoking. The studies may be designed by experts, but experts make mistakes too. Finding the weak points is more an art than a science, and often depends on information outside the study.

2. THE CLOFIBRATE TRIAL

The Coronary Drug Project was a randomized, controlled double-blind experiment, whose objective was to evaluate five drugs for the prevention of heart attacks. The subjects were middle-aged men with heart trouble. Of the 8,341 subjects, 5,552 were assigned at random to the drug groups and 2,789 to the control group. The drugs and the placebo (lactose) were administered in identical capsules. The patients were followed for 5 years.

One of the drugs on test was clofibrate, which reduces the levels of cholesterol in the blood. Unfortunately, this treatment did not save any lives. About 20% of the clofibrate group died over the period of followup, compared to 21% of the control group. A possible reason for this failure was suggested—many subjects in the clofibrate group did not take their medicine.

Subjects who took more than 80% of their prescribed medicine (or placebo) were called "adherers" to the protocol. For the clofibrate group, the 5-year mortality rate among the adherers was only 15%, compared to 25% among the non-adherers (table 1). This looks like strong evidence for the effectiveness of the drug. However, caution is in order. This particular comparison is observational not experimental—even though the data were collected while an experiment was going on. After all, the investigators did not decide who would adhere to protocol and who would not. The subjects decided.

Table 1. The clofibrate trial. Numbers of subjects, and percentages who died during 5 years of followup. Adherers take 80% or more of prescription.

	Clofibrate		*Placebo*	
	Number	*Deaths*	*Number*	*Deaths*
Adherers	708	15%	1,813	15%
Non-adherers	357	25%	882	28%
Total group	1,103	20%	2,789	21%

Note: Data on adherence missing for 38 subjects in the clofibrate group and 94 in the placebo group. Deaths from all causes.
Source: The Coronary Drug Project Research Group, "Influence of adherence to treatment and response of cholesterol on mortality in the Coronary Drug Project," *New England Journal of Medicine* vol. 303 (1980) pp. 1038–41.

Maybe adherers were different from non-adherers in other ways, besides the amount of the drug they took. To find out, the investigators compared adherers and non-adherers in the control group. Remember, the experiment was double-blind. The controls did not know whether they were taking an active drug or the placebo; neither did the subjects in the clofibrate group. The psychological basis for adherence was the same in both groups.

In the control group too, the adherers did better. Only 15% of them died during the 5-year period, compared to 28% among the non-adherers. The conclusions:

(i) Clofibrate does not have an effect.
(ii) Adherers are different from non-adherers.

Probably, adherers are more concerned with their health and take better care of themselves in general. That would explain why they took their capsules and why they lived longer. Observational comparisons can be quite misleading. The investigators in the clofibrate trial were unusually careful, and they found out what was wrong with comparing adherers to non-adherers.[2]

"TO ADHERE OR NOT TO ADHERE, THAT IS THE QUESTION."

3. MORE EXAMPLES

Example 1. "*Pellagra* was first observed in Europe in the eighteenth century by a Spanish physician, Gaspar Casal, who found that it was an important cause of ill-health, disability, and premature death among the very poor inhabitants of the Asturias. In the ensuing years, numerous . . . authors described the same condition in northern Italian peasants, particularly those from the plain of Lombardy. By the beginning of the nineteenth century, pellagra had spread across Europe, like a belt, causing the progressive physical and mental deterioration of thousands of people in southwestern France, in Austria, in Rumania, and in the domains of the Turkish Empire. Outside Europe, pellagra was recognized in Egypt and South Africa, and by the first decade of the twentieth century it was rampant in the United States, especially in the south . . . ."[3]

Pellagra seemed to hit some villages much more than others. Even within affected villages, many households were spared; but some had pellagra cases year after year. Sanitary conditions in diseased households were primitive; flies were everywhere. One blood-sucking fly (*Simulium*) had the same geographical range as pellagra, at least in Europe; and the fly was most active in the spring, just when most pellagra cases developed. Many epidemiologists concluded the disease was infectious, and—like malaria, yellow fever, or typhus—was transmitted from one person to another by insects. Was this conclusion justified?

Discussion. Starting around 1914, the American epidemiologist Joseph Goldberger showed by a series of observational studies and experiments that pellagra is caused by a bad diet, and is not infectious. The disease can be prevented or cured by foods rich in what Goldberger called the P-P (pellagra-preventive) factor. Since 1940, most of the flour sold in the United States is enriched with the P-P factor, among other vitamins; the P-P factor is called "niacin" on the label.

Niacin occurs naturally in meat, milk, eggs, some vegetables, and certain grains. Corn, however, contains relatively little niacin. In the pellagra areas, the poor ate corn—and not much else. Some villages and some households were poorer than others, and had even more restricted diets. That is why they were harder hit by the disease. The flies were a marker of poverty, not a cause of pellagra. Association is not the same as causation.

Example 2. Cervical cancer and circumcision. For many years, cervical cancer was one of the most common cancers among women. Many epidemiologists worked on identifying the causes of this disease. They found that in several different countries, cervical cancer was quite rare among Jews. They also found the disease to be very unusual among Moslems. In the 1950s, several investigators wrote papers concluding that circumcision of the males was the protective factor. Was this conclusion justified?

Discussion. There are differences between Jews or Moslems and members of other communities, besides circumcision. It turns out that cervical cancer is a sexually transmitted disease, spread by contact. Current research suggests that certain strains of HPV (human papilloma virus) are the causal agents. Some women are more active sexually than others, and have more partners; they are more likely to be exposed to the viruses causing the disease. That seems to be what makes the rate of cervical cancer higher for some groups of women. Early studies did not pay attention to this confounding variable, and reached the wrong conclusions.[4] (Cancer takes a long time to develop; sexual behavior in the 1930s or 1940s was the issue.)

Example 3. Ultrasound and low birthweight. Human babies can now be examined in the womb using ultrasound. Several experiments on lab animals have shown that ultrasound examinations can cause low birthweight. If this is true for humans, there are grounds for concern. Investigators ran an observational study to find out, at the Johns Hopkins hospital in Baltimore.

Of course, babies exposed to ultrasound differed from unexposed babies in many ways besides exposure; this was an observational study. The investigators found a number of confounding variables and adjusted for them. Even so, there was an association. Babies exposed to ultrasound in the womb had lower birthweight, on average, than babies who were not exposed. Is this evidence that ultrasound causes lower birthweight?

Discussion. Obstetricians suggest ultrasound examinations when something seems to be wrong. The investigators concluded that the ultrasound exams and low birthweights had a common cause—problem pregnancies. Later, a randomized controlled experiment was done to get more definite evidence. If anything, ultrasound was protective.[5]

Example 4. The Samaritans and suicide. Over the period 1964–70, the suicide rate in England fell by about one-third. During this period, a volunteer welfare organization called "The Samaritans" was expanding rapidly. One investigator thought that the Samaritans were responsible for the decline in suicides. He did an observational study to prove it. This study was based on 15 pairs of towns. To control for confounding, the towns in a pair were matched on the variables regarded as important. One town in each pair had a branch of the Samaritans; the other did not. On the whole, the towns with the Samaritans had lower suicide rates. So the Samaritans prevented suicides. Or did they?

Discussion. A second investigator replicated the study, with a bigger sample and more careful matching. He found no effect. Furthermore, the suicide rate was stable in the 1970s (after the first investigator had published his paper) although the Samaritans continued to expand. The decline in suicide rates in the 1960s is better explained by a shift from coal gas to natural gas for heating and cooking. Natural gas is less toxic. In fact, about one-third of suicides in the early 1960s were by gas. At the end of the decade, there were practically no such cases, explaining the decline in suicides. The switch to natural gas was complete, so the suicide rate by gas couldn't decline much further. Finally, the suicide rate by methods other than gas was nearly constant over the 1960s—despite the Samaritans. The Samaritans were a good organization, but they do not seem to have had much effect on the suicide rate. And observational studies, no matter how carefully done, are not experiments.[6]

4. SEX BIAS IN GRADUATE ADMISSIONS

To review briefly, one source of trouble in observational studies is that subjects differ among themselves in crucial ways besides the treatment. Sometimes these differences can be adjusted for, by comparing smaller and more homogeneous subgroups. Statisticians call this technique *controlling for* the confounding factor—the third sense of the word *control.*

An observational study on sex bias in admissions was done by the Graduate Division at the University of California, Berkeley.[7] During the study period, there were 8,442 men who applied for admission to graduate school and 4,321 women. About 44% of the men and 35% of the women were admitted. Taking percents adjusts for the difference in numbers of male and female applicants: 44 out of every 100 men were admitted, and 35 out of every 100 women.

Assuming that the men and women were on the whole equally well qualified (and there is no evidence to the contrary), the difference in admission rates looks like a strong piece of evidence to show that men and women are treated differently in the admissions procedure. The university seems to prefer men, 44 to 35.

Each major did its own admissions to graduate work. By looking at them separately, the university should have been able to identify the ones which discriminated against the women. At that point, a puzzle appeared. Major by major, there did not seem to be any bias against women. Some majors favored men, but others favored women. On the whole, if there was any bias, it ran against the men. What was going on?

"YES, ON THE SURFACE IT WOULD APPEAR TO BE SEX-BIAS BUT LET US ASK THE FOLLOWING QUESTIONS..."

Over a hundred majors were involved. However, the six largest majors together accounted for over one-third of the total number of applicants to the campus. And the pattern for these majors was typical of the whole campus. Table 2 shows the number of male and female applicants, and the percentage admitted, for each of these majors.

Table 2. Admissions data for the graduate programs in the six largest majors at University of California, Berkeley.

	Men		*Women*	
Major	*Number of applicants*	*Percent admitted*	*Number of applicants*	*Percent admitted*
A	825	62	108	82
B	560	63	25	68
C	325	37	593	34
D	417	33	375	35
E	191	28	393	24
F	373	6	341	7

Note: University policy does not allow these majors to be identified by name.
Source: The Graduate Division, University of California, Berkeley.

In each major, the percentage of female applicants who were admitted is roughly equal to the percentage for male applicants. The only exception is major A, which appears to discriminate against men. It admitted 82% of the women but only 62% of the men. The department that looks most biased against women is E. It admitted 28% of the men and 24% of the women. This difference only amounts to 4 percentage points. However, when all six majors are taken together, they admitted 44% of the male applicants, and only 30% of the females. The difference is 14 percentage points.

This seems paradoxical, but here is the explanation.

- The first two majors were easy to get into. Over 50% of the men applied to these two majors.
- The other four majors were much harder to get into. Over 90% of the women applied to these four majors.

The men were applying to the easy majors, the women to the harder ones. There was an effect due to the choice of major, confounded with the effect due to sex. When the choice of major is controlled for, as in table 2, there is little difference in the admissions rates for men or women. The statistical lesson: relationships between percentages in subgroups (for instance, admissions rates for men and women in each department separately) can be reversed when the subgroups are combined. This is called *Simpson's paradox.*[8]

Technical note. Table 2 is hard to read because it compares twelve admissions rates. A statistician might summarize table 2 by computing one overall admissions rate for men and another for women, but adjusting for the sex difference in application rates. The procedure would be to take some kind of average admission rate separately for the men and women. An ordinary average ignores the differences in size among the departments. Instead, a *weighted average* of the admission rates could be used, the weights being the total number of applicants (male and female) to each department; see table 3.

Table 3. Total number of applicants, from table 2.

Major	*Total number of applicants*
A	933
B	585
C	918
D	792
E	584
F	714
	4,526

The weighted average admission rate for men is

$$\frac{.62\times 933 + .63\times 585 + .37\times 918 + .33\times 792 + .28\times 584 + .06\times 714}{4,526}$$

This works out to 39%. Similarly, the weighted average admission rate for the women is

$$\frac{.82\times 933 + .68\times 585 + .34\times 918 + .35\times 792 + .24\times 584 + .07\times 714}{4{,}526}$$

This works out to 43%. In these formulas, the weights are the same for the men and women; they are the totals from table 3. The admission rates are different for men and women; they are the rates from table 2. The final comparison: the weighted average admission rate for men is 39%, while the weighted average admission rate for women is 43%. The weighted averages control for the confounding factor—choice of major. These averages suggest that if anything, the admissions process is biased against the men.

5. CONFOUNDING

Hidden confounders are a major problem in observational studies. As discussed in section 1, epidemiologists found an association between exposure (smoking) and disease (lung cancer): heavy smokers get lung cancer at higher rates than light smokers; light smokers get the disease at higher rates than nonsmokers. According to the epidemiologists, the association comes about because smoking causes lung cancer. However, some statisticians—including Sir R. A. Fisher—thought the association could be explained by confounding.

Confounders have to be associated with (i) the disease and (ii) the exposure. For example, suppose there is a gene which increases the risk of lung cancer. Now, if the gene also gets people to smoke, it meets both the tests for a confounder. This gene would create an association between smoking and lung cancer. The idea is a bit subtle: a gene that causes cancer but is unrelated to smoking is not a confounder and is sideways to the argument, because it does not account for the facts—the association between smoking and cancer.[9] Fisher's "constitutional hypothesis" explained the association on the basis of genetic confounding; nowadays, there is evidence from twin studies to refute this hypothesis (review exercise 11, chapter 15).

> Confounding means a difference between the treatment and control groups—other than the treatment—which affects the responses being studied. A confounder is a third variable, associated with exposure and with disease.

Exercise Set A

1. In the U.S. in 2000, there were 2.4 million deaths from all causes, compared to 1.9 million in 1970—a 25% increase.[10] True or false, and explain: the data show that the public's health got worse over the period 1970–2000.

2. Data from the Salk vaccine field trial suggest that in 1954, the school districts in the NFIP trial and in the randomized controlled experiment had similar exposures to the polio virus.
   (a) The data also show that children in the two vaccine groups (for the randomized controlled experiment and the NFIP design) came from families with similar incomes and educational backgrounds. Which two numbers in table 1 (p. 6) confirm this finding?
   (b) The data show that children in the two no-consent groups had similar family backgrounds. Which pair of numbers in the table confirm this finding?
   (c) The data show that children in the two control groups had different family backgrounds. Which pair of numbers in the table confirm this finding?
   (d) In the NFIP study, neither the control group nor the no-consent group got the vaccine. Yet the no-consent group had a lower rate of polio. Why?
   (e) To show that the vaccine works, someone wants to compare the 44/100,000 in the NFIP study with the 25/100,000 in the vaccine group. What's wrong with this idea?

3. Polio is an infectious disease; for example, it seemed to spread when children went swimming together. The NFIP study was not done blind: could that bias the results? Discuss briefly.

4. The Salk vaccine field trials were conducted only in certain experimental areas (school districts), selected by the Public Health Service in consultation with local officials.[11] In these areas, there were about 3 million children in grades 1, 2, or 3; and there were about 11 million children in those grades in the United States. In the experimental areas, the incidence of polio was about 25% higher than in the rest of the country. Did the Salk vaccine field trials cause children to get polio instead of preventing it? Answer yes or no, and explain briefly.

5. Linus Pauling thought that vitamin C prevents colds, and cures them too. Thomas Chalmers and associates did a randomized controlled double-blind experiment to find out.[12] The subjects were 311 volunteers at the National Institutes of Health. These subjects were assigned at random to 1 of 4 groups:

Group	*Prevention*	*Therapy*
1	placebo	placebo
2	vitamin C	placebo
3	placebo	vitamin C
4	vitamin C	vitamin C

   All subjects were given six capsules a day for prevention, and an additional six capsules a day for therapy if they came down with a cold. However, in group 1 both sets of capsules just contained the placebo (lactose). In group 2, the prevention capsules had vitamin C while the therapy capsules were filled with the placebo. Group 3 was the reverse. In group 4, all the capsules were filled with vitamin C.

   There was quite a high dropout rate during the trial. And this rate was significantly higher in the first 3 groups than in the 4th. The investigators noticed this, and found the reason. As it turned out, many of the subjects broke the blind. (That is quite easy to do; you just open a capsule and taste the contents; vitamin C—ascorbic acid—is sour, lactose is not.) Subjects who were getting the placebo were more likely to drop out.

The investigators analyzed the data for the subjects who remained blinded, and vitamin C had no effect. Among those who broke the blind, groups 2 and 4 had the fewest colds; groups 3 and 4 had the shortest colds. How do you interpret these results?

6. (Hypothetical.) One of the other drugs in the Coronary Drug Project (section 2) was nicotinic acid.[13] Suppose the results on nicotinic acid were as reported below. Something looks wrong. What, and why?

	Nicotinic acid		*Placebo*	
	Number	*Deaths*	*Number*	*Deaths*
Adherers	558	13%	1,813	15%
Non-adherers	487	26%	882	28%
Total group	1,045	19%	2,695	19%

7. (Hypothetical.) In a clinical trial, data collection usually starts at "baseline," when the subjects are recruited into the trial but before they are assigned to treatment or control. Data collection continues until the end of followup. Two clinical trials on prevention of heart attacks report baseline data on smoking, shown below. In one of these trials, the randomization did not work. Which one, and why?

		Number of persons	*Percent who smoked*
(i)	Treatment	1,012	49.3%
	Control	997	69.0%
(ii)	Treatment	995	59.3%
	Control	1,017	59.0%

8. Some studies find an association between liver cancer and smoking. However, alcohol consumption is a confounding variable. This means—
   (i) Alcohol causes liver cancer.
   (ii) Drinking is associated with smoking, and alcohol causes liver cancer.

   Choose one option, and explain briefly.

9. Breast cancer is one of the most common malignancies among women in the U.S. If it is detected early enough—before the cancer spreads—chances of successful treatment are much better. Do screening programs speed up detection by enough to matter?

   The first large-scale trial was run by the Health Insurance Plan of Greater New York, starting in 1963. The subjects (all members of the plan) were 62,000 women age 40 to 64. These women were divided at random into two equal groups. In the treatment group, women were encouraged to come in for annual screening, including examination by a doctor and X-rays. About 20,200 women in the treatment group did come in for the screening; but 10,800 refused. The control group was offered usual health care. All the women were followed for many years.

Results for the first 5 years are shown in the table below.[14] ("HIP" is the usual abbreviation for the Health Insurance Plan.)

Deaths in the first five years of the HIP screening trial, by cause. Rates per 1,000 women.

		Cause of Death			
		Breast cancer		*All other*	
		Number	*Rate*	*Number*	*Rate*
Treatment group					
Examined	20,200	23	1.1	428	21
Refused	10,800	16	1.5	409	38
Total	31,000	39	1.3	837	27
Control group	31,000	63	2.0	879	28

Epidemiologists who worked on the study found that (i) screening had little impact on diseases other than breast cancer; (ii) poorer women were less likely to accept screening than richer ones; and (iii) most diseases fall more heavily on the poor than the rich.

(a) Does screening save lives? Which numbers in the table prove your point?
(b) Why is the death rate from all other causes in the whole treatment group ("examined" and "refused" combined) about the same as the rate in the control group?
(c) Breast cancer (like polio, but unlike most other diseases) affects the rich more than the poor. Which numbers in the table confirm this association between breast cancer and income?
(d) The death rate (from all causes) among women who accepted screening is about half the death rate among women who refused. Did screening cut the death rate in half? If not, what explains the difference in death rates?

10. (This continues exercise 9.)

(a) To show that screening reduces the risk from breast cancer, someone wants to compare 1.1 and 1.5. Is this a good comparison? Is it biased against screening? For screening?
(b) Someone claims that encouraging women to come in for breast cancer screening increases their health consciousness, so these women take better care of themselves and live longer for that reason. Is the table consistent or inconsistent with the claim?
(c) In the first year of the HIP trial, 67 breast cancers were detected in the "examined" group, 12 in the "refused" group, and 58 in the control group. True or false, and explain briefly: screening causes breast cancer.

11. Cervical cancer is more common among women who have been exposed to the herpes virus, according to many observational studies.[15] Is it fair to conclude that the virus causes cervical cancer?

12. Physical exercise is considered to increase the risk of spontaneous abortion. Furthermore, women who have had a spontaneous abortion are more likely to have another. One observational study finds that women who exercise regularly have fewer spontaneous abortions than other women.[16] Can you explain the findings of this study?

13. A hypothetical university has two departments, A and B. There are 2,000 male applicants, of whom half apply to each department. There are 1,100 female applicants: 100 apply to department A and 1,000 to department B. Department A admits 60% of the men who apply and 60% of the women. Department B admits 30% of the men who apply and 30% of the women. "For each department, the percentage of men admitted equals the percentage of women admitted; this must be so for both departments together." True or false, and explain briefly.

Exercises 14 and 15 are designed as warm-ups for the next chapter. Do not use a calculator when working them. Just remember that "%" means "per hundred." For example, 41 people out of 398 is just about 10%. The reason: 41 out of 398 is like 40 out of 400, that's 10 out of 100, and that's 10%.

14. Say whether each of the following is about 1%, 10%, 25%, or 50%—
    - (a) 39 out of 398 (b) 99 out of 407
    - (c) 57 out of 209 (d) 99 out of 197

15. Among beginning statistics students in one university, 46 students out of 446 reported family incomes ranging from $40,000 to $50,000 a year.
    - (a) About what percentage had family incomes in the range $40,000 to $50,000 a year?
    - (b) Guess the percentage that had family incomes in the range $45,000 to $46,000 a year.
    - (c) Guess the percentage that had family incomes in the range $46,000 to $47,000 a year.
    - (d) Guess the percentage that had family incomes in the range $47,000 to $49,000 a year.

The answers to these exercises are on pp. A43–45.

6. REVIEW EXERCISES

Review exercises may cover material from previous chapters.

1. The Federal Bureau of Investigation reports state-level and national data on crimes.[17]
    - (a) An investigator compares the incidence of crime in Minnesota and in Michigan. In 2001, there were 3,584 crimes in Minnesota, compared to 4,082 in Michigan. He concludes that Minnesotans are more law-abiding. After all, Michigan includes the big bad city of Detroit. What do you say?
    - (b) An investigator compares the incidence of crime in the U.S. in 1991 and 2001. In 1991, there were 28,000 crimes, compared to 22,000 in 2001. She concludes that the U.S. became more law-abiding over that time period. What do you say?

2. The National Highway and Traffic Safety Administration analyzed thefts of new cars in 2002, as well as sales figures for that year.[18]
    - (a) There were 99 Corvettes stolen, and 26 Infiniti Q45 sedans. Should you conclude that American thieves prefer American cars? Or is something missing from the equation?

(b) There were 50 BMW 7-series cars stolen, compared to 146 in the 3-series. Should you conclude that thieves prefer smaller cars, which are more economical to run and easier to park? Or is something missing from the equation?
(c) There were 429 Liberty Jeeps stolen, compared to 207,991 sold, for a rate of 2 per 100,000. True or false and explain: the rate is low because the denominator is large.

3. From table 1 in chapter 1 (p. 6), those children whose parents refused to participate in the randomized controlled Salk trial got polio at the rate of 46 per 100,000. On the other hand, those children whose parents consented to participation got polio at the slightly higher rate of 49 per 100,000 in the treatment group and control group taken together. Suppose that this field trial was repeated the following year. On the basis of the figures, some parents refused to allow their children to participate in the experiment and be exposed to this higher risk of polio. Were they right? Answer yes or no, and explain briefly.

4. The Public Health Service studied the effects of smoking on health, in a large sample of representative households.[19] For men and for women in each age group, those who had never smoked were on average somewhat healthier than the current smokers, but the current smokers were on average much healthier than those who had recently stopped smoking.
   (a) Why did they study men and women and the different age groups separately?
   (b) The lesson seems to be that you shouldn't start smoking, but once you've started, don't stop. Comment briefly.

5. There is a rare neurological disease (idiopathic hypoguesia) that makes food taste bad. It is sometimes treated with zinc sulfate. One group of investigators did two randomized controlled experiments to test this treatment. In the first trial, the subjects did not know whether they were being given the zinc sulfate or a placebo. However, the doctors doing the evaluations did know. In this trial, patients on zinc sulfate improved significantly; the placebo group showed little improvement. The second trial was run double-blind: neither the subjects nor the doctors doing the evaluation were told who had been given the drug or the placebo. In the second trial, zinc sulfate had no effect.[20] Should zinc sulfate be given to treat the disease? Answer yes or no, and explain briefly.

6. (Continues the previous exercise.) The second trial used what is called a "crossover" design. The subjects were assigned at random to one of four groups:

placebo	placebo
placebo	zinc
zinc	placebo
zinc	zinc

In the first group, the subjects stayed on the placebo through the whole experiment. In the second group, subjects began with the placebo, but halfway through the experiment they were switched to zinc sulfate. Similarly, in the third group, subjects began on zinc sulfate but were switched to placebo. In the last group, they stayed on zinc sulfate. Subjects knew the design of the study, but were not told the group to which they were assigned.

Some subjects did not improve during the first half of the experiment. In each of the four groups, these subjects showed some improvement (on average) during the second half of the experiment. How can this be explained?

7. According to a study done at Kaiser Permanente in Walnut Creek, California, users of oral contraceptives have a higher rate of cervical cancer than non-users, even after adjusting for age, education, and marital status. Investigators concluded that the pill causes cervical cancer.[21]

   (a) Is this a controlled experiment or an observational study?
   (b) Why did the investigators adjust for age? education? marital status?
   (c) Women using the pill were likely to differ from non-users on another factor which affects the risk of cervical cancer. What factor is that?
   (d) Were the conclusions of the study justified by the data? Answer yes or no, and explain briefly.

8. Ads for ADT Security Systems claim[22]

   > When you go on vacation, burglars go to work.... According to FBI statistics, over 25% of home burglaries occur between Memorial Day and Labor Day.

   Do the statistics prove that burglars go to work when other people go on vacation? Answer yes or no, and explain briefly.

9. People who get lots of vitamins by eating five or more servings of fresh fruit and vegetables each day (especially "cruciferous" vegetables like broccoli) have much lower death rates from colon cancer and lung cancer, according to many observational studies. These studies were so encouraging that two randomized controlled experiments were done. The treatment groups were given large doses of vitamin supplements, while people in the control groups just ate their usual diet. One experiment looked at colon cancer; the other, at lung cancer.

   The first experiment found no difference in the death rate from colon cancer between the treatment group and the control group. The second experiment found that beta carotene (as a diet supplement) increased the death rate from lung cancer.[23] True or false, and explain:

   (a) The experiments confirmed the results of the observational studies.
   (b) The observational studies could easily have reached the wrong conclusions, due to confounding—people who eat lots of fruit and vegetables have lifestyles that are different in many other ways too.
   (c) The experiments could easily have reached the wrong conclusions, due to confounding—people who eat lots of fruit and vegetables have lifestyles that are different in many other ways too.

10. A study of young children found that those with more body fat tended to have more "controlling" mothers; the *San Francisco Chronicle* concluded that "Parents of Fat Kids Should Lighten Up."[24]

    (a) Was this an observational study or a randomized controlled experiment?
    (b) Did the study find an association between mother's behavior and her child's level of body fat?
    (c) If controlling behavior by the mother causes children to eat more, would that explain an association between controlling behavior by the mother and her child's level of body fat?
    (d) Suppose there is a gene which causes obesity. Would that explain the association?
    (e) Can you think of another way to explain the association?
    (f) Do the data support the *Chronicle*'s advice on child-rearing?

    Discuss briefly.

11. California is evaluating a new program to rehabilitate prisoners before their release; the object is to reduce the recidivism rate—the percentage who will be back in prison within two years of release. The program involves several months of "boot camp"—military-style basic training with very strict discipline. Admission to the program is voluntary. According to a prison spokesman, "Those who complete boot camp are less likely to return to prison than other inmates."[25]

    (a) What is the treatment group in the prison spokesman's comparison? the control group?
    (b) Is the prison spokesman's comparison based on an observational study or a randomized controlled experiment?
    (c) True or false: the data show that boot camp worked.

    Explain your answers.

12. (Hypothetical.) A study is carried out to determine the effect of party affiliation on voting behavior in a certain city. The city is divided up into wards. In each ward, the percentage of registered Democrats who vote is higher than the percentage of registered Republicans who vote. True or false: for the city as a whole, the percentage of registered Democrats who vote must be higher than the percentage of registered Republicans who vote. If true, why? If false, give an example.

7. SUMMARY AND OVERVIEW

1. In an *observational study*, the investigators do not assign the subjects to treatment or control. Some of the subjects have the condition whose effects are being studied; this is the treatment group. The other subjects are the controls. For example, in a study on smoking, the smokers form the treatment group and the non-smokers are the controls.

2. Observational studies can establish *association*: one thing is linked to another. Association may point to causation: if exposure causes disease, then people who are exposed should be sicker than similar people who are not exposed. But association does not prove causation.

3. In an observational study, the effects of treatment may be confounded with the effects of factors that got the subjects into treatment or control in the first place. Observational studies can be quite misleading about cause-and-effect relationships, because of confounding. A *confounder* is a third variable, associated with exposure and with disease.

4. When looking at a study, ask the following questions. Was there any control group at all? Were historical controls used, or contemporaneous controls? How were subjects assigned to treatment—through a process under the control of the investigator (a controlled experiment), or a process outside the control of the investigator (an observational study)? If a controlled experiment, was the assignment made using a chance mechanism (randomized controlled), or did assignment depend on the judgment of the investigator?

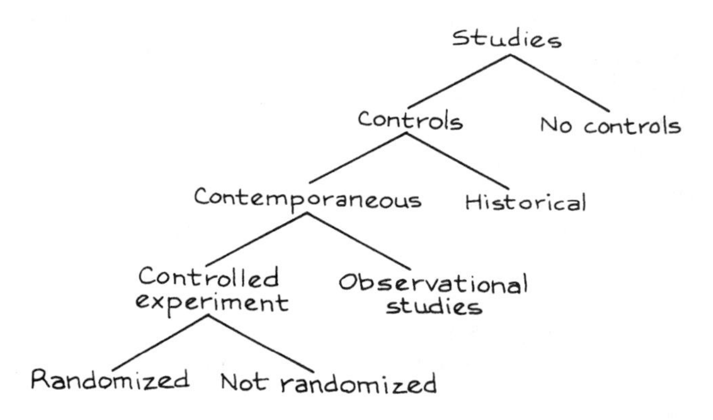

5. With observational studies, and with nonrandomized controlled experiments, try to find out how the subjects came to be in treatment or in control. Are the groups comparable? different? What factors are confounded with treatment? What adjustments were made to take care of confounding? Were they sensible?

6. In an observational study, a confounding factor can sometimes be *controlled for*, by comparing smaller groups which are relatively homogeneous with respect to the factor.

7. Study design is a central issue in applied statistics. Chapter 1 introduced the idea of randomized experiments, and chapter 2 draws the contrast with observational studies. The great weakness of observational studies is confounding; randomized experiments minimize this problem. Statistical inference from randomized experiments will be discussed in chapter 27.

PART II

Descriptive Statistics

3

The Histogram

Grown-ups love figures. When you tell them that you have made a new friend, they never ask you any questions about essential matters. They never say to you, "What does his voice sound like? What games does he love best? Does he collect butterflies?" Instead, they demand: "How old is he? How many brothers has he? How much does he weigh? How much money does his father make?" Only from these figures do they think they have learned anything about him.

—*The Little Prince*[1]

1. INTRODUCTION

In the U.S., how are incomes distributed? How much worse off are minority groups? Some information is provided by government statistics, obtained from the Current Population Survey. Each month, interviewers talk to a representative cross section of about 50,000 American families (for details, see part VI). In March, these families are asked to report their incomes for the previous year. We are going to look at the results for 1973. These data have to be summarized—nobody wants to look at 50,000 numbers. To summarize data, statisticians often use a graph called a *histogram* (figure 1 on the next page).

This section explains how to read histograms. First of all, there is no vertical scale: unlike most other graphs, a histogram does not need a vertical scale. Now look at the horizontal scale. This shows income in thousands of dollars. The graph itself is just a set of blocks. The bottom edge of the first block covers the range from \$0 to \$1,000, the bottom edge of the second goes from \$1,000 to \$2,000;

Figure 1. A histogram. This graph shows the distribution of families by income in the U.S. in 1973.

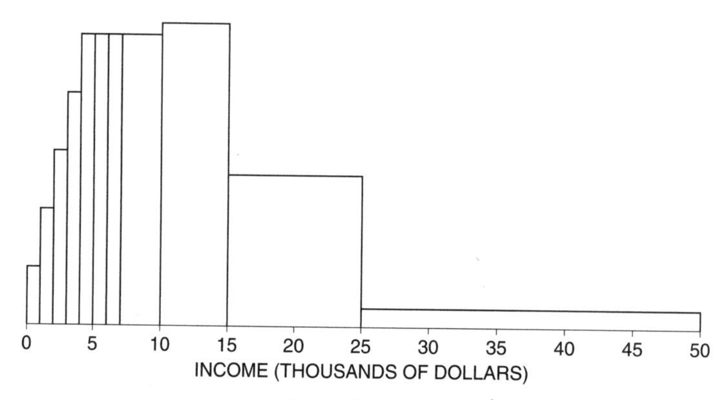

Source: Current Population Survey.[2]

and so on until the last block, which covers the range from $25,000 to $50,000. These ranges are called *class intervals*. The graph is drawn so the area of a block is proportional to the number of families with incomes in the corresponding class interval.

To see how the blocks work, look more closely at figure 1. About what percentage of the families earned between $10,000 and $15,000? The block over this interval amounts to something like one-fourth of the total area. So about one-fourth, or 25%, of the families had incomes in that range.

Take another example. Were there more families with incomes between $10,000 and $15,000, or with incomes between $15,000 and $25,000? The block over the first interval is taller, but the block over the second interval is wider. The areas of the two blocks are about the same, so the percentage of families earning $10,000 to $15,000 is about the same as the percentage earning $15,000 to $25,000.

For a last example, take the percentage of families with incomes under $7,000. Is this closest to 10%, 25%, or 50%? By eye, the area under the histogram between $0 and $7,000 is about a quarter of the total area, so the percentage is closest to 25%.

In a histogram, the areas of the blocks represent percentages.

The horizontal axis in figure 1 stops at $50,000. What about the families earning more than that? The histogram simply ignores them. In 1973, only 1% of American families had incomes above that level: most are represented in the figure.

At this point, a good way to learn more about histograms is to do some exercises. Figure 2 shows the same histogram as figure 1, but with a vertical scale supplied. This scale will be useful in working exercise 1. Exercise 8 compares the income data for 1973 and 2004.

Figure 2. The histogram from figure 1, with a vertical scale supplied.

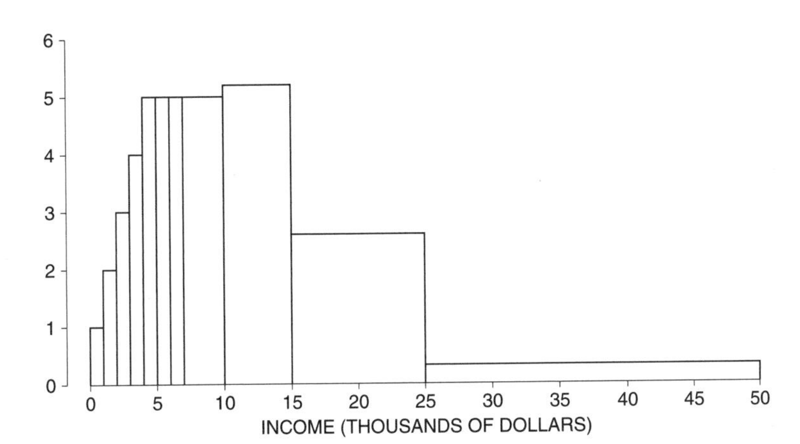

Exercise Set A

1. About 1% of the families in figure 2 had incomes between \$0 and \$1,000. Estimate the percentage who had incomes—
   (a) between \$1,000 and \$2,000
   (b) between \$2,000 and \$3,000
   (c) between \$3,000 and \$4,000
   (d) between \$4,000 and \$5,000
   (e) between \$4,000 and \$7,000
   (f) between \$7,000 and \$10,000

2. In figure 2, were there more families earning between \$10,000 and \$11,000 or between \$15,000 and \$16,000? Or were the numbers about the same? Make your best guess.

3. The histogram below shows the distribution of final scores in a certain class.
   (a) Which block represents the people who scored between 60 and 80?
   (b) Ten percent scored between 20 and 40. About what percentage scored between 40 and 60?
   (c) About what percentage scored over 60?

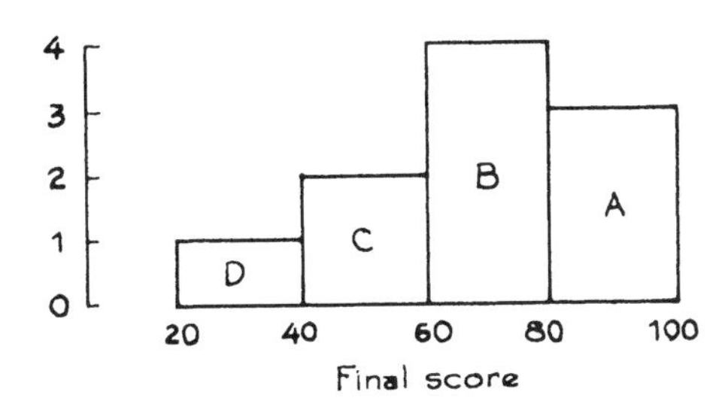

4. Below are sketches of histograms for test scores in three different classes. The scores range from 0 to 100; a passing score was 50. For each class, was the percent who passed about 50%, well over 50%, or well under 50%?

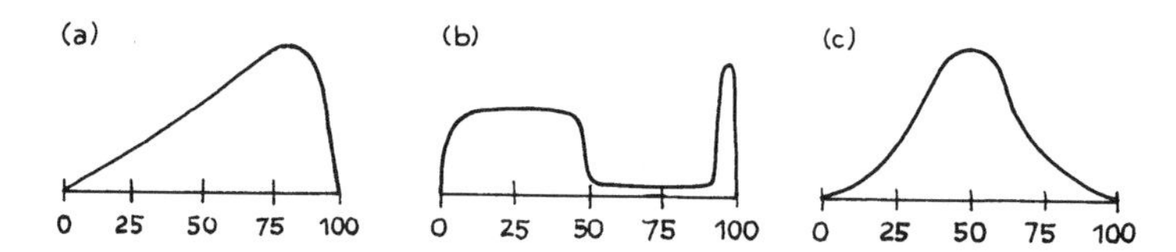

5. One class in exercise 4 had two quite distinct groups of students, with one group doing rather poorly on the test, and the other group doing very well. Which class was it?

6. In class (b) of exercise 4, were there more people with scores in the range 40–50 or 90–100?

7. An investigator collects data on hourly wage rates for three groups of people. Those in group B earn about twice as much as those in group A. Those in group C earn about $10 an hour more than those in group A. Which histogram belongs to which group? (The histograms don't show wages above $50 an hour.)

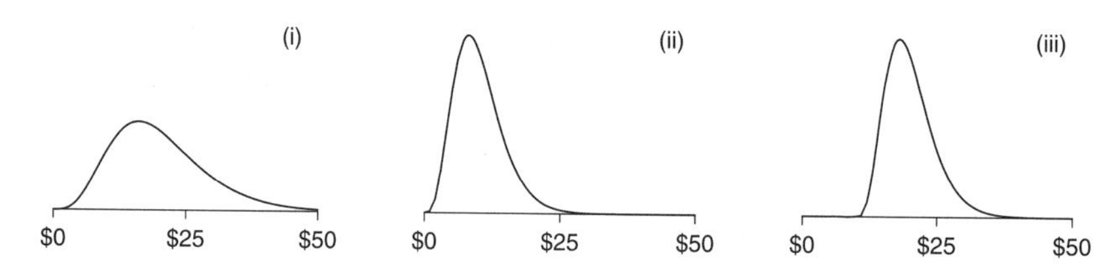

8. The figure below compares the histograms for family incomes in the U.S. in 1973 and in 2004. It looks as if family income went up by a factor of 4 over 30 years. Or did it? Discuss briefly.

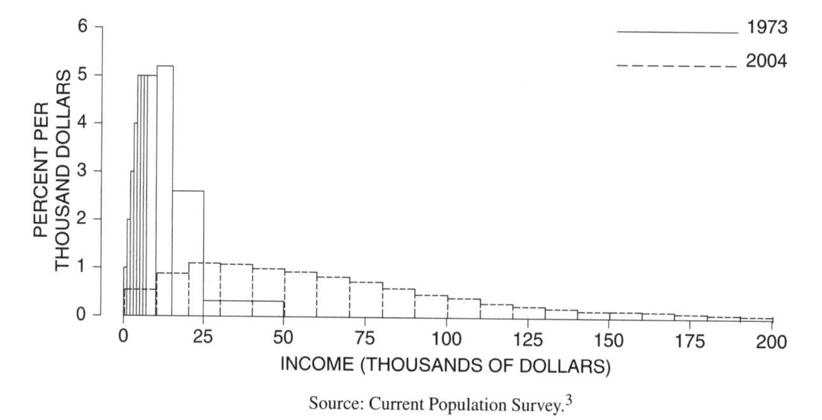

Source: Current Population Survey.[3]

The answers to these exercises are on pp. A45–46.

2. DRAWING A HISTOGRAM

This section explains how to draw a histogram. The method is not difficult, but there are a couple of wrong turns to avoid. The starting point in drawing a histogram is a *distribution table*, which shows the percentage of families with incomes in each class interval (table 1). These percentages are found by going back to the original data—on the 50,000 families—and counting. Nowadays this sort of work is done by computer, and in fact table 1 was drawn up with the help of a computer at the Bureau of the Census.

The computer has to be told what to do with families that fall right on the boundary between two class intervals. This is called an *endpoint convention*. The convention followed in table 1 is indicated by the caption. The left endpoint is included in the class interval, the right endpoint is excluded. In the first line of the table, for example, \$0 is included and \$1,000 is excluded. This interval has the families that earn \$0 or more, but less than \$1,000. A family that earns \$1,000 exactly goes in the next interval.

Table 1. Distribution of families by income in the U.S. in 1973. Class intervals include the left endpoint, but not the right endpoint.

Income level	*Percent*
\$0–\$1,000	1
\$1,000–\$2,000	2
\$2,000–\$3,000	3
\$3,000–\$4,000	4
\$4,000–\$5,000	5
\$5,000–\$6,000	5
\$6,000–\$7,000	5
\$7,000–\$10,000	15
\$10,000–\$15,000	26
\$15,000–\$25,000	26
\$25,000–\$50,000	8
\$50,000 and over	1

Note: Percents do not add to 100%, due to rounding.
Source: Current Population Survey.[4]

The first step in drawing a histogram is to put down a horizontal axis. For the income histogram, some people get

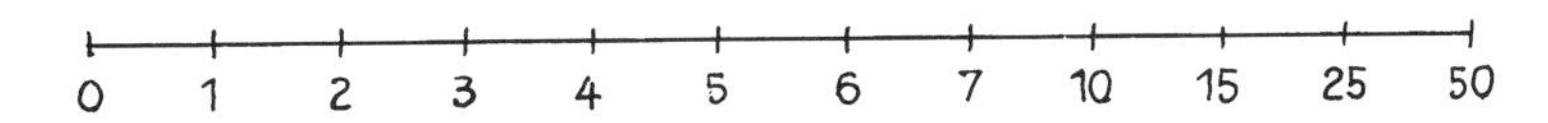

That is a mistake. The interval from \$7,000 to \$10,000 is three times as long as the interval from \$6,000 to \$7,000. So the horizontal axis should look like this:

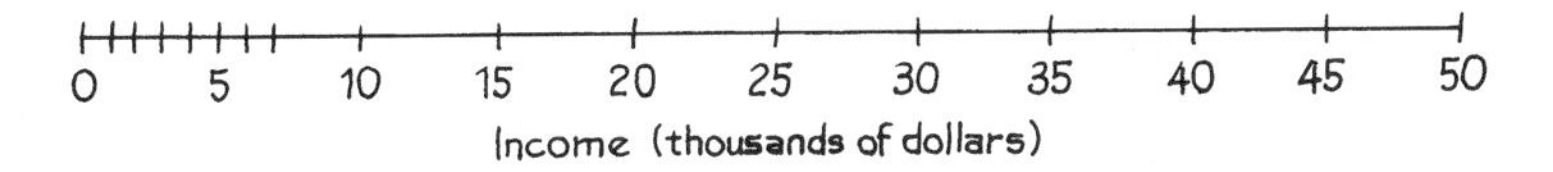

The next step is to draw the blocks. It's tempting to make their heights equal to the percents in the table. Figure 3 shows what happens if you make that mistake. The graph gives much too rosy a picture of the income distribution. For example, figure 3 says there were many more families with incomes over \$25,000 than under \$7,000. The U.S. was a rich country in 1973, but not that rich.

Figure 3. Don't plot the percents.

The source of the trouble is that some class intervals are longer than others, so the percents in table 1 are not on a par with one another. The 8% who earn \$25,000 to \$50,000, for instance, are spread over a much larger range of incomes than the 15% who earn \$7,000 to \$10,000. Plotting percents directly ignores this, and makes the blocks over the longer class intervals too big.

There is a simple way to compensate for the different lengths of the class intervals—use thousand-dollar intervals as a common unit. For example, the class interval from \$7,000 to \$10,000 contains three of these intervals: \$7,000 to \$8,000, \$8,000 to \$9,000, and \$9,000 to \$10,000. From table 1, 15% of the families had incomes in the whole interval. Within each of the thousand-dollar sub-intervals, there will only be about 5% of the families. This 5, not the 15, is what should be plotted above the interval \$7,000 to \$10,000.

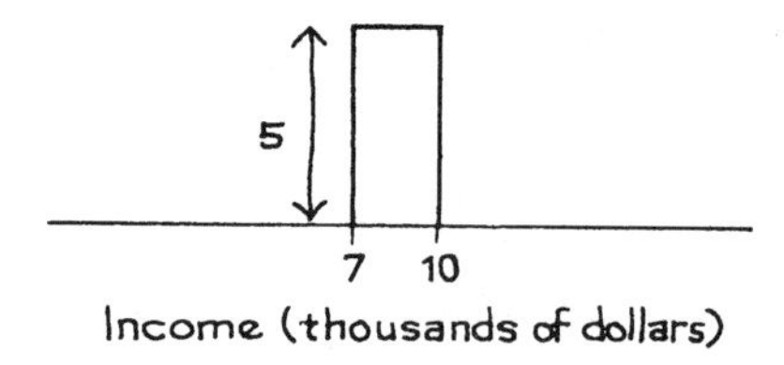

For a second example, take the interval from \$10,000 to \$15,000. This contains 5 of the thousand-dollar intervals. According to table 1, 26% of the families had incomes in the whole interval. Within each of the 5 smaller intervals there will be about 5.2% of the families: $26/5 = 5.2$. The height of the block over the interval \$10,000 to \$15,000 is 5.2.

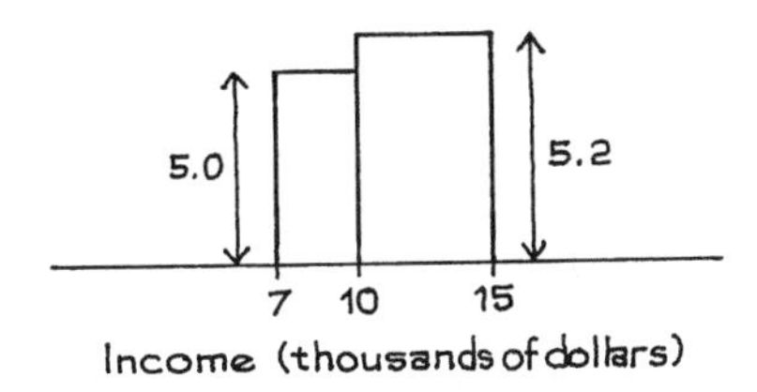

The work is done for two of the lines in table 1. To complete the histogram, do the same thing for the rest of the class intervals. Figure 4 (below) is the result.

> To figure out the height of a block over a class interval, divide the percentage by the length of the interval.

That way, the area of the block equals the percentage of families in the class interval. The histogram represents the distribution as if the percent is spread evenly over the class interval. Often, this is a good first approximation.

Figure 4. Distribution of families by income in the U.S. in 1973.

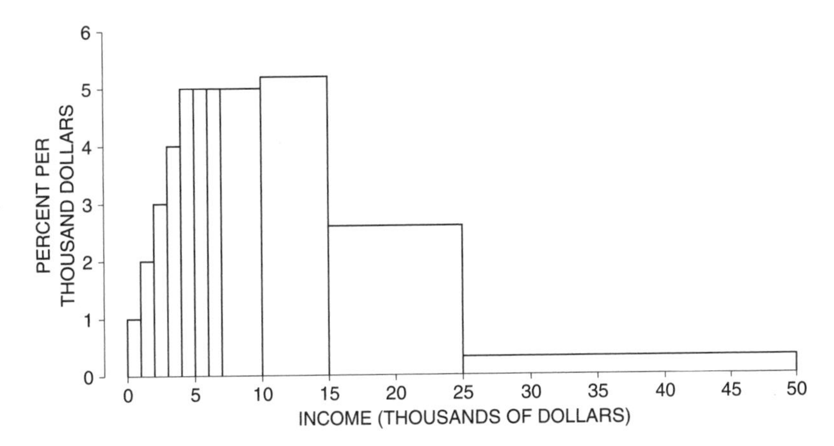

The procedure is straightforward, but the units on the vertical scale are a little complicated. For instance, to get the height of the block over the interval \$7,000 to \$10,000, you divide 15 percent by 3 thousand dollars. So the units for the answer are percent per thousand dollars. Think about the "per" just as you would when reading that there are 50,000 people per square mile in Tokyo: in each square mile of the city, there are about 50,000 people. It is the same with histograms. The height of the block over the interval \$7,000 to \$10,000 is 5% per thousand dollars: in each thousand-dollar interval between \$7,000 and \$10,000, there are about 5% of the families. Figure 4 shows the complete histogram with these units on the vertical scale.

Exercise Set B

1. The table below gives the distribution of educational level for persons age 25 and over in the U.S. in 1960, 1970, and 1991. ("Educational level" means the number of years of schooling completed.) The class intervals include the left endpoint, but not the right; for example, from the second line of the table, in 1960 about 14% of the people had completed 5–8 years of schooling, 8 not included; in 1991, about 4% of the people were in this category. Draw a histogram for the 1991 data. You can interpret "16 or more" as 16–17 years of schooling; not many people completed more than 16 years of school, especially in 1960 and 1970. Why does your histogram have spikes at 8, 12, and 16 years of schooling?

Educational level (years of schooling)	*1960*	*1970*	*1991*
0–5	8	6	2
5–8	14	10	4
8–9	18	13	4
9–12	19	19	11
12–13	25	31	39
13–16	9	11	18
16 or more	8	11	21

Source: Statistical Abstract, 1988, Table 202; 1992, Table 220.

2. Redraw the histogram for the 1991 data, combining the first two class intervals into one (0–8 years, with 6% of the people). Does this change the histogram much?

3. Draw the histogram for the 1970 data, and compare it to the 1991 histogram. What happened to the educational level of the population between 1970 and 1991—did it go up, go down, or stay about the same?

4. What happened to the educational level from 1960 to 1970?

The answers to these exercises are on p. A46.

3. THE DENSITY SCALE

When reading areas off a histogram, it is convenient to have a vertical scale. The income histogram in the previous section was drawn using the *density scale*.[5] The unit on the horizontal axis was $1,000 of family income, and the vertical axis showed the percentage of families per $1,000 of income. Figure 5 is another example of a histogram with a density scale. This is a histogram for educational level of persons age 25 and over in the U.S. in 1991. "Educational level" means years of schooling completed; kindergarten doesn't count.

The endpoint convention followed in this histogram is a bit fussy. The block over the interval 8–9 years, for example, represents all the people who finished eighth grade, but not ninth grade; people who dropped out part way through ninth

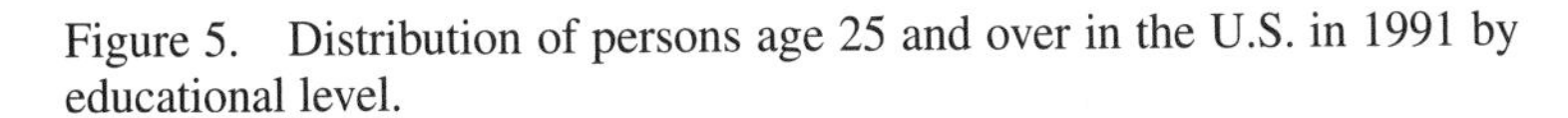

Figure 5. Distribution of persons age 25 and over in the U.S. in 1991 by educational level.

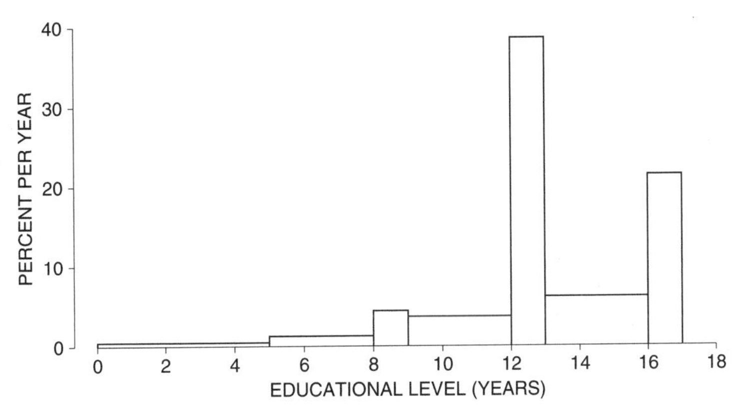

Source: *Statistical Abstract*, 1992, Table 220.

grade are included. The units on the horizontal axis of the histogram are years, so the units on the vertical axis are percent per year. For instance, the height of the histogram over the interval 13–16 years is 6% per year. In other words, about 6% of the population finished the first year of college, another 6% finished the second year, and another 6% finished the third year.

Section 1 described how area in a histogram represents percent. If one block covers a larger area than another, it represents a larger percent of the cases. What does the height of a block represent? Look at the horizontal axis in figure 5. Imagine the people lined up on this axis, with each person stationed at his or her educational level. Some parts of the axis—years—will be more crowded than others. The height of the histogram shows the crowding.

The histogram is highest over the interval 12–13 years, so the crowding is greatest there. This interval has all the people with high-school degrees. (Some people in this interval may have gone on to college, but they did not even finish the first year.) There are two other peaks, a small one at 8–9 years (finishing middle school) and a big one at 16–17 years—finishing college. The peaks show how people tend to stop their schooling at one of the three possible graduations rather than dropping out in between.

At first, it may be difficult to keep apart the notion of the crowding in an interval, represented by the height of the block, and the number in an interval, represented by the area of the block. An example will help. Look at the blocks over the intervals 8–9 years and 9–12 years in figure 5. The first block is a little taller, so this interval is a little more crowded. However, the block over 9–12 years has a much larger area, so this interval has many more people. Of course, there is more room in the second interval—it's 3 times as long. The two intervals are like the Netherlands and the U.S. The Netherlands is more crowded, but the U.S. has more people.

> In a histogram, the height of a block represents crowding—percentage per horizontal unit.

By contrast, the area of the block represents the percentage of cases in the corresponding class interval (section 1).

Once you learn how to use it, the density scale can be quite helpful. For example, take the interval from 9 to 12 years in figure 5—the people who got through their first year of high school but didn't graduate. The height of the block over this interval is nearly 4% per year. In other words, each of the three one-year intervals 9–10, 10–11, and 11–12 holds nearly 4% of the people. So the whole three-year interval must hold nearly $3 \times 4\% = 12\%$ of the people. Nearly 12% of the population age 25 and over got through at least one year of high school, but failed to graduate.

Example 1. The sketch below shows one block of the family-income histogram for a certain city. About what percent of the families in the city had incomes between \$15,000 and \$25,000?

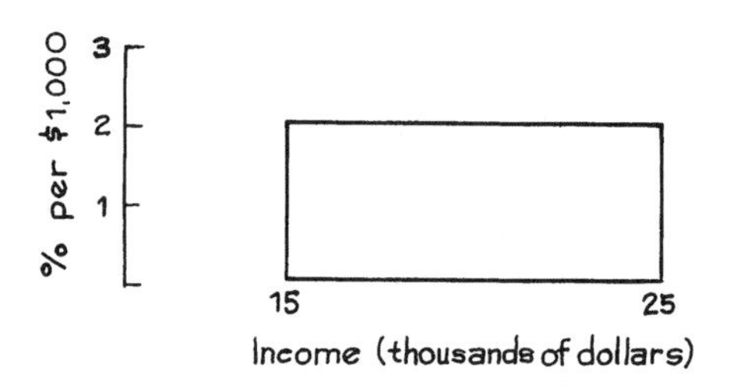

Solution. The height of the block is 2% per thousand dollars. Each thousand-dollar interval between \$15,000 and \$25,000 contains about 2% of the families in the city. There are 10 of these thousand-dollar intervals between \$15,000 and \$25,000. The answer is $10 \times 2\% = 20\%$. About 20% of the families in the city had incomes between \$15,000 and \$25,000.

The example shows that with the density scale, the areas of the blocks come out in percent. The horizontal units—thousands of dollars—cancel:

$$2\% \text{ per thousand dollars} \times 10 \text{ thousand dollars} = 20\%.$$

Example 2. Someone has sketched a histogram for the weights of some people, using the density scale. What's wrong?

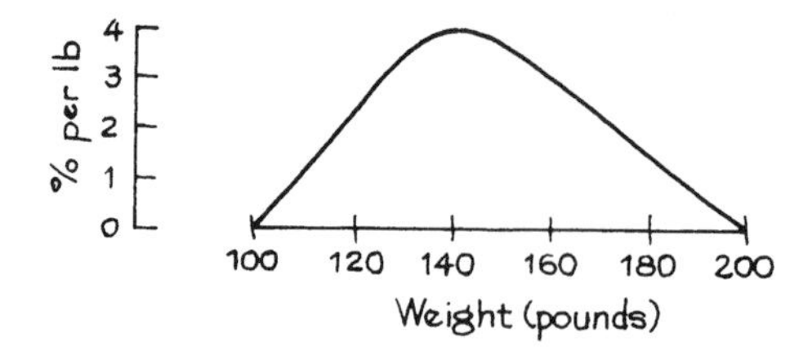

Solution. The total area is 200%, and should only be 100%. The area can be calculated as follows. The histogram is almost a triangle, whose height is 4% per pound and whose base is 200 lb − 100 lb = 100 lb. The area is

$$\frac{1}{2} \times \text{base} \times \text{height} = \frac{1}{2} \times 100 \text{ lb} \times 4\% \text{ per lb} = 200\%.$$

> With the density scale on the vertical axis, the areas of the blocks come out in percent. The area under the histogram over an interval equals the percentage of cases in that interval.[6] The total area under the histogram is 100%.

Since 1991, the educational level in the U.S. has continued to increase. Then, 21% of the population had a bachelor's degree or better (the "population" means people age 25 and over). In 2005, the corresponding figure was 28%.

Exercise Set C

1. A histogram of monthly wages for part-time employees is shown below (densities are marked in parentheses). Nobody earned more than $1,000 a month. The block over the class interval from $200 to $500 is missing. How tall must it be?

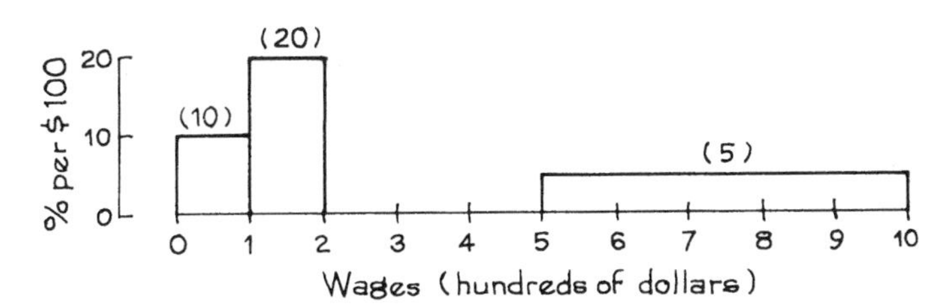

2. Three people plot histograms for the weights of subjects in a study, using the density scale. Only one is right. Which one, and why?

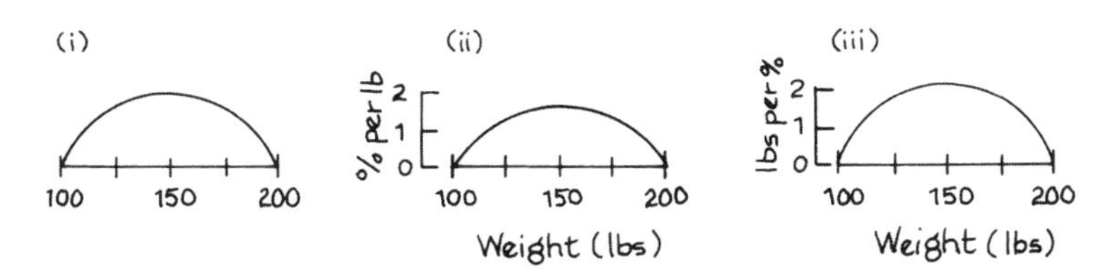

3. An investigator draws a histogram for some height data, using the metric system. She is working in centimeters (cm). The vertical axis shows density, and the top of the vertical axis is 10 percent per cm. Now she wants to convert to millimeters (mm). There are 10 millimeters to the centimeter. On the horizontal axis, she has to change 175 cm to ______ mm, and 200 cm to ______ mm. On the vertical axis, she has to change 10 percent per cm to ______ percent per mm, and 5 percent per cm to ______ percent per mm.

4. In a Public Health Service study, a histogram was plotted showing the number of cigarettes per day smoked by each subject (male current smokers), as shown below.[7] The density is marked in parentheses. The class intervals include the right endpoint, not the left.

   (a) The percentage who smoked 10 cigarettes or less per day is around

   1.5% 15% 30% 50%

   (b) The percentage who smoked more than a pack a day, but not more than 2 packs, is around

   1.5% 15% 30% 50%

   (There are 20 cigarettes in a pack.)

   (c) The percent who smoked more than a pack a day is around

   1.5% 15% 30% 50%

   (d) The percent who smoked more than 3 packs a day is around

   0.25 of 1% 0.5 of 1% 10%

   (e) The percent who smoked 15 cigarettes per day is around

   0.35 of 1% 0.5 of 1% 1.5% 3.5% 10%

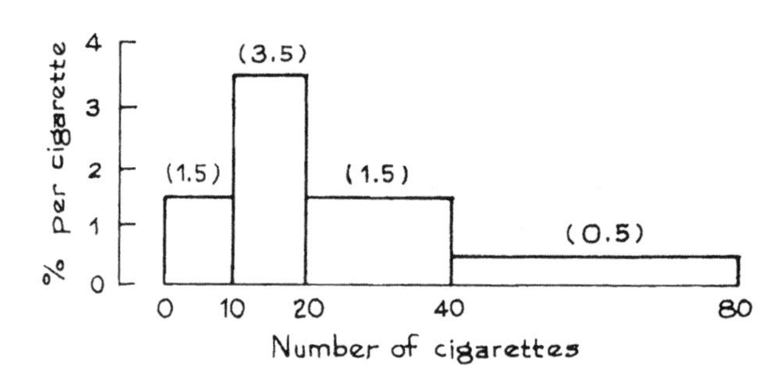

The answers to these exercises are on p. A46.

4. VARIABLES

The Current Population Survey covers many other variables besides income. A *variable* is a characteristic which changes from person to person in a study. Interviewers for the survey use a battery of questions: How old are you? How many people are there in your family? What is your family's total income? Are you married? Do you have a job? The corresponding variables would be: age, family size, family income, marital status, and employment status. Some questions are answered by giving a number: the corresponding variables are *quantitative*. Age, family size, and family income are examples of quantitative variables. Some questions are answered with a descriptive word or phrase, and the corresponding variables are *qualitative*: examples are marital status (single, married, widowed, divorced, separated) and employment status (employed, unemployed, not in the labor force).

Quantitative variables may be *discrete* or *continuous*. This is not a hard-and-fast distinction, but it is a useful one.[8] For a discrete variable, the values can only differ by fixed amounts. Family size is discrete. Two families can differ in size by 0 or 1 or 2, and so on. Nothing in between is possible. Age, on the other hand, is a continuous variable. This doesn't refer to the fact that a person is continuously getting older; it just means that the difference in age between two people can be arbitrarily small—a year, a month, a day, an hour, ... Finally, the terms *qualitative*, *quantitative*, *discrete*, and *continuous* are also used to describe data—qualitative data are collected on a qualitative variable, and so on.

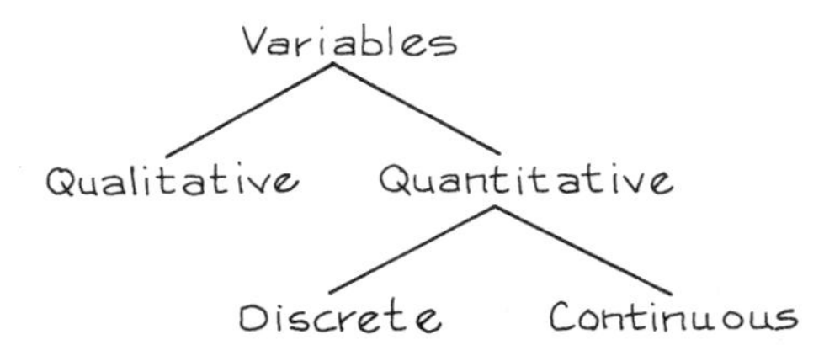

Section 2 showed how to plot a histogram starting with a distribution table. Often the starting point is the raw data—a list of cases (individuals, families, schools, etc.) and the corresponding values of the variable. In order to draw the histogram, a distribution table must be prepared. The first step is to choose the class intervals. With too many or too few, the histogram will not be informative. There is no rule, it is a matter of judgment or trial and error. It is common to start with ten or fifteen class intervals and work from there. In this book, the class intervals will always be given.[9]

When plotting a histogram for a continuous variable, investigators also have to decide on the endpoint convention—what to do with cases that fall right on the boundary. With a discrete variable, there is a convention which gets around this nuisance: center the class intervals at the possible values. For instance, family size can be 2 or 3 or 4, and so on. (The Census does not recognize one person as a family.) The corresponding class intervals in the distribution table would be

Center	*Class interval*
2	1.5 to 2.5
3	2.5 to 3.5
4	3.5 to 4.5
.	.
.	.
.	.

Since a family cannot have 2.5 members, there is no problem with endpoints. Figure 6 (on the next page) shows the histogram for family size. The bars seem to stop at 8; that is because there are so few families with 9 or more people.

Figure 6. Histogram showing distribution of families by size in 2005. With a discrete variable, the class intervals are centered at the possible values.

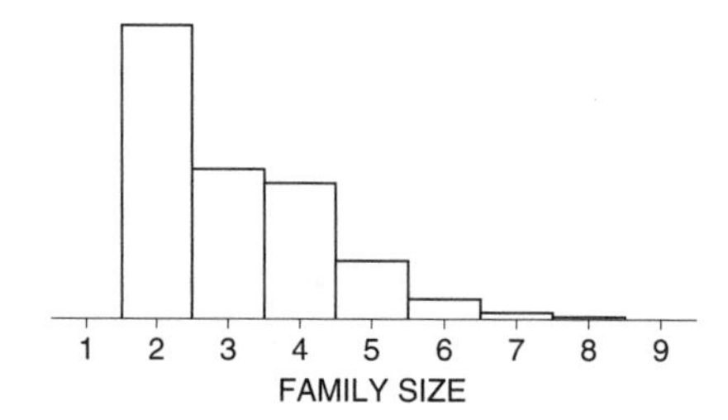

Source: March 2005 Current Population Survey; CD-ROM supplied by the Bureau of the Census.

Exercise Set D

1. Classify each of the following variables as qualitative or quantitative; if quantitative, as discrete or continuous.
   (a) occupation (b) region of residence (c) weight
   (d) height (e) number of automobiles owned

2. In the March Current Population Survey, women are asked how many children they have. Results are shown below for women age 25–39, by educational level.
   (a) Is the number of children discrete or continuous?
   (b) Draw histograms for these data. (You may take "5 or more" as 5—very few women had more than 5 children.)
   (c) What do you conclude?

Distribution of women age 25–39 by educational level and number of children (percent).

Number of children	*Women who are high-school graduates*	*Women with college degrees*
0	30.2	47.9
1	21.8	19.4
2	28.4	22.7
3	13.7	8.0
4	4.4	1.5
5 or more	1.5	0.5

Note: High-school graduates with no further education. College degrees at the level of a B.A. or B.Sc. Own, never-married children under the age of 18. Percents may not add to 100%, due to rounding.
Source: March 2005 Current Population Survey; CD-ROM supplied by the Bureau of the Census.

The answers to these exercises are on p. A47.

5. CONTROLLING FOR A VARIABLE

In the 1960s, many women began using oral contraceptives, "the pill." Since the pill alters the body's hormone balance, it is important to see what the side effects are. Research on this question is carried out by the Contraceptive Drug Study at the Kaiser Clinic in Walnut Creek, California. Over 20,000 women in the Walnut Creek area belong to the Kaiser Foundation Health Plan, paying a monthly insurance fee and getting medical services from Kaiser. One of these services is a routine checkup called the "multiphasic." During the period 1969–1971, about 17,500 women age 17–58 took the multiphasic and became subjects for the Drug Study. Investigators compared the multiphasic results for two different groups of women:

- "users" who take the pill (the treatment group);
- "non-users" who don't take the pill (the control group).

Figure 7. The effect of the pill. The top panel shows histograms for the systolic blood pressures of the 1,747 users and the 3,040 non-users age 25–34 in the Contraceptive Drug Study. The bottom panel shows the histogram for the non-users shifted to the right by 5 mm.

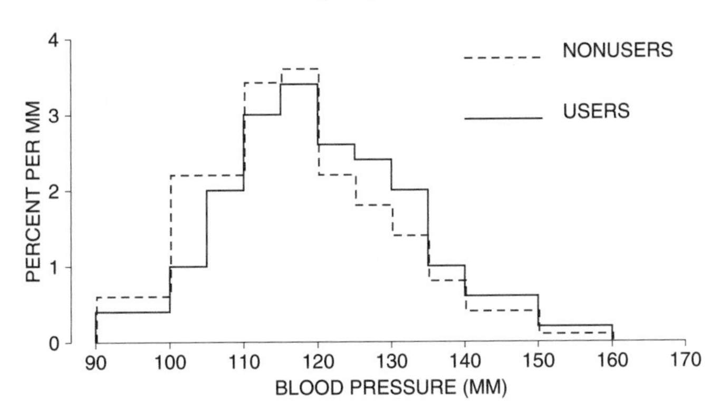

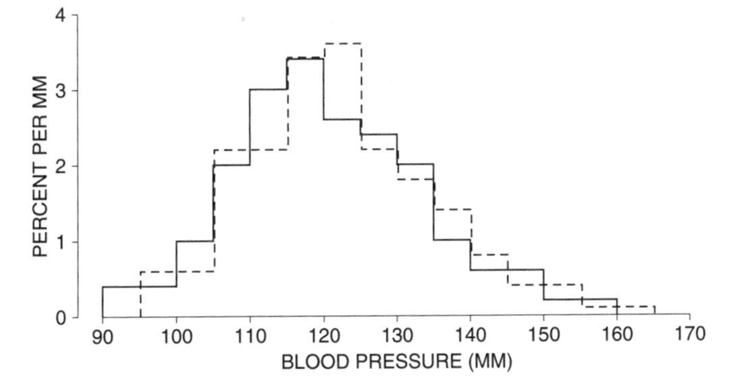

This is an observational study. It is the women who decided whether to take the pill or not. The investigators just watched what happens.

One issue was the effect of the pill on blood pressure. It might seem natural to compare the blood pressures for the users and non-users. However, this could be misleading. Blood pressure tends to go up with age, and the non-users were on the whole older than the users. For example, about 70% of the non-users were over 30, compared to 50% of the users. The effect of age is confounded with the effect of the pill. To make the full effect of the pill visible, it is necessary to make a separate comparison for each age group: this controls for age.[10] We will look only at the women age 25–34. Figure 7 shows the histograms for the users and non-users in this age group. (Blood pressure is measured relative to the length of a column of mercury; the units are "mm," that is, millimeters.)

The two histograms in the top panel of figure 7 have very similar shapes. However, the user histogram is higher to the right of 120 mm, lower to the left. High blood pressure (above 120 mm) is more prevalent among users, low blood pressure less prevalent. Now imagine that 5 mm were added to the blood pressure of each non-user. That would shift their histogram 5 mm to the right, as shown in the bottom panel of figure 7. In the bottom panel, the two histograms match up quite well. As far as the histograms are concerned, it is as if using the pill adds about 5 mm to the blood pressure of each woman.

This conclusion must be treated with caution. The results of the Contraceptive Drug Study suggest that if a woman goes on the pill, her blood pressure will go up by around 5 mm. But the proof is not complete. It cannot be, because of the design. The Drug Study is an observational study, not a controlled experiment. Part I showed that observational studies can be misleading about cause-and-effect relationships. There could be some factor other than the pill or age, as yet unidentified, which is affecting the blood pressures. For the Drug Study, this is a bit farfetched. The physiological mechanism by which the pill affects blood pressure is well established. The Drug Study data show the size of the effect.

Exercise Set E

1. As a sideline, the Drug Study compared blood pressures for women having different numbers of children. Below are sketches of the histograms for women with 2 or 4 children. Which group has higher blood pressure? Does having children cause the blood pressures of the mothers to change? Or could the change be due to some other factor, whose effects are confounded with the effect of having children?

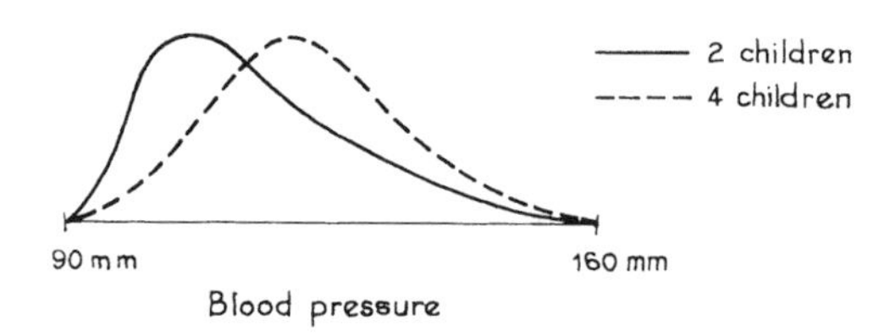

2. (Hypothetical.) The sketches on the next page show results from two other studies

of the pill, for women age 25–29. In one study, the pill adds about 10 mm to blood pressures; in the other, the pill adds about 10%. Which is which, and why?

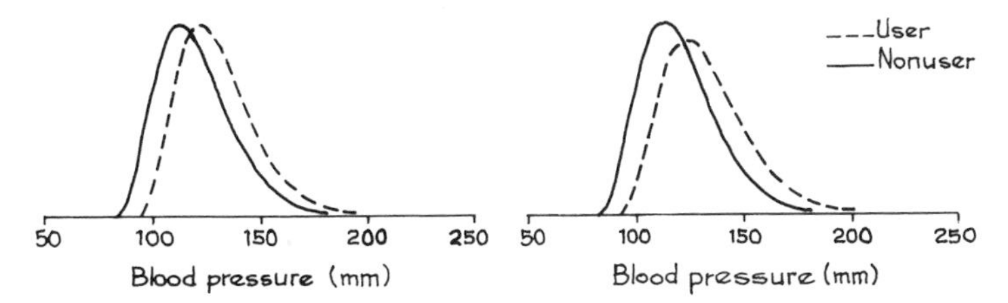

The answers to these exercises are on p. A47.

6. CROSS-TABULATION

The previous section explained how to control for the effect of age: it was a matter of doing the comparison separately for each age group. The comparison was made graphically, through the histograms in figure 7. Some investigators prefer to make the comparison in tabular form, using what is called a *cross-tab* (short for *cross-tabulation*). A cross-tab for blood pressure by age and pill use is shown in table 2. Such tables are a bit imposing, and the eye naturally tends to skip over

Table 2. Systolic blood pressure by age and pill use, for women in the Contraceptive Drug Study, excluding those who were pregnant or taking hormonal medication other than the pill. Class intervals include the left endpoint, but not the right. – means negligible. Table entries are in percent; columns may not add to 100 due to rounding.

Blood pressure (millimeters)	*Age 17–24*		*Age 25–34*		*Age 35–44*		*Age 45–58*	
	Non-users	*Users*	*Non-users*	*Users*	*Non-users*	*Users*	*Non-users*	*Users*
	(%)	(%)	(%)	(%)	(%)	(%)	(%)	(%)
under 90	–	1	1	–	1	1	1	–
90–95	1	–	1	–	2	1	1	1
95–100	3	1	5	4	5	4	4	2
100–105	10	6	11	5	9	5	6	4
105–110	11	9	11	10	11	7	7	7
110–115	15	12	17	15	15	12	11	10
115–120	20	16	18	17	16	14	12	9
120–125	13	14	11	13	9	11	9	8
125–130	10	14	9	12	10	11	11	11
130–135	8	12	7	10	8	10	10	9
135–140	4	6	4	5	5	7	8	8
140–145	3	4	2	4	4	6	7	9
145–150	2	2	2	2	2	5	7	9
150–155	–	1	1	1	1	3	2	4
155–160	–	–	–	1	1	1	1	3
160 and over	–	–	–	–	1	2	2	5
Total percent	100	98	100	99	100	100	99	99
Total number	1,206	1,024	3,040	1,747	3,494	1,028	2,172	437

them until some of the numbers are needed. However, all the cross-tab amounts to is a distribution table for blood pressures, made separately for users and non-users in each age group.

Look at the columns for the age group 17–24. There were 1,206 non-users and 1,024 users. About 1% of the users had blood pressure below 90 mm; the corresponding percentage of non-users was negligible—that is what the dash means. To see the effect of the pill on the blood pressures of women age 17–24, it is a matter of looking at the percents in the columns for non-users and users in the age group 17–24. To see the effect of age, look first at the non-users column in each age group and see how the percents shift toward the high blood pressures as age goes up. Then do the same thing for the users.

Exercise Set F

1. Use table 2 to answer the following questions.
   (a) What percentage of users age 17–24 have blood pressures of 140 mm or more?
   (b) What percentage of non-users age 17–24 have blood pressures of 140 mm or more?
   (c) What do you conclude?
2. Draw histograms for the blood pressures of the users and non-users age 17–24. What do you conclude?
3. Compare the histograms of blood pressures for non-users age 17–24 and for non-users age 25–34. What do you conclude?

The answers to these exercises are on p. A47.

7. SELECTIVE BREEDING

In 1927, the psychologist Charles Spearman published *The Abilities of Man*, his theory of human intelligence. Briefly, Spearman held that test scores of intellectual abilities (like reading comprehension, arithmetic, or spatial perception) were weighted sums of two independent components: a general intelligence factor which Spearman called "g," and an ability factor specific to each test. This theory attracted a great deal of attention.

As part of his Ph.D. research in the psychology department at Berkeley, Robert Tryon decided to check the theory on an animal population, where it is simpler to control extraneous variables.[11] Tryon used rats, which are easy to breed in the laboratory. To test their intelligence, he put the rats into a maze. When they ran the maze, the rats made errors by going into blind alleys. The test consisted of 19 runs through the maze; the animal's "intelligence score" was the total number of errors it made. So the bright rats are the ones with low scores, the dulls are the ones with high scores. Tryon started out with 142 rats, and the distribution of their intelligence scores is sketched in figure 8.

Figure 8. Tryon's experiment. Distribution of intelligence in the original population.

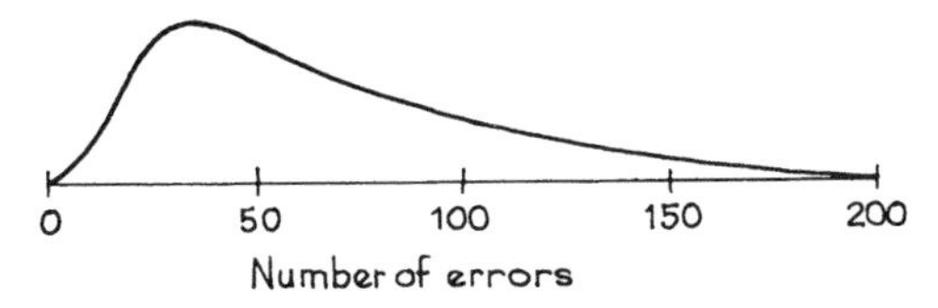

The next step in the experiment was to breed for intelligence. In each generation, the "maze-bright" rats (the ones making only a small number of errors) were bred with each other. Similarly, the "maze-dull" animals (with high scores) were bred together. Seven generations later, Tryon had 85 rats in the maze-bright strain, and 68 in the maze-dull strain. There was a clear separation in scores. Figure 9 shows the distribution of intelligence for the two groups, and the histograms barely overlap. (In fact, Tryon went on with selective breeding past the seventh generation, but didn't get much more separation in scores.)

Figure 9. Tryon's experiment. After seven generations of selective breeding, there is a clear separation into "maze-bright" and "maze-dull" strains.

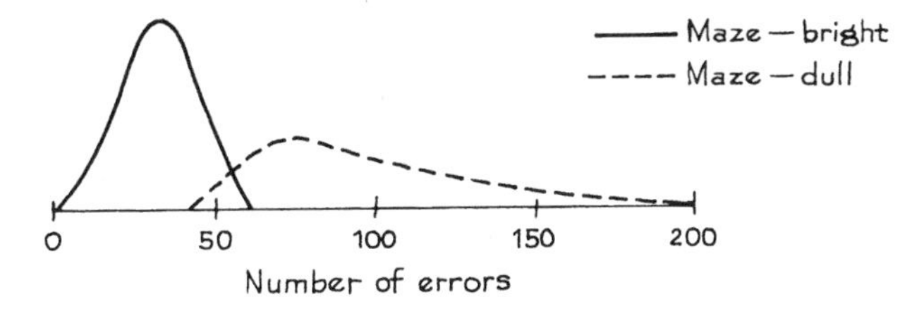

The two strains created by Tryon were used for many other experiments by the Berkeley psychology department. Generations later, rats from the maze-bright population continued to outperform the dulls at maze-running. So Tryon managed to breed for a mental ability—evidence that some mental abilities are at least in part genetically determined. What did the experiment say about Spearman's theory? Tryon found that the maze-bright rats did no better than the maze-dulls on other tests of animal intelligence, such as discriminating between geometric shapes, or between intensities of light. This was evidence against Spearman's theory of a general intelligence factor (at least for rats). On the other hand, Tryon did find intriguing psychological differences between the two rat populations. The "brights" seemed to be unsociable introverts, well adjusted to life in the maze, but neurotic in their relationships with other rats. The "dulls" were quite the opposite.

8. REVIEW EXERCISES

Review exercises may cover material from previous chapters.

1. The figure below shows a histogram for the heights of a representative sample of men. The shaded area represents the percentage of men whose heights were between ______ and ______. Fill in the blanks.

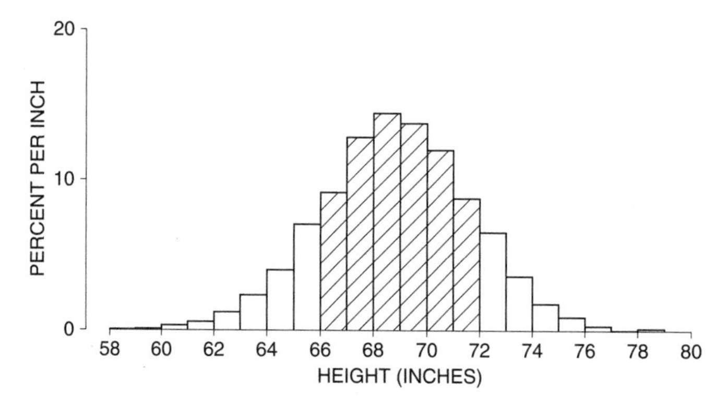

Source: Data tape supplied by the Inter-University Consortium for Political and Social Research.

2. The age distribution of people in the U.S. in 2004 is shown below. Draw the histogram. (The class intervals include the left endpoint, not the right; for instance, on the second line of the table, 14% of the people were age 5 years or more but had not yet turned 15. The interval for "75 and over" can be ended at 85. Men and women are combined in the data.) Use your histogram to answer the following questions.

   (a) Are there more children age 1, or elders age 71?
   (b) Are there more 21-year-olds, or 61-year-olds?
   (c) Are there more people age 0–4, or 65–69?
   (d) The percentage of people age 35 and over is around 25%, 50%, or 75%?

Age	*Percent of population*	*Age*	*Percent of population*
0–5	7	35–45	15
5–15	14	45–55	14
15–20	7	55–65	10
20–25	7	65–75	6
25–30	7	75 and over	6
30–35	7		

Source: *Statistical Abstract*, 2006, Table 11.

3. The American Housing Survey is done every year by the Bureau of the Census. Data from the 2003 survey can be used to find the distribution of occupied housing units (this includes apartments) by number of rooms. Results for the whole U.S. are shown below, separately for "owner-occupied" and "renter-

occupied" units. Draw a histogram for each of the two distributions. (You may assume that "10 or more" means 10 or 11; very few units have more than 11 rooms.)

(a) The owner-occupied percents add up to 100.2% while the renter-occupied percents add up to 100.0%. Why?
(b) The percentage of one-room units is much smaller for owner-occupied housing. Is that because there are so many more owner-occupied units in total? Answer yes or no, and explain briefly.
(c) Which are larger, on the whole: the owner-occupied units or the renter-occupied units?

Number of rooms in unit	*Owner-occupied (percent)*	*Renter-occupied (percent)*
1	0.0	1.0
2	0.1	2.8
3	1.4	22.7
4	9.7	34.5
5	23.3	22.6
6	26.4	10.4
7	17.5	3.6
8	10.4	1.2
9	5.0	0.5
10 or more	6.4	0.7
Total	100.2	100.0
Number	72.2 million	33.6 million

Source: www.census.gov/hhres/www/housing/ahs/nationaldata.html

4. The figure below is a histogram showing the distribution of blood pressure for all 14,148 women in the Drug Study (section 5). Use the histogram to answer the following questions:

(a) Is the percentage of women with blood pressures above 130 mm around 25%, 50%, or 75%?
(b) Is the percentage of women with blood pressures between 90 mm and 160 mm around 1%, 50%, or 99%?
(c) In which interval are there more women: 135–140 mm or 140–150 mm?

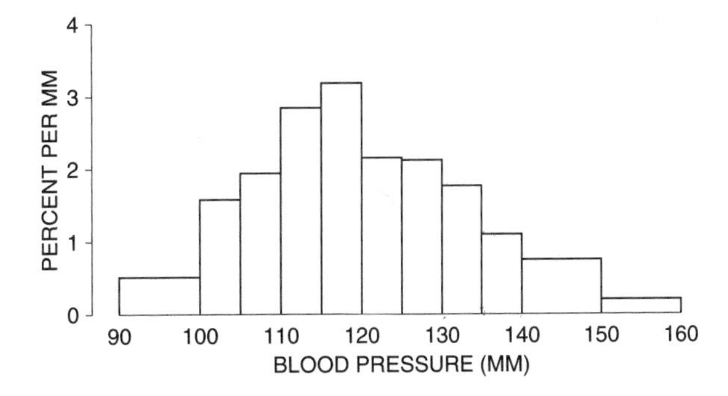

(d) Which interval is more crowded: 135–140 mm or 140–150 mm?
(e) On the interval 125–130 mm, the height of the histogram is about 2.1% per mm. What percentage of the women had blood pressures in this class interval?
(f) Which interval has more women: 97–98 mm or 102–103 mm?
(g) Which is the most crowded millimeter of all?

5. Someone has sketched one block of a family-income histogram for a wealthy suburb. About what percentage of the families in this suburb had incomes between \$90,000 and \$100,000 a year?

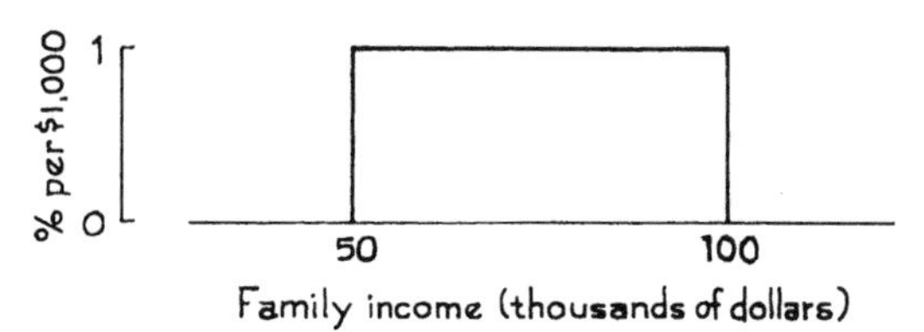

6. (Hypothetical.) In one study, 100 people had their heights measured to the nearest eighth of an inch. A histogram for the results is shown below. Two of the following lists have this histogram. Which ones, and why?

(i) 25 people, 67 inches tall; 50 people, 68 inches tall; 25 people, 69 inches tall.
(ii) 10 people, $66\frac{3}{4}$ inches tall; 15 people, $67\frac{1}{4}$ inches tall; 50 people, 68 inches tall; 25 people, 69 inches tall.
(iii) 30 people, 67 inches tall; 40 people, 68 inches tall; 30 people, 69 inches tall.

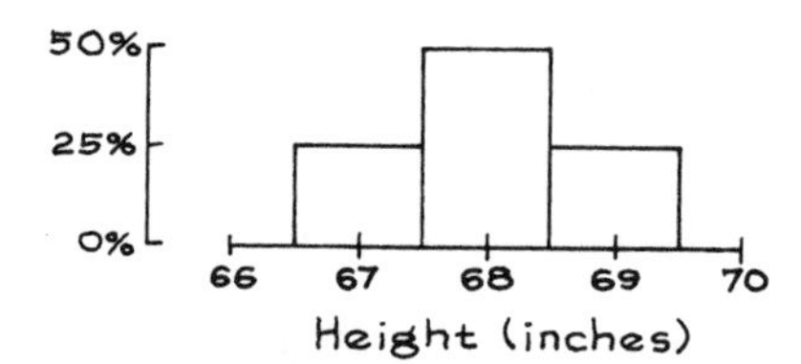

7. Two histograms are sketched below. One shows the distribution of age at death from natural causes (heart disease, cancer, and so forth). The other shows age at death from trauma (accident, murder, suicide). Which is which, and why?

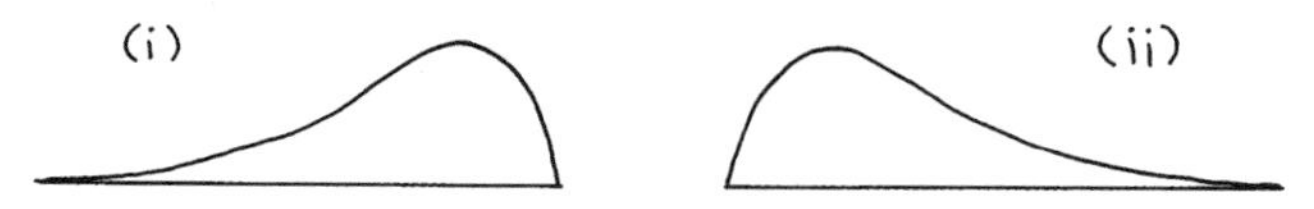

8. The figure on the next page (adapted from the *San Francisco Chronicle*, May 18, 1992) shows the distribution of American families by income. Ranges include

the left endpoint but not the right. For example, 3.7% of the families had incomes in the range \$0–\$4,999, 5.8% had incomes in the range \$5,000–\$9,999, and so forth. True or false, and explain:

(a) Although American families are not spread evenly over the whole income range, the families that earn between \$10,000 and \$35,000 are spread fairly evenly over that range.
(b) The families that earn between \$35,000 and \$75,000 are spread fairly evenly over that range.
(c) The graph is a histogram.

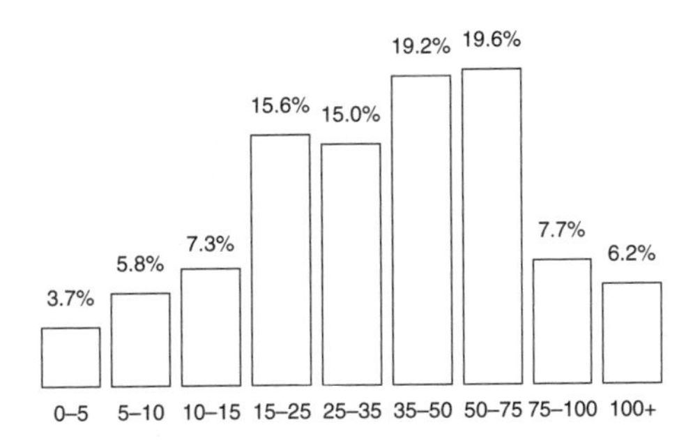

9. In a survey carried out at the University of California, Berkeley, a sample of students were interviewed and asked what their grade-point average was. A histogram of the results is shown below. (GPA ranges from 0 to 4, and 2 is a bare pass.)

(a) True or false: more students reported a GPA in the range 2.0 to 2.1 than in the range 1.5 to 1.6.
(b) True or false: more students reported a GPA in the range 2.0 to 2.1 than in the range 2.5 to 2.6.
(c) What accounts for the spike at 2?

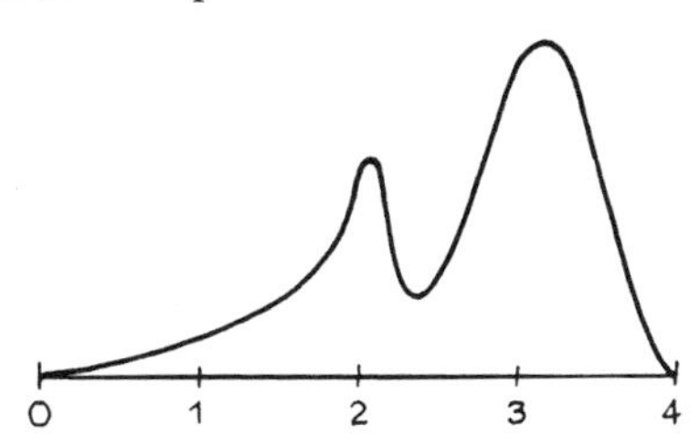

10. The table on the next page shows the distribution of adults by the last digit of their age, as reported in the Census of 1880 and the Census of 1970.[12] You might expect each of the ten possible digits to turn up for 10% of the people, but this is not the case. For example, in 1880, 16.8% of all persons reported an age ending in 0—like 30 or 40 or 50. In 1970, this percentage was only 10.6%.

(a) Draw histograms for these two distributions.

(b) In 1880, there was a strong preference for the digits 0 and 5. How can this be explained?
(c) In 1970, the preference was much weaker. How can this be explained?
(d) Are even digits more popular, or odd ones, in 1880? 1970?

Digit	*1880*	*1970*
0	16.8	10.6
1	6.7	9.9
2	9.4	10.0
3	8.6	9.6
4	8.8	9.8
5	13.4	10.0
6	9.4	9.9
7	8.5	10.2
8	10.2	10.0
9	8.2	10.1

Source: United States Census.

11. In the Sanitary District of Chicago, operating engineers are hired on the basis of a competitive civil-service examination. In 1966, there were 223 applicants for 15 jobs. The exam was held on March 12; the test scores are shown below, arranged in increasing order. The height of each bar in the histogram (top of next page) shows the number of people with the corresponding score. The examiners were charged with rigging the exam.[13] Why?

26 27 27 27 27 29 30 30 30 30 31 31 31 32 32
33 33 33 33 33 34 34 34 35 35 36 36 36 37 37
37 37 37 37 37 39 39 39 39 39 39 39 40 41 42
42 42 42 42 43 43 43 43 43 43 43 43 44 44 44
44 44 44 45 45 45 45 45 45 45 46 46 46 46 46

46 47 47 47 47 47 47 48 48 48 48 48 48 48 48
49 49 49 49 50 50 51 51 51 51 51 52 52 52 52
52 53 53 53 53 53 54 54 54 54 54 55 55 55 56
56 56 56 56 57 57 57 57 58 58 58 58 58 58 58
58 59 59 59 59 60 60 60 60 60 60 61 61 61 61

61 61 62 62 62 63 63 64 65 66 66 66 67 67 67
67 68 68 69 69 69 69 69 69 69 69 71 71 72 73
74 74 74 75 75 76 76 78 80 80 80 80 81 81 81
82 82 83 83 83 83 84 84 84 84 84 84 84 90 90
90 91 91 91 92 92 92 93 93 93 93 95 95

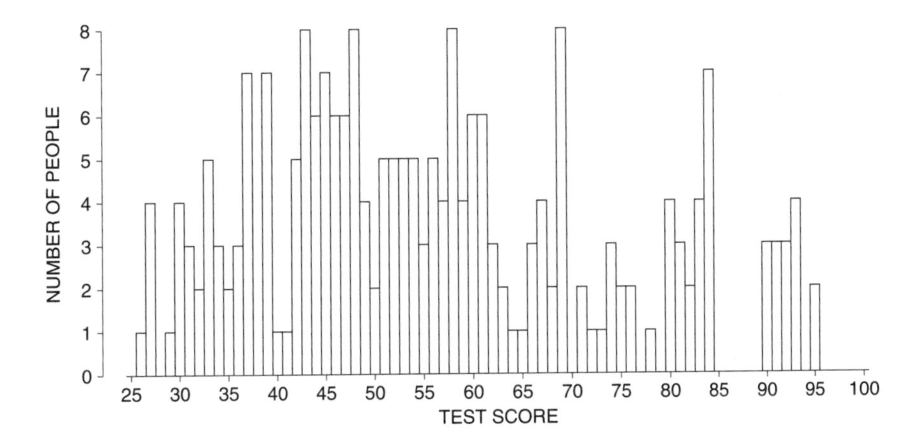

12. The late 1960s and early 1970s were years of turmoil in the U.S. Psychologists thought that rioting was related (among other things) to temperature, with hotter weather making people more aggressive.[14] Two investigators, however, argued that "the frequency of riots should increase with temperature through the mid-80s but then go down sharply with increases in temperature beyond this level."

    To support their theory, they collected data on 102 riots over the period 1967–71, including the temperature in the city where the riot took place. They plotted a histogram for the distribution of riots by temperature (a sketch is shown below). There is a definite peak around 85°. True or false, and explain: the histogram shows that higher temperatures prevent riots.

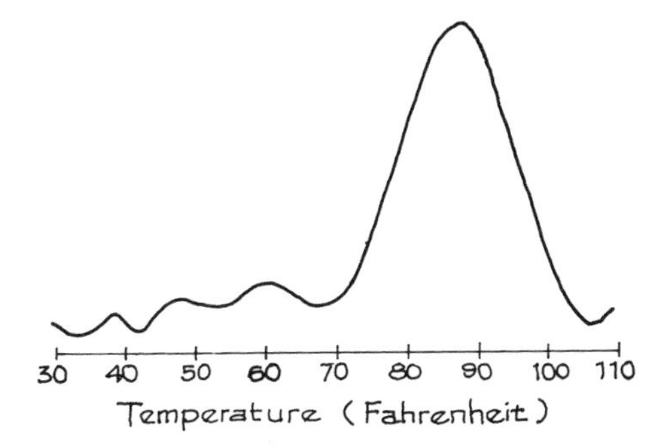

9. SUMMARY

1. A *histogram* represents percents by area. It consists of a set of blocks. The area of each block represents the percentage of cases in the corresponding *class interval.*

2. With the *density scale*, the height of each block equals the percentage of cases in the corresponding class interval, divided by the length of that interval.

3. With the density scale, area comes out in percent, and the total area is 100%. The area under the histogram between two values gives the percentage of cases falling in that interval.

4. A *variable* is a characteristic of the subjects in a study. It can be either *qualitative* or *quantitative.* A quantitative variable can be either *discrete* or *continuous.*

5. A confounding factor is sometimes controlled for by *cross-tabulation.*

4

The Average and the Standard Deviation

It is difficult to understand why statisticians commonly limit their enquiries to Averages, and do not revel in more comprehensive views. Their souls seem as dull to the charm of variety as that of the native of one of our flat English counties, whose retrospect of Switzerland was that, if its mountains could be thrown into its lakes, two nuisances would be got rid of at once.

—SIR FRANCIS GALTON (ENGLAND, 1822–1911)[1]

1. INTRODUCTION

A histogram can be used to summarize large amounts of data. Often, an even more drastic summary is possible, giving just the center of the histogram and the spread around the center. ("Center" and "spread" are ordinary words here, without any special technical meaning.) Two histograms are sketched in figure 1 on the next page. The center and spread are shown. Both histograms have the same center, but the second one is more spread out—there is more area farther away from the center. For statistical work, precise definitions have to be given, and there are several ways to go about this. The *average* is often used to find the center, and so is the *median*.[2] The *standard deviation* measures spread around the average; the *interquartile range* is another measure of spread.

The histograms in figure 1 can be summarized by the center and the spread. However, things do not always work out so well. For instance, figure 2 gives the distribution of elevation over the earth's surface. Elevation is shown along the

Figure 1. Center and spread. The centers of the two histograms are the same, but the second histogram is more spread out.

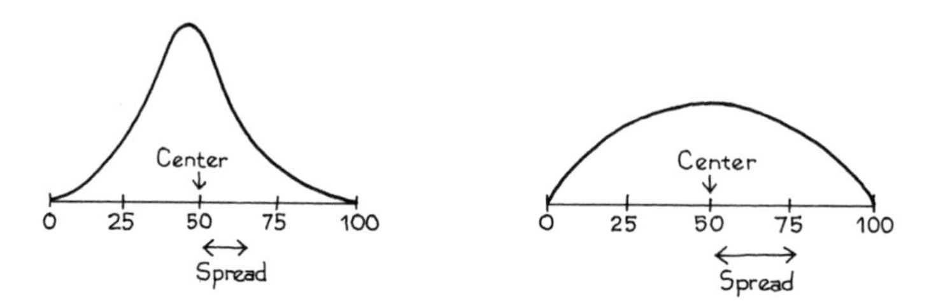

horizontal axis, in miles above (+) or below (–) sea level. The area under the histogram between two elevations gives the percentage of the earth's surface area between those elevations. There are clear peaks in this histogram. Most of the surface area is taken up by the sea floors, around 3 miles below sea level; or the continental plains, around sea level. Reporting only the center and spread of this histogram would miss the two peaks.[3]

Figure 2. Distribution of the surface area of the earth by elevation above (+) or below (–) sea level.

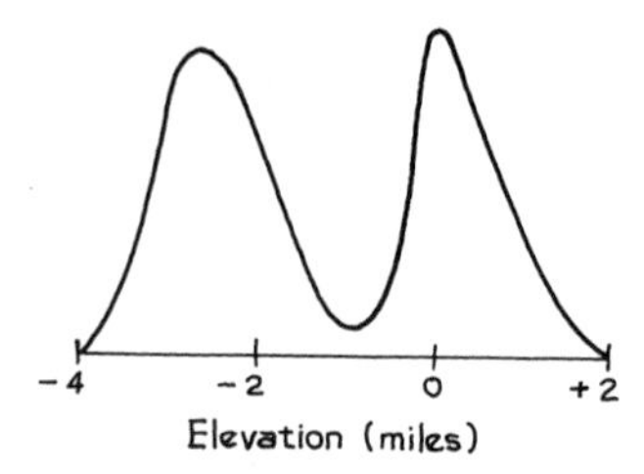

2. THE AVERAGE

The object of this section is to review the average; the difference between *cross-sectional* and *longitudinal* surveys will also be discussed. The context is HANES—the Health and Nutrition Examination Survey, in which the Public Health Service examines a representative cross section of Americans. This survey has been done at irregular intervals since 1959 (when it was called the Health Examination Survey). The objective is to get baseline data about—

- demographic variables, like age, education, and income;
- physiological variables like height, weight, blood pressure, and serum cholesterol levels;
- dietary habits;
- prevalence of diseases.

Subsequent analysis focuses on the interrelationships among the variables, and has some impact on health policy.[4]

The HANES2 sample was taken during the period 1976–80. Before looking at the data, let's make a quick review of averages.

> The average of a list of numbers equals their sum, divided by how many there are.

For instance, the list 9, 1, 2, 2, 0 has 5 entries, the first being 9. The average of the list is

$$\frac{9+1+2+2+0}{5} = \frac{14}{5} = 2.8$$

Let's get back to HANES. What did the men and women in the sample (age 18–74) look like?

- The average height of the men was 5 feet 9 inches, and their average weight was 171 pounds.
- The average height of the women was 5 feet 3.5 inches, and their average weight was 146 pounds.

They're pretty chubby.

What's happened since 1980? The survey was done again in 2003–04 (HANES5). Average heights went up by a fraction of an inch, while weights went up by nearly 20 pounds—both for men and for women.

Figure 3 shows the averages for men and women, and for each age group; averages are joined by straight lines. From HANES2 to HANES5, average heights went up a little in each group—but average weights went up a lot. This could become a serious public-health problem, because excess weight is associated with many diseases, including heart disease, cancer, and diabetes.

Figure 3. Age-specific average heights and weights for men and women 18–74 in the HANES sample. The panel on the left shows height, the panel on the right shows weight.

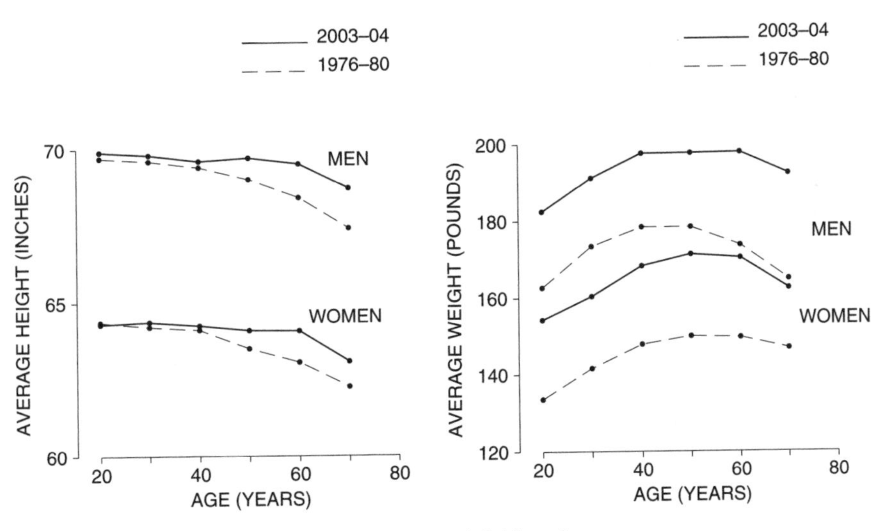

Source: www.cdc.gov/nchs/nhanes.htm

The average is a powerful way of summarizing data—many histograms are compressed into the four curves. But this compression is achieved only by smoothing away individual differences. For instance, in 2003–04, the average height of the men age 18–24 was 5 feet 10 inches. But 15% of them were taller than 6 feet 1 inch; another 15% were shorter than 5 feet 6 inches. This diversity is hidden by the average.

For a moment, we return to design issues (chapter 2). In the 1976–80 data, the average height of men appears to decrease after age 20, dropping about two inches in 50 years. Similarly for women. Should you conclude that an average person got shorter at this rate? Not really. HANES is *cross-sectional*, not *longitudinal*. In a cross-sectional study, different subjects are compared to each other at one point in time. In a longitudinal study, subjects are followed over time, and compared with themselves at different points in time. The people age 18–24 in figure 3 are completely different from those age 65–74. The first group was born a lot later than the second.

There is evidence to suggest that, over time, Americans have been getting taller. This is called the *secular trend* in height, and its effect is confounded with the effect of age in figure 3. Most of the two-inch drop in height seems to be due to the secular trend. The people age 65–74 were born around 50 years before those age 18–24, and are an inch or two shorter for that reason.[5] On the other hand, the secular trend has slowed down. (Reasons are unclear.) Average heights only increased a little from 1976–80 to 2003–04. The slowing also explains why the height curves for 2003–04 are flatter than the curves for 1976–80.

> If a study draws conclusions about the effects of age, find out whether the data are cross-sectional or longitudinal.

Exercise Set A

1. (a) The numbers 3 and 5 are marked by crosses on the horizontal line below. Find the average of these two numbers and mark it by an arrow.

(b) Repeat (a) for the list 3, 5, 5.

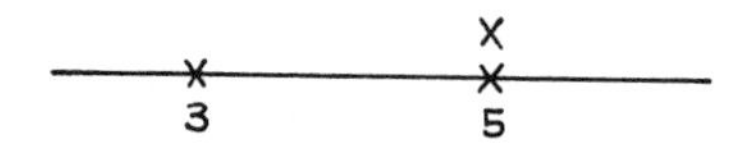

(c) Two numbers are shown below by crosses on a horizontal axis. Draw an arrow pointing to their average.

2. A list has 10 entries. Each entry is either 1 or 2 or 3. What must the list be if the average is 1? If the average is 3? Can the average be 4?

3. Which of the following two lists has a bigger average? Or are they the same? Try to answer without doing any arithmetic.

   (i) 10, 7, 8, 3, 5, 9 (ii) 10, 7, 8, 3, 5, 9, 11

4. Ten people in a room have an average height of 5 feet 6 inches. An 11th person, who is 6 feet 5 inches tall, enters the room. Find the average height of all 11 people.

5. Twenty-one people in a room have an average height of 5 feet 6 inches. A 22nd person, who is 6 feet 5 inches tall, enters the room. Find the average height of all 22 people. Compare with exercise 4.

6. Twenty-one people in a room have an average height of 5 feet 6 inches. A 22nd person enters the room. How tall would he have to be to raise the average height by 1 inch?

7. In figure 2, are the Rocky Mountains plotted near the left end of the axis, the middle, or the right end? What about Kansas? What about the trenches in the sea floor, like the Marianas trench?

8. Diastolic blood pressure is considered a better indicator of heart trouble than systolic pressure. The figure below shows age-specific average diastolic blood pressure for the men age 20 and over in HANES5 (2003–04).[6] True or false: the data show that as men age, their diastolic blood pressure increases until age 45 or so, and then decreases. If false, how do you explain the pattern in the graph? (Blood pressure is measured in "mm," that is, millimeters of mercury.)

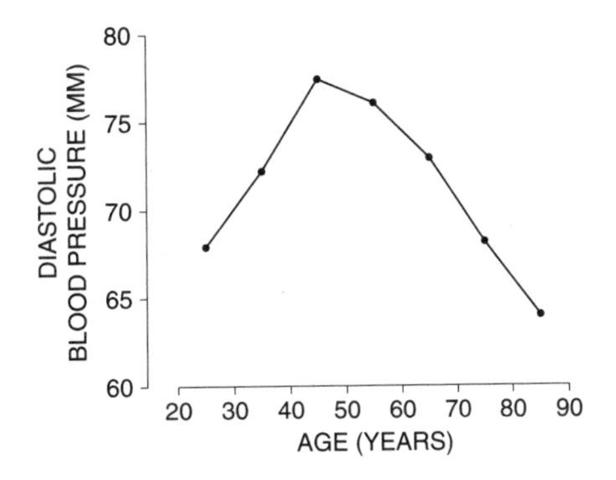

9. Average hourly earnings are computed each month by the Bureau of Labor Statistics using payroll data from commercial establishments. The Bureau figures the total wages paid out (to nonsupervisory personnel), and divides by the total hours worked. During recessions, average hourly earnings typically go up. When the recession ends, average hourly earnings often start going down. How can this be?

The answers to these exercises are on pp. A47–48.

3. THE AVERAGE AND THE HISTOGRAM

This section will indicate how the average and the median are related to histograms. To begin with an example, there were 2,696 women age 18 and over in HANES5 (2003–04). Their average weight was 164 pounds. It is natural to guess

Figure 4. Histogram for the weights of the 2,696 women in the HANES5 sample. The average is marked by a vertical line. Only 41% of the women were above average in weight.

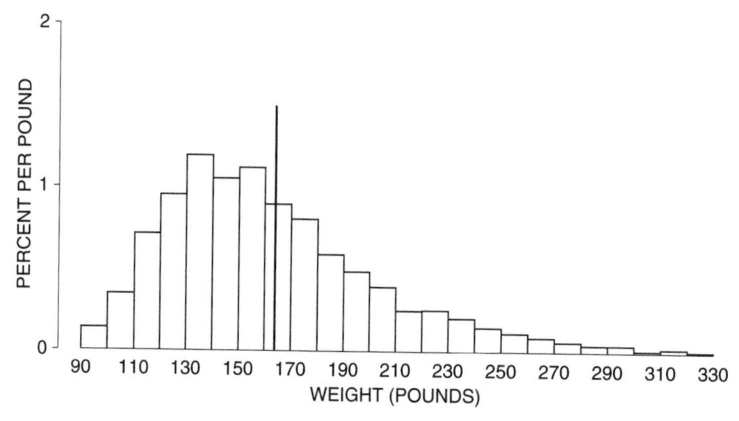

Source: www.cdc.gov/nchs/nhanes.htm.

that 50% of them were above average in weight, and 50% were below average. However, this guess is somewhat off. In fact, only 41% were above average, and 59% were below average. Figure 4 shows a histogram for the data: the average is marked by a vertical line. In other situations, the percentages can be even farther from 50%.

How is this possible? To find out, it is easiest to start with some hypothetical data—the list 1, 2, 2, 3. The histogram for this list (figure 5) is symmetric about the value 2. And the average equals 2. If the histogram is symmetric around a value, that value equals the average. Furthermore, half the area under the histogram lies to the left of that value, and half to the right. (What does symmetry mean? Imagine drawing a vertical line through the center of the histogram and folding the histogram in half around that line: the two halves should match up.)

Figure 5. Histogram for the list 1, 2, 2, 3. The histogram is symmetric around 2, the average: 50% of the area is to the left of 2, and 50% is to the right.

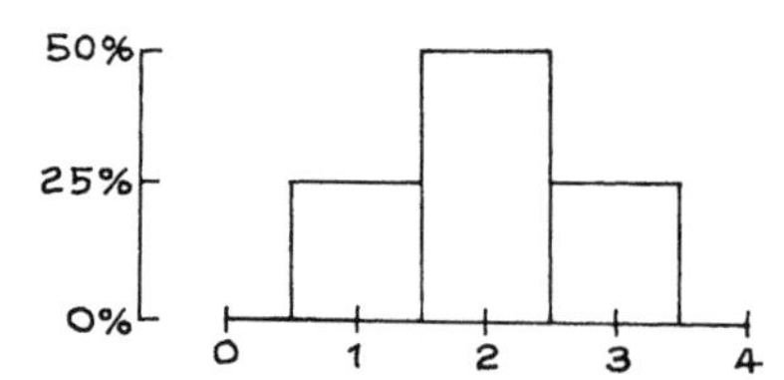

What happens when the value 3 on the list 1, 2, 2, 3 is increased, say to 5 or 7? As shown in figure 6, the rectangle over that value moves off to the right, destroying the symmetry. The average for each histogram is marked with an arrow, and the arrow shifts to the right following the rectangle. To see why, imagine the histogram is made out of wooden blocks attached to a stiff, weightless board. Put the histogram across a taut wire, as illustrated in the bottom panel of figure 6. The histogram will balance at the average.[7] A small area far away from the average can balance a large area close to the average, because areas are weighted by their distance from the balance point.

Figure 6. The average. The top panel shows three histograms; the averages are marked by arrows. As the shaded box moves to the right, it pulls the average along with it. The area to the left of the average gets up to 75%. The bottom panel shows the same three histograms made out of wooden blocks attached to a stiff, weightless board. The histograms balance when supported at the average.

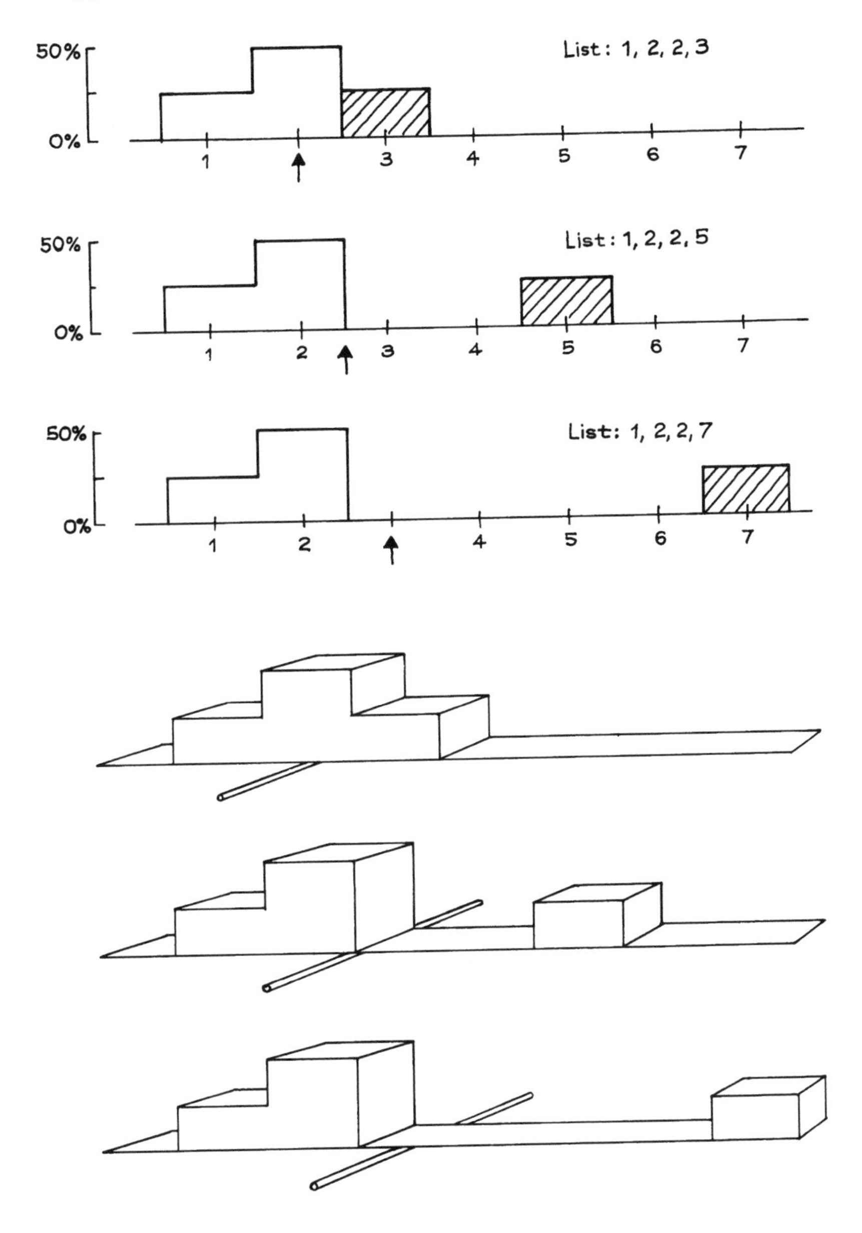

A histogram balances when supported at the average.

A small child sits farther away from the center of a seesaw in order to balance a large child sitting closer to the center. Blocks in a histogram work the same way. That is why the percentage of cases on either side of the average can differ from 50%.

The *median* of a histogram is the value with half the area to the left and half to the right. For all three histograms in figure 6, the median is 2. With the second and third histograms, the area to the right of the median is far away by comparison with the area to the left. Consequently, if you tried to balance one of those histograms at the median, it would tip to the right. More generally, the average is to the right of the median whenever the histogram has a long right-hand tail, as in figure 7. The weight histogram (figure 4 on p. 62) had an average of 164 lbs and a median of 155 lbs. The long right-hand tail is what made the average bigger than the median.

Figure 7. The tails of a histogram.

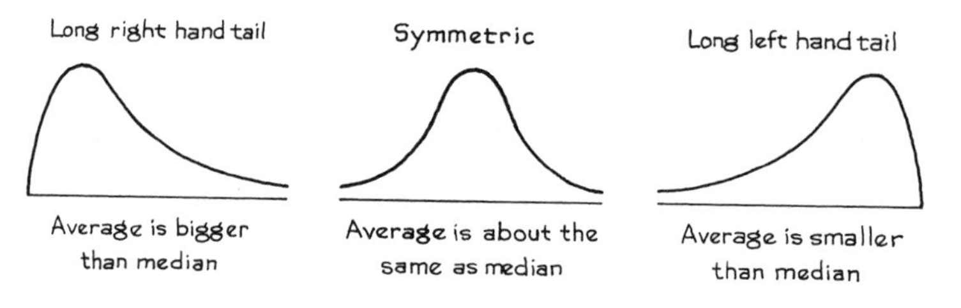

For another example, median family income in the U.S. in 2004 was about \$54,000. The income histogram has a long right-hand tail, and the average was higher—\$60,000.[8] When dealing with long-tailed distributions, statisticians might use the median rather than the average, if the average pays too much attention to the extreme tail of the distribution. We return to this point in the next chapter.

Exercise Set B

1. Below are sketches of histograms for three lists. Fill in the blank for each list: the average is around ______. Options: 25, 40, 50, 60, 75.

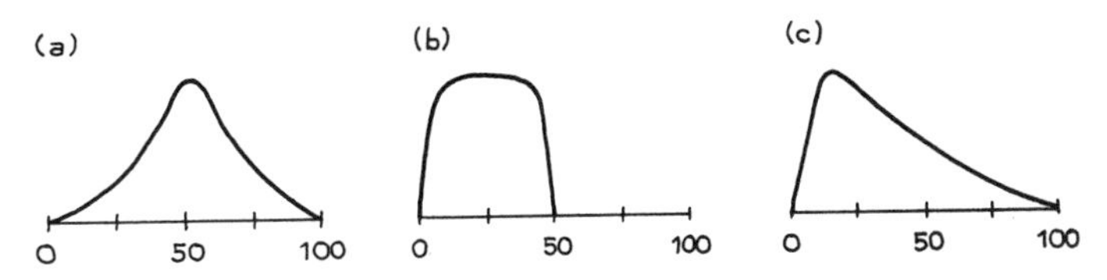

2. For each histogram in exercise 1, is the median equal to the average? or is it to the left? to the right?
3. Look back at the cigarette histogram on p. 42. The median is around ______. Fill in the blank. Options: 10, 20, 30, 40
4. For this cigarette histogram, is the average around 15, 20, or 25?
5. For registered students at universities in the U.S., which is larger: average age or median age?
6. For each of the following lists of numbers, say whether the entries are on the whole around 1, 5, or 10 in size. No arithmetic is needed.

   (a) 1.3, 0.9, 1.2, 0.8 (b) 13, 9, 12, 8
   (c) 7, 3, 6, 4 (d) 7, −3, −6, 4

The answers to these exercises are on pp. A48–49.

Technical note. The median of a list is defined so that half or more of the entries are at the median or bigger, and half or more are at the median or smaller. This will be illustrated on 4 lists—

(a) 1, 5, 7
(b) 1, 2, 5, 7
(c) 1, 2, 2, 7, 8
(d) 8, −3, 5, 0, 1, 4, −1

For list (a), the median is 5: two entries out of the three are 5 or more, and two are 5 or less. For list (b), any value between 2 and 5 is a median; if pressed, most statisticians would choose 3.5 (which is halfway between 2 and 5) as "the" median. For list (c), the median is 2: four entries out of five are 2 or more, and three are 2 or less. To find the median of list (d), arrange it in increasing order:

$$-3, \quad -1, \quad 0, \quad 1, \quad 4, \quad 5, \quad 8$$

There are seven entries on this list: four are 1 or more, and four are 1 or less. So, 1 is the median.

4. THE ROOT-MEAN-SQUARE

The next main topic in the chapter is the *standard deviation*, which is used to measure spread. This section presents a mathematical preliminary, illustrated on the list

$$0, \quad 5, \quad -8, \quad 7, \quad -3$$

How big are these five numbers? The average is 0.2, but this is a poor measure of size. It only means that to a large extent, the positives cancel the negatives. The simplest way around the problem would be to wipe out the signs and then take the average. However, statisticians do something else: they apply the *root-mean-square* operation to the list. The phrase "root-mean-square" says how to do the arithmetic, provided you remember to read it backwards:

- SQUARE all the entries, getting rid of the signs.
- Take the MEAN (average) of the squares.
- Take the square ROOT of the mean.

This can be expressed as an equation, with root-mean-square abbreviated to r.m.s.

$$\text{r.m.s. size of a list} = \sqrt{\text{average of (entries}^2\text{)}}.$$

Example 1. Find the average, the average neglecting signs, and the r.m.s. size of the list 0, 5, −8, 7, −3.

Solution.

$$\text{average} = \frac{0+5-8+7-3}{5} = 0.2$$

$$\text{average neglecting signs} = \frac{0+5+8+7+3}{5} = 4.6$$

$$\text{r.m.s. size} = \sqrt{\frac{0^2+5^2+(-8)^2+7^2+(-3)^2}{5}} = \sqrt{29.4} \approx 5.4$$

The r.m.s. size is a little bigger than the average neglecting signs. It always turns out like that—except in the trivial case when all the entries are the same size. The root and the square do not cancel, due to the intervening operation of taking the mean. (The "≈" means "nearly equal:" some rounding has been done.)

There doesn't seem to be much to choose between the 5.4 and the 4.6 as a measure of the overall size for the list in the example. Statisticians use the r.m.s. size because it fits in better with the algebra that they have to do.[9] Whether this explanation is appealing or not, don't worry. Everyone is suspicious of the r.m.s. at first, and gets used to it very quickly.

Exercise Set C

1. (a) Find the average and the r.m.s. size of the numbers on the list
   1, −3, 5, −6, 3.

   (b) Do the same for the list −11, 8, −9, −3, 15.

2. Guess whether the r.m.s. size of each of the following lists of numbers is around 1, 10, or 20. No arithmetic is required.
   (a) 1, 5, −7, 8, −10, 9, −6, 5, 12, −17
   (b) 22, −18, −33, 7, 31, −12, 1, 24, −6, −16
   (c) 1, 2, 0, 0, −1, 0, 0, −3, 0, 1

3. (a) Find the r.m.s. size of the list 7, 7, 7, 7.

   (b) Repeat, for the list 7, −7, 7, −7.

4. Each of the numbers 103, 96, 101, 104 is almost 100 but is off by some amount. Find the r.m.s. size of the amounts off.

5. The list 103, 96, 101, 104 has an average. Find it. Each number in the list is off the average by some amount. Find the r.m.s. size of the amounts off.

6. A computer is programmed to predict test scores, compare them with actual scores, and find the r.m.s. size of the prediction errors. Glancing at the printout, you see the r.m.s. size of the prediction errors is 3.6, and the following results for the first ten students:

predicted score:	90	90	87	80	42	70	67	60	83	94
actual score:	88	70	81	85	63	77	66	49	71	69

   Does the printout seem reasonable, or is something wrong with the computer?

The answers to these exercises are on p. A49.

5. THE STANDARD DEVIATION

As the quote at the beginning of the chapter suggests, it is often helpful to think of the way a list of numbers spreads out around the average. This spread is usually measured by a quantity called the *standard deviation*, or SD. The SD measures the size of deviations from the average: it is a sort of average deviation. The program is to interpret the SD in the context of real data, and then see how to calculate it.

There were 2,696 women age 18 and over in the HANES5 sample. The average height of these women was about 63.5 inches, and the SD was close to 3 inches. The average tells us that most of the women were somewhere around 63.5 inches tall. But there were deviations from the average. Some of the women were taller than average, some shorter. How big were these deviations? That is where the SD comes in.

> The SD says how far away numbers on a list are from their average. Most entries on the list will be somewhere around one SD away from the average. Very few will be more than two or three SDs away.

The SD of 3 inches says that many of the women differed from the average height by 1 or 2 or 3 inches: 1 inch is a third of an SD, and 3 inches is an SD. Few women differed from the average height by more than 6 inches (two SDs).

There is a rule of thumb which makes this idea more quantitative, and which applies to many data sets.

> Roughly 68% of the entries on a list (two in three) are within one SD of the average, the other 32% are further away. Roughly 95% (19 in 20) are within two SDs of the average, the other 5% are further away. This is so for many lists, but not all.

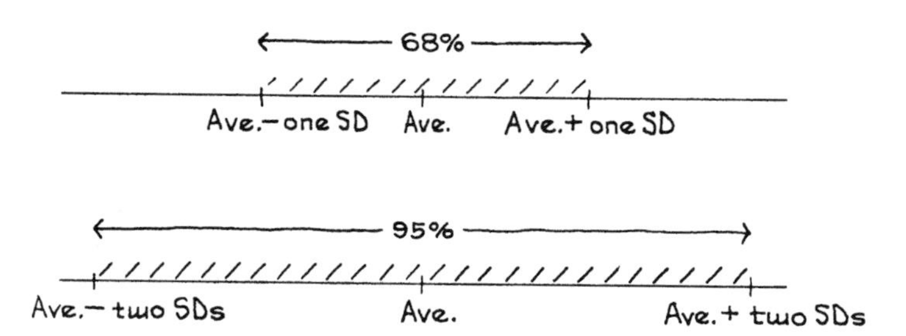

Figure 8 shows the histogram for the heights of women age 18 and over in HANES5. The average is marked by a vertical line, and the region within one SD of the average is shaded. This shaded area represents the women who differed from average height by one SD or less. The area is about 72%. About 72% of the women differed from the average height by one SD or less.

Figure 8. The SD and the histogram. Heights of 2,696 women age 18 and over in HANES5. The average of 63.5 inches is marked by a vertical line. The region within one SD of the average is shaded: 72% of the women differed from average by one SD (3 inches) or less.

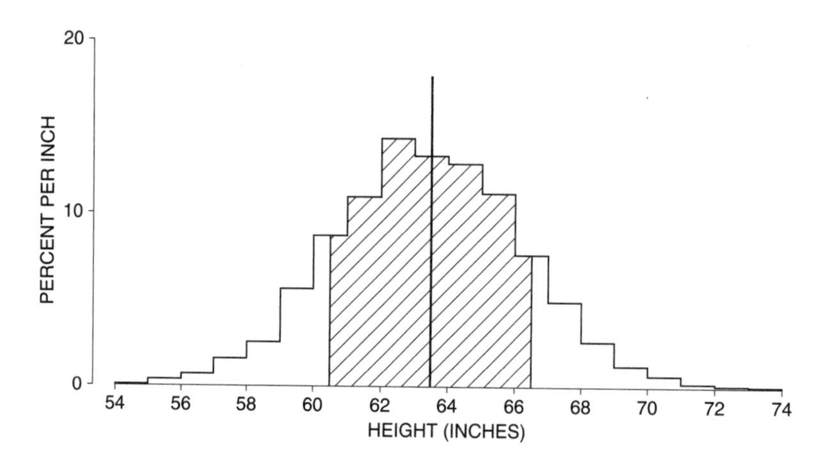

Figure 9 shows the same histogram. Now the area within two SDs of average is shaded. This shaded area represents the women who differed from average height by two SDs or less. The area is about 97%. About 97% of the women differed from the average height by two SDs or less.

Figure 9. The SD and the histogram. Heights of 2,696 women age 18 and over in HANES5. The average of 63.5 inches is marked by a vertical line. The region within two SDs of the average is shaded: 97% of the women differed from average by two SDs (6 inches) or less.

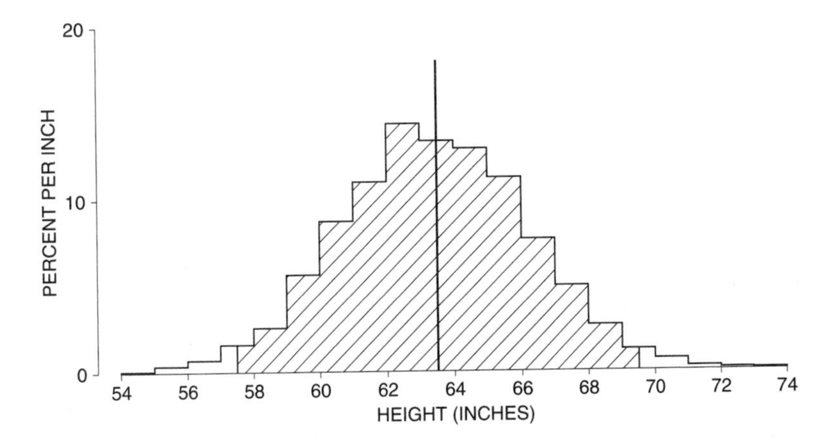

To sum up, about 72% of the women differed from average by one SD or less, and 97% differed from average by two SDs or less. There was only one woman in the sample who was more than three SDs away from the average, and none were more than four SDs away. For this data set, the 68%–95% rule works quite well. Where do the 68% and 95% come from? See chapter 5.[10]

About two-thirds of the HANES women differed from the average by less than one SD.

Exercise Set D

1. The Public Health Service found that for boys age 11 in HANES2, the average height was 146 cm and the SD was 8 cm. Fill in the blanks.

   (a) One boy was 170 cm tall. He was above average, by ______ SDs.
   (b) Another boy was 148 cm tall. He was above average, by ______ SDs.
   (c) A third boy was 1.5 SDs below average height. He was ______ cm tall.
   (d) If a boy was within 2.25 SDs of average height, the shortest he could have been is ______ cm and the tallest is ______ cm.

2. This continues exercise 1.

   (a) Here are the heights of four boys: 150 cm, 130 cm, 165 cm, 140 cm. Match the heights with the descriptions. A description may be used twice.

   unusually short   about average   unusually tall

   (b) About what percentage of boys age 11 in the study had heights between 138 cm and 154 cm? Between 130 and 162 cm?

3. Each of the following lists has an average of 50. For which one is the spread of the numbers around the average biggest? smallest?

   (i) 0, 20, 40, 50, 60, 80, 100
   (ii) 0, 48, 49, 50, 51, 52, 100
   (iii) 0, 1, 2, 50, 98, 99, 100

4. Each of the following lists has an average of 50. For each one, guess whether the SD is around 1, 2, or 10. (This does not require any arithmetic.)

   (a) 49, 51, 49, 51, 49, 51, 49, 51, 49, 51
   (b) 48, 52, 48, 52, 48, 52, 48, 52, 48, 52
   (c) 48, 51, 49, 52, 47, 52, 46, 51, 53, 51
   (d) 54, 49, 46, 49, 51, 53, 50, 50, 49, 49
   (e) 60, 36, 31, 50, 48, 50, 54, 56, 62, 53

5. The SD for the ages of the people in the HANES5 sample is around ______. Fill in the blank, using one of the options below. Explain briefly. (This survey was discussed in section 2; the age range was 0–85 years.)

   5 years   25 years   50 years

6. Below are sketches of histograms for three lists. Match the sketch with the description. Some descriptions will be left over. Give your reasoning in each case.

   (i) ave $\approx$ 3.5, SD $\approx$ 1
   (ii) ave $\approx$ 3.5, SD $\approx$ 0.5
   (iii) ave $\approx$ 3.5, SD $\approx$ 2
   (iv) ave $\approx$ 2.5, SD $\approx$ 1
   (v) ave $\approx$ 2.5, SD $\approx$ 0.5
   (vi) ave $\approx$ 4.5, SD $\approx$ 0.5

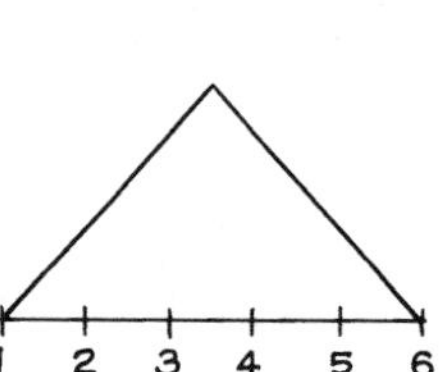

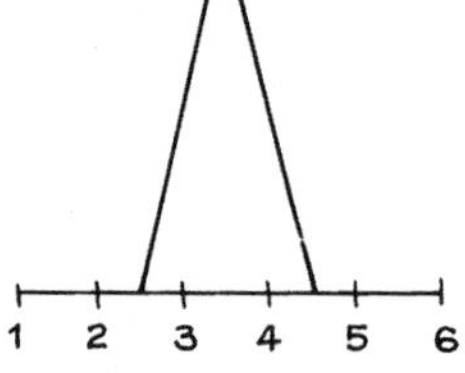

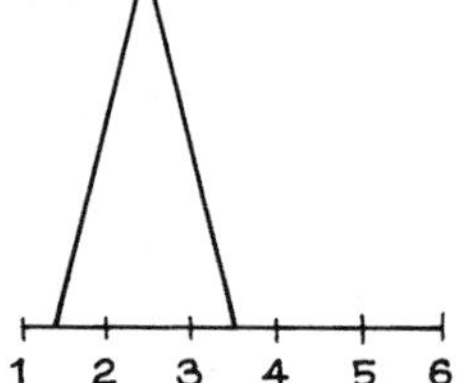

7. (Hypothetical). In a clinical trial, data collection usually starts at "baseline," when the subjects are recruited into the trial but before they are randomized to treatment or control. Data collection continues until the end of followup. Two clinical trials on prevention of heart attacks report baseline data on weight, shown below. In one of these trials, the randomization did not work. Which one, and why?

		Number of persons	*Average weight*	*SD*
(i)	Treatment	1,012	185 lb	25 lb
	Control	997	143 lb	26 lb
(ii)	Treatment	995	166 lb	27 lb
	Control	1,017	163 lb	25 lb

8. One investigator takes a sample of 100 men age 18–24 in a certain town. Another takes a sample of 1,000 such men.
   (a) Which investigator will get a bigger average for the heights of the men in his sample? or should the averages be about the same?
   (b) Which investigator will get a bigger SD for the heights of the men in his sample? or should the SDs be about the same?
   (c) Which investigator is likely to get the tallest of the sample men? or are the chances about the same for both investigators?
   (d) Which investigator is likely to get the shortest of the sample men? or are the chances about the same for both investigators?

9. The men in the HANES5 sample had an average height of 69 inches, and the SD was 3 inches. Tomorrow, one of these men will be chosen at random. You have to guess his height. What should you guess? You have about 1 chance in 3 to be off by more than ______. Fill in the blank. Options: 1/2 inch, 3 inches, 5 inches.

10. As in exercise 9, but tomorrow a whole series of men will be chosen at random. After each man appears, his actual height will be compared with your guess to see how far off you were. The r.m.s. size of the amounts off should be ______. Fill in the blank. (Hint: Look at the bottom of this page.)

The answers to these exercises are on pp. A49–50.

6. COMPUTING THE STANDARD DEVIATION

To find the standard deviation of a list, take the entries one at a time. Each deviates from the average by some amount, perhaps 0:

$$\text{deviation from average} = \text{entry} - \text{average}.$$

The SD is the r.m.s. size of these deviations. (Reminder: "r.m.s." means root-mean-square. See p. 66.)

SD = r.m.s. deviation from average.

Example 2. Find the SD of the list 20, 10, 15, 15.

Solution. The first step is to find the average:

$$\text{average} = \frac{20 + 10 + 15 + 15}{4} = 15.$$

The second step is to find the deviations from the average: just subtract the average from each entry. The deviations are

$$5 \quad -5 \quad 0 \quad 0$$

The last step is to find the r.m.s. size of the deviations:

$$\begin{aligned} \text{SD} &= \sqrt{\frac{5^2 + (-5)^2 + 0^2 + 0^2}{4}} \\ &= \sqrt{\frac{25 + 25 + 0 + 0}{4}} \\ &= \sqrt{\frac{50}{4}} = \sqrt{12.5} \approx 3.5 \end{aligned}$$

This completes the calculation.

The SD comes out in the same units as the data. For example, suppose heights are measured in inches. The intermediate squaring step in the procedure changes the units to inches squared, but the square root returns the answer to the original units.[11] Do not confuse the SD of a list with its r.m.s. size. The SD is the r.m.s., not of the original numbers on the list, but of their deviations from average.

Exercise Set E

1. Guess which of the following two lists has the larger SD. Check your guess by computing the SD for both lists.
   (i) 9, 9, 10, 10, 10, 12
   (ii) 7, 8, 10, 11, 11, 13

2. Someone is telling you how to calculate the SD of the list 1, 2, 3, 4, 5:

   The average is 3, so the deviations from average are

   $$-2 \quad -1 \quad 0 \quad 1 \quad 2$$

   Drop the signs. The average deviation is

   $$\frac{2 + 1 + 0 + 1 + 2}{5} = 1.2$$

   And that's the SD.

   Is this right? Answer yes or no, and explain briefly.

3. Someone is telling you how to calculate the SD of the list 1, 2, 3, 4, 5:

   The average is 3, so the deviations from average are

   $$-2 \quad -1 \quad 0 \quad 1 \quad 2$$

   The 0 doesn't count, so the r.m.s. deviation is

   $$\sqrt{\frac{4+1+1+4}{4}} = 1.6$$

   And that's the SD.

   Is this right? Answer yes or no, and explain briefly.

4. Three instructors are comparing scores on their finals; each had 99 students. In class A, one student got 1 point, another got 99 points, and the rest got 50 points. In class B, 49 students got a score of 1, one student got a score of 50, and 49 students got a score of 99. In class C, one student got a score of 1, one student got a score of 2, one student got a score of 3, and so forth, all the way through 99.
   (a) Which class had the biggest average? or are they the same?
   (b) Which class had the biggest SD? or are they the same?
   (c) Which class had the biggest range? or are they the same?

5. (a) For each list below, work out the average, the deviations from average, and the SD.
      (i) 1, 3, 4, 5, 7
      (ii) 6, 8, 9, 10, 12
   (b) How is list (ii) related to list (i)? How does this relationship carry over to the average? the deviations from the average? the SD?

6. Repeat exercise 5 for the following two lists:
   (i) 1, 3, 4, 5, 7
   (ii) 3, 9, 12, 15, 21

7. Repeat exercise 5 for the following two lists:
   (i) 5, −4, 3, −1, 7
   (ii) −5, 4, −3, 1, −7

8. (a) The Governor of California proposes to give all state employees a flat raise of $250 a month. What would this do to the average monthly salary of state employees? to the SD?
   (b) What would a 5% increase in the salaries, across the board, do to the average monthly salary? to the SD?

9. What is the r.m.s. size of the list 17, 17, 17, 17, 17? the SD?

10. For the list 107, 98, 93, 101, 104, which is smaller—the r.m.s. size or the SD? No arithmetic is needed.

11. Can the SD ever be negative?

12. For a list of positive numbers, can the SD ever be larger than the average?

The answers to these exercises are on pp. A50–51.

Technical note. There is an alternative way to compute the SD, which is more efficient in some cases:[12]

$$\text{SD} = \sqrt{\text{average of (entries}^2) - (\text{average of entries})^2}.$$

7. USING A STATISTICAL CALCULATOR

Most statistical calculators produce not the SD, but the slightly larger number SD^+. (The distinction between SD and SD^+ will be explained more carefully in section 6 of chapter 26.) To find out what your machine is doing, put in the list -1, 1. If the machine gives you 1, it's working out the SD. If it gives you 1.41 . . . , it's working out the SD^+. If you're getting the SD^+ and you want the SD, you have to multiply by a conversion factor. This depends on the number of entries on the list. With 10 entries, the conversion factor is $\sqrt{9/10}$. With 20 entries, it is $\sqrt{19/20}$. In general,

$$\text{SD} = \sqrt{\frac{\text{number of entries} - \text{one}}{\text{number of entries}}} \times \text{SD}^+$$

8. REVIEW EXERCISES

Review exercises may cover material from previous chapters.

1. (a) Find the average and SD of the list 41, 48, 50, 50, 54, 57.
   (b) Which numbers on the list are within 0.5 SDs of average? within 1.5 SDs of average?

2. (a) Both of the following lists have the same average of 50. Which one has the smaller SD, and why? No computations are necessary.
   (i) 50, 40, 60, 30, 70, 25, 75
   (ii) 50, 40, 60, 30, 70, 25, 75, 50, 50, 50
   (b) Repeat, for the following two lists.
   (i) 50, 40, 60, 30, 70, 25, 75
   (ii) 50, 40, 60, 30, 70, 25, 75, 99, 1

3. Here is a list of numbers:

   0.7 1.6 9.8 3.2 5.4 0.8 7.7 6.3 2.2 4.1
   8.1 6.5 3.7 0.6 6.9 9.9 8.8 3.1 5.7 9.1

   (a) Without doing any arithmetic, guess whether the average is around 1, 5, or 10.
   (b) Without doing any arithmetic, guess whether the SD is around 1, 3, or 6.

4. For persons age 25 and over in the U.S., would the average or the median be higher for income? for years of schooling completed?

5. For the men age 18–24 in HANES5, the average systolic blood pressure was 116 mm and the SD was 11 mm.[13] Say whether each of the following blood pressures is unusually high, unusually low, or about average:

   80 mm   115 mm   120 mm   210 mm

6. Below are sketches of histograms for three lists.
   (a) In scrambled order, the averages are 40, 50, 60. Match the histograms with the averages.
   (b) Match the histogram with the description:
      the median is less than the average
      the median is about equal to the average
      the median is bigger than the average
   (c) Is the SD of histogram (iii) around 5, 15, or 50?
   (d) True or false, and explain: the SD for histogram (i) is a lot smaller than that for histogram (iii).

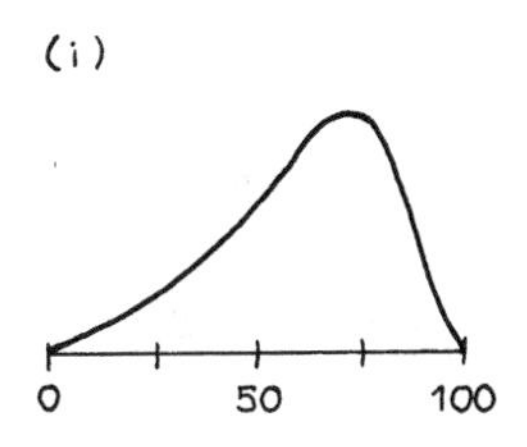

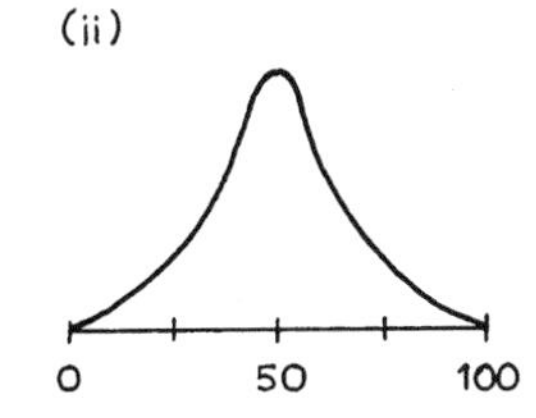

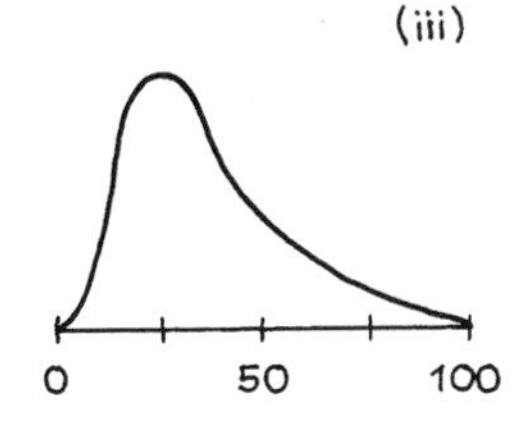

7. A study on college students found that the men had an average weight of about 66 kg and an SD of about 9 kg. The women had an average weight of about 55 kg and an SD of 9 kg.
   (a) Find the averages and SDs, in pounds (1 kg = 2.2 lb).
   (b) Just roughly, what percentage of the men weighed between 57 kg and 75 kg?
   (c) If you took the men and women together, would the SD of their weights be smaller than 9 kg, just about 9 kg, or bigger than 9 kg? Why?

8. In the HANES5 sample, the average height of the boys was 137 cm at age 9 and 151 cm at age 11. At age 11, the average height of all the children was 151 cm.[14]
   (a) On the average, are boys taller than girls at age 11?
   (b) Guess the average height of the 10-year-old boys.

9. An investigator has a computer file showing family incomes for 1,000 subjects in a certain study. These range from $5,800 a year to $98,600 a year. By accident, the highest income in the file gets changed to $986,000.
   (a) Does this affect the average? If so, by how much?
   (b) Does this affect the median? If so, by how much?

10. Incoming students at a certain law school have an average LSAT (Law School Aptitude Test) score of 163 and an SD of 8. Tomorrow, one of these students

will be picked at random. You have to guess the score now; the guess will be compared with the actual score, to see how far off it is. Each point off will cost a dollar. (For example, if the guess is 158 and the score is really 151, you will have to pay $7.)

(a) Is the best guess 150, 163, or 170?
(b) You have about 1 chance in 3 to lose more than ______. Fill in the blank. Options: $1, $8, $20.

(LSAT scores range from 120 to 180; the average across all test-takers is about 150 and the SD is about 9. The test is re-normed from time to time, the data are for 2005.)

11. As in exercise 10, but a whole series of students are chosen. The r.m.s. size of your losses should be around ______. Fill in the blank.

12. Many observers think there is a permanent underclass in American society—most of those in poverty typically remain poor from year to year. Over the period 1970–2000, the percentage of the American population in poverty each year has been remarkably stable, at 12% or so. Income figures for each year were taken from the March Current Population Survey of that year; the cutoff for poverty was based on official government definitions.[15]

To what extent do these data support the theory of the permanent underclass? Discuss briefly.

9. SUMMARY

1. A typical list of numbers can be summarized by its *average* and *standard deviation* (SD).

2. Average of a list $= \dfrac{\text{sum of entries}}{\text{number of entries}}$.

3. The average locates the center of a histogram, in the sense that the histogram balances when supported at the average.

Drawing by Dana Fradon; © 1976 The New Yorker Magazine, Inc.

4. Half the area under a histogram lies to the left of the *median*, and half to the right. The median is another way to locate the center of a histogram.

5. The *r.m.s. size* of a list measures how big the entries are, neglecting signs.

6. r.m.s. size of a list $= \sqrt{\text{average of (entries}^2\text{)}}$.

7. The SD measures distance from the average. Each number on a list is off the average by some amount. The SD is a sort of average size for these amounts off. More technically, the SD is the r.m.s. size of the deviations from the average.

8. Roughly 68% of the entries on a list of numbers are within one SD of the average, and about 95% are within two SDs of the average. This is so for many lists, but not all.

9. If a study draws conclusions about the effects of age, find out whether the data are cross-sectional or longitudinal.

5

The Normal Approximation for Data

1. THE NORMAL CURVE

The normal curve was discovered around 1720 by Abraham de Moivre, while he was developing the mathematics of chance. (His work will be discussed again in parts IV and V.) Around 1870, the Belgian mathematician Adolph Quetelet had the idea of using the curve as an ideal histogram, to which histograms for data could be compared.

The normal curve has a formidable-looking equation:

$$y = \frac{100\%}{\sqrt{2\pi}} e^{-x^2/2}, \quad \text{where } e = 2.71828\ldots.$$

This equation involves three of the most famous numbers in the history of mathematics: $\sqrt{2}$, π, and e. This is just to show off a little. You will find it is easy to work with the normal curve through diagrams and tables, without ever using the equation. A graph of the curve is shown in figure 1.

Figure 1. The normal curve.

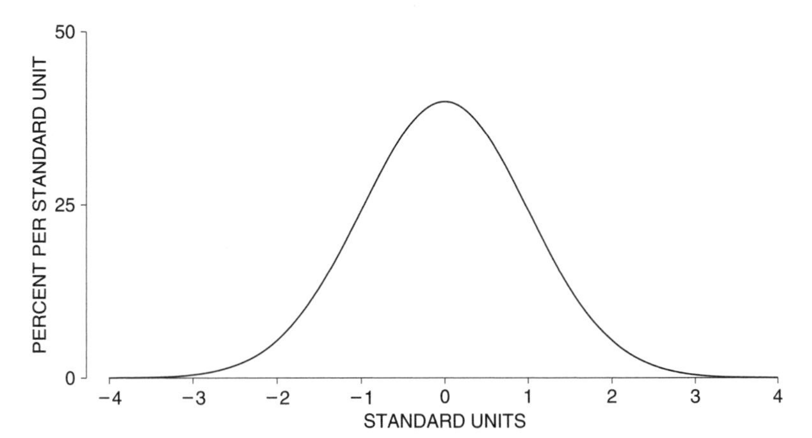

Several features of this graph will be important. First, the graph is symmetric about 0: the part of the curve to the right of 0 is a mirror image of the part to the left. Next, the total area under the curve equals 100%. (Areas come out in percent, because the vertical axis uses the density scale.) Finally, the curve is always above the horizontal axis. It appears to stop between 3 and 4, but that's only because the curve gets so low there. Only about 6/100,000 of the area is outside the interval from −4 to 4.

It will be helpful to find areas under the normal curve between specified values. For instance,

- the area under the normal curve between −1 and +1 is about 68%;
- the area under the normal curve between −2 and +2 is about 95%;
- the area under the normal curve between −3 and +3 is about 99.7%.

Finding these areas is a matter of looking things up in a table, or pushing a button on the right kind of calculator; the table will be explained in section 2.

Many histograms for data are similar in shape to the normal curve, provided they are drawn to the same scale. Making the horizontal scales match up involves *standard units*.[1]

> A value is converted to standard units by seeing how many SDs it is above or below the average.

Values above the average are given a plus sign; values below the average get a minus sign. The horizontal axis of figure 1 is in standard units.

For instance, take the women age 18 and over in the HANES5 sample. Their average height was 63.5 inches; the SD was 3 inches. One of these women was 69.5 inches tall. What was her height in standard units? Our subject was 6 inches taller than average, and 6 inches is 2 SDs. In standard units, her height was +2.

Example 1. For women age 18 and over in the HANES5 sample—

(a) Convert the following to standard units:
(i) 66.5 inches (ii) 57.5 inches (iii) 64 inches (iv) 63.5 inches

(b) Find the height which is −1.2 in standard units.

Solution. Part (a). For (i), 66.5 inches is 3 inches above the average. That is 1 SD above the average. In standard units, 66.5 inches is +1. For (ii), 57.5 inches is 6 inches below the average. That is 2 SDs below average. In standard units, 57.5 inches is −2. For (iii), 64 inches is 0.5 inches above average. That is $0.5/3 \approx 0.17$ SDs. The answer is 0.17. For (iv), 63.5 inches is the average. So, 63.5 inches is 0 SDs away from average. The answer is 0. (Reminder: "≈" means "nearly equal.")

Part (b). The height is 1.2 SDs below the average, and 1.2×3 inches = 3.6 inches. The height is

$$63.5 \text{ inches} - 3.6 \text{ inches} = 59.9 \text{ inches}.$$

That is the answer.

Standard units are used in figure 2. In this figure, the histogram for the heights of the women age 18 and over in the HANES5 sample is compared to the normal curve. The horizontal axis for the histogram is in inches; the horizontal axis for the normal curve is in standard units. The two match up as indicated in example 1. For instance, 66.5 inches is directly above +1, and 57.5 inches is directly above −2.

There are also two vertical axes in figure 2. The histogram is drawn relative to the inside one, in percent per inch. The normal curve is drawn relative to the outside one, in percent per standard unit. To see how the scales match up, take the top value on each axis: 60% per standard unit matches 20% per inch because there are 3 inches to the standard unit. Spreading 60% over an SD is the same as spreading 60% over 3 inches, and that comes to 20% per inch—

$$\begin{aligned} 60\% \text{ per standard unit} &= 60\% \text{ per 3 inches} \\ &= 60\% \div 3 \text{ inches} = 20\% \text{ per inch.} \end{aligned}$$

Similarly, 30% per standard unit matches 10% per inch. Any other pair of values can be dealt with in the same way.

The last chapter said that for many lists, roughly 68% of the entries are within one SD of average. This is the range

$$\text{average} - \text{SD} \quad \text{to} \quad \text{average} + \text{SD}.$$

Figure 2. A histogram for heights of women compared to the normal curve. The area under the histogram between 60.5 inches and 66.5 inches (the percentage of women within one SD of average with respect to height) is about equal to the area between −1 and +1 under the curve—68%.

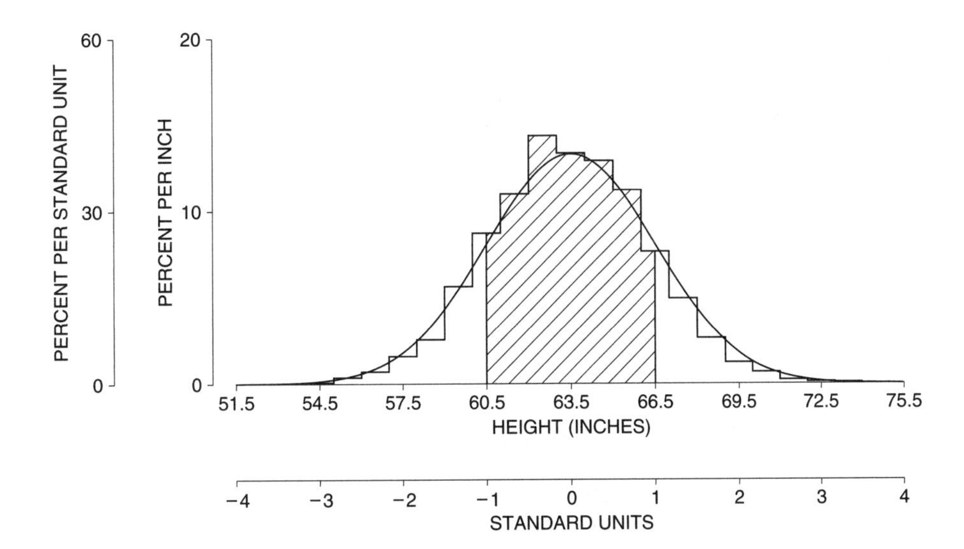

To see where the 68% comes from, look at figure 2. The percentage of women whose heights are within one SD of average equals the area under the histogram within one SD of average. This area is shaded in figure 2. The histogram follows the normal curve fairly well. Parts of it are higher than the curve, and parts of it are lower. But the highs balance out the lows. And the shaded area under the histogram is about the same as the area under the curve. The area under the normal curve between −1 and +1 is 68%. That is where the 68% comes from.

For many lists, roughly 95% of the entries are within 2 SDs of average. This is the range

$$\text{average} - 2\,\text{SDs} \quad \text{to} \quad \text{average} + 2\,\text{SDs}.$$

The reasoning is similar. If the histogram follows the normal curve, the area under the histogram will be about the same as the area under the curve. And the area under the curve between −2 and +2 is 95%:

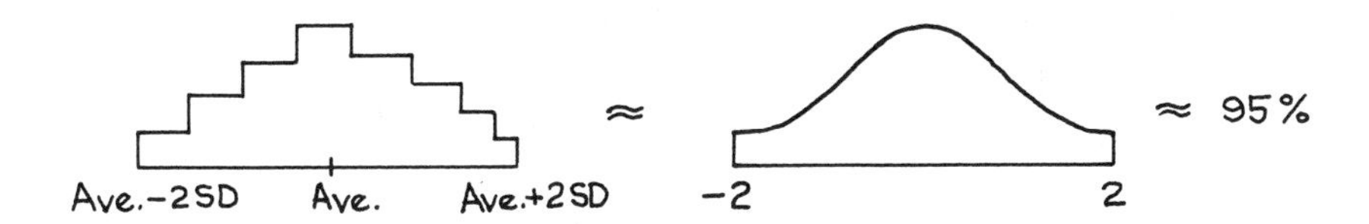

The normal curve can be used to estimate the percentage of entries in an interval, as follows.[2] First, convert the interval to standard units; second, find the corresponding area under the normal curve. The method for getting areas will be explained in section 2. Finally, section 3 will put the two steps together. The whole procedure is called the *normal approximation*. The approximation consists in replacing the original histogram by the normal curve before finding the area.

Exercise Set A

1. On a certain exam, the average of the scores was 50 and the SD was 10.
   (a) Convert each of the following scores to standard units: 60, 45, 75.
   (b) Find the scores which in standard units are: 0, +1.5, −2.8.

2. (a) Convert each entry on the following list to standard units (that is, using the average and SD of the list): 13, 9, 11, 7, 10.
   (b) Find the average and SD of the converted list.

The answers to these exercises are on p. A51.

2. FINDING AREAS UNDER THE NORMAL CURVE

At the end of the book, there is a table giving areas under the normal curve (p. A104). For example, to find the area under the normal curve between −1.20 and 1.20, go to 1.20 in the column marked z and read off the entry in the column marked *Area*. This is about 77%, so the area under the normal curve between −1.20 and 1.20 is about 77%.

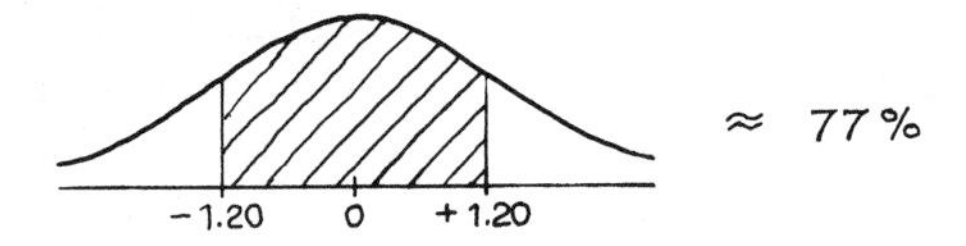

But you are also going to want to find other areas:

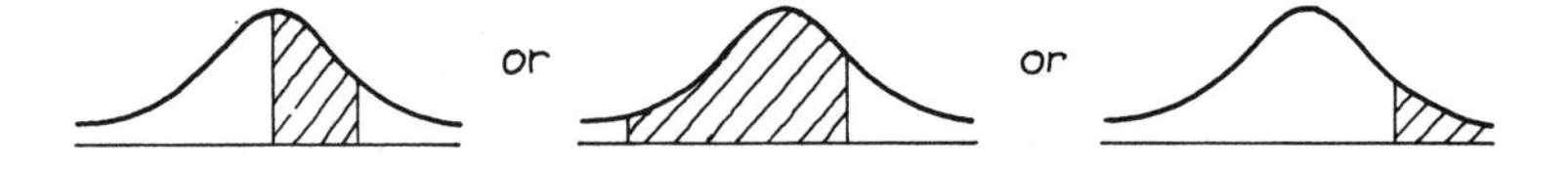

The method for finding such areas is indicated by example.

Example 2. Find the area between 0 and 1 under the normal curve.

Solution. First make a sketch of the normal curve, and then shade in the area to be found.

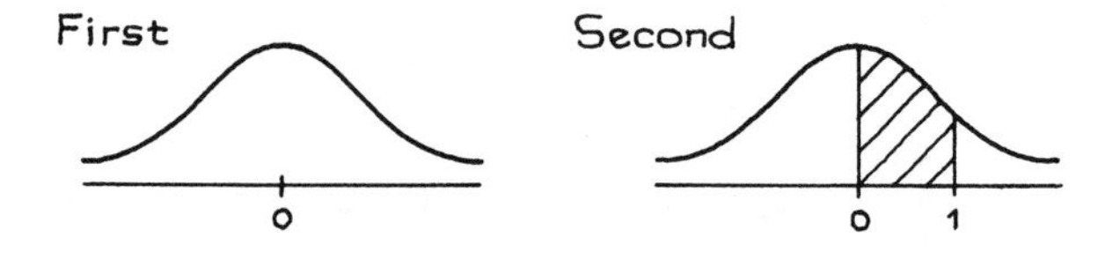

The table will give you the area between -1 and $+1$. This is about 68%. By symmetry, the area between 0 and 1 is half the area between -1 and $+1$, that is,

$$\frac{1}{2} \times 68\% = 34\%$$

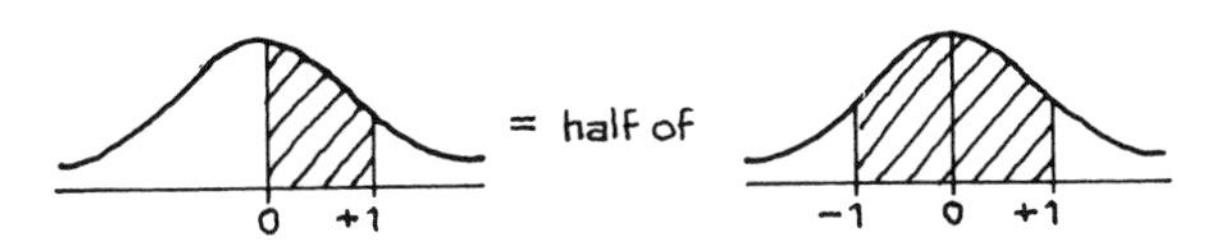

Example 3. Find the area between 0 and 2 under the normal curve.

Solution. This isn't double the area between 0 and 1 because the normal curve isn't a rectangle.

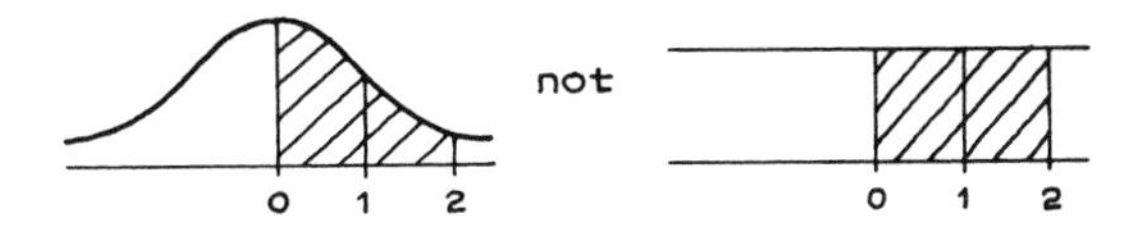

The procedure is the same as in example 2. The area between -2 and 2 can be found from the table. It is about 95%. The area between 0 and 2 is half that, by symmetry:

$$\frac{1}{2} \times 95\% \approx 48\%.$$

Example 4. Find the area between -2 and 1 under the normal curve.

Solution. The area between -2 and 1 can be broken down into two other areas—

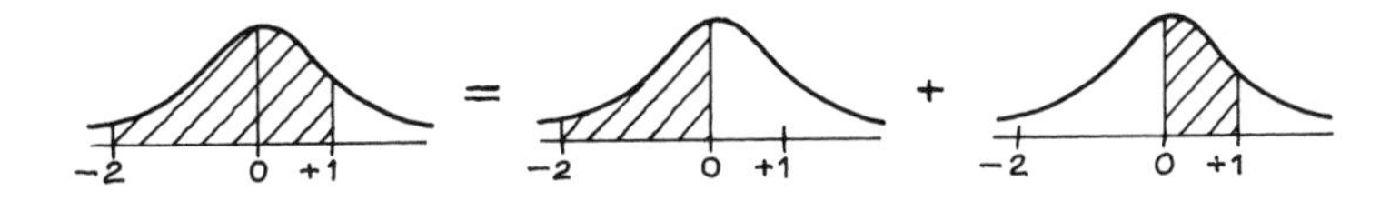

The area between -2 and 0 is the same as the area between 0 and 2, by symmetry, and is about 48% (example 3). The area between 0 and 1 is about 34% (example 2). The area between -2 and 1 is about

$$48\% + 34\% = 82\%.$$

Example 5. Find the area to the right of 1 under the normal curve.

Solution. The table gives the area between -1 and 1, which is 68%. The area outside this interval is 32%.

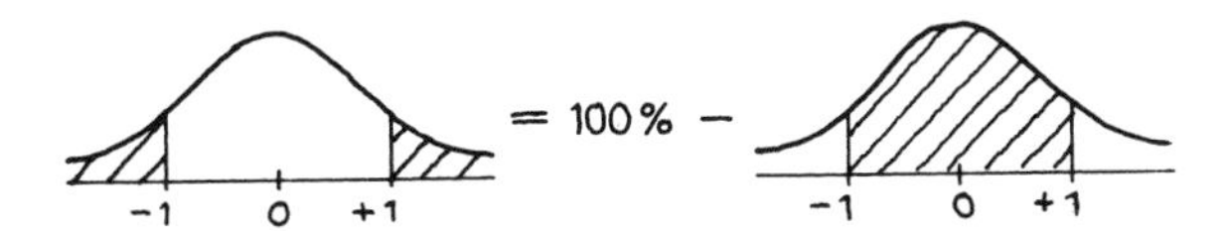

By symmetry, the area to the right of 1 is half this, or 16%.

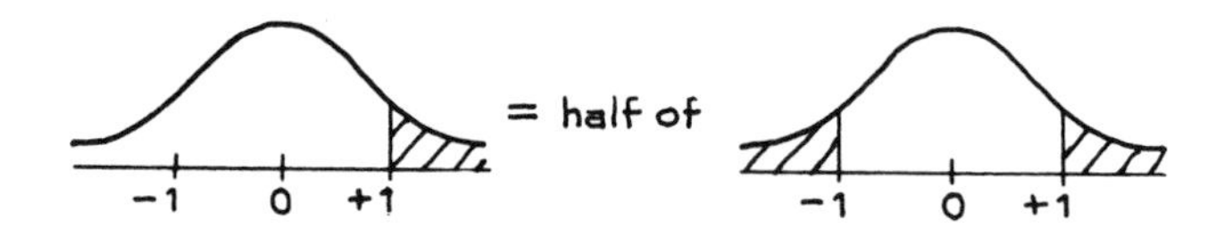

Example 6. Find the area to the left of 2 under the normal curve.

Solution. The area to the left of 2 is the sum of the area to the left of 0, and the area between 0 and 2.

The area to the left of 0 is half the total area, by symmetry:

$$\frac{1}{2} \times 100\% = 50\%$$

The area between 0 and 2 is about 48%. The sum is $50\% + 48\% = 98\%$.

Example 7. Find the area between 1 and 2 under the normal curve.

Solution.

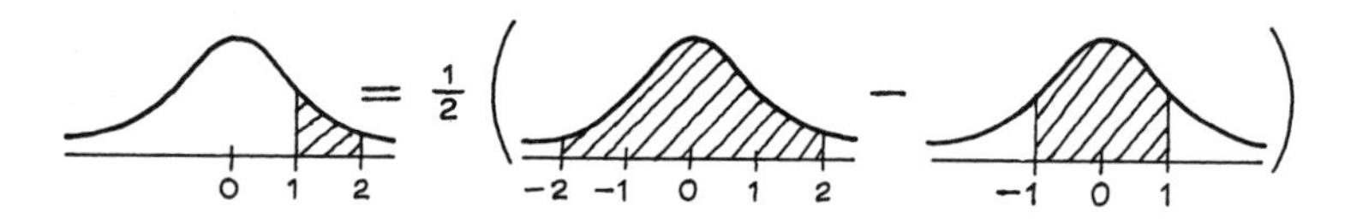

The area between −2 and 2 is about 95%; the area between −1 and 1 is about 68%. Half the difference is

$$\frac{1}{2} \times (95\% - 68\%) = \frac{1}{2} \times 27\% \approx 14\%.$$

There is no set procedure to use in solving this sort of problem. It is a matter of drawing pictures which relate the area you want to areas that can be read from the table.

Exercise Set B

1. Find the area under the normal curve—

   (a) to the right of 1.25
   (b) to the left of −0.40
   (c) to the left of 0.80
   (d) between 0.40 and 1.30
   (e) between −0.30 and 0.90
   (f) outside −1.5 to 1.5

2. Fill in the blanks:
   (a) The area between ± ______ under the normal curve equals 68%.
   (b) The area between ± ______ under the normal curve equals 75%.

3. The normal curve is sketched below; solve for z.

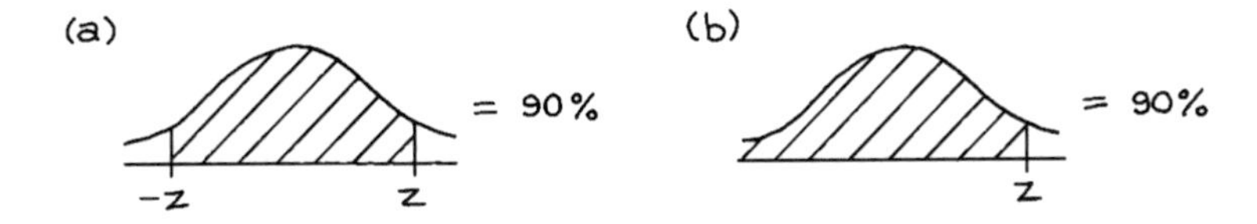

4. A certain curve (not the normal) is sketched below. The total area under it is 100%, and the area between 0 and 1 is 39%.
   (a) If possible, find the area to the right of 1.
   (b) If possible, find the area between 0 and 0.5.

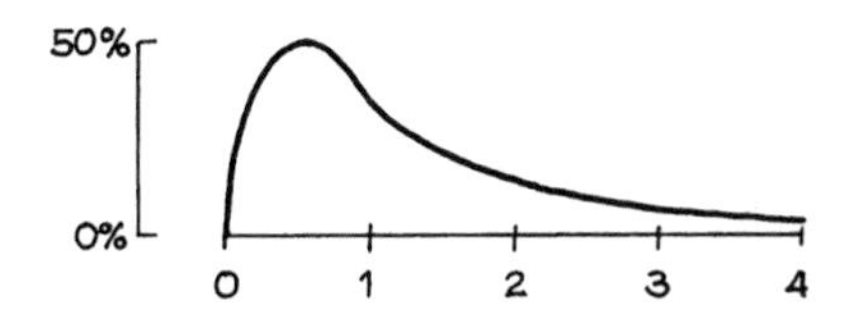

5. A certain curve (not the normal) is sketched below. It is symmetric around 0, and the total area under it is 100%. The area between −1 and 1 is 58%.
   (a) If possible, find the area between 0 and 1.
   (b) If possible, find the area to the right of 1.
   (c) If possible, find the area to the right of 2.

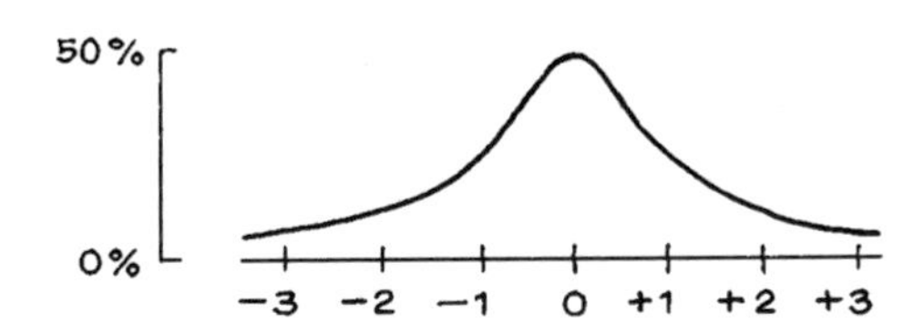

The answers to these exercises are on p. A51.

3. THE NORMAL APPROXIMATION FOR DATA

The method for the normal approximation will be explained here by example. The diagrams look so simple that you may not think they are worth drawing. However, it is easy to lose track of the area that is wanted. Please draw the diagrams.

Example 8. The heights of the men age 18 and over in HANES5 averaged 69 inches; the SD was 3 inches. Use the normal curve to estimate the percentage of these men with heights between 63 inches and 72 inches.

Solution. The percentage is given by the area under the height histogram, between 63 inches and 72 inches.

Step 1. Draw a number line and shade the interval.

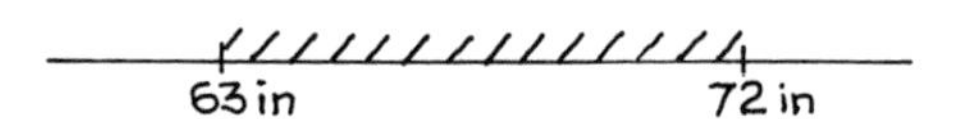

Step 2. Mark the average on the line and convert to standard units.

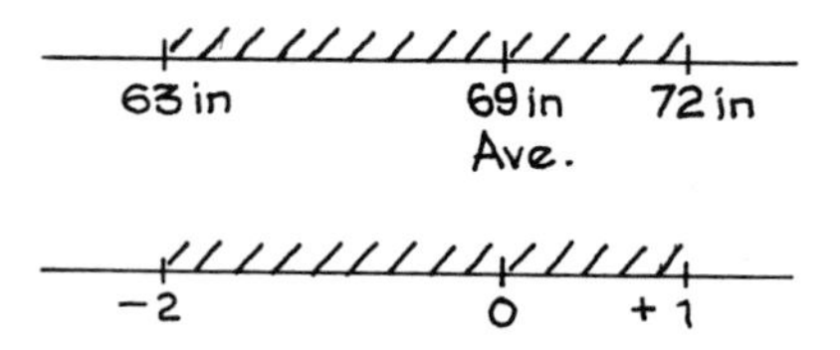

Step 3. Sketch in the normal curve, and find the area above the shaded standard-units interval obtained in step 2. The percentage is approximately equal to the shaded area, which is almost 82%.

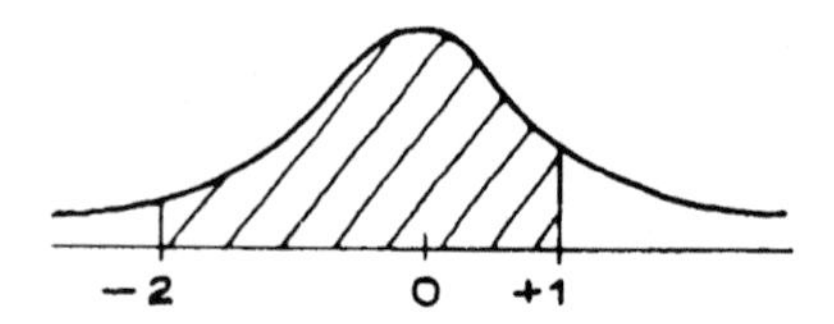

Using the normal curve, we estimate that about 82% of the heights were between 63 inches and 72 inches. This is only an approximation, but it is pretty good: 81% of the men were in that range. Figure 3 shows the approximation.

Figure 3. The normal approximation consists in replacing the original histogram by the normal curve before computing areas.

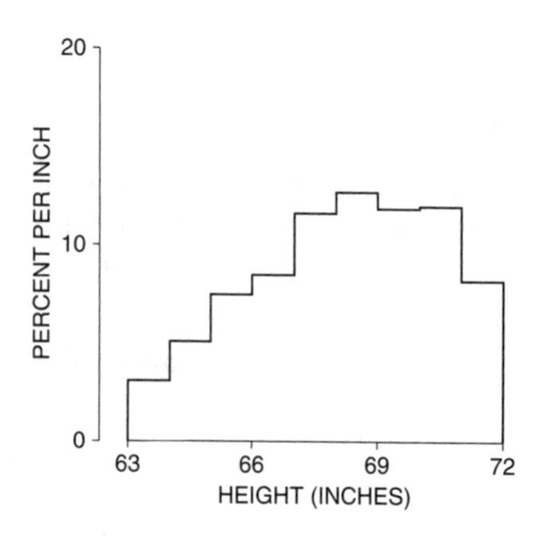

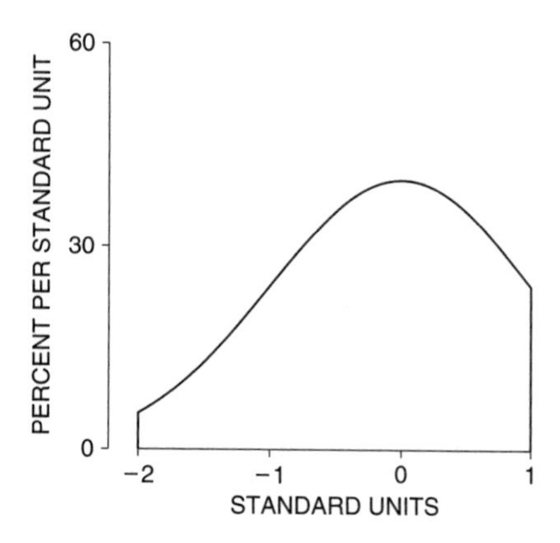

Example 9. The heights of the women age 18 and over in HANES5 averaged 63.5 inches; the SD was 3 inches. Use the normal curve to estimate the percentage with heights above 59 inches.

Solution. A height of 59 inches is 1.5 SDs below average:

$$(59 - 63.5)/3 = -1.5.$$

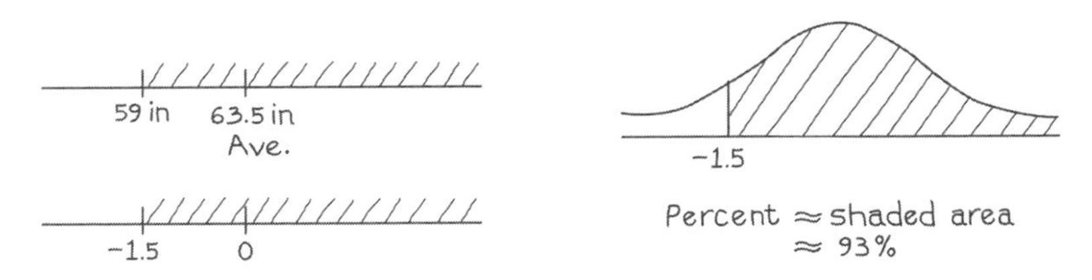

Using the normal curve, we estimate that 93% of the women were more than 59 inches in height. This estimate is about right: 96% of the women were taller than 59 inches.

It is a remarkable fact that many histograms follow the normal curve. (The story continues in part V.) For such histograms, the average and SD are good summary statistics. If a histogram follows the normal curve, it looks something like the sketch in figure 4. The average pins down the center, and the SD gives the spread. That is nearly all there is to say about the histogram—if its shape is like the normal curve. Many other histograms, however, do not follow the normal curve. In such cases, the average and SD are poor summary statistics. More about this in the next section.

Figure 4. The average and SD. By locating the center and measuring the spread around the center, the average and SD summarize a histogram which follows the normal curve.

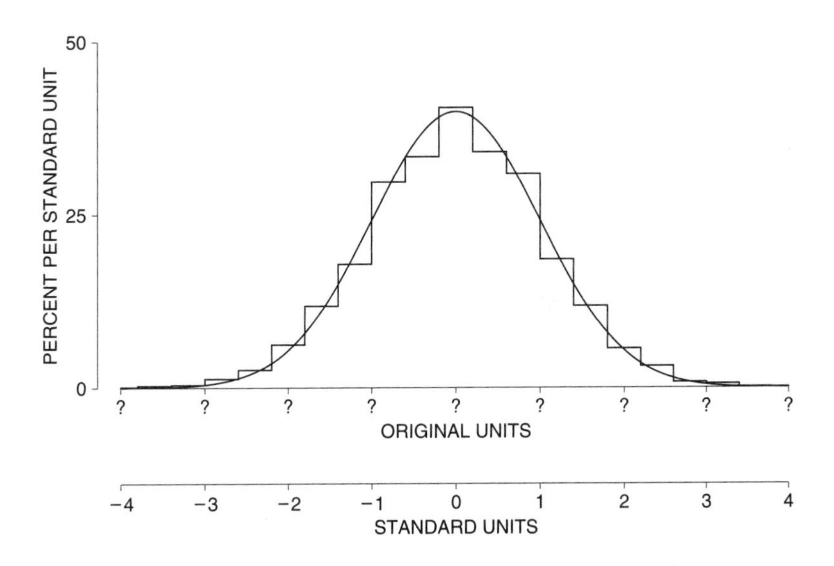

Exercise Set C

1. For the women age 18–24 in HANES2, the average height was about 64.3 inches; the SD was about 2.6 inches. Using the normal curve, estimate the percentage of women with heights—
   (a) below 66 inches.
   (b) between 60 inches and 66 inches.
   (c) above 72 inches.

2. In a law school class, the entering students averaged about 160 on the LSAT; the SD was about 8. The histogram of LSAT scores followed the normal curve reasonably well. (LSAT scores range from 120 to 180; among all test-takers, the average is around 150 and the SD is around 9.)
   (a) About what percentage of the class scored below 166?
   (b) One student was 0.5 SDs above average on the LSAT. About what percentage of the students had lower scores than he did?

3. In figure 2 (p. 81), the percentage of women with heights between 61 inches and 66 inches is exactly equal to the area between 61 inches and 66 inches under the ______ and approximately equal to the area under the ______. Options: normal curve, histogram.

The answers to these exercises are on pp. A51–52.

4. PERCENTILES

The average and SD can be used to summarize data following the normal curve. They are less satisfactory for other kinds of data. Take the distribution of family income in the U.S. in 2004, shown in figure 5.

Figure 5. Distribution of families by income: the U.S. in 2004.

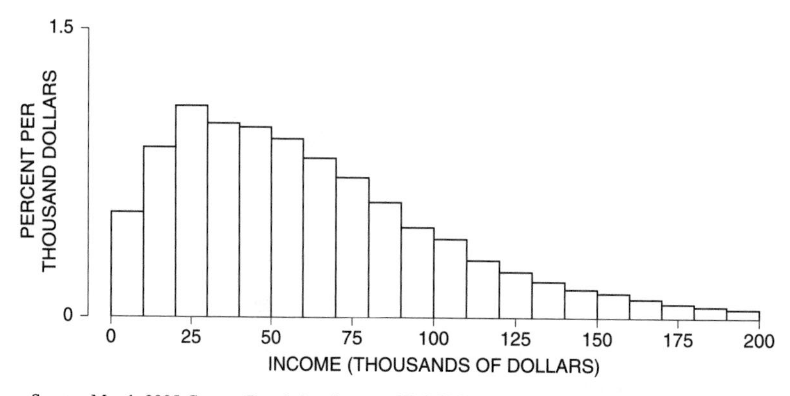

Source: March 2005 Current Population Survey; CD-ROM supplied by the Bureau of the Census. Primary families.

The average income for the families in figure 5 was about \$60,000; the SD was about \$40,000.[3] So the normal approximation suggests that about 7% of these families had negative incomes:

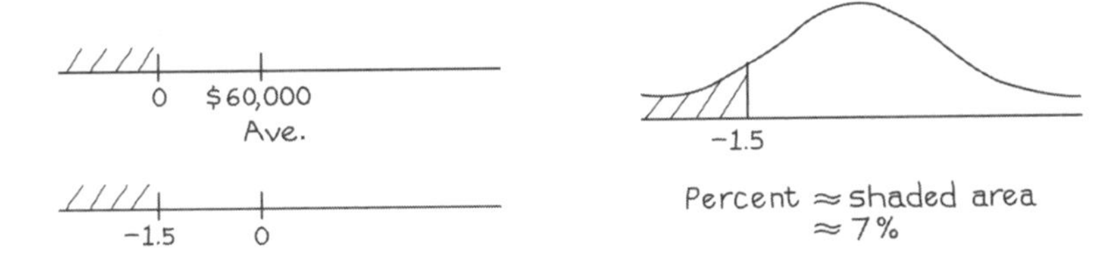

The reason for this blunder: the histogram in figure 5 does not follow the normal curve at all well, it has a long right-hand tail. To summarize such histograms, statisticians often use *percentiles* (table 1).

Table 1. Selected percentiles for family income in the U.S. in 2004.

1	$0
10	$15,000
25	$29,000
50	$54,000
75	$90,000
90	$135,000
99	$430,000

Source: March 2005 Current Population Survey; CD-ROM supplied by the Bureau of the Census. Primary families.

The 1st percentile of the income distribution was $0, meaning that about 1% of the families had incomes of $0 or less, and about 99% had incomes above that level. (Mainly, the families with no income were retired or not working for some other reason.) The 10th percentile was $15,000: about 10% of the families had incomes below that level, and 90% were above. The 50th percentile is just the median (chapter 4).

By definition, the *interquartile range* equals

$$\text{75th percentile} - \text{25th percentile}.$$

This is sometimes used as a measure of spread, when the distribution has a long tail. For table 1, the interquartile range is $61,000.

For reasons of their own, statisticians call de Moivre's curve "normal." This gives the impression that other curves are abnormal. Not so. Many histograms follow the normal curve very well, and many others—like the income histogram—do not. Later in the book, we will present a mathematical theory which helps explain when histograms should follow the normal curve.

Exercise Set D

1. Fill in the blanks, using the options below.
   (a) The percentage of families in table 1 with incomes below $90,000 was about ______.
   (b) About 25% of the families in table 1 had incomes below ______.
   (c) The percentage of families in table 1 with incomes between $15,000 and $125,000 was about ______.

   5%   10%   25%   60%   75%   95% $29,000 $90,000

2. In 2004, a family with an income of $9,000 was at the ______ th percentile of the income distribution, while a family that made $174,000 was at the ______ th percentile. Options: 5, 95.

3. Is the 25th percentile for the distribution of family income in 1973 around $7,000, $10,000, or $25,000? (See table 1 on p. 35.)

4. Skinfold thickness is used to measure body fat. A histogram for skinfold thickness is shown below; the units on the horizontal axis are millimeters (mm). The 25th percentile of skinfold thickness is ______ 25 mm. Fill in the blank, using one of the phrases below. Or can this be determined from the figure?

   quite a bit smaller than
   around
   quite a bit bigger than

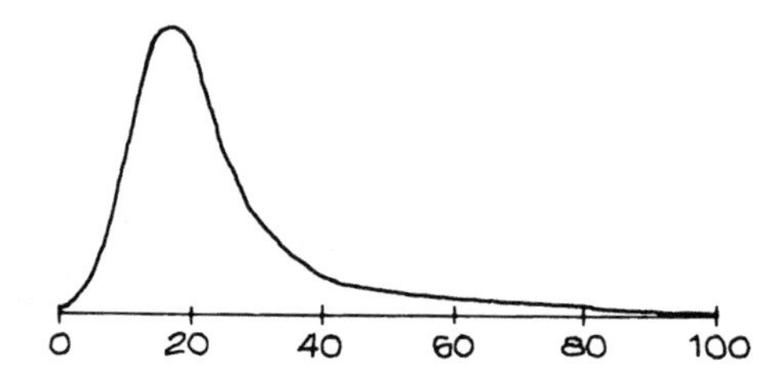

5. A histogram is sketched below.
   (a) How is it different from the normal curve?
   (b) Is the interquartile range around 15, 25, or 50?

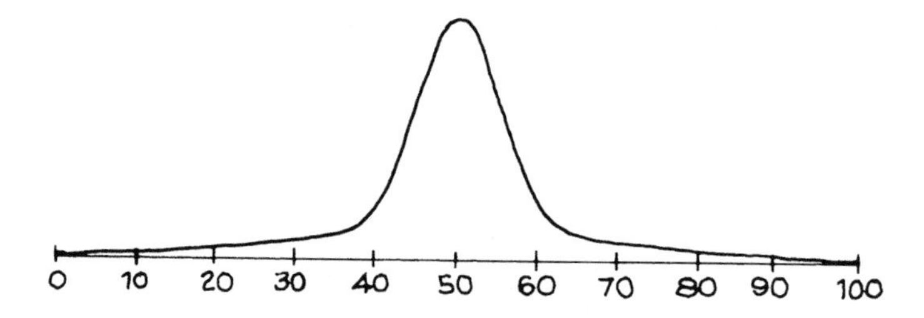

The answers to these exercises are on p. A52.

5. PERCENTILES AND THE NORMAL CURVE

When a histogram does follow the normal curve, the table can be used to estimate its percentiles. The method is indicated by example.

Example 10. Among all applicants to a certain university one year, the Math SAT scores averaged 535, the SD was 100, and the scores followed the normal curve. Estimate the 95th percentile of the score distribution.

Solution. This score is above average, by some number of SDs. We need to find that number, call it z. There is an equation for z:

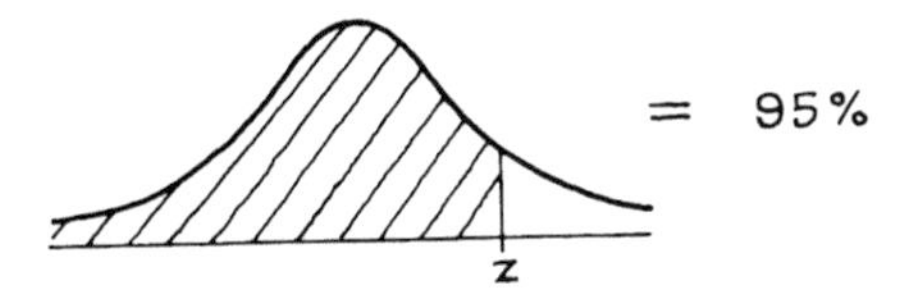

The normal table cannot be used directly, because it gives the area between $-z$ and z rather than the area to the left of z.

The area to the right of our z is 5%, so the area to the left of $-z$ is 5% too. Then the area between $-z$ and z must be $100\% - 5\% - 5\% = 90\%$.

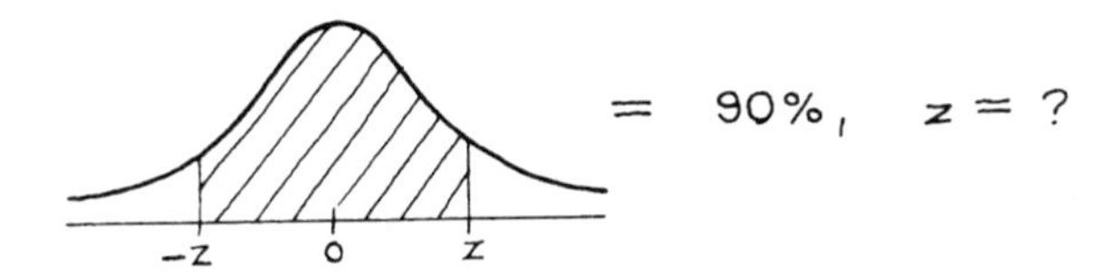

From the table, $z \approx 1.65$. You have to score 1.65 SDs above average to be in the 95th percentile of the Math SAT. Translated back to points, this score is above average by $1.65 \times 100 = 165$ points. The 95th percentile of the score distribution is $535 + 165 = 700$.

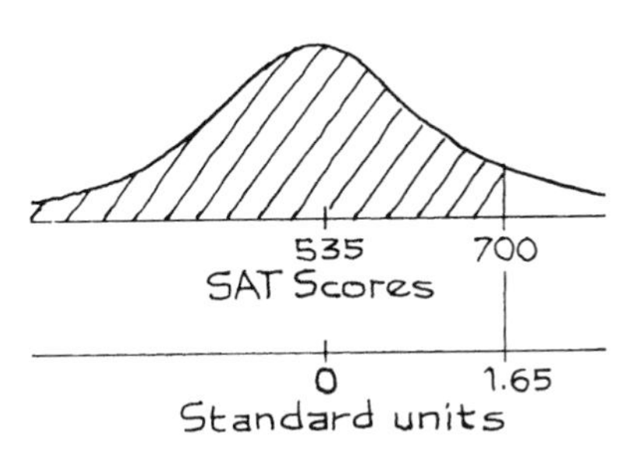

The terminology is a little confusing. A *percentile* is a score: in example 10, the 95th percentile is a score of 700. A *percentile rank*, however, is a percent: if you score 700, your percentile rank is 95%. There is even a third way to say the same thing: a score of 700 puts you at the 95th percentile of the score distribution.

Exercise Set E

1. At the university in example 10, one applicant scored 750 on the Math SAT. She was at the ______ percentile of the score distribution.

2. For the university in example 10, estimate the 80th percentile of the Math SAT scores.

3. For Berkeley freshmen, the average GPA (grade point average) is around 3.0; the SD is about 0.5. The histogram follows the normal curve. Estimate the 30th percentile of the GPA distribution.

The answers to these exercises are on p. A52.

6. CHANGE OF SCALE

If you add the same number to every entry on a list, that number just gets added to the average; the SD does not change. (The deviations from the average do not change, because the added constant just cancels.) Furthermore, if you multiply every entry on a list by the same number, the average and the SD simply get multiplied by that number. There is one exception: if that constant multiplier is negative, wipe out its sign before applying it to the SD. Exercises 5–8 on p. 73 illustrated these ideas.

Example 11.

(a) Find the average and SD of the list 1, 3, 4, 5, 7.

(b) Take the list in part (a), multiply each entry by 3 and then add 7, to get the list 10, 16, 19, 22, 28. Find the average and SD of this new list.

Solution. Part (a). The average is 4. So the deviations from average are $-3, -1, 0, 1, 3$. The SD is 2.

Part (b). The average is $3 \times 4 + 7 = 19$, the SD is $3 \times 2 = 6$. (Of course, you can work these numbers out directly.)

Example 12. Convert the following lists to standard units:

(a) 1, 3, 4, 5, 7

(b) 10, 16, 19, 22, 28

(These are the two lists in the previous example.)

Solution. Part (a). The average is 4, and the deviations from average are $-3, -1, 0, 1, 3$. The SD is 2. Divide by 2 to get the list in standard units:

$$-1.5 \quad -0.5 \quad 0 \quad 0.5 \quad 1.5$$

Part (b). Now the average is 19, and the deviations from average are $-9, -3, 0, 3, 9$. The SD is 6. Divide by 6 to get the list in standard units:

$$-1.5 \quad -0.5 \quad 0 \quad 0.5 \quad 1.5$$

The two lists are the same in standard units.

List (b) comes from list (a) by changing the scale: multiply by 3, add 7. The 7 washes out when computing the deviations from average. The 3 washes out when dividing by the SD—because the SD got multiplied by 3 along with all the deviations. That is why the lists are the same in standard units. To summarize:

(i) Adding the same number to every entry on a list adds that constant to the average; the SD does not change.

(ii) Multiplying every entry on a list by the same positive number multiplies the average and the SD by that constant.

(iii) These changes of scale do not change the standard units.

Conversion of temperature from Fahrenheit to Celsius is a practical example:

$$\text{C}^\circ = \frac{5}{9}(\text{F}^\circ - 32^\circ)$$

Statisticians call this a *change of scale*, because it is only the units that change. (What happens if you multiply all the numbers on a list by the same negative constant? In standard units, that just reverses all the signs.)

Exercise Set F

1. A group of people have an average temperature of 98.6 degrees Fahrenheit, with an SD of 0.3 degrees.
   (a) Translate these results into degrees Celsius.
   (b) Someone's temperature is 1.5 SDs above average on the Fahrenheit scale. Convert this temperature to standard units, for an investigator who is using the Celsius scale.

The answers to these exercises are on p. A52.

7. REVIEW EXERCISES

Review exercises may cover material from previous chapters.

1. The following list of test scores has an average of 50 and an SD of 10:

   39 41 47 58 65 37 37 49 56 59 62 36 48
   52 64 29 44 47 49 52 53 54 72 50 50

   (a) Use the normal approximation to estimate the number of scores within 1.25 SDs of the average.
   (b) How many scores really were within 1.25 SDs of the average?

2. You are looking at a computer printout of 100 test scores, which have been converted to standard units. The first 10 entries are

   −6.2 3.5 1.2 −0.13 4.3 −5.1 −7.2 −11.3 1.8 6.3

   Does the printout look reasonable, or is something wrong with the computer?

3. From the mid-1960s to the early 1990s, there was a slow but steady decline in SAT scores. For example, take the Verbal SAT. The average in 1967 was about 543; by 1994, the average was down to about 499. However, the SD stayed close to 110. The drop in averages has a large effect on the tails of the distribution.

   (a) Estimate the percentage of students scoring over 700 in 1967.
   (b) Estimate the percentage of students scoring over 700 in 1994.

   You may assume that the histograms follow the normal curve.

   Comments. SAT scores range from 200 to 800. It does not seem that the SAT was getting harder. Most of the decline in the 1960s is thought to result from changes in the population of students taking the test. The decline in the 1970s cannot be explained that way. From 1994 to 2005, scores generally increased. The test was re-normalized in 1996, which complicates the interpretation; the averages mentioned above were converted to the new scale.[4]

4. On the Math SAT, men have a distinct edge. In 2005, for instance, the men averaged about 538, and the women averaged about 504.

   (a) Estimate the percentage of men getting over 700 on this test in 2005.
   (b) Estimate the percentage of women getting over 700 on this test in 2005.

   You may assume (i) the histograms followed the normal curve, and (ii) both SDs were about 120.[4]

5. In HANES5, the men age 18 and over had an average height of 69 inches and an SD of 3 inches. The histogram is shown below, with a normal curve. The percentage of men with heights between 66 inches and 72 inches is exactly equal to the area between ___(a)___ and ___(b)___ under the ___(c)___. This percentage is approximately equal to the area between ___(d)___ and ___(e)___ under the ___(f)___. Fill in the blanks. For (a), (b), (d) and (e), your options are

   66 inches   72 inches   −1   +1

   For (c) and (f), your options are: normal curve, histogram

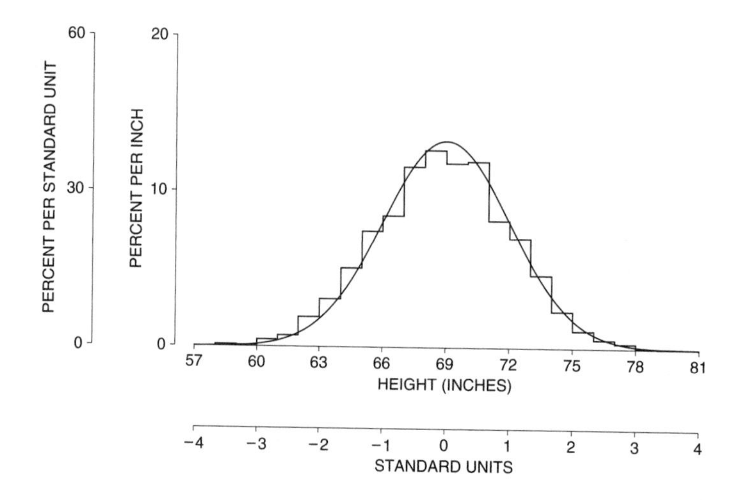

6. Among applicants to one law school, the average LSAT score was about 169, the SD was about 9, and the highest score was 178. Did the LSAT scores follow the normal curve?

7. Among freshmen at a certain university, scores on the Math SAT followed the normal curve, with an average of 550 and an SD of 100. Fill in the blanks; explain briefly.
   (a) A student who scored 400 on the Math SAT was at the ______ th percentile of the score distribution.
   (b) To be at the 75th percentile of the distribution, a student needed a score of about ______ points on the Math SAT.

8. True or false, and explain briefly—
   (a) If you add 7 to each entry on a list, that adds 7 to the average.
   (b) If you add 7 to each entry on a list, that adds 7 to the SD.
   (c) If you double each entry on a list, that doubles the average.
   (d) If you double each entry on a list, that doubles the SD.
   (e) If you change the sign of each entry on a list, that changes the sign of the average.
   (f) If you change the sign of each entry on a list, that changes the sign of the SD.

9. Which of the following are true? false? Explain or give examples.
   (a) The median and the average of any list are always close together.
   (b) Half of a list is always below average.
   (c) With a large, representative sample, the histogram is bound to follow the normal curve quite closely.
   (d) If two lists of numbers have exactly the same average of 50 and the same SD of 10, then the percentage of entries between 40 and 60 must be exactly the same for both lists.

10. For women age 25–34 with full time jobs, the average income in 2004 was \$32,000. The SD was \$26,000, and 1/4 of 1% had incomes above \$150,000. Was the percentage with incomes in the range from \$32,000 to \$150,000 about 40%, 50%, or 60%? Choose one option and explain briefly.[5]

11. One term, about 700 Statistics 2 students at the University of California, Berkeley, were asked how many college mathematics courses they had taken, other than Statistics 2. The average number of courses was about 1.1; the SD was about 1.5. Would the histogram for the data look like (i), (ii), or (iii)? Why?

12. In 2005, the average score on the Math SAT was about 520. However, among students who took a subject-matter test, the average score on the Math SAT was about 624.[6] What accounts for the difference?

8. SUMMARY

1. The *normal curve* is symmetric about 0, and the total area under it is 100%.

2. *Standard units* say how many SDs a value is, above (+) or below (−) the average.

3. Many histograms have roughly the same shape as the normal curve.

4. If a list of numbers follows the normal curve, the percentage of entries falling in a given interval can be estimated by converting the interval to standard units, and then finding the corresponding area under the normal curve. This procedure is called the *normal approximation*.

5. A histogram which follows the normal curve can be reconstructed fairly well from its average and SD. In such cases, the average and SD are good summary statistics.

6. All histograms, whether or not they follow the normal curve, can be summarized using *percentiles*.

7. If you add the same number to every entry on a list, that constant just gets added to the average; the SD does not change. If you multiply every entry on a list by the same positive number, the average and the SD just get multiplied by that constant. (If the constant is negative, wipe out the sign before multiplying the SD.)

"LOOK, FRED! THIS SEEMS TO BE THE SAME THING, SUMMARIZED."

6

Measurement Error

Jesus: I am come to bear witness unto the truth.
Pilate: What is truth?

1. INTRODUCTION

In an ideal world, if the same thing is measured several times, the same result would be obtained each time. In practice, there are differences. Each result is thrown off by chance error, and the error changes from measurement to measurement. One of the earliest scientists to deal with this problem was Tycho Brahé (1546–1601), the Danish astronomer. But it was probably noticed first in the market place, as merchants weighed out spices and measured off lengths of silk.

There are several questions about chance errors. Where do they come from? How big are they likely to be? How much is likely to cancel out in the average? The first question has a short answer: in most cases, nobody knows. The second question will be dealt with later in this chapter, and the third will be answered in part VII.

2. CHANCE ERROR

This section will discuss chance errors in precision weighing done at the National Bureau of Standards.[1] First, a brief explanation of standard weights. Stores weigh merchandise on scales. The scales are checked periodically by county weights-and-measures officials, using county standard weights. The county standards too must be *calibrated* (checked against external standards) periodically. This is done at the state level. And state standards are calibrated against national standards, by the National Bureau of Standards in Washington, D.C.

This chain of comparisons ends at the International Prototype Kilogram (for short, The Kilogram), a platinum-iridium weight held at the International Bureau of Weights and Measures near Paris. By international treaty—The Treaty of the Meter, 1875—"one kilogram" was defined to be the weight of this object under standard conditions.[2] All other weights are determined relative to The Kilogram. For instance, something weighs a pound if it weighs just a bit less than half as much as The Kilogram. More precisely,

$$\text{The Pound} = 0.4539237 \text{ of The Kilogram.}$$

To say that a package of butter weighs a pound means that it has been connected by some long and complicated series of comparisons to The Kilogram in Paris, and weighs 0.4539237 times as much.

Each country that signed the Treaty of the Meter got a national prototype kilogram, whose exact weight had been determined as accurately as possible relative to The Kilogram. These prototypes were distributed by lot, and the United States got Kilogram #20. The values of all the U.S. national standards are determined relative to K_{20}.

In the U.S., accuracy in weighing at the supermarket ultimately depends on the accuracy of the calibration work done at the Bureau. One basic issue is reproducibility: if a measurement is repeated, how much will it change? The Bureau gets at this issue by making repeated measurements on some of their own weights. We will discuss the results for one such weight, called NB 10 because it is owned by the National Bureau and its nominal value is 10 grams—the weight of two nickels. (A package of butter has a "nominal" weight of 1 pound; the exact weight will be a little different—chance error in butter; similarly, the people who manufactured NB 10 tried to make it weigh 10 grams, and missed by a little.)

NB 10 was acquired by the Bureau around 1940, and they've weighed it many times since then. We are going to look at 100 of these weighings. These measurements were made in the same room, on the same apparatus, by the same technicians. Every effort was made to follow the same procedure each time. All the factors known to affect the results, like air pressure or temperature, were kept as constant as possible.

The first five weighings in the series were

9.999591 grams
9.999600 grams
9.999594 grams
9.999601 grams
9.999598 grams

At first glance, these numbers all seem to be the same. But look more closely. It is only the first 4 digits that are solid, at 9.999. The last 3 digits are shaky, they change from measurement to measurement. This is chance error at work.[3]

NB 10 does weigh a bit less than 10 grams. Instead of writing out the 9.999 each time, the Bureau just reports the amount by which NB10 fell short of 10 grams. For the first weighing, this was

0.000409 grams.

The 0's are distracting, so the Bureau works not in grams but in micrograms: a *microgram* is the millionth part of a gram. In these units, the first five measurements on NB 10 are easier to read. They are

409 400 406 399 402.

All 100 measurements are shown in table 1. Look down the table. You can see that the results run around 400 micrograms, but some are more, some are less. The smallest is 375 micrograms (#94); the largest is 437 micrograms (#86). And there is a lot of variability in between. To keep things in perspective, one microgram is the weight of a large speck of dust; 400 micrograms is the weight of a grain or two of salt. This really is precision weighing!

Even so, the different measurements can't all be right. The exact amount by which NB 10 falls short of 10 grams is very unlikely to equal the first number

Table 1. One hundred measurements on NB 10. Almer and Jones, National Bureau of Standards. Units are micrograms below 10 grams.

No.	*Result*	*No.*	*Result*	*No.*	*Result*	*No.*	*Result*
1	409	26	397	51	404	76	404
2	400	27	407	52	406	77	401
3	406	28	401	53	407	78	404
4	399	29	399	54	405	79	408
5	402	30	401	55	411	80	406
6	406	31	403	56	410	81	408
7	401	32	400	57	410	82	406
8	403	33	410	58	410	83	401
9	401	34	401	59	401	84	412
10	403	35	407	60	402	85	393
11	398	36	423	61	404	86	437
12	403	37	406	62	405	87	418
13	407	38	406	63	392	88	415
14	402	39	402	64	407	89	404
15	401	40	405	65	406	90	401
16	399	41	405	66	404	91	401
17	400	42	409	67	403	92	407
18	401	43	399	68	408	93	412
19	405	44	402	69	404	94	375
20	402	45	407	70	407	95	409
21	408	46	406	71	412	96	406
22	399	47	413	72	406	97	398
23	399	48	409	73	409	98	406
24	402	49	404	74	400	99	403
25	399	50	402	75	408	100	404

in the table, or the second, or any of them. Despite the effort of making these 100 measurements, the exact weight of NB 10 remains unknown and perhaps unknowable.

Why does the Bureau bother to weigh the same weight over and over again? One of the objectives is quality control. If the measurements on NB 10 jump from 400 micrograms below 10 grams to 500 micrograms above 10 grams, something has gone wrong and needs to be fixed. (For this reason, NB 10 is called a *check weight*; it is used to check the weighing process.)

To see another use for repeated measurements, imagine that a scientific laboratory sends a nominal 10-gram weight off to the Bureau for calibration. One measurement can't be the last word, because of chance error. The lab will want to know how big this chance error is likely to be. There is a direct way to find out: send the same weight back for a second weighing. If the two results differ by a few micrograms, the chance error in each one is only likely to be a few micrograms in size. On the other hand, if the two results differ by several hundred micrograms, each measurement is likely to be off by several hundred micrograms. The repeated weighings on NB 10 save everybody the bother of sending in weights more than once. There is no need to ask for replicate calibrations because the Bureau has already done the work.

> No matter how carefully it was made, a measurement could have come out a bit differently. If the measurement is repeated, it will come out a bit differently. By how much? The best way to answer this question is to replicate the measurement.

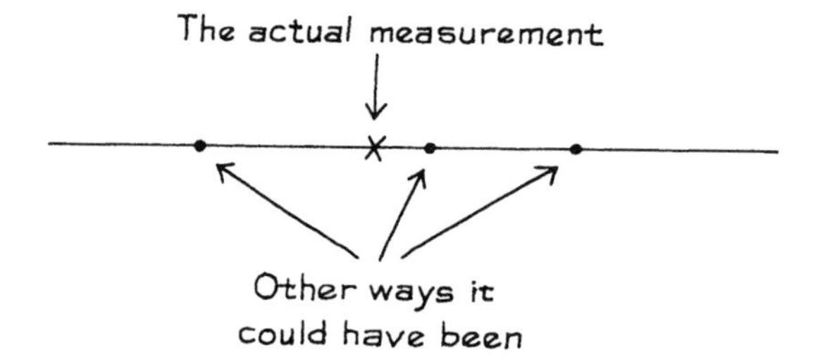

The SD of the 100 measurements in table 1 is just over 6 micrograms. The SD tells you that each measurement on NB 10 was thrown off by a chance error something like 6 micrograms in size. Chance errors around 2 or 5 or 10 micrograms in size were fairly common. Chance errors around 50 or 100 micrograms must have been extremely rare. The conclusion: in calibrating other 10-gram weights by the same process, the chance errors should be something like 6 micrograms in size.

> The SD of a series of repeated measurements estimates the likely size of the chance error in a single measurement.

There is an equation which helps explain the idea:

$$\text{individual measurement} = \text{exact value} + \text{chance error}.$$

The chance error throws each individual measurement off the exact value by an amount which changes from measurement to measurement. The variability in repeated measurements reflects the variability in the chance errors, and both are gauged by the SD of the data. Mathematically, the SD of the chance errors must equal the SD of the measurements: adding the exact value is just a change of scale (pp. 92–93).

To go at this more slowly, the average of all 100 measurements reported in table 1 was 405 micrograms below 10 grams. This is very likely to be close to the exact weight of NB 10. The first measurement in table 1 differed from the average by 4 micrograms:

$$409 - 405 = 4.$$

This measurement must have differed from the exact weight by nearly 4 micrograms. The chance error was nearly 4 micrograms. The second measurement was below average by 5 micrograms; the chance error must have been around −5 micrograms. The typical deviation from average was around 6 micrograms in size, because the SD was 6 micrograms. Therefore, the typical chance error must have been something like 6 micrograms in size.

Of course, the average of all 100 measurements (405 micrograms below 10 grams) is itself only an estimate for the exact weight of NB10. This estimate too must be off by some infinitesimal chance error. Chapter 24 will explain how to figure the likely size of the chance error in this sort of average.

Figure 1. The U.S. national prototype kilogram, K_{20}.

Source: National Institute of Science and Technology.

3. OUTLIERS

How well do the measurements reported in table 1 fit the normal curve? The answer is, not very well. Measurement #36 is 3 SDs away from the average; #86 and #94 are 5 SDs away—minor miracles. Such extreme measurements are called *outliers*. They do not result from blunders. As far as the Bureau could tell, nothing went wrong when these 3 observations were made. However, the 3 outliers inflate the SD. Consequently, the percentage of results falling closer to the average than one SD is 86%—quite a bit larger than the 68% predicted by the normal curve.

When the 3 outliers are discarded, the remaining 97 measurements average out to 404 micrograms below 10 grams, with an SD of only 4 micrograms. The average doesn't change much, but the SD drops by about 30%. As figure 2 shows,

Figure 2. Outliers. The top panel shows the histogram for all 100 measurements on NB 10; a normal curve is drawn for comparison. The curve does not fit well. The second panel shows the data with 3 outliers removed. The curve fits better. Most of the data follow the normal curve, but a few measurements are much further away from average than the curve suggests.

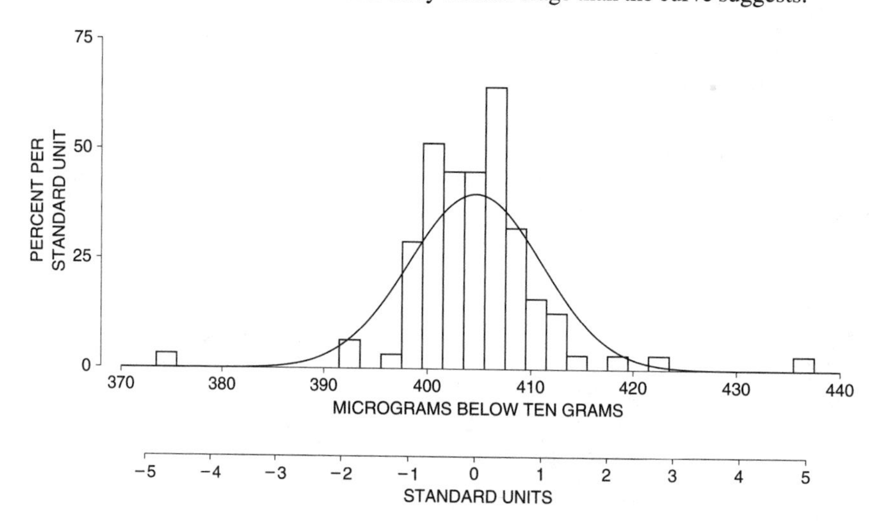

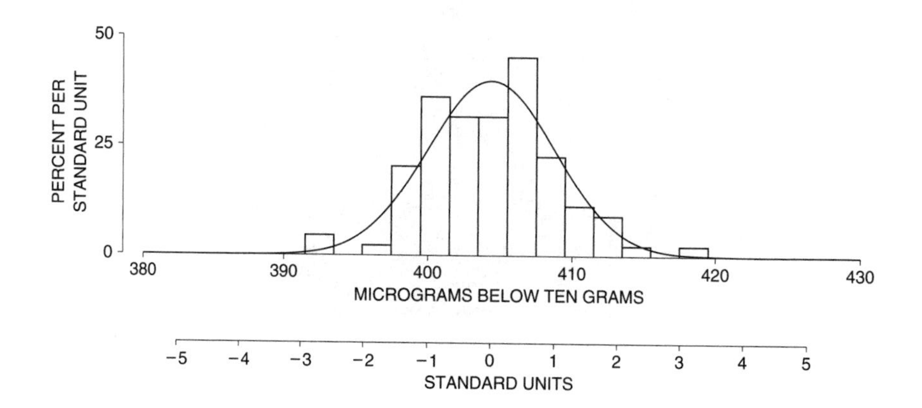

the remaining 97 measurements come closer to the normal curve. In sum, most of the data have an SD of about 4 micrograms. But a few of the measurements are quite a bit further away from the average than the SD would suggest. The overall SD of 6 micrograms is a compromise between the SD of the main part of the histogram—4 micrograms—and the outliers.

In careful measurement work, a small percentage of outliers is expected. The only unusual aspect of the NB 10 data is that the outliers are reported. Here is what the Bureau has to say about *not* reporting outliers.[4] For official prose, the tone is quite stern.

> A major difficulty in the application of statistical methods to the analysis of measurement data is that of obtaining suitable collections of data. The problem is more often associated with conscious, or perhaps unconscious, attempts to make a particular process perform as one would like it to perform rather than accepting the actual performance.... Rejection of data on the basis of arbitrary performance limits severely distorts the estimate of real process variability. Such procedures defeat the purpose of the... program. Realistic performance parameters require the acceptance of all data that cannot be rejected for cause.

There is a hard choice to make when investigators see an outlier. Either they ignore it, or they have to concede that their measurements don't follow the normal curve. The prestige of the curve is so high that the first choice is the usual one—a triumph of theory over experience.

4. BIAS

Suppose a butcher weighs a steak with his thumb on the scale. That causes an error in the measurement, but little has been left to chance. Take another example. Suppose a fabric store uses a cloth tape measure which has stretched from 36 inches to 37 inches in length. Every "yard" of cloth they sell to a customer has an extra inch tacked onto it. This isn't a chance error, because it always works for the customer. The butcher's thumb and the stretched tape are two examples of *bias*, or *systematic error.*

> Bias affects all measurements the same way, pushing them in the same direction. Chance errors change from measurement to measurement, sometimes up and sometimes down.

The basic equation has to be modified when each measurement is thrown off by bias as well as chance error:

$$\text{individual measurement} = \text{exact value} + \text{bias} + \text{chance error}.$$

If there is no bias in a measurement procedure, the long-run average of repeated measurements should give the exact value of the thing being measured: the chance errors should cancel out. However, when bias is present, the long-run average will itself be either too high or too low.

Usually, bias cannot be detected just by looking at the measurements themselves. Instead, the measurements have to be compared to an external standard or to theoretical predictions. In the U.S., all weight measurements depend on the connection between K_{20} and The Kilogram. These two weights have been compared a number of times, and it is estimated that K_{20} is a tiny bit lighter than The Kilogram—by 19 parts in a billion. All weight calculations at the Bureau are revised upward by 19 parts in a billion, to compensate. However, this factor itself is likely to be just a shade off: it too was the result of some measurement process. All weights measured in the U.S. are systematically off, by the same (tiny) percentage. This is another example of bias, but not one to worry about.

5. REVIEW EXERCISES

1. True or false, and explain: "An experienced scientist who is using the best equipment available only needs to measure things once—provided he doesn't make a mistake. After all, if he measures the same thing twice, he'll get the same results both times."

2. A carpenter is using a tape measure to get the length of a board.
   (a) What are some possible sources of bias?
   (b) Which is more subject to bias, a steel tape or a cloth tape?
   (c) Would the bias in a cloth tape change over time?

3. True or false, and explain.
   (a) Bias is a kind of chance error.
   (b) Chance error is a kind of bias.
   (c) Measurements are usually affected by both bias and chance error.

4. You send a yardstick to a local laboratory for calibration, asking that the procedure be repeated three times. They report the following values:

   35.96 inches   36.01 inches   36.03 inches

   If you send the yardstick back for a fourth calibration, you would expect to get 36 inches give or take—

   .01 inches or so   .03 inches or so   .06 inches or so

5. Nineteen students in a beginning statistics course were asked to measure the thickness of a table top, using a vernier gauge reading to 0.001 of an inch. Each person made two measurements, shown at the top of the next page. (The units are inches; for instance, the first person got 1.317 and 1.320 for the two measurements.)
   (a) Did the students work independently of one another?
   (b) Some friends of yours do not believe in chance error. How could you use these data to convince them?

Person	Measurements (inches) 1st	2nd
1	1.317	1.320
2	13.26	13.25
3	1.316	1.335
4	1.316	1.328
5	1.318	1.324
6	1.329	1.326
7	1.332	1.334
8	1.342	1.328
9	1.337	1.342
10	13.26	13.25

Person	Measurements (inches) 1st	2nd
11	1.333	1.334
12	1.315	1.317
13	1.316	1.318
14	1.321	1.319
15	1.337	1.343
16	1.349	1.336
17	1.320	1.336
18	1.342	1.340
19	1.317	1.318

6. SPECIAL REVIEW EXERCISES

These exercises cover all of parts I and II.

1. In one course, a histogram for the scores on the final looked like the sketch below. True or false: because this isn't like the normal curve, there must have been something wrong with the test. Explain.

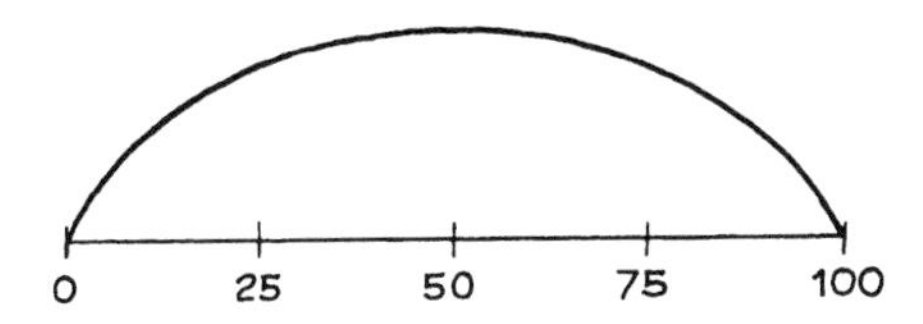

2. Fill in the blanks, using the options below, and give examples to show that you picked the right answers.
   (a) The SD of a list is 0. This means ______ .
   (b) The r.m.s. size of a list is 0. This means ______ .

   Options:
   (i) there are no numbers on the list
   (ii) all the numbers on the list are the same
   (iii) all the numbers on the list are 0
   (iv) the average of the list is 0

3. A personality test is administered to a large group of subjects. Five scores are shown below, in original units and in standard units. Fill in the blanks.

79	64	52	72	___
1.8	0.8	___	___	−1.4

4. Among first-year students at a certain university, scores on the Verbal SAT follow the normal curve; the average is around 550 and the SD is about 100.
   (a) What percentage of these students have scores in the range 400 to 700?

(b) There were about 1,000 students with scores in the range 450–650 on the Verbal SAT. About ______ of them had scores in the range 500 to 600. Fill in the blank; explain briefly.

5. In Cycle III of the Health Examination Survey (like HANES, but done in 1966–70), there were 6,672 subjects. The sex of each subject was recorded at two different stages of the survey. In 17 cases, there was a discrepancy: the subject was recorded as male at one interview, female at the other. How would you account for this?

6. Among entering students at a certain college, the men averaged 650 on the Math SAT, and their SD was 125. The women averaged 600, but had the same SD of 125. There were 500 men in the class, and 500 women.

   (a) For the men and the women together, the average Math SAT score was ______.

   (b) For the men and the women together, was the SD of Math SAT scores less than 125, just about 125, or more than 125?

7. Repeat exercise 6, when there are 600 men in the class, and 400 women. (The separate averages and SDs for the men and women stay the same.)

8. Table 1 on p. 99 reported 100 measurements on the weight of NB 10; the top panel in figure 2 on p. 102 shows the histogram. The average was 405 micrograms, and the SD was 6 micrograms. If you used the normal approximation to estimate how many of these measurements were in the range 400 to 406 micrograms, would your answer be too low, too high, or about right? Why?

9. A teaching assistant gives a quiz to his section. There are 10 questions on the quiz and no part credit is given. After grading the papers, the TA writes down for each student the number of questions the student got right and the number wrong. The average number of right answers is 6.4 with an SD of 2.0. The average number of wrong answers is ______ with an SD of ______. Fill in the blanks—or do you need the data? Explain briefly.

10. A large, representative sample of Americans was studied by the Public Health Service, in the Health and Nutrition Examination Survey (HANES2).[5] The percentage of respondents who were left-handed decreased steadily with age, from 10% at 20 years to 4% at 70. "The data show that many people change from left-handed to right-handed as they get older." True or false? Why? If false, how do you explain the pattern in the data?

11. For a certain group of women, the 25th percentile of height is 62.2 inches and the 75th percentile is 65.8 inches. The histogram follows the normal curve. Find the 90th percentile of the height distribution.

12. In March, the Current Population Survey asks a large, representative sample of Americans to say what their incomes were during the previous year.[6] A histogram for family income in 2004 is shown at the top of the next page. (Class intervals include the left endpoint but not the right.) From $15,000 and on to the right, the blocks alternate regularly from high to low. Why is that?

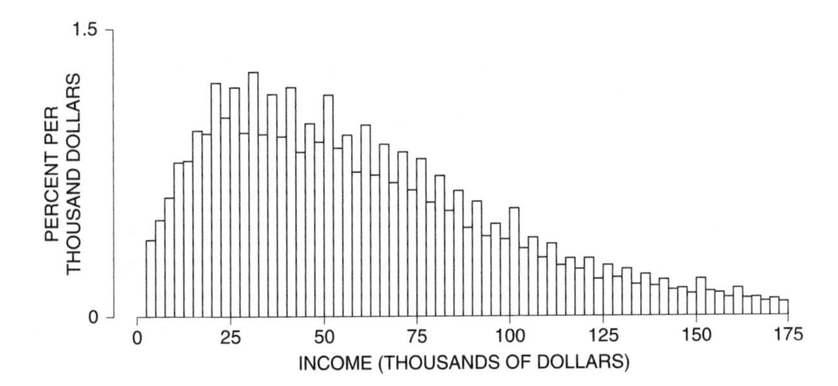

13. To measure the effect of exercise on the risk of heart disease, investigators compared the incidence of this disease for two large groups of London Transport Authority busmen—drivers and conductors. The conductors got a lot more exercise as they walked around all day collecting fares.

    The age distributions for the two groups were very similar, and all the subjects had been on the same job for 10 years or more. The incidence of heart disease was substantially lower among the conductors, and the investigators concluded that exercise prevents heart disease.

    Other investigators were skeptical. They went back and found that London Transport Authority had issued uniforms to drivers and conductors at the time of hire; a record had been kept of the sizes.[7]

    (a) Why does it matter that the age distributions of the two groups were similar?
    (b) Why does it matter that all the subjects had been on the job for 10 years or more?
    (c) Why did the first group of investigators compare the conductors to drivers, not to London Transport Authority executive staff?
    (d) Why might the second group of investigators have been skeptical?
    (e) What would you do with the sizes of the uniforms?

14. Breast cancer is one of the most common malignancies among women in Canada and the U.S. If it is detected early enough—before the cancer spreads—chances of successful treatment are much better. Do screening programs speed up detection by enough to matter? Many studies have examined this question.

    The Canadian National Breast Cancer Study was a randomized controlled experiment on mammography, that is, x-ray screening for breast cancer. The study found no benefit from screening. (The benefit was measured by comparing death rates from breast cancer in the treatment and control groups.)

    Dr. Daniel Kopans argued that the randomization was not done properly: instead of following instructions, nurses assigned high risk women to the treatment group.[8] Would this bias the study? If so, would the bias make the benefit from screening look bigger or smaller than it really is? Explain your answer.

15. In some jurisdictions, there are "pretrial conferences," where the judge confers with the opposing lawyers to settle the case or at least to define the issues before trial. Observational data suggest that pretrial conferences promote settlements and speed up trials, but there were doubts.

    In New Jersey courts, pretrial conferences were mandatory. However, an experiment was done in 7 counties. During a six-month period, 2,954 personal injury cases (mainly automobile accidents) were assigned at random to treatment or control. For the 1,495 control cases (group A), pretrial conferences remained mandatory. For the 1,459 treatment cases, the conferences were made optional—either lawyer could request one. Among the treatment cases, 701 opted for a pretrial conference (group C), and 758 did not (group B).

    The investigator who analyzed the data looked to see whether pretrial conferences encouraged cases to settle before reaching trial; or, if they went to trial, whether the conferences shortened the amount of trial time. (This matters, because trial time is very expensive.)

    The investigator reported the main results as follows; tabular material is quoted from his report.[9]

    (i) Pretrial conferences had no impact on settlement; the same percentage go to trial in group B as in group A + C.

Percentage of cases reaching trial

	Group B	*Group A + C*
Reached trial	22%	23%
Number of cases	701	2,079

    (ii) Pretrial conferences do not shorten trial time; the percentage of short trials is highest in cases that refused pretrial conferences.

Distribution of trial time among cases that go to trial

	Group B	*Group A*	*Group C*
Trial time (in hours)			
1. 5 or less	43%	34%	28%
2. Over 5 to 10	35%	41%	39%
3. Over 10	22%	26%	33%
Number of cases	63	176	70

    Comment briefly on the analysis.

7. SUMMARY AND OVERVIEW

1. No matter how carefully it was made, a measurement could have turned out a bit differently. This reflects *chance error*. Before investigators rely on a measurement, they should estimate the likely size of the chance error. The best way to do that: *replicate* the measurement.

2. The likely size of the chance error in a single measurement can be estimated by the SD of a sequence of repeated measurements made under the same conditions.

3. *Bias*, or *systematic error*, causes measurements to be systematically too high or systematically too low. The equation is

$$\text{individual measurement} = \text{exact value} + \text{bias} + \text{chance error.}$$

The chance error changes from measurement to measurement, but the bias stays the same. Bias cannot be estimated just by repeating the measurements.

4. Even in careful measurement work, a small percentage of *outliers* can be expected.

5. The average and SD can be strongly influenced by outliers. Then the histogram will not follow the normal curve at all well.

6. This part of the book introduced two basic descriptive statistics, the average and the standard deviation; histograms were used to summarize data. For many data sets, the histogram follows the normal curve. Chapter 6 illustrates these ideas on measurement data. Later in the book, histograms will be used for probability distributions, and statistical inference will be based on the normal curve. This is legitimate when the probability histograms follow the curve—the topic of chapter 18.

7

Plotting Points and Lines

Q. What did the dot say to the line?
A. Come to the point.

1. READING POINTS OFF A GRAPH

This chapter reviews some of the ideas about plotting points and lines which will be used in part III. You can either read this chapter now, or return to it if you run into difficulty in part III. If you read the chapter now, the first four sections are the most important; the last section is more difficult.

Figure 1 shows a horizontal axis (the x-axis) and a vertical axis (the y-axis). The point shown in the figure has an x-coordinate of 3, because it is in line with 3 on the x-axis. It has a y-coordinate of 2, because it is in line with 2 on the y-axis. This point is written $x = 3$, $y = 2$. Sometimes, it is abbreviated even more, to $(3, 2)$. The point shown in figure 2 is $(-2, -1)$: it is directly below -2 on the x-axis, and directly to the left of -1 on the y-axis.

Figure 1. Figure 2.

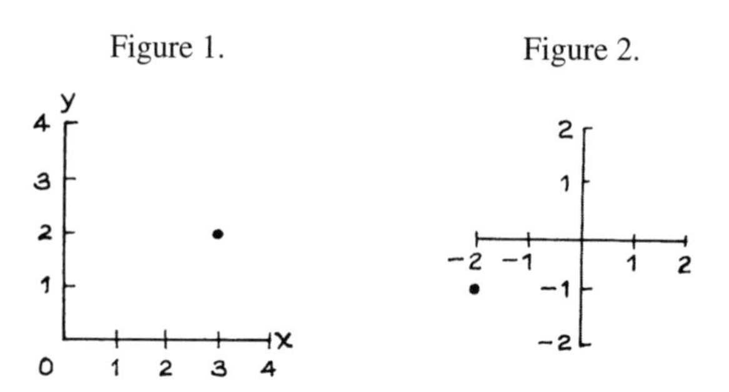

The idea of representing points by pairs of numbers is due to René Descartes (France, 1596–1650). In his honor, the x- and y-coordinates are often called "cartesian coordinates."

Exercise Set A

1. Figure 3 shows five points. Write down the x-coordinate and y-coordinate for each point.
2. As you move from point A to point B in figure 3, your x-coordinate goes up by ______; your y-coordinate goes up by ______.
3. One point in figure 3 has a y-coordinate 1 bigger than the y-coordinate of point E. Which point is that?

Figure 3.

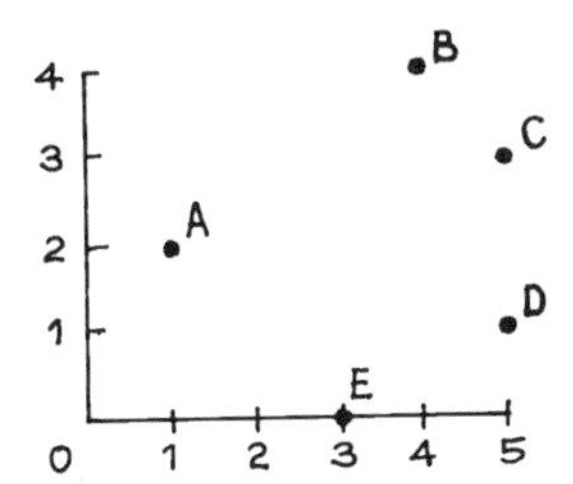

The answers to these exercises are on p. A52.

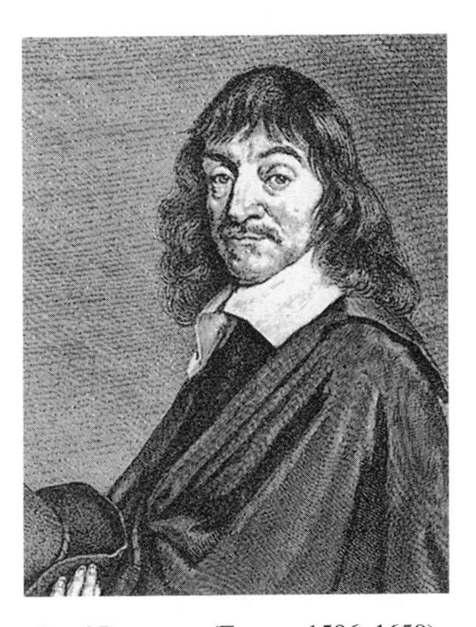

René Descartes (France, 1596–1650)

Wolff-Leavenworth Collection, courtesy of the Syracuse University Art Collection.

2. PLOTTING POINTS

Figure 4 shows a pair of axes. To plot the point (2, 1), find the 2 on the x-axis. The point will be directly above this, as in figure 5. Find the 1 on the y-axis, the point will be directly to the right of this, as in figure 6.

Figure 4.

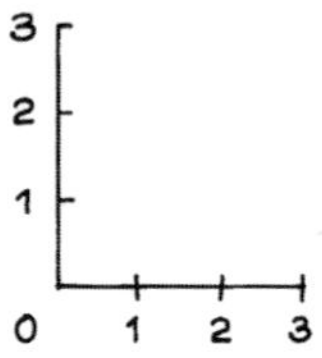

Figure 5.

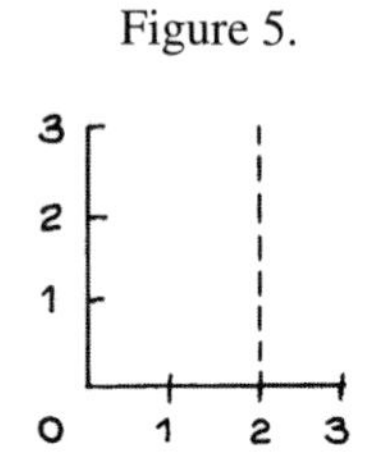

Figure 6.

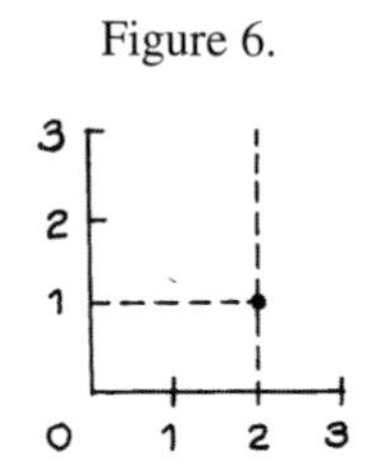

Exercise Set B

1. Draw a pair of axes and plot each of the following points:

   (1, 1)   (2, 2)   (3, 3)   (4, 4)

   What can you say about them?

2. Three out of the following four points lie on a line. Which is the maverick? Is it above or below the line?

   (0, 0)   (0.5, 0.5)   (1, 2)   (2.5, 2.5)

3. The table below shows four points. In each case, the y-coordinate is computed from the x-coordinate by the rule $y = 2x + 1$. Fill in the blanks, then plot the four points. What can you say about them?

x	y
1	3
2	5
3	–
4	–

4. Figure 7 below shows a shaded region. Which of the following two points is in the region: (1, 2) or (2, 1)?

5. Do the same for figure 8.

6. Do the same for figure 9.

Figure 7.

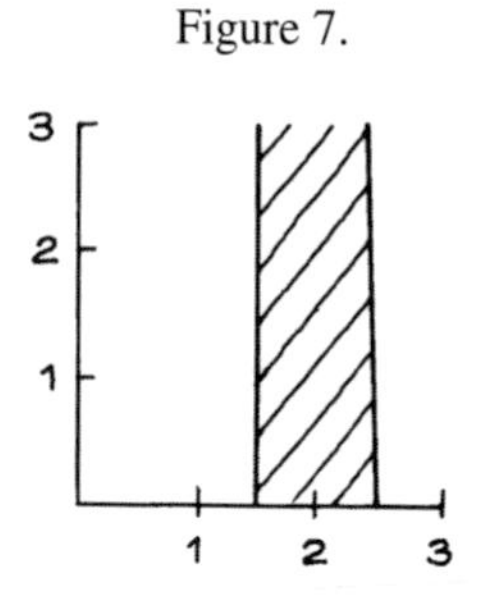

Figure 8.

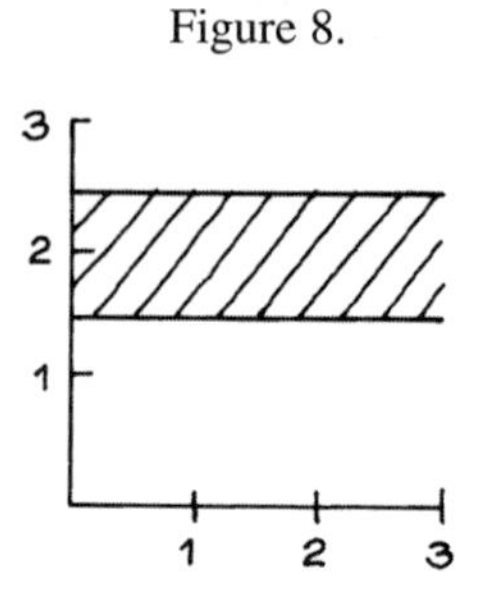

Figure 9.

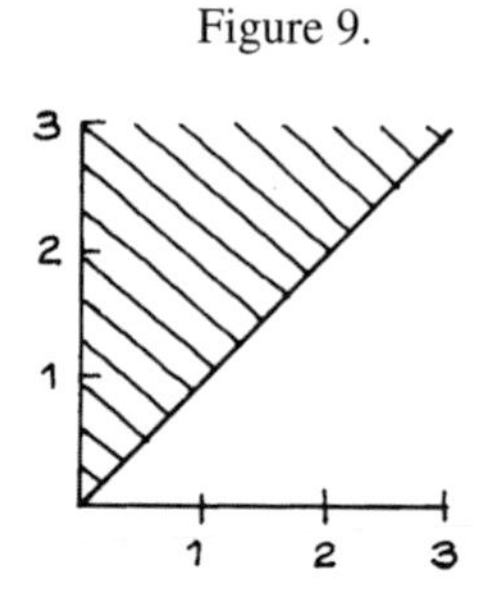

The answers to these exercises are on p. A53.

3. SLOPE AND INTERCEPT

Figure 10 shows a line. Take any point on the line—for instance, point A. Now move up the line to any other point—for instance, point B. Your x-coordinate has increased by some amount, called the *run*. In this case, the run was 2. At the same time your y-coordinate has increased by some other amount, called the *rise*. In this case, the rise was 1. Notice that in this case, the rise was half the run. Whatever two points you take on this line, the rise will be half the run. The ratio rise/run is called the *slope* of the line:

$$\text{slope} = \text{rise/run}.$$

The slope is the rate at which y increases with x, along the line. To interpret it another way, imagine the line as a road going up a hill. The slope measures the steepness of the grade. For the line in figure 10, the grade is 1 in 2—quite steep for a road. In figure 11, the slope of the line is 0. In figure 12, the slope is -1. If the slope is positive, the line is going uphill. If the slope is 0, the line is horizontal. If the slope is negative, the line is going downhill.

Figure 10. Slope is 1/2.

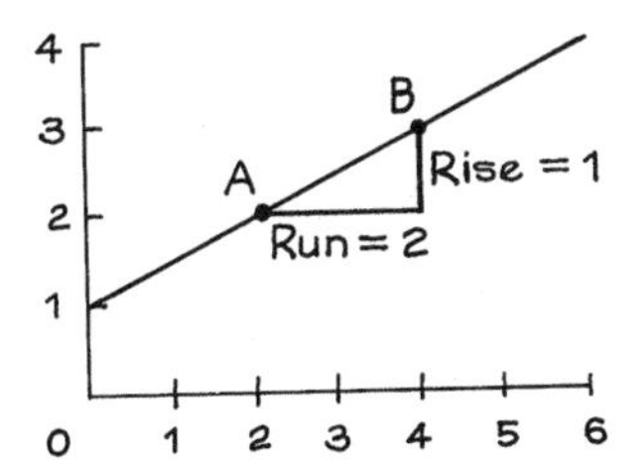

Figure 11. Slope is 0.

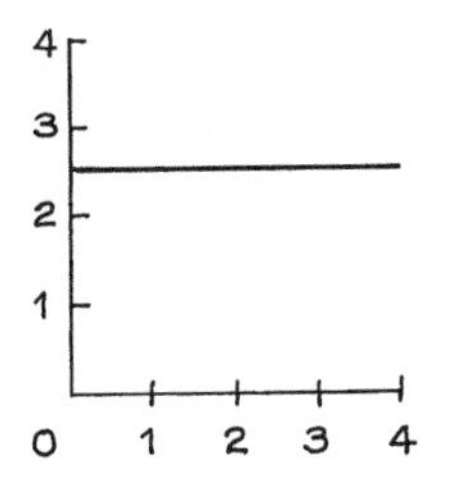

Figure 12. Slope is -1.

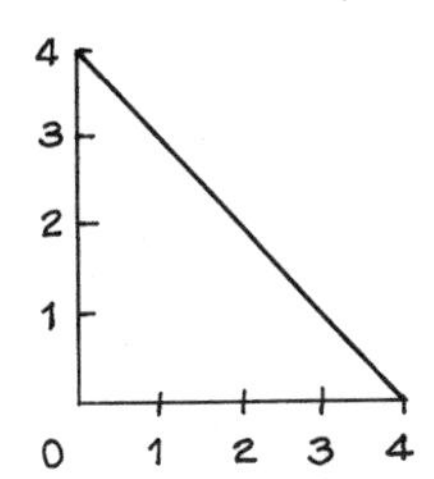

The *intercept* of a line is its height at $x = 0$. Usually, the axes cross at 0. Then, the intercept is where the line crosses the y-axis. In figure 13, the intercept is 2. Sometimes, the axes don't cross at 0, and then you have to be a little bit careful. In figure 14, the axes cross at (1, 1). The intercept of the line in figure 14 is 0—that would be its height at $x = 0$.

Often, the axes of a graph show units. For example, in figure 15 the units for the x-axis are inches, the units for the y-axis are degrees celsius. Then the slope and intercept have units too. In figure 15, the slope of the line is 2.5 degrees per inch; the intercept is -5 degrees.

Figure 13.

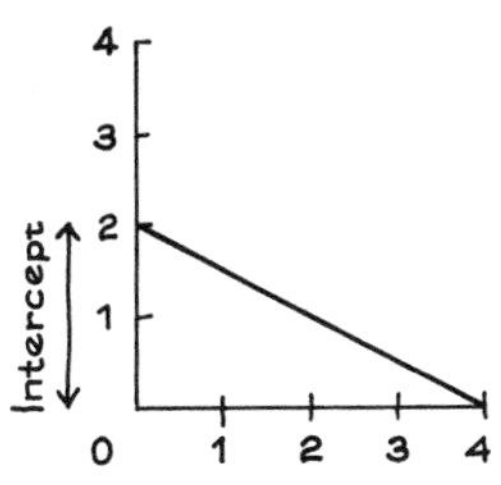

Figure 14.

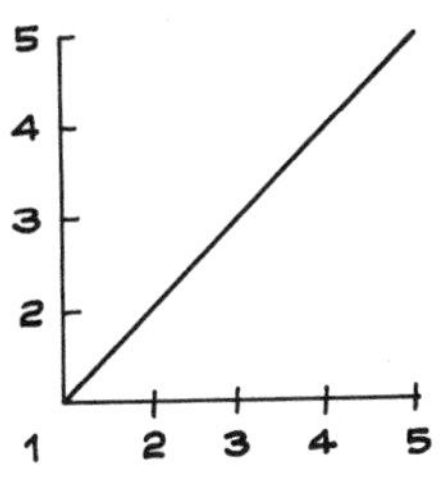

Figure 15.

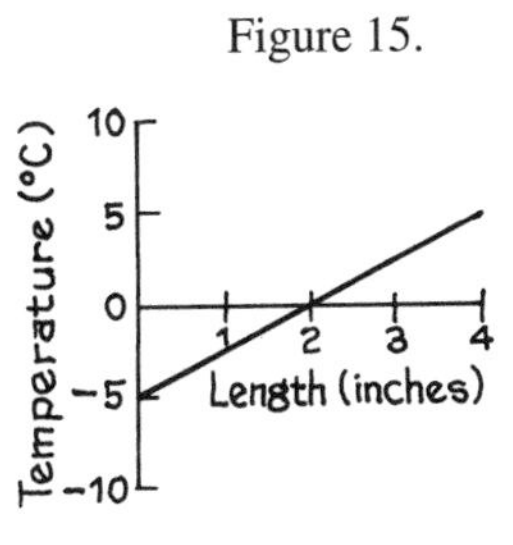

Exercise Set C

1. Figures 16 to 18 show lines. For each line, find the slope and intercept. Note: the axes do not cross at 0 in each case.

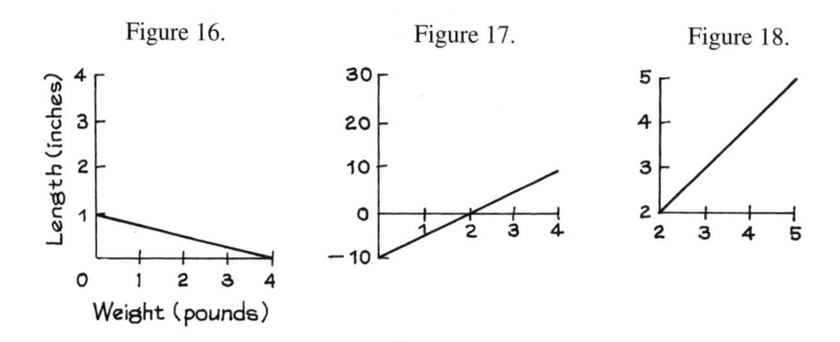

The answers to these exercises are on p. A53.

4. PLOTTING LINES

Example 1. Plot the line which passes through the point (2, 1) and has slope 1/2.

Solution. First draw a pair of axes and plot the given point (2, 1), as in figure 19. Then move any convenient distance off directly to the right from the given point: figure 20 shows a run of 3. Make a construction point at this new location. Since the line slopes up, it passes above the construction point. How far? That is, how much will the line rise in a run of 3? The answer is given by the slope. The line is rising at the rate of half a vertical unit per horizontal unit, and in this case there is a run of 3 horizontal units, so the rise is $3 \times 1/2 = 1.5$:

$$\text{rise} = \text{run} \times \text{slope}.$$

Make a vertical move of 1.5 from the construction point, and mark a point at this third location, as in figure 21. This third point is on the line. Put a ruler down and join it to the given point (2, 1).

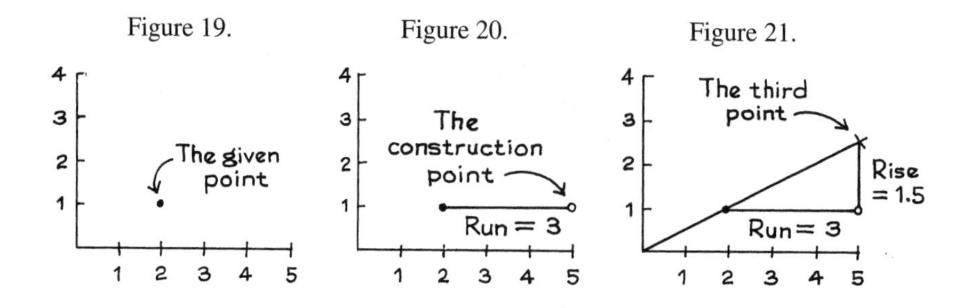

Exercise Set D

1. Draw lines through the point (2, 1) with the following slopes:

   (a) $+1$ (b) -1 (c) 0

2. Start at the point (2, 1) in figure 21. If you move over 2 and up 1, will you be on the line, above the line, or below the line?
3. The same, but move over 4 and up 2.
4. The same but move over 6 and up 5.
5. Draw the line with intercept 2 and slope -1. Hint: this line goes through the point (0, 2).
6. Draw the line with intercept 2 and slope 1.

The answers to these exercises are on p. A54.

5. THE ALGEBRAIC EQUATION FOR A LINE

Example 2. Here is a rule for computing the y-coordinate of a point from its x-coordinate: $y = \frac{1}{2}x + 1$. The table below shows the points with x-coordinates of 1, 2, 3, 4. Plot the points. Do they fall on a line? If so, find the slope and intercept of this line.

Solution. The points are plotted in figure 22. They do fall on a line. Any point whose y-coordinate is related to its x-coordinate by the same equation $y = \frac{1}{2}x + 1$ will fall on the same line. This line is called the *graph* of the equation. The slope of the line is $\frac{1}{2}$, the coefficient of x in the equation. The intercept is 1, the constant term in the equation.

Figure 22.

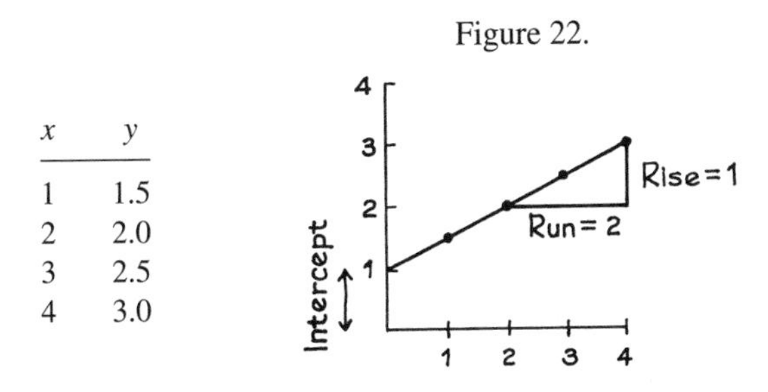

x	y
1	1.5
2	2.0
3	2.5
4	3.0

> The graph of the equation $y = mx + b$ is a straight line, with slope m and intercept b.

Example 3. Figure 23 shows a line. What is the equation of this line? What is the height of this line at $x = 1$?

Solution. This line has slope -1 and intercept 4. Therefore, its equation is $y = -x + 4$. Substituting $x = 1$ gives $y = 3$; so the height of the line is 3 when x is 1.

Figure 23.

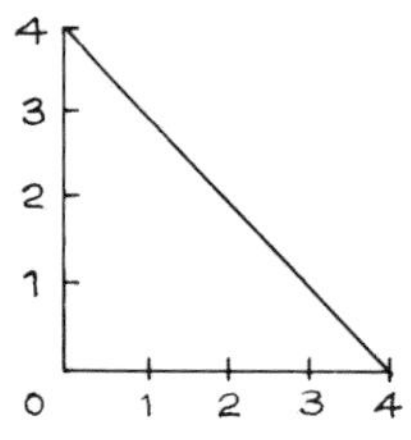

Figure 24.

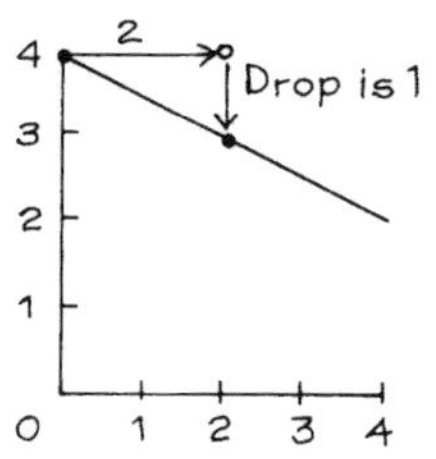

Figure 25.

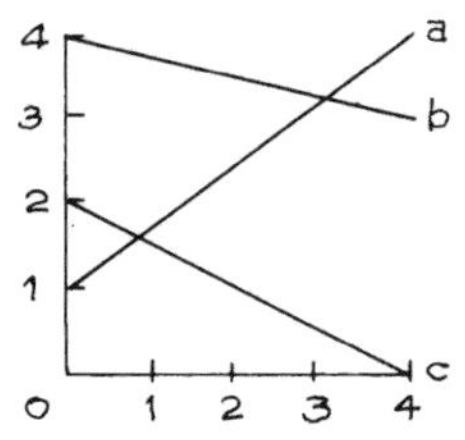

Example 4. Plot the line whose equation is $y = -\frac{1}{2}x + 4$.

Solution. The intercept of this line is 4. Plot the point (0, 4) as in figure 24. The line must go through this point. Make any convenient horizontal move—say 2. The slope is $-1/2$, so the line must drop 1. Mark the point which is 2 over and 1 down from the first point in figure 24. Then join these two points by a straight line.

Exercise Set E

1. Plot the graphs of the following equations:

   (a) $y = 2x + 1$ (b) $y = \frac{1}{2}x + 2$

   In each case, say what the slope and intercept are, and give the height of the line at $x = 2$.

2. Figure 25 shows three lines. Match the lines with the equations:

$$y = \tfrac{3}{4}x + 1 \qquad y = -\tfrac{1}{4}x + 4 \qquad y = -\tfrac{1}{2}x + 2$$

3. Plot four different points whose y-coordinates are double their x-coordinates. Do these points lie on a line? If so, what is the equation of the line?

4. Plot the points (1, 1), (2, 2), (3, 3), and (4, 4) on the same graph. These points all lie on a line. What is the equation of this line?

5. For each of the following points, say whether it is on the line of exercise 4, or above, or below:

   (a) (0, 0) (b) (1.5, 2.5) (c) (2.5, 1.5)

6. True or false:

   (a) If y is bigger than x, then the point (x, y) is above the line of exercise 4.
   (b) If $y = x$, then the point (x, y) is on the line of exercise 4.
   (c) If y is smaller than x, then the point (x, y) is below the line of exercise 4.

The answers to these exercises are on pp. A54–55.

PART III

Correlation and Regression

8

Correlation

Like father, like son.

1. THE SCATTER DIAGRAM

The methods discussed in part II are good for dealing with one variable at a time. Other methods are needed for studying the relationship between two variables.[1] Sir Francis Galton (England, 1822–1911) made some progress on this front while he was thinking about the degree to which children resemble their parents. Statisticians in Victorian England were fascinated by the idea of quantifying hereditary influences and gathered huge amounts of data in pursuit of this goal. We are going to look at the results of a study carried out by Galton's disciple Karl Pearson (England, 1857–1936).[2]

As part of the study, Pearson measured the heights of 1,078 fathers, and their sons at maturity. A list of 1,078 pairs of heights would be hard to grasp. But the relationship between the two variables—father's height and son's height—can be brought out in a *scatter diagram* (figure 1 on the next page). Each dot on the diagram represents one father-son pair. The x-coordinate of the dot, measured along the horizontal axis, gives the height of the father. The y-coordinate of the dot, along the vertical axis, gives the height of the son.

Figure 1. Scatter diagram for heights of 1,078 fathers and sons. Shows positive association between son's height and father's height. Families where the height of the son equals the height of the father are plotted along the 45-degree line $y = x$. Families where the father is 72 inches tall (to the nearest inch) are plotted in the vertical strip.

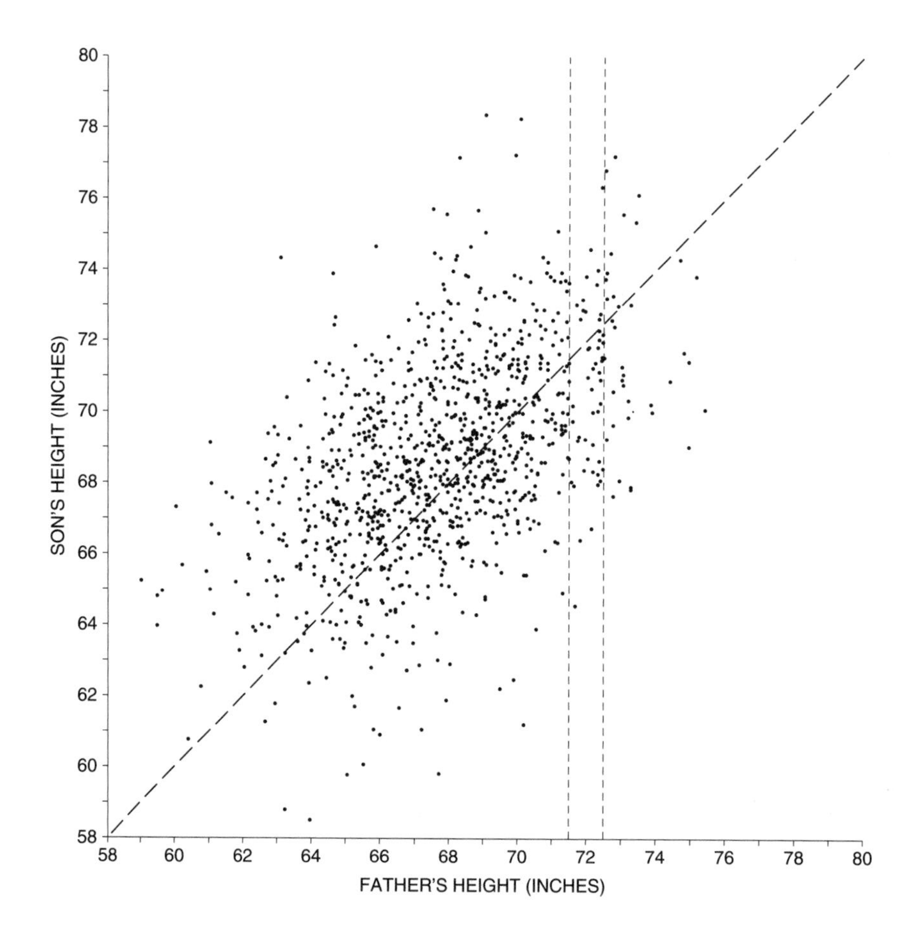

Figure 2a illustrates the mechanics of plotting scatter diagrams. (Chapter 7 has details.) The scatter diagram in figure 1 is a cloud shaped something like a football, with points straggling off the edges. When making a rough sketch of such a scatter diagram, it is only necessary to show the main oval portion—figure 2b.

The swarm of points in figure 1 slopes upward to the right, the y-coordinates of the points tending to increase with their x-coordinates. A statistician might say there is a *positive association* between the heights of fathers and sons. As a rule, the taller fathers have taller sons. This confirms the obvious. Now look at the 45-degree line in figure 1. This line corresponds to the families where son's height equals father's height. Along the line, for example, if the father is 72 inches tall then the son is 72 inches tall; if the father is 64 inches tall, the son is too; and so

Figure 2a. A point on a scatter diagram.

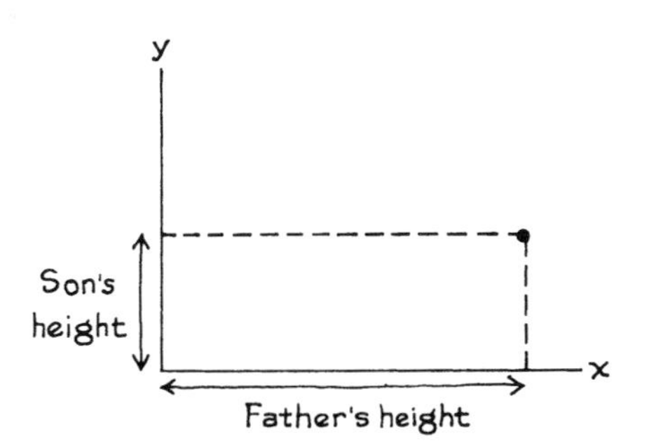

Figure 2b. Rough sketch.

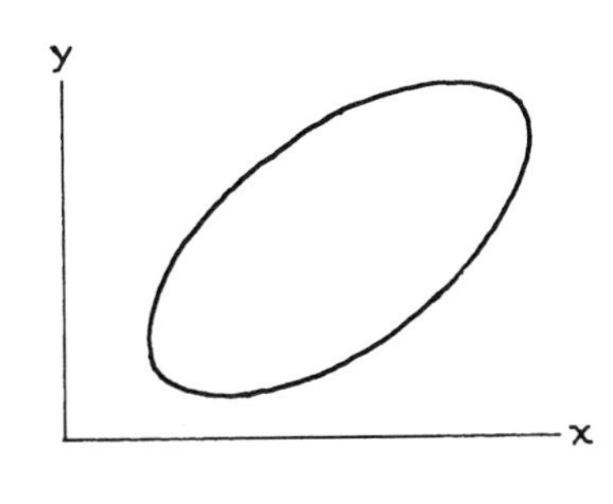

forth. Similarly, if a son's height is close to his father's height, then their point on the scatter diagram will be close to the line, like the points in figure 3.

There is a lot more spread around the 45-degree line in the actual scatter diagram than in figure 3. This spread shows the weakness of the relationship between father's height and son's height. For instance, suppose you have to guess the height of a son. How much help does the father's height give you? In figure 1, the dots in the chimney represent all the father-son pairs where the father is 72 inches tall to the nearest inch (father's height between 71.5 inches and 72.5 inches, where the dashed vertical lines cross the x-axis). There is still a lot of variability in the heights of the sons, as indicated by the vertical scatter in the chimney. Even if you know the father's height, there is still a lot of room for error in trying to guess the height of his son.

> If there is a strong association between two variables, then knowing one helps a lot in predicting the other. But when there is a weak association, information about one variable does not help much in guessing the other.

Figure 3. Son's height close to father's height.

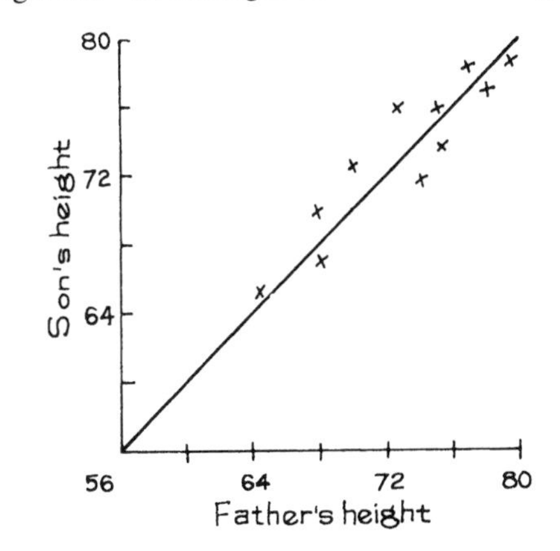

Sir Francis Galton (England, 1822–1911)

Source: *Biometrika* (November, 1903).

In social science studies of the relationship between two variables, it is usual to label one as *independent* and the other as *dependent*. Ordinarily, the independent variable is thought to influence the dependent variable, rather than the other way around. In figure 1, father's height is taken as the independent variable and plotted along the x-axis: father's height influences son's height. However, there is nothing to stop an investigator from using son's height as the independent variable. This choice might be appropriate, for example, if the problem is to guess a father's height from his son's height.

Before going on, it would be a good idea to work the exercises of this section. They are easy, and they will really help you understand the rest of this chapter. If you have trouble with them, review chapter 7.

Exercise Set A

1. Use figure 1 (p. 120) to answer the following questions:
   (a) What is the height of the shortest father? of his son?
   (b) What is the height of tallest father? of his son?
   (c) Take the families where the father was 72 inches tall, to the nearest inch. How tall was the tallest son? the shortest son?
   (d) How many families are there where the sons are more than 78 inches tall? How tall are the fathers?
   (e) Was the average height of the fathers around 64, 68, or 72 inches?
   (f) Was the SD of the fathers' heights around 3, 6, or 9 inches?

2. Below is the scatter diagram for a certain data set. Fill in the blanks.

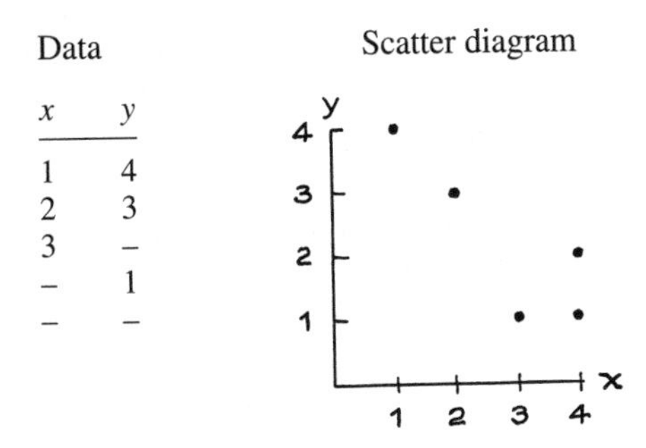
Data

x	y
1	4
2	3
3	–
–	1
–	–

3. Below is a scatter diagram for some hypothetical data.
   (a) Is the average of the x-values around 1, 1.5, or 2?
   (b) Is the SD of the x-values around 0.1, 0.5, or 1?
   (c) Is the average of the y-values around 1, 1.5, or 2?
   (d) Is the SD of the y-values around 0.5, 1.5, or 3?

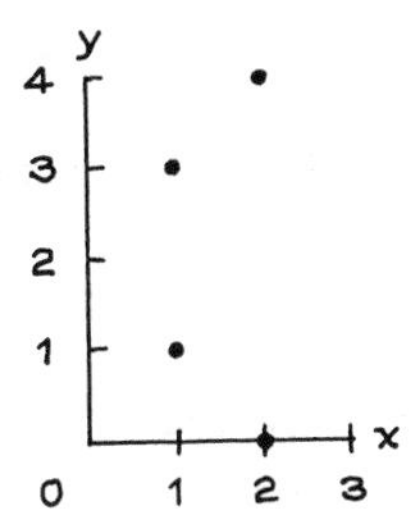

4. Draw the scatter diagram for each of the following hypothetical data sets. The variable labeled "x" should be plotted along the x-axis, the one labeled "y" along the y-axis. Mark each axis fully. In some cases, you will have to plot the same point more than once. The number of times such a multiple point appears can be indicated next to the point, as in the diagram below; please follow this convention.

(a)

x	y
1	2
3	1
2	3
1	2

(b)

x	y
3	5
1	4
3	1
2	3
1	4
4	1

Scatter diagram

y

3 •

x

5. Students named A, B, C, D, E, F, G, H, I, and J took a midterm and a final in a certain course. A scatter diagram for the scores is shown on the next page.
   (a) Which students scored the same on the midterm as on the final?
   (b) Which students scored higher on the final?

(c) Was the average score on the final around 25, 50, or 75?

(d) Was the SD of the scores on the final around 10, 25, or 50?

(e) For the students who scored over 50 on the midterm, was the average score on the final around 30, 50, or 70?

(f) True or false: on the whole, students who did well on the midterm also did well on the final.

(g) True or false: there is strong positive association between midterm scores and final scores.

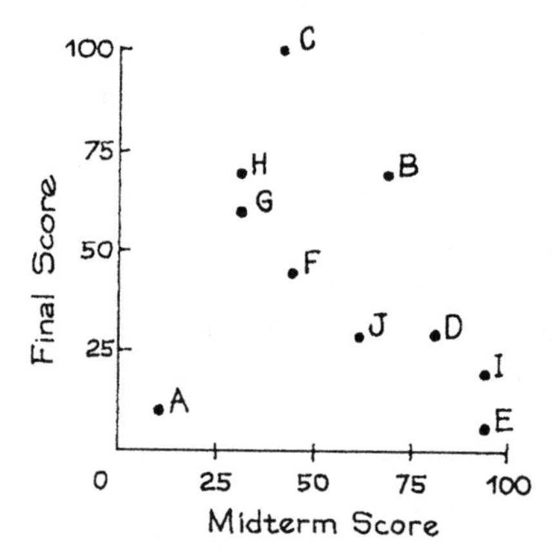

6. The scatter diagram below shows scores on the midterm and final in a certain course.

(a) Was the average midterm score around 25, 50, or 75?

(b) Was the SD of the midterm scores around 5, 10, or 20?

(c) Was the SD of the final scores around 5, 10, or 20?

(d) Which exam was harder—the midterm or the final?

(e) Was there more spread in the midterm scores, or the final scores?

(f) True or false: there was a strong positive association between midterm scores and final scores.

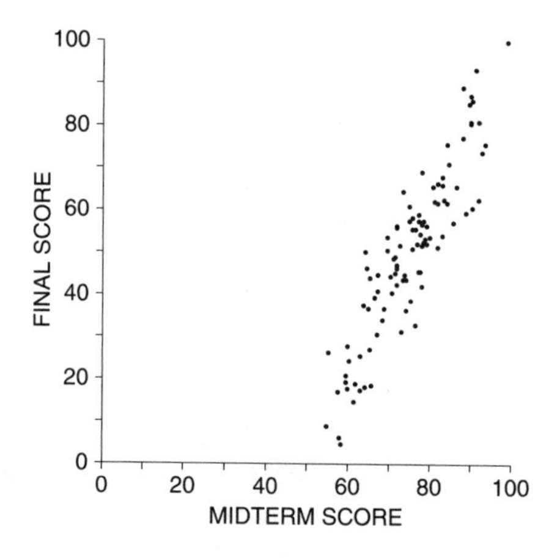

The answers to these exercises are on pp. A55–56.

2. THE CORRELATION COEFFICIENT

Suppose you are looking at the relationship between two variables, and have already plotted the scatter diagram. The graph is a football-shaped cloud of points. How can it be summarized? The first step would be to mark a point showing the average of the x-values and the average of the y-values (figure 4a). This is the *point of averages*, which locates the center of the cloud.[3] The next step would be to measure the spread of the cloud from side to side. This can be done using the SD of the x-values—the horizontal SD. Most of the points will be within 2 horizontal SDs on either side of the point of averages (figure 4b). In the same way, the SD of the y-values—the vertical SD—can be used to measure the spread of the cloud from top to bottom. Most of the points will be within 2 vertical SDs above or below the point of averages (figure 4c).

Figure 4. Summarizing a scatter diagram.

(a) The point of averages (b) The horizontal SD (c) The vertical SD

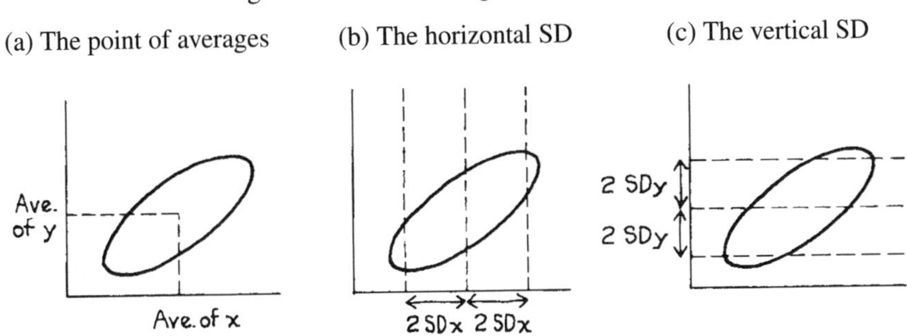

So far, the summary statistics are

- average of x-values, SD of x-values,
- average of y-values, SD of y-values.

These statistics tell us the center of the cloud, and how spread out it is, both horizontally and vertically. But there is still something missing—the strength of the association between the two variables. Look at the scatter diagrams in figure 5.

Figure 5. Summarizing a scatter diagram. The correlation coefficient measures clustering around a line.

(a) Correlation near 1 means tight clustering.

(b) Correlation near 0 means loose clustering.

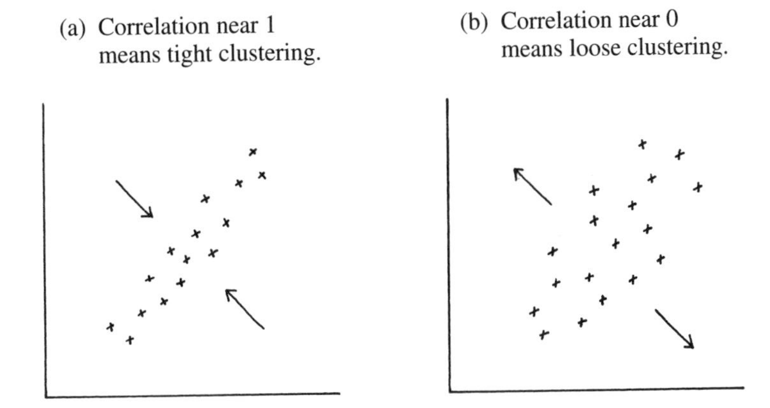

Both clouds have the same center and show the same spread, horizontally and vertically. However, the points in the first cloud are tightly clustered around a line: there is a strong linear association between the two variables. In the second cloud, the clustering is much looser. The strength of the association is different in the two diagrams. To measure the association, one more summary statistic is needed—the *correlation coefficient*. This coefficient is usually abbreviated as r, for no good reason (although there are two r's in "correlation").

> The correlation coefficient is a measure of linear association, or clustering around a line. The relationship between two variables can be summarized by
>
> - the average of the x-values, the SD of the x-values,
> - the average of the y-values, the SD of the y-values,
> - the correlation coefficient r.

The formula for computing r will be presented in section 4, but right now we want to focus on the graphical interpretation. Figure 6 shows six scatter diagrams for hypothetical data, each with 50 points. The diagrams were generated by computer. In all six pictures, the average is 3 and the SD is 1 for x and for y. The computer has printed the value of the correlation coefficient over each diagram. The one at the top left shows a correlation of 0. The cloud is completely formless. As x increases, y shows no tendency to increase or decrease: it just straggles around.

The next scatter diagram has $r = 0.40$; a linear pattern is beginning to emerge. The next one has $r = 0.60$, with a stronger linear pattern. And so on, through the last one. The closer r is to 1, the stronger is the linear association between the variables, and the more tightly clustered are the points around a line. A correlation of 1, which does not appear in the figure, is often referred to as a *perfect correlation*—all the points lie exactly on a line, so there is a perfect linear relationship between the variables. Correlations are always 1 or less.

The correlation between the heights of identical twins is around 0.95.[4] The lower right scatter diagram in figure 6 has a correlation coefficient of 0.95. A scatter diagram for the twins would look about the same. Identical twins are like each other in height, and their points on a scatter diagram are fairly close to the line $y = x$. However, such twins do not have exactly the same height. That is what the scatter around the 45-degree line shows.

For another example, in the U.S. in 2005, the correlation between income and education was 0.07 for men age 18–24, rising to 0.43 for men age 55–64.[5] As the scatter diagrams in figure 6 indicate, the relationship between income and education is stronger for the older men, but it is still quite rough. Weak associations are common in social science studies, 0.3 to 0.7 being the usual range for r in many fields.

A word of warning: $r = 0.80$ does not mean that 80% of the points are tightly clustered around a line, nor does it indicate twice as much linearity as $r = 0.40$. Right now, there is no direct way to interpret the exact numerical value of the correlation coefficient; that will be done in chapters 10 and 11.

Figure 6. The correlation coefficient—six positive values. The diagrams are scaled so that the average equals 3 and the SD equals 1, horizontally and vertically; there are 50 points in each diagram. Clustering is measured by the correlation coefficient.

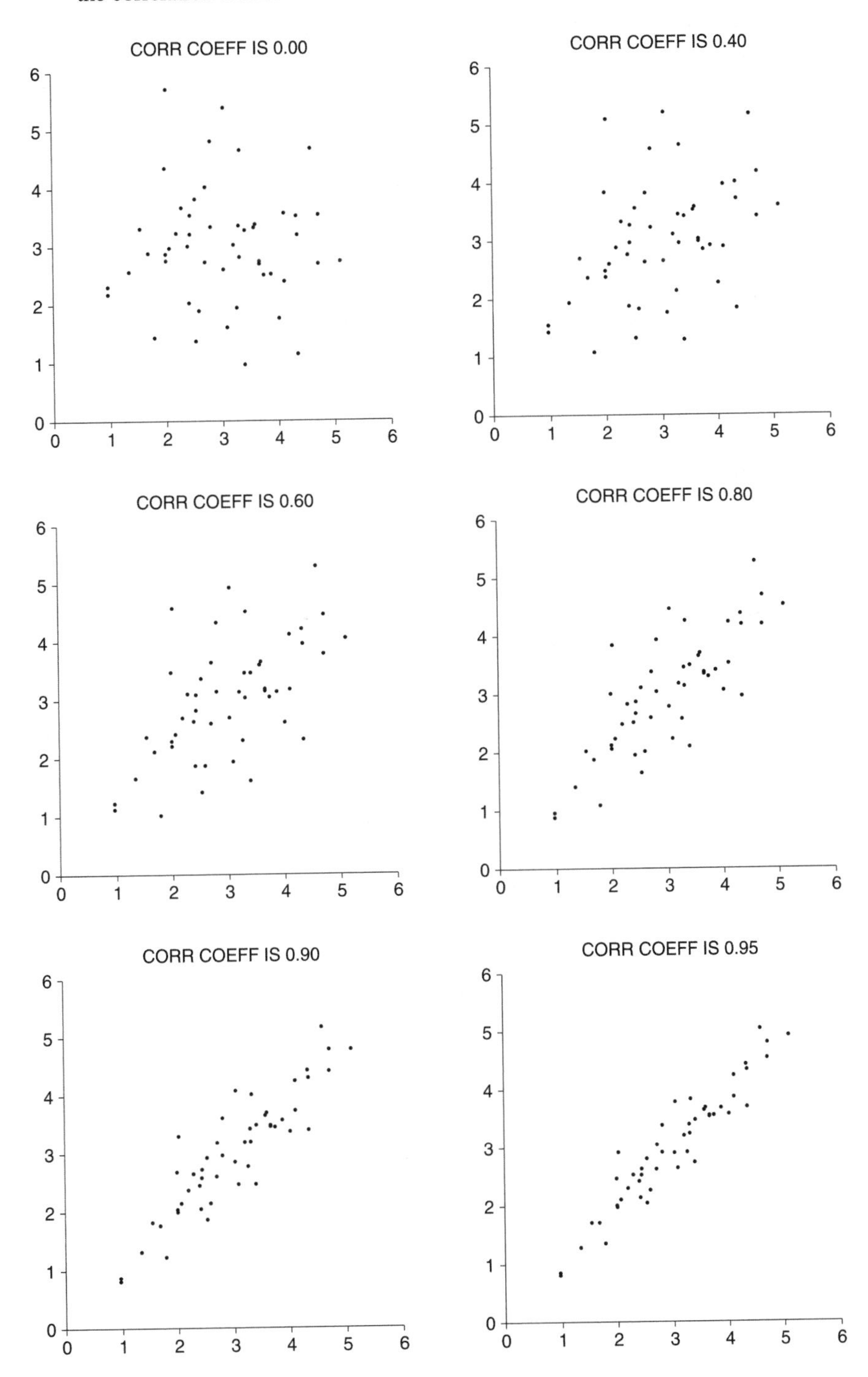

So far, only positive association has been discussed. Negative association is indicated by a negative sign in the correlation coefficient. Figure 7 shows six more scatter diagrams for hypothetical data, each with 50 points. They are scaled just like figure 6, each variable having an average of 3 and an SD of 1.

A correlation of −0.90, for instance, indicates the same degree of clustering as one of +0.90. With the negative sign, the clustering is around a line which slopes down; with a positive sign, the line slopes up. For women age 25–39 in the U.S. in 2005, the correlation between education and number of children was about −0.2, a weak negative association.[6] A perfect negative correlation of −1 indicates that all the points lie on a line which slopes down.

> Correlations are always between −1 and 1, but can take any value in between. A positive correlation means that the cloud slopes up; as one variable increases, so does the other. A negative correlation means that the cloud slopes down; as one variable increases, the other decreases.

In a real data set, both SDs will be positive. As a technical matter, if either SD is zero, there is no good way to define the correlation coefficient.

Exercise Set B

1. (a) Would the correlation between the age of a second-hand car and its price be positive or negative? Why? (Antiques are not included.)
   (b) What about the correlation between weight and miles per gallon?

2. For each scatter diagram below:
   (a) The average of x is around
   1.0 1.5 2.0 2.5 3.0 3.5 4.0
   (b) Same, for y.
   (c) The SD of x is around
   0.25 0.5 1.0 1.5
   (d) Same, for y.
   (e) Is the correlation positive, negative, or 0?

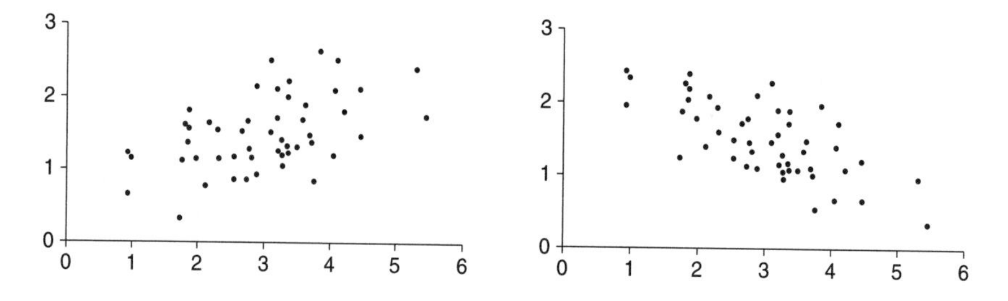

3. For which of the diagrams in the previous exercise is the correlation closer to 0, forgetting about signs?

Figure 7. The correlation coefficient—six negative values. The diagrams are scaled so the average equals 3 and the SD equals 1, horizontally and vertically; there are 50 points in each diagram. Clustering is measured by the correlation coefficient.

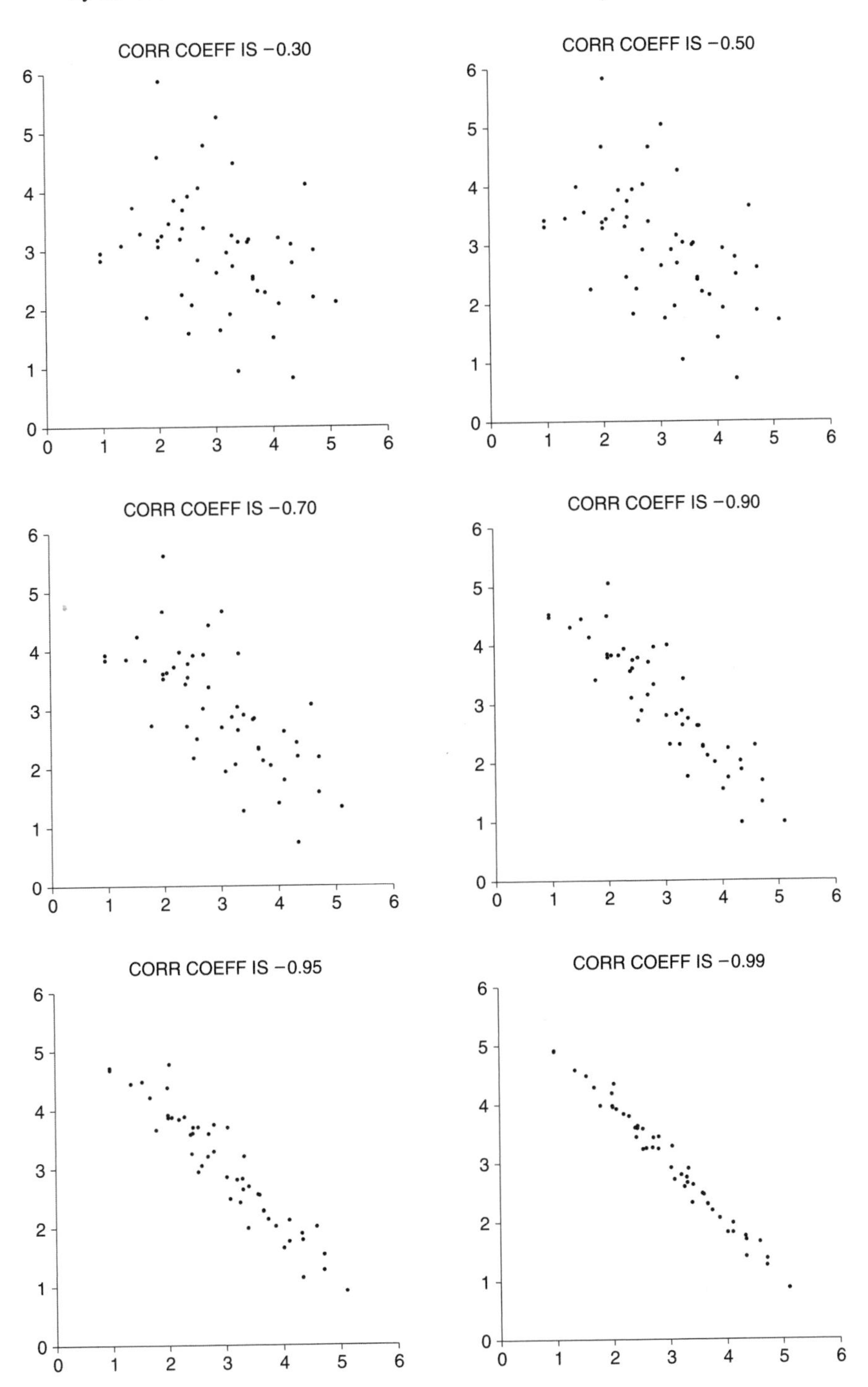

4. In figure 1, is the correlation between the heights of the fathers and sons around −0.3, 0, 0.5, or 0.8?

5. In figure 1, if you took only the fathers who were taller than 6 feet, and their sons, would the correlation between the heights be around −0.3, 0, 0.5 or 0.8?

6. (a) If women always married men who were five years older, the correlation between the ages of husbands and wives would be ______. Choose one of the options below, and explain.
   (b) The correlation between the ages of husbands and wives in the U.S. is ______. Choose one option, and explain.

   exactly −1 close to −1 close to 0 close to 1 exactly 1

7. Investigators are studying registered students at the University of California. The students fill out questionnaires giving their year of birth, age (in years), age of mother, and so forth. Fill in the blanks, using the options given below, and explain briefly.
   (a) The correlation between student's age and year of birth is ______.
   (b) The correlation between student's age and mother's age is ______.

   −1 nearly −1 somewhat negative
   0 somewhat positive nearly 1 1

8. Investigators take a sample of DINKS (dual-income families—where husband and wife both work—and no kids). The investigators have data on the husband's income and the wife's income. By definition,

   family income = husband's income + wife's income.

   The average family income was around $85,000, and 10% of the couples had family income in the range $80,000–$90,000. Fill in the blanks, using the options given below, and explain briefly.
   (a) The correlation between wife's income and family income is ______.
   (b) Among couples whose family income is in the range $80,000–$90,000, the correlation between wife's income and husband's income is ______.

   −1 nearly −1 somewhat negative
   0 somewhat positive nearly 1 1

9. True or false, and explain: if the correlation coefficient is 0.90, then 90% of the points are highly correlated.

The answers to these exercises are on p. A56.

3. THE SD LINE

The points in a scatter diagram generally seem to cluster around the *SD line*. This line goes through the point of averages; and it goes through all the points which are an equal number of SDs away from the average, for both variables. For example, take a scatter diagram showing heights and weights. Someone who happened to be 1 SD above average in height and also 1 SD above average in weight would be plotted on the SD line. But a person who is 1 SD above average in height and 0.5 SDs above average in weight would be off the line. Similarly, a person who is 2 SDs below average in height and also 2 SDs below average in weight would be on the line. Someone who is 2 SDs below average in height and 2.5 SDs below average in weight would be off the line.

Figure 8 shows how to plot the SD line on a graph. The line goes through the point of averages, and climbs at the rate of one vertical SD for each horizontal SD. More technically, the slope is the ratio.

$$(\text{SD of } y)/(\text{SD of } x).$$

This is for positive correlations. When the correlation coefficient is negative, the SD line goes down; the slope is[7]

$$-(\text{SD of } y)/(\text{SD of } x).$$

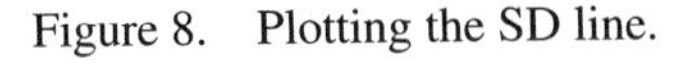
Figure 8. Plotting the SD line.

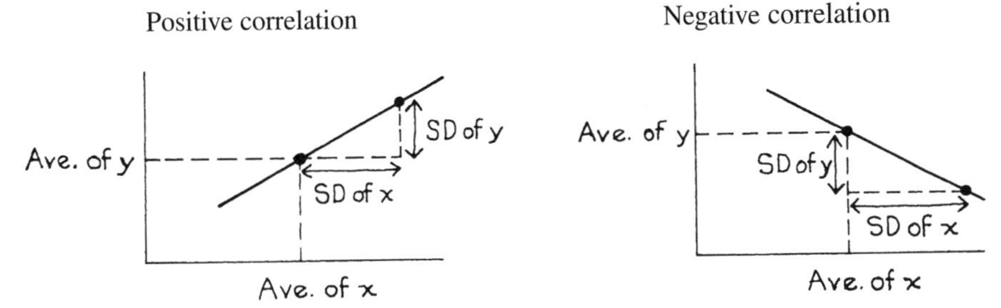

Exercise Set C

1. True or false:
   (a) The SD line always goes through the point of averages.
   (b) The SD line always goes through the point (0, 0).

2. For the scatter diagram shown below, say whether it is the solid line or the dashed line which is the SD line.

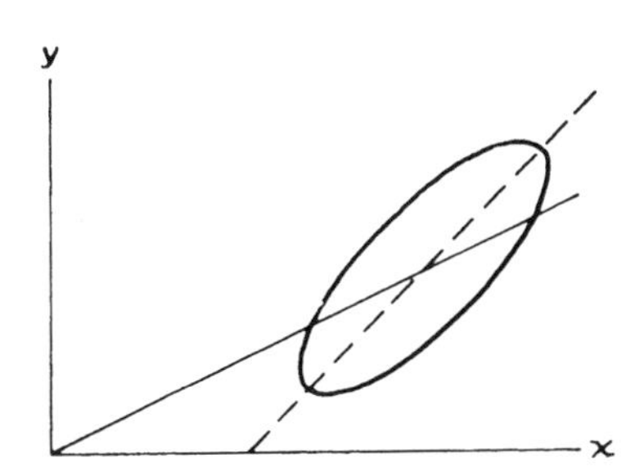

3. One study on male college students found their average height to be 69 inches, with an SD of 3 inches. Their average weight was 140 pounds, with an SD of 20 pounds. And the correlation was 0.60. If one of these people is 72 inches tall, how heavy would he have to be to fall on the SD line?

4. Using the same data as in exercise 3, say whether each of the following students was on the SD line:
   (a) height 75 inches, weight 180 pounds
   (b) height 66 inches, weight 130 pounds
   (c) height 66 inches, weight 120 pounds

The answers to these exercises are on p. A57.

4. COMPUTING THE CORRELATION COEFFICIENT

Here is the procedure for computing the correlation coefficient.

> Convert each variable to standard units. The average of the products gives the correlation coefficient.

(Standard units were discussed on pp. 79–80.) This procedure can be given as a formula, where x stands for the first variable, y for the second variable, and r for the correlation coefficient:

$$r = \text{average of } (x \text{ in standard units}) \times (y \text{ in standard units}).$$

Example 1. Compute r for the hypothetical data in table 1.

Table 1. Data.

x	y
1	5
3	9
4	7
5	1
7	13

Note. The first row of table 1 represents two measurements on one subject in the study; the two numbers are the x- and y-coordinates of the corresponding point on the scatter diagram. Similarly for the other rows. The pairing matters: r is defined only when you have two variables, and both are measured for every subject in the study.

Solution. The work can be laid out as in table 2.

Step 1. Convert the x-values to standard units, as in chapter 5. This is quite a lot of work. First, you have to find the average and SD of the x-values:

$$\text{average of } x\text{-values} = 4, \quad \text{SD} = 2.$$

Then, you have to subtract the average from each x-value, and divide by the SD:

$$\frac{1-4}{2} = -1.5 \quad \frac{3-4}{2} = -0.5 \quad \frac{4-4}{2} = 0 \quad \frac{5-4}{2} = 0.5 \quad \frac{7-4}{2} = 1.5$$

Table 2. Computing r.

x	y	*x in standard units*	*y in standard units*	*Product*
1	5	−1.5	−0.5	0.75
3	9	−0.5	0.5	−0.25
4	7	0.0	0.0	0.00
5	1	0.5	−1.5	−0.75
7	13	1.5	1.5	2.25

The results go into the third column of table 2. The numbers tell you how far above or below average the x-values are, in terms of the SD. For instance, the value 1 is 1.5 SDs below average.

Step 2. Convert the y-values to standard units; the results go into the fourth column of the table. That finishes the worst of the arithmetic.

Step 3. For each row of the table, work out the product

$$(x \text{ in standard units}) \times (y \text{ in standard units})$$

The products go into the last column of the table.

Step 4. Take the average of the products:

$$r = \text{average of } (x \text{ in standard units}) \times (y \text{ in standard units})$$
$$= \frac{0.75 - 0.25 + 0.00 - 0.75 + 2.25}{5} = 0.40$$

This completes the solution. If you plot a scatter diagram for the data (figure 9a), the points slope up but are only loosely clustered.

Why does r work as a measure of association? In figure 9a, the products are marked at the corresponding dots. Horizontal and vertical lines are drawn through the point of averages, dividing the scatter diagram into four quadrants. If a point is in the lower left quadrant, both variables are below average and are negative in

Figure 9. How the correlation coefficient works.

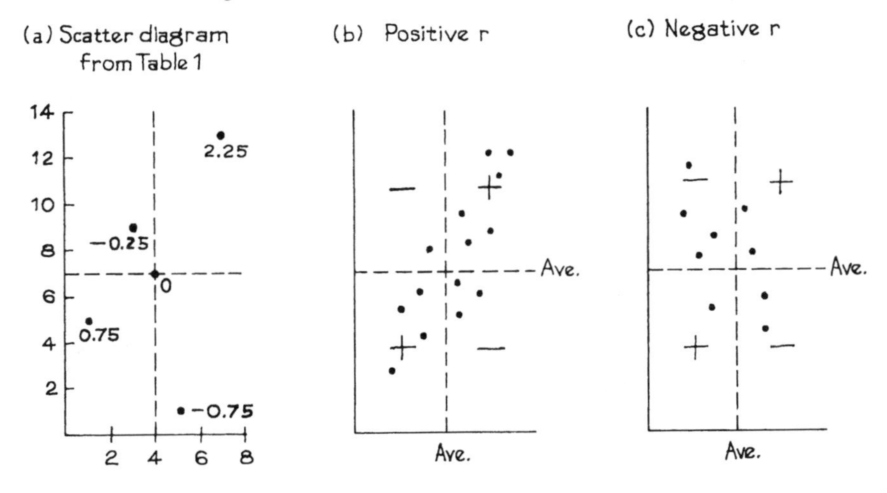

standard units; the product of two negatives is positive. In the upper right quadrant, the product of two positives is positive. In the remaining two quadrants, the product of a positive and a negative is negative. The average of all these products is the correlation coefficient. If r is positive, then points in the two positive quadrants will predominate, as in figure 9b. If r is negative, points in the two negative quadrants will predominate, as in figure 9c.

Exercise Set D

1. For each of the data sets shown below, calculate r.

(a)		(b)		(c)	
x	y	x	y	x	y
1	6	1	2	1	7
2	7	2	1	2	6
3	5	3	4	3	5
4	4	4	3	4	4
5	3	5	7	5	3
6	1	6	5	6	2
7	2	7	6	7	1

2. Find the scatter diagram in figure 6 (p. 127) with a correlation of 0.95. In this diagram, the percentage of points where both variables are simultaneously above average is around

5% 25% 50% 75% 95%.

3. Repeat exercise 2, for a correlation of 0.00.

4. Using figure 7, repeat exercise 2 for a correlation of −0.95.

The answers to these exercises are on p. A57.

Technical note. There is another way to compute r, which is sometimes useful:[8]

$$r = \frac{\text{cov}(x, y)}{(\text{SD of } x) \times (\text{SD of } y)}$$

where

$$\text{cov}(x, y) = (\text{average of products } xy) - (\text{ave of } x) \times (\text{ave of } y).$$

5. REVIEW EXERCISES

Review exercises may cover material from previous chapters.

1. A study of the IQs of husbands and wives obtained the following results:

for husbands, average IQ = 100, SD = 15
for wives, average IQ = 100, SD = 15
$r = 0.6$

One of the following is a scatter diagram for the data. Which one? Say briefly why you reject the others.

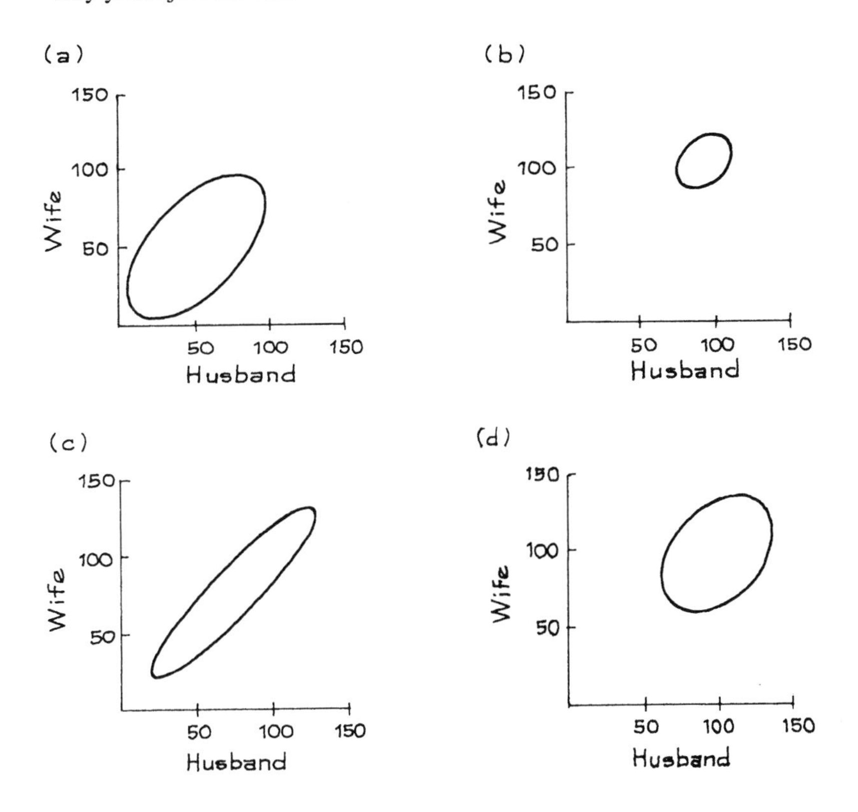

2. (a) For a representative sample of cars, would the correlation between the age of the car and its gasoline economy (miles per gallon) be positive or negative?
   (b) The correlation between gasoline economy and income of owner turns out to be positive.[9] How do you account for this positive association?

3. Suppose men always married women who were exactly 8% shorter. What would the correlation between their heights be?

4. Is the correlation between the heights of husbands and wives in the U.S. around −0.9, −0.3, 0.3, or 0.9? Explain briefly.

5. Three data sets are collected, and the correlation coefficient is computed in each case. The variables are
   (i) grade point average in freshman year and in sophomore year
   (ii) grade point average in freshman year and in senior year
   (iii) length and weight of two-by-four boards

   Possible values for correlation coefficients are

   −0.50   0.0   0.30   0.60   0.95

   Match the correlations with the data sets; two will be left over. Explain your choices.

6. In one class, the correlation between scores on the final and the midterm was 0.50, while the correlation between the scores on the final and the homework was 0.25. True or false, and explain: the relationship between the final scores and the midterm scores is twice as linear as the relationship between the final scores and the homework scores.

7. The figure below has six scatter diagrams for hypothetical data. The correlation coefficients, in scrambled order, are:

   −0.85 −0.38 −1.00 0.06 0.97 0.62

   Match the scatter diagrams with the correlation coefficients.

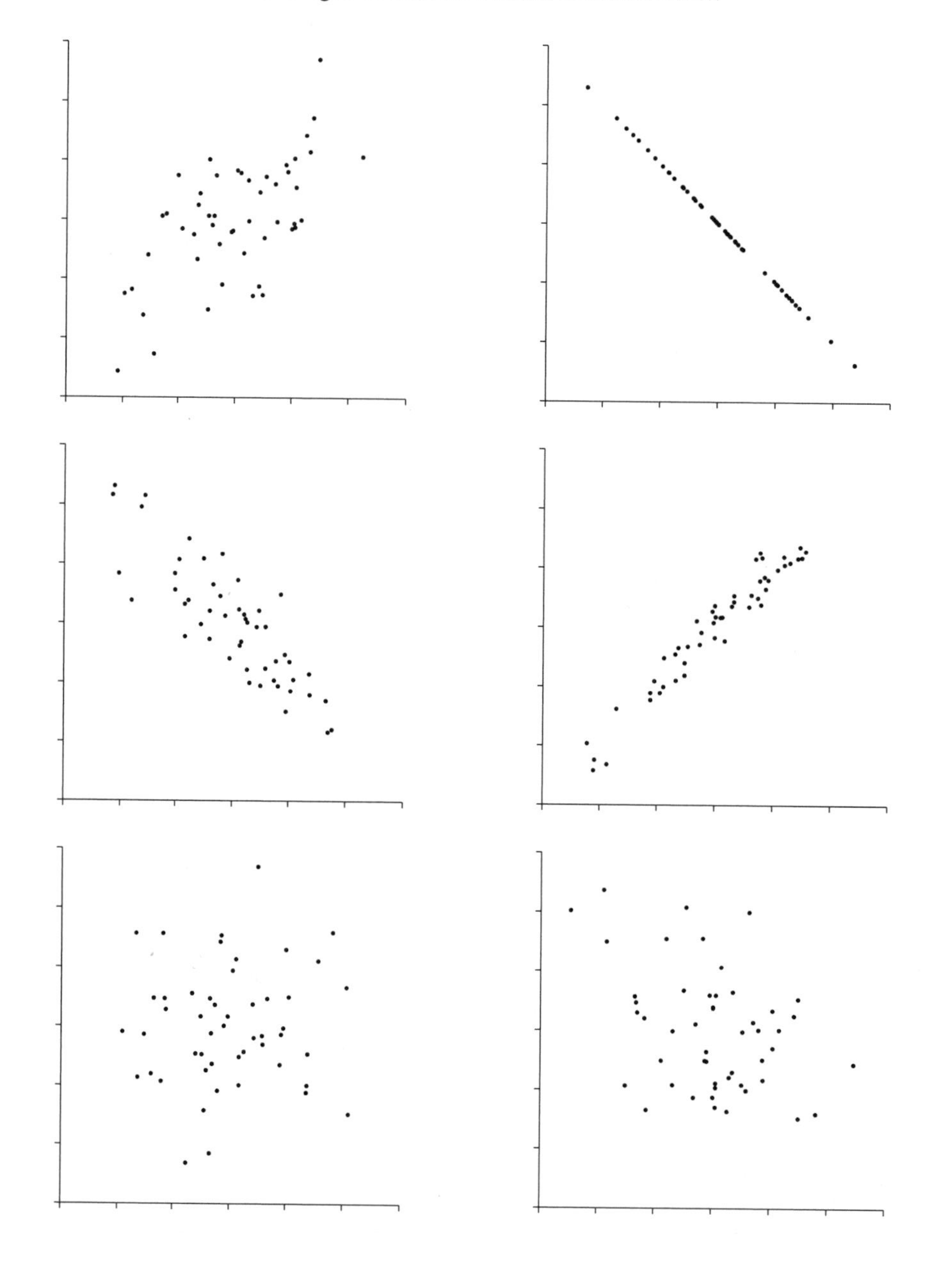

8. A longitudinal study of human growth was started in 1929 at the Berkeley Institute of Human Development.[10] The scatter diagram below shows the heights of 64 boys, measured at ages 4 and 18.
   (a) The average height at age 4 is around
   38 inches 42 inches 44 inches
   (b) The SD of height at age 18 is around
   0.5 inches 1.0 inches 2.5 inches
   (c) The correlation coefficient is around
   0.50 0.80 0.95
   (d) Which is the SD line—solid or dashed?

   Explain your answers.

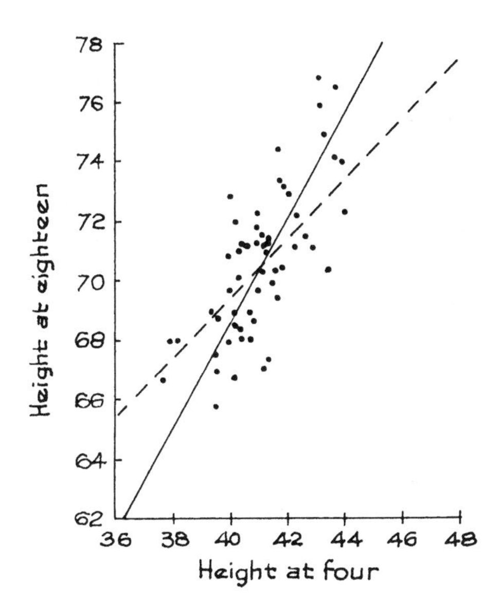

9. Find the correlation coefficient for each of the three data sets shown below.

(a)

x	y
1	5
1	3
1	5
1	7
2	3
2	3
2	1
3	1
3	1
4	1

(b)

x	y
1	1
1	2
1	1
1	3
2	1
2	4
2	1
3	2
3	2
4	3

(c)

x	y
1	2
1	2
1	2
1	2
2	4
2	4
2	4
3	6
3	6
4	8

10. In a large psychology study, each subject took two IQ tests (form L and form M of the Stanford-Binet). A scatter diagram for the test scores is sketched at the top of the next page. You are trying to predict the score on

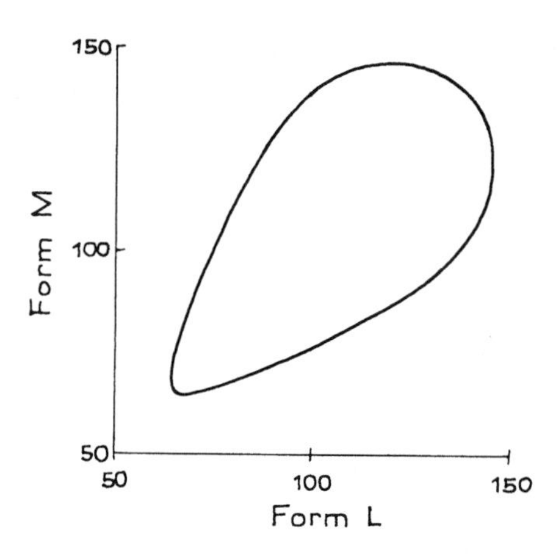

form M from the score on form L. Each prediction is off by some amount. On the whole, will these prediction errors be smaller when the score on form L is 75, or 125? or is it about the same for both?

11. A teaching assistant gives a quiz with 10 questions and no part credit. After grading the papers, the TA writes down for each student the number of questions the student got right and the number wrong. The average number of right answers is 6.4 with an SD of 2.0; the average number of wrong answers is 3.6 with the same SD of 2.0. The correlation between the number of right answers and the number of wrongs is

    0 $\quad -0.50 \quad +0.50 \quad -1 \quad +1$ can't tell without the data

    Explain.

Figure for exercise 12

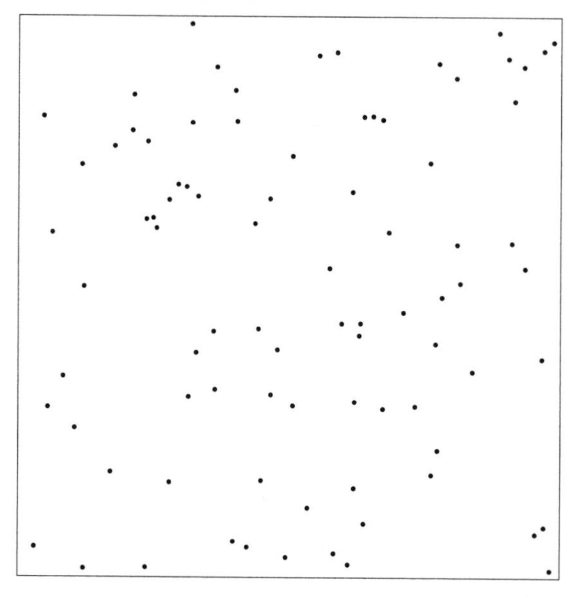

12. Fifteen students in an elementary statistics course at U.C. Berkeley were asked to count the dots in a figure like the one at the bottom of the previous page; there were 85 dots in the figure. The counts are shown in the table below. Make a scatter diagram for the counts. Represent each student by one point on your diagram, showing the first and second count. Label both your axes fully. Choose the scale so you can see the pattern in the points. Use your scatter diagram to answer the following questions:
    (a) Did the students work independently?
    (b) True or false: those students who counted high the first time also tended to be high the second time.

The two counts

1st	*2nd*
91	85
81	83
86	85
83	84
85	85
85	84
85	89
84	83
91	82
91	82
91	82
85	85
85	85
87	85
90	85

6. SUMMARY

1. The relationship between two variables can be represented by a *scatter diagram*. When the scatter diagram is tightly clustered around a line, there is a strong *linear association* between the variables.

2. A scatter diagram can be summarized by means of five statistics:

- the average of the x-values, the SD of the x-values,
- the average of the y-values, the SD of the y-values,
- the *correlation coefficient* r.

3. Positive association (a cloud which slopes up) is indicated by a plus-sign in the correlation coefficient. Negative association (a cloud which slopes down) is indicated by a minus-sign.

4. In a series of scatter diagrams with the same SDs, as r gets closer to ± 1, the points cluster more tightly around a line.

5. The correlation coefficient ranges from -1 (when all the points lie on a line which slopes down), to $+1$ (when all the points lie on a line which slopes up).

6. The *SD line* goes through the point of averages. When r is positive, the slope of the line is

$$(\text{SD of } y)/(\text{SD of } x).$$

When r is negative, the slope is

$$-(\text{SD of } y)/(\text{SD of } x).$$

7. To calculate the correlation coefficient, convert each variable to standard units, and then take the average product.

9

More about Correlation

"Very true," said the Duchess: "flamingoes and mustard both bite. And the moral of that is—'Birds of a feather flock together.'"
"Only mustard isn't a bird," Alice remarked.
"Right, as usual," said the Duchess: "what a clear way you have of putting things!"
—*Alice in Wonderland*

1. FEATURES OF THE CORRELATION COEFFICIENT

The correlation coefficient is a pure number. Why? Because the first step in computing r is a conversion to standard units. Original units—like inches for height data or degrees for temperature data—cancel out. In a similar way, r is not affected if you multiply all the values of one variable by the same positive number, or if you add the same number to all the values of one variable. (As a statistician might say, r is not affected by *changes of scale*; see pp. 92–93.)

For example, if you multiply each value of x by 3, then the average gets multiplied by 3. All the deviations from average get multiplied by 3 as well, and so does the SD. This common factor cancels in the conversion to standard units. So r stays the same. For another example, suppose you add 7 to each value of x. Then the average of x goes up by 7 too. However, the deviations from average do not change. Neither does r.

Figure 1 (on the next page) shows the correlation between daily maximum temperatures at New York and Boston. There is a dot in the diagram for each day of June 2005. The temperature in New York that day is plotted on the horizontal axis; the Boston temperature, on the vertical. The left hand panel does it in degrees

Fahrenheit, and $r = 0.5081$. The right hand panel does it in degrees Celsius, and r stays the same.[1] The conversion from Fahrenheit to Celsius is just a change of scale, which does not affect the correlation.

Figure 1. Daily maximum temperatures. New York and Boston, June 2005. The left hand panel plots the data in degrees Fahrenheit; the right hand panel, in degrees Celsius. This does not change r.

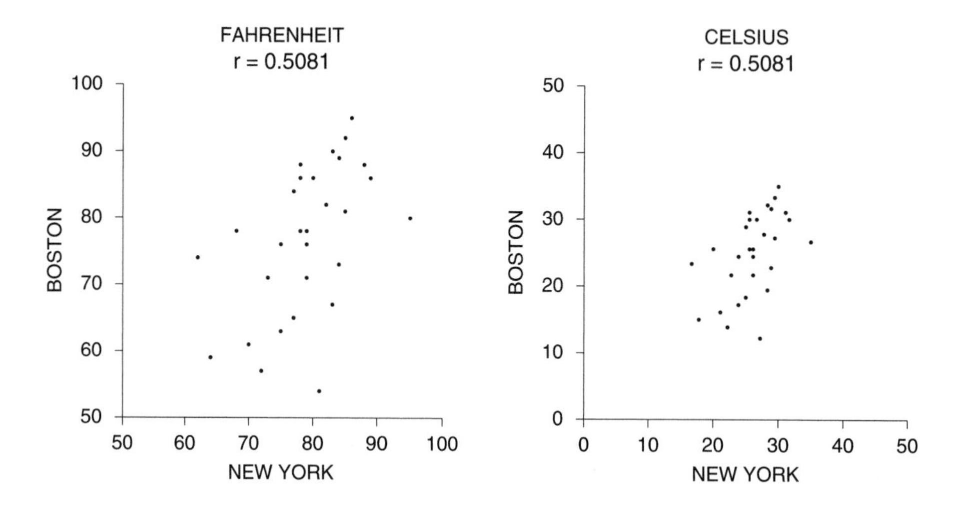

Another feature: The correlation between x and y is the same as the correlation between y and x. For example, the left hand panel in figure 2 is a scatter diagram for temperature data at New York in June 2005. The minimum tempera-

Figure 2. Daily temperatures. New York, June 2005.

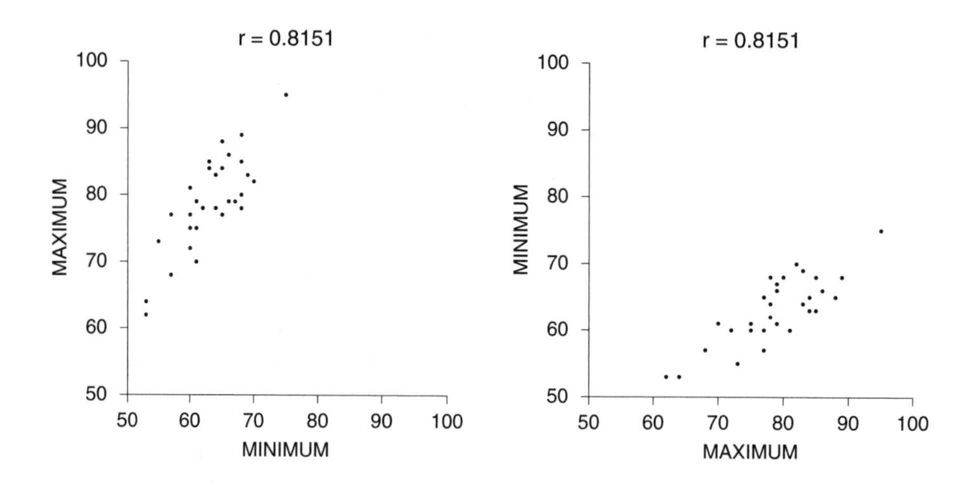

ture each day is plotted on the horizontal axis; the maximum, on the vertical. The correlation between the minimum and the maximum temperature is 0.8151. The right hand panel shows exactly the same data. This time, the minimum is plotted on the vertical instead of the horizontal. The pictures look different because the points are reflected around the diagonal. But r stays the same. Switching the order of the variables does not affect r. Why? Remember, r is the average of the products after conversion to standard units. Products do not depend on the order of the factors ($a \times b = b \times a$). It may be surprising that the correlation is only 0.8151, but the weather is full of surprises.

> The correlation coefficient is a pure number, without units. It is not affected by
>
> - interchanging the two variables,
> - adding the same number to all the values of one variable,
> - multiplying all the values of one variable by the same positive number.

Exercise Set A

1. (a) In June 2005, which city was warmer—Boston or New York? Or were they about the same?
   (b) In the left hand panel of figure 2, all the dots are above the 45-degree line. Why?

2. A small data set is shown below; $r \approx 0.76$. If you switch the two columns, does this change r? Explain or calculate.

x	y
1	2
2	3
3	1
4	5
5	6

3. As in exercise 2, but you add 3 to each value of y instead of interchanging the columns.

4. As in exercise 2, but you double each value of x.

5. As in exercise 2, but you interchange the last two values (5 and 6) for y.

6. Suppose the correlation between x and y is 0.73.
   (a) Does the scatter diagram slope up or down?
   (b) If you multiply all the values of y by -1, would the new scatter diagram slope up or down?
   (c) If you multiply all the values of y by -1, what happens to the correlation?

7. Two different investigators are working on a growth study. The first measures the heights of 100 children, in inches. The second prefers the metric system, and changes the results to centimeters (multiplying by the conversion factor 2.54 centimeters per inch). A scatter diagram is plotted, showing for each child its height in inches on the horizontal axis, and height in centimeters on the vertical axis.
   (a) If no mistakes are made in the conversion, what is the correlation?
   (b) What happens to r if mistakes are made in the arithmetic?
   (c) What happens to r if the second investigator goes out and measures the same children again, using metric equipment?

8. In figure 1 on p. 120, the correlation is 0.5. Suppose we plot on the horizontal axis the height of the paternal grandfather (not the father); the height of the son is still plotted on the vertical axis. Would the correlation be more or less than 0.5?

9. Two weathermen compute the correlation between daily maximum temperatures for Washington and Boston. One does it for June; the other does it for the whole year. Who gets the bigger correlation? ("Washington" is the city, not the state.)

10. Six data sets are shown below. In (i), the correlation is 0.8571, and in (ii) the correlation is 0.7857. Find the correlations for the remaining data sets. No arithmetic is necessary.

(i)		(ii)		(iii)		(iv)		(v)		(vi)	
x	y	x	y	x	y	x	y	x	y	x	y
1	2	1	2	2	1	2	2	1	4	0	6
2	3	2	3	3	2	3	3	2	6	1	9
3	1	3	1	1	3	4	1	3	2	2	3
4	4	4	4	4	4	5	4	4	8	3	12
5	6	5	6	6	5	6	6	5	12	4	18
6	5	6	7	7	6	7	5	6	10	5	21
7	7	7	5	5	7	8	7	7	14	6	15

The answers to these exercises are on pp. A57–58.

2. CHANGING SDs

The appearance of a scatter diagram depends on the SDs. For instance, look at figure 3. In both diagrams, r is 0.70. However, the top one looks more tightly clustered around the SD line. That is because its SDs are smaller. The formula for r involves converting the variables to standard units: deviations from average are divided by the SD. So, r measures clustering not in absolute terms but in relative terms—relative to the SDs.

To interpret a correlation coefficient graphically, draw the scatter diagram in your mind's eye so the vertical SD covers the same distance on the page as the vertical SDs in figure 6 on p. 127; and likewise for the horizontal SD. If r for your scatter diagram is 0.40, it will probably show about the same amount of clustering around the diagonal as the one with an r of 0.40 in the figure at the top right. If r is 0.90, it will look like the diagram in the figure at the bottom left. In general, your scatter diagram will match the one that has a similar value for r.

Figure 3. The effect of changing SDs. The two scatter diagrams have the same correlation coefficient of 0.70. The top diagram looks more tightly clustered around the SD line because its SDs are smaller.

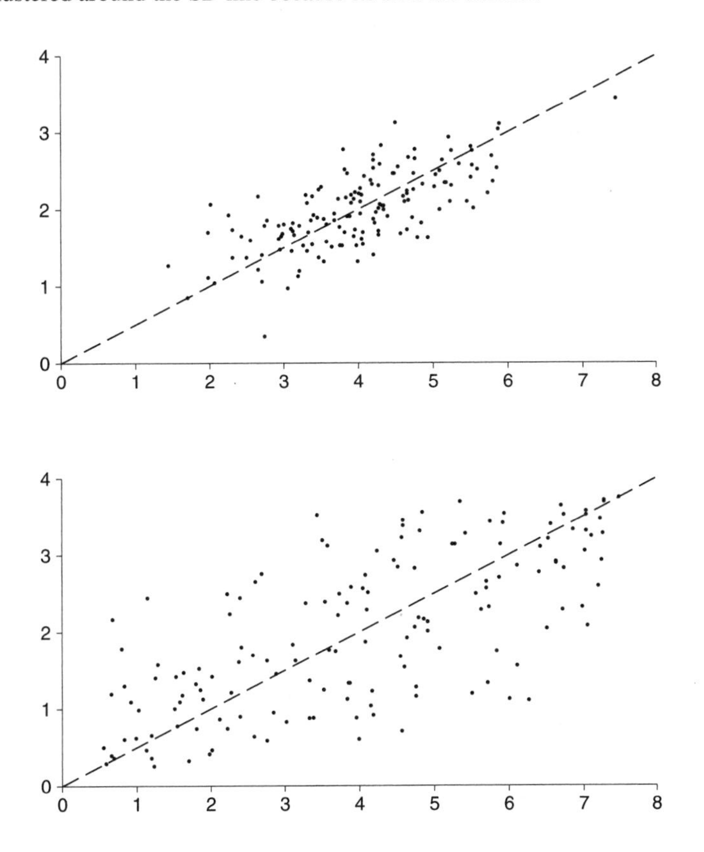

Exercise Set B

1. In the figure below, 6 scatter diagrams are plotted on the same pair of axes; in the first, the points are marked "a"; in the second, "b"; and so forth. For each of the 6 diagrams taken on its own, the correlation is around 0.6. Now take all the points together. For the combined diagram, is the correlation around 0.0, 0.6, or 0.9?

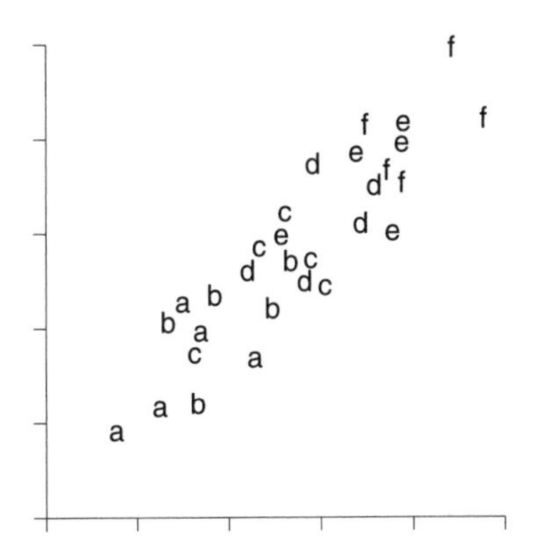

2. The National Health and Nutrition Examination Survey (p. 58) also covers children. In HANES2, at each age from 6 to 11, the correlation between height and weight was just about 0.67. For all the children together, would the correlation between height and weight be just about 0.67, somewhat more than 0.67, or somewhat less than 0.67? Choose one option and explain.

3. Below are three scatter diagrams. Do they have the same correlation? Try to answer without calculating.

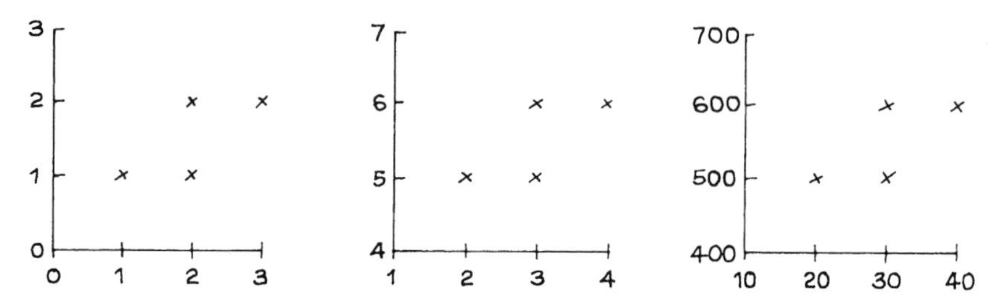

4. Someone hands you the scatter diagram shown below, but has forgotten to label the axes. Can you still calculate r? If so, what is it? Or do you need the labels?

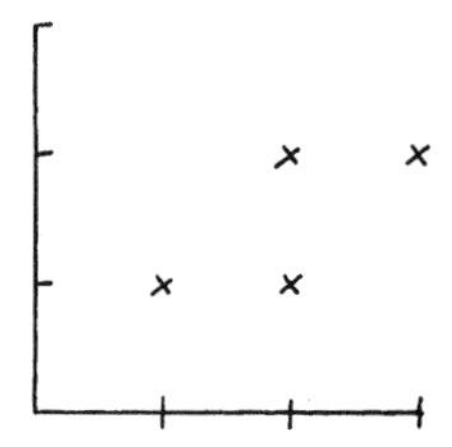

The answers to these exercises are on p. A58.

Technical notes. (i) If r is close to 1, then a typical point is only a small fraction of a vertical SD above or below the SD line. If r is close to 0, then a typical point is above or below the line by an amount roughly comparable in size to the vertical SD: see figure 4. (The "vertical SD" is the SD of the variable plotted on the y-axis.)

Figure 4. The correlation coefficient. As r gets close to 1, the distance of a typical point above or below the SD line becomes a small fraction of the vertical SD.

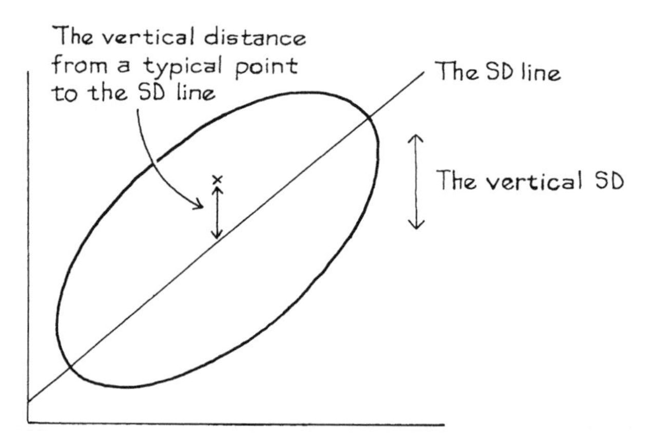

(ii) The connection between the correlation coefficient and the typical distance above or below the SD line can be expressed mathematically, as follows. The r.m.s. vertical distance to the SD line equals

$$\sqrt{2(1 - |r|)} \times \text{the vertical SD}$$

Take, for example, a correlation of 0.95. Then

$$\sqrt{2(1 - |r|)} = \sqrt{0.1} \approx 0.3$$

So the spread around the SD line is about 30% of a vertical SD. That is why a scatter diagram with $r = 0.95$ shows a fair amount of spread around the line (figure 6 on p. 127). There are similar formulas for the horizontal direction.

3. SOME EXCEPTIONAL CASES

The correlation coefficient is useful for football-shaped scatter diagrams. For other diagrams, r can be misleading. Outliers and non-linearity are problem cases. In figure 5a, the dots show a perfect correlation of 1. The outlier, marked by a cross, brings the correlation down almost to 0. Figure 5a should not be summarized using r. Some people get carried away in pursuit of outliers. However, in any scatter diagram there will be some points more or less detached from the main part of the cloud. These points should be rejected only if there is good reason to do so.

Figure 5. The correlation coefficient can be misleading in the presence of outliers or non-linear association.

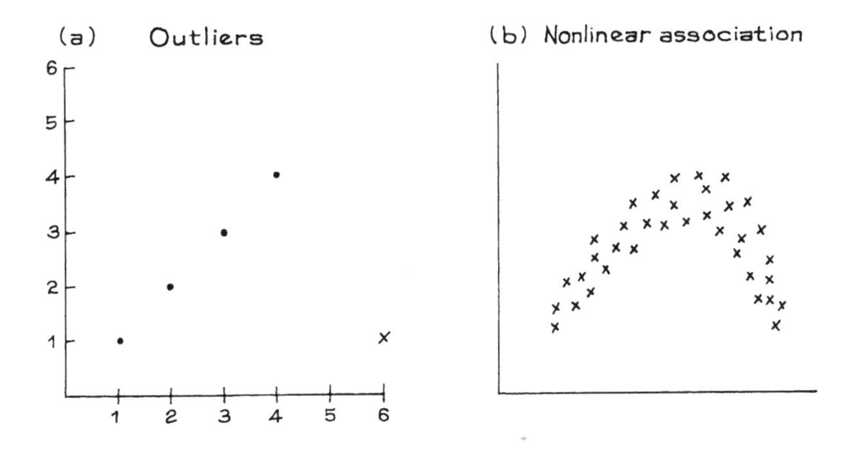

In figure 5b, the correlation coefficient is close to 0, even though the points show a strong association. The reason is that the graph does not look at all like a straight line: as x increases, y rises then falls. This pattern is shown by the association between weight and age for adult men (figure 3 on p. 59). Again, such data should not be summarized using r—the pattern gets lost.

> r measures linear association, not association in general.

Exercise Set C

1. Which of the following three scatter diagrams should be summarized by r?

2. A class of 15 students happens to include 5 basketball players. True or false, and explain: the relationship between heights and weights for this class should be summarized using r.

3. A circle of diameter d has area $\frac{1}{4}\pi d^2$. An investigator plots a scatter diagram of area against diameter for a sample of circles with different diameters. (The diagram is shown below.) The correlation coefficient is ______. Fill in the blank, and explain. Options:

-1 nearly -1 nearly 0 nearly 1 1

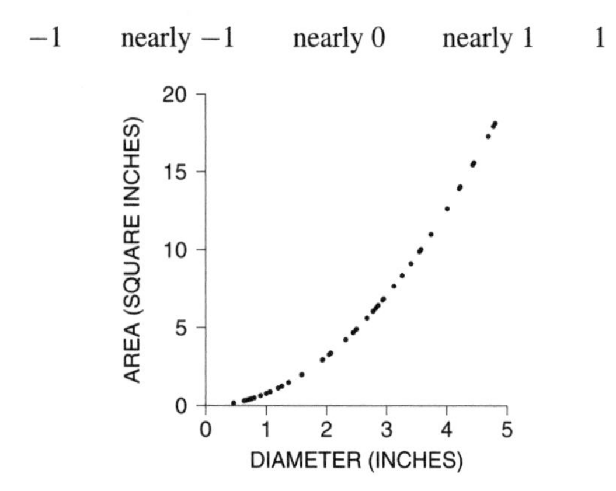

4. For a certain data set, $r = 0.57$. Say whether each of the following statements is true or false, and explain briefly; if you need more information, say what you need, and why.
   (a) There are no outliers.
   (b) There is a non-linear association.

The answers to these exercises are on p. A58.

4. ECOLOGICAL CORRELATIONS

In 1955, Sir Richard Doll published a landmark article on the relationship between cigarette smoking and lung cancer.[2] One piece of evidence was a scatter diagram showing the relationship between the rate of cigarette smoking (per capita) and the rate of deaths from lung cancer in eleven countries. The correlation between these eleven pairs of rates was 0.7, and this was taken as showing the strength of the relationship between smoking and cancer. However, it is not countries which smoke and get cancer, but people. To measure the strength of the relationship for people, it is necessary to have data relating smoking and cancer for individuals rather than countries. Such studies are available, and show that smoking does indeed cause cancer.

The statistical point: correlations based on rates or averages can be misleading. Here is another example. From Current Population Survey data for 2005, you can compute the correlation between income and education for men age 25–64 in the United States: $r \approx 0.42$. For each state (and D.C.), you can compute average educational level and average income. Finally, you can compute the correlation between the 51 pairs of averages: $r \approx 0.70$. If you used the correlation for the states to estimate the correlation for the individuals, you would be way off. The reason is that within each state, there is a lot of spread around the averages. Replacing the states by their averages eliminates the spread, and gives a misleading impression of tight clustering. Figure 6 shows the effect for three states.[3]

Ecological correlations are based on rates or averages. They are often used in political science and sociology. And they tend to overstate the strength of an association. So watch out.

Figure 6. Ecological correlations (based on rates or averages) are usually too big. The panel on the left represents income and education for individuals in three states, labeled A, B, C. Each individual is marked by a letter showing state of residence. The correlation is moderate. The panel on the right shows the averages for each state. The correlation between the averages is almost 1.

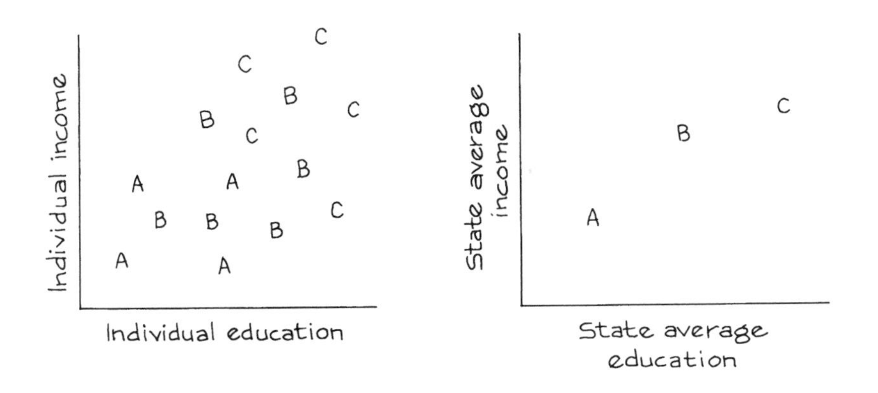

Exercise Set D

1. The table at the top of the next page is adapted from Doll and shows per capita consumption of cigarettes in various countries in 1930, and the death rates from lung cancer for men in 1950. (In 1930, hardly any women smoked; and a long period of time is needed for the effects of smoking to show up.)

Country	*Cigarette consumption*	*Deaths per million*
Australia	480	180
Canada	500	150
Denmark	380	170
Finland	1,100	350
Great Britain	1,100	460
Iceland	230	60
Netherlands	490	240
Norway	250	90
Sweden	300	110
Switzerland	510	250
U.S.	1,300	200

(a) Plot a scatter diagram for these data.

(b) True or false: the higher cigarette consumption was in 1930 in one of these countries, on the whole the higher the death rate from lung cancer in 1950. Or can this be determined from the data?

(c) True or false: death rates from lung cancer tend to be higher among those persons who smoke more. Or can this be determined from the data?

2. A sociologist is studying the relationship between suicide and literacy in nineteenth-century Italy.[4] He has data for each province, showing the percentage of literates and the suicide rate in that province. The correlation is 0.6. Does this give a fair estimate of the strength of the association between literacy and suicide?

The answers to these exercises are on p. A59.

5. ASSOCIATION IS NOT CAUSATION

For school children, shoe size is strongly correlated with reading skills. However, learning new words does not make the feet get bigger. Instead, there is a third factor involved—age. As children get older, they learn to read better and they outgrow their shoes. (According to the statistical jargon of chapter 2, age is a confounder.) In the example, the confounder was easy to spot. Often, this is not so easy. And the arithmetic of the correlation coefficient does not protect you against third factors.[5]

> Correlation measures association. But association is not the same as causation.

Example 1. Education and unemployment. During the Great Depression of 1929–1933, better-educated people tended to have shorter spells of unemployment. Does education protect you against unemployment?

Discussion. Perhaps, but the data were observational. As it turned out, age was a confounding variable. The younger people were better educated, because the educational level had been going up over time. (It still is.) Given a choice in hiring, employers seemed to prefer younger job-seekers. Controlling for age made the effect of education on unemployment much weaker.[6]

Example 2. Range and duration of species. Does natural selection operate at the level of species? This is a question of some interest for paleontologists. David Jablonski argues that geographical range is a heritable characteristic of species: a species with a wide range survives longer, because if a disaster strikes in one place, the species stays alive at other places.

One piece of evidence is a scatter diagram (figure 7). Ninety-nine species of gastropods (slugs, snails, etc.) are represented in the diagram. The duration of the species—its lifetime, in millions of years—is plotted on the vertical axis; its range is on the horizontal, in kilometers. Both variables are determined from the fossil record. There is a good positive association: r is about 0.64. (The cloud looks formless, but that is because of a few straggling points at the bottom right and the top left.) Does a wide geographical range promote survival of the species?

Figure 7. Duration of species in millions of years plotted against geographical range in kilometers, for 99 species of gastropods. Several species can be plotted at the same point; the number of such species is indicated next to the point.

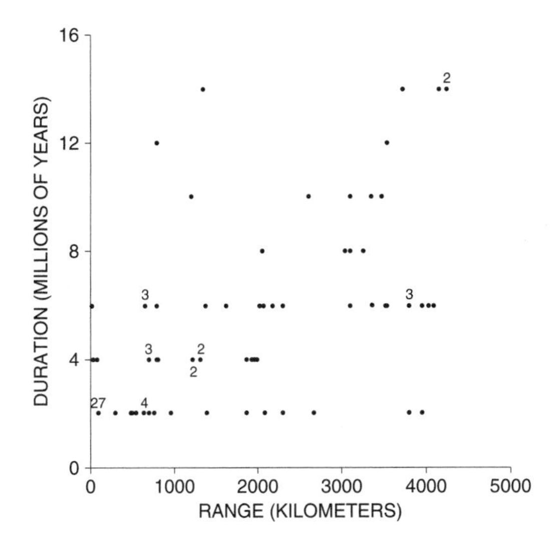

Discussion. A wide range may cause a long lifetime. Or, a long lifetime may cause a wide range. Or, there may be something else going on. Jablonski had his eye on the first possibility. The second one is unlikely, because other evidence suggests that species achieve their ranges very soon after they emerge. But what about the third explanation? Michael Russell and David Lindberg point out that species with a wide geographical range have more chances to be preserved in the fossil record, which can create the appearance of a long lifetime. If so, figure 7 is a statistical artifact.[7] Association is not causation.

Example 3. Fat in the diet and cancer. In countries where people eat lots of fat—like the U.S.—rates of breast cancer and colon cancer are high. See figure 8 for data on breast cancer. This correlation is often used to argue that fat in the diet causes cancer. How good is the evidence?

Figure 8. Death rates from breast cancer plotted against fat in the diet, for a sample of countries.

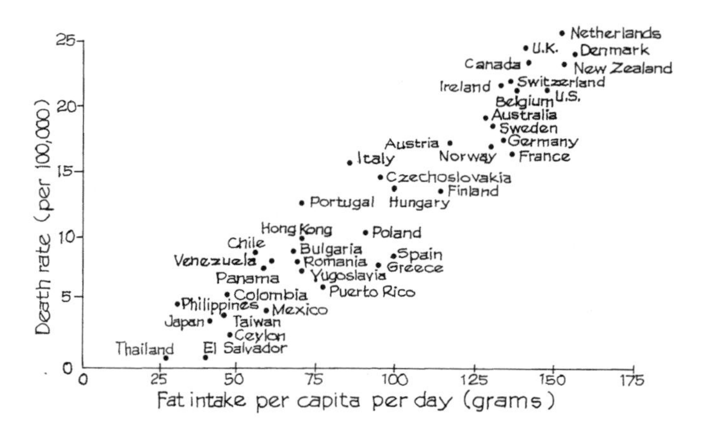

Note: Age standardized.
Source: K. Carroll, "Experimental evidence of dietary factors and hormone-dependent cancers," *Cancer Research* vol. 35 (1975) p. 3379. Copyright by *Cancer Research*. Reproduced by permission.

Discussion. If fat in the diet causes cancer, then the points in the diagram should slope up, other things being equal. So the diagram is some evidence for the theory. But the evidence is quite weak, because other things aren't equal. For example, the countries with lots of fat in the diet also have lots of sugar. A plot of breast cancer rates against sugar consumption would look just like figure 8, and nobody thinks that sugar causes breast cancer. As it turns out, fat and sugar are relatively expensive. In rich countries, people can afford to eat fat and sugar rather than starchier grain products. Some aspects of the diet in these countries, or other factors in the life-style, probably do cause certain kinds of cancer—and protect against other kinds. So far, epidemiologists can identify only a few of these factors with any real confidence.[8]

Exercise Set E

1. The scatter diagram in figure 7 shows stripes. Why?
2. Is the correlation in figure 8 ecological? How is that relevant to the argument?
3. The correlation between height and weight among men age 18–74 in the U.S. is about 0.40. Say whether each conclusion below follows from the data; explain your answer.
   (a) Taller men tend to be heavier.
   (b) The correlation between weight and height for men age 18–74 is about 0.40.
   (c) Heavier men tend to be taller.

(d) If someone eats more and puts on 10 pounds, he is likely to get somewhat taller.

4. Studies find a negative correlation between hours spent watching television and scores on reading tests.[9] Does watching television make people less able to read? Discuss briefly.

5. Many studies have found an association between cigarette smoking and heart disease. One study found an association between coffee drinking and heart disease.[10] Should you conclude that coffee drinking causes heart disease? Or can you explain the association between coffee drinking and heart disease in some other way?

6. Many economists believe that there is trade-off between unemployment and inflation: low rates of unemployment will cause high rates of inflation, while higher rates of unemployment will reduce the rate of inflation. The relationship between the two variables is shown below for the U.S. in the decade 1960–69. There is one point for each year, with the rate of unemployment that year shown on the x-axis, and the rate of inflation shown on the y-axis. The points fall very close to a smooth curve known as the *Phillips Curve*. Is this an observational study or a controlled experiment? If you plotted the points for the 1970s or the 1950s, would you expect them to fall along the curve?

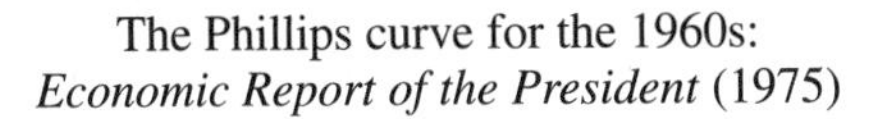
The Phillips curve for the 1960s:
Economic Report of the President (1975)

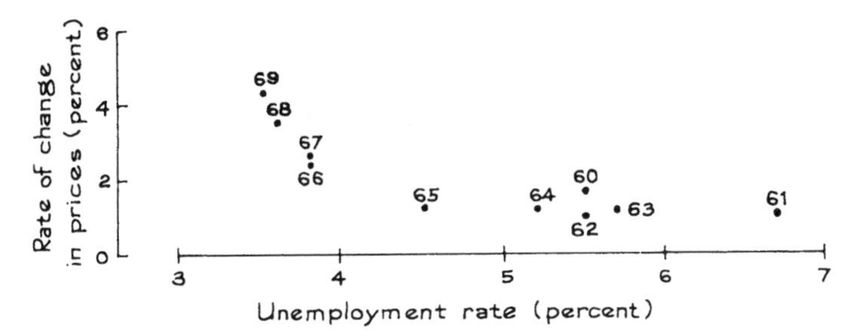

The answers to these exercises are on p. A59.

6. REVIEW EXERCISES

Review exercises may cover material from previous chapters.

1. When studying one variable, you can use a graph called a ______. When studying the relationship between two variables, you can use a graph called a ______.

2. True or false, and explain briefly:
   (a) If the correlation coefficient is -0.80, below-average values of the dependent variable are associated with below-average values of the independent variable.
   (b) If y is usually less than x, the correlation coefficient between x and y will be negative.

3. In each case, say which correlation is higher, and explain briefly. (Data are from a longitudinal study of growth.)
   (a) Height at age 4 and height at age 18, height at age 16 and height at age 18.
   (b) Height at age 4 and height at age 18, weight at age 4 and weight at age 18.
   (c) Height and weight at age 4, height and weight at age 18.

4. An investigator collected data on heights and weights of college students; results can be summarized as follows.

	Average	*SD*
Men's height	70 inches	3 inches
Men's weight	144 pounds	21 pounds
Women's height	64 inches	3 inches
Women's weight	120 pounds	21 pounds

The correlation coefficient between height and weight for the men was about 0.60; for the women, it was about the same. If you take the men and women together, the correlation between height and weight would be ______.

just about 0.60 somewhat lower somewhat higher

Choose one option, and explain briefly.

5. A number is missing in each of the data sets below. If possible, fill in the blank to make r equal to 1. If this is not possible, say why not.

(a)

x	y
1	1
2	3
2	3
4	–

(b)

x	y
1	1
2	3
3	4
4	–

6. A computer program prints out r for the two data sets shown below. Is the program working correctly? Answer yes or no, and explain briefly.

(i)

x	y
1	2
2	1
3	4
4	3
5	7
6	5
7	6

$r = 0.8214$

(ii)

x	y
1	5
2	4
3	7
4	6
5	10
6	8
7	9

$r = 0.7619$

7. In 1910, Hiram Johnson entered the California gubernatorial primaries. For each county, data are available to show the percentage of native-born Americans in that county, as well as the percentage of the vote for Johnson. A

political scientist calculated the correlation between these percentages.[11] It is 0.5. Is this a fair measure of the extent to which "Johnson received native, as opposed to immigrant, support?" Answer yes or no, and explain briefly.

8. For women age 25 and over in the U.S. in 2005, the relationship between age and educational level (years of schooling completed) can be summarized as follows:[12]

$$\text{average age} \approx 50 \text{ years}, \quad \text{SD} \approx 16 \text{ years}$$
$$\text{average ed. level} \approx 13.2 \text{ years}, \quad \text{SD} \approx 3.0 \text{ years}, \quad r \approx -0.20$$

True or false, and explain: as you get older, you become less educated. If this statement is false, what accounts for the negative correlation?

9. At the University of California, Berkeley, Statistics 2 is a large lecture course with small discussion sections led by teaching assistants. As part of a study, at the second-to-last lecture one term, the students were asked to fill out anonymous questionnaires rating the effectiveness of their teaching assistants (by name), and the course, on the scale

1	2	3	4	5
poor	fair	good	very good	excellent

The following statistics were computed.

- The average rating of the assistant by the students in each section.
- The average rating of the course by the students in each section.
- The average score on the final for the students in each section.

Results are shown below (sections are identified by letter). Draw a scatter diagram for each pair of variables—there are three pairs—and find the correlations.

Section	*Ave. rating of assistant*	*Ave. rating of course*	*Ave. score on final*
A	3.3	3.5	70
B	2.9	3.2	64
C	4.1	3.1	47
D	3.3	3.3	63
E	2.7	2.8	69
F	3.4	3.5	69
G	2.8	3.6	69
H	2.1	2.8	63
I	3.7	2.8	53
J	3.2	3.3	65
K	2.4	3.3	64

The data are section averages. Since the questionnaires were anonymous, it was not possible to link up student ratings with scores on an individual basis. Student ability may be a confounding factor. However, controlling for pretest results turned out to make no difference in the analysis.[13] Each assistant taught one section. True or false, and explain:

(a) On the average, those sections that liked their TA more did better on the final.
(b) There was almost no relationship between the section's average rating of the assistant and the section's average rating of the course.
(c) There was almost no relationship between the section's average rating of the course and the section's average score on the final.

10. In a study of 2005 Math SAT scores, the Educational Testing Service computed the average score for each of the 51 states, and the percentage of the high-school seniors in that state who took the test.[14] (For these purposes, D.C. counts as a state.) The correlation between these two variables was equal to −0.84.
    (a) True or false: test scores tend to be lower in the states where a higher percentage of the students take the test. If true, how do you explain this? If false, what accounts for the negative correlation?
    (b) In Connecticut, the average score was only 517. But in Iowa, the average was 608. True or false, and explain: the data show that on average, the schools in Iowa are doing a better job at teaching math than the schools in Connecticut.

11. As part of the study described in exercise 10, the Educational Testing Service computed the average Verbal SAT score for each state, as well as the average Math SAT score for each state. (Again, D.C. counts as a state.) The correlation between these 51 pairs of averages was 0.97. Would the correlation between the Math SAT and the Verbal SAT—computed from the data on all the individuals who took the tests—be larger than 0.97, about 0.97, or less than 0.97? Explain briefly.

12. Shown below is a scatter diagram for educational levels (years of schooling completed) of husbands and wives in South Carolina, from the March 2005 Current Population Survey.
    (a) The points make vertical and horizontal stripes. Why?

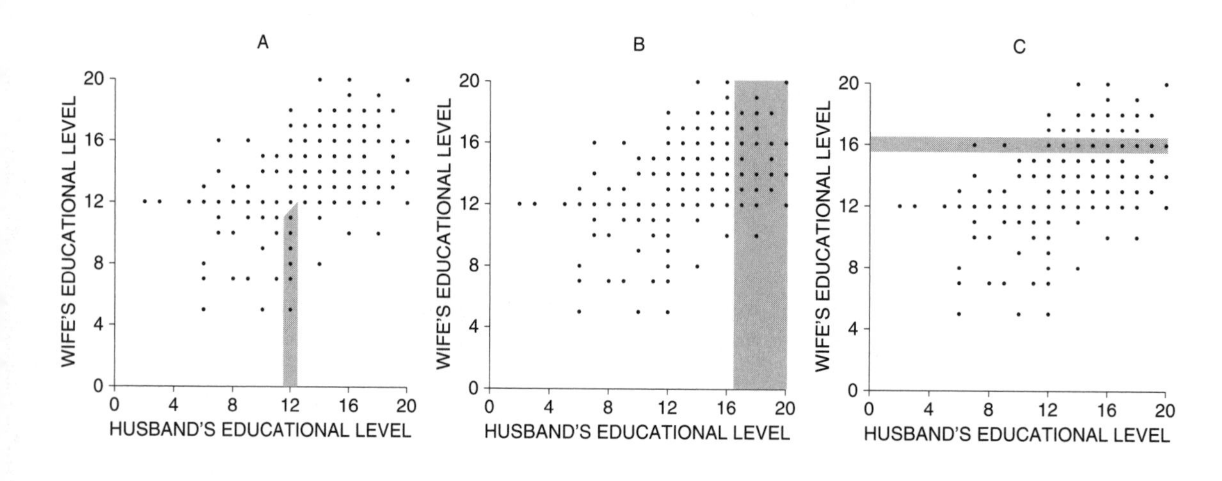

(b) There were 530 couples in the sample, and there is a dot for each couple. But if you count, there are only 104 dots in the scatter diagram. How can that be? Explain briefly.
(c) Three areas are shaded. Match the area with the description. (One description will be left over.)
    (i) Wife completed 16 years of schooling.
    (ii) Wife completed more years of schooling than husband.
    (iii) Husband completed more than 16 years of schooling.
    (iv) Husband completed 12 years of schooling and wife completed fewer years of schooling than husband.

7. SUMMARY

1. The correlation coefficient is a pure number, without units. It is not affected by

- interchanging the two variables,
- adding the same number to all the values of one variable,
- multiplying all the values of one variable by the same positive number.

2. The correlation coefficient measures clustering around a line, relative to the SDs.

3. The correlation coefficient can be misleading in the presence of outliers or non-linear association. Whenever possible, look at the scatter diagram to check for these problems.

4. *Ecological* correlations, which are based on rates or averages, tend to overstate the strength of associations for individuals.

5. Correlation measures association. But association does not necessarily show causation. It may only show that both variables are simultaneously influenced by some third variable.

10

Regression

You've got to draw the line somewhere.

1. INTRODUCTION

The regression method describes how one variable depends on another. For example, take height and weight. We have data for 471 men age 18–24 (from the Health and Nutrition Examination Survey—HANES5; see p. 58). In round numbers the average height of these men was 70 inches, and their overall average weight was 180 pounds. Naturally, the taller men weighed more. How much of an increase in weight is associated with a unit increase in height? To get started, look at the scatter diagram (figure 1 on the next page). Height is plotted on the horizontal axis, and weight on the vertical. The summary statistics are[1]

average height $\approx$ 70 inches, SD $\approx$ 3 inches
average weight $\approx$ 180 pounds, SD $\approx$ 45 pounds, $r \approx 0.40$

The scales on the vertical and horizontal axes have been chosen so that one SD of height and one SD of weight cover the same distance on the page. This makes the SD line (dashed) rise at 45 degrees across the page. There is a fair amount of scatter around the line: r is only 0.40.

The vertical strip in figure 1 shows the men who were one SD above average in height (to the nearest inch). The men who were also one SD above average in weight would be plotted on the SD line. However, most of the points in the strip are well below the SD line. In other words, most of the men who were one SD above average in height were quite a bit less than one SD above average in

Figure 1. Scatter diagram. Each point shows the height and weight for one of the 471 men age 18–24 in HANES5. The vertical strip represents men who are about one SD above average in height. Those who are also one SD above average in weight would be plotted along the dashed SD line. Most of the men in the strip are below the SD line: they are only part of an SD above average in weight. The solid regression line estimates average weight at each height.

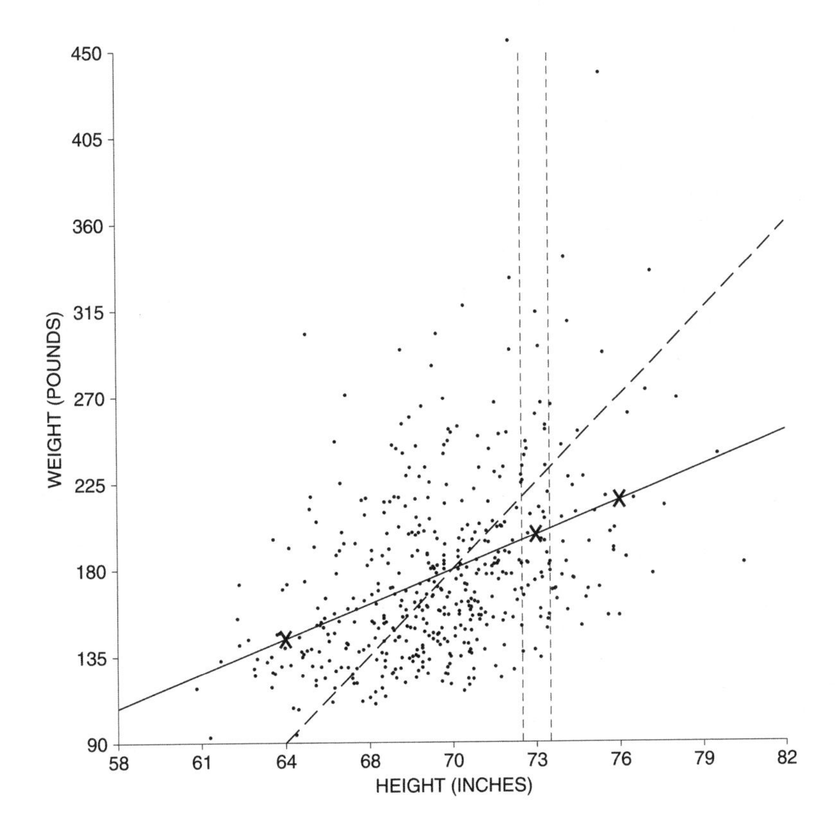

weight. The average weight of these men is only part of an SD above the overall average weight. This is where the correlation of 0.40 comes in. Associated with an increase of one SD in height there is an increase of only 0.40 SDs in weight, on the average.

To be more specific, take the men who are one SD above average in height:

$$\text{average height} + \text{SD of height} = 70 \text{ in} + 3 \text{ in} = 73 \text{ in}.$$

Their average weight will be above the overall average by 0.40 SDs of weight. Translated back to pounds, that's

$$0.40 \times 45 \text{ lb} = 18 \text{ lb}.$$

So, the average weight of these men is around

$$180 \text{ lb} + 18 \text{ lb} = 198 \text{ lb}.$$

The point (73 inches, 198 pounds) is marked by a cross in figure 1.

What about the men who are 2 SDs above average in height? Now

$$\text{average height} + 2 \text{ SD of height} = 70 \text{ in} + 2 \times 3 \text{ in} = 76 \text{ in}.$$

The average weight of this second group of men should be above the overall average by $0.40 \times 2 = 0.80$ SDs of weight. That's 0.80×45 lb $= 36$ lb. So their average is around 180 lb + 36 lb = 216 lb. The point (76 inches, 216 pounds) is also marked by a cross in figure 1.

What about the men who are 2 SDs below average in height? Their height equals

$$\text{average height} - 2 \text{ SD of height} = 70 \text{ in} - 2 \times 3 \text{ in} = 64 \text{ in}.$$

Their average weight is below the overall average by $0.40 \times 2 = 0.80$ SDs of weight. That's 0.80×45 lb $= 36$ lb. The average weight of this third group is around 180 lb − 36 lb = 144 lb. The point (64 inches, 144 pounds) is marked by a third cross in figure 1.

All the points (height, estimate for average weight) fall on the solid line shown in figure 1. This is the *regression line*. The line goes through the point of averages: men of average height should also be of average weight.

> The regression line for y on x estimates the average value for y corresponding to each value of x.

Along the regression line, associated with each increase of one SD in height there is an increase of only 0.40 SDs in weight. To be more specific, imagine grouping the men by height. There is a group which is average in height, another group which is one SD above average in height, and so on. From each group to the next, the average weight also goes up, but only by around 0.40 SDs. Remember where the 0.40 comes from. It is the correlation between height and weight.

This way of using the correlation coefficient to estimate the average value of y for each value of x is called the *regression method*. The method can be stated as follows.

> Associated with each increase of one SD in x there is an increase of only r SDs in y, on the average.

Two different SDs are involved here: the SD of x, to gauge changes in x; and the SD of y, to gauge changes in y. It is easy to get carried away by the rhythm: if x goes up by one SD, so does y. But that's wrong. On the average, y only goes up by r SDs (figure 2, next page).

Why is r the right factor? Three cases are easy to see directly. First, suppose r is 0. Then there is no association between x and y. So a one-SD increase in x is accompanied by a zero-SD increase in y, on the average. Second, suppose r is 1. Then all the points lie on the SD line: a one-SD increase in x is accompanied by a one-SD increase in y. Third, suppose r is -1. The argument is the same, except

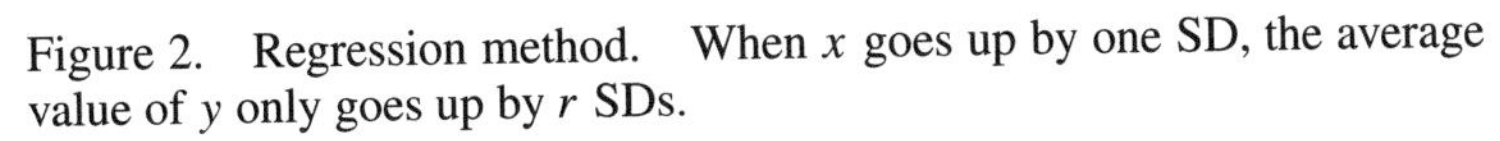
Figure 2. Regression method. When x goes up by one SD, the average value of y only goes up by r SDs.

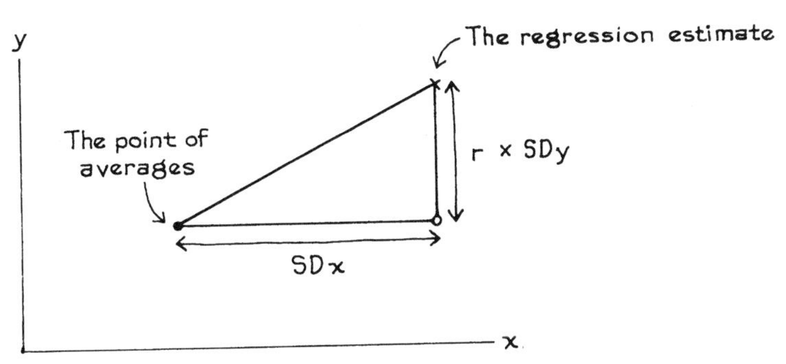

the line slopes down. With in-between values of r, a complicated mathematical argument is needed—but r is the factor to use.

Exercise Set A

1. In a certain class, midterm scores average out to 60 with an SD of 15, as do scores on the final. The correlation between midterm scores and final scores is about 0.50. Estimate the average final score for the students whose midterm scores were

   (a) 75 (b) 30 (c) 60

   Plot your regression estimates, as in figure 1.

2. For the men age 18 and over in HANES5,

   average height $\approx$ 69 inches, SD $\approx$ 3 inches
   average weight $\approx$ 190 pounds, SD $\approx$ 42 pounds, $r \approx 0.41$

   Estimate the average weight of the men whose heights were

   (a) 69 inches (b) 66 inches (c) 24 inches (d) 0 inches

   Comment on your answers to (c) and (d).

3. The men age 45–74 in HANES5 had an average height of 69 inches, equal to the overall average height (exercise 2). True or false, and explain: their average weight should be around 190 pounds, that being the overall average weight.

4. For women age 25–34 in the U.S. in 2005, with full-time jobs, the relationship between education (years of schooling completed) and personal income can be summarized as follows:[2]

   average education $\approx$ 14 years, SD $\approx$ 2.4 years
   average income $\approx$ \$32,000, SD $\approx$ \$26,000, $r \approx 0.34$

   Estimate the average income of those women who have finished high school but have not gone on to college (so they have 12 years of education).

5. Suppose $r = -1$. Can you explain why a one-SD increase in x is matched by a one-SD decrease in y?

The answers to these exercises are on pp. A59–60.

2. THE GRAPH OF AVERAGES

Figure 3 is the *graph of averages* for the heights and weights of the men age 18–24 in the HANES5 sample.[3] The graph shows the average weight for men at each height, and is close to a straight line in the middle—where most of the people are. But at the ends, the graph is quite bumpy. For instance, the men who were 78 inches tall (to the nearest inch) had an average weight of 241 pounds. This is represented by the point (78 inches, 241 pounds) in the figure. The men who were 80 inches tall averaged 211 pounds in weight. This is noticeably less than the average for the men who were 78 inches tall. The taller men weighed less than the shorter men. Chance variation is at work. The men were chosen for the sample at random. By the luck of the draw, the 78-inch men were too heavy, and the 80-inch men weren't heavy enough. Of course, there were only 2 men in each group, as indicated by the little numbers above or below the dots. The regression line smooths away this kind of chance variation.

> The regression line is a smoothed version of the graph of averages. If the graph of averages follows a straight line, that line is the regression line.

Figure 3. The graph of averages. Shows average weight at each height for the 471 men age 18–24 in the HANES5 sample. The regression line smooths this graph.

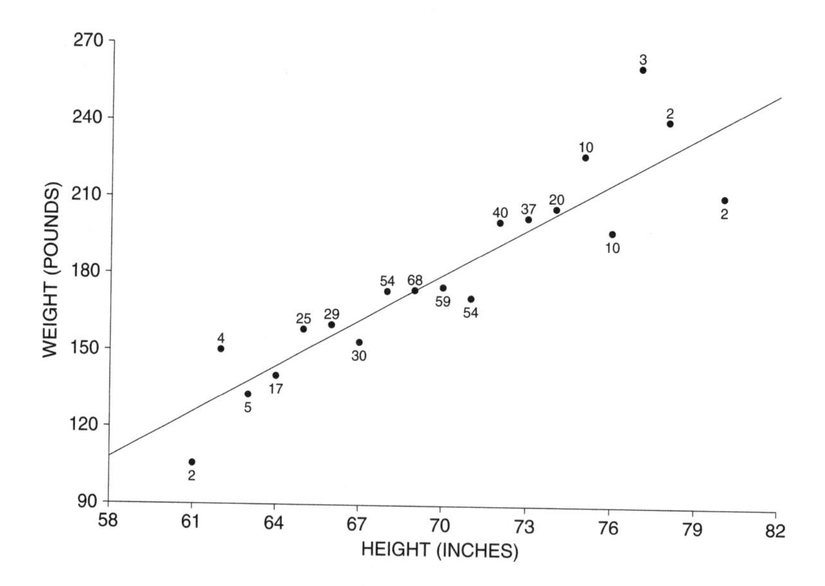

In some situations, the regression line smooths away too much. If there is a non-linear association between the two variables, as in figure 4 on the next page, the regression line will pass it by. Then, it is better to use the graph of averages. (Non-linearity came up for the correlation coefficient, section 3 of chapter 9; also see pp. 59 and 61 for data where the graph of averages is non-linear.)

Figure 4. Non-linear association. Regression lines should not be used when there is a non-linear association between the variables.

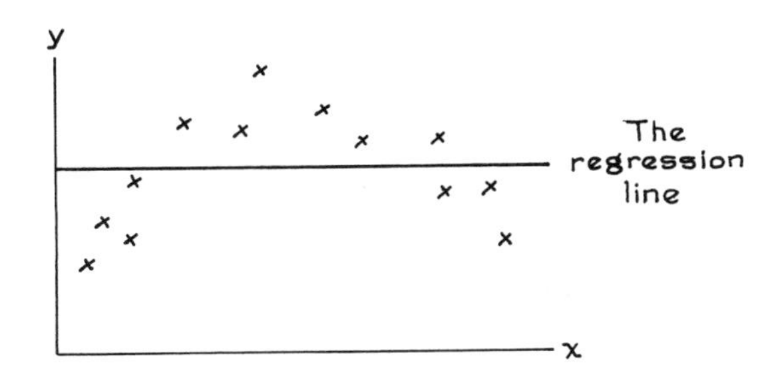

Exercise Set B

1. The figure below is based on a representative sample of married couples in New York. The graph shows the average income of the wives, given their husband's income. With 102 couples, the husband's income was in the range \$1–\$5,000; for those couples, the wife's income averaged \$15,390, as indicated by the point (\$2,500, \$15,390). With 58 couples, the husband's income was in the range \$5,001–\$10,000; for those couples, the wife's income averaged \$18,645, as indicated by the point (\$7,500, \$18,645). And so forth. The regression line is plotted too.[4]
   - (a) True or false: there is a positive association between husband's income and wife's income. If true, how would you explain the association?
   - (b) Why is the dot at \$127,500 so far below the regression line?
   - (c) If you use the regression line to estimate wife's income from husband's income, would your estimates generally be a little too high, a little too low, or just about right—for the couples in the sample with husband's income in the range \$65,000–\$80,000?

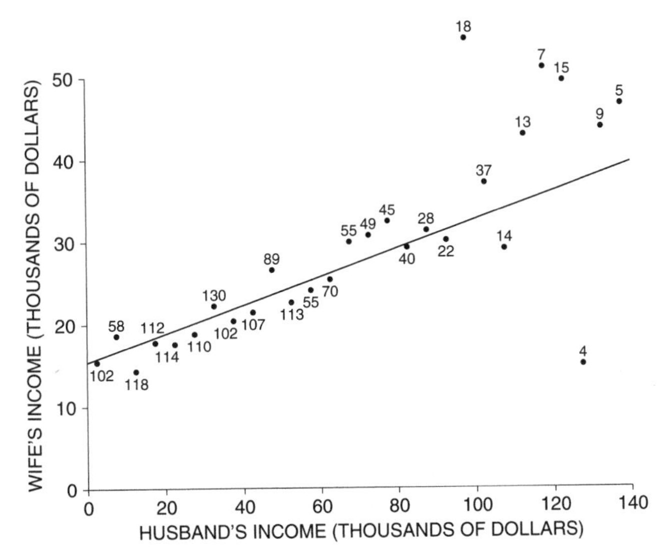

Source: March 2005 Current Population Survey; CD-ROM supplied by the Bureau of the Census.

2. Trace the diagram below on a piece of paper, and make a cross at the average for each of the vertical strips; one of them has already been done. Then draw the regression line for y on x. (The SD line is dashed.)

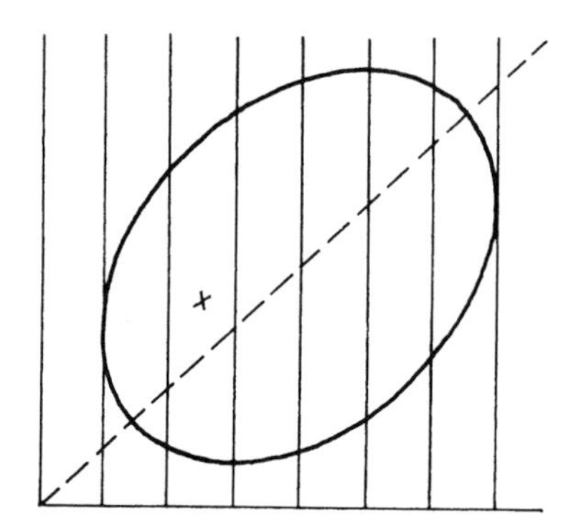

3. Below are four scatter diagrams, each with a solid line and a dashed line. For each diagram, say which is the SD line and which is the regression line for y on x.

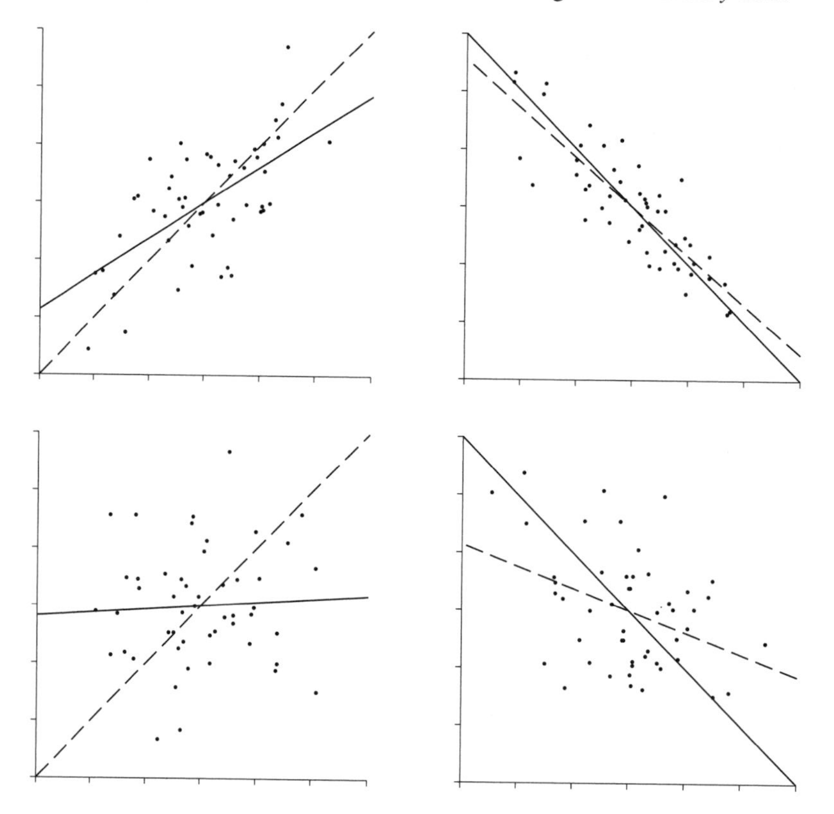

4. At the top of the next page are some hypothetical data sets. For each one, draw the scatter diagram, plot the graph of averages, and draw the regression line for y on x. Please do not do any calculations: make the best guess you can.

(a)

x	y
1	0
1	6
2	5
3	6
3	8

(b)

x	y
0	0
0	2
1	2

(c)

x	y
0	0
1	1
2	4

(d)

x	y
0	2
1	3
2	0
2	4
3	1
4	2

The answers to these exercises are on pp. A61–62.

Technical note. In general, the regression line fitted to the graph of averages, with each point weighted according to the number of cases it represents, coincides with the regression line fitted to the original scatter diagram. This is exact when points with different x-coordinates are kept separate in the graph of averages; otherwise, it is a good approximation.

3. THE REGRESSION METHOD FOR INDIVIDUALS

For the men age 18–24 in HANES5, the relationship between height and weight can be summarized as follows:

$$\text{average height} \approx 70 \text{ inches}, \quad \text{SD} \approx 3 \text{ inches}$$
$$\text{average weight} \approx 180 \text{ pounds}, \quad \text{SD} \approx 45 \text{ pounds}, \quad r \approx 0.40$$

Suppose one of these men is picked at random, and you have to guess his weight without being told anything about him. The best guess is the overall average weight, 180 pounds. Next, you are told the man's height: 73 inches, for example. This man is tall, and likely to be heavier than average. Your best guess for his weight is the average for all the 73-inch men in the study. This new average can be estimated by the regression method, as 198 pounds (p. 159). The rule: if you have to predict one variable from another, use the new average. In many cases, the regression method gives a sensible way of estimating the new average. Of course, if there is a non-linear association between the variables, the regression method would not apply.

Example 1. A university has made a statistical analysis of the relationship between Math SAT scores (ranging from 200 to 800) and first-year GPAs (ranging from 0 to 4.0), for students who complete the first year. The results:

$$\text{average SAT score} = 550, \quad \text{SD} = 80$$
$$\text{average first-year GPA} = 2.6, \quad \text{SD} = 0.6, \quad r = 0.4$$

The scatter diagram is football-shaped. A student is chosen at random, and has an SAT of 650. Predict this individual's first-year GPA.

Solution. This student is $100/80 = 1.25$ SDs above average on the SAT. The regression estimate for first-year GPA is, above average by $0.4 \times 1.25 = 0.5$ SDs. That's $0.5 \times 0.6 = 0.3$ GPA points. The predicted GPA is $2.6 + 0.3 = 2.9$.

The logic: for all students with an SAT of around 650, the average first-year GPA is about 2.9, by the regression method. That is why we predict a first-year GPA of 2.9 for this individual.

Usually, investigators work out regression estimates from a study, and then extrapolate: they use the estimates on new subjects. In many cases this makes sense, provided the subjects in the survey are representative of the people about whom the inferences are going to be made. But you have to think about the issue each time. The mathematics of the regression method will not protect you. In example 1, the university only has experience with the students it admits. There could be a problem in using the regression procedure on students who are quite different from that group. (Admissions officers typically do extrapolate, from admitted students to students who are denied admission.)

Now, another use for the regression method—to predict *percentile ranks.* If your percentile rank on a test is 90%, you did very well: only 10% of the class scored higher, the other 90% scored lower. A percentile rank of 25% is not so good: 75% of the class scored higher, the other 25% scored lower (p. 91).

Example 2. (This continues example 1.) Suppose the percentile rank of one student on the SAT is 90%, among the first-year students. Predict his percentile rank on first-year GPA. The scatter diagram is football-shaped. In particular, the SAT scores and GPAs follow the normal curve.

Solution. We are going to use the regression method. This student is above the average on the SAT. By how many SDs? Because SAT scores follow the normal curve, his percentile rank has this information—in disguise (section 5 of chapter 5):

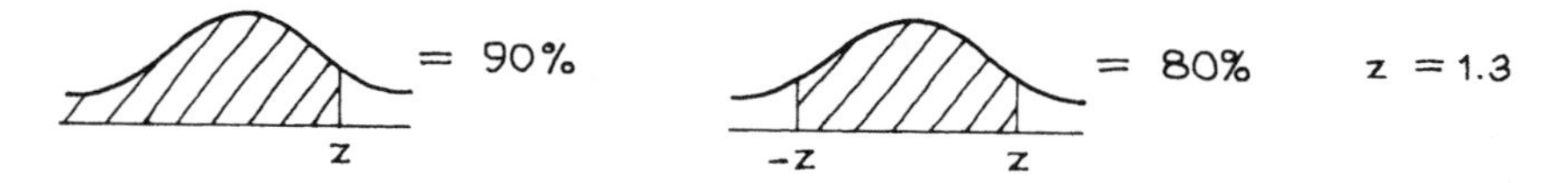

This student scored 1.3 SDs above average on the SAT. The regression method predicts he will be $0.4 \times 1.3 \approx 0.5$ SDs above average on first-year GPA. Finally, this can be translated back into a percentile rank:

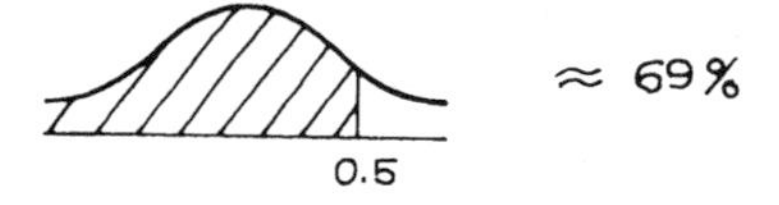

That is the answer. The percentile rank on first-year GPA is predicted as 69%.

In solving this problem, the averages and SDs of the two variables were never used. All that mattered was r. Basically, this is because the whole problem was worked in standard units. The percentile ranks give you the standard units.

The student in example 2 was compared with his class in two different competitions, the SAT and the first-year exams. He did very well on the SAT, scoring at the 90th percentile. But the regression estimate only puts him at the 69th percentile on the first year exams; still above average, but not as much. On the other hand, for poor students—say at the 10th percentile of the SAT—the regression method predicts an improvement. It will put them at the 31st percentile on the first-year tests. This is still below average, but closer.

To go at this more carefully, take all the people at the 90th percentile on the SAT—good students. Some of them will move up on the first-year tests, some will move down. On the average, however, this group moves down. For comparison, take all the people at the 10th percentile of the SAT—poor students. Again, some will do better on the first-year tests, others worse. On the average, however, this group moves up. That is what the regression method is telling us.

Initially, many people would predict a first-year rank equal to the SAT rank. This is not a good strategy. To see why, imagine that you had to predict a student's rank in a mathematics class. In the absence of other information, the safest guess is to put her at the median. However, if you knew that this student was very good in physics, you would probably put her well above the median in mathematics. After all, there is a strong correlation between physics and mathematics. On the other hand, if all you knew was her rank in a pottery class, that would not help very much in guessing the mathematics rank. The median looks good: there is not much correlation between pottery and mathematics.

Now, back to the problem of predicting first-year rank from SAT rank. If the two sets of scores are perfectly correlated, first-year rank will be equal to SAT rank. At the other extreme, if the correlation is zero, SAT rank does not help at all in predicting first-year rank. The correlation is somewhere between the two extremes, so we have to predict a rank on the first-year tests somewhere between the SAT rank and the median. The regression method tells us where.

Exercise Set C

1. In a certain class, midterm scores average out to 60 with an SD of 15, as do scores on the final. The correlation between midterm scores and final scores is about 0.50. The scatter diagram is football-shaped. Predict the final score for a student whose midterm score is

   (a) 75 (b) 30 (c) 60 (d) unknown

   Compare your answers to exercise 1 on p. 161.

2. For the first-year students at a certain university, the correlation between SAT scores and first-year GPA was 0.60. The scatter diagram is football-shaped. Predict the percentile rank on the first-year GPA for a student whose percentile rank on the SAT was

   (a) 90% (b) 30% (c) 50% (d) unknown

   Compare your answer to (a) with example 2.

3. The scatter diagram below shows the scores on the midterm and final in a certain course. Three lines are drawn across the diagram.
   (a) People who have the same percentile rank on both tests are plotted along one of these lines. Which one, and why?
   (b) One of these lines would be used to predict final score from midterm score. Which one, and why?

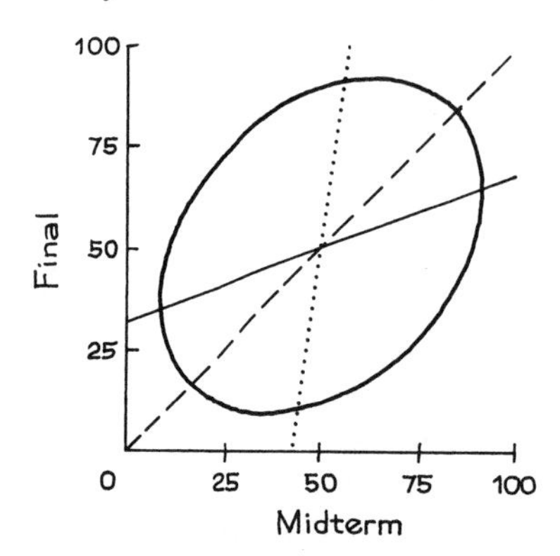

4. The scatter diagram below shows ages of husbands and wives in Tennessee. (Data are from the March 2005 Current Population Survey.)
   (a) Why are there no dots in the lower left hand corner of the diagram?
   (b) Why does the diagram show vertical and horizontal stripes?

5. For the men age 18 and over in the HANES5 sample, the correlation between height and weight was 0.41; the SD of height was about 3 inches and the SD of weight was about 42 pounds. The men age 55–64 averaged about half an inch shorter than the men age 18–24. True or false, and explain: since half an inch is $1/6 \approx 0.17$ SDs of height, the men age 55–64 must have averaged about $0.41 \times 0.17 \times 42 \approx 3$ pounds lighter than the men age 18–24.

The answers to these exercises are on p. A62.

Technical note. The method discussed in example 2 is for median ranks. To see why, assume normality and $r = 0.4$. Of students at the 90th percentile on the SAT (relative to their classmates), about half will rank above the 69th percentile on first-year GPA, and half will rank below. The procedure for estimating average ranks is harder.

4. THE REGRESSION FALLACY

A preschool program tries to boost children's IQs. Children are tested when they enter the program (the pre-test), and again when they leave (the post-test). On both occasions, the scores average out to nearly 100, and the SD is about 15. The program seems to have no effect. A closer look at the data, however, shows something very surprising. The children who were below average on the pre-test had an average gain of about 5 IQ points at the post-test. Conversely, those children who were above average on the pre-test had an average loss of about 5 points. What does this prove? Does the program operate to equalize intelligence? Perhaps when the brighter children play with the duller ones, the difference between the two groups tends to be diminished. Is this desirable or undesirable?

These speculations may be interesting, but the sad fact is that nothing much is going on, good or bad. Here is why. The children cannot be expected to score exactly the same on the two tests. There will be differences between the two scores. Nobody would think these differences mattered, or needed any explanation. But they make the scatter diagram for the test scores spread out around the SD line into that familiar football-shaped cloud. The spread around the line makes the bottom group come up and the top group come down. There is nothing else to it.

> In virtually all test-retest situations, the bottom group on the first test will on average show some improvement on the second test—and the top group will on average fall back. This is the *regression effect*.

Thinking that the regression effect must be due to something important, not just the spread around the line, is the *regression fallacy*.

We are now going to see why the regression effect appears whenever there is spread around the SD line. This effect was first noticed by Galton in his study of family resemblances, so that is the context for the discussion. But the reasoning is general. Figure 5 shows a scatter diagram for the heights of 1,078 pairs of fathers and sons, as discussed in chapter 8. The summary statistics are[5]

average height of fathers $\approx$ 68 inches, SD $\approx$ 2.7 inches
average height of sons $\approx$ 69 inches, SD $\approx$ 2.7 inches, $r \approx 0.5$

The sons average 1 inch taller than the fathers. On this basis, it is natural to guess that a 72-inch father should have a 73-inch son; similarly, a 64-inch father should have a 65-inch son; and so on. Such fathers and sons are plotted along the dashed line in figure 5. Of course, not many families are going to be right on the line. In fact, there is a lot of spread around the line. Some of the sons are taller than their fathers; others are shorter.

Take the fathers who are 72 inches tall, to the nearest inch. The corresponding families are plotted in the vertical strip over 72 inches in figure 5, and there is quite a range in the sons' heights. Some of the points are above the dashed line: the son is taller than 73 inches. But most of the points are below the dashed line: the son is shorter than 73 inches. All in all, the sons of the 72-inch fathers only average 71 inches in height. With tall fathers (high score on first test), on the average the sons are shorter (score on second test drops).

Now look at the points in the vertical strip over 64 inches, representing the families where the father is 64 inches tall, to the nearest inch. The height of the dashed line there is 65 inches, representing a son who is 1 inch taller than his 64-inch father. Some of the points fall below the dashed line, but most are above, and the sons of the 64-inch fathers average 67 inches in height. With short fathers (low score on first test), on the average the sons are taller (score on second test goes up). The aristocratic Galton termed this "regression to mediocrity."

The dashed line in figure 5 goes through the point corresponding to an average father of height 68 inches, and his average son of height 69 inches. Along the dashed line, each one-SD increase in father's height is matched by a one-SD increase in son's height. These two facts make it the SD line. The cloud is symmetric around the SD line, but the strip at 72 inches is not. The strip only contains points with unusually big x-coordinates. And most of the points in this strip fall below the SD line. Conversely, the strip at 64 inches only contains points with unusually small x-coordinates. Most of the points in this strip fall above the SD line. The hidden imbalance is always there in football-shaped clouds. The graphical explanation for the regression effect may not seem very romantic. But then, statistics isn't known as a romantic subject.

Figure 5 also shows the regression line for the son's height on father's height. This solid line rises less steeply than the dashed SD line, and it picks off the center of each vertical strip of dots—the average y-value in the strip. For instance, take the fathers who are 72 inches tall. They are 4 inches above average in height:

Figure 5. The regression effect. If a son is 1 inch taller than his father, the family is plotted along the dashed line. The points in the strip over 72 inches correspond to the families where the father is 72 inches tall, to the nearest inch; most of these points are below the dashed line. The points in the strip over 64 inches correspond to families where the father is 64 inches tall, to the nearest inch; most of these points are above the dashed line. The solid regression line picks off the centers of all the vertical strips, and is flatter than the dashed line.

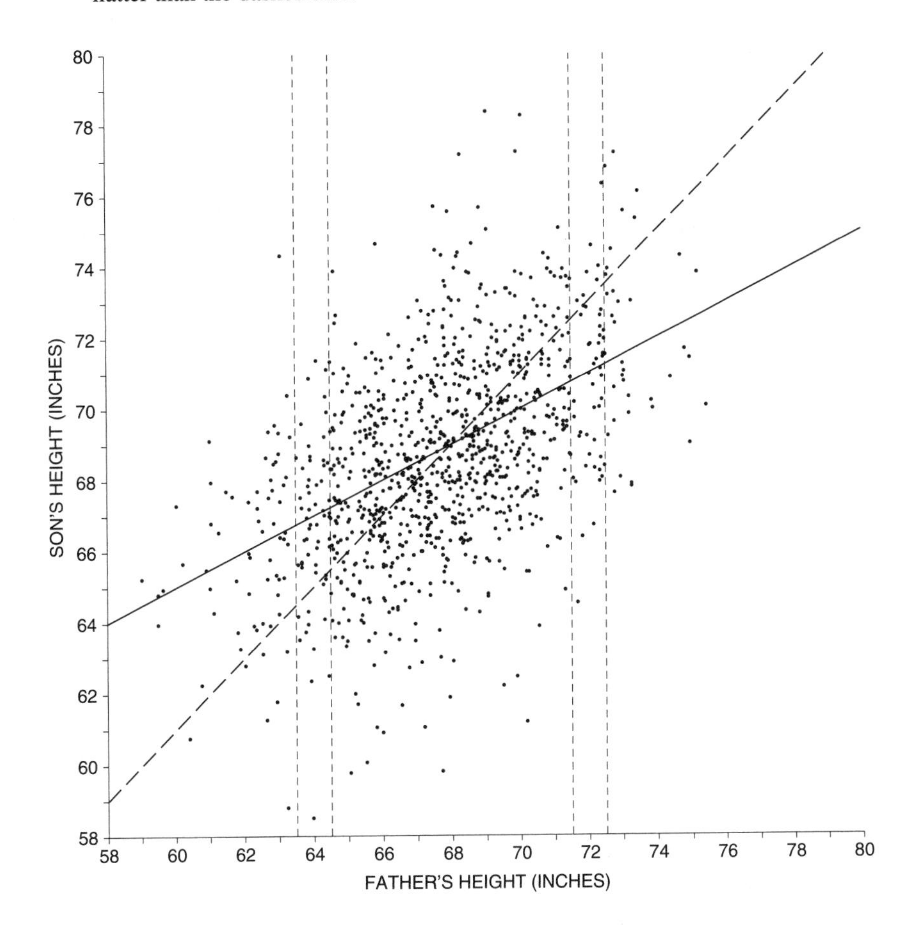

4 inches/2.7 inches $\approx$ 1.5 SDs. The regression line says their sons should be taller than average, by about

$$r \times 1.5 \text{ SDs} = 0.75 \text{ SDs} \approx 2 \text{ inches.}$$

The overall average height for sons is 69 inches, so the regression estimate for the average height of these sons is 71 inches—dead on.

Figure 6 shows the regression effect at its starkest, without the cloud. The dashed SD line rises at a 45 degree angle. The dots show the average height of the sons corresponding to each value of father's height. These dots are the centers of the vertical strips in figure 5. The dots rise less steeply than the SD line—the regression effect. On the whole, the dots are halfway between the SD line and the horizontal line through the point of averages. That is because the correlation coefficient is one half. Each one-SD increase in father's height is accompanied by a half-SD increase in son's height, not a one-SD increase. The solid regression line goes up at the half-to-one rate, and tracks the graph of averages quite well indeed.

Figure 6. The regression effect. The SD line is dashed, the regression line is solid. The dots show the average height of the sons, for each value of father's height. They rise less steeply than the SD line. This is the regression effect. The regression line follows the dots.

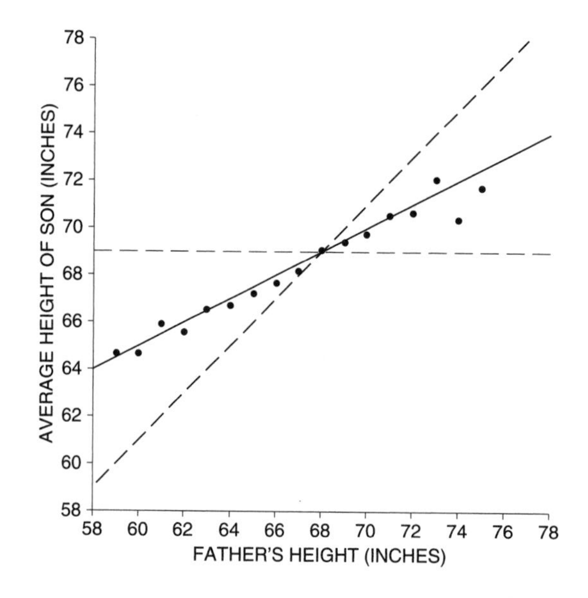

At first glance, the scatter diagram in figure 5 is rather chaotic. It was a stroke of genius on Galton's part to see a straight line in the chaos. Since Galton's time, many other investigators have found that the averages in their scatter diagrams followed straight lines too. That is why the regression line is so useful.

Now, a look behind the scenes: the regression effect can be understood a little better in some cases, for instance, in the context of a repeated IQ test. The basic fact is that the two scores are apt to be different. The difference can be explained in terms of chance variability. Each person may be lucky or unlucky on the first test. But if the score on the first test is very high, that suggests the person was lucky on that occasion, implying that the score on the second test will probably be lower. (You wouldn't say, "He scored very high, must have had bad luck that day.") On the other hand, if the score on the first test was very low, the person was probably unlucky to some extent on that occasion and will do better next time.

Here is a crude model for the test-retest situation, which brings the explanation into sharper focus. The basic equation is

$$\text{observed test score} = \text{true score} + \text{chance error}.$$

Assume that the distribution of true scores in the population follows the normal curve, with an average of 100 and an SD of 15. Suppose too that the chance error is as likely to be positive as negative, and tends to be about 5 points in size. Someone who has a true score of 135 is just as likely to score 130 as 140 on the test. Someone with a true score of 145 is just as likely to score 140 as 150. Of course, the chance error could also be ± 4, or ± 6, and so forth: any symmetric pair of values can be dealt with in a similar way.

Figure 7. A model for the regression effect.

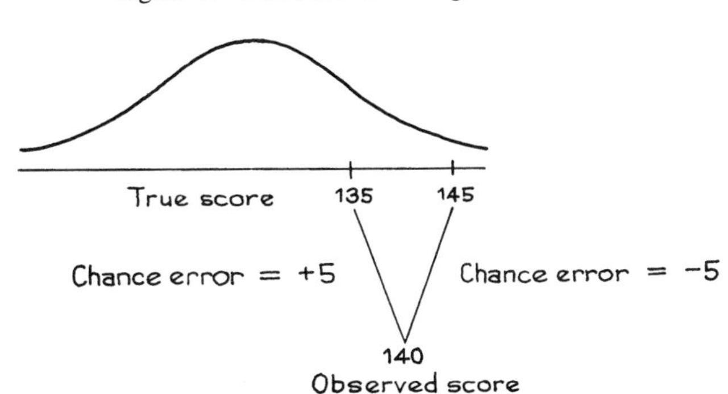

Take the people who scored 140 on the first test. There are two alternative explanations for this observed score:

- true score below 140, with a positive chance error;
- true score above 140, with a negative chance error.

The first explanation is more likely. For instance, more people have true scores of 135 than 145, as figure 7 shows.

The model accounts for the regression effect. If someone scores above average on the first test, the true score is probably a bit lower than the observed score. If this person takes the test again, we predict that the second score will be a bit lower than the first score. On the other hand, if a person scores below average on the first test, we estimate that the true score is a bit higher than the observed score, and our prediction for the second score will be a bit higher than the first score.

Exercise Set D

1. As part of their training, air force pilots make two practice landings with instructors, and are rated on performance. The instructors discuss the ratings with the pilots after each landing. Statistical analysis shows that pilots who make poor landings the first time tend to do better the second time. Conversely, pilots who make good landings the first time tend to do worse the second time. The conclusion: criticism helps the pilots while praise makes them do worse. As a result, instructors were ordered to criticize all landings, good or bad. Was this warranted by the facts? Answer yes or no, and explain briefly.[6]

2. An instructor standardizes her midterm and final each semester so the class average is 50 and the SD is 10 on both tests. The correlation between the tests is around 0.50. One semester, she took all the students who scored below 30 at the midterm, and gave them special tutoring. They all scored above 50 on the final. Can this be explained by the regression effect? Answer yes or no, and explain briefly.

3. In the data set of figures 5 and 6, are the sons of the 61-inch fathers taller on the average than the sons of the 62-inch fathers, or shorter? What is the explanation?

The answers to these exercises are on pp. A62–63.

5. THERE ARE TWO REGRESSION LINES

In fact, two regression lines can be drawn across a scatter diagram. For example, a height-weight scatter diagram is sketched in figure 8. The left hand panel shows the regression line for weight on height. This picks off the centers of the vertical strips, and estimates the average weight for each height. The right hand panel shows the regression line for height on weight. This picks off the centers of the horizontal strips, and estimates the average height for each weight. In both panels, the regression line is solid and the SD line is dashed. The regression of weight on height seems more natural for most purposes, but the other line may come in handy too.

Figure 8. The left hand panel shows the regression of weight on height; the right hand panel, height on weight. The SD line is dashed.

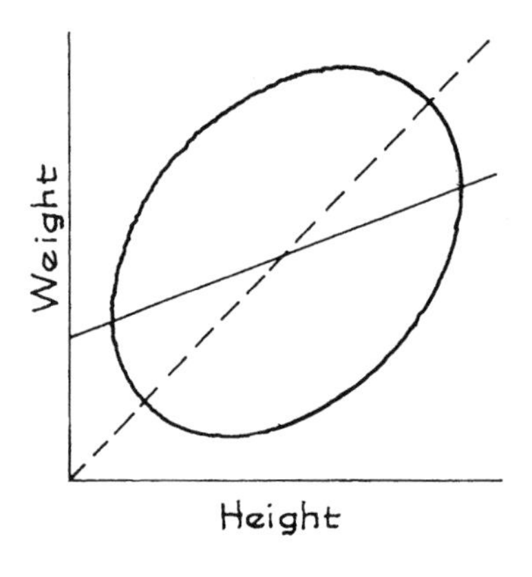

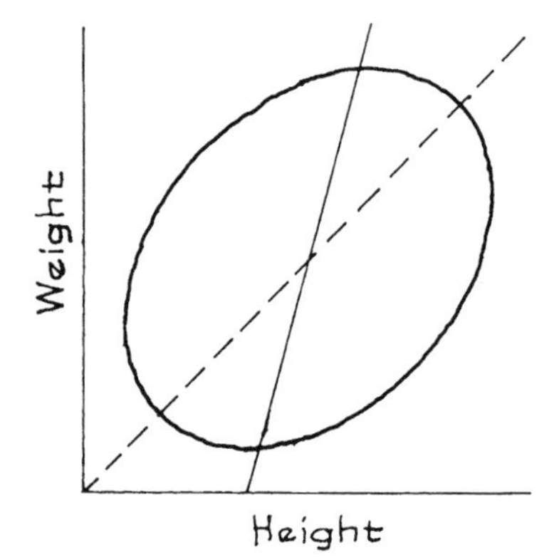

Example 3. IQ scores are scaled to have an average of about 100, and an SD of about 15, both for men and for women. The correlation between the IQs of husbands and wives is about 0.50. A large study of families found that the men whose IQ was 140 had wives whose IQ averaged 120. Look at the wives in the study whose IQ was 120. Should the average IQ of their husbands be greater than 120? Answer yes or no, and explain briefly.

Solution. No, the average IQ of their husbands will be around 110. See figure 9. The families where the husband has an IQ of 140 are shown in the vertical strip. The average y-coordinate in this strip is 120. The families where the wife has an IQ of 120 are shown in the horizontal strip. This is a completely different set of families. The average x-coordinate for points in the horizontal strip is about 110. Remember, there are two regression lines. One line is for predicting the wife's IQ from her husband's IQ. The other line is for predicting the husband's IQ from his wife's.

Figure 9. The two regression lines.

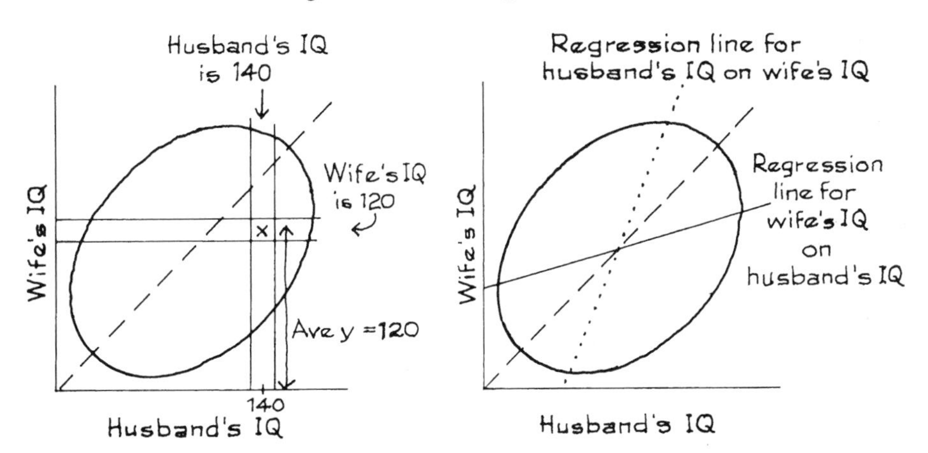

Exercise Set E

1. For the men age 18–24 in the HANES5 sample, the ones who were 63 inches tall averaged 138 pounds in weight. True or false, and explain: the ones who weighed 138 pounds must have averaged 63 inches in height.
2. In Pearson's study, the sons of the 72-inch fathers only averaged 71 inches in height. True or false: if you take the 71-inch sons, their fathers will average about 72 inches in height. Explain briefly.
3. In example 2 (p. 166), the regression method predicted that a student at the 90th percentile on the SAT would only be at the 69th percentile on first-year GPA. True or false, and explain: a student at the 69th percentile on first-year GPA should be at the 90th percentile on the SAT.

The answers to these exercises are on p. A63.

6. REVIEW EXERCISES

Review exercises may cover material from previous chapters.

1. Shown below is a scatter diagram for Math and Verbal SAT scores for graduating seniors at a certain high school. Three areas are shaded. Match the area with the description. (One description will be left over.)
   (i) Total score (Math + Verbal) is below 800.
   (ii) Total score (Math + Verbal) is around 800.
   (iii) Math score is about equal to Verbal score.
   (iv) Math score is less than Verbal score.

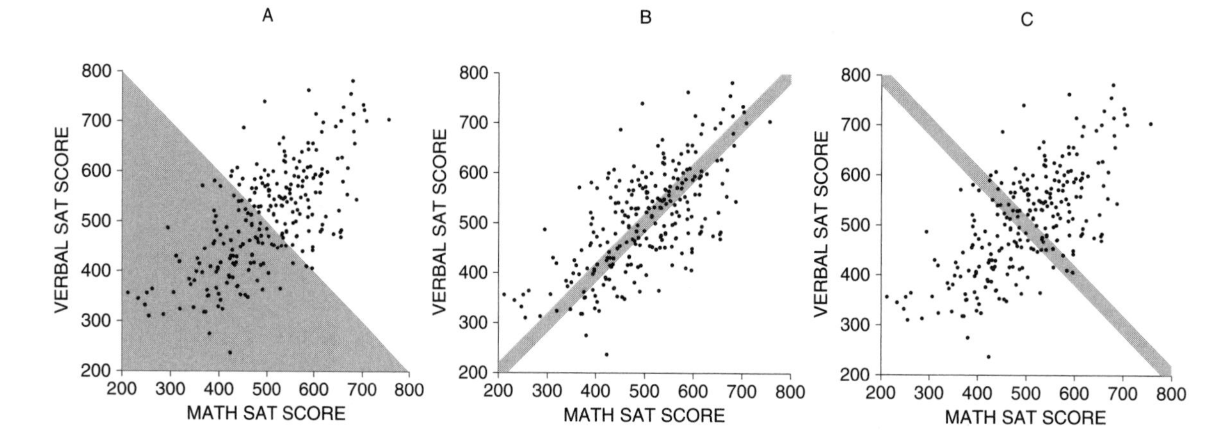

2. In a study of the stability of IQ scores, a large group of individuals is tested once at age 18 and again at age 35. The following results are obtained.

   age 18: average score $\approx$ 100, SD $\approx$ 15
   age 35: average score $\approx$ 100, SD $\approx$ 15, $r \approx 0.80$

   (a) Estimate the average score at age 35 for all the individuals who scored 115 at age 18.
   (b) Predict the score at age 35 for an individual who scored 115 at age 18.

3. Pearson and Lee obtained the following results in a study of about 1,000 families:

   average height of husband $\approx$ 68 inches, SD $\approx$ 2.7 inches
   average height of wife $\approx$ 63 inches, SD $\approx$ 2.5 inches, $r \approx 0.25$

   Predict the height of a wife when the height of her husband is
   (a) 72 inches (b) 64 inches (c) 68 inches (d) unknown

4. In one study, the correlation between the educational level of husbands and wives in a certain town was about 0.50; both averaged 12 years of schooling completed, with an SD of 3 years.[7]

(a) Predict the educational level of a woman whose husband has completed 18 years of schooling.
(b) Predict the educational level of a man whose wife has completed 15 years of schooling.
(c) Apparently, well-educated men marry women who are less well educated than themselves. But the women marry men with even less education. How is this possible?

5. An investigator measuring various characteristics of a large group of athletes found that the correlation between the weight of an athlete and the amount of weight that athlete could lift was 0.60. True or false, and explain:
   (a) On the average, an athlete can lift 60% of his body weight.
   (b) If an athlete gains 10 pounds, he can expect to lift an additional 6 pounds.
   (c) The more an athlete weighs, on the average the more he can lift.
   (d) The more an athlete can lift, on the average the more he weighs.
   (e) 60% of an athlete's lifting ability can be attributed to his weight alone.

6. Three lines are drawn across the scatter diagram below. One is the SD line, one is the regression line for y on x, and one is the regression line for x on y. Which is which? Why? (The "regression line for y on x" is used to predict y from x.)

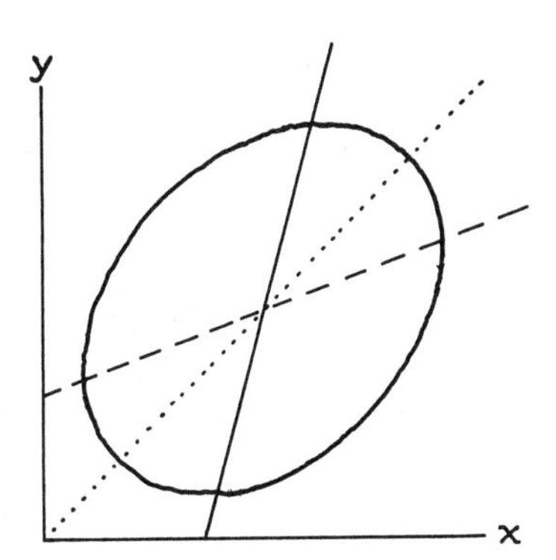

7. A doctor is in the habit of measuring blood pressures twice. She notices that patients who are unusually high on the first reading tend to have somewhat lower second readings. She concludes that patients are more relaxed on the second reading. A colleague disagrees, pointing out that the patients who are unusually low on the first reading tend to have somewhat higher second readings, suggesting they get more nervous. Which doctor is right? Or perhaps both are wrong? Explain briefly.

8. A large study was made on the blood-pressure problem discussed in the previous exercise. It found that first readings average 130 mm, and second readings average 120 mm; both SDs were about 15 mm. Does this support either doctor's argument? Or is it the regression effect? Explain.

9. In a large statistics class, the correlation between midterm scores and final scores is found to be nearly 0.50, every term. The scatter diagrams are football-shaped. Predict the percentile rank on the final for a student whose percentile rank on the midterm is
   (a) 5% (b) 80% (c) 50% (d) unknown

10. True or false: A student who is at the 40th percentile of first-year GPAs is also likely to be at the 40th percentile of second-year GPAs. Explain briefly. (The scatter diagram is football-shaped.)

7. SUMMARY

1. Associated with an increase of one SD in x, there is an increase of only r SDs in y, on the average. Plotting these *regression estimates* gives the *regression line* for y on x.

2. The *graph of averages* is often close to a straight line, but may be a little bumpy. The regression line smooths out the bumps. If the graph of averages is a straight line, then it coincides with the regression line. If the graph of averages has a strong non-linear pattern, regression may be inappropriate.

3. The regression line can be used to make predictions for individuals. But if you have to extrapolate far from the data, or to a different group of subjects, be careful.

4. In a typical test-retest situation, the subjects get different scores on the two tests. Take the bottom group on the first test. Some improve on the second test, others do worse. On average, the bottom group shows an improvement. Now, the top group: some do better the second time, others fall back. On average, the top group does worse the second time. This is the *regression effect*, and it happens whenever the scatter diagram spreads out around the SD line into a football-shaped cloud of points.

5. The *regression fallacy* consists in thinking that the regression effect must be due to something other than spread around the SD line.

6. There are two regression lines that can be drawn on a scatter diagram. One predicts y from x; the other predicts x from y.

11

The R.M.S. Error for Regression

Such are the formal mathematical consequences of normal correlation. Much biometric material certainly shows a general agreement with the features to be expected on this assumption: although I am not aware that the question has been subjected to any sufficiently critical enquiry. Approximate agreement is perhaps all that is needed to justify the use of the correlation as a quantity descriptive of the population; its efficacy in this respect is undoubted, and it is not improbable that in some cases it affords, in conjunction with the means and variances, a complete description of the simultaneous variation of the variates.

—SIR R. A. FISHER (ENGLAND, 1890–1962)[1]

1. INTRODUCTION

The regression method can be used to predict y from x. However, actual values differ from predictions. By how much? The object of this section is to measure the overall size of the differences using the r.m.s. error. For example, take the heights and weights of the 471 men age 18–24 in the HANES5 sample (section 1 of chapter 10). The summary statistics:

average height $\approx$ 70 inches, SD $\approx$ 3 inches
average weight $\approx$ 180 pounds, SD $\approx$ 45 pounds, $r \approx 0.40$

To review briefly, given a man's height, his weight is predicted by the average weight for all the men with that height. The average can be estimated by the regression method. Figure 1 shows the regression line. Person A on the diagram is about 72 inches tall. The regression estimate for average weight at this height is

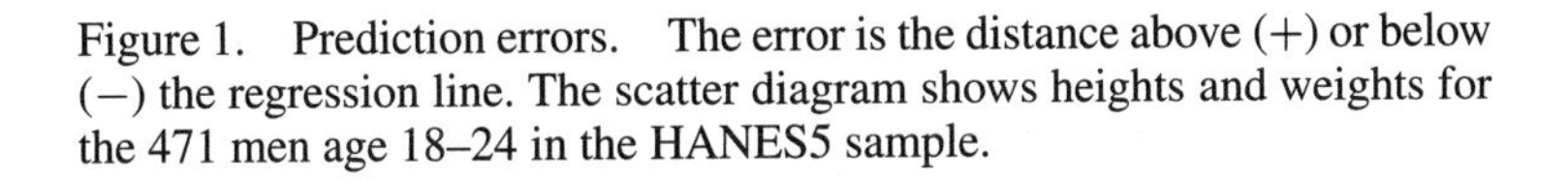

Figure 1. Prediction errors. The error is the distance above (+) or below (−) the regression line. The scatter diagram shows heights and weights for the 471 men age 18–24 in the HANES5 sample.

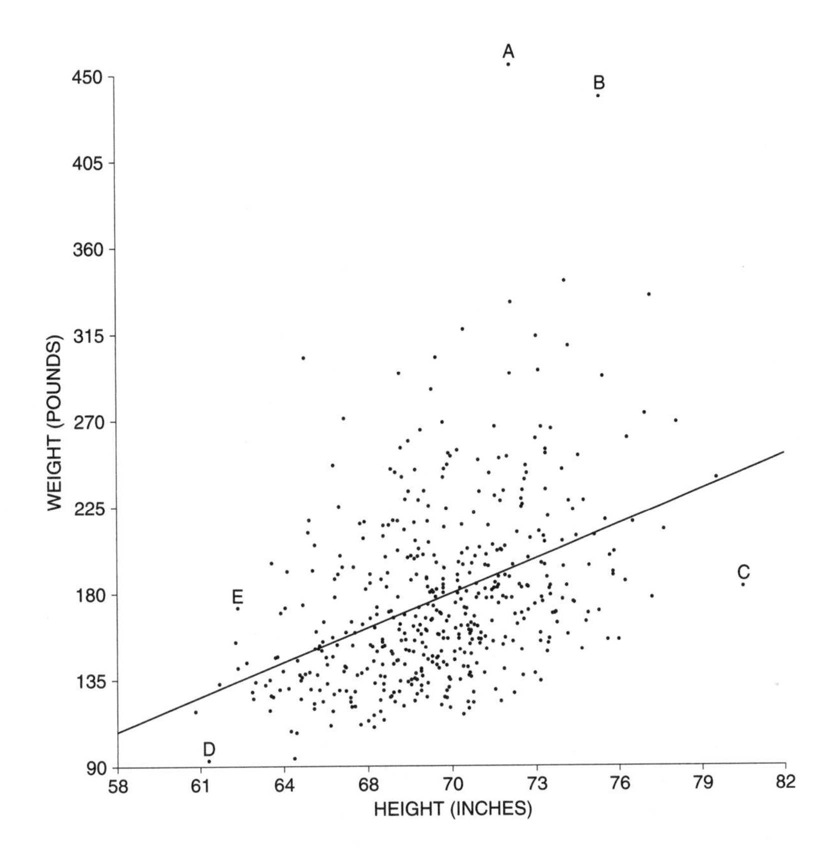

192 pounds (section 1 of chapter 10). However, A's actual weight is 456 pounds. The prediction is off, by 264 pounds:

$$\begin{aligned}\text{error} &= \text{actual weight} - \text{predicted weight}\\ &= 456\text{ lb} - 192\text{ lb} = 264\text{ lb.}\end{aligned}$$

In the diagram, the prediction error is the vertical distance of A above the regression line.

Person C on the diagram is 80.5 inches tall and weighs 183 pounds. The regression line predicts his weight as 243 pounds. So there is a prediction error of 183 lb − 243 lb = −60 lb. In the diagram, this error is represented by the vertical distance of C below the regression line.

> The distance of a point above (+) or below (−) the regression line is
>
> $$\text{error} = \text{actual} - \text{predicted.}$$

Figure 2. Prediction error equals vertical distance from the line.

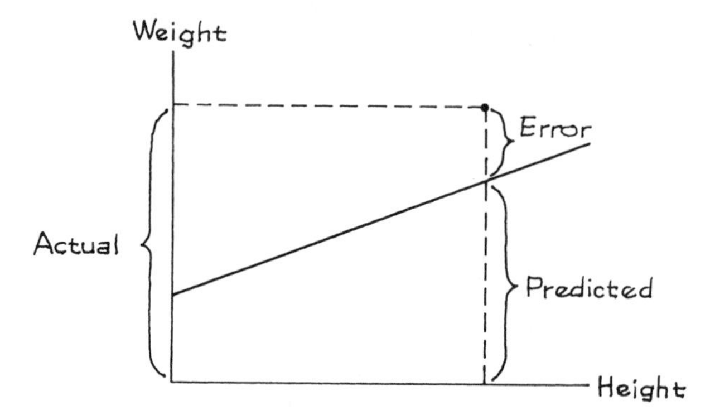

Figure 2 shows the connection between prediction errors and distances from the line. The overall size of these errors is measured by taking their root-mean-square (p. 66). The result is called the *r.m.s. error of the regression line*.

Go back to figure 1. Each of the 471 points in the scatter diagram is some vertical distance above or below the regression line, corresponding to a prediction error made by the line. The r.m.s. error of the regression line for predicting weight from height is

$$\sqrt{\frac{(\text{error \#1})^2 + (\text{error \#2})^2 + \cdots + (\text{error \#471})^2}{471}}$$

This looks painful, but the answer is about 41 pounds. (A short-cut through the arithmetic will be presented in the next section.)

The r.m.s. error has a graphical interpretation: a typical point in figure 1 is above or below the regression line by something like 41 pounds. Since the line is predicting weight from height, we conclude that for typical men in the study, actual weight differs from predicted weight by around 41 pounds or so.

The r.m.s. error for regression says how far typical points are above or below the regression line.

The r.m.s. error is to the regression line as the SD is to the average. For instance, about 68% of the points on a scatter diagram will be within one r.m.s. error of the regression line; about 95% of them will be within two r.m.s. errors. This rule of thumb holds for many data sets, but not all; it is illustrated in figure 3.

What about the height-weight data? The computer found that the predictions were right to within one r.m.s. error (41 pounds) for 340 out of 471 men, or 72% of them. The rule of thumb doesn't look bad at all. The predictions were right to

Figure 3. Rule of thumb. About 68% of the points on a scatter diagram fall inside the strip whose edges are parallel to the regression line, and one r.m.s. error away (up or down). About 95% of the points are in the wider strip whose edges are parallel to the regression line, and twice the r.m.s. error away.

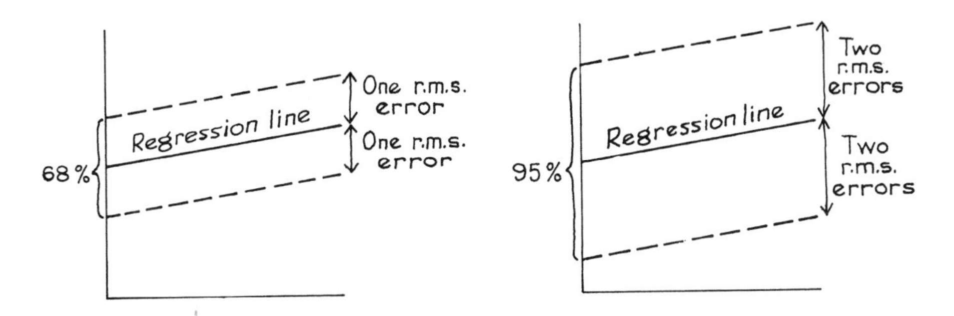

within two r.m.s. errors (82 pounds) for 451 out of the 471 men, which is 96%. This is even better for the rule of thumb.

Soon, we will compare the r.m.s. error for regression to the r.m.s. error for a baseline prediction method. The baseline method just ignores the x-values and uses the average value of y to predict y. With this method, the predictions fall along a horizontal line through the average of y.

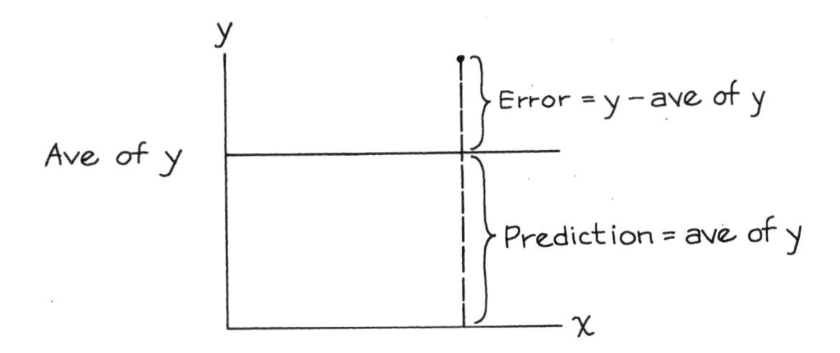

Graphically, the prediction errors for the second method are the vertical distances above and below this horizontal line, as shown by the sketch. Numerically, the errors are the deviations from the average of y. So the r.m.s. error for the second method is the SD of y: remember, the SD is the r.m.s. of the deviations from average.

> The SD of y says how far typical points are above or below a horizontal line through the average of y. In other words, the SD of y is the r.m.s. error for the baseline method—predicting y by its average, just ignoring the x-values.

Exercise Set A

1. Look at figure 1, then fill in the blanks: person B is ______ and ______, while D is ______ and ______. Options: short, tall, skinny, chubby.

2. Look at figure 1, then say whether each statement is true or false:

   (a) E is above average in weight.
   (b) E is above average in weight, for men of his height.

3. A regression line is fitted to a small data set. For each subject, the table shows the actual value of y and the predicted value from the regression line. (The value of x is not shown.) Compute the prediction errors, and the r.m.s. error of the regression line.

Actual value of y	*Predicted value of y*
57	64
63	62
43	40
51	52
49	45

4. Below are three scatter diagrams. The regression line has been drawn across each one, by eye. In each case, guess whether the r.m.s. error is 0.2, or 1, or 5.

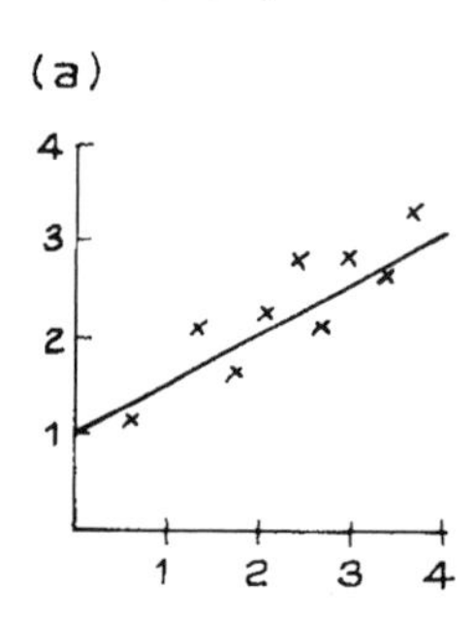

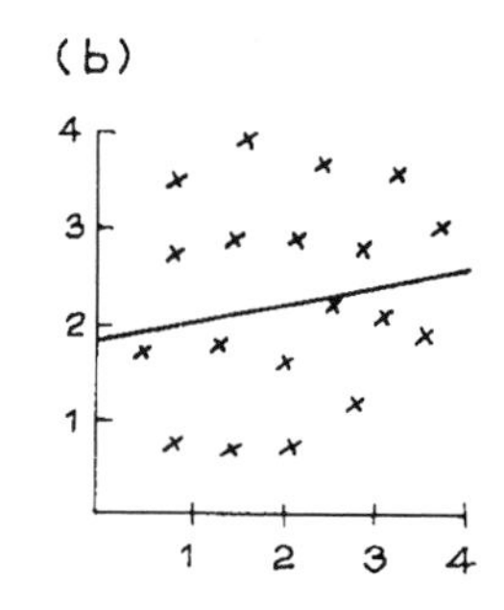

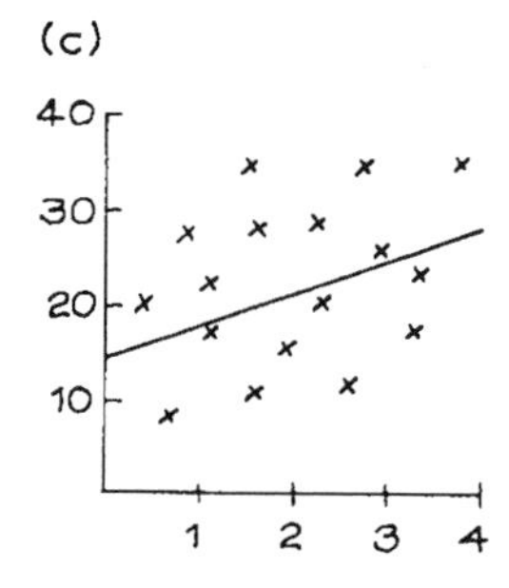

5. A regression line for predicting income has an r.m.s. error of $2,000. It predicts someone's income as $20,000. This is likely to be right give or take: a few hundred dollars, a few thousand dollars, ten or twenty thousand dollars.

6. An admissions officer is trying to choose between two methods of predicting first-year scores. One method has an r.m.s. error of 12. The other has an r.m.s. error of 7. Other things being equal, which should he choose? Why?

7. A regression line for predicting test scores has an r.m.s. error of 8 points.

   (a) About 68% of the time, the predictions will be right to within ______ points.
   (b) About 95% of the time, the predictions will be right to within ______ points.

8. The scatter diagram on the next page shows incomes for a sample of 168 working couples in Louisiana. Summary statistics are as follows:

   average husband's income = $45,000, SD = $25,000
   average wife's income = $28,000, SD = $20,000

   (a) If you predict wife's income as $28,000, ignoring husband's income, your r.m.s. error will be ______.

(b) All the predictions are on one of the lines in the diagram. Which one? Explain your answer.

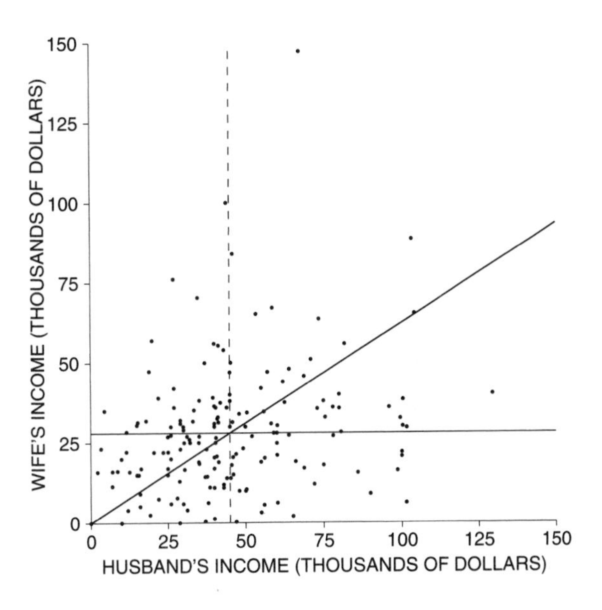

The answers to these exercises are on pp. A63–64.

2. COMPUTING THE R.M.S. ERROR

The r.m.s. error for the regression line measures distances above or below the regression line (left-hand panel of figure 4). The right-hand panel of figure 4 shows another line, namely, the horizontal line through the average of y. The r.m.s. error for that line is just the SD of y, as discussed on p. 183.

Figure 4. The r.m.s. error of the regression line, and the SD of y.

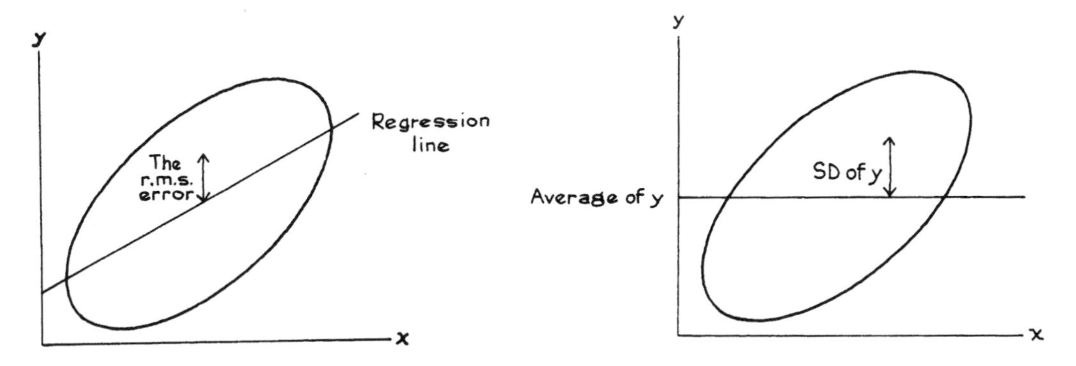

The r.m.s. error for the regression line will be smaller than the SD of y, because the regression line gets closer to the points than the horizontal line. The r.m.s. will be smaller by the factor $\sqrt{1 - r^2}$.

The r.m.s. error for the regression line of y on x can be figured as

$$\sqrt{1-r^2} \times \text{the SD of } y.$$

Which SD goes into the formula? The SD of the variable being predicted. If you are predicting weight from height, use the SD of weight. The r.m.s. error has to come out in pounds, not inches. If you are predicting income from education, use the SD of income. The r.m.s. error has to come out in dollars, not years.

The units for the r.m.s. error are the same as the units for the variable being predicted.

In the height-weight scatter diagram (figure 1), there were 471 prediction errors, one for each man. Finding the root-mean-square of these 471 errors looked like a lot of work. But the factor $\sqrt{1-r^2}$ gives you a shortcut through the arithmetic. The r.m.s. error of the regression line for predicting weight from height equals

$$\sqrt{1-r^2} \times \text{SD of weight} = \sqrt{1-0.40^2} \times 45 \text{ lb} \approx 41 \text{ lb}.$$

The r.m.s. error isn't much smaller than the SD of weight, because weight is not that well correlated with height: $r \approx 0.40$. Knowing a man's height does not help so much in predicting his weight.

The formula is hard to prove without algebra. But three special cases are easy to see. First, suppose $r = 1$. Then all the points lie on a straight line which slopes up. The regression line goes through all the points on the scatter diagram, and all the prediction errors are 0. So the r.m.s. error should be 0. And that is what the formula says. The factor works out to

$$\sqrt{1-r^2} = \sqrt{1-1^2} = \sqrt{1-1} = 0.$$

The case $r = -1$ is the same, except that the line slopes down. The r.m.s. error should still be 0, and the factor is

$$\sqrt{1-r^2} = \sqrt{1-(-1)^2} = \sqrt{1-1} = 0.$$

The third case is $r = 0$. Then there is no linear relationship between the variables. So the regression line does not help in predicting y, and its r.m.s. error should equal the SD. The factor is

$$\sqrt{1-r^2} = \sqrt{1-0^2} = \sqrt{1-0} = 1.$$

The r.m.s. error measures spread around the regression line in absolute terms: pounds, dollars, and so on. The correlation coefficient, on the other hand, measures spread relative to the SD, and has no units. The r.m.s. error is connected to the SD through the correlation coefficient. This is the third time that r comes into the story.

- r describes the clustering of the points around a line, relative to the SDs (chapter 8).
- r says how the average value of y depends on x—associated with each one-SD increase in x there is an increase of only r SDs in y, on the average (chapter 10).
- r determines the accuracy of the regression predictions, through the formula for r.m.s. error.

A cautionary note. If you extrapolate beyond the data, or use the line to make estimates for people who are different from the subjects in the study, the r.m.s. error cannot tell you how far off you are likely to be. That is beyond the power of mathematics.

Exercise Set B

1. A law school finds the following relationship between LSAT scores and first-year scores:

$$\begin{aligned} \text{average LSAT score} &= 165, \quad \text{SD} = 5 \\ \text{average first-year score} &= 65, \quad \text{SD} = 10, \quad r = 0.6 \end{aligned}$$

The admissions officer uses the regression line to predict first-year scores from LSAT scores. The r.m.s. error of the line is ______. Options:

$$5 \qquad 10 \qquad \sqrt{1 - 0.6^2} \times 5 \qquad \sqrt{1 - 0.6^2} \times 10$$

2. (This continues exercise 1.)
   (a) One of these students is chosen at random; you have to guess his first-year score, without being told his LSAT score. How would you do this?
   (b) Your r.m.s. error would be ______. Options:

$$5 \qquad 10 \qquad \sqrt{1 - 0.6^2} \times 5 \qquad \sqrt{1 - 0.6^2} \times 10$$

   (c) Repeat parts (a) and (b), if you are allowed to use his LSAT score.

3. At a certain college, first-year GPAs average about 3.0, with an SD of about 0.5; they are correlated about 0.6 with high-school GPA. Person A predicts first- year GPAs just using the average. Person B predicts first-year GPAs by regression, using the high-school GPAs. Which person makes the smaller r.m.s. error? Smaller by what factor?

The answers to these exercises are on *p. A64.*

3. PLOTTING THE RESIDUALS

Prediction errors are often called *residuals*. Statisticians recommend graphing the residuals. The method is indicated by figure 5 on the next page. Each point on the scatter diagram is transferred to a second diagram, called the *residual plot*, in the following way.The x-coordinate is left alone. But the y-coordinate is replaced by the residual at the point—the distance above (+) or below (−)

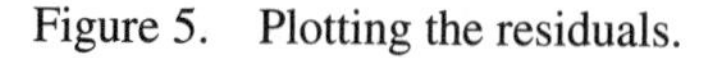
Figure 5. Plotting the residuals.

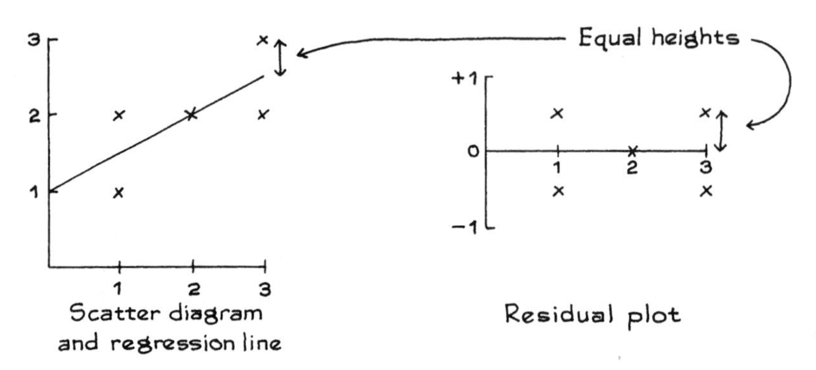

the regression line. Figure 6 shows the residual plot for the height-weight scatter diagram of figure 1. Figures 5 and 6 suggest that the positive residuals balance out the negative ones. Mathematically, the residuals from the regression line must average out to 0. The figures show something else too. As you look across the residual plot, there is no systematic tendency for the points to drift up (or down). Basically, the reason is that all the trend up or down has been taken out of the residuals, and has been absorbed into the regression line.

> The residuals average out to 0; and the regression line for the residual plot is horizontal.

Figure 6. A residual plot. The scatter diagram at the left shows the heights and weights of the 471 men age 18–24 in the HANES5 sample, with the regression line. The residual plot is shown at the right. There is no trend or pattern in the residuals.

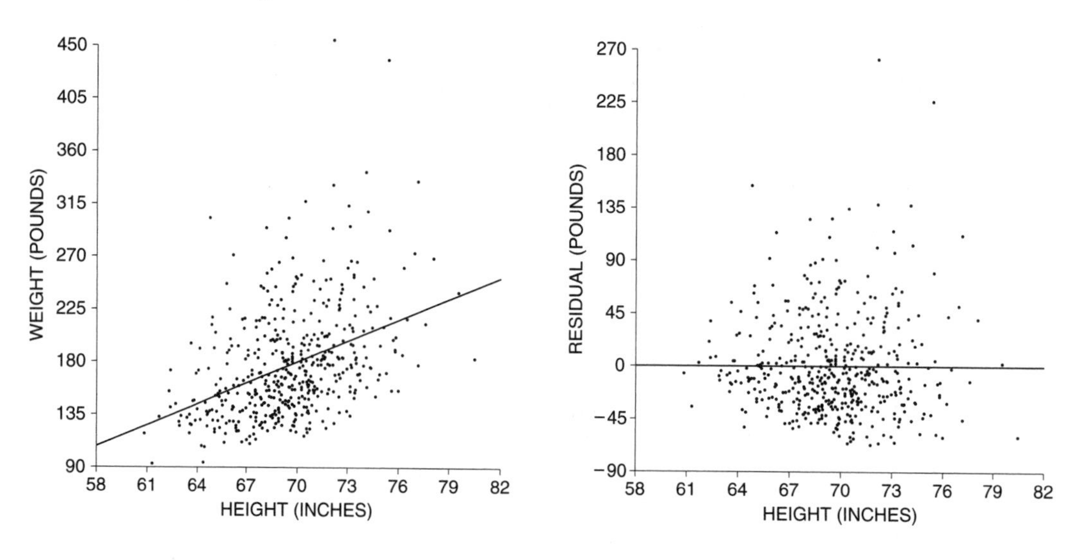

The residual plot in figure 6 shows no pattern. By comparison, figure 7 shows a residual plot (for hypothetical data) with a strong pattern. With this kind of pattern, it is probably a mistake to use a regression line. Often, you can spot non-linearities by looking at the scatter diagram. However, the residual plot may give a more sensitive test—because the vertical scale can be made big enough so things can be examined carefully. Residual plots are useful diagnostics in *multiple regression*; for example, in predicting first-year GPA from SAT scores and high-school GPA.[2] (Multiple regression is discussed in section 3 of chapter 12.)

Figure 7. A residual plot with a strong pattern. It may have been a mistake to fit the regression line.

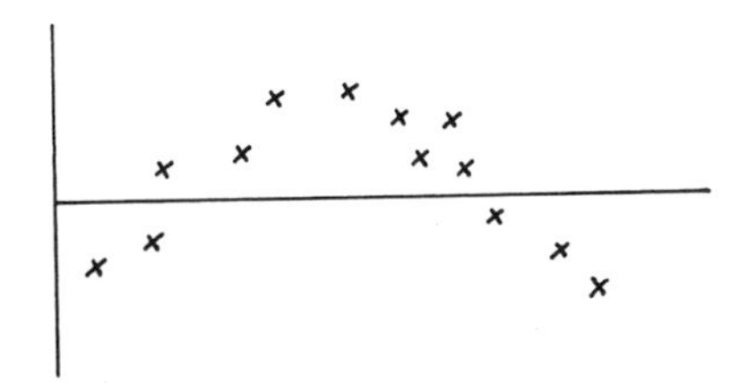

Exercise Set C

1. Several different regression lines are used to predict the price of a stock (from different independent variables). Histograms for the residuals from each line are sketched below. Match the description with the histogram:

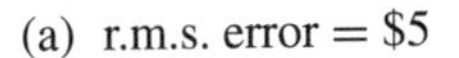

   (a) r.m.s. error = $5 (b) r.m.s. error = $15 (c) something's wrong

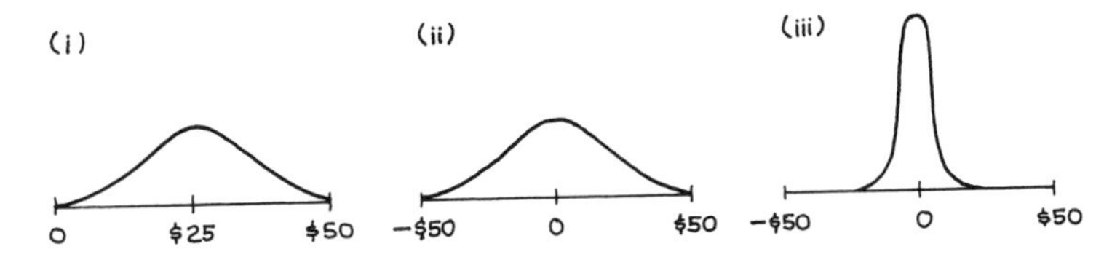

2. Several regression lines are used to predict the monthly salaries in a certain company, from different independent variables. Residual plots from each regression are shown below. Match the description with the plot. Explain. (You may use the same description more than once.)

   (a) r.m.s. error = $1,000 (b) r.m.s. error = $5,000 (c) something's wrong

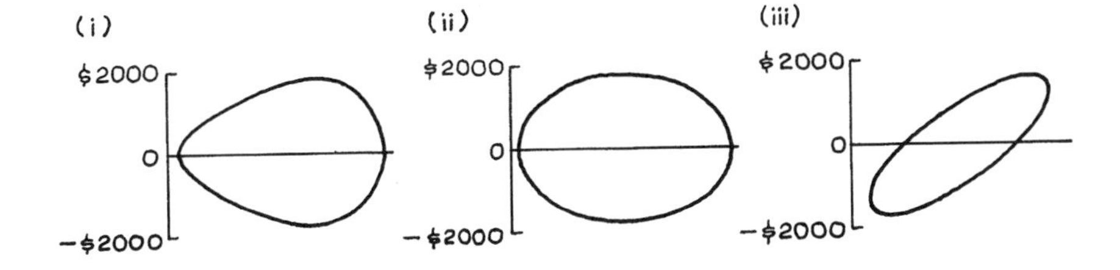

3. Look at the figure below.
   (a) Is the SD of y about 0.6, 1.0, or 2.0?
   (b) Is the SD of the residuals about 0.6, 1.0, or 2.0?
   (c) Take the points in the scatter diagram whose x-coordinates are between 4.5 and 5.5. Is the SD of their y-coordinates about 0.6, 1.0, or 2.0?

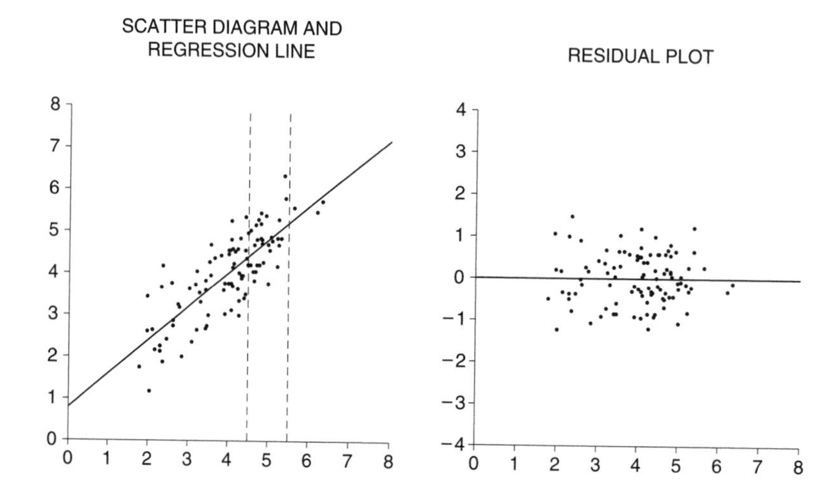

The answers to these exercises are on p. A64.

4. LOOKING AT VERTICAL STRIPS

Figure 8 repeats the scatter diagram for the heights of the 1,078 fathers and sons in Pearson's study (section 1 of chapter 8). The families where the father is 64 inches tall, to the nearest inch, are plotted in the vertical strip on the left. A histogram for son's heights in these families is shown at the bottom of the figure (solid line). The families with 72-inch fathers are plotted in the vertical strip on the right. A histogram for the heights of those sons is shown too (dashed line). The dashed histogram is farther to the right than the solid one: on the average, the taller fathers do have taller sons. However, both histograms have similar shapes, and just about the same amount of spread.[3]

When all the vertical strips in a scatter diagram show similar amounts of spread, the diagram is said to be *homoscedastic*. The scatter diagram in figure 8 is homoscedastic. The range of sons' heights for given father's height is greater in the middle of the picture, but that is only because there are more families in the middle of things than at the extremes. The SD of sons' height for given father's height is pretty much the same from one end of the picture to the other. *Homo* means "same," *scedastic* means "scatter." *Homoscedasticity* is a terrible word, but statisticians insist on it: we prefer "football-shaped."[4]

When the scatter diagram is football-shaped, the prediction errors are similar all along the regression line. In figure 8, the regression line for predicting son's height from father's height had an r.m.s. error of 2.3 inches. If the father is 64 inches tall, the prediction for the son's height is 67 inches, and this is likely to be off by 2.3 inches or so. If the father is 72 inches tall, the prediction for the son's height is 71 inches, and this is likely to be off by the same amount, 2.3 inches or so.[5]

Figure 8. Homoscedastic scatter diagram. Heights of fathers and sons. Families with 64-inch fathers are plotted in the solid vertical strip: the solid histogram is for the heights of those sons. Families with 72-inch fathers are plotted in the dashed vertical strip; the dashed histogram is for the heights of those sons. The two histograms have similar shapes, and their SDs are nearly the same.

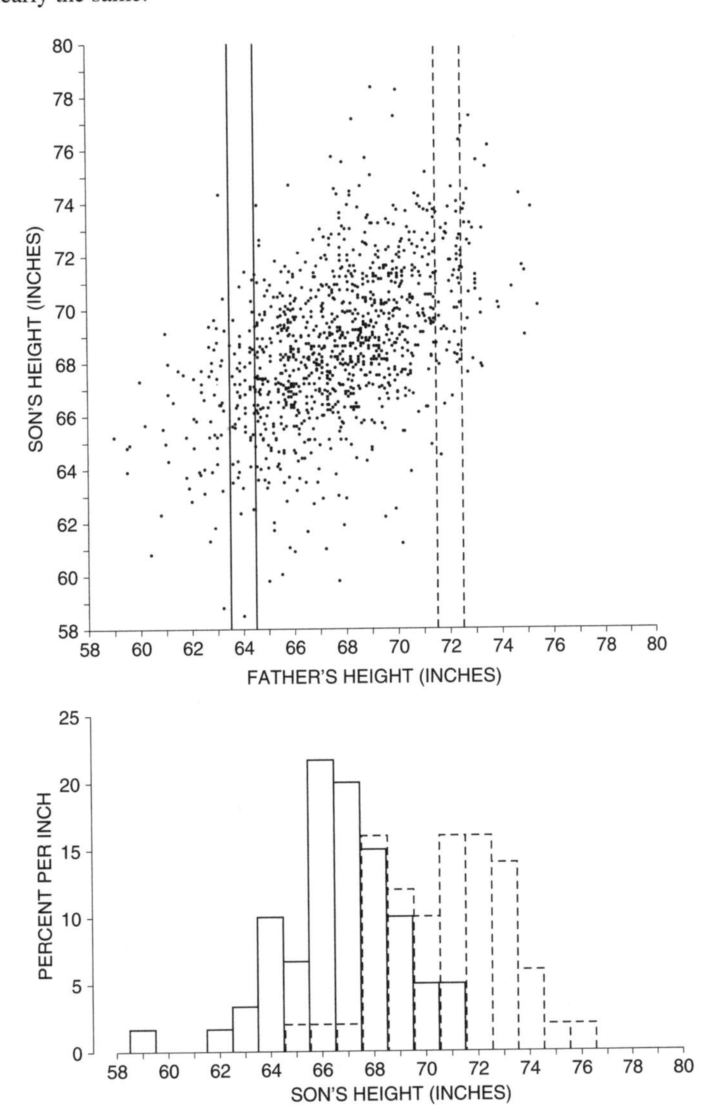

By comparison, figure 9 shows the *heteroscedastic* scatter diagram of income against education (*hetero* means "different"). As education goes up, average income goes up, and so does the spread in income. When the scatter diagram is heteroscedastic, the regression method is off by different amounts in different parts of the scatter diagram. In figure 9, the r.m.s. error of the regression line is about \$19,000. However, it is quite a bit harder to predict the incomes of the highly educated people. With 8 years of schooling, the prediction errors are something like \$6,000. At 12 years, the errors go up to \$15,000 or so. At 16 years, the errors go up even more, to \$27,000 or so. In this case, the r.m.s. error of the regression line gives a sort of average error—across all the different x-values.

Figure 9. Heteroscedastic scatter diagram. Income and education (years of schooling completed) for a sample of 570 California women age 25–29 in 2005.[6] The regression line is shown too.

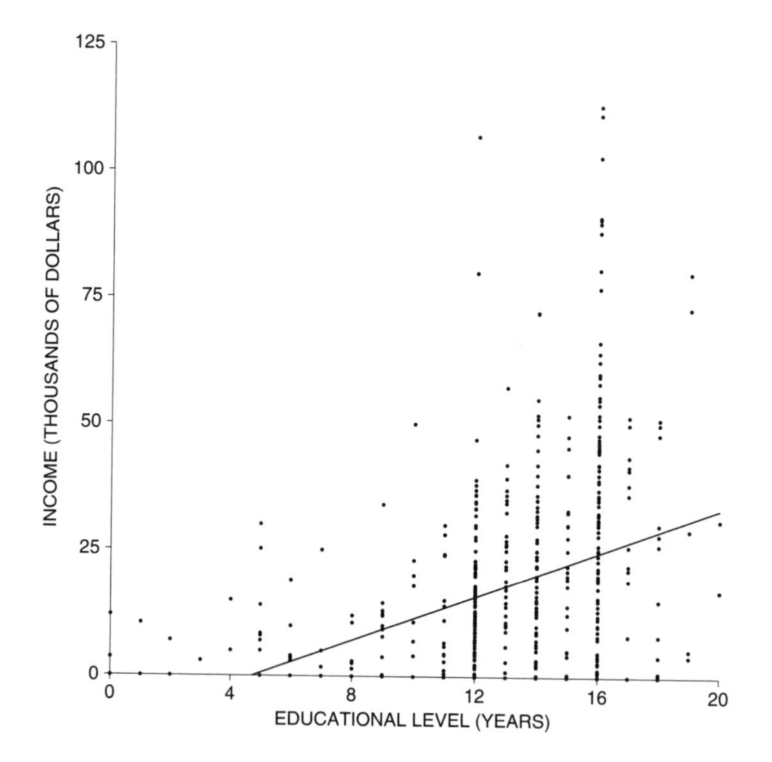

> Suppose that a scatter diagram is football-shaped. Take the points in a narrow vertical strip. They will be off the regression line (up or down) by amounts similar in size to the r.m.s. error. If the diagram is heteroscedastic, the r.m.s. error should not be used for individual strips.

Exercise Set D

1. In 1937, the Stanford-Binet IQ test was restandardized with two forms (L and M). A large number of subjects took both tests. The results can be summarized as follows:

   Form L average $\approx$ 100, SD $\approx$ 15
   Form M average $\approx$ 100, SD $\approx$ 15, $r \approx 0.80$

   (a) True or false, and explain: the regression line for predicting the score on form M from the score on form L has an r.m.s. error of about 9 points.
   (b) Suppose the scatter diagram looks like (i) below. If someone scores 130 on form L, the regression method predicts 124 for the score on form M. True or false, and explain: this prediction is likely to be off by 9 points or so.
   (c) Repeat, if the scatter diagram looks like (ii).

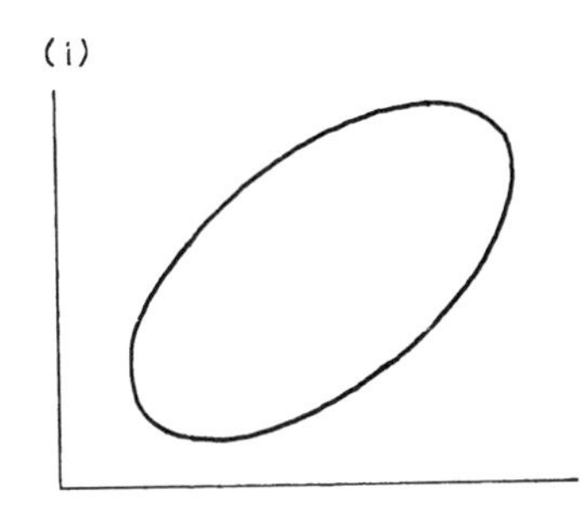

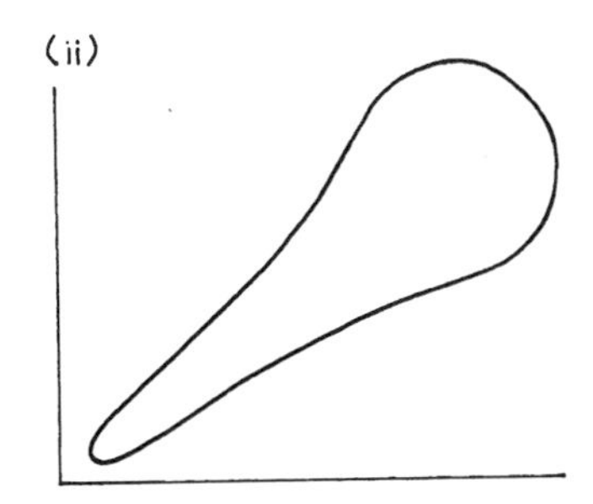

2. The data in figure 8 can be summarized as follows:

   average height of fathers $\approx$ 68 inches, SD $\approx$ 2.7 inches
   average height of sons $\approx$ 69 inches, SD $\approx$ 2.7 inches, $r \approx 0.5$

   (a) Find the r.m.s. error of the regression line for predicting son's height from father's height.
   (b) If a father is 72 inches tall, predict his son's height.
   (c) This prediction is likely to be off by ______ inches or so. If more information is needed, say what it is, and why.
   (d) Repeat parts (b) and (c), if the father is 66 inches tall.

3. The data in figure 9 can be summarized as follows:

   average education $\approx$ 13.0 years, SD $\approx$ 3.4 years
   average income $\approx$ \$18,000, SD $\approx$ \$20,000, $r \approx 0.37$

   (a) Find the r.m.s. error of the regression line for predicting income from education.
   (b) Predict the income of a woman with 16 years of education.
   (c) This prediction is likely to be off by \$______ or so. If more information is needed, say what it is, and why.
   (d) Repeat parts (b) and (c), for a woman with 8 years of education.

4. The figure below is a scatter diagram for the ages of husbands and wives in Indiana. Data are from the March 2005 Current Population Survey.[7] The vertical strip represents the families where the ______ is between ______ and ______ years of age.

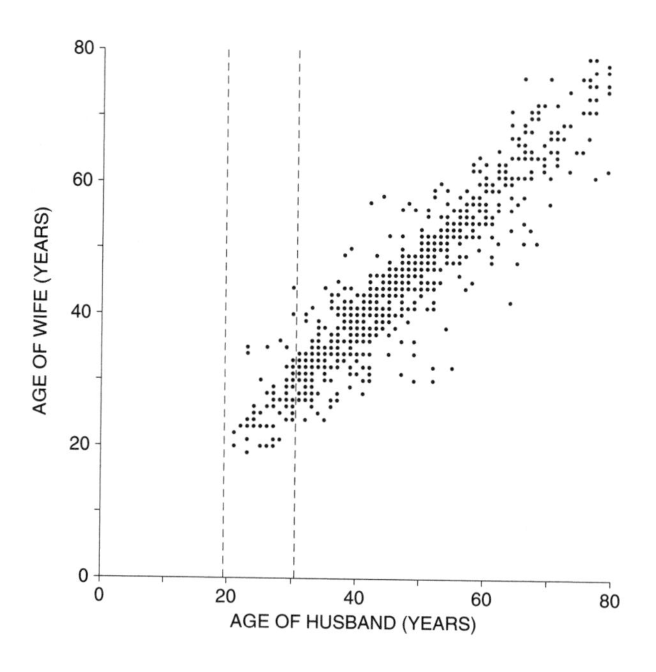

5. (Continues exercise 4.) Fill in the blanks, using the options given below.

.25 .5 .95 1 5 15 25 50

   (a) The average age for all the husbands is about ______; the SD is about ______.
   (b) The average age for all the wives is about ______; the SD is about ______.
   (c) The correlation between the ages of all the husbands and wives is about ______.
   (d) Among families plotted in the vertical strip, the average age for the wives is about ______; the SD is about ______.
   (e) Among families plotted in the vertical strip, the correlation between the ages of the husbands and wives is about ______.

6. (Continues exercises 4 and 5.)
   (a) The SD is computed for the ages of—
      (i) all the wives, and
      (ii) the wives whose husbands are 20–30 years old.

   Which SD is bigger? Or are the SDs about the same?
   (b) The SD is computed for the ages of—
      (i) all the wives, and
      (ii) the wives whose husbands were born in March.

   Which SD is bigger? Or are the SDs about the same?

7. In one study of identical male twins, the average height was found to be about 68 inches, with an SD of about 3 inches. The correlation between the heights of the twins was about 0.95, and the scatter diagram was football-shaped.
   (a) You have to guess the height of one of these twins, without any further information. What method would you use?
   (b) Find the r.m.s. error for the method in (a).
   (c) One twin of the pair is standing in front of you. You have to guess the height of the other twin. What method would you use? (For instance, suppose the twin you see is 6 feet 6 inches.)
   (d) Find the r.m.s. error for the method in (c).

The answers to these exercises are on pp. A64–65.

5. USING THE NORMAL CURVE INSIDE A VERTICAL STRIP

Often, it is possible to use the normal approximation when working inside a vertical strip. For this to be legitimate, the scatter diagram has to be football-shaped, with the dots thickly scattered in the center of the picture and fading off toward the edges. Figure 8 is a good example. On the other hand, if the scatter diagram is heteroscedastic (figure 9), or shows a non-linear pattern (figure 7), do not use the method of this section. With the height-weight data in figure 6, the normal curve would not work especially well either: the cloud isn't football-shaped, it is stretched out on top and squeezed in at the bottom.

Example 1. A law school finds the following relationship between LSAT scores and first-year scores (for students who finish the first year):

$$\text{average LSAT score} = 162, \quad \text{SD} = 6$$
$$\text{average first-year score} = 68, \quad \text{SD} = 10, \quad r = 0.60$$

The scatter diagram is football-shaped.

(a) About what percentage of the students had first-year scores over 75?
(b) Of the students who scored 165 on the LSAT, about what percentage had first-year scores over 75?

Solution. Part (a). This is a straightforward normal approximation problem. The LSAT results and r have nothing to do with it.

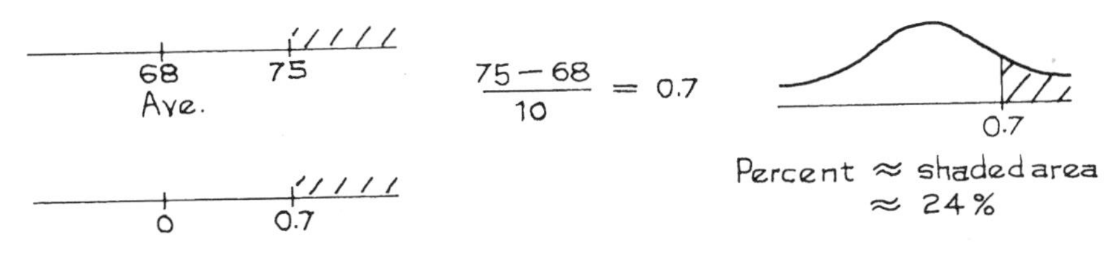

Part (b). This is a new problem. It is about a special group of students—those who scored 165 on the LSAT. These students are all in the same vertical

Figure 10. A football-shaped scatter diagram. Take the points inside a narrow vertical strip. Their y-values are a new data set. The new average is given by the regression method. The new SD is given by the r.m.s. error of the regression line. Inside the strip, a typical y-value is around the new average—give or take the new SD.

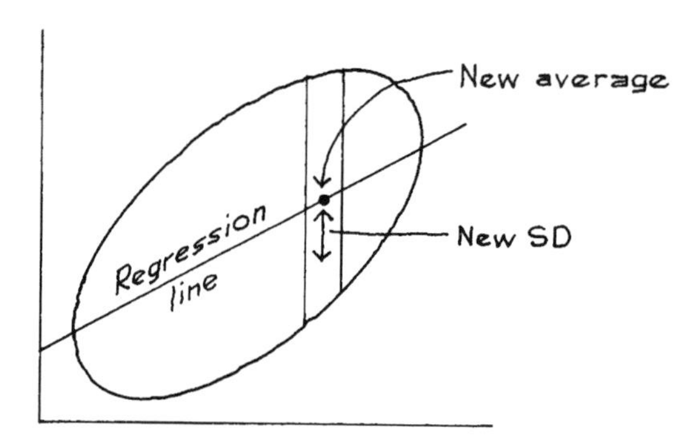

strip (figure 10). Their first-year scores are a new data set. To do the normal approximation, you need the average and the SD of this new data set.

The new average. The students who scored 165 on the LSAT are better than average. As a group, they will do better than average in the first year of law school—although there is a fair amount of spread (vertical scatter inside the strip). The group average can be estimated by the regression method: 165 is 0.5 SDs above average, so the group will score above average in the first year, by about $r \times 0.5 = 0.6 \times 0.5 = 0.3$ SDs. This is $0.3 \times 10 = 3$ points. The new average is $68 + 3 = 71$.

The new SD. The students who scored 165 on the LSAT are a smaller and more homogeneous group. So the SD of their first-year scores is less than 10 points. How much less? Since the diagram is football-shaped, the scatter around the regression line is about the same in each vertical strip, and is given by the r.m.s. error for the regression line (section 4). The new SD is

$$\sqrt{1 - r^2} \times \text{SD of } y = \sqrt{1 - 0.6^2} \times 10 = 8 \text{ points.}$$

(We are predicting first-year scores from LSAT scores, so the error is in first-year points: 10 goes into the formula, not 6.) A typical student who scored around 165 on the LSAT will have a first-year score of about 71, give or take 8 or so. The new average is 71, and the new SD is 8.

The normal approximation is the last step. This is done as usual, but is based on the new average and the new SD.

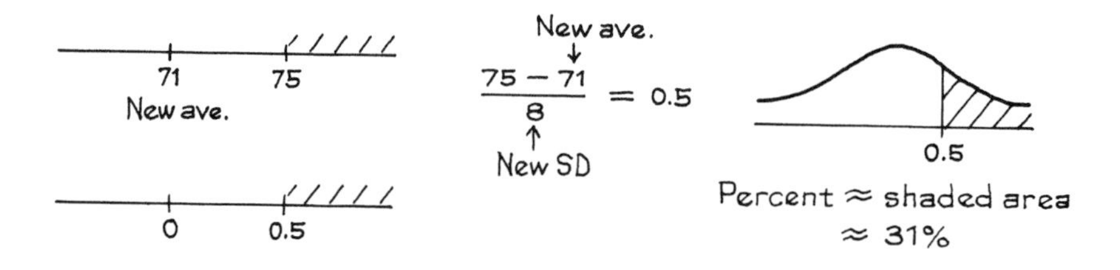

Why is the new SD smaller? Look at figure 10: there is less vertical scatter in the strip than in the whole diagram. Also see exercises 4–6 on p. 194.

> Suppose that a scatter diagram is football-shaped. Take the points in narrow vertical strip. Their y-values are a new data set. The new average is estimated by the regression method. The new SD is about equal to the r.m.s. error for the regression line.

The normal approximation can be done as usual, based on the new average and the new SD.

Technical note. What can you do with non-linear or heteroscedastic data? Often a transformation will help—for example, taking logarithms. The left hand panel in figure 11 shows a scatter diagram for Secchi depth (a measure of water clarity) versus total chlorophyll concentration (a measure of algae in the water).[8] The data are non-linear and heteroscedastic. The right hand panel shows the same data, after taking logs: the diagram is more like a football.

Figure 11. Left-hand panel: scatter diagram for Secchi depth versus total chlorophyll concentration. (Units for chlorophyll concentration are ppb, or parts per billion in the water.) Right-hand panel: data have been transformed by taking logarithms to base 10.

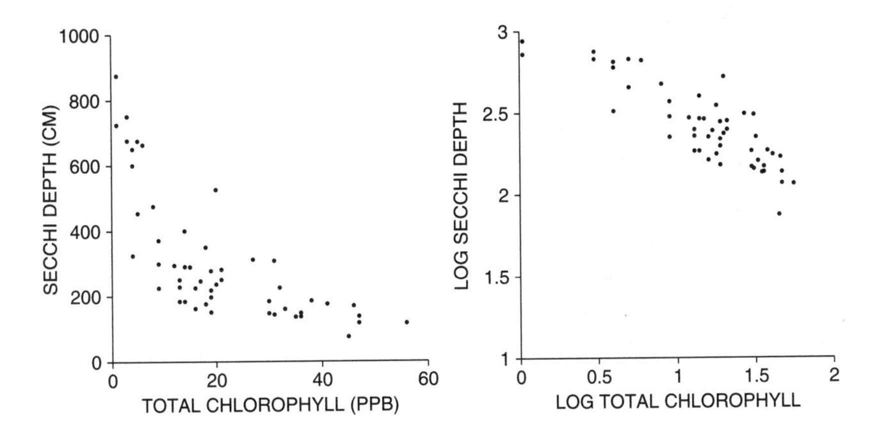

Exercise Set E

1. Pearson and Lee obtained the following results for about 1,000 families:

$$\text{average height of husband} \approx 68 \text{ inches}, \quad \text{SD} \approx 2.7 \text{ inches}$$
$$\text{average height of wife} \approx 63 \text{ inches}, \quad \text{SD} \approx 2.5 \text{ inches}, \quad r \approx 0.25$$

   (a) What percentage of the women were over 5 feet 8 inches?
   (b) Of the women who were married to men of height 6 feet, what percentage were over 5 feet 8 inches?

2. From the same study:

$$\begin{aligned} \text{average height of father} &\approx 68 \text{ inches}, \quad \text{SD} \approx 2.7 \text{ inches} \\ \text{average height of son} &\approx 69 \text{ inches}, \quad \text{SD} \approx 2.7 \text{ inches}, \quad r \approx 0.50 \end{aligned}$$

(a) What percentage of the sons were over 6 feet tall?
(b) What percentage of the 6-foot fathers had sons over 6 feet tall?

3. From the same study:

$$\begin{aligned} \text{average height of men} &\approx 68 \text{ inches}, \quad \text{SD} \approx 2.7 \text{ inches} \\ \text{average forearm length} &\approx 18 \text{ inches}, \quad \text{SD} \approx 1 \text{ inch}, \quad r \approx 0.80 \end{aligned}$$

(a) What percentage of men have forearms which are 18 inches long, to the nearest inch?
(b) Of the men who are 68 inches tall, what percentage have forearms which are 18 inches long, to the nearest inch?

The answers to these exercises are on p. A65.

6. REVIEW EXERCISES

Review exercises may cover material from previous chapters.

1. The r.m.s. error of the regression line for predicting y from x is ______.

(i) SD of y	(iv) $r \times$ SD of x
(ii) SD of x	(v) $\sqrt{1-r^2} \times$ SD of y
(iii) $r \times$ SD of y	(vi) $\sqrt{1-r^2} \times$ SD of x

2. A computer program is developed to predict the GPA of college freshmen from their high-school GPAs. This program is tried out on a class whose college GPAs are known. The r.m.s. error is 3.12. Is anything wrong? Answer yes or no, and explain.

3. Tuddenham and Snyder obtained the following results for 66 California boys at ages 6 and 18 (the scatter diagram is football-shaped):[9]

$$\begin{aligned} \text{average height at 6} &\approx 3 \text{ feet 10 inches}, \quad \text{SD} \approx 1.7 \text{ inches}, \\ \text{average height at 18} &\approx 5 \text{ feet 10 inches}, \quad \text{SD} \approx 2.5 \text{ inches}, \quad r \approx 0.80 \end{aligned}$$

(a) Find the r.m.s. error for the regression prediction of height at 18 from height at 6.
(b) Find the r.m.s. error for the regression prediction of height at 6 from height at 18.

4. A statistical analysis was made of the midterm and final scores in a large course, with the following results:

$$\begin{aligned} \text{average midterm score} &\approx 50, \quad \text{SD} \approx 25 \\ \text{average final score} &\approx 55, \quad \text{SD} \approx 15, \quad r \approx 0.60 \end{aligned}$$

The scatter diagram was football-shaped. For each student, the final score was predicted from the midterm score using the regression line.

(a) For about $1/3$ of the students, the prediction for the final score was off by more than ______ points. Options: 6, 9, 12, 15, 25.
(b) Predict the final score for a student whose midterm score was 80.
(c) This prediction is likely to be off by ______ points or so. Options: 6, 9, 12, 15, 25.

Explain your answers.

5. Use the data in exercise 4 to answer the following questions.

    (a) About what percentage of students scored over 80 on the final?
    (b) Of the students who scored 80 on the midterm, about what percentage scored over 80 on the final?

    Explain your answers.

6. In a study of high-school students, a positive correlation was found between hours spent per week doing homework, and scores on standardized achievement tests. The investigators concluded that doing homework helps prepare students for these tests. Does the conclusion follow from the data? Answer yes or no, and explain briefly.

7. The freshmen at a large university are required to take a battery of aptitude tests. Students who score high on the mathematics test also tend to score high on the physics test. On both tests, the average score is 60; the SDs are the same too. The scatter diagram is football-shaped. Of the students who scored about 75 on the mathematics test:

    (i) just about half scored over 75 on the physics test.
    (ii) more than half scored over 75 on the physics test.
    (iii) less than half scored over 75 on the physics test.

    Choose one option and explain.

8. The bends are caused by rapid changes in air pressure, resulting in the formation of nitrogen bubbles in the blood. The symptoms are acute pain, and sometimes paralysis leading to death. In World War II, pilots got the bends during certain battle maneuvers. It was feasible to simulate these conditions in a pressure chamber. As a result, pilot trainees were tested under these conditions once, at the beginning of their training. If they got the bends (only mild cases were induced), they were excluded from the training on the grounds that they were more likely to get the bends under battle conditions. This procedure was severely criticized by the statistician Joe Berkson, and he persuaded the Air Force to replicate the test—that is, repeat it several times for each trainee.

    (a) Why might Berkson have suggested this?
    (b) Give another example where replication is helpful.

9. Every year, baseball's major leagues honor their outstanding first-year players with the title "Rookie of the Year." The overall batting average for the Rookies of the Year is around .290, far above the major league batting average of .260. However, Rookies of the Year don't do so well in their second year—their

overall second-season batting average is only .275. Baseball writers call this "sophomore slump," the idea being that star players get distracted by outside activities like product endorsements and television appearances. Do the data support the idea of the sophomore slump? Answer yes or no, and explain briefly.[10]

10. A study was made of the relationship between stock prices on the last trading day of 2005 and the last trading day of 2006. A formula was developed to predict the 2006 price from the 2005 price, using data on 100 stocks. An analyst is now reviewing the results. Data are shown below for five out of the 100 stocks; prices are in dollars. Was the regression method used to predict the 2006 price from the 2005 price? Answer yes or no and explain. If you need more information, explain why.

Stock	*2005 price* *actual*	*2006 price* *predicted*	*actual*
A	10	8	8
B	10	8	3
C	12	13	17
D	14	12	6
E	15	20	27

11. The figure below is a scatter plot of income against education, for a representative sample of men age 25–29 in Texas. Or is something wrong? Explain briefly. ("Educational level" means years of schooling completed, not counting kindergarten.)

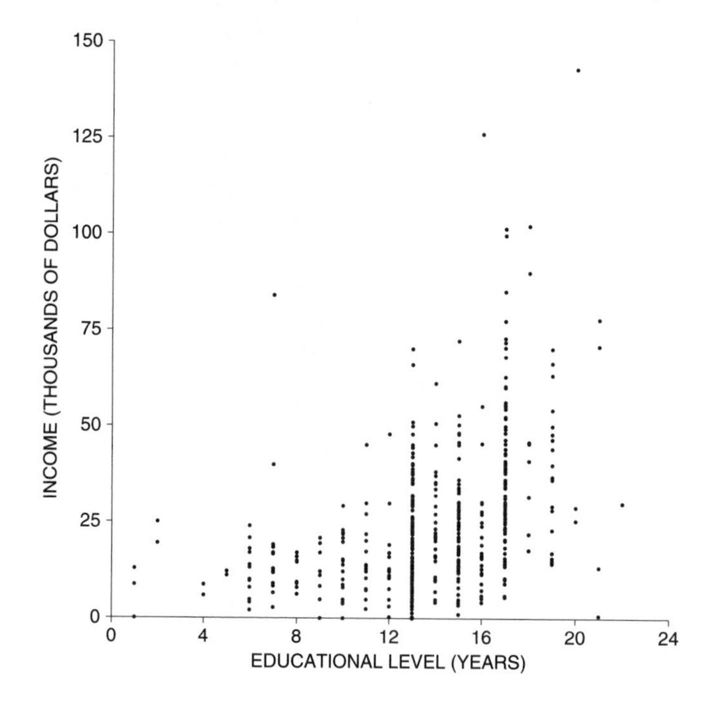

12. For the men age 25–34 in HANES5, the relationship between education (years of schooling completed) and systolic blood pressure can be summarized as follows.[11]

$$\text{average education} \approx 13 \text{ years}, \quad \text{SD} \approx 3 \text{ years}$$
$$\text{average blood pressure} \approx 119 \text{ mm}, \quad \text{SD} \approx 11 \text{ mm}, \quad r \approx -0.1$$

One man in the sample had 20 years of education, and his blood pressure was 118 mm. True or false, and explain: compared to other men at his educational level, his blood pressure was a bit on the high side.

7. SUMMARY

1. When the regression line is used to predict y from x, the difference between the actual value and the predicted value is a *residual*, or prediction error.

2. In a scatter diagram, the vertical distance of a point above or below the regression line is the graphical counterpart of the prediction error made by the regression method.

3. The *r.m.s. error* of the regression line is the root-mean-square of the residuals. This measures the accuracy of the regression predictions. The predictions are off by amounts similar in size to the r.m.s. error. For many scatter diagrams, about 68% of the predictions will be right to within one r.m.s. error. About 95% will be right to within two r.m.s. errors.

4. The SD of y is equal to the r.m.s. error of a horizontal line through the average of y. The r.m.s. error of the regression line is smaller, by the factor $\sqrt{1 - r^2}$. Therefore, the r.m.s. error for the regression line of y on x can be figured as

$$\sqrt{1 - r^2} \times \text{the SD of } y.$$

5. After carrying out a regression, statisticians often graph the residuals. If the *residual plot* shows a pattern, the regression may not have been appropriate.

6. When all the vertical strips in a scatter diagram show similar amounts of spread, the diagram is *homoscedastic*: the prediction errors are similar in size all along the regression line. When the scatter diagram is *heteroscedastic*, the prediction errors are different in different parts of the scatter diagram. Football-shaped diagrams are homoscedastic.

7. Suppose that a scatter diagram is football-shaped. Take the points inside a narrow vertical strip. Their y-values are a new data set. The new average is estimated by the regression method. The new SD is about equal to the r.m.s. error for the regression line. And the normal approximation can be done as usual, based on the new average and the new SD.

12

The Regression Line

The estimation of a magnitude using an observation subject to a larger or smaller error can be compared not inappropriately to a game of chance in which one can only lose and never win and in which each possible error corresponds to a loss . . . . However, what specific loss we should ascribe to any specific error is by no means clear of itself. In fact, the determination of this loss depends at least in part on our judgment . . . . Among the infinite variety of possible functions the one that is simplest seems to have the advantage and this is unquestionably the square . . . . Laplace treated the problem in a similar fashion, but he chose the size of the error as the measure of loss. However, unless we are mistaken this choice is surely not less arbitrary than ours.

—C. F. GAUSS (GERMANY, 1777–1855)[1]

1. SLOPE AND INTERCEPT

Does education pay? Figure 1 shows the relationship between income and education, for a sample of 562 California men age 25–29 in 2005. The summary statistics:[2]

$$\text{average education} \approx 12.5 \text{ years}, \quad \text{SD} \approx 3 \text{ years}$$
$$\text{average income} \approx \$30{,}000, \quad \text{SD} \approx \$24{,}000, \quad r \approx 0.25$$

The regression estimates for average income at each educational level fall along the regression line shown in the figure. The line slopes up, showing that on the average, income does go up with education.

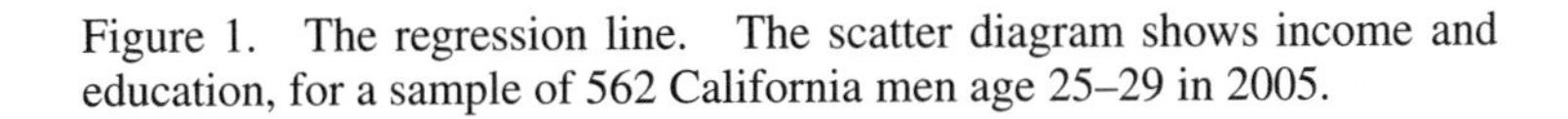
Figure 1. The regression line. The scatter diagram shows income and education, for a sample of 562 California men age 25–29 in 2005.

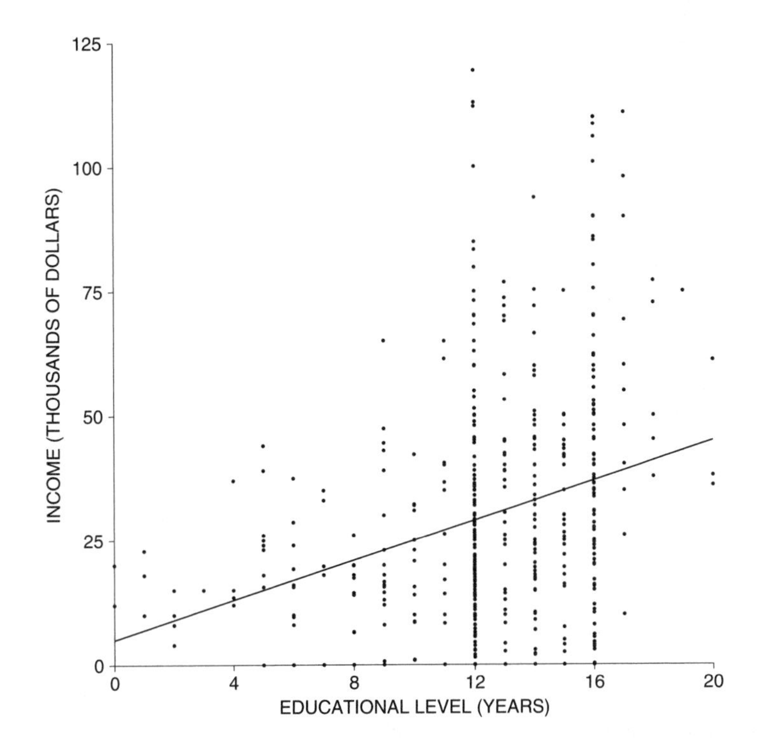

Any line can be described in terms of its slope and intercept (chapter 7). The y-intercept is the height of the line when x is 0. And the slope is the rate at which y increases, per unit increase in x. Slope and intercept are illustrated in figure 2.

Figure 2. Slope and intercept.

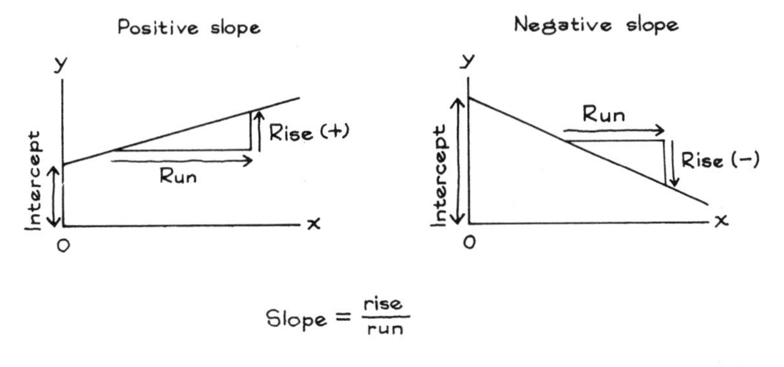

How do you get the slope of the regression line? Take the income-education example. Associated with an increase of one SD in education, there is an increase of r SDs in income. On this basis, 3 extra years of education are worth an extra

Figure 3. Finding the slope and intercept of the regression line.

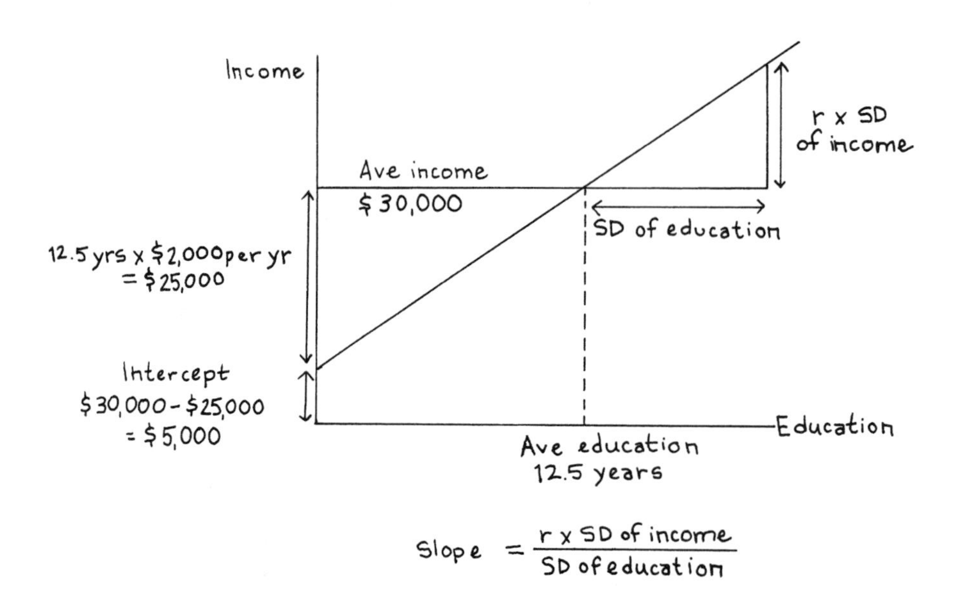

0.25 × \$24,000 = \$6,000 of income, on the average. So each extra year is worth \$6,000/3 = \$2,000. The slope of the regression line is \$2,000 per year.

The intercept of the regression line is the height when $x = 0$, corresponding to men with 0 years of education. These men are 12.5 years below average in education. Each year costs \$2,000—that is what the slope says. A man with no education should have an income which is below average by

$$12.5 \text{ years} \times \$2{,}000 \text{ per year} = \$25{,}000.$$

His income should be \$30,000 − \$25,000 = \$5,000. That is the intercept (figure 3): the predicted value of y when $x = 0$. Zero years of education may sound extreme, but there were four men who reported having no education; their points are in the lower left corner of figure 1. (Three of these men reported incomes of \$20,000, so their points plot one on top of the other.)

> Associated with a unit increase in x there is some average change in y. The slope of the regression line estimates this change. The formula for the slope is
>
> $$\frac{r \times \text{SD of } y}{\text{SD of } x}$$
>
> The intercept of the regression line is just the predicted value for y when x is 0.

The equation of a line can be written in terms of the slope and intercept:

$$y = \text{slope} \times x + \text{intercept}.$$

The equation for the regression line is called (not surprisingly) the *regression equation.* In figure 3, the regression equation is

$$\text{predicted income} = \$2{,}000 \text{ per year} \times \text{education} + \$5{,}000.$$

There is nothing new here. The regression equation is just a way of predicting y from x by the regression method. However, social scientists often report the regression equation; the slope and intercept can be interesting in their own right.

Example 1. Education and income for 570 California women age 25–29 are shown in figure 9 on p. 192. The summary statistics are:[3]

$$\begin{aligned} \text{average education} &\approx 13.0 \text{ years}, \quad \text{SD} \approx 3.4 \text{ years} \\ \text{average income} &\approx \$18{,}000, \quad \text{SD} \approx \$20{,}000, \quad r \approx 0.37 \end{aligned}$$

(a) Find the regression equation for predicting income from education.

(b) Use the equation to predict the income of a woman whose educational level is: 8 years, 12 years, 16 years.

Solution. Part (a). The first step is to find the slope. In a run of one SD of education, the regression line rises r SDs of income. So

$$\text{slope} = \frac{0.37 \times \$20{,}000}{3.4 \text{ years}} \approx \$2{,}176 \text{ per year.}$$

On the average, each extra year of schooling is worth an extra \$2,176 of income; each year less of schooling costs \$2,176 of income. (Income has such a large SD because the distribution has a long right hand tail.)

The next step is to find the intercept. That is the height of the regression line at $x = 0$—in other words, the predicted income for a woman with no education. She is 13 years below average. Her income should be below average by

$$13 \text{ years} \times \$2{,}176 \text{ per year} = \$28{,}288.$$

Her predicted income is

$$\$18{,}000 - \$28{,}288 = -\$10{,}288.$$

That is the intercept: the prediction for y when $x = 0$. The regression equation is

$$\text{predicted income} = \$2{,}176 \text{ per year} \times \text{education} - \$10{,}288.$$

The regression line becomes unreliable when you are far from the center of the data, so a negative intercept is not too disturbing.

Part (b). Substitute 8 years for education, to get

$$\$2{,}176 \text{ per year} \times 8 \text{ years} - \$10{,}288 = \$7{,}120.$$

Substitute 12 years for education:

$$\$2,176 \text{ per year} \times 12 \text{ years} - \$10,288 = \$15,824.$$

Substitute 16 years:

$$\$2,176 \text{ per year} \times 16 \text{ years} - \$10,288 = \$24,528.$$

This completes the solution. Despite the negative intercept, the predictions are quite reasonable for most of the women.

In example 1, the slope is $2,176 per year. Associated with each extra year of education, there is an increase of $2,176 in income, on the average. The phrase "associated with" sounds like it is talking around some difficulty, and here is the issue. Are income differences caused by differences in educational level, or do both reflect the common influence of some third variable? The phrase "associated with" was invented to let statisticians talk about regressions without having to commit themselves on this sort of point.

The slope is often used to predict how y will respond, if someone intervenes and changes x. This is legitimate when the data come from a controlled experiment. With observational studies, the inference is often shaky—because of confounding. Look at example 1. On the average, the women who finished high school (12 years of education) earned about $9,000 more than women who just finished middle school (8 years).

Now, take a representative group of women with 8 years of education. If the government intervened and sent them on to get high-school degrees, the slope suggests that their incomes would go up by an average of $4 \times \$2,176 \approx \$9,000$. However, example 1 is based on survey data rather than a controlled experiment. One group of women in the survey had 8 years of education. Another, separate, group had 12 years. The two groups were different with respect to many factors besides education—like intelligence, ambition, and family background.

The effects of these factors are confounded with the effect of education, and go into the slope. Sending people off to high school probably would make their incomes go up, but not by the full $9,000. To measure the impact on incomes, it might be necessary to run a controlled experiment. (Many investigators would use a technique called *multiple regression*; more about this in section 3.[4])

With an observational study, the slope and intercept of the regression line are only descriptive statistics. They say how the average value of one variable is related to values of another variable in the population being observed. The slope cannot be relied on to predict how y would respond if you intervene to change the value of x.

> If you run an observational study, the regression line only describes the data that you see. The line cannot be relied on for predicting the results of interventions.

There is another assumption that we have been making throughout this section: that the average of y depends linearly on x. If the relationship is non-linear, the regression line may be quite misleading—whether the data come from an experiment or an observational study.[5]

Exercise Set A

1. For the men in figure 1, the regression equation for predicting average income from education is

$$\text{predicted income} = \$2{,}000 \text{ per year} \times \text{education} + \$5{,}000.$$

   Predict the income for one of these men who has

   (a) 8 years of schooling—elementary education
   (b) 12 years of schooling—a high-school diploma
   (c) 16 years of schooling—a college degree.

2. The International Rice Research Institute in the Philippines developed the hybrid rice IR 8, setting off "the green revolution" in tropical agriculture. Among other things, they made a thorough study of the effects of fertilizer on rice yields. These experiments involved a number of experimental plots (of about 20 square yards in size). Each plot was planted with IR 8, and fertilized with some amount of nitrogen chosen by the investigators. (The amounts ranged from 0 to about a pound.) When the rice was harvested, the yield was measured and related to the amount of nitrogen applied. In one such experiment, the correlation between rice yield and nitrogen was 0.95, and the regression equation was[6]

$$\text{predicted rice yield} = (20 \text{ oz rice per oz nitrogen}) \times (\text{nitrogen}) + 240 \text{ oz}.$$

   (a) An unfertilized plot can be expected to produce around ______ of rice.
   (b) Each extra ounce of nitrogen fertilizer can be expected to increase the rice yield by ______ .
   (c) Predict the rice yield when the amount of fertilizer is

   3 ounces of nitrogen   4 ounces of nitrogen

   (d) Was this an observational study or a controlled experiment?
   (e) In fact, fertilizer was applied only in the following amounts: 0 ounces, 4 ounces, 8 ounces, 12 ounces, 16 ounces. Would you trust the prediction for 3 ounces of nitrogen, even though this particular amount was never applied?
   (f) Would you trust the prediction for 100 ounces of nitrogen?

3. Summary statistics for heights of fathers and sons are on p. 170.

   (a) Find the regression equation for predicting the height of a son from the height of his father.
   (b) Find the regression equation for predicting the height of a father from the height of his son.

4. An expert witness offers testimony that[7]

   > Regression is a substitute for controlled experiments. It provides a precise estimate of the effect of one variable on another.

   Comment briefly.

The answers to these exercises are on pp. A65–66.

2. THE METHOD OF LEAST SQUARES

Chapter 10 discussed regression from one point of view, and section 1 went over the same ground using the regression equation. This section is a third pass at the same topic, from yet another perspective. (For statisticians, regression is an important technique.) Sometimes the points on a scatter diagram seem to be following a line. The problem discussed in this section is how to find the line which best fits the points. Usually, this involves a compromise: moving the line closer to some points will increase its distance from others. To resolve the conflict, two steps are necessary. First, define an average distance from the line to all the points. Second, move the line around until this average distance is as small as possible.

To be more specific, suppose the line will be used to predict y from x. Then, the error made at each point is the vertical distance from the point to the line. In statistics, the usual way to define the average distance is by taking the root-mean-square of the errors. This measure of average distance is called the *r.m.s. error of the line*. (It was first proposed by Gauss; see the chapter opening quote.)

The second problem, how to move the line around to minimize the r.m.s. error, was also solved by Gauss.

> Among all lines, the one that makes the smallest r.m.s. error in predicting y from x is the regression line.

For this reason, the regression line is often called the *least squares line*: the errors are squared to compute the r.m.s. error, and the regression line makes the r.m.s. error as small as possible. (The r.m.s. error of the regression line was discussed in section 1 of chapter 11.)

Now, an example. Robert Hooke (England, 1653–1703) was able to determine the relationship between the length of a spring and the load placed on it. He just hung weights of different sizes on the end of a spring, and watched what happened. When he increased the load, the spring got longer. When he reduced the load, the spring got shorter. And the relationship was linear.

Let b be the length of the spring with no load. A weight of x kilograms is attached to the end of the spring. As illustrated in figure 4, the spring stretches to

Figure 4. Hooke's law: the stretch is proportional to the load.

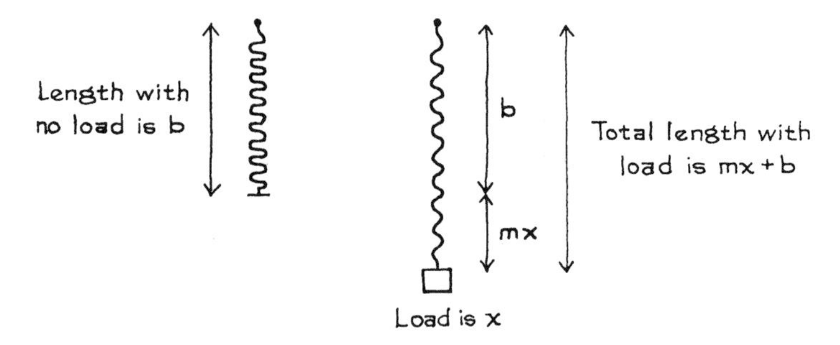

a new length. According to Hooke's law, the amount of stretch is proportional to the weight x. The new length of the spring is

$$y = mx + b.$$

In this equation, m and b are constants which depend on the spring. Their values are unknown, and have to be estimated using experimental data.

Table 1 shows the results of an experiment on Hooke's law, done in a physics class at the University of California, Berkeley. Different weights were hung on the end of a long piano wire.[8] The first column shows the load. The second column shows the measured length. With 20 pounds of load, this "spring" only stretched about 0.2 inch (10 kg $\approx$ 22 lb, 0.5 cm $\approx$ 0.2 in). Piano wire is not very stretchy.

Table 1. Data on Hooke's law.

Weight (kg)	*Length (cm)*
0	439.00
2	439.12
4	439.21
6	439.31
8	439.40
10	439.50

The correlation coefficient for the data in table 1 is 0.999, very close to 1 indeed. So the points almost form a straight line (figure 5), just as Hooke's law predicts. The minor deviations from linearity are probably due to measurement error; neither the weights nor the lengths have been measured with perfect accuracy. (Nothing ever is.)

Figure 5. Scatter diagram for table 1.

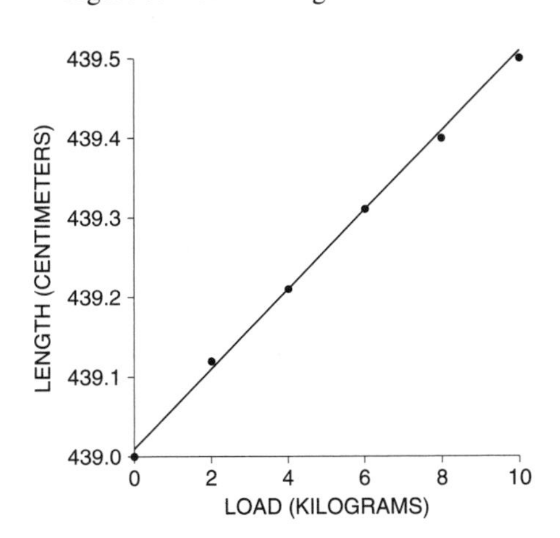

Our goal is to estimate m and b in the equation of Hooke's law for the piano wire:

$$y = mx + b.$$

The graph of this equation is a perfect straight line. If the points in figure 5 happened to fall exactly on some line, the slope of that line would estimate m, and its intercept would estimate b. However, the points do not line up perfectly. Many different lines could be drawn across the scatter diagram, each having a slightly different slope and intercept.

Which line should be used? Hooke's equation predicts length from weight. As discussed above, it is natural to choose m and b so as to minimize the r.m.s. error: this is the *method of least squares*. The line $y = mx + b$ which does the job is the regression line.[9] In other words, m in Hooke's law should be estimated as the slope of the regression line, and b as its intercept. These are called the *least squares estimates*, because they minimize root-mean-square error. If you do the arithmetic,

$$m \approx 0.05 \text{ cm per kg and } b \approx 439.01 \text{ cm}$$

The length of the spring under no load is estimated as 439.01 cm. And each kilogram of load causes the spring to stretch by about 0.05 cm. There is no need to hedge, because the estimates are based on a controlled experiment. The investigator puts the weights on, and the wire stretches. Take the weights off, and the wire comes back to its original length. This can be repeated as often as you want. There is no question here about what is causing what; the language of "association" is not needed. Of course, even Hooke's law has its limits: beyond some point, the spring will break. *Extrapolating beyond the data is risky.*

The method of least squares and the regression method involve the same mathematics; but the contexts may be different. In some fields, investigators talk about "least squares" when they are estimating parameters—unknown constants of nature like m and b in Hooke's law. In other fields, investigators talk about regression when they are studying the relationship between two variables, like income and education, using non-experimental data.

A technical point: The least squares estimate for the length of the spring under no load was 439.01 cm. This is a tiny bit longer than the measured length at no load (439.00 cm). A statistician might trust the least squares estimate over the measurement. Why? Because the least squares estimate takes advantage of all six measurements, not just one: some of the measurement error is likely to cancel out. Of course, the six measurements are tied together by a good theory—Hooke's law. Without the theory, the least squares estimate wouldn't be worth much.

Exercise Set B

1. For the men age 25–34 in the HANES2 sample (p. 58), the regression equation for predicting height from education is[10]

   $$\text{predicted height} = (0.25 \text{ inches per year}) \times (\text{education}) + 66.75 \text{ inches}$$

   Predict the height of a man with 12 years of education; with 16 years of education. Does going to college increase a man's height? Explain.

2. For the data in table 1 (p. 209), the regression equation for predicting length from weight is

$$\text{predicted length} = (0.05 \text{ cm per kg}) \times (\text{weight}) + 439.01 \text{ cm}$$

Predict the length of the wire when the weight is 3 kg; 5 kg. Does putting more weight on the spring make it longer? Explain.

3. A study is made of Math and Verbal SAT scores for the entering class at a certain college. The summary statistics:

$$\begin{aligned} \text{average M-SAT} &= 560, \quad \text{SD} = 120 \\ \text{average V-SAT} &= 540, \quad \text{SD} = 110, \quad r = 0.66 \end{aligned}$$

The investigator uses the SD line to predict V-SAT score from M-SAT score.

(a) If a student scores 680 on the M-SAT, the predicted V-SAT score is ______.
(b) If a student scores 560 on the M-SAT, the predicted V-SAT score is ______.
(c) The investigator's r.m.s. error is ______ $\sqrt{1 - 0.66^2} \times 110$. Options:

greater than   equal to   less than

If more information is needed, say what you need, and why.

4. Repeat exercise 3, if the investigator always predicts a V-SAT of 540.

5. Exercise 3 describes one way to predict V-SAT from M-SAT; exercise 4 describes a second way; and regression is a third way. Which way will have the smallest r.m.s. error?

The answers to these exercises are on p. A66.

3. DOES THE REGRESSION MAKE SENSE?

A regression line can be put down on any scatter diagram. However, there are two questions to ask: First, was there a non-linear association between the variables? If so, the regression line may be quite misleading (p. 163). Even if the association looks linear, there is a second question: Did the regression make sense? The second question is harder. Answering it requires some understanding of the mechanism which produced the data. If this mechanism is not understood, fitting a line can be intellectually disastrous.

To make up an example, suppose an investigator does not know the formula for the area of a rectangle. He thinks area ought to depend on perimeter. Taking an empirical approach, he draws 20 typical rectangles, measuring the area and the perimeter for each one. The correlation coefficient turns out to be 0.98—almost as good as Hooke's law. The investigator concludes that he is really on to something. His regression equation is

$$\text{area} = (1.60 \text{ inches}) \times (\text{perimeter}) - 10.51 \text{ square inches}$$

(Area is measured in square inches and perimeter in inches.)

Figure 6. Scatter diagram of area against perimeter for 20 rectangles; the regression line is shown too.

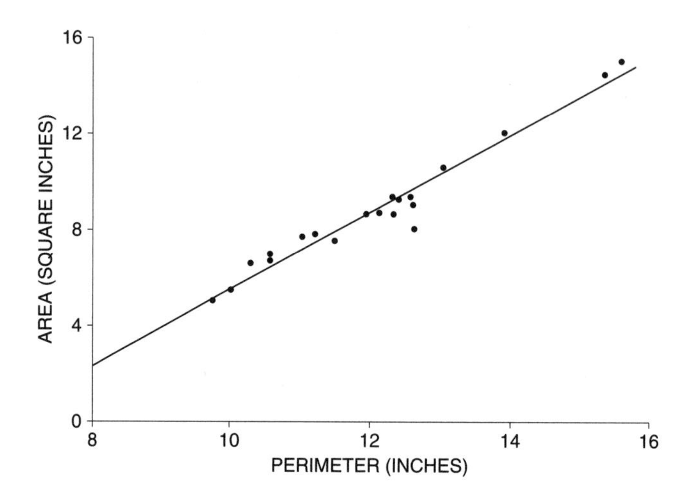

There is a scatter diagram in figure 6, with one dot for each rectangle; the regression line is plotted too. The rectangles themselves are shown in figure 7. The arithmetic is all in order, but the regression is silly. The investigator should have looked at two other variables, length and width. These two variables determine both area and perimeter:

$$\text{area} = \text{length} \times \text{width}, \qquad \text{perimeter} = 2(\text{length} + \text{width})$$

Our straw-man investigator would never find this out by doing regressions.

When looking at a regression study, ask yourself whether it is more like Hooke's law, or more like area and perimeter. Of course, the area-perimeter example is hypothetical. But many investigators do fit lines to data without facing the issues. That can make a lot of trouble.[11]

Technical note. Example 1 in section 1 presented a regression equation for predicting income from education. This is a good way to describe the relationship between income and education. But it may not be legitimate to interpret the slope as the effect on income if you intervene to change education. The problem—the effects of other variables may be confounded with the effects of education.

Many investigators would use multiple regression to control for other variables. For instance, they might develop some measure for the socioeconomic status of parents, and fit a multiple regression equation of the form

$$y = a + b \times E + c \times S,$$

where

$$y = \text{predicted income}, \quad E = \text{educational level},$$
$$S = \text{measure of parental status}.$$

Figure 7. The 20 rectangles themselves.

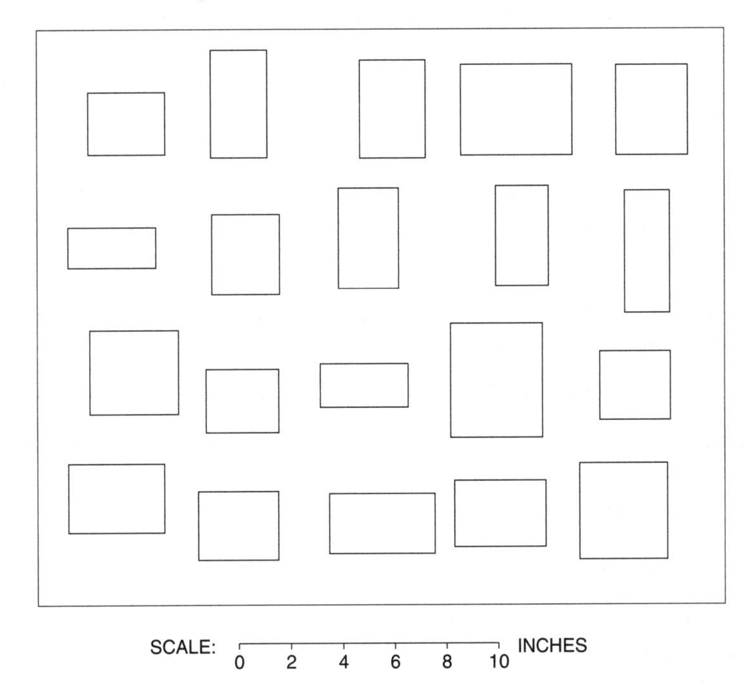

The coefficient b would be interpreted as showing the effect of education, controlling for the effect of parental status.

This might give sensible results. But it can equally well produce nonsense. Take the hypothetical investigator who was working on the area of rectangles. He could decide to control for the shape of the rectangles by multiple regression, using the length of the diagonal to measure shape. (Of course, this isn't a good measure of shape, but nobody knows how to measure status very well either.) The investigator would fit a multiple regression equation of the form

$$\text{predicted area} = a + b \times \text{perimeter} + c \times \text{diagonal}.$$

He might tell himself that b measures the effect of perimeter, controlling for the effect of shape. As a result, he would be even more confused than before. The perimeter and the diagonal do determine the area, but only by a non-linear formula. Multiple regression is a powerful technique, but it is not a substitute for understanding.

4. REVIEW EXERCISES

Review exercises may cover material from previous chapters.

1. Find the regression equation for predicting final score from midterm score, based on the following information:

$$\text{average midterm score} = 70, \quad \text{SD} = 10$$
$$\text{average final score} = 55, \quad \text{SD} = 20, \quad r = 0.60$$

2. For women age 25–34 in the HANES5 sample, the relationship between height and income can be summarized as follows:[12]

$$\text{average height} \approx 64 \text{ inches}, \quad \text{SD} \approx 2.5 \text{ inches}$$
$$\text{average income} \approx \$21{,}000, \quad \text{SD} \approx \$20{,}000, \quad r \approx 0.2$$

What is the regression equation for predicting income from height? What does the equation tell you?

3. For men age 18–24 in the HANES5 sample, the regression equation for predicting height from weight is

$$\text{predicted height} = (0.0267 \text{ inches per pound}) \times (\text{weight}) + 65.2 \text{ inches}$$

(Height is measured in inches and weight in pounds.) If someone puts on 20 pounds, will he get taller by

$$20 \text{ pounds} \times 0.0267 \text{ inches per pound} \approx 0.5 \text{ inches?}$$

If not, what does the slope mean?

4. (a) Is the r.m.s. error of the line below around 0.1, 0.3, or 1?
(b) Is it the regression line?

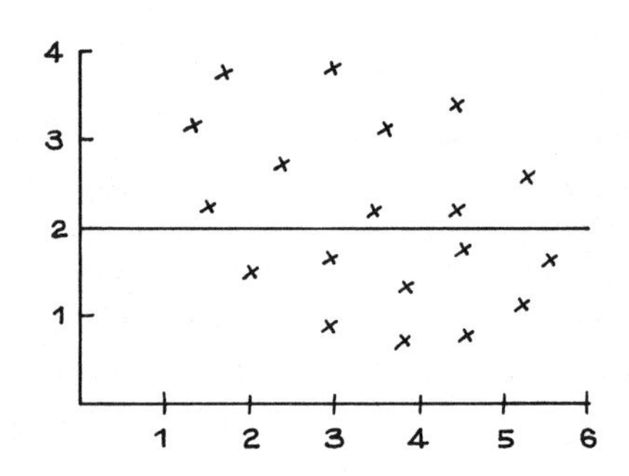

5. A study is made of working couples. The regression equation for predicting wife's income from husband's income is

$$\text{wife's income} = 0.1667 \times \text{husband's income} + \$24{,}000.$$

Another investigator solves this equation for husband's income, and gets

$$\text{husband's income} = 6 \times \text{wife's income} - \$144{,}000.$$

True or false, and explain: the second investigator has found the regression equation for predicting husband's income from wife's income. If you want to compute anything,

$$\text{husband's average income} = \$54{,}000, \quad \text{SD} = \$39{,}000$$
$$\text{wife's average income} = \$33{,}000, \quad \text{SD} = \$26{,}000, \quad r = 0.25$$

6. (Continues exercise 5.) The couples in the previous exercise are followed for a year. Suppose everyone's income goes up by 10%. Find the new regression line for predicting wife's income from husband's income.

7. A statistician is doing a study on a group of undergraduates. On average, these students drink 4 beers a month, with an SD of 8. They eat 4 pizzas a month, with an SD of 4. There is some positive association between beer and pizza, and the regression equation is[13]

$$\text{predicted number of beers} = \underline{\qquad\quad} \times \text{number of pizzas} + 2.$$

However, the statistician lost the data and forgot the slope of the equation. (Perhaps he had too much beer and pizza.) Can you help him remember the slope? Explain.

8. An investigator wants to use a straight line to predict IQ from lead levels in the blood, for a representative group of children aged 5–9.[14] There is a weak positive association in the data. True or false, and explain—
   (a) He can use many different lines.
   (b) He has to use the regression line.
   (c) Only the regression line has an r.m.s. error.
   (d) Any line he uses will have an r.m.s. error.
   (e) Among all lines, the regression line has the smallest r.m.s. error.

9. In a large study (hypothetical) of the relationship between parental income and the IQs of their children, the following results were obtained:

$$\begin{array}{rll} \text{average income} \approx \$90{,}000, & \text{SD} \approx \$45{,}000 & \\ \text{average IQ} \approx 100, & \text{SD} \approx 15, & r \approx 0.50 \end{array}$$

For each income group ($0–$9999, $10,000–$19,999, $20,000–$29,999, etc.), the average IQ of children with parental income in that group was calculated and then plotted above the midpoint of the group ($5,000, $15,000, $25,000, etc.). It was found that the points on this graph followed a straight line very closely. The slope of this line (in IQ points per dollar) would be about:

6,000 3,000 1,500 500 1/500 1/1,500 1/3,000 1/6,000
can't say from the information given

Explain briefly.

10. One child in the study referred to in exercise 9 had an IQ of 110, but the information about his parents' income was lost. At $150,000 the height of the line plotted in exercise 9 corresponds to an IQ of 110. Is $150,000 a good estimate for the parents' income? Or is the estimate likely to be too high? too low? Explain.

11. (Hypothetical.) A congressional report is discussing the relationship between income of parents and educational attainment of their daughters. Data are

from a sample of families with daughters age 18–24. Average parental income is $79,300; average educational attainment of the daughters is 12.7 years of schooling completed; the correlation is 0.37.

The regression line for predicting daughter's education from parental income is reported as $y = mx + b$, with x = parental income (dollars), y = predicted education (years), $m = 0.00000925$ years per dollar, and $b = 10.3$ years:

$$\text{predicted education} = 0.00000925 \times \text{income} + 10.3$$

Is anything wrong? Or do you need more information to decide? Explain briefly.

12. Many epidemiologists think that a high level of salt in the diet causes high blood pressure. INTERSALT is often cited to support this view. INTERSALT was a large study done at 52 centers in 32 countries.[15] Each center recruited 200 subjects in 8 age- and sex-groups. Salt intake was measured, as well as blood pressure and several possible confounding variables. After adjusting for age, sex, and the other confounding variables, the authors found a significant association between high salt intake and high blood pressure. However, a more detailed analysis showed that in 25 of the centers, there was a positive association between blood pressure and salt; in the other 27, the association was negative. Do the data support the theory that high levels of salt cause high blood pressure? Answer yes or no, and explain briefly.

5. SUMMARY AND OVERVIEW

1. The regression line can be specified by two descriptive statistics: the *slope* and the *intercept*.

2. The slope of the regression line for y on x is the average change in y, per unit change in x. This equals

$$r \times \text{SD of } y/\text{SD of } x.$$

3. The intercept of the regression line equals the regression estimate for y, when x is 0.

4. The equation of the regression line for y on x is

$$y = \text{slope} \times x + \text{intercept}.$$

5. The equation can be used to make all the regression predictions, by substitution.

6. Among all lines, the regression line for y on x makes the smallest r.m.s. error in predicting y from x. For that reason, the regression line is often called the *least squares line*.

7. Sometimes, two quantities are thought to be connected by a linear relationship (for example, length and weight in Hooke's law). The statistical problem is to estimate the slope and intercept of the line. The *least squares estimates* are the slope and intercept of the regression line.

8. In this part of the book, scatter diagrams are used to graph the association between two variables. If the scatter diagram is football-shaped, it can be summarized by the average and SD of the two variables, with r to measure the strength of the association.

9. How does the average of one variable depend on the values of the other variable? The regression line can be used to answer that question.

10. With a controlled experiment, the slope can tell you the average change in y that would be caused by a change in x. With an observational study, however, the slope cannot be relied on to predict the results of interventions. It takes a lot of hard work to draw causal inferences from observational data, with or without regression.

11. If the average of y depends on x in a non-linear way, the regression line can be quite misleading.

PART IV

Probability

13

What Are the Chances?

In the long run, we are all dead.
—JOHN MAYNARD KEYNES (ENGLAND, 1883–1946)

1. INTRODUCTION

People talk loosely about chance all the time, without doing any harm. What are the chances of getting a job? of meeting someone? of rain tomorrow? But for scientific purposes, it is necessary to give the word *chance* a definite, clear interpretation. This turns out to be hard, and mathematicians have struggled with the job for centuries. They have developed some careful and rigorous theories, but these theories cover just a small range of the cases where people ordinarily speak of chance. This book will present the *frequency theory*, which works best for processes which can be repeated over and over again, independently and under the same conditions.[1] Many games fall into this category, and the frequency theory was originally developed to solve gambling problems. One of the great early masters was Abraham de Moivre, a French Protestant who fled to England to avoid religious persecution. Part of the dedication to his book, *The Doctrine of Chances*, is reproduced in figure 1 on the next page.[2]

Figure 1. De Moivre's dedication to *The Doctrine of Chances.*

To the Right Honorable the
Lord CARPENTER.

My Lord,

There are many people in the World who are prepossessed with an Opinion, that the Doctrine of Chances has a Tendency to promote Play; but they soon will be undeceived, if they think fit to look into the general Design of this Book; in the mean time it will not be improper to inform them, that your Lordship is pleased to espouse the Patronage of this second Edition; which your strict Probity, and the distinguished Character you bear in the World, would not have permitted, were not their Apprehensions altogether groundless.

Your Lordship does easily perceive, that this Doctrine is so far from encouraging Play, that it is rather a Guard against it, by setting in a clear light, the Advantages and Disadvantages of those Games wherein Chance is concerned....

Another use to be made of this Doctrine of Chances is that it may serve in conjunction with the other parts of the Mathematicks, as a fit Introduction to the Art of Reasoning: it being known by experience that nothing can contribute more to the attaining of that Art, than the consideration of a long Train of Consequences, rightly deduced from undoubted Principles, of which this Book affords many Examples.

One simple game of chance involves betting on the toss of a coin. The process of tossing the coin can be repeated over and over again, independently and under the same conditions. The chance of getting heads is 50%: in the long run, heads will turn up about 50% of the time.

Take another example. A die (plural, "dice") is a cube with six faces, labelled

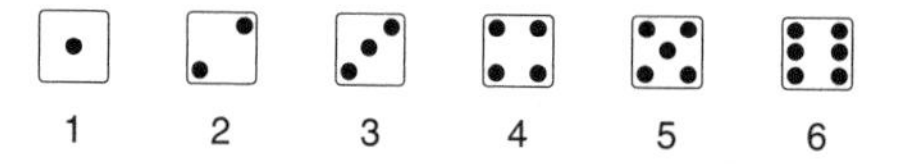

When the die is rolled, the faces are equally likely to turn up. The chance of getting an ace—⚀—is 1 in 6, or $16\frac{2}{3}\%$. The interpretation: if the die is rolled over and over again, repeating the basic chance process under the same conditions, in the long run an ace will show about $16\frac{2}{3}\%$ of the time.

> The chance of something gives the percentage of time it is expected to happen, when the basic process is done over and over again, independently and under the same conditions.

If something is impossible, it happens 0% of the time. At the other extreme, if something is sure to happen, then it happens 100% of the time. All chances are between these two extremes.

> Chances are between 0% and 100%.

Here is another basic fact. Suppose you are playing a game, and have a 45% chance to win. In other words, you expect to win about 45% of the time. So you must expect to lose the other 55% of the time.

> The chance of something equals 100% minus the chance of the opposite thing.

Abraham de Moivre (England, 1667–1754)
Etching by Faber. Copyright © British Museum.

Example 1. A box contains red marbles and blue marbles. One marble is drawn at random from the box (each marble has an equal chance to be drawn). If it is red, you win \$1. If it is blue, you win nothing. You can choose between two boxes:

- box A contains 3 red marbles and 2 blue ones;
- box B contains 30 red marbles and 20 blue ones.

Which box offers a better chance of winning, or are they the same?

Solution. Some people prefer box A, because it has fewer blue marbles. Others prefer B, because it has more red marbles. Both views are wrong. The two boxes offer the same chance of winning, 3 in 5. To see why, imagine drawing many times at random from box A (replacing the marble after each draw, so as not to change the conditions of the experiment). In the long run each of the

5 marbles will appear about 1 time in 5. So the red marbles will turn up about 3/5 of the time. With box A, your chance of drawing a red marble is 3/5, that is, 60%.

Now imagine drawing many times at random with replacement from box B. Each of the 50 marbles will turn up about 1 time in 50. But now there are 30 red marbles. With box B, your chance of winning is $30/50 = 3/5 = 60\%$, just as for box A. What counts is the ratio

$$\frac{\text{number of red marbles}}{\text{total number of marbles}}.$$

The ratio is the same in both boxes. De Moivre's solution for this example is given in figure 2.

Figure 2. De Moivre's solution.

The Probability of an Event is greater or less, according to the number of Chances by which it may happen, compared with the whole number of Chances by which it may either happen or fail.

Wherefore, if we constitute a Fraction whereof the Numerator be the number of Chances whereby an Event may happen, and the Denominator the number of all the Chances whereby it may either happen or fail, that Fraction will be a proper designation of the Probability of it happening. Thus if an Event has 3 Chances to happen, and 2 to fail, the Fraction 3/5 will fitly represent the Probability of its happening, and may be taken as the measure of it.

The same things may be said of the Probability of failing, which will likewise be measured by a Fraction, whose Numerator is the number of Chances whereby it may fail, and the Denominator the whole number of Chances, both for its happening and failing; thus the Probability of the failing of that Event which has 2 Chances to fail and 3 to happen will be measured by the Fraction 2/5.

The Fractions which represent the Probabilities of happening and failing, being added together, their Sum will always be equal to Unity, since the Sum of their Numerators will be equal to their common Denominator: now it being a certainty that an Event will either happen or fail, it follows that Certainty, which may be conceived under the notion of an infinitely great degree of Probability, is fitly represented by Unity. [By "Unity," de Moivre means the number 1.]

These things will easily be apprehended, if it be considered that the word Probability includes a double Idea: first, of the number of Chances whereby an Event may happen; secondly, of the number of Chances whereby it may either happen or fail.

Many problems, like example 1, take the form of drawing at random from a box. A typical instruction is,

Draw two tickets at random WITH replacement from the box

[1] [2] [3]

This asks you to imagine the following process: shake the box, draw out one ticket at random (equal chance for all three tickets), make a note of the number on it, put it back in the box, shake the box again, draw a second ticket at random (equal chance for all three tickets), make a note of the number on it, and put the second ticket back in the box. The contrast is with the instruction,

Draw two tickets at random WITHOUT replacement from the box

[1] [2] [3]

The second instruction asks you to imagine the following process: shake the box, draw out one ticket at random (equal chance for all three tickets), set it aside, draw out a second ticket at random (equal chance for the two tickets left in the box). See figure 3.

Figure 3. The difference between drawing with or without replacement. Two draws are made at random from the box [1] [2] [3]. Suppose the first draw is [3].

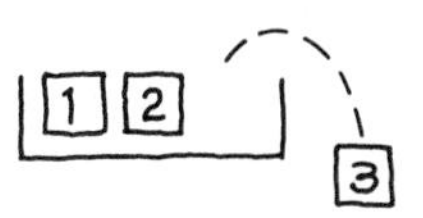

WITH replacement . . . the second draw is from

WITHOUT replacement . . . the second draw is from

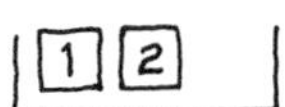

> When you draw at random, all the tickets in the box have the same chance to be picked.

Exercise Set A

1. A computer is programmed to compute various chances. Match the numerical answers with the verbal descriptions (which may be used more than once).

Numerical answer	*Verbal description*
(a) −50%	(i) This is as likely to happen as not.
(b) 0%	(ii) This is very likely to happen, but it's not certain.
(c) 10%	(iii) This won't happen.
(d) 50%	(iv) This may happen, but it's not likely.
(e) 90%	(v) This will happen, for sure.
(f) 100%	(vi) There's a bug in the program.
(g) 200%	

2. A coin will be tossed 1,000 times. About how many heads are expected?
3. A die will be rolled 6,000 times. About how many aces are expected?
4. In five-card draw poker, the chance of being dealt a full house (one pair and three of a kind) is 0.14 of 1%. If 10,000 hands are dealt, about how many will be a full house?
5. One hundred tickets will be drawn at random with replacement from one of the two boxes shown below. On each draw, you will be paid the amount shown on the ticket, in dollars. Which box is better and why?

(i) | [1] [2] | (ii) | [1] [3] |

The answers to these exercises are on p. A66.

2. CONDITIONAL PROBABILITIES

This section introduces conditional probabilities. The examples involve cards. A deck of cards has 4 suits: clubs, diamonds, hearts, spades. There are 13 cards in each suit: 2 through 10, jack, queen, king, ace. So there are $4 \times 13 = 52$ cards in the deck.

Example 2. A deck of cards is shuffled and the top two cards are put on a table, face down. You win $1 if the second card is the queen of hearts.

(a) What is your chance of winning the dollar?

(b) You turn over the first card. It is the seven of clubs. Now what is your chance of winning?

Solution. Part (a). The bet is about the second card, not the first. Initially, this will seem a little strange. Some illustrations may help.

- If the first card is the two of spades and the second is the queen of hearts, you win.
- If the first card is the jack of clubs and the second is the queen of hearts, you win.
- If the first card is the seven of clubs and the second is the king of hearts, you lose.

The bet can be settled without even looking at the first card. The second card is all you need to know.

The chance of winning is 1/52. To see why, think about shuffling the deck. That brings the cards into random order. The queen of hearts has to wind up somewhere. There are 52 possible positions, and they are all equally likely. So there is 1 chance in 52 for her to wind up as the second card in the deck—and bring you the dollar.

Part (b). There are 51 cards left. They are in random order, and the queen of hearts is one of them. She has 1 chance in 51 to be on the table. Your chance goes up a little, to 1/51. That is the answer.

The 1/51 in part (b) is a *conditional* chance. The problem puts a condition on the first card: it has to be the seven of clubs. A mathematician might talk about the conditional probability that the second card is the queen of hearts *given* the first card is the seven of clubs. To emphasize the contrast, the 1/52 in part (a) is called an *unconditional* chance: the problem puts no conditions on the first card.

Exercise Set B

1. Two tickets are drawn at random without replacement from the box | 1 2 3 4 |.
   (a) What is the chance that the second ticket is 4?
   (b) What is the chance that the second ticket is 4, given the first is 2?
2. Repeat exercise 1, if the draws are made with replacement.
3. A penny is tossed 5 times.
   (a) Find the chance that the 5th toss is a head.
   (b) Find the chance that the 5th toss is a head, given the first 4 are tails.
4. Five cards are dealt off the top of a well-shuffled deck.
   (a) Find the chance that the 5th card is the queen of spades.
   (b) Find the chance that the 5th card is the queen of spades, given that the first 4 cards are hearts.

The answers to these exercises are on pp. A66–67.

Technical notes. (i) Mathematicians write the probability for the second card to be the queen of hearts as follows:

P(2nd card is queen of hearts).

The "P" is short for "probability."

(ii) The conditional probability for the second card to be the queen of hearts, given the first was the seven of clubs, is written as follows:

P(2nd card is queen of hearts | 1st card is seven of clubs).

The vertical bar is read "given."

3. THE MULTIPLICATION RULE

This section will show how to figure the chance that two events happen, by multiplying probabilities.

Example 3. A box has three tickets, colored red, white and blue.

| R | W | B |

Two tickets will be drawn at random without replacement. What is the chance of drawing the red ticket and then the white?

Solution. Imagine a large group of people. Each of these people holds a box | R | W | B | and draws two tickets at random without replacement. About one third of the people get [R] on the first draw, and are left with

| W | B |

On the second draw, about half of these people will get [W]. The fraction who draw [R] [W] is therefore

$$\frac{1}{2} \text{ of } \frac{1}{3} = \frac{1}{2} \times \frac{1}{3} = \frac{1}{6}.$$

The chance is 1 in 6, or $16\frac{2}{3}\%$.

For instance, suppose you start with 600 people. About 200 of them will get [R] on the first draw. Of these 200 people, about 100 will get [W] on the second draw. So $100/600 = 1/6$ of the people draw the red ticket first and then the white one. In figure 4, the people who draw [R] [W] are at the top left.

Statisticians usually multiply the chances in reverse order:

$$\frac{1}{3} \times \frac{1}{2} = \frac{1}{6}.$$

The reason: $1/3$ refers to the first draw, and $1/2$ to the second.

Figure 4. The multiplication rule. Each stick figure corresponds to 100 people.

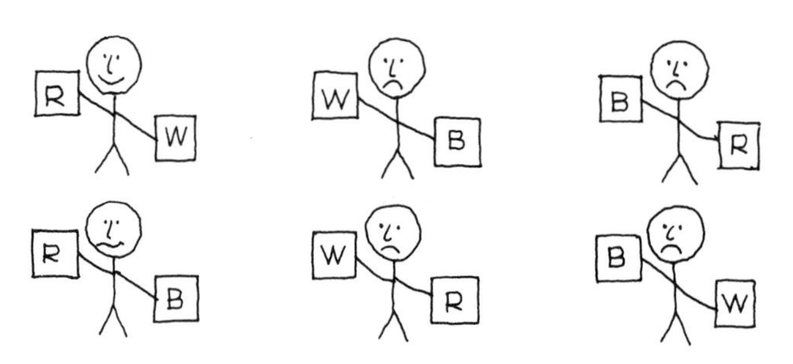

The method in example 3 is called the multiplication rule.

> *Multiplication Rule.* The chance that two things will both happen equals the chance that the first will happen, multiplied by the chance that the second will happen given the first has happened.

Example 4. Two cards will be dealt off the top of a well-shuffled deck. What is the chance that the first card will be the seven of clubs and the second card will be the queen of hearts?

Solution. This is like example 3, with a much bigger box. The chance that the first card will be the seven of clubs is 1/52. Given that the first card was the seven of clubs, the chance that the second card will be the queen of hearts is 1/51. The chance of getting both cards is

$$\frac{1}{52} \times \frac{1}{51} = \frac{1}{2{,}652}.$$

This is a small chance: about 4 in 10,000, or 0.04 of 1%.

Example 5. A deck of cards is shuffled, and two cards are dealt. What is the chance that both are aces?

Solution. The chance that the first card is an ace equals 4/52. Given that the first card is an ace, there are 3 aces among the 51 remaining cards. So the chance of a second ace equals 3/51. The chance that both cards are aces equals

$$\frac{4}{52} \times \frac{3}{51} = \frac{12}{2{,}652}.$$

This is about 1 in 200, or 1/2 of 1%.

Example 6. A coin is tossed twice. What is the chance of a head followed by a tail?

Solution. The chance of a head on the first toss equals 1/2. No matter how the first toss turns out, the chance of tails on the second toss equals 1/2. So the chance of heads followed by tails equals

$$\frac{1}{2} \times \frac{1}{2} = \frac{1}{4}.$$

Exercise Set C

1. A deck is shuffled and two cards are dealt.
   (a) Find the chance that the second card is a heart given the first card is a heart.
   (b) Find the chance that the first card is a heart and the second card is a heart.

2. A die is rolled three times.

   (a) Find the chance that the first roll is an ace ⚀.
   (b) Find the chance that the first roll is an ace ⚀, the second roll is a deuce ⚁, and the third roll is a trey ⚂.

3. A deck is shuffled and three cards are dealt.

   (a) Find the chance that the first card is a king.
   (b) Find the chance that the first card is a king, the second is a queen, and the third is a jack.

4. A die will be rolled six times. You have a choice—

   (i) to win \$1 if at least one ace shows
   (ii) to win \$1 if an ace shows on all the rolls

   Which option offers the better chance of winning? Or are they the same? Explain.

5. Someone works example 2(a) on p. 226 this way:

   > For me to win, the queen can't be the first card dealt (51 chances in 52) and she must be the second card (1 chance in 51), so the answer is
   >
   > $$\frac{51}{52} \times \frac{1}{51} = \frac{1}{52}.$$

   Is the multiplication legitimate? Why?

6. "A cat-o'nine-______ can be used to punish ______ of state, but this is seldom done." A coin is tossed twice, to fill in the blanks. What is the chance of the coin getting it right?

7. A coin is tossed 3 times.

   (a) What is the chance of getting 3 heads?
   (b) What is the chance of not getting 3 heads?
   (c) What is the chance of getting at least 1 tail?
   (d) What is the chance of getting at least 1 head?

The answers to these exercises are on p. A67.

4. INDEPENDENCE

This section introduces the idea of independence, which will be used many times in the rest of the book.

> Two things are *independent* if the chances for the second given the first are the same, no matter how the first one turns out. Otherwise, the two things are *dependent*.

Example 7. Someone is going to toss a coin twice. If the coin lands heads on the second toss, you win a dollar.

(a) If the first toss is heads, what is your chance of winning the dollar?
(b) If the first toss is tails, what is your chance of winning the dollar?

(c) Are the tosses independent?

Solution. If the first toss is heads, there is a 50% chance to get heads the second time. If the first toss is tails, the chance is still 50%. The chances for the second toss stay the same, however the first toss turns out. That is independence.

Example 8. Two draws will be made at random with replacement from

| [1] [1] [2] [2] [3] |

(a) Suppose the first draw is [1]. What is the chance of getting a [2] on the second draw?
(b) Suppose the first draw is [2]. What is the chance of getting [2] on the second draw?
(c) Are the draws independent?

Solution. Whether the first draw is [1] or [2] or anything else, the chance of getting [2] on the second draw stays the same—two in five, or 40%. The reason: the first ticket is replaced, so the second draw is always made from the same box | [1] [1] [2] [2] [3] |. The draws are independent.

Example 9. As in example 8, but the draws are made without replacement.

Solution. If the first draw turns out to be [1] then the second draw is from the box | [1] [2] [2] [3] |. The chance for the second draw to be [2] is 50%. On the other hand, if the first draw turns out to be [2], then the second draw is from the box | [1] [1] [2] [3] |. Now there is only a 25% chance for the second to be [2]. The draws are dependent.

> When drawing at random with replacement, the draws are independent. Without replacement, the draws are dependent.

What does independence of the draws mean? To answer this question, think about bets which can be settled on one draw: for instance, that the draw will be 3 or more. Then the conditional chance of winning the bet must stay the same, no matter how the other draws turn out.

Example 10. A box has three tickets, colored red, white, and blue.

| [R] [W] [B] |

Two tickets will be drawn at random with replacement. What is the chance of drawing the red ticket and then the white?

Solution. The draws are independent, so the chance is

$$\frac{1}{3} \times \frac{1}{3} = \frac{1}{9}.$$

Compare this with example 3. The answers are different. Independence matters. And it's easier this time: you don't need to work out conditional probabilities.

> If two things are independent, the chance that both will happen equals the product of their unconditional probabilities. This is a special case of the multiplication rule (p. 229).

Exercise Set D

1. For each of the following boxes, say whether color and number are dependent or independent.

   (a) | [1] [2] [2] **[1]** **[2]** **[2]** | (c) | [1] [2] [3] **[1]** **[2]** **[2]** |

   (b) | [1] [2] *[1]* *[2]* **[1]** **[2]** |

2. (a) In the box shown below, each ticket has two numbers.

   | [1|2] [1|3] [4|2] [4|3] |

   (For instance, the first number on [4|2] is 4 and the second is 2.) A ticket is drawn at random. Are the two numbers dependent or independent?

   (b) Repeat, for the box

   | [1|2] [1|3] [1|3] [4|2] [4|3] [4|3] |

   (c) Repeat, for the box

   | [1|2] [1|3] [1|3] [4|2] [4|2] [4|3] |

3. Every week you buy a ticket in a lottery that offers one chance in a million of winning. What is the chance that you never win, even if you keep this up for ten years?

4. Two draws are made at random without replacement from the box | [1] [2] [3] [4] |. The first ticket is lost, and nobody knows what was written on it. True or false, and explain: the two draws are independent.

5. Suppose that in a certain class, there are
   - 80% men and 20% women;
   - 15% freshmen and 85% sophomores.

   (a) The percentage of sophomore women in the class can be as small as ______.
   (b) This percentage can be as large as ______.

6. One student is chosen at random from the class described in the previous exercise.

   (a) The chance of getting a sophomore woman can be as small as ______.
   (b) This chance can be as large as ______.

7. In 2002, about 50.9% of the population of the United States was female. Also, 1.6% of the population was age 85 and over.[3] True or false, and explain: the percentage of the population consisting of women age 85 and over is

$$50.9\% \text{ of } 1.6\% = 0.509 \times 1.6\% \approx 0.8 \text{ of } 1\%$$

8. (Hard.) In a certain psychology experiment, each subject is presented with three ordinary playing cards, face down. The subject takes one of these cards. The subject also takes one card at random from a separate, full deck of playing cards. If the two cards are from the same suit, the subject wins a prize. What is the chance of winning? If more information is needed, explain what you need, and why.

The answers to these exercises are on pp. A67–68.

5. THE COLLINS CASE

People v. Collins is a law case in which there was a major statistical issue. A black man and a white woman were charged with robbery. The facts were described by the court as follows.[4]

> On June 18, 1964, about 11:30 A.M. Mrs. Juanita Brooks, who had been shopping, was walking home along an alley in the San Pedro area of the City of Los Angeles. She was pulling behind her a wicker basket carryall containing groceries and had her purse on top of the packages. She was using a cane. As she stooped down to pick up an empty carton, she was suddenly pushed to the ground by a person whom she neither saw nor heard approach. She was stunned by the fall and felt some pain. She managed to look up and saw a young woman running from the scene. According to Mrs. Brooks the latter appeared to weigh about 145 pounds, was wearing "something dark," and had hair "between a dark blond and a light blond," but lighter than the color of defendant Janet Collins' hair as it appeared at trial. Immediately after the incident, Mrs. Brooks discovered that her purse, containing between $35 and $40, was missing.
>
> About the same time as the robbery, John Bass, who lived on the street at the end of the alley, was in front of his house watering his lawn. His attention was attracted by "a lot of crying and screaming" coming from the alley. As he looked in that direction, he saw a woman run out of the alley and enter a yellow automobile parked across the street from him. He was unable to give the make of the car. The car started off immediately and pulled wide around another parked vehicle so that in the narrow street it passed within six feet of Bass. The latter then saw that it was being driven by a male Negro, wearing a mustache and beard. At the trial Bass identified defendant as the driver of the yellow automobile. However, an attempt was made to impeach his identification by his admission that at the preliminary hearing he testified to an uncertain identification at the police lineup shortly after the attack on Mrs. Brooks, when defendant was beardless.
>
> In his testimony Bass described the woman who ran from the alley as a Caucasian, slightly over five feet tall, of ordinary build, with her hair in a dark blond ponytail, and wearing dark clothing. He further testified that her ponytail was "just like" one which Janet had in a police photograph taken on June 22, 1964.

The prosecutor then had a mathematics instructor at a local state college explain the multiplication rule, without paying much attention to independence, or the distinction between conditional and unconditional probabilities. After this testimony, the prosecution assumed the following chances:

Yellow automobile	1/10	Woman with blond hair	1/3
Man with mustache	1/4	Black man with beard	1/10
Woman with ponytail	1/10	Interracial couple in car	1/1,000

When multiplied together, these come to 1 in 12,000,000. According to the prosecution, this procedure gave the chance "that any [other] couple possessed the distinctive characteristics of the defendants." If no other couple possessed these characteristics, the defendants were guilty. The jury convicted. On appeal, the Supreme Court of California reversed the verdict. It found no evidence to support the assumed values for the six chances. Furthermore, these were presented as unconditional probabilities. The basis for multiplying them, as the mathematics instructor should have explained, was independence. And there was no evidence to support that assumption either. On the contrary, some factors were clearly dependent—like "Black man with beard" and "interracial couple in car."

> Blindly multiplying chances can make real trouble. Check for independence, or use conditional probabilities.

There is another objection to the prosecutor's reasoning. Probability calculations like the multiplication rule were developed for dealing with games of chance, where the basic process can be repeated independently and under the same conditions. The prosecutor was trying to apply this theory to a unique event: something that either happened—or didn't happen—on June 18, 1964, at 11:30 A.M. What does chance mean, in this new context? It was up to the prosecutor to answer this question, and to show that the theory applied to his situation.[5]

In the 1990s, DNA evidence began to be used for identification of criminals: the idea is to match a suspect's DNA with DNA left at the scene of the crime—for instance, in bloodstains. Matching is done on a set of characteristics of DNA. The technical issues are similar to those raised by the Collins case: Can you estimate the fraction of the population with a given characteristic? Are those characteristics independent? Is the lab work accurate? Many experts believe that such questions have satisfactory answers, others are quite skeptical.[6]

6. REVIEW EXERCISES

When a die is rolled, each of the six faces is equally likely to come up. A deck of cards has 4 suits (clubs, diamonds, hearts, spades) with 13 cards in each suit—2, 3, ..., 10, jack, queen, king, ace. *See pp. 222 and 226.*

1. True or false, and explain:
   (a) If something has probability 1,000%, it is sure to happen.
   (b) If something has probability 90%, it can be expected to happen about nine times as often as its opposite.

2. Two cards will be dealt off the top of a well-shuffled deck. You have a choice:
   (i) To win $1 if the first is a king.

(ii) To win \$1 if the first is a king and the second is a queen.

Which option is better? Or are they equivalent? Explain briefly.

3. Four cards will be dealt off the top of a well-shuffled deck. There are two options:

    (i) To win \$1 if the first card is a club and the second is a diamond and the third is a heart and the fourth is a spade.
    (ii) To win \$1 if the four cards are of four different suits.

    Which option is better? Or are they the same? Explain.

4. A poker hand is dealt. Find the chance that the first four cards are aces and the fifth is a king.

5. One ticket will be drawn at random from the box below. Are color and number independent? Explain.

6. A deck of cards is shuffled and the top two cards are placed face down on a table. True or false, and explain:

    (a) There is 1 chance in 52 for the first card to be the ace of clubs.
    (b) There is 1 chance in 52 for the second card to be the ace of diamonds.
    (c) The chance of getting the ace of clubs and then the ace of diamonds is $1/52 \times 1/52$.

7. A coin is tossed six times. Two possible sequences of results are

    (i) H T T H T H   (ii) H H H H H H

    (The coin must land H or T in the order given; H = heads, T = tails.) Which of the following is correct? Explain.[7]

    (a) Sequence (i) is more likely.
    (b) Sequence (ii) is more likely.
    (c) Both sequences are equally likely.

8. A die is rolled four times. What is the chance that—

    (a) all the rolls show 3 or more spots?
    (b) none of the rolls show 3 or more spots?
    (c) not all the rolls show 3 or more spots?

9. A die is rolled 10 times. Find the chance of—

    (a) getting 10 sixes.
    (b) not getting 10 sixes.
    (c) all the rolls showing 5 spots or less.

10. Which of the two options is better, or are they the same? Explain briefly.

    (i) You toss a coin 100 times. On each toss, if the coin lands heads, you win \$1. If it lands tails, you lose \$1.
    (ii) You draw 100 times at random with replacement from | [1] [0] |. On each draw, you are paid (in dollars) the number on the ticket.

11. In the box shown below, each ticket should have two numbers:

| 1 | 1 2 | 1 2 | 1 3 | 3 1 | 3 2 | 3 | 3 |

A ticket will be drawn at random. Can you fill in the blanks so the two numbers are independent?

12. You are thinking about playing a lottery. The rules: you buy a ticket, choose 3 different numbers from 1 to 100, and write them on the ticket. The lottery has a box with 100 balls numbered from 1 through 100. Three balls are drawn at random without replacement. If the numbers on these balls are the same as the numbers on your ticket, you win. (Order doesn't matter.) If you decide to play, what is your chance of winning?

7. SUMMARY

1. The *frequency theory* of chance applies most directly to chance processes which can be repeated over and over again, independently and under the same conditions.

2. The chance of something gives the percentage of times the thing is expected to happen, when the basic process is repeated over and over again.

3. Chances are between 0% and 100%. Impossibility is represented by 0%, certainty by 100%.

4. The chance of something equals 100% minus the chance of the opposite thing.

5. The chance that two things will both happen equals the chance that the first will happen, multiplied by the *conditional* chance that the second will happen given that the first has happened. This is the *multiplication rule.*

6. Two things are *independent* if the chances for the second one stay the same no matter how the first one turns out.

7. If two things are independent, the chance that both will happen equals the product of their unconditional chances. This is a special case of the multiplication rule.

8. When you draw at random, all the tickets in the box have the same chance to be picked. Draws made at random with replacement are independent. Without replacement, the draws are dependent.

9. Blindly multiplying chances can make real trouble. Check for independence, or use conditional chances.

10. The mathematical theory of chance only applies in some situations. Using it elsewhere can lead to ridiculous results.

14

More about Chance

Some of the Problems about Chance having a great appearance of Simplicity, the Mind is easily drawn into a belief, that their Solution may be attained by the meer Strength of natural good Sense; which generally proving otherwise and the Mistakes occasioned thereby being not unfrequent, 'tis presumed that a Book of this Kind, which teaches to distinguish Truth from what seems so nearly to resemble it, will be looked upon as a help to good Reasoning.

—ABRAHAM DE MOIVRE (ENGLAND, 1667–1754)[1]

1. LISTING THE WAYS

A *probabilist* is a mathematician who specializes in computing the probabilities of complex events. In the twentieth century, two of the leading probabilists were A. N. Kolmogorov (Russia, 1903–1987) and P. Lévy (France, 1886–1971). The techniques they developed are beyond the scope of this book, but we can look at more basic methods, developed by earlier mathematicians.

When trying to figure chances, it is sometimes very helpful to list all the possible ways that a chance process can turn out. If this is too hard, writing down a few typical ones is a good start.

Example 1. Two dice are thrown. What is the chance of getting a total of 2 spots?

Solution. The chance process here consists of throwing the two dice. What matters is the number of spots shown by each die. To keep the dice separate, imagine that one is white and the other black. One way for the dice to fall is

This means the white die showed 2 spots, and the black die showed 3. The total number of spots is 5.

How many ways are there for the two dice to fall? To begin with, the white die can fall in any one of 6 ways:

When the white die shows , say, there are still 6 possible ways for the black die to fall:

We now have 6 of the possible ways that the two dice can fall. These ways are shown in the first row of figure 1. Similarly, the second row shows another 6 ways for the dice to fall, with the white die showing . And so on. The figure shows there are $6 \times 6 = 36$ possible ways for the dice to fall. They are all equally likely, so each has 1 chance in 36. There is only one way to get a total of 2 spots: . The chance is $1/36$. That is the answer.

There may be several methods for answering questions about chance. In figure 1, for example, the chance for each of the 36 outcomes can also be worked out using the multiplication rule: $1/6 \times 1/6 = 1/36$.

Example 2. A pair of dice are thrown. What is the chance of getting a total of 4 spots?

Solution. Look at figure 1. There are 3 ways to get a total of four spots:

The chance is 3 in 36. That is the answer.

What about three dice? A three-dimensional picture like figure 1 would be a bit much to absorb, but similar reasoning can be used. In the seventeenth century, Italian gamblers used to bet on the total number of spots rolled with three dice. They believed that the chance of rolling a total of 9 ought to equal the chance of

Figure 1. Throwing a pair of dice. There are 36 ways for the dice to fall, shown in the body of the diagram; all are equally likely.

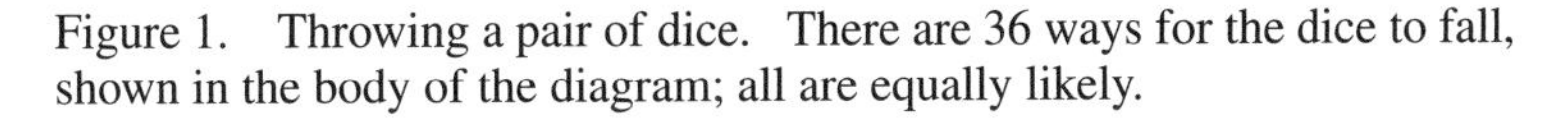

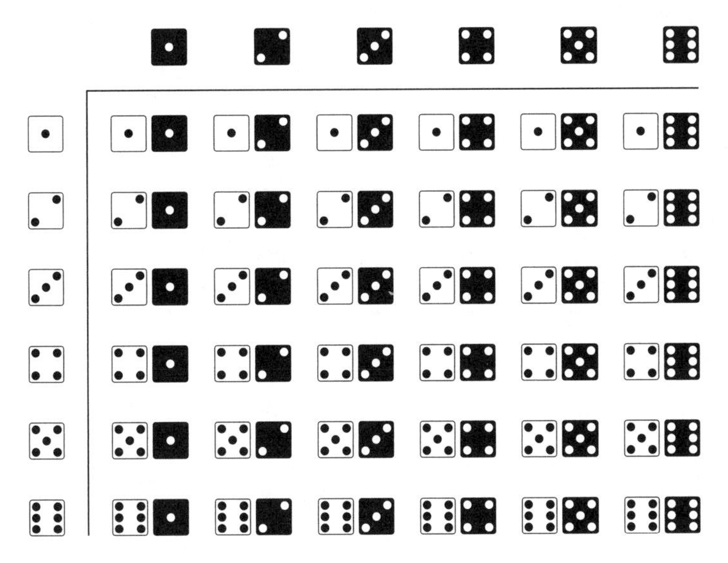

rolling a total of 10. For instance, they said, one combination with a total of 9 spots is

1 spot on one die, 2 spots on another die, 6 spots on the third die.

This can be abbreviated as "1 2 6." There are altogether six combinations for 9:

1 2 6   1 3 5   1 4 4   2 3 4   2 2 5   3 3 3

Similarly, they found six combinations for 10:

1 4 5   1 3 6   2 2 6   2 3 5   2 4 4   3 3 4

Thus, argued the gamblers, 9 and 10 should by rights have the same chance. However, experience showed that 10 came up a bit more often than 9.

They asked Galileo for help, and he reasoned as follows. Color one of the dice white, another one grey, and another one black—so they can be kept apart. This won't affect the chances. How many ways can the three dice fall? The white die can land in 6 ways. Corresponding to each of them, the grey die can land in 6 ways, making 6×6 possibilities. Corresponding to each of these possibilities, there are still 6 for the black die. Altogether, there are $6 \times 6 \times 6 = 6^3$ ways for three dice to land. (With 4 dice, there would be 6^4; with 5 dice, 6^5 and so on.)

Now $6^3 = 216$ is a lot of ways for three dice to fall. But Galileo sat down and listed them. Then he went through his list and counted the ones with a total of 9 spots. He found 25. And he found 27 ways to get a total of 10 spots. He concluded that the chance of rolling 9 is $25/216 \approx 11.6\%$, while the chance of rolling 10 is $27/216 = 12.5\%$.

The gamblers made a basic error: they didn't get down to the different ways for the dice to land. For instance, the triplet 3 3 3 for 9 only corresponds to one way for the dice to land:

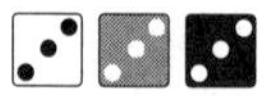

But the triplet 3 3 4 for 10 corresponds to three ways for the dice to land:

The gamblers' argument is corrected in table 1.

Table 1. The chance of getting 9 or 10 spots with three dice.

Triplets for 9	*Number of ways to roll each triplet*	*Triplets for 10*	*Number of ways to roll each triplet*
1 2 6	6	1 4 5	6
1 3 5	6	1 3 6	6
1 4 4	3	2 2 6	3
2 3 4	6	2 3 5	6
2 2 5	3	2 4 4	3
3 3 3	1	3 3 4	3
Total	25	Total	27

Galileo (Italy, 1564–1642)

Wolff-Leavenworth Collection, courtesy of the Syracuse University Art Collection.

Exercise Set A

1. Look at figure 1 and make a list of the ways to roll a total of 5 spots. What is the chance of throwing a total of 5 spots with two dice?
2. Two draws are made at random with replacement from the box | 1 2 3 4 5 |.

Draw a picture like figure 1 to represent all possible results. How many are there? What is the chance that the sum of the two draws turns out to equal 6?

3. A pair of dice is thrown 1,000 times. What total should appear most often? What totals should appear least often?

4. (a) In the box shown below, each ticket has two numbers.

| 1 2 | 1 3 | 3 1 | 3 2 |

(For instance, the first number on | 3 1 | is 3 and the second is 1.) A ticket is drawn at random. Find the chance that the sum of the two numbers is 4.

(b) Repeat, for the box

| 1 2 | 1 3 | 1 3 | 3 2 | 3 3 | 3 3 |

(c) Repeat, for the box

| 1 2 | 1 3 | 1 3 | 3 1 | 3 2 | 3 3 |

The answers to these exercises are on p. A68.

2. THE ADDITION RULE

This section is about the chance that at least one of two specified things will happen: either the first happens, or the second, or both. The possibility of both happening turns out to be a complication, which can sometimes be ruled out.

> Two things are *mutually exclusive* when the occurrence of one prevents the occurrence of the other: one excludes the other.

Example 3. A card is dealt off the top of a well-shuffled deck. The card might be a heart. Or, it might be a spade. Are these two possibilities mutually exclusive?

Solution. If the card is a heart, it can't be a spade. These two possibilities are mutually exclusive.

We can now state a general principle for figuring chances. It is called the addition rule.

> *Addition Rule.* To find the chance that at least one of two things will happen, check to see if they are mutually exclusive. If they are, add the chances.

Example 4. A card is dealt off the top of a well-shuffled deck. There is 1 chance in 4 for it to be a heart. There is 1 chance in 4 for it to be a spade. What is the chance for it to be in a major suit (hearts or spades)?

Solution. The question asks for the chance that one of the following two things will happen:

- the card is a heart;
- the card is a spade.

As in example 3, if the card is a heart then it can't be a spade: these are mutually exclusive events. So it is legitimate to add the chances. The chance of getting a card in a major suit is $1/4 + 1/4 = 1/2$. (A check on the reasoning: there are 13 hearts and 13 spades, so $26/52 = 1/2$ of the cards in the deck are in a major suit.)

Example 5. Someone throws a pair of dice. True or false: the chance of getting at least one ace is $1/6 + 1/6 = 1/3$.

Solution. This is false. Imagine one of the dice is white, the other black.

The question asks for the chance that one of the following two things will happen:

- the white die lands ace ⚀;
- the black die lands ace ⚀.

A white ace does not prevent a black ace. These two events are not mutually exclusive, so the addition rule does not apply. Adding the chances gives the wrong answer.

Look at figure 1. There are 6 ways for the white die to show ⚀. There are 6 ways for the black die to show ⚀. But the number of ways to get at least one ace is not $6 + 6$. Addition double counts the outcome ⚀ ⚀ at the top left corner. The chance of getting at least one ace is

$$(6 + 6 - 1)/36 = 11/36, \quad \text{not} \quad (6 + 6)/36 = 12/36 = 1/3.$$

> If you want to find the chance that at least one event occurs, and the events are not mutually exclusive, do not add the chances: the sum will be too big.

Blindly adding chances can give the wrong answer, by double counting the chance that two things happen. With mutually exclusive events, there is no double counting: that is why the addition rule works.

Exercise Set B

1. Fifty children went to a party where cookies and ice cream were served: 12 children took cookies; 17 took ice cream. True or false: 29 children must have had cookies or ice cream. Explain briefly.

2. There are 20 dots in the diagram below, and 3 circles. The circles are labeled A, B, and C. One of the dots will be chosen at random.
   (a) What is the probability that the dot falls inside circle A?
   (b) What is the probability that the dot falls inside circle B?
   (c) What is the probability that the dot falls inside circle C?
   (d) What is the probability that the dot falls inside at least one of the circles?

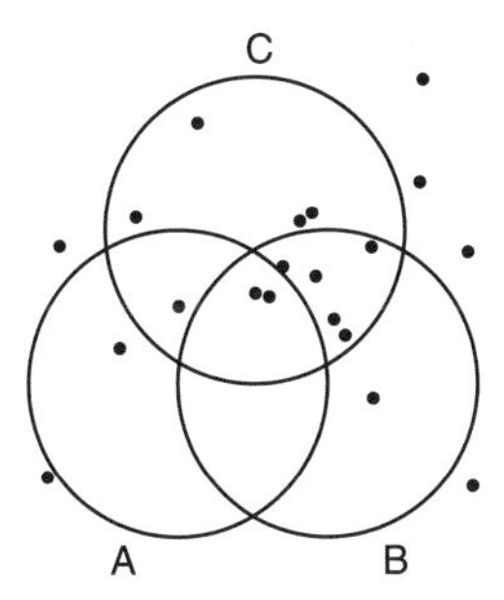

3. Two cards are dealt off the top of a well-shuffled deck. You have a choice:
   (i) to win $1 if the first card is an ace or the second card is an ace;
   (ii) to win $1 if at least one of the two cards is an ace.

   Which option is better? or are they the same? Explain briefly.

4. Two dice will be rolled. The chance that the first one lands ⚀ is $1/6$. The chance that the second one lands ⚁ is $1/6$. True or false: the chance that the first one lands ⚀ or the second one lands ⚁ equals $1/6 + 1/6$. Explain briefly.

5. A box contains 10 tickets numbered 1 through 10. Five draws will be made at random with replacement from this box. True or false: there are 5 chances in 10 of getting $\boxed{7}$ at least once. Explain briefly.

6. A number is drawn at random from a box. There is a 20% chance for it to be 10 or less. There is a 10% chance for it to be 50 or more. True or false: the chance of getting a number between 10 and 50 (exclusive) is 70%. Explain briefly.

The answers to these exercises are on pp. A68–69.

3. TWO FAQs (FREQUENTLY ASKED QUESTIONS)

- What's the difference between mutually exclusive and independent?
- When do I add and when do I multiply?

"Mutually exclusive" is one idea; independence is another. Both ideas apply to pairs of events, and say something about how the events are related. However, the relationships are quite different.

- Two events are mutually exclusive if the occurrence of one prevents the other from happening.
- Two events are independent if the occurrence of one does not change the chances for the other.

The addition rule, like the multiplication rule, is a way of combining chances. However, the two rules solve different problems (pp. 229 and 241).

- The addition rule finds the chance that at least one of two things happens.
- The multiplication rule finds the chance that two things both happen.

So, the first step in deciding whether to add or to multiply is to read the question: Do you want to know P(A or B), P(A and B), or something else entirely? But there is also a second step—because the rules apply only if the events are related in the right way.

- Adding the probabilities of two events requires them to be mutually exclusive.[2]
- Multiplying the unconditional probabilities of two events requires them to be independent. (For dependent events, the multiplication rule uses conditional probabilities.)

Example 6. A die is rolled 6 times; a deck of cards is shuffled.

(a) The chance that the first roll is an ace or the last roll is an ace equals ______.

(b) The chance that the first roll is an ace and the last roll is an ace equals ______.

(c) The chance that the top card is the ace of spades or the bottom card is the ace of spades equals ______.

(d) The chance that the top card is the ace of spades and the bottom card is the ace of spades equals ______.

Options for parts (a) and (b):

(i) $\frac{1}{6} + \frac{1}{6}$ (ii) $\frac{1}{6} \times \frac{1}{6}$ (iii) neither of these

Options for parts (c) and (d):

(i) $\frac{1}{52} + \frac{1}{52}$ (ii) $\frac{1}{52} \times \frac{1}{52}$ (iii) neither of these

Solution. Part (a). You want the chance that at least one of the two things will happen, so the addition rule looks relevant. However, the two things are not mutually exclusive. Do not use the addition rule, it will give the wrong answer (example 5). If you can't add, maybe you can multiply? The two events are independent, but you do not want the chance that both happen. Do not use the multiplication rule either, it too will give the wrong answer. Choose option (iii).

Part (b). You want the chance that both events happen, and they are independent. Now is the time to multiply. Choose option (ii).

Part (c). The chance the top card is the ace of spades equals $1/52$. The chance that the bottom card is the ace of spades—computed before looking at any of the cards (example 2 on p. 226) also equals $1/52$. The two events are mutually exclusive; you want the chance that at least one of the two will occur. This is when the addition rule shines. Choose (i).

Part (d). The two events are mutually exclusive, but you do not want the chance that at least one of the two will occur. Therefore, do not use the addition rule, it will give the wrong answer. You want the chance that both things happen, so multiplication may be relevant. However, the events are dependent. Do not multiply the unconditional probabilities, you will get the wrong answer. Choose (iii). (The chance is 0: the ace of spades cannot turn up in both places.)

As example 6 indicates, you may not be able either to add or to multiply. Then more thinking is needed. (The cartoon is trying to tell you something.) The next section gives an example—The Paradox of the Chevalier de Méré.

Technical notes. The chance of two aces is $1/36$, so the chance in example 6(a) can be figured as

$$\frac{1}{6} + \frac{1}{6} - \frac{1}{36} = \frac{11}{36}$$

However, if the die is rolled 3 times, the chance of getting at least one ace is not

$$\frac{1}{6} + \frac{1}{6} + \frac{1}{6} - \left(\frac{1}{6}\right)^3$$

Think about 12 rolls! This sort of problem will be solved in the next section.

In example 6(d), the multiplication rule can be used, with conditional probabilities—although this is very fussy. The chance that the top card is the ace of spades equals $1/52$. Given that the top card is the ace of spades, the conditional chance that the bottom card is the ace of spades equals 0. The chance that both things happen equals $1/52 \times 0 = 0$.

Exercise Set C

1. A large group of people are competing for all-expense-paid weekends in Philadelphia. The Master of Ceremonies gives each contestant a well-shuffled deck of cards. The contestant deals two cards off the top of the deck, and wins a weekend in Philadelphia if the first card is the ace of hearts or the second card is the king of hearts.
    (a) All the contestants whose first card was the ace of hearts are asked to step forward. What fraction of the contestants do so?
    (b) The contestants return to their original places. Then, the ones who got the king of hearts for their second card are asked to step forward. What fraction of the contestants do so?
    (c) Do any of the contestants step forward twice?
    (d) True or false, and explain: the chance of winning a weekend in Philadelphia is $1/52 + 1/52$.

"PHILADELPHIA, PLEASE."

2. A large group of people are competing for all-expense-paid weekends in Philadelphia. The Master of Ceremonies gives each contestant a well-shuffled deck of cards. The contestant deals two cards off the top of the deck, and wins a weekend in Philadelphia if the first card is the ace of hearts or the second card is the ace of hearts. (This is like exercise 1, but the winning cards are a little different.)
   (a) All the contestants whose first card was the ace of hearts are asked to step forward. What fraction of the contestants do so?
   (b) The contestants return to their original places. Then, the ones who got the ace of hearts for their second card are asked to step forward. What fraction of the contestants do so?
   (c) Do any of the contestants step forward twice?
   (d) True or false, and explain: the chance of winning a weekend in Philadelphia is $1/52 + 1/52$.

3. A deck of cards is shuffled. True or false, and explain briefly:
   (a) The chance that the top card is the jack of clubs equals $1/52$.
   (b) The chance that the bottom card is the jack of diamonds equals $1/52$.
   (c) The chance that the top card is the jack of clubs or the bottom card is the jack of diamonds equals $2/52$.
   (d) The chance that the top card is the jack of clubs or the bottom card is the jack of clubs equals $2/52$.
   (e) The chance that the top card is the jack of clubs and the bottom card is the jack of diamonds equals $1/52 \times 1/52$.
   (f) The chance that the top card is the jack of clubs and the bottom card is the jack of clubs equals $1/52 \times 1/52$.

4. The unconditional probability of event A is $1/2$. The unconditional probability of event B is $1/3$. Say whether each of the following is true or false, and explain briefly.
   (a) The chance that A and B both happen must be $1/2 \times 1/3 = 1/6$.
   (b) If A and B are independent, the chance that they both happen must be $1/2 \times 1/3 = 1/6$.
   (c) If A and B are mutually exclusive, the chance that they both happen must be $1/2 \times 1/3 = 1/6$.
   (d) The chance that at least one of A or B happens must be $1/2 + 1/3 = 5/6$.
   (e) If A and B are independent, the chance that at least one of them happens must be $1/2 + 1/3 = 5/6$.
   (f) If A and B are mutually exclusive, the chance that at least one of them happens must be $1/2 + 1/3 = 5/6$.

5. Two cards are dealt off the top of a well-shuffled deck.
   (a) Find the chance that the second card is an ace.
   (b) Find the chance that the second card is an ace, given the first card is a king.
   (c) Find the chance that the first card is a king and the second card is an ace.

The answers to these exercises are on pp. A69–70.

4. THE PARADOX OF THE CHEVALIER DE MÉRÉ

In the seventeenth century, French gamblers used to bet on the event that with 4 rolls of a die, at least one ace would turn up: an ace is ⚀. In another game, they bet on the event that with 24 rolls of a pair of dice, at least one double-ace would turn up: a double-ace is a pair of dice which show ⚀⚀.

The Chevalier de Méré, a French nobleman of the period, thought the two events were equally likely. He reasoned this way about the first game:

- In one roll of a die, I have $1/6$ of a chance to get an ace.
- So in 4 rolls, I have $4 \times 1/6 = 2/3$ of a chance to get at least one ace.

His reasoning for the second game was similar:

- In one roll of a pair of dice, I have $1/36$ of a chance to get a double-ace.
- So in 24 rolls, I must have $24 \times 1/36 = 2/3$ of a chance to get at least one double-ace.

By this argument, both chances were the same, namely $2/3$. However, the gamblers found that the first event was a bit more likely than the second. This contradiction became known as the *Paradox of the Chevalier de Méré.*

De Méré asked the philosopher Blaise Pascal about the problem, and Pascal solved it with the help of his friend, Pierre de Fermat. Fermat was a judge and a member of parliament, who is remembered today for the mathematical research he did after hours. Fermat saw that de Méré was adding chances for events that were not mutually exclusive. In fact, pushing de Méré's argument a little further, it shows the chance of getting an ace in 6 rolls of a die to be $6/6$, or 100%. Something had to be wrong.

The question is how to calculate the chances correctly. Pascal and Fermat solved this problem, with a typically indirect piece of mathematical reasoning—

Blaise Pascal (France, 1623–1662)

Wolff-Leavenworth Collection, courtesy of the Syracuse University Art Collection.

Pierre de Fermat (France, 1601–1665)

From the *Oeuvres Complètes*

the kind that always leaves non-mathematicians feeling a bit cheated. Of course, a direct attack like Galileo's (section 1) could easily bog down. With 4 rolls of a die, there are $6^4 = 1{,}296$ outcomes to worry about. With 24 rolls of a pair of dice, there are $36^{24} \approx 2.2 \times 10^{37}$ outcomes.

The conversation between Pascal and Fermat is lost to history, but here is a reconstruction.[3]

Pascal. Let's look at the first game first.

Fermat. Bon. The chance of winning is hard to compute, so let's work out the chance of the opposite event—losing. Then

$$\text{chance of winning} = 100\% - \text{chance of losing}.$$

Pascal. D'accord. The gambler loses when none of the four rolls shows an ace. But how do you work out the chances?

Fermat. It does look complicated. Let's start with one roll. What's the chance that the first roll doesn't show an ace?

Pascal. It has to show something from 2 through 6, so the chance is $5/6$.

Fermat. C'est ça. Now, what's the chance that the first two rolls don't show aces?

Pascal. We can use the multiplication rule. The chance that the first roll doesn't give an ace and the second doesn't give an ace equals $5/6 \times 5/6 = (5/6)^2$. After all, the rolls are independent, n'est-ce pas?

Fermat. What about 3 rolls?

Pascal. It looks like $5/6 \times 5/6 \times 5/6 = (5/6)^3$.

Fermat. Oui. Now what about 4 rolls?

Pascal. Must be $(5/6)^4$.

Fermat. Sans doute, and that's about 0.482, or 48.2%.

Pascal. So there is a 48.2% chance of losing. Now

$$\begin{aligned}\text{chance of winning} &= 100\% - \text{chance of losing}\\ &= 100\% - 48.2\% = 51.8\%.\end{aligned}$$

Fermat That settles the first game. The chance of winning is a little over 50%. Now what about the second?

Pascal Eh bien, in one roll of a pair of dice, there is 1 chance in 36 of getting a double-ace, and 35 chances in 36 of not getting a double-ace. By the multiplication rule, in 24 rolls of a pair of dice the chance of getting no double-aces must be

$$(35/36)^{24}.$$

Fermat Entendu. That's about 50.9%. So we have the chance of losing. Now

$$\begin{aligned}\text{chance of winning} &= 100\% - \text{chance of losing}\\ &= 100\% - 50.9\% = 49.1\%.\end{aligned}$$

Pascal Le résultat is a bit less than 50%. Voilà. That's why you win the second game a bit less frequently than the first. But you have to roll a lot of dice to see the difference.

> If the chance of an event is hard to find, try to find the chance of the opposite event. Then subtract from 100%. (See p. 223.) This is useful when the chance of the opposite event is easier to compute.

Exercise Set D

1. A die is rolled three times. You bet $1 on some proposition. Below is a list of 6 bets, and then a list of 3 outcomes. For each bet, find all the outcomes where you win. For instance, with (a), you win on (i) only.

   Bets

   (a) all aces
   (b) at least one ace
   (c) no aces
   (d) not all aces
   (e) 1st roll is an ace, or 2nd roll is an ace, or 3rd roll is an ace
   (f) 1st roll is an ace, and 2nd roll is an ace, and 3rd roll is an ace

   Outcomes

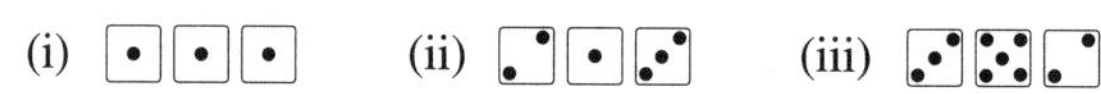

2. In exercise 1, which is a better bet—(a) or (f)? Or are they same? What about (b) and (e)? What about (c) and (d)? (You do not need to compute the chances.)

3. A box contains four tickets, one marked with a star, and the other three blank:

   Two draws are made at random with replacement from this box.

   (a) What is the chance of getting a blank ticket on the first draw?
   (b) What is the chance of getting a blank ticket on the second draw?
   (c) What is the chance of getting a blank ticket on the first draw and a blank ticket on the second draw?
   (d) What is the chance of not getting the star in the two draws?
   (e) What is the chance of getting the star at least once in the two draws?

4. (a) A die is rolled 3 times. What is the chance of getting at least one ace?
   (b) Same, with 6 rolls.
   (c) Same, with 12 rolls.

5. A pair of dice is rolled 36 times. What is the chance of getting at least one double-ace?

6. According to de Moivre, in eighteenth-century England people played a game similar to modern roulette. It was called "Royal Oak." There were 32 "points" or num-

bered pockets on a table. A ball was thrown in such a way that it landed in each pocket with an equal chance, 1 in 32.

If you bet 1 pound on a point and it came up, you got your stake back, together with winnings of 27 pounds. If your point didn't come up, you lost your pound. The players (or "Adventurers," as de Moivre called them) complained that the game was unfair, and they should have won 31 pounds if their point came up. (They were right; section 1 of chapter 17.) De Moivre continues:

> The Master of the Ball maintained they had no reason to complain; since he would undertake that any particular point of the Ball should come up in Two-and-Twenty Throws: of this he would offer to lay a Wager, and actually laid it when required.The seeming contradiction between the Odds of One-and-Thirty to One, and Twenty-two Throws for any [point] to come up, so perplexed the Adventurers, that they begun to think the Advantage was on their side: for which reason they played on and continued to lose. [Two-and-Twenty is 22, One-and-Thirty is 31.]

What is the chance that the point 17, say, will come up in Two-and-Twenty Throws? (The Master of the Ball laid this wager at even money, so if the chance is over 50%, he shows a profit here too.)

7. In his novel *Bomber*, Len Deighton argues that a World War II pilot had a 2% chance of being shot down on each mission. So in 50 missions he is "mathematically certain" to be shot down: $50 \times 2\% = 100\%$. Is this a good argument?

   Hint: To make chance calculations, you have to see how the situation is like a game of chance. The analogy here is getting the card "survive" every time, if you draw 50 times at random with replacement from the box

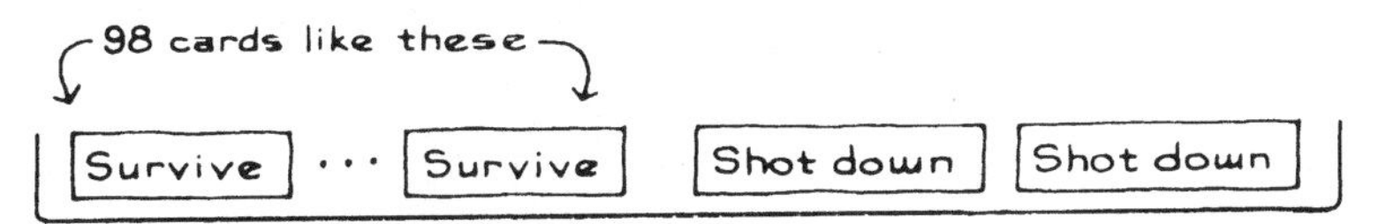

The answers to these exercises are on p. A70.

5. ARE REAL DICE FAIR?

According to Galileo (section 1), when a die is rolled it is equally likely to show any of its 6 faces. Galileo was thinking of an ideal die which is perfectly symmetric. This is like ignoring friction in the study of physics: the results are only a first approximation. What does Galileo's calculation say about real dice?

- For real dice, the 216 possible ways three dice can land are close to being equally likely.
- If these ways were equally likely, the chance of rolling a total of 9 spots would be exactly 25 in 216.
- So for real dice, the chance of rolling a total of 9 spots is just about 25 in 216.

For loaded dice, the calculations would be badly off. But ordinary dice, coins, and the like are very close to fair—in the sense that all the outcomes are equally likely. Of course, you have to put some effort into shaking the dice or flipping the coins. And the games of chance based on these fair mechanisms may be quite unfair (chapter 17).

In a similar way, if you are told that a ticket is drawn at random, you should assume that each ticket in the box is equally likely to be drawn. If the tickets are close to the same size, shape, and texture, and the box is well shaken, this is quite a reasonable approximation.

6. REVIEW EXERCISES

Review exercises may cover material from previous chapters.

When a die is rolled, each of the 6 faces is equally likely to come up. A deck of cards has 4 suits (clubs, diamonds, hearts, spades) with 13 cards in each suit— 2, 3, . . . , 10, jack, queen, king, ace. See pp. 222 and 226.

1. A pair of dice are thrown.
   (a) Find the chance that both dice show 3 spots.
   (b) Find the chance that both dice show the same number of spots.
2. In the game of Monopoly, a player rolls two dice, counts the total number of spots, and moves that many squares. Find the chance that the player moves 11 squares (no more and no less).
3. True or false, and explain:
   (a) If a die is rolled three times, the chance of getting at least one ace is $1/6 + 1/6 + 1/6 = 1/2$.
   (b) If a coin is tossed twice, the chance of getting at least one head is 100%.
4. Two cards will be dealt off the top of a well-shuffled deck. You have a choice:

(i) to win $1 if at least one of the two cards is a queen.
(ii) to win $1 if the first is a queen.

Which option is better? Or are they equivalent? Explain.

5. The chance of A is 1/3; the chance of B is 1/10. True or false, and explain:
   (a) If A and B are independent, they must also be mutually exclusive.
   (b) If A and B are mutually exclusive, they cannot be independent.

6. One event has chance 1/2, another has chance 1/3. Fill in the blanks, using one phrase from each pair below, to make up two true sentences. Write out both sentences.

   "If you want to find the chance that __(i)__ will happen, check to see if they are __(ii)__. If so, you can __(iii)__ the chances."

   (i) at least one of the two events, both events
   (ii) independent, mutually exclusive
   (iii) add, multiply

7. Four draws are going to be made at random with replacement from the box | [1] [2] [2] [3] [3] |. Find the chance that [2] is drawn at least once.

8. Repeat exercise 7, if the draws are made at random without replacement.

9. One ticket will be drawn at random from each of the two boxes shown below:

   (A) | [1] [2] [3] | (B) | [1] [2] [3] [4] |

   Find the chance that:
   (a) The number drawn from A is larger than the one from B.
   (b) The number drawn from A equals the one from B.
   (c) The number drawn from A is smaller than the one from B.

10. There are two options:
    (i) A die will be rolled 60 times. Each time it shows an ace or a six, you win $1; on the other rolls, you win nothing.
    (ii) Sixty draws will be made at random with replacement from the box | [1] [1] [1] [0] [0] [0] |. On each draw, you will be paid the amount shown on the ticket, in dollars.

    Which option is better? or are they the same? Explain briefly.

11. Three cards are dealt from a well-shuffled deck.
    (a) Find the chance that all of the cards are diamonds.
    (b) Find the chance that none of the cards are diamonds.
    (c) Find the chance that the cards are not all diamonds.

12. A coin is tossed 10 times. True or false, and explain:
    (a) The chance of getting 10 heads in a row is 1/1,024.
    (b) Given that the first 9 tosses were heads, the chance of getting 10 heads in a row is 1/2.

Exercises 13 and 14 are more difficult.

13. A box contains 2 red marbles and 98 blue ones. Draws are made at random with replacement. In ______ draws from the box, there is better than a 50% chance for a red marble to appear at least once. Fill in the blank with the smallest number that makes the statement true. (You will need a calculator.)

14. In Lotto 6-53, there is a box with 53 balls, numbered from 1 to 53. Six balls are drawn at random without replacement from the box. You win the grand prize if the numbers on your lottery ticket are the same as the numbers on the six balls; order does not matter.

    Person A bought two tickets, with the following numbers:

Ticket #1	5	12	21	30	42	51
Ticket #2	5	12	23	30	42	49

    Person B bought two tickets, with the following numbers:

Ticket #1	7	11	25	28	34	50
Ticket #2	9	14	20	22	37	45

    Which person has the better chance of winning? Or are their chances the same? Explain briefly.

7. SUMMARY

1. When figuring chances, one helpful strategy is to write down a complete list of all the possible ways that the chance process can turn out. If this is too hard, at least write down a few typical ways, and count how many ways there are in total.

2. The chance that at least one of two things will happen equals the sum of the individual chances, provided the things are mutually exclusive. Otherwise, adding the chances will give the wrong answer—double counting.

3. If you are having trouble working out the chance of an event, try to figure out the chance of its opposite; then subtract from 100%.

15

The Binomial Formula

Man is a reed, but a reed that thinks.
—BLAISE PASCAL (FRANCE, 1623–1662)

1. INTRODUCTION

This chapter explains how to answer questions like the following.

- A coin is tossed four times. What is the chance of getting exactly one head?
- A die is rolled ten times. What is the chance of getting exactly three aces?
- A box contains one red marble and nine green ones. Five draws are made at random with replacement. What is the chance that exactly two draws will be red?

These problems are all similar, and can be solved using the *binomial coefficients*, discovered by Pascal and Newton.[1] The method will be illustrated on the marbles.

The problem is to find the chance of getting two reds (no more and no less) in five draws from the box; so the other three draws must be green. One way this can happen is that the first two draws are red and the final three are green. With R for red and G for green, this possibility can be written

R R G G G

Of course, there are many other ways to get two reds. For example, the second and the fifth draws might be red, while all the rest are green:

G R G G R

Isaac Newton (England, 1642–1727).

From the Warden Collection; engraved by W.T. Fry after a painting by G. Kneller.

To solve the problem, we must find all the possible ways, calculate the chance of each, and then use the addition rule to add up the chances. The first task seems formidable, so we postpone it for a moment and turn to the second one.

The chance of the pattern R R G G G is

$$\frac{1}{10} \times \frac{1}{10} \times \frac{9}{10} \times \frac{9}{10} \times \frac{9}{10} = \left(\frac{1}{10}\right)^2 \left(\frac{9}{10}\right)^3$$

This follows from the multiplication rule: on each draw, the chance of red is $1/10$, the chance of green is $9/10$.

Similarly, the chance of the pattern G R G G R equals

$$\frac{9}{10} \times \frac{1}{10} \times \frac{9}{10} \times \frac{9}{10} \times \frac{1}{10} = \left(\frac{1}{10}\right)^2 \left(\frac{9}{10}\right)^3$$

The pattern G R G G R has the same chance as the pattern R R G G G. In fact, each pattern with 2 reds and 3 greens has the same chance, $(1/10)^2(9/10)^3$, since the 2 reds will contribute $(1/10)^2$ to the product and the 3 greens will contribute $(9/10)^3$. The sum of the chances of all the patterns, therefore, equals the number of patterns times the common chance.

How many patterns are there? Each pattern is specified by writing down in a row 2 R's and 3 G's, in some order. The number of patterns is given by the *binomial coefficient,*

$$\frac{5 \times 4 \times 3 \times 2 \times 1}{(2 \times 1) \times (3 \times 2 \times 1)} = 10$$

In other words, there are 10 different patterns with 2 R's and 3 G's. So the chance of drawing exactly 2 reds is

$$10 \times \left(\frac{1}{10}\right)^2 \left(\frac{9}{10}\right)^3 \approx 7\%$$

Binomial coefficients look messy. Mathematicians get around this by introducing convenient notation. They use an exclamation mark (!) to indicate the result of multiplying together a number and all the numbers which come before it. For example,

$$\begin{aligned} 1! &= 1 \\ 2! &= 2 \times 1 = 2 \\ 3! &= 3 \times 2 \times 1 = 6 \\ 4! &= 4 \times 3 \times 2 \times 1 = 24 \end{aligned}$$

And so on. The exclamation mark is read "factorial," so that $4! = 24$ is read "four-factorial equals twenty-four." Now the binomial coefficient is easier to read:

$$\frac{5!}{2!\,3!}$$

Remember what the formula represents—the number of different ways of arranging 2 R's and 3 G's in a row.

The 5 in the numerator of the formula is the sum of 2 and 3 in the denominator. Binomial coefficients always take this form. For example, the number of ways to arrange four R's and one G in a row is

$$\frac{5!}{4!\,1!} = 5$$

The patterns are

R R R R G   R R R G R   R R G R R   R G R R R   G R R R R

How many ways are there to arrange five R's and zero G's in a row? There is only one way, R R R R R. Applying the formula mechanically gives

$$\frac{5!}{5!\,0!}$$

But we have not yet said what 0! means. It is a convention of mathematics that $0! = 1$. With this convention, the binomial coefficient does equal 1.

Binomial coefficients and factorials get very large very quickly. For instance, the number of ways to arrange 10 R's and 10 G's in a row is given by the binomial coefficient

$$\frac{20!}{10!\,10!} = 184{,}756$$

However, there was a lot of cancellation going on: $10! = 3{,}628{,}800$; and $20! \approx 2 \times 10^{18}$, or 2 followed by 18 zeros. (A trillion is 1 followed by 12 zeros.)

Exercise Set A

1. Find the number of different ways of arranging one R and three G's in a row. Write out all the patterns.

2. Find the number of different ways of arranging two R's and two G's in a row. Write out all the patterns.

3. A box contains one red ball and five green ones. Four draws are made at random with replacement from the box. Find the chance that—
   (a) a red ball is never drawn
   (b) a red ball appears exactly once
   (c) a red ball appears exactly twice
   (d) a red ball appears exactly three times
   (e) a red ball appears on all the draws
   (f) a red ball appears at least twice

4. A die is rolled four times. Find the chance that—
   (a) an ace (one dot) never appears
   (b) an ace appears exactly once
   (c) an ace appears exactly twice

5. A coin is tossed 10 times. Find the chance of getting exactly 5 heads. Find the chance of obtaining between 4 and 6 heads inclusive.

6. It is claimed that a vitamin supplement helps kangaroos learn to run a special maze with high walls. To test whether this is true, 20 kangaroos are divided up into 10 pairs. In each pair, one kangaroo is selected at random to receive the vitamin supplement; the other is fed a normal diet. The kangaroos are then timed as they learn to run the maze. In 7 of the 10 pairs, the treated kangaroo learns to run the maze more quickly than its untreated partner. If in fact the vitamin supplement has

"Did you bring the vitamins ?"

no effect, so that each animal of the pair is equally likely to be the quicker, what is the probability that 7 or more of the treated animals would learn the maze more quickly than their untreated partners, just by chance?

The answers to these exercises are on pp. A70–71.

2. THE BINOMIAL FORMULA

The reasoning of section 1 is summarized in the *binomial formula*. Suppose a chance process is carried out as a sequence of trials. An example would be rolling a die 10 times, where each roll counts as a trial. There is an event of interest which may or may not occur at each trial: the die may or may not land ace. The problem is to calculate the chance that the event will occur a specified number of times.

The chance that an event will occur exactly k times out of n is given by the binomial formula

$$\frac{n!}{k!\,(n-k)!}\,p^k(1-p)^{n-k}$$

In this formula, n is the number of trials, k is the number of times the event is to occur, and p is the probability that the event will occur on any particular trial. The assumptions:

- The value of n must be fixed in advance.
- p must be the same from trial to trial.
- The trials must be independent.

The formula starts with the binomial coefficient (p. 256),

$$\frac{n!}{k!\,(n-k)!}$$

Remember, this is the number of ways to arrange n objects in a row, when k are alike of one kind and $n - k$ are alike of another (for instance, red and green marbles).

Example 1. A die is rolled 10 times. What is the chance of getting exactly 2 aces?

Solution. The number of trials is fixed in advance. It is 10. So $n = 10$. The event of interest is rolling an ace. The probability of rolling an ace is the same from trial to trial. It is $1/6$. So $p = 1/6$. The trials are independent. The binomial formula can be used, and the answer is

$$\frac{10!}{2!\,8!}\left(\frac{1}{6}\right)^2\left(\frac{5}{6}\right)^8 \approx 29\%$$

Example 2. A die is rolled until it first lands six. If this can be done using the binomial formula, find the chance of getting 2 aces. If not, why not?

Solution. The number of trials is not fixed in advance. It could be 1, if the die lands six right away. Or it could be 2, if the die lands five then six. Or it could be 3. And so forth. The binomial formula does not apply.

Example 3. Ten draws are made at random with replacement from the box | [1] [1] [2] [3] [4] [5] |. However, just before the last draw is made, whatever else has gone on, the ticket [5] is removed from the box. True or false: the chance of drawing exactly two [1]'s is

$$\frac{10!}{2!\,8!}\left(\frac{2}{6}\right)^2\left(\frac{4}{6}\right)^8$$

Solution. In this example, n is fixed in advance and the trials are independent. However, p changes at the last trial from $2/6$ to $2/5$. So the binomial formula does not apply, and the statement is false.

Example 4. Four draws are made at random without replacement from the box in example 3. True or false: the chance of drawing exactly two [1]'s is

$$\frac{4!}{2!\,2!}\left(\frac{2}{6}\right)^2\left(\frac{4}{6}\right)^2$$

Solution. The trials are dependent, so the binomial formula does not apply.

Technical notes. (i) To work out the chance in example 4, take a pattern with exactly two 1's, like 1 1 N N, where N means "not 1." The chance of getting 1 1 N N equals

$$\frac{2}{6} \times \frac{1}{5} \times \frac{4}{4} \times \frac{3}{3} = \frac{1}{15}$$

Surprisingly, the chance is the same for all such patterns. How many patterns have exactly two 1's? The answer is

$$\frac{4!}{2!\,2!} = 6$$

So the chance of getting exactly two 1's is

$$6 \times \frac{1}{15} = \frac{2}{5}$$

(ii) Mathematicians usually write $\binom{n}{k}$ for the binomial coefficient:

$$\binom{n}{k} = \frac{n!}{k!\,(n-k)!}$$

They read $\binom{n}{k}$ as "n choose k," the idea being that the formula gives the number of ways to choose k things out of n. Older books write the binomial coefficient as ${}_nC_k$ or ${}^nC_k$, the "number of combinations of n things taken k at a time."

3. REVIEW EXERCISES

Review exercises may cover material from previous chapters.

1. A die will be rolled 6 times. What is the chance of obtaining exactly 1 ace?

2. A die will be rolled 10 times. The chance it never lands six can be found by one of the following calculations. Which one, and why?

   (i) $\left(\frac{1}{6}\right)^{10}$ (ii) $1 - \left(\frac{1}{6}\right)^{10}$ (iii) $\left(\frac{5}{6}\right)^{10}$ (iv) $1 - \left(\frac{5}{6}\right)^{10}$

3. Of families with 4 children, what proportion have more girls than boys? You may assume that the sex of a child is determined as if by drawing at random with replacement from[2]

   [M] [F]   M = male, F = female

4. A box contains 8 red marbles and 3 green ones. Six draws are made at random without replacement. True or false: the chance that the 3 green marbles are drawn equals

$$\frac{6!}{3!\,3!}\left(\frac{8}{11}\right)^3\left(\frac{3}{11}\right)^3$$

   Explain briefly.

5. There are 8 people in a club.[3] One person makes up a list of all the possible committees with 2 members. Another person makes up a list of all the possible committees with 5 members. True or false: the second list is longer than the first. Explain briefly.

6. There are 8 people in a club. One person makes up a list of all the possible committees with 2 members. Another person makes up a list of all the possible committees with 6 members. True or false: the second list is longer than the first. Explain briefly.

7. A box contains one red marble and nine green ones. Five draws are made at random with replacement. The chance that exactly two draws will be red is

$$10 \times \left(\frac{1}{10}\right)^2\left(\frac{9}{10}\right)^3$$

   Is the addition rule used in deriving this formula? Answer yes or no, and explain carefully.

8. A coin will be tossed 10 times. Find the chance that there will be exactly 2 heads among the first 5 tosses, and exactly 4 heads among the last 5 tosses.

9. For each question (a–e) below, choose one of the answers (i–viii); explain your choice.

   Questions

   A deck of cards is shuffled. What is the chance that—

   (a) the top card is the king of spades and the bottom card is the queen of spades?
   (b) the top card is the king of spades and the bottom card is the king of spades?
   (c) the top card is the king of spades or the bottom card is the king of spades?
   (d) the top card is the king of spades or the bottom card is the queen of spades?
   (e) of the top and bottom cards, one is the king of spades and the other is the queen of spades?

   Answers

   (i) $1/52 \times 1/51$
   (ii) $1/52 + 1/51$
   (iii) $1/52 \times 1/52$
   (iv) $1/52 + 1/52$
   (v) $1 - (1/52 \times 1/51)$
   (vi) $1 - (1/52 \times 1/52)$
   (vii) $2/52 \times 1/51$
   (viii) None of the above

10. A box contains 3 red tickets and 2 green ones. Five draws will be made at random. You win $1 if 3 of the draws are red and 2 are green. Would you prefer the draws to be made with or without replacement? Why?

11. It is now generally accepted that cigarette smoking causes heart disease, lung cancer, and many other diseases. However, in the 1950s, this idea was controversial. There was a strong association between smoking and ill-health, but association is not causation. R. A. Fisher advanced the "constitutional hypothesis:" there is some genetic factor that disposes you both to smoke and to die.

    To refute Fisher's idea, the epidemiologists used twin studies. They identified sets of smoking-discordant monozygotic twin pairs. ("Monozygotic" twins come from one egg and have identical genetic makeup; "smoking-discordant" means that one twin smokes, the other doesn't.) Now there is a race. Which twin dies first, the smoker or the non-smoker? Data from a Finnish twin study are shown at the top of the next page.[4]

Data from the Finnish twin study

	Smokers	*Non-smokers*
All causes	17	5
Coronary heart disease	9	0
Lung cancer	2	0

According to the first line of the table, there were 22 smoking-discordant monozygotic twin pairs where at least one twin of the pair died. In 17 cases, the smoker died first; in 5 cases, the non-smoker died first. According to the second line, there were 9 pairs where at least one twin died of coronary heart disease; in all 9 cases, the smoker died first. According to the last line, there were 2 pairs where at least one twin died of lung cancer, and in both pairs the smoker won the race to death. (Lung cancer is a rare disease, even among smokers.)

For parts (a–c), suppose that each twin in the pair is equally likely to die first, so the number of pairs in which the smoker dies first is like the number of heads in coin-tossing.

(a) On this basis, what is the chance of having 17 or more pairs out of 22 where the smoker dies first?
(b) Repeat the test in part (a), for the 9 deaths from coronary heart disease.
(c) Repeat the test in part (a), for the 2 deaths from lung cancer.
(d) Can the difference between the death rates for smoking and non-smoking twins be explained by
   (i) chance?
   (ii) genetics?
   (iii) health effects of smoking?

4. SPECIAL REVIEW EXERCISES

These exercises cover all of parts I–IV.

1. In the U.S. in 1990, 20,273 people were murdered, compared to 16,848 in 1970—nearly a 20% increase. "These figures show that the U.S. became a more violent society over the period 1970–1990." True or false, and explain briefly.[5]

2. A leading cause of death in the U.S. is coronary artery disease—a breakdown of the main arteries to the heart. The disease can be treated with coronary bypass surgery (section 3 of chapter 1). In one of the first trials, Dr. Daniel Ullyot and associates performed coronary bypass surgery on a test group of patients: 98% survived 3 years or more. The conventional treatment used drugs and special diets to reduce blood pressure and eliminate fatty deposits in the arteries. According to previous studies, only 68% of the patients getting the conventional treatment survived 3 years or more. [Exercise continues. . . .]

A newspaper article described Ullyot's results as "spectacular," because the survival rate among Ullyot's patients was much higher than the survival rate in previous studies.[6]

(a) Did Ullyot's study have contemporaneous controls? If not, what patients were used as the comparison group?
(b) Was the newspaper article's enthusiasm justified by the study? Discuss briefly.

3. Susan Bouman was denied a promotion to sergeant in the Los Angeles County Sheriff's Department after she took a competitive exam for the position. She filed suit in federal court in April 1980, claiming that the exam was discriminatory.[7] Data for 1975 and 1977 are shown below. In 1975, the pass rate for women was 10/79 = 12.7%; for men, it was 250/1,312 = 19.1%. The "selection ratio" was 12.7/19.1 = 66.5%: in other words, the women's pass rate was only 66.5% of the men's pass rate. For 1977, the selection ratio was 67.1%. Selection ratios below 80% are generally regarded by the Equal Opportunity Employment Commission as showing "adverse impact" on a "protected group."

Results can also be analyzed by "pooling" data for the two years—just adding up the numbers. For the two years combined, there were 102 + 79 = 181 women applicants of whom 10 + 18 = 28 passed, and so forth. True or false, and explain: "The selection ratio was 66.5% in 1975 and 67.1% in 1977; therefore, the selection ratio for the pooled data must be between 66.5% and 67.1%."

	Women	*Men*
1975		
Applicants	79	1,312
Passed the exam	10	250
1977		
Applicants	102	1,259
Passed the exam	18	331

4. Three people have tried to sketch the histogram for blood pressures of the subjects in a certain study, using the density scale. Only one is right. Which one, and why?

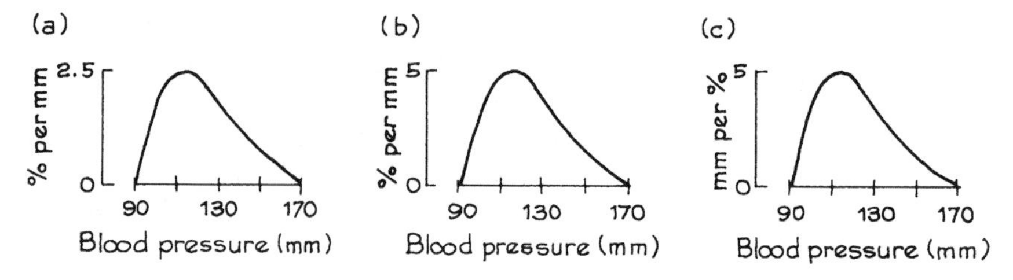

5. A study is made of the age at entrance of college freshmen.[8] Is the SD about 1 month, 1 year, or 5 years. Why?

6. A study is based on a representative sample of men age 25–64 in 2005, who were working full time. The figure below plots average income for each age group.[9] True or false, and explain: the data show that on average, if a man keeps working, his income will increase until age 50 or so, then stabilize. If false, how do you account for the pattern in the data?

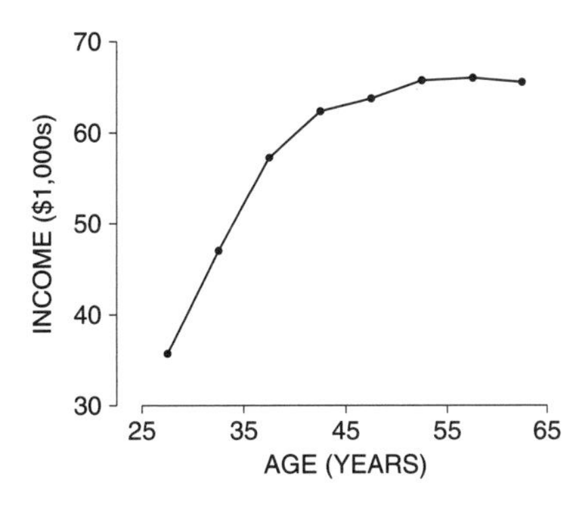

Source: March 2005 Current Population Survey; CD-ROM supplied by the Bureau of the Census.

7. True or false, and explain: for the histogram below, the 60th percentile is equal to twice the 30th percentile. (You may assume the distribution is uniform on each class interval.)

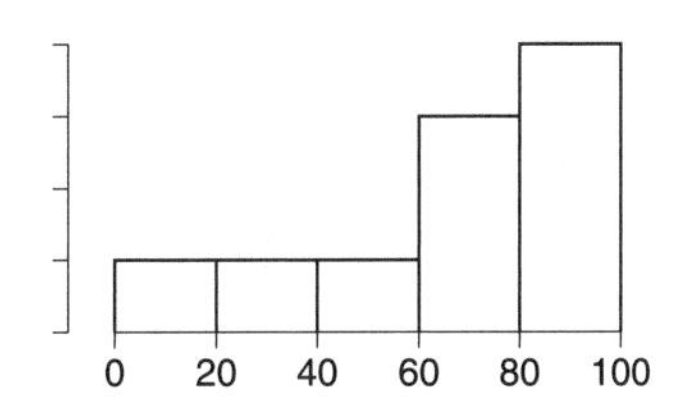

8. True or false, and explain. (You don't need to compute the average or the SD of the lists.)

   (a) The following two lists are the same, when converted to standard units:

(i)	1	3	4	7	9	9	9	21	32
(ii)	3	7	9	15	19	19	19	43	65

   (b) The following two lists are the same, when converted to standard units:

(i)	1	3	4	7	9	9	9	21	32
(ii)	−1	−5	−7	−13	−17	−17	−17	−41	−63

9. In a large class, the average score on the final was 50 out of 100, and the SD was 20. The scores followed the normal curve.

   (a) Two brothers took the final. One placed at the 70th percentile and the other was at the 80th percentile. How many points separated them?

   (b) Two sisters took the final. One placed at the 80th percentile and the other was at the 90th percentile. How many points separated them?

10. The figure below is a scatter plot of income against education (years of schooling completed) for a representative sample of men age 25–34 in Kansas. Or is something wrong? Explain briefly.

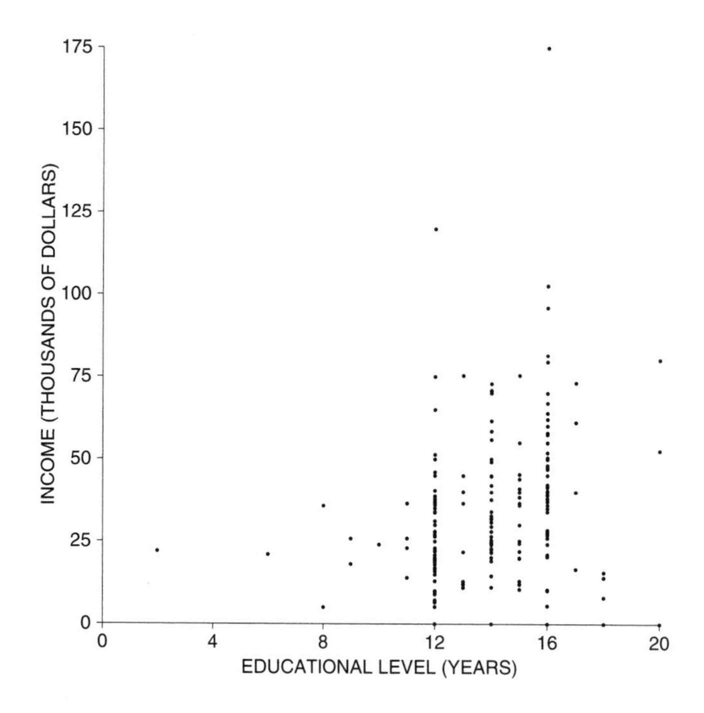

11. (a) Find the correlation coefficient for the data set in table (i) below.
(b) If possible, fill in the blanks in table (ii) below so the correlation coefficient is 1. If this is not possible, explain why not.

(i)

x	y
4	7
5	0
7	9
8	9
8	13
10	16

(ii)

x	y
___	7
5	___
7	9
8	9
8	13
10	___

12. In a study of Danish draftees, T. W. Teasdale and associates found a positive correlation between near-sightedness and intelligence.[10] True or false, and explain:

(a) Draftees who were more near-sighted were also more intelligent, on average.
(b) Draftees who were more intelligent were also more near-sighted, on average.
(c) The data show that near-sightedness causes intelligence.
(d) The data show that intelligence causes near-sightedness.

13. For each diagram below, say whether r is nearly -1, 0, or 1. Explain briefly.

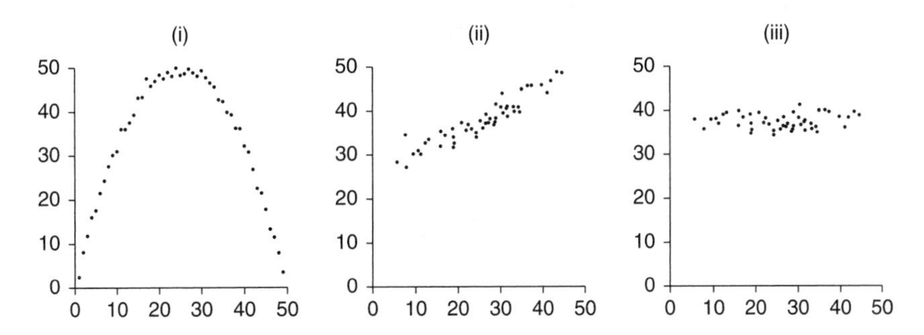

14. The figure below shows a scatter diagram for test scores. Verbal SAT is plotted on the vertical axis and Math SAT on the horizontal. Three lines are drawn across the diagram. Match the line with the description (one description will be left over). Explain briefly.

    (i) estimated average score on V-SAT for given score on M-SAT
    (ii) estimated average score on M-SAT for given score on V-SAT
    (iii) nearly equal percentile ranks on both tests
    (iv) total score on the two tests is about 1,100

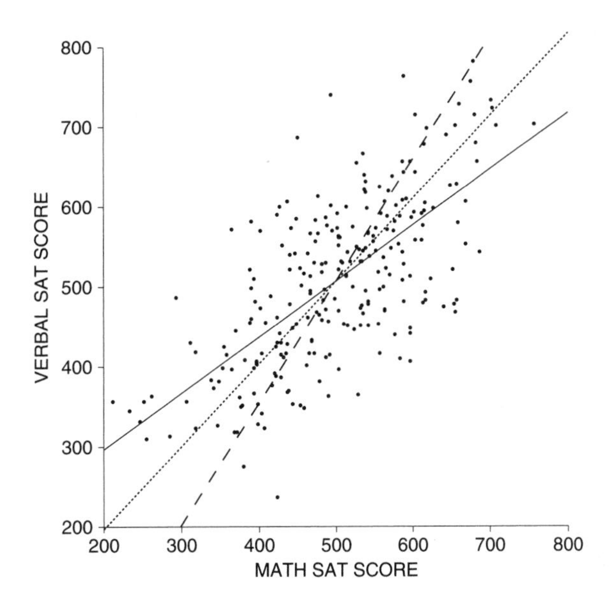

15. At a certain law school, first-year scores average 65 and the SD is 12. The correlation between LSAT scores and first-year scores is 0.55. The scatter diagram is football-shaped. The dean's office uses regression to predict first-year scores from LSAT scores. About what percent of the students do better than predicted, by 10 points or more? Explain your answer. If you need more information, say what you need and why.

16. The great prime ministers of France generally served under kings who were less talented. Similarly, the great kings typically had prime ministers who were not as great. Is this a fact of French history, or of statistics? Discuss briefly.

17. In a large class, the midterm had an average of 50 points with an SD of 22. The final scores averaged out to 60 with an SD of 20. The correlation between midterm and final scores was 0.60. The scatter diagram was football-shaped. Of the students who scored around 50 on the midterm, about what percentage were in the top 25% of the class on the final?

18. One ticket is drawn at random from each of the two boxes below:

(A) | 1 2 3 4 5 | (B) | 1 2 3 4 5 6 |

Find the chance that—
   (a) One of the numbers is 2 and the other is 5.
   (b) The sum of the numbers is 7.
   (c) One number is bigger than twice the other.

19. There are 52 cards in a deck, and 13 of them are hearts.
   (a) Four cards are dealt, one at a time, off the top of a well-shuffled deck. What is the chance that a heart turns up on the fourth card, but not before? Explain briefly.
   (b) A deck of cards is shuffled. You have to deal one card at a time until a heart turns up. You have dealt 3 cards, and still have not seen a heart. What is the chance of getting a heart on the 4th card? Explain briefly.

20. A coin is tossed 10 times. Find the chance of getting 7 heads and 3 tails.

5. SUMMARY AND OVERVIEW

1. The *binomial coefficient* $\frac{n!}{k!\,(n-k)!}$ gives the number of ways to arrange n objects in a row, when k are alike of one kind and $n - k$ are alike of another kind (for instance, red and blue marbles).

2. The chance that an event will occur exactly k times out of n is given by the *binomial formula*

$$\frac{n!}{k!\,(n-k)!}\,p^k(1-p)^{n-k}$$

In this formula, n is the number of trials, k is the number of times the event is to occur, and p is the probability that the event will occur on any particular trial. The assumptions:

- The value of n must be fixed in advance.
- p must be the same from trial to trial.
- The trials must be independent.

3. This part of the book defined conditional probabilities, independence, and the multiplication rule. The addition rule was introduced for mutually exclusive events.

4. The binomial formula is an application of the multiplication rule combined with the addition rule.

5. Independence is the basis for the statistical theory to be developed in part V, and the crucial assumption behind many of the procedures to be discussed in parts VI–VIII.

PART V

Chance Variability

16

The Law of Averages

The roulette wheel has neither conscience nor memory.
—JOSEPH BERTRAND (FRENCH MATHEMATICIAN, 1822–1900)

1. WHAT DOES THE LAW OF AVERAGES SAY?

A coin lands heads with chance 50%. After many tosses, the number of heads should equal the number of tails: isn't that the law of averages? John Kerrich, a South African mathematician, found out the hard way. He was visiting Copenhagen when World War II broke out. Two days before he was scheduled to fly to England, the Germans invaded Denmark. Kerrich spent the rest of the war interned at a camp in Jutland. To pass the time he carried out a series of experiments in probability theory.[1] One experiment involved tossing a coin 10,000 times. With his permission, some of the results are summarized in table 1 and figure 1 (pp. 274–275 below). What do these results say about the law of averages? To find out, let's pretend that at the end of World War II, Kerrich was invited to demonstrate the law of averages to the King of Denmark. He is discussing the invitation with his assistant.

Assistant. So you're going to tell the king about the law of averages.

Kerrich. Right.

Assistant. What's to tell? I mean, everyone knows about the law of averages, don't they?

Kerrich. OK. Tell me what the law of averages says.

Assistant. Well, suppose you're tossing a coin. If you get a lot of heads, then tails start coming up. Or if you get too many tails, the chance for heads goes up. In the long run, the number of heads and the number of tails even out.

Kerrich. It's not true.

Assistant. What do you mean, it's not true?

Kerrich. I mean, what you said is all wrong. First of all, with a fair coin the chance for heads stays at 50%, no matter what happens. Whether there are two heads in a row or twenty, the chance of getting a head next time is still 50%.

Assistant. I don't believe it.

Kerrich. All right. Take a run of four heads, for example. I went through the record of my first 2,000 tosses. In 130 cases, the coin landed heads four times in a row; 69 of these runs were followed by a head, and only 61 by a tail. A run of heads just doesn't make tails more likely next time.

Assistant. You're always telling me these things I don't believe. What are you going to tell the king?

Kerrich. Well, I tossed the coin 10,000 times, and I got about 5,000 heads. The exact number was 5,067. The difference of 67 is less than 1% of the number of tosses. I have the record here in table 1.

Assistant. Yes, but 67 heads is a lot of heads. The king won't be impressed, if that's the best the law of averages can do.

Kerrich. What do you suggest?

Table 1. John Kerrich's coin-tossing experiment. The first column shows the number of tosses. The second shows the number of heads. The third shows the difference

number of heads − half the number of tosses.

Number of tosses	*Number of heads*	*Difference*	*Number of tosses*	*Number of heads*	*Difference*
10	4	−1	600	312	12
20	10	0	700	368	18
30	17	2	800	413	13
40	21	1	900	458	8
50	25	0	1,000	502	2
60	29	−1	2,000	1,013	13
70	32	−3	3,000	1,510	10
80	35	−5	4,000	2,029	29
90	40	−5	5,000	2,533	33
100	44	−6	6,000	3,009	9
200	98	−2	7,000	3,516	16
300	146	−4	8,000	4,034	34
400	199	−1	9,000	4,538	38
500	255	5	10,000	5,067	67

Assistant. Toss the coin another 10,000 times. With 20,000 tosses, the number of heads should be quite a bit closer to the expected number. After all, eventually the number of heads and the number of tails have to even out, right?

Kerrich. You said that before, and it's wrong. Look at table 1. In 1,000 tosses, the difference between the number of heads and the expected number was 2. With 2,000 tosses, the difference went up to 13.

Assistant. That was just a fluke. By toss 3,000, the difference was only 10.

Kerrich. That's just another fluke. At toss 4,000, the difference was 29. At 5,000, it was 33. Sure, it dropped back to 9 at toss 6,000, but look at figure 1. The chance error is climbing pretty steadily from 1,000 to 10,000 tosses, and it's going straight up at the end.

Assistant. So where's the law of averages?

Kerrich. With a large number of tosses, the size of the difference between the number of heads and the expected number is likely to be quite large in absolute terms. But compared to the number of tosses, the difference is likely to be quite small. That's the law of averages. Just like I said, 67 is only a small fraction of 10,000.

Assistant. I don't understand.

Kerrich. Look. In 10,000 tosses you expect to get 5,000 heads, right?

Assistant. Right.

Kerrich. But not exactly. You only expect to get around 5,000 heads. I mean, you could just as well get 5,001 or 4,998 or 5,007. The amount off 5,000 is what we call "chance error."

Figure 1. Kerrich's coin-tossing experiment. The "chance error" is

number of heads − half the number of tosses.

This difference is plotted against the number of tosses. As the number of tosses goes up, the size of the chance error tends to go up. The horizontal axis is not to scale and the curve is drawn by linear interpolation.

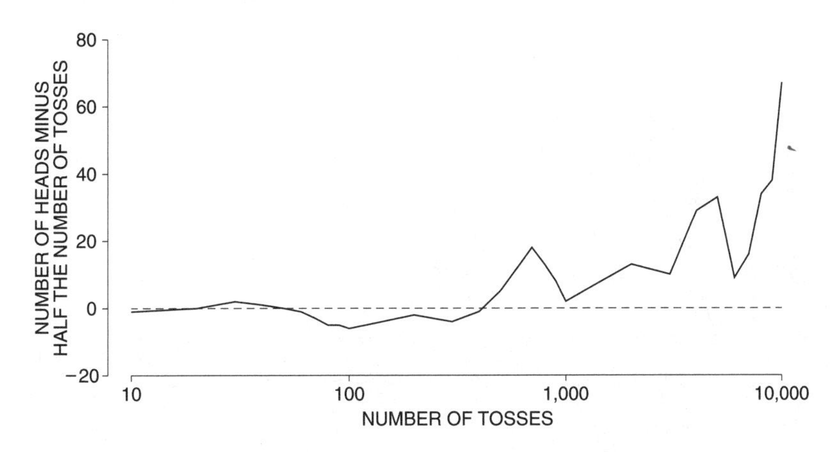

Assistant. Can you be more specific?

Kerrich. Let me write an equation:

$$\text{number of heads} = \text{half the number of tosses} + \text{chance error.}$$

This error is likely to be large in absolute terms, but small compared to the number of tosses. Look at figure 2. That's the law of averages, right there.

Assistant. Hmmm. But what would happen if you tossed the coin another 10,000 times. Then you'd have 20,000 tosses to work with.

Kerrich. The chance error would go up, but not by a factor of two. In absolute terms, the chance error gets bigger.[2] But as a percentage of the number of tosses, it gets smaller.

Assistant. Tell me again what the law of averages says.

Kerrich. The number of heads will be around half the number of tosses, but it will be off by some amount—chance error. As the number of tosses goes up, the chance error gets bigger in absolute terms. Compared to the number of tosses, it gets smaller.

Assistant. Can you give me some idea of how big the chance error is likely to be?

Kerrich. Well, with 100 tosses, the chance error is likely to be around 5 in size. With 10,000 tosses, the chance error is likely to be around 50 in size. Multiplying the number of tosses by 100 only multiplies the likely size of the chance error by $\sqrt{100} = 10$.

Assistant. What you're saying is that as the number of tosses goes up, the difference between the number of heads and half the number of tosses gets

Figure 2. The chance error expressed as a percentage of the number of tosses. When the number of tosses goes up, this percentage goes down: the chance error gets smaller relative to the number of tosses. The horizontal axis is not to scale and the curve is drawn by linear interpolation.

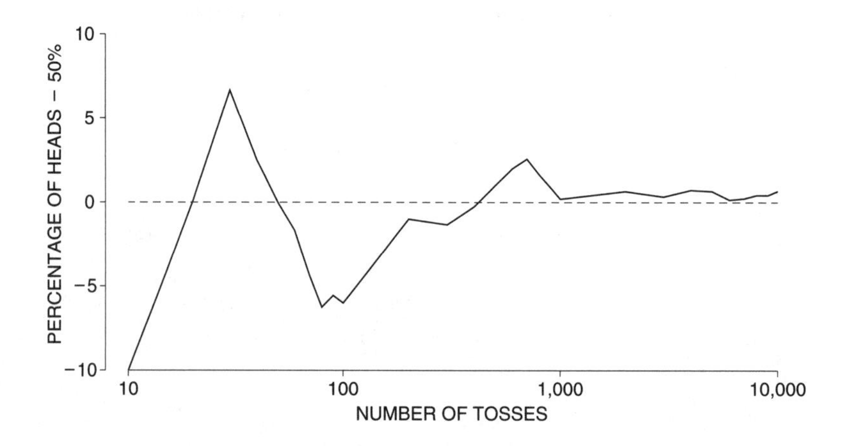

bigger; but the difference between the percentage of heads and 50% gets smaller.

Kerrich. That's it.

Exercise Set A

1. A machine has been designed to toss a coin automatically and keep track of the number of heads. After 1,000 tosses, it has 550 heads. Express the chance error both in absolute terms and as a percentage of the number of tosses.

2. After 1,000,000 tosses, the machine in exercise 1 has 501,000 heads. Express the chance error in the same two ways.

3. A coin is tossed 100 times, landing heads 53 times. However, the last seven tosses are all heads. True or false: the chance that the next toss will be heads is somewhat less than 50%. Explain.

4. (a) A coin is tossed, and you win a dollar if there are more than 60% heads. Which is better: 10 tosses or 100? Explain.
   (b) As in (a), but you win the dollar if there are more than 40% heads.
   (c) As in (a), but you win the dollar if there are between 40% and 60% heads.
   (d) As in (a), but you win the dollar if there are exactly 50% heads.

5. With a Nevada roulette wheel, there are 18 chances in 38 that the ball will land in a red pocket. A wheel is going to be spun many times. There are two choices:
   (i) 38 spins, and you win a dollar if the ball lands in a red pocket 20 or more times.
   (ii) 76 spins, and you win a dollar if the ball lands in a red pocket 40 or more times.

   Which is better? Or are they the same? Explain.

The next three exercises involve drawing at random from a box. This was described in section 1 of chapter 13 and is reviewed in section 3 below.

6. A box contains 20% red marbles and 80% blue marbles. A thousand marbles are drawn at random with replacement. One of the following statements is true. Which one, and why?

   (i) Exactly 200 marbles are going to be red.
   (ii) About 200 marbles are going to be red, give or take a dozen or so.

7. Repeat exercise 6, if the draws are made at random without replacement and the box contains 50,000 marbles.

8. One hundred tickets will be drawn at random with replacement from one of the two boxes shown below. On each draw, you will be paid the amount shown on the ticket, in dollars. (If a negative number is drawn, that amount will be taken away from you.) Which box is better? Or are they the same?

   (i) | −1 | −1 | 1 | 1 | (ii) | −1 | 1 |

9. (Hard.) Look at figure 1. If Kerrich kept on tossing, would the graph ever get negative?

The answers to these exercises are on pp. A71–72.

2. CHANCE PROCESSES

Kerrich's assistant was struggling with the problem of chance variability. He came to see that when a coin is tossed a large number of times, the actual number of heads is likely to differ from the expected number. But he didn't know how big a difference to anticipate. A method for calculating the likely size of the difference will be presented in the next chapter. This method works in many different situations. For example, it can be used to see how much money the house should expect to win at roulette (chapter 17) or how accurate a sample survey is likely to be (chapter 21).

What is the common element? All these problems are about chance processes.[3] Take the number of heads in Kerrich's experiment. Chance comes in with each toss of the coin. If you repeat the experiment, the tosses turn out differently, and so does the number of heads. Second example: the amount of money won or lost at roulette. Spinning the wheel is a chance process, and the amounts won or lost depend on the outcome. Spin again, and winners become losers. A final example: the percentage of Democrats in a random sample of voters. A chance process is used to draw the sample. So the number of Democrats in the sample is determined by the luck of the draw. Take another sample, and the percentages would change.

To what extent are the numbers influenced by chance? This sort of question must be faced over and over again in statistics. A general strategy will be presented in the next few chapters. The two main ideas:

- Find an analogy between the process being studied (sampling voters in the poll example) and drawing numbers at random from a box.

- Connect the variability you want to know about (for example, in the estimate for the Democratic vote) with the chance variability in the sum of the numbers drawn from the box.

The analogy between a chance process and drawing from a box is called a *box model*. The point is that the chance variability in the sum of numbers drawn from a box will be easy to analyze mathematically. More complicated processes can then be dealt with through the analogy.

3. THE SUM OF DRAWS

The object of this section is to illustrate the following process. There is a box of tickets. Each ticket has a number written on it. Then some tickets are drawn at random from the box, and the numbers on these tickets are added up. For example, take the box

[1] [2] [3] [4] [5] [6]

Imagine drawing twice at random with replacement from this box. You shake the box to mix up the tickets, pick one ticket at random, make a note of the number on it, put it back in the box. Then you shake the box again, and make a second draw at random. The phrase "with replacement" reminds you to put the ticket back in the box before drawing again. Putting the tickets back enables you to draw over and over again, under the same conditions. (Drawing with and without replacement was discussed in section 1 of chapter 13.)

Having drawn twice at random with replacement, you add up the two numbers. For example, the first draw might be [3] and the second [5]. Then the sum of the draws is 8. Or the first draw might be [3] and the second [3] too, so the sum of the draws is 6. There are many other possibilities. The sum is subject to chance variability. If the draws turn out one way, the sum is one thing; if they turn out differently, the sum is different too.

At first, this example may seem artificial. But it is just like a turn at Monopoly—you roll a pair of dice, add up the two numbers, and move that many squares. Rolling a die is just like picking a number from the box.

Next, imagine taking 25 draws from the same box

[1] [2] [3] [4] [5] [6]

Of course, the draws must be made with replacement. About how big is their sum going to be? The most direct way to find out is by experiment. We programmed the computer to make the draws.[4] It got 3 on the first draw, 2 on the second, 4 on the third. Here they all are:

3 2 4 6 2   3 5 4 4 2   3 6 4 1 2   4 1 5 5 6   2 2 2 5 5

The sum of these 25 draws is 88.

Of course, if the draws had been different, their sum would have been different. So we had the computer repeat the whole process ten times. Each time, it made 25 draws at random with replacement from the box, and took their sum. The results:

88 84 80 90 83 78 95 94 80 89

Chance variability is easy to see. The first sum is 88, the second drops to 84, the third drops even more to 80. The values range from a low of 78 to a high of 95.

In principle, the sum could have been as small as $25 \times 1 = 25$, or as large as $25 \times 6 = 150$. But in fact, the ten observed values are all between 75 and 100. Would this keep up with more repetitions? Just what is the chance that the sum turns out to be between 75 and 100? That kind of problem will be solved in the next two chapters.

The *sum of the draws* from a box is shorthand for the process discussed in this section:

- Draw tickets at random from a box.
- Add up the numbers on the tickets.[5]

Exercise Set B

1. One hundred draws are made at random with replacement from the box | [1] [2] |. Forty-seven draws turn out to be [1], and the remaining 53 are [2]. How much is the sum?

2. One hundred draws are made at random with replacement from the box | [1] [2] |.
   (a) How small can the sum be? How large?
   (b) How many times do you expect the ticket [1] to turn up? The ticket [2]?
   (c) About how much do you expect the sum to be?

3. One hundred draws are made at random with replacement from the box | [1] [2] [9] |.
   (a) How small can the sum be? How large?
   (b) About how much do you expect the sum to be?

4. One hundred draws will be made at random with replacement from one of the following boxes. Your job is to guess what the sum will be, and you win $1 if you are right to within 10. In each case, what would you guess? Which box is best? Worst?

   (i) | [1] [9] | (ii) | [4] [6] | (iii) | [5] [5] |

5. One ticket will be drawn at random from the box

   | [1] [2] [3] [4] [5] [6] [7] [8] [9] [10] |

What is the chance that it will be 1? That it will be 3 or less? 4 or more?

6. Fifty draws will be made at random with replacement from one of the two boxes shown below. On each draw, you will be paid in dollars the amount shown on the ticket; if a negative number is drawn, that amount will be taken away from you. Which box is better? Or are they the same? Explain.

(i) [−1] [2] (ii) [−1] [−1] [2]

7. You gamble four times at a casino. You win $4 on the first play, lose $2 on the second, win $5 on the third, lose $3 on the fourth. Which of the following calculations tells how much you come out ahead? (More than one may be correct.)
   (i) $\$4 + \$5 - (\$2 + \$3)$
   (ii) $\$4 + (-\$2) + \$5 + (-\$3)$
   (iii) $\$4 + \$2 + \$5 - \$3$
   (iv) $-\$4 + \$2 + \$5 + \$3$

The answers to these exercises are on p. A72.

4. MAKING A BOX MODEL

The object of this section is to make some box models, as practice for later. The sum of the draws from the box turns out to be the key ingredient for many statistical procedures, so keep your eye on the sum. There are three questions to answer when making a box model:

- What numbers go into the box?
- How many of each kind?
- How many draws?

The purpose of a box model is to analyze chance variability, which can be seen in its starkest form at any gambling casino. So this section will focus on box models for roulette. A Nevada roulette wheel has 38 pockets. One is numbered 0, another is numbered 00, and the rest are numbered from 1 through 36. The croupier spins the wheel, and throws a ball onto the wheel. The ball is equally likely to land in any one of the 38 pockets. Before it lands, bets can be placed on the table (figure 3 on the next page).

One bet is *red or black*. Except for 0 and 00, which are colored green, the numbers on the roulette wheel alternate red and black. If you bet a dollar on red, say, and a red number comes up, you get the dollar back together with another dollar in winnings. If a black or green number comes up, the croupier smiles and rakes in your dollar.

Suppose you are at the Golden Nugget in Las Vegas. You have just put a dollar on red, and the croupier spins the wheel. It may seem hard to figure your chances, but a box model will help. What numbers go into the box? You will either win a dollar or lose a dollar. So the tickets must show either +$1 or −$1.

The second question is, how many of each kind? You win if one of the 18 red numbers comes up, and lose if one of the 18 black numbers comes up. But you also lose if 0 or 00 come up. And that is where the house gets its edge. Your chance of winning is only 18 in 38, and the chance of losing is 20 in 38. So there are 18 $\boxed{+\$1}$'s and 20 $\boxed{-\$1}$'s. The box is

$$\boxed{18 \text{ tickets } \boxed{+\$1} \quad 20 \text{ tickets } \boxed{-\$1}}$$

As far as the chances are concerned, betting a dollar on red is just like drawing a ticket at random from the box. The great advantage of the box model is that all

Figure 3. A Nevada roulette table.

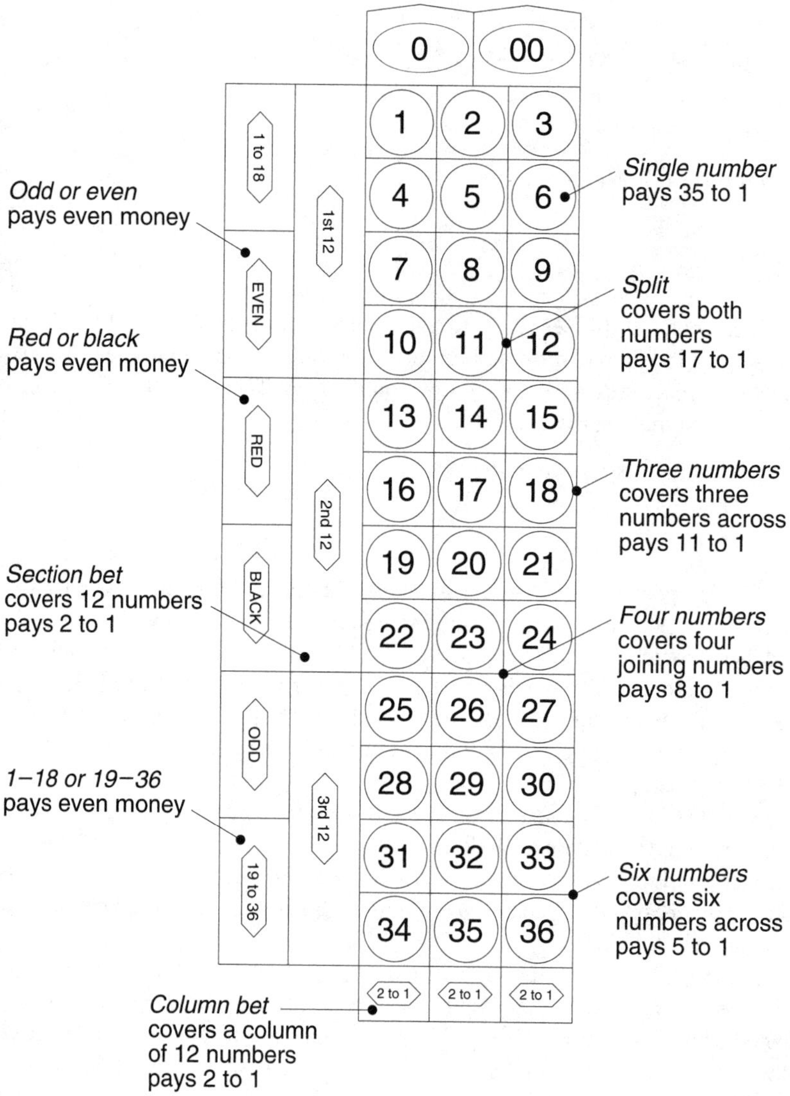

Roulette is a pleasant, relaxed, and highly comfortable way to lose your money.

—JIMMY THE GREEK

the irrelevant details—the wheel, the table, and the croupier's smile—have been stripped away. And you can see the cruel reality: you have 18 tickets, they have 20.

That does one play. But suppose you play roulette ten times, betting a dollar on red each time. What is likely to happen then? You will end up ahead or behind by some amount. This amount is called your *net gain*. The net gain is positive if you come out ahead, negative if you come out behind.

To figure the chances, the net gain has to be connected to the box. On each play, you win or lose some amount. These ten win-lose numbers are like ten draws from the box, made at random with replacement. (Replacing the tickets keeps the chances on each draw the same as the chances for the wheel.) The net gain—the total amount won or lost—is just the sum of these ten win-lose numbers. Your net gain in ten plays is like the sum of ten draws made at random with replacement from the box

| 18 tickets [+$1] 20 tickets [−$1] |

This is our first model, so it is a good idea to look at it more closely. Suppose, for instance, that the ten plays came out this way:

R R R B G R R B B R

(R means red, B means black, and G means green—the house numbers 0 and 00). Table 2 below shows the ten corresponding win-lose numbers, and the net gain.

Table 2. The net gain. This is the cumulative sum of the win-lose numbers.

Plays	R	R	R	B	G	R	R	B	B	R
Win-lose numbers	+1	+1	+1	−1	−1	+1	+1	−1	−1	+1
Net gain	1	2	3	2	1	2	3	2	1	2

Follow the net gain along. When you get a red, the win-lose number is +1, and the net gain goes up by 1. When you get a black or a green, the win-lose number is −1, and the net gain goes down by 1. The net gain is just the sum of the win-lose numbers, and these are like the draws from the box. That is why the net gain is like the sum of draws from the box. This game had a happy ending: you came out ahead $2. To see what would happen if you kept on playing, read the next chapter.

Example 1. If you bet a dollar on a single number at Nevada roulette, and that number comes up, you get the $1 back together with winnings of $35. If any other number comes up, you lose the dollar. Gamblers say that a single number *pays 35 to 1*. Suppose you play roulette 100 times, betting a dollar on the number 17 each time. Your net gain is like the sum of ______ draws made at random with replacement from the box ______. Fill in the blanks.

Solution. What numbers go into the box? To answer this question, think about one play of the game. You put a dollar chip on 17. If the ball drops into the pocket 17, you'll be up $35. If it drops into any other pocket, you'll be down $1. So the box has to contain the tickets [$35] and [−$1].

The tickets in the box show the various amounts that can be won or lost on a single play.

How many tickets of each kind? Keep thinking about one play. You have only 1 chance in 38 of winning, so the chance of drawing [$35] has to be 1 in 38. You have 37 chances in 38 of losing, so the chance of drawing [−$1] has to be 37 in 38. The box is

1 ticket [+$35] 37 tickets [−$1]

The chance of drawing any particular number from the box must equal the chance of winning that amount on a single play. ("Winning" a negative amount is the mathematical equivalent of what most people call losing.)

How many draws? You are playing 100 times. The number of draws has to be 100. Tickets must be replaced after each draw, so as not to change the odds.

The number of draws equals the number of plays.

So, the net gain in 100 plays is like the sum of 100 draws made at random with replacement from the box

1 ticket [+$35] 37 tickets [−$1]

This completes the solution.

Exercise Set C

1. Consider the following three situations.
   (i) A box contains one ticket marked "0" and nine marked "1." A ticket is drawn at random. If it shows "1" you win a panda bear.
   (ii) A box contains ten tickets marked "0" and ninety marked "1." One ticket is drawn at random. If it shows "1" you win the panda.
   (iii) A box contains one ticket marked "0" and nine marked "1." Ten draws are made at random with replacement. If the sum of the draws equals 10, you win the panda.

   Assume you want the panda. Which is better—(i) or (ii)? Or are they the same? What about (i) and (iii)?

2. A gambler is going to play roulette 25 times, putting a dollar on a *split* each time. (A split is two adjacent numbers, like 11 and 12 in figure 3 on p. 282.) If either

number comes up, the gambler gets the dollar back, together with winnings of $17. If neither number comes up, he loses the dollar. So a split pays 17 to 1, and there are 2 chances in 38 to win. The gambler's net gain in the 25 plays is like the sum of 25 draws made from one of the following boxes. Which one, and why?

(i) | [0] [00] 36 tickets numbered [1] through [36] |

(ii) | [$17] [$17] 34 tickets [−$1] |

(iii) | [$17] [$17] 36 tickets [−$1] |

3. In one version of chuck-a-luck, 3 dice are rolled out of a cage. You can bet that all 3 show six. The house pays 36 to 1, and the bettor has 1 chance in 216 to win. Suppose you make this bet 10 times, staking $1 each time. Your net gain is like the sum of ______ draws made at random with replacement from the box ______. Fill in the blanks.

The answers to these exercises are on p. A72.

5. REVIEW EXERCISES

1. A box contains 10,000 tickets: 4,000 [0]'s and 6,000 [1]'s. And 10,000 draws will be made at random with replacement from this box. Which of the following best describes the situation, and why?
   (i) The number of 1's will be 6,000 exactly.
   (ii) The number of 1's is very likely to equal 6,000, but there is also some small chance that it will not be equal to 6,000.
   (iii) The number of 1's is likely to be different from 6,000, but the difference is likely to be small compared to 10,000.

2. Repeat exercise 1 for 10,000 draws made at random without replacement from the box.

3. A gambler loses ten times running at roulette. He decides to continue playing because he is due for a win, by the law of averages. A bystander advises him to quit, on the grounds that his luck is cold. Who is right? Or are both of them wrong?

4. (a) A die will be rolled some number of times, and you win $1 if it shows an ace ([•]) more than 20% of the time. Which is better: 60 rolls, or 600 rolls? Explain.
   (b) As in (a), but you win the dollar if the percentage of aces is more than 15%.
   (c) As in (a), but you win the dollar if the percentage of aces is between 15% and 20%.
   (d) As in (a), but you win the dollar if the percentage of aces is exactly $16\frac{2}{3}\%$.

5. True or false: if a coin is tossed 100 times, it is not likely that the number of heads will be exactly 50, but it is likely that the percentage of heads will be exactly 50%. Explain.

6. According to genetic theory, there is very close to an even chance that both children in a two-child family will be of the same sex. Here are two possibilities.

   (i) 15 couples have two children each. In 10 or more of these families, it will turn out that both children are of the same sex.
   (ii) 30 couples have two children each. In 20 or more of these families, it will turn out that both children are of the same sex.

   Which possibility is more likely, and why?

7. A quiz has 25 multiple choice questions. Each question has 5 possible answers, one of which is correct. A correct answer is worth 4 points, but a point is taken off for each incorrect answer. A student answers all the questions by guessing at random. The score will be like the sum of ______ draws from the box ______. Fill in the first blank with a number and the second with a box of tickets. Explain your answers.

8. A gambler will play roulette 50 times, betting a dollar on four joining numbers each time (like 23, 24, 26, 27 in figure 3, p. 282). If one of these four numbers comes up, she gets the dollar back, together with winnings of $8. If any other number comes up, she loses the dollar. So this bet pays 8 to 1, and there are 4 chances in 38 of winning. Her net gain in 50 plays is like the sum of ______ draws from the box ______. Fill in the blanks; explain.

9. A box contains red and blue marbles; there are more red marbles than blue ones. Marbles are drawn one at a time from the box, at random with replacement. You win a dollar if a red marble is drawn more often than a blue one.[6] There are two choices:

   (A) 100 draws are made from the box.
   (B) 200 draws are made from the box.

   Choose one of the four options below; explain your answer.

   (i) A gives a better chance of winning.
   (ii) B gives a better chance of winning.
   (iii) A and B give the same chance of winning.
   (iv) Can't tell without more information.

10. Two hundred draws will be made at random with replacement from the box [-3 | -2 | -1 | 0 | 1 | 2 | 3].

    (a) If the sum of the 200 numbers drawn is 30, what is their average?
    (b) If the sum of the 200 numbers drawn is −20, what is their average?
    (c) In general, how can you figure the average of the 200 draws, if you are told their sum?
    (d) There are two alternatives:
        (i) winning $1 if the sum of the 200 numbers drawn is between −5 and +5.
        (ii) winning $1 if the average of the 200 numbers drawn is between −0.025 and +0.025.

    Which is better, or are they the same? Explain.

6. SUMMARY

1. There is *chance error* in the number of heads:

number of heads = half the number of tosses + chance error.

The error is likely to be large in absolute terms, but small relative to the number of tosses. That is the *law of averages*.

2. The law of averages can be stated in percentage terms. With a large number of tosses, the percentage of heads is likely to be close to 50%, although it is not likely to be exactly equal to 50%.

3. The law of averages does not work by changing the chances. For example, after a run of heads in coin tossing, a head is still just as likely as a tail.

4. A complicated chance process for generating a number can often be modeled by drawing from a box. The sum of the draws is a key ingredient.

5. The basic questions to ask when making a box model:

- Which numbers go into the box?
- How many of each kind?
- How many draws?

6. For gambling problems in which the same bet is made several times, a box model can be set up as follows:

- The tickets in the box show the amounts that can be won (+) or lost (−) on each play.
- The chance of drawing any particular value from the box equals the chance of winning that amount on a single play.
- The number of draws equals the number of plays.

Then, the *net gain* is like the sum of the draws from the box.

Drawing by Dana Fradon; © 1976 The New Yorker Magazine, Inc.

17

The Expected Value and Standard Error

If you believe in miracles, head for the Keno lounge.
—JIMMY THE GREEK

1. THE EXPECTED VALUE

A chance process is running. It delivers a number. Then another. And another. You are about to drown in random output. But mathematicians have found a little order in this chaos. The numbers delivered by the process vary around the *expected value*, the amounts off being similar in size to the *standard error*. To be more specific, imagine generating a number through the following chance process: count the number of heads in 100 tosses of a coin. You might get 57 heads. This is 7 above the expected value of 50, so the chance error is +7. If you made another 100 tosses, you would get a different number of heads, perhaps 46. The chance error would be −4. A third repetition might generate still another number, say 47; and the chance error would be −3. Your numbers will be off 50 by chance amounts similar in size to the standard error, which is 5 (section 5 below).

The formulas for the expected value and standard error depend on the chance process which generates the number. This chapter deals with the sum of draws from a box, and the formula for the expected value will be introduced with an example: the sum of 100 draws made at random with replacement from the box

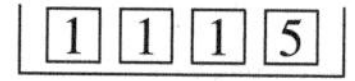

About how large should this sum be? To answer this question, think how the draws should turn out. There are four tickets in the box, so [5] should come up on around one-fourth of the draws, and [1] on three-fourths. With 100 draws, you can expect to get around twenty-five [5]'s, and seventy-five [1]'s. The sum of the draws should be around

$$25 \times 5 + 75 \times 1 = 200.$$

That is the expected value.

The formula for the expected value is a short-cut. It has two ingredients:

- the number of draws;
- the average of the numbers in the box, abbreviated to "average of box."

> The expected value for the sum of draws made at random with replacement from a box equals
>
> (number of draws) × (average of box).

To see the logic behind the formula, go back to the example. The average of the box is

$$\frac{1+1+1+5}{4} = 2.$$

On the average, each draw adds around 2 to the sum. With 100 draws, the sum must be around $100 \times 2 = 200$.

Example 1. Suppose you are going to Las Vegas to play Keno. Your favorite bet is a dollar on a single number. When you win, they give you the dollar back and two dollars more. When you lose, they keep the dollar. There is 1 chance in 4 to win.[1] About how much should you expect to win (or lose) in 100 plays, if you make this bet on each play?

Solution. The first step is to write down a box model. On each play, your net gain either goes up by \$2 or goes down by \$1. There is 1 chance in 4 to go up; there are 3 chances in 4 to go down. So your net gain after 100 plays is like the sum of 100 draws at random with replacement from the box

[[\$2] [−\$1] [−\$1] [−\$1]]

The average of this box is

$$\frac{\$2 - \$1 - \$1 - \$1}{4} = -\$0.25$$

On the average, each play costs you a quarter. In 100 plays, you can expect to lose around \$25. This is the answer. If you continued on, in 1,000 plays you should expect to lose around \$250. The more you play, the more you lose. Perhaps you should look for another game.

Exercise Set A

1. Find the expected value for the sum of 100 draws at random with replacement from the box—

   (a) | 0 | 1 | 1 | 6 | (b) | −2 | −1 | 0 | 2 |

   (c) | −2 | −1 | 3 | (d) | 0 | 1 | 1 |

2. Find the expected number of squares moved on the first play in Monopoly (p. 279).

3. Someone is going to play roulette 100 times, betting a dollar on the number 17 each time. Find the expected value for the net gain. (See pp. 283–284.)

4. You are going to play roulette 100 times, staking \$1 on red-or-black each time. Find the expected value for your net gain. (This bet pays even money, and you have 18 chances in 38 of winning; figure 3 on p. 282.)

5. Repeat exercise 4 for 1,000 plays.

6. A game is *fair* if the expected value for the net gain equals 0: on the average, players neither win nor lose. A generous casino would offer a bit more than \$1 in winnings if a player staked \$1 on red-and-black in roulette and won. How much should they pay to make it a fair game? (Hint: Let x stand for what they should pay. The box has 18 tickets | x | and 20 tickets | −\$1 |. Write down the formula for the expected value in terms of x and set it equal to 0.)

7. If an Adventurer at the Game of the Royal Oak staked 1 pound on a point and won, how much should the Master of the Ball have paid him, for the Game to be fair? (The rules are explained in exercise 6 on pp. 250–251.)

The answers to these exercises are on pp. A72–73.

2. THE STANDARD ERROR

Suppose 25 draws are made at random with replacement from the box

| 0 | 2 | 3 | 4 | 6 |

(There is nothing special about the numbers in the box; they were chosen to make later calculations come out evenly.) Each of the five tickets should appear on about one-fifth of the draws, that is, 5 times. So the sum should be around

$$5 \times 0 + 5 \times 2 + 5 \times 3 + 5 \times 4 + 5 \times 6 = 75.$$

That is the expected value for the sum. Of course, each ticket won't appear on exactly one-fifth of the draws, just as Kerrich didn't get heads on exactly half the tosses. The sum will be off the expected value by a chance error:

$$\text{sum} = \text{expected value} + \text{chance error}.$$

The chance error is the amount above (+) or below (−) the expected value. For example, if the sum is 70, the chance error is −5.

How big is the chance error likely to be? The answer is given by the *standard error*, usually abbreviated to SE.

> A sum is likely to be around its expected value, but to be off by a chance error similar in size to the standard error.

There is a formula to use in computing the SE for a sum of draws made at random with replacement from a box. It is called the square root law, because it involves the square root of the number of draws. The statistical procedures in the rest of the book depend on this formula.[2]

> *The square root law.* When drawing at random with replacement from a box of numbered tickets, the standard error for the sum of the draws is
>
> $$\sqrt{\text{number of draws}} \times (\text{SD of box}).$$

The formula has two ingredients: the square root of the number of draws, and the SD of the list of numbers in the box (abbreviated to "SD of the box"). The SD measures the spread among the numbers in the box. If there is a lot of spread in the box, the SD is big, and it is hard to predict how the draws will turn out. So the standard error must be big too. Now for the number of draws. The sum of two draws is more variable than a single draw. The sum of 100 draws is still more variable. Each draw adds some extra variability to the sum, because you don't know how it is going to turn out. As the number of draws goes up, the sum gets harder to predict, the chance errors get bigger, and so does the standard error. However, the standard error goes up slowly, by a factor equal to the square root of the number of draws. For instance, the sum of 100 draws is only $\sqrt{100} = 10$ times as variable as a single draw.

The SD and the SE are different.[3] The SD applies to spread in lists of numbers. It is worked out using the method explained on p. 71. By contrast, the SE applies to chance variability—for instance, in the sum of the draws.

The SD is for a list

1 2 3 4 5 6

The SE is for a chance process

At the beginning of the section, we looked at the sum of 25 draws made at random with replacement from the box

[0] [2] [3] [4] [6]

The expected value for this sum is 75. The sum will be around 75, but will be off by a chance error. How big is the chance error likely to be? To find out, calculate the standard error. The average of the numbers in the box is 3. The deviations from the average are

$$-3 \quad -1 \quad 0 \quad 1 \quad 3$$

The SD of the box is

$$\sqrt{\frac{(-3)^2 + (-1)^2 + 0^2 + 1^2 + 3^2}{5}} = \sqrt{\frac{9+1+0+1+9}{5}}$$
$$= \sqrt{\frac{20}{5}} = 2.$$

This measures the variability in the box. According to the square root law, the sum of 25 draws is more variable, by the factor $\sqrt{25} = 5$. The SE for the sum of 25 draws is $5 \times 2 = 10$. In other words, the likely size of the chance error is 10. And the sum of the draws should be around 75, give or take 10 or so. In general, the sum is likely to be around its expected value, give or take a standard error or so.

To show what this means empirically, we had the computer programmed to draw 25 times at random with replacement from the box | [0] [2] [3] [4] [6] |. It got

00440 43262 20262 64263 03640

The sum of these 25 draws is 71. This is 4 below the expected value, so the chance error is −4. The computer drew another 25 times and took the sum, getting 76. The chance error was +1. The third sum was 86, with a chance error of +11. In fact, we had the computer generate 100 sums, shown in table 1. These numbers are all around 75, the expected value. They are off by chance errors similar in size to 10, the standard error.

> The sum of the draws is likely to be around ______, give or take ______ or so. The expected value for the sum fills in the first blank. The SE for the sum fills in the second blank.

Some terminology: the number 71 in table 1 is an *observed value* for the sum of the draws; the 76 is another observed value. All told, the table has 100 observed values for the sum. These observed values differ from the expected value of 75. The difference is chance error. For example, the chance error in 71 is −4, because $71 - 75 = -4$. The chance error in 76 is +1, because $76 - 75 = 1$. And so forth.

The observed values in table 1 show remarkably little spread around the expected value. In principle, they could be as small as 0, or as large as $25 \times 6 = 150$. However, all but one of them are between 50 and 100, that is, within 2.5 SEs of the expected value.

> Observed values are rarely more than 2 or 3 SEs away from the expected value.

Table 1. Computer simulation: the sum of 25 draws made at random with replacement from the box | 0 2 3 4 6 |.

Repetition	*Sum*	*Repetition*	*Sum*	*Repetition*	*Sum*	*Repetition*	*Sum*	*Repetition*	*Sum*
1	71	21	80	41	64	61	64	81	60
2	76	22	77	42	65	62	70	82	67
3	86	23	70	43	88	63	65	83	82
4	78	24	71	44	77	64	78	84	85
5	88	25	79	45	82	65	64	85	77
6	67	26	56	46	73	66	77	86	79
7	76	27	79	47	92	67	81	87	82
8	59	28	65	48	75	68	72	88	88
9	59	29	72	49	57	69	66	89	76
10	75	30	73	50	68	70	74	90	75
11	76	31	78	51	80	71	70	91	77
12	66	32	75	52	70	72	76	92	66
13	76	33	89	53	90	73	80	93	69
14	84	34	77	54	76	74	70	94	86
15	58	35	81	55	77	75	56	95	81
16	60	36	68	56	65	76	49	96	90
17	79	37	70	57	67	77	60	97	74
18	78	38	86	58	60	78	98	98	72
19	66	39	70	59	74	79	81	99	57
20	71	40	71	60	83	80	72	100	62

Exercise Set B

1. One hundred draws are going to be made at random with replacement from the box | 1 2 3 4 5 6 7 |.
   (a) Find the expected value and standard error for the sum.
   (b) The sum of the draws will be around ______, give or take ______ or so.
   (c) Suppose you had to guess what the sum was going to be. What would you guess? Would you expect to be off by around 2, 4, or 20?

2. You gamble 100 times on the toss of a coin. If it lands heads, you win \$1. If it lands tails, you lose \$1. Your net gain will be around ______, give or take ______ or so. Fill in the blanks, using the options

   −\$10 −\$5 \$0 +\$5 +\$10

3. The expected value for a sum is 50, with an SE of 5. The chance process generating the sum is repeated ten times. Which is the sequence of observed values?
   (i) 51, 57, 48, 52, 57, 61, 58, 41, 53, 48
   (ii) 51, 49, 50, 52, 48, 47, 53, 50, 49, 47
   (iii) 45, 50, 55, 45, 50, 55, 45, 50, 55, 45

4. Fifty draws are made at random with replacement from the box | 1 2 3 4 5 |; the sum of the draws turns out to be 157. The expected value for the sum is ______, the observed value is ______, the chance error is ______, and the standard error is ______. Fill in the blanks, and explain briefly.

5. Tickets are drawn at random with replacement from a box of numbered tickets. The sum of 25 draws has expected value equal to 50, and the SE is 10. If possible, find the expected value and SE for the sum of 100 draws. Or do you need more information?

6. One hundred draws are going to be made at random with replacement from the box | [0] [2] [3] [4] [6] |. True or false and explain.

   (a) The expected value for the sum of the draws is 300.
   (b) The expected value for the sum of the draws is 300, give or take 20 or so.
   (c) The sum of the draws will be 300.
   (d) The sum of the draws will be around 300, give or take 20 or so.

7. In the simulation for table 1 (p. 293), if the computer kept on running, do you think it would eventually generate a sum more than 3 SEs away from the expected value? Explain.

The answers to these exercises are on p. A73.

3. USING THE NORMAL CURVE

A large number of draws will be made at random with replacement from a box. What is the chance that the sum of the draws will be in a given range? Mathematicians discovered the normal curve while trying to solve problems of this kind. The logic behind the curve will be discussed in the next chapter. The object of this section is only to sketch the method, which applies whenever the number of draws is reasonably large. Basically, it is a matter of converting to standard units (using the expected value and standard error) and then working out areas under the curve, just as in chapter 5.

Now for an example. Suppose the computer is programmed to take the sum of 25 draws made at random with replacement from the magic box

| [0] [2] [3] [4] [6] |

It prints out the result, repeating the process over and over again. About what percentage of the observed values should be between between 50 and 100?

Each sum will be somewhere on the horizontal axis between 0 and $25 \times 6 = 150$.

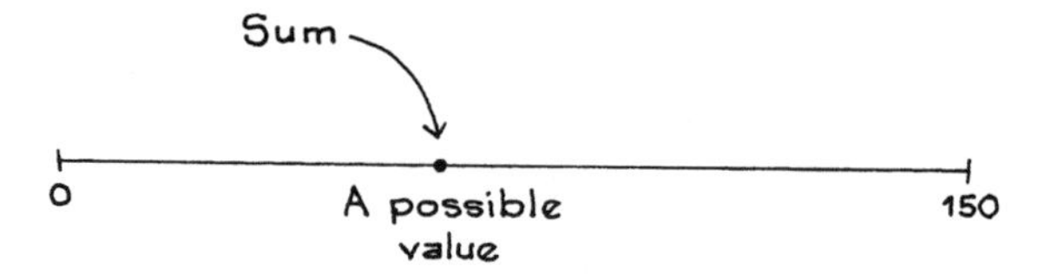

The problem is asking for the chance that the sum will turn out to be between 50 and 100.

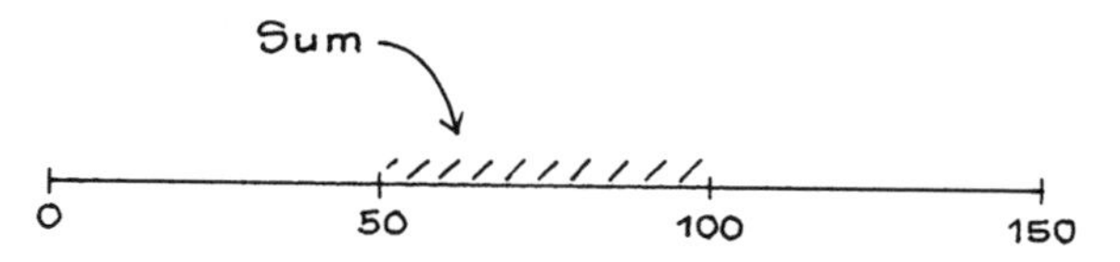

To find the chance, convert to standard units and use the normal curve. Standard units say how many SEs a number is away from the expected value.[4] In the example, 100 becomes 2.5 in standard units. The reason: the expected value for the sum is 75 and the SE is 10, so 100 is 2.5 SEs above the expected value. Similarly, 50 becomes −2.5.

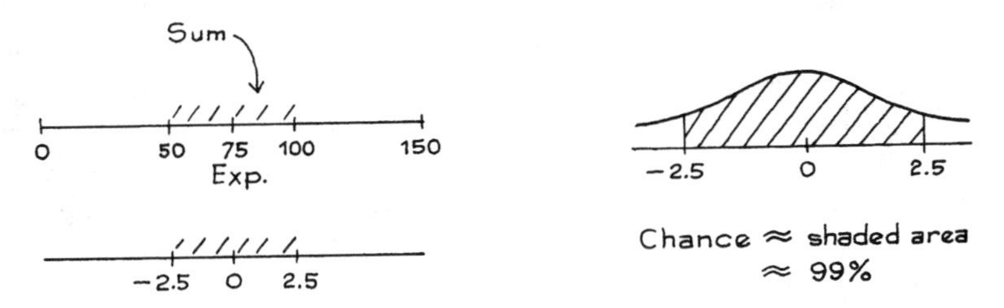

The interval from 50 to 100 is the interval within 2.5 SEs of the expected value, so the sum should be there about 99% of the time.

That finishes the calculation. Now for some data. Table 1 above reported 100 observed values for the sum: about 99 of them should be in the interval from 50 to 100, and in fact 99 of them are. To take some less extreme ranges, about 68% of the observed values should be in the interval from $75 - 10$ to $75 + 10$. In fact, 73 are. Finally, about 95% of the observed values in table 1 should be in the range 75 ± 20, and 98 of them are. The theory looks pretty good. (Ranges include endpoints; ± is read "plus-or-minus.")

Example 2. In a month, there are 10,000 independent plays on a roulette wheel in a certain casino. To keep things simple, suppose the gamblers only stake \$1 on red at each play. Estimate the chance that the house will win more than \$250 from these plays.[5] (Red-or-black pays even money, and the house has 20 chances in 38 to win.)

Solution. The problem asks for the chance that the net gain of the house will be more than \$250.

Net gain

\$250

The box model is the first thing. The box is

20 tickets [+\$1] 18 tickets [−\$1]

The net gain for the house is like the sum of 10,000 draws from this box.

The expected value for the net gain is the number of draws times the average of the numbers in the box. The average is

$$\frac{\overbrace{\$1 + \cdots + \$1}^{20 \text{ tickets}} \overbrace{- \$1 - \cdots - \$1}^{18 \text{ tickets}}}{38} = \frac{\$20 - \$18}{38} = \frac{\$2}{38} \approx \$0.05$$

On the average, each draw adds around \$0.05 to the sum. The sum of 10,000 draws has an expected value of 10,000 $\times$ \$0.05 = \$500. The house averages about a nickel on each play, so in 10,000 plays it can expect to win around \$500. (The gambler and the house are on opposite sides of the box: 20 tickets are good for the house, and 18 are good for the gambler; see pp. 281–283.)

Finding the SE for the net gain comes next. This requires the SD of the numbers in the box. The deviations from average are all just about \$1, because the average is close to \$0. So the SD of the box is about \$1. This \$1 measures the variability in the box. According to the square root law, the sum of 10,000 draws is more variable, by the factor $\sqrt{10{,}000} = 100$. The SE for the sum of 10,000 draws is 100 $\times$ \$1 = \$100. The house can expect to win around \$500, give or take \$100 or so.

Now the normal curve can be used.

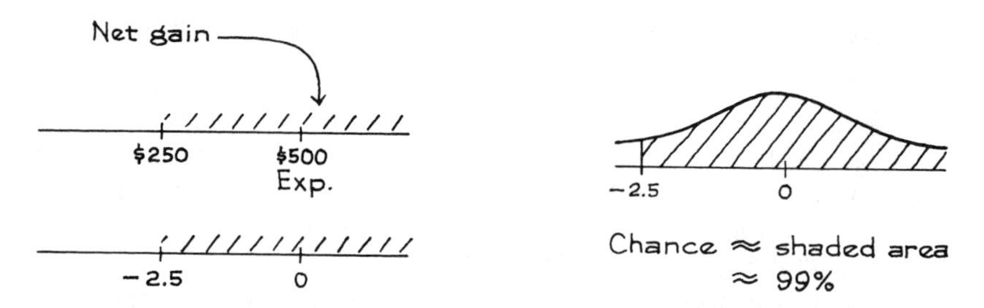

This completes the solution. The key idea: the net gain is like the sum of the draws from a box; that provided a logical basis for the square root law.

The house has about a 99% chance to win more than \$250. This may not seem like much, but you have to remember that the house owns many wheels, there often is a crowd of gamblers playing on each spin of each wheel, and a lot of bets are over a dollar. The house can expect to win about 5% of the money that crosses the table, and the square root law virtually eliminates the risk. For instance, suppose the house runs 25 wheels. To be very conservative, suppose each wheel operates under the conditions of example 2. With these assumptions, the casino's expected winnings go up by a full factor of 25, to 25 $\times$ \$500 = \$12,500. But their standard error only goes up by the factor $\sqrt{25} = 5$, to \$500. Now the casino can be virtually certain—99%—of winning at least \$11,000. For the casino, roulette is a volume business, just like groceries are for Safeway.

Exercise Set C

1. One hundred draws will be made at random with replacement from the box | [1] [1] [2] [2] [2] [4] |.

   (a) The smallest the sum can be is ______, the largest is ______.
   (b) The sum of the draws will be around ______, give or take ______ or so.
   (c) The chance that the sum will be bigger than 250 is almost ______%.

2. One hundred draws will be made at random with replacement from the box | 1 3 3 9 |.
   (a) How large can the sum be? How small?
   (b) How likely is the sum to be in the range from 370 to 430?

3. You can draw either 10 times or 100 times at random with replacement from the box | −1 1 |. How many times should you draw—
   (a) To win $1 when the sum is 5 or more, and nothing otherwise?
   (b) To win $1 when the sum is −5 or less, and nothing otherwise?
   (c) To win $1 when the sum is between −5 and 5, and nothing otherwise?

   No calculations are needed, but explain your reasoning.

4. There are two options:
   (i) One hundred draws will be made at random with replacement from the box | 1 1 5 7 8 8 |.
   (ii) Twenty-five draws will be made at random with replacement from the box | 14 17 21 23 25 |.

   Which is better, if the payoff is—
   (a) $1 when the sum is 550 or more, and nothing otherwise?
   (b) $1 when the sum is 450 or less, and nothing otherwise?
   (c) $1 when the sum is between 450 and 550, and nothing otherwise?

5. Suppose that in one week at a certain casino, there are 25,000 independent plays at roulette. On each play, the gamblers stake $1 on red. Is the chance that the casino will win more than $1,000 from these 25,000 plays closest to 2%, 50%, or 98%? Explain briefly.

6. Suppose that one person stakes $25,000 on one play at red-or-black in roulette. Is the chance that the casino will win more than $1,000 from this play closest to 2%, 50%, or 98%? Explain briefly.

7. A gambler plays once at roulette, staking $1,000 on each number (including 0 and 00). So this person has staked $38,000 in all. What will happen? Explain briefly.

8. A box contains 10 tickets. Each ticket is marked with a whole number between −5 and 5. The numbers are not all the same; their average equals 0. There are two choices:
   (A) 100 draws are made from the box, and you win $1 if the sum is between −15 and 15.
   (B) 200 draws are made from the box, and you win $1 if the sum is between −30 and 30.

   Choose one of the four options below; explain your answer.[6]
   (i) A gives a better chance of winning.
   (ii) B gives a better chance of winning.
   (iii) A and B give the same chance of winning.
   (iv) Can't tell without the SD of the box.

The answers to these exercises are on p. A74.

4. A SHORT-CUT

Finding SDs can be painful, but there is a short-cut for lists with only two different numbers, a big one and a small one.[7] (Each number can be repeated several times.)

> When a list has only two different numbers ("big" and "small"), the SD equals
>
> $$\left(\text{big number} - \text{small number}\right) \times \sqrt{\text{fraction with big number} \times \text{fraction with small number}}$$

For example, take the list 5, 1, 1, 1. The short-cut can be used because there are only two different numbers, 5 and 1. The SD is

$$(5 - 1) \times \sqrt{\frac{1}{4} \times \frac{3}{4}} \approx 1.73$$

The short-cut involves much less arithmetic than finding the root-mean-square of the deviations from average (p. 71), and gives exactly the same answer. The short-cut is helpful in many gambling problems (and in other contexts too).

Example 3. A gambler plays roulette 100 times, staking $1 on the number 10 each time. The bet pays 35 to 1, and the gambler has 1 chance in 38 to win. Fill in the blanks: the gambler will win $______, give or take $______ or so.

Solution. The first thing to do is to make a box model for the net gain. (See example 1 on pp. 283–284.) The gambler's net gain is like the sum of 100 draws made at random with replacement from

| 1 ticket [+$35] 37 tickets [−$1] |

What is the expected net gain? This is 100 times the average of the box. The average of the numbers in the box is their total, divided by 38. The winning ticket contributes $35 to the total, while the 37 losing tickets take away $37 in all. So the average is

$$\frac{\$35 - \$37}{38} = \frac{-\$2}{38} \approx -\$0.05$$

In 100 plays, the expected net gain is

$$100 \times (-\$0.05) = -\$5$$

In other words, the gambler expects to lose about $5 in 100 plays.

The next step is to find the SE for the sum of the draws: this is $\sqrt{100}$ times the SD of the box. The short-cut can be used, and the SD of the box equals

$$[\$35 - (-\$1)] \times \sqrt{\frac{1}{38} \times \frac{37}{38}} \approx \$36 \times 0.16 \approx \$5.76$$

The SE for the sum of the draws is $\sqrt{100} \times \$5.76 \approx \$58$.

The gambler will lose about \$5, give or take \$58 or so. This completes the solution. The large SE gives the gambler a reasonable chance of winning, and that is the attraction. Of course, on average the gambler loses; and the SE also means that the gambler can lose a bundle.

Exercise Set D

1. Does the formula give the SD of the list? Explain.

List	*Formula*
(a) 7, 7, 7, −2, −2	$5 \times \sqrt{3/5 \times 2/5}$
(b) 0, 0, 0, 0, 5	$5 \times \sqrt{1/5 \times 4/5}$
(c) 0, 0, 1	$\sqrt{2/3 \times 1/3}$
(d) 2, 2, 3, 4, 4, 4	$2 \times \sqrt{1/6 \times 2/6 \times 3/6}$

2. Suppose a gambler bets a dollar on a single number at Keno (example 1 on p. 289). In 100 plays, the gambler's net gain will be \$______, give or take \$______ or so.

3. At Nevada roulette tables, the "house special" is a bet on the numbers 0, 00, 1, 2, 3. The bet pays 6 to 1, and there are 5 chances in 38 to win.
   (a) For all other bets at Nevada roulette tables, the house expects to make about 5 cents out of every dollar put on the table. How much does it expect to make per dollar on the house special?
   (b) Someone plays roulette 100 times, betting a dollar on the house special each time. Estimate the chance that this person comes out ahead.

4. A gambler plays roulette 100 times. There are two possibilities:
   (i) Betting \$1 on a section each time (see figure 3 on p. 282).
   (ii) Betting \$1 on red each time.

   A section bet pays 2 to 1, and there are 12 chances in 38 to win. Red pays even money, and there are 18 chances in 38 to win. True or false, and explain:
   (a) The chance of coming out ahead is the same with (i) and (ii).
   (b) The chance of winning more than \$10 is bigger with (i).
   (c) The chance of losing more than \$10 is bigger with (i).

The answers to these exercises are on pp. A74–75.

5. CLASSIFYING AND COUNTING

Some chance processes involve counting. The square root law can be used to get the standard error for a count, but the box model has to be set up correctly. The next example will show how to do this.

Example 4. A die is rolled 60 times.

(a) The total number of spots should be around ______, give or take ______ or so.

(b) The number of 6's should be around ______, give or take ______ or so.

By way of illustration, table 2 shows the results of throwing a die 60 times: the first throw was a 4, the second was a 5, and so on.

Table 2. Sixty throws of a die.

4 5 5 2 4	5 3 2 6 3	5 4 6 2 6	4 4 2 5 6
1 5 3 1 2	2 1 2 5 3	3 6 6 1 1	5 1 6 1 2
4 4 2 1 4	4 5 2 6 3	2 4 6 1 6	4 6 1 5 2

Solution. *Part (a)* is familiar. It involves adding. Each throw contributes some number of spots, and we add these numbers up. The total number of spots in 60 throws of the die is like the sum of 60 draws from the box

$$\boxed{1}\ \boxed{2}\ \boxed{3}\ \boxed{4}\ \boxed{5}\ \boxed{6}$$

The average of this box is 3.5 and the SD is 1.71. The expected value for the sum is $60 \times 3.5 = 210$; the SE for the sum is $\sqrt{60} \times 1.71 \approx 13$. The total number of spots will be around 210, give or take 13 or so. In fact, the sum of the numbers in table 2 is 212. The sum was off its expected value by around one-sixth of an SE.

Part (b). Filling in the first blank is easy. Each of the six faces should come up on about one-sixth of the throws, so the expected value for the number of 6's is $60 \times 1/6 = 10$. The second blank is harder. We need a new kind of box because the sum of the draws from $\boxed{1}\ \boxed{2}\ \boxed{3}\ \boxed{4}\ \boxed{5}\ \boxed{6}$ is no longer relevant. Instead of being added, each throw of the die is classified: is it a 6, or not? (There are only two classes here, 6's on one hand, everything else on the other.) Then, the number of 6's is counted up.

The point to notice is that on each throw, the number of 6's either goes up by 1, or stays the same:

- 1 is added to the count if the throw is 6;
- 0 is added to the count if the throw is anything else.

The count has 1 chance in 6 to go up by one, and 5 chances in 6 to stay the same. Therefore, on each draw, the sum must have 1 chance in 6 to go up by one, and 5 chances in 6 to stay the same. The right box to use is

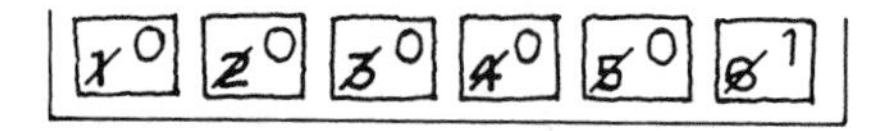

As far as the chances are concerned, the number of 6's in 60 throws of the die is just like the sum of 60 draws from the new box. This puts us in a position to use the square root law.

The new box has five $\boxed{0}$'s and a $\boxed{1}$. The SD is $\sqrt{1/6 \times 5/6} \approx 0.37$, by the short-cut method. And the SE for the sum of the draws is $\sqrt{60} \times 0.37 \approx 3$. In 60 throws of a die, the number of 6's will be around 10, give or take 3 or so. In fact, in table 2 there were eleven 6's. The observed number of 6's was off its expected value by a third of an SE. This completes the example. It's the old story, for a new box.

This example makes a general point. Although they may look quite different, many problems about chance processes can be solved in the same way. In these problems, some tickets are drawn at random from a box. An operation is performed on the draws, and the problem asks for the chance that the result will be in a given interval. In this chapter, there are two possible operations on the draws:

- adding,
- classifying and counting.

The message is that both operations can be treated the same way—provided you change the box.

> If you have to classify and count the draws, put 0's and 1's on the tickets. Mark 1 on the tickets that count for you, 0 on the others.

For adding up the draws, the box is

$\boxed{1}\,\boxed{2}\,\boxed{3}\,\boxed{4}\,\boxed{5}\,\boxed{6}$

For counting 6's, the box is

$\boxed{0}\,\boxed{0}\,\boxed{0}\,\boxed{0}\,\boxed{0}\,\boxed{1}$

Remember to change the tickets!

Example 5. A coin will be tossed 100 times. Find the expected value and standard error for the number of heads. Estimate the chance of getting between 40 and 60 heads.

Solution. The first thing is to make a box model. The problem involves classifying the tosses as heads or tails, and then counting the number of heads. So there should be only 0's and 1's in the box. The chances are 50–50 for heads, so the box should be $\boxed{0}\,\boxed{1}$. The number of heads in 100 tosses of a coin is like the sum of 100 draws made at random with replacement from the box $\boxed{0}\,\boxed{1}$. (The coin is even simpler than the die in example 4: each toss either pushes the number of heads up by 1 or leaves it alone, with a 50–50 chance; likewise, each draw from the box either pushes the sum up by 1 or leaves it alone, with the same 50–50 chance.) This completes the model.

Since the number of heads is like the sum of the draws, the square root law can be used. The SD of the box is 1/2. So the SE for the sum of 100 draws is $\sqrt{100} \times 1/2 = 5$. The number of heads will be around 50, give or take 5 or so.

The range from 40 to 60 heads represents the expected value, give or take 2 SEs. And the chance is around 95%. This completes the solution.

To interpret this 95% chance, imagine counting the number of heads in 100 tosses of a coin. You might get 44 heads. Toss again: you might get 54 heads. A third time, the number would change once more, perhaps to 48 heads. And so on. In the long run, about 95% of these counts would come out in the range from 40 to 60. John Kerrich actually did this experiment. Table 3 shows the results, with Kerrich's 10,000 tosses broken down into successive groups of one hundred. In fact, 95 out of 100 groups had 40 to 60 heads (inclusive). The theory looks good.

Table 3. Kerrich's coin tossing experiment, showing the number of heads he got in each successive group of 100 tosses.

Group of 100 tosses	*No. of heads*	*Group of 100 tosses*	*No. of heads*	*Group of 100 tosses*	*No. of heads*	*Group of 100 tosses*	*No. of heads*
1–100	44	2,501–2,600	44	5,001–5,100	42	7,501–7,600	48
101–200	54	2,601–2,700	34	5,101–5,200	68	7,601–7,700	43
201–300	48	2,701–2,800	59	5,201–5,300	45	7,701–7,800	58
301–400	53	2,801–2,900	50	5,301–5,400	37	7,801–7,900	57
401–500	56	2,901–3,000	51	5,401–5,500	47	7,901–8,000	48
501–600	57	3,001–3,100	51	5,501–5,600	52	8,001–8,100	45
601–700	56	3,101–3,200	48	5,601–5,700	51	8,101–8,200	50
701–800	45	3,201–3,300	56	5,701–5,800	49	8,201–8,300	53
801–900	45	3,301–3,400	57	5,801–5,900	48	8,301–8,400	46
901–1,000	44	3,401–3,500	50	5,901–6,000	37	8,401–8,500	56
1,001–1,100	40	3,501–3,600	54	6,001–6,100	47	8,501–8,600	58
1,101–1,200	54	3,601–3,700	47	6,101–6,200	52	8,601–8,700	54
1,201–1,300	53	3,701–3,800	53	6,201–6,300	45	8,701–8,800	49
1,301–1,400	55	3,801–3,900	50	6,301–6,400	48	8,801–8,900	48
1,401–1,500	52	3,901–4,000	53	6,401–6,500	44	8,901–9,000	45
1,501–1,600	54	4,001–4,100	52	6,501–6,600	51	9,001–9,100	55
1,601–1,700	58	4,101–4,200	54	6,601–6,700	55	9,101–9,200	51
1,701–1,800	50	4,201–4,300	55	6,701–6,800	53	9,201–9,300	48
1,801–1,900	53	4,301–4,400	52	6,801–6,900	52	9,301–9,400	56
1,901–2,000	42	4,401–4,500	51	6,901–7,000	60	9,401–9,500	55
2,001–2,100	56	4,501–4,600	53	7,001–7,100	50	9,501–9,600	55
2,101–2,200	53	4,601–4,700	54	7,101–7,200	57	9,601–9,700	50
2,201–2,300	53	4,701–4,800	47	7,201–7,300	49	9,701–9,800	48
2,301–2,400	45	4,801–4,900	42	7,301–7,400	46	9,801–9,900	59
2,401–2,500	52	4,901–5,000	44	7,401–7,500	62	9,901–10,000	52

It is time to connect the square root law and the law of averages. Suppose a coin is tossed a large number of times. Then heads will come up on about half the tosses:

$$\text{number of heads} = \text{half the number of tosses} + \text{chance error.}$$

How big is the chance error likely to be? At first, Kerrich's assistant thought it would be very small. The record showed him to be wrong. As Kerrich kept tossing the coin, the chance error grew in absolute terms but shrank relative to the number of tosses, just as the mathematics predicts. (See figures 1 and 2, pp. 275–276.)

According to the square root law, the likely size of the chance error is $\sqrt{\text{number of tosses}} \times 1/2$. For instance, with 10,000 tosses the standard error is $\sqrt{10{,}000} \times 1/2 = 50$. When the number of tosses goes up to 1,000,000, the standard error goes up too, but only to 500—because of the square root. As the number of tosses goes up, the SE for the number of heads gets bigger and bigger in absolute terms, but smaller and smaller relative to the number of tosses. That is why the percentage of heads gets closer and closer to 50%. The square root law is the mathematical explanation for the law of averages.

Exercise Set E

1. A coin is tossed 16 times.
   (a) The number of heads is like the sum of 16 draws made at random with replacement from one of the following boxes. Which one and why?

   (i) | head | tail | (ii) | 0 | 1 | (iii) | 0 | 1 | 1 |

   (b) The number of heads will be around ______, give or take ______ or so.

2. One hundred draws are made at random with replacement from | 1 | 2 | 3 | 4 | 5 |. What is the chance of getting between 8 and 32 tickets marked "5"?

3. According to the simplest genetic model, the sex of a child is determined at random, as if by drawing a ticket at random from the box

   | male | female |

   What is the chance that of the next 2,500 births (not counting twins or other multiple births), more than 1,275 will be females?

4. This exercise and the next are based on Kerrich's coin-tossing experiment (table 3, p. 302). For example, in tosses 1–100, the observed number of heads was 44, the expected number was 50, so the chance error was $44 - 50 = -6$. Fill in the blanks.

Group of 100 tosses	*Observed value*	*Expected value*	*Chance error*	*Standard error*
1–100	44	50	−6	—
101–200	54	50	—	—
201–300	48	—	—	—
301–400	—	—	—	—

5. How many of the counts in table 3 on p. 302 should be in the range 45 to 55? How many are? (Endpoints included.)

6. (a) A coin is tossed 10,000 times. What is the chance that the number of heads will be in the range 4,850 to 5,150?
   (b) A coin is tossed 1,000,000 times. What is the chance that the number of heads will be in the range 498,500 to 501,500?

7. Fifty draws are made at random with replacement from the box | [0] [0] [1] [1] [1] |; there are 33 [1]'s among the draws. The expected number of [1]'s is ______, the observed number is ______, the chance error is ______, and the SE is ______.

8. A computer program is written to do the following job. There is a box with ten blank tickets. You tell the program what numbers to write on the tickets, and how many draws to make. Then, the computer will draw that many tickets at random with replacement from the box, add them up, and print out the sum—but not the draws. This program does not know anything about coin tossing. Still, you can use it to simulate the number of heads in 1,000 tosses of a coin. How?

9. A die is rolled 100 times. Someone figures the expected number of aces as $100 \times 1/6 = 16.67$, and the SE as $\sqrt{100} \times \sqrt{1/6 \times 5/6} \approx 3.73$. (An ace is [•].) Is this right? Answer yes or no, and explain.

The answers to these exercises are on p. A75.

6. REVIEW EXERCISES

1. One hundred draws will be made at random with replacement from the box | [1] [6] [7] [9] [9] [10] |.

   (a) How small can the sum of the draws be? How large?
   (b) The sum is between 650 and 750 with a chance of about

   1% 10% 50% 90% 99%

   Explain.

2. A gambler plays roulette 100 times, betting a dollar on a column each time. The bet pays 2 to 1, and there are 12 chances in 38 to win. Fill in the blanks; show work.

   (a) In 100 plays, the gambler's net gain will be around $______, give or take $______ or so.
   (b) In 100 plays, the gambler should win ______ times, give or take ______ or so.
   (c) How does the column bet compare with betting on a single number at Keno (example 1 on p. 289)?

3. Match the lists with the SDs. Explain your reasoning

(a) $1, -2, -2$	(i) $\sqrt{1/3 \times 2/3}$
(b) $15, 15, 16$	(ii) $2 \times \sqrt{1/3 \times 2/3}$
(c) $-1, -1, -1, 1$	(iii) $3 \times \sqrt{1/3 \times 2/3}$
(d) $0, 0, 0, 1$	(iv) $\sqrt{1/4 \times 3/4}$
(e) $0, 0, 2$	(v) $2 \times \sqrt{1/4 \times 3/4}$

4. A large group of people get together. Each one rolls a die 180 times, and counts the number of [•]'s. About what percentage of these people should get counts in the range 15 to 45?

5. A die will be thrown some number of times, and the object is to guess the total number of spots. There is a one-dollar penalty for each spot that the guess is off. For instance, if you guess 200 and the total is 215, you lose $15. Which do you prefer: 50 throws, or 100? Explain.

6. One hundred draws are made at random with replacement from the box | [1] [1] [2] [3] |. The draws come out as follows: 45 [1]'s, 23 [2]'s, 32 [3]'s. For each number below, find the phrase which describes it.

Number	*Phrase*
12	observed value for the sum of the draws
45	observed value for the number of 3's
187	observed value for the number of 1's
25	expected value for the sum of the draws
50	expected value for the number of 3's
175	expected value for the number of 1's
5	chance error in the sum of the draws
32	standard error for the number of 1's

7. One hundred draws are made at random with replacement from the box | [1] [2] [3] [4] [5] [6] |.
   (a) If the sum of the draws is 321, what is their average?
   (b) If the average of the draws is 3.78, what is the sum?
   (c) Estimate the chance that the average of the draws is between 3 and 4.

8. A coin is tossed 100 times.
   (a) The difference "number of heads – number of tails" is like the sum of 100 draws from one of the following boxes. Which one, and why?

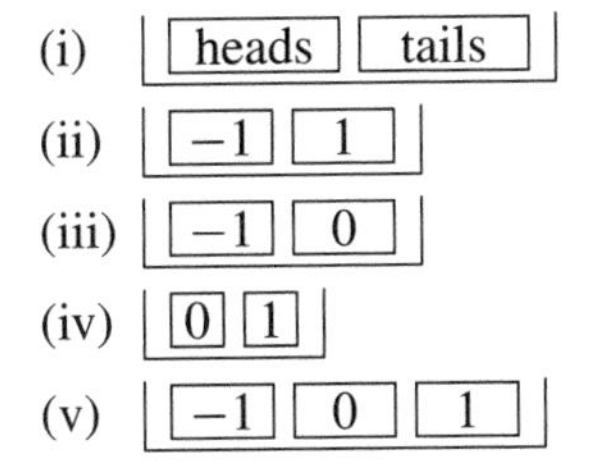

   (b) Find the expected value and standard error for the difference.

9. A gambler plays roulette 1,000 times. There are two possibilities:
   (i) Betting $1 on a column each time.
   (ii) Betting $1 on a number each time.

   A column pays 2 to 1, and there are 12 chances in 38 to win; a number pays 35 to 1, and there is 1 chance in 38 to win. True or false and explain:

(a) The chance of coming out ahead is the same with (i) and (ii).
(b) The chance of winning more than $100 is bigger with (ii).
(c) The chance of losing more than $100 is bigger with (ii).

10. A box contains numbered tickets. Draws are made at random with replacement from the box. Below are three statements about this particular box; (i) and (ii) are true. Is (iii) true or false? Explain.
    (i) For a certain number of draws, the expected value for the sum of the draws equals 400.
    (ii) For that same number of draws, there is about a 75% chance that the sum will be between 350 and 450.
    (iii) For twice that number of draws, there is about a 75% chance that the sum will be between 700 and 900.

11. One hundred draws are made at random with replacement from the box | -2 | -1 | 0 | 1 | 3 |. The sum of the positive numbers will be around ______, give or take ______ or so.

12. One hundred draws are made at random with replacement from the box | 1 | 2 | 3 | 4 | 5 | 6 | 7 |.
    (a) The sum of the draws is 431. The expected value for the sum of the draws is ______, the observed value is ______, the chance error is ______, and the standard error is ______.
    (b) The sum of the draws is 386. The expected value for the sum of the draws is ______, the observed value is ______, the chance error is ______, and the standard error is ______.
    (c) The sum of the draws is 417. The expected value for the sum of the draws is ______, the observed value is ______, the chance error is ______, and the standard error is ______.

13. A letter is drawn 1,000 times, at random, from the word A R A B I A. There are two offers.
    (A) You win a dollar if the number of A's among the draws is 10 or more above the expected number.
    (B) You win a dollar if the number of B's among the draws is 10 or more above the expected number.

    Choose one option and explain.
    (i) A gives a better chance of winning than B.
    (ii) A and B give the same chance of winning.
    (iii) B gives better chance of winning than A.
    (iv) There is not enough information to decide.

14. In roulette, once in a while, someone will bet $1 on red; and, at the same time, someone else will bet $1 on black (p. 282). Suppose this pair of bets is made 100 times in the course of an evening.
    (a) The house will make money on ______ of the 100 pairs of bets, give or take ______ or so.

(b) The net gain for the house from the 100 pairs of bets will be around ______ give or take ______ or so.

7. POSTSCRIPT

The exercises of this chapter teach a melancholy lesson. The more you gamble, the more you lose. The basic reason is that all the bets are unfair, in the sense that your expected net gain is negative. So the law of averages works for the house, not for you. Of course, this chapter only discussed simple strategies, and gamblers have evolved complicated systems for betting on roulette, craps, and the like. But it is a theorem of mathematics that no system for compounding unfair bets can ever make your expected net gain positive. In proving this theorem, only two assumptions are needed: (i) you aren't clairvoyant, and (ii) your financial resources are finite. The game of blackjack is unusual. Under some circumstances there are bets with a positive expected net gain.[8] As a result, people have won a lot of money on blackjack. However, the casinos change the rules to make this harder and harder.

8. SUMMARY

1. An *observed value* should be somewhere around the *expected value*; the difference is chance error. The likely size of the chance error is given by the *standard error*. For instance, the sum of the draws from a box will be around the expected value, give or take a standard error or so.

2. When drawing at random with replacement from a box of numbered tickets, each draw adds to the sum an amount which is around the average of the box. So the expected value for the sum is

$$\text{(number of draws)} \times \text{(average of box)}.$$

3. When drawing at random with replacement from a box of numbered tickets,

$$\text{SE for sum} = \sqrt{\text{number of draws}} \times \text{(SD of box)}.$$

This is the *square root law.*

4. When a list has only two different numbers ("big" and "small"), the SD can be figured by a short-cut method:

$$\left(\begin{array}{c}\text{big}\\ \text{number}\end{array} - \begin{array}{c}\text{small}\\ \text{number}\end{array}\right) \times \sqrt{\begin{array}{c}\text{fraction with}\\ \text{big number}\end{array} \times \begin{array}{c}\text{fraction with}\\ \text{small number}\end{array}}$$

5. If you have to classify and count the draws, remember to put 1 on the tickets that count for you, 0 on the others.

6. Provided the number of draws is sufficiently large, the normal curve can be used to figure chances for the sum of the draws.

18

The Normal Approximation for Probability Histograms

Everybody believes in the [normal approximation], the experimenters because they think it is a mathematical theorem, the mathematicians because they think it is an experimental fact.

—G. LIPPMANN (FRENCH PHYSICIST, 1845–1921)

1. INTRODUCTION

According to the law of averages, when a coin is tossed a large number of times, the percentage of heads will be close to 50%. Around 1700, the Swiss mathematician James Bernoulli put this on a rigorous mathematical footing. Twenty years later, Abraham de Moivre made a substantial improvement on Bernoulli's work, by showing how to compute the chance that the percentage of heads will fall in any given interval around 50%. The computation is not exact, but the approximation gets better and better as the number of tosses goes up. (De Moivre's work was discussed before, in chapter 13.)

Bernoulli and de Moivre both made the same assumptions about the coin: the tosses are independent, and on each toss the coin is as likely to land heads as tails. From these assumptions, it follows that the coin is as likely to land in any specific pattern of heads and tails as in any other. What Bernoulli did was to show that for most patterns, about 50% of the entries are heads.

You can see this starting to happen even with 5 tosses. Imagine tossing the coin 5 times, and keeping a record of how it lands on each toss. There is one possible pattern with 5 heads: H H H H H. How many patterns are there with four heads? The answer is 5:

T H H H H   H T H H H   H H T H H   H H H T H   H H H H T

The pattern T H H H H, for instance, means that the coin landed tails on the first toss, then gave four straight heads. Table 1 shows how many patterns there are, for any given number of heads. With 5 tosses, there are altogether $2^5 = 32$ possible patterns in which the coin can land. And 20 patterns out of the 32 have nearly half heads (two or three out of five).

Table 1. The number of patterns corresponding to a given number of heads, in 5 tosses of a coin.

Number of heads	*Number of patterns*
zero	1
one	5
two	10
three	10
four	5
five	1

De Moivre managed to count, to within a small margin of error, the number of patterns having a given number of heads—for any number of tosses. With 100 tosses, the number of patterns he had to think about is 2^{100}. This is quite a large number. If you tried to write all these patterns out, it might be possible to get a hundred of them on a page the size of this one. By the time you finished writing, you would have enough books to fill a shelf reaching from the earth to the farthest known star.

Still and all, mathematicians have a formula for the number of patterns with exactly 50 heads:

$$\frac{100!}{50! \times 50!} = \frac{100 \times 99 \times \cdots \times 51}{50 \times 49 \times \cdots \times 1}.$$

(Binomial coefficients are covered in chapter 15; they won't really matter here.)

The formula was of no immediate help to de Moivre, because the arithmetic is nearly impossible to do by hand. By calculator,[1]

$$\frac{100 \times 99 \times \cdots \times 51}{50 \times 49 \times \cdots \times 1} \approx 1.01 \times 10^{29}.$$

Similarly, the total number of patterns is $2^{100} \approx 1.27 \times 10^{30}$. So the chance of getting exactly 50 heads in 100 tosses of a coin is

$$\frac{\text{number of patterns with 50 heads}}{\text{total number of patterns}} \approx \frac{1.01 \times 10^{29}}{1.27 \times 10^{30}} \approx 0.08 = 8\%.$$

Of course, de Moivre did not have anything like a modern calculator available. He needed a mathematical way of estimating the binomial coefficients, without having to work the arithmetic out. And he found a way to do it (though the approximation is usually credited to another mathematician, James Stirling). De Moivre's procedure led him to the normal curve. For example, he found that the chance of getting exactly 50 heads in 100 tosses of a coin was about equal to the area under the normal curve between −0.1 and +0.1. In fact, he was able to prove that the whole *probability histogram* for the number of heads is close to the normal curve when the number of tosses is large. Modern researchers have extended this result to the sum of draws made at random from any box of tickets. The details of de Moivre's argument are too complicated to go into here—but we can present his idea graphically, using a computer to draw the pictures.[2]

2. PROBABILITY HISTOGRAMS

When a chance process generates a number, the expected value and standard error are a guide to where that number will be. But the probability histogram gives a complete picture.

> A probability histogram is a new kind of graph. This graph represents chance, not data.

Here is an example. Gamblers playing craps bet on the total number of spots shown by a pair of dice. (The numbers range from 2 through 12.) So the odds depend on the chance of rolling each possible total. To find the chances, a casino might hire someone to throw a pair of dice. This experiment was simulated on the computer; results for the first 100 throws are shown in table 2.

Table 2. Rolling a pair of dice. The computer simulated rolling a pair of dice, and finding the total number of spots. It repeated this process 10,000 times. The first 100 repetitions are shown in the table.

Repetition	*Total*	*Repetition*	*Total*	*Repetition*	*Total*	*Repetition*	*Total*	*Repetition*	*Total*
1	8	21	10	41	8	61	8	81	11
2	9	22	4	42	10	62	5	82	9
3	7	23	8	43	6	63	3	83	7
4	10	24	7	44	3	64	11	84	4
5	9	25	7	45	4	65	9	85	7
6	5	26	3	46	8	66	4	86	4
7	5	27	8	47	4	67	12	87	7
8	4	28	8	48	4	68	7	88	6
9	4	29	12	49	5	69	10	89	7
10	4	30	2	50	4	70	4	90	11
11	10	31	11	51	11	71	7	91	6
12	8	32	12	52	8	72	4	92	11
13	3	33	12	53	10	73	7	93	8
14	11	34	7	54	9	74	9	94	8
15	7	35	7	55	10	75	9	95	7
16	8	36	6	56	12	76	11	96	9
17	9	37	6	57	7	77	6	97	10
18	8	38	2	58	6	78	9	98	5
19	6	39	6	59	7	79	9	99	7
20	8	40	3	60	7	80	7	100	7

The top panel in figure 1 shows the histogram for the data in table 2. The total of 7 came up 20 times, so the rectangle over 7 has an area of 20%, and similarly for the other possible totals. The next panel shows the empirical histogram for the first 1,000 repetitions, and the third is for all 10,000. These empirical his-

Figure 1. Empirical histograms converging to a probability histogram. The computer simulated rolling a pair of dice and finding the total number of spots. It repeated the process 100 times, and made a histogram for the 100 numbers (top panel). This is an empirical histogram—based on observations. The second panel is for 1,000 repetitions, the third panel for 10,000. (Each repetition involves rolling a pair of dice.) The bottom panel is the ideal or probability histogram for the total number of spots when a pair of dice are rolled.

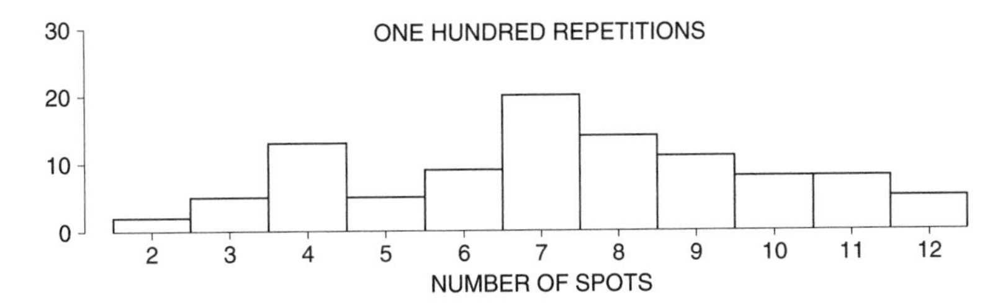

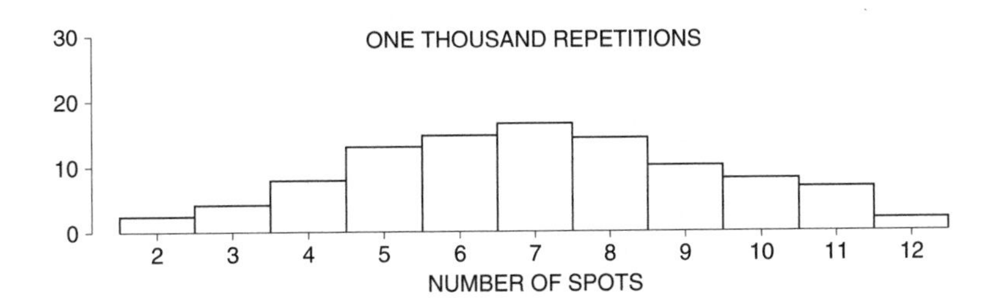

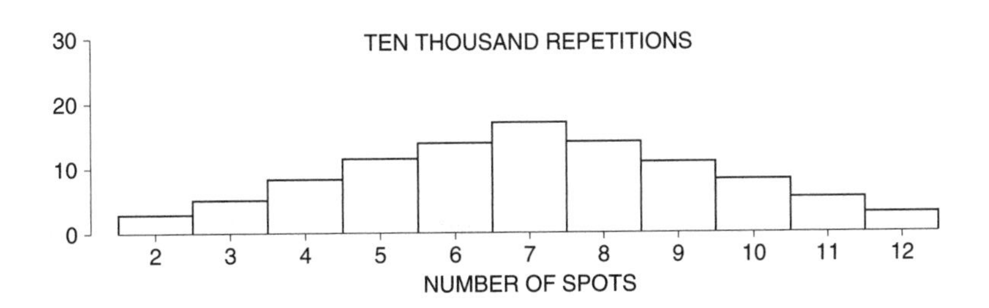

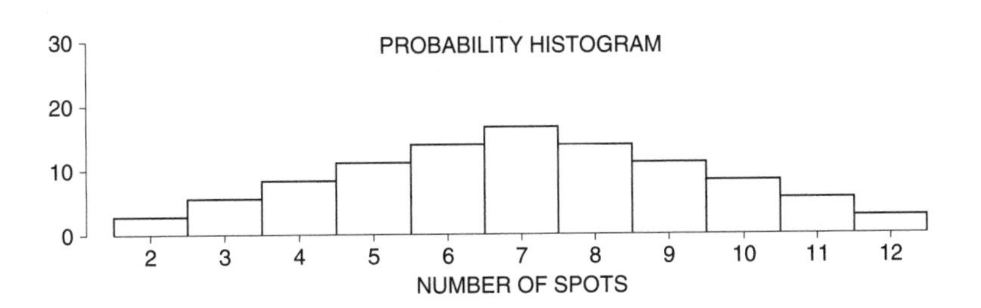

tograms converge to the ideal probability histogram shown in the bottom panel of the figure. (*Empirical* means "experimentally observed," *converge* means "gets closer and closer to.")

Of course, this probability histogram can be computed using a theoretical argument. As shown in chapter 14, there are 6 chances in 36 of rolling a 7. That's $16\frac{2}{3}\%$. Consequently, the area of the rectangle over 7 in the probability histogram equals $16\frac{2}{3}\%$. Similarly for the other rectangles.

A probability histogram represents chance by area.

The probability histogram (bottom panel, figure 1) is made up of rectangles. The base of each rectangle is centered at a possible value for the sum of the draws, and the area of the rectangle equals the chance of getting that value.[3] The total area of the histogram is 100%.

For another example, look at the product of the numbers on a pair of dice, instead of the sum. The computer was programmed to repeat the following chance process over and over again: roll a pair of dice and take the product of the numbers. The top panel of figure 2 gives the empirical histogram for 100 repetitions. The product 10 came up 4 times, so the area of the rectangle over 10 equals 4%. Other values are done the same way. The second panel gives the empirical histogram for 1,000 repetitions; the third, for 10,000. (Each repetition involves rolling a pair of dice and taking the product.) The last panel shows the probability histogram. The empirical histogram for 10,000 repetitions looks almost exactly like the probability histogram.

Figure 2 is very different from figure 1: the new histograms have gaps. To see why, it helps to think about the possible values of the product. The smallest value is 1, if both dice show ⚀; the biggest is 36, if both show ⚅. But there is no way to get the product 7. There is no rectangle over 7, because the chance is zero. For the same reason, there is no rectangle over 11. All the gaps can be explained in this way.

Exercise Set A

1. The figure below is a probability histogram for the sum of 25 draws from the box | [1] [2] [3] [4] [5] |. The shaded area represents the chance that the sum will be between _____ and _____ (inclusive).

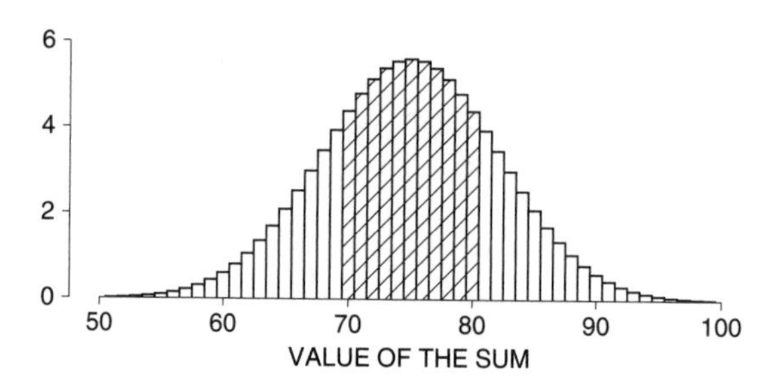

Figure 2. Empirical histograms converging to a probability histogram. The computer simulated rolling a pair of dice and taking the product of the two numbers. It repeated the process 100 times, and made a histogram for the 100 products (top panel). This is an empirical histogram—based on observations. The second panel is for 1,000 repetitions, the third panel for 10,000. (Each repetition involves rolling a pair of dice.) The bottom panel is the ideal or probability histogram for the product of the two numbers that come up when a pair of dice are rolled.

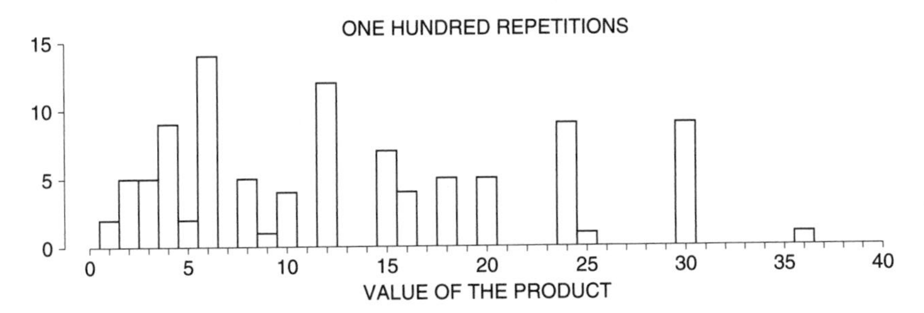

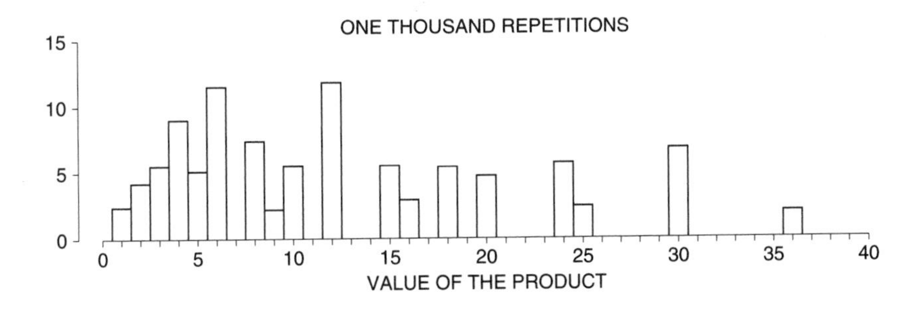

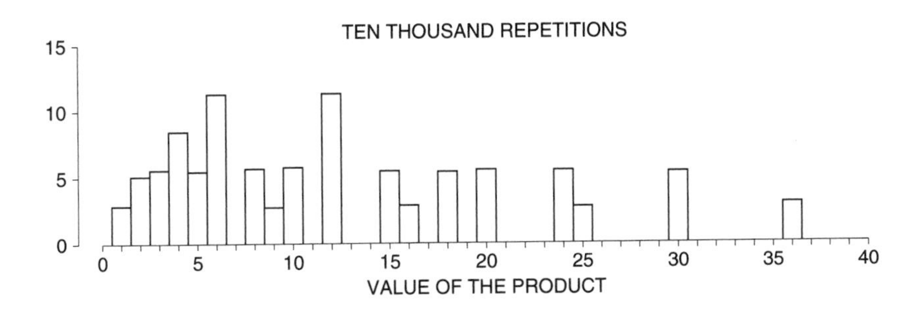

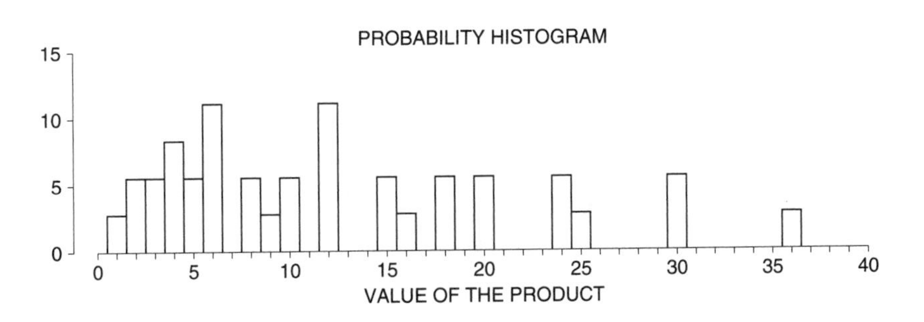

2. The bottom panel of figure 1 (p. 311) shows the probability histogram for the total number of spots when a pair of dice are rolled.
   (a) The chance that the total number of spots will be between 7 and 10 (inclusive) equals the area under the histogram between ______ and ______.
   (b) The chance that the total number of spots will be 7 equals the area under the histogram between ______ and ______.

3. This exercise—like exercise 2—refers to figure 1 on p. 311.
   (a) If a pair of dice are rolled, the total number of spots is most likely to be ______.
   (b) In 1,000 rolls of the pair of dice, which total came up most often?
   (c) In the top panel of figure 1, the rectangle over 4 is bigger than the rectangle over 5. Is this because 4 is more likely than 5? Explain.
   (d) Look at the top panel of the figure. The area of the rectangle above 8 represents—
      (i) the chance of getting a total of 8 spots when a pair of dice are rolled.
      (ii) the chance of getting a total of 8 spots when 100 dice are rolled.
      (iii) the percentage of times the total of 8 comes up in table 2.

   Choose one option, and explain.

4. Figure 2 on p. 313 is about the product of the numbers on a pair of dice.
   (a) If the dice land ⚀⚂, what is the product? If they land ⚁⚂?
   (b) "2 is as likely a value for the product as 3." Which panel should you look at to check this statement? Is it true?
   (c) In 1,000 rolls, which value appeared more often for the product: 2 or 3? Explain.
   (d) None of the histograms has a rectangle above 14. Why?
   (e) In the bottom panel of figure 2, the area of the rectangle above 6 is 11.1%. What does this 11.1% represent?

5. The figure below shows the probability histograms for the sum of 25 draws made at random with replacement from boxes (i) and (ii). Which histogram goes with which box? Explain.

   (i) [0] [1] [2] (ii) [0] [1] [2] [3] [4]

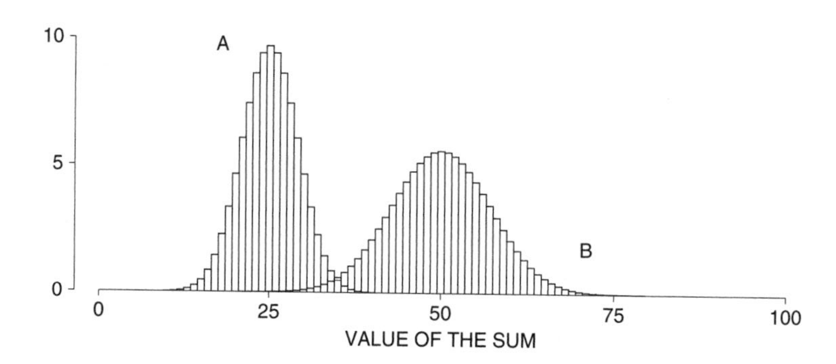

6. The figure at the top of the next page is the probability histogram for the sum of 25 draws made at random with replacement from a box. True or false: the shaded area represents the percentage of times you draw a number between 5 and 10 inclusive.

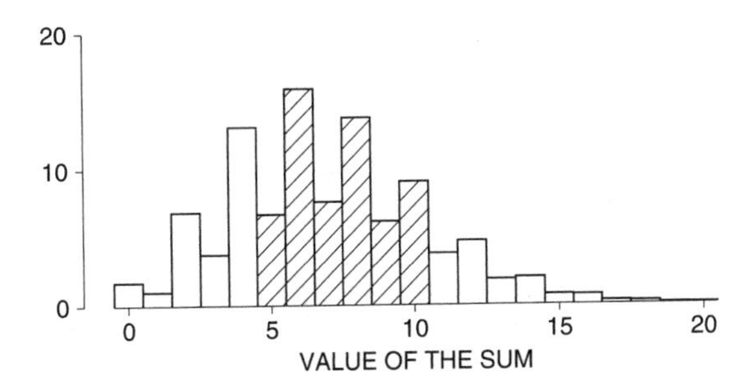

The answers to these exercises are on p. A76.

3. PROBABILITY HISTOGRAMS AND THE NORMAL CURVE

The object of this section is to show how the probability histogram for the number of heads gets close to the normal curve when the number of tosses becomes large. For instance, suppose the coin is tossed 100 times. The probability histogram for the number of heads is a bit jagged, but follows the normal curve quite well (figure 3).

The figure has two horizontal axes. The probability histogram is drawn relative to the upper axis, showing the number of heads. The normal curve is drawn relative to the lower axis, showing standard units. The expected number of heads is 50, and the SE is 5. So 50 on the number-of-heads axis corresponds to 0 on the standard-units axis, 55 corresponds to +1, and so on.

There are also two vertical axes in the figure. The probability histogram is drawn relative to the inside one, showing percent per head. The normal curve is drawn relative to the outside one, showing percent per standard unit. To see how the scales match up, take the top value on each axis. Why does 50% per standard unit match up with 10% per head? The SE is 5, so there are 5 heads to the standard unit. And $50/5 = 10$. Any other pair of values can be dealt with in the same way. (Also see p. 80 on data histograms.)

Figure 3. The probability histogram for the number of heads in 100 tosses of a coin, compared to the normal curve. The curve is drawn on the standard-units scale for the histogram.

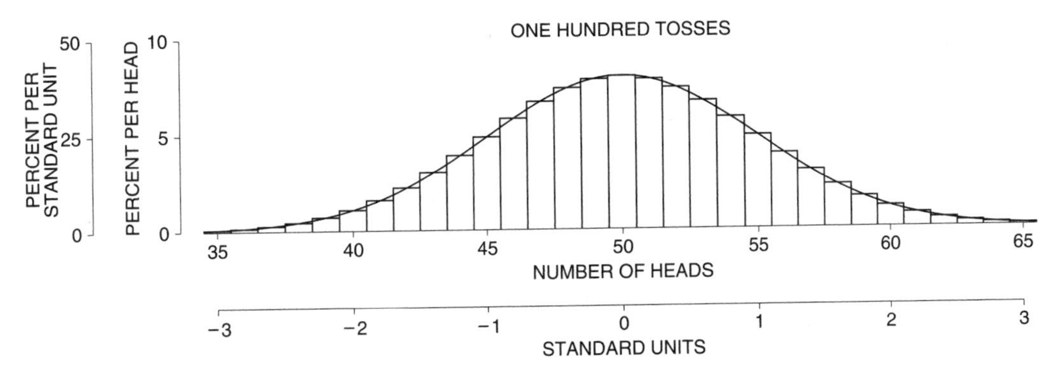

Figure 4 shows probability histograms for the number of heads in 100, 400, and 900 tosses of a coin. With 100 tosses, the histogram follows the curve but is more jagged. With 900 tosses, the histogram is practically the same as the curve. In the early eighteenth century, de Moivre proved this convergence had to take place, by pure mathematical reasoning.

Figure 4. The normal approximation. Probability histograms are shown for the number of heads in 100, 400, and 900 tosses of a coin. The normal curve is shown for comparison. The histograms follow the curve better and better as the number of tosses goes up.

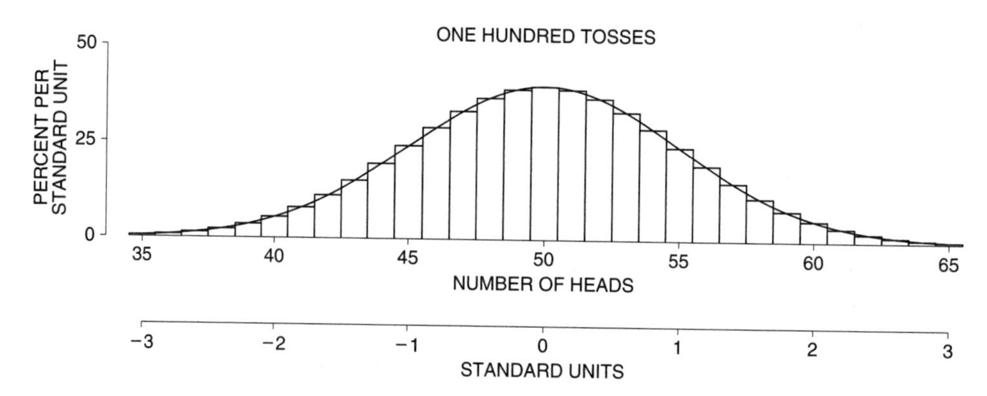

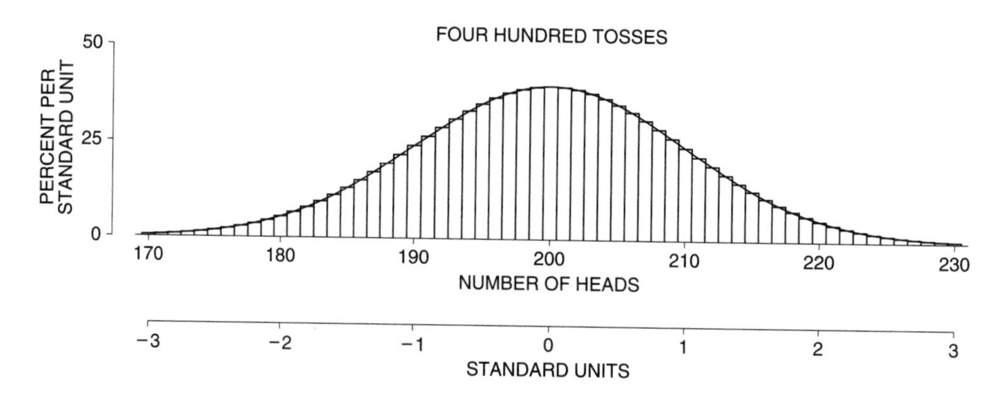

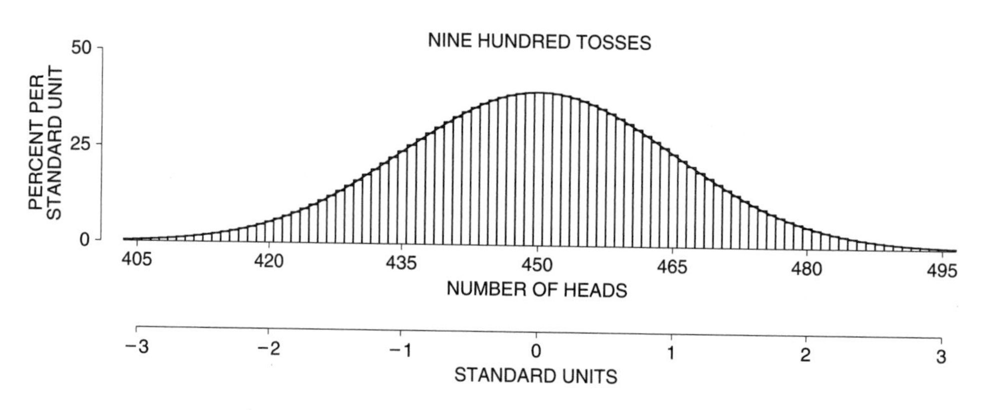

4. THE NORMAL APPROXIMATION

The normal curve has already been used in chapter 17 to figure chances. This section will explain the logic. It will also present a technique for taking care of endpoints, which should be used when the number of tosses is small or extra accuracy is wanted.

Example 1. A coin will be tossed 100 times. Estimate the chance of getting—

(a) exactly 50 heads.
(b) between 45 and 55 heads inclusive.
(c) between 45 and 55 heads exclusive.

Solution. The expected number of heads is 50 and the standard error is 5, as shown on p. 301.

Part (a). Look at figure 3 (p. 315). The chance of getting exactly 50 heads equals the area of the rectangle over 50. The base of this rectangle goes from 49.5 to 50.5 on the number-of-heads scale. In standard units, the base of the rectangle goes from −0.1 to 0.1:

$$\frac{49.5 - 50}{5} = -0.1, \qquad \frac{50.5 - 50}{5} = 0.1$$

But the histogram and the normal curve almost coincide. So the area of the rectangle is nearly equal to the area between −0.1 and 0.1 under the curve.

(The exact chance is 7.96%, to two decimals; the approximation is excellent.[4])

Part (b). The chance of getting between 45 and 55 heads inclusive equals the area of the eleven rectangles over the values 45 through 55 in figure 3. That is the area under the histogram between 44.5 and 55.5 on the number-of-heads scale, which correspond to −1.1 and 1.1 on the standard-units scale. Because the histogram follows the normal curve so closely, this area is almost equal to the area under the curve.

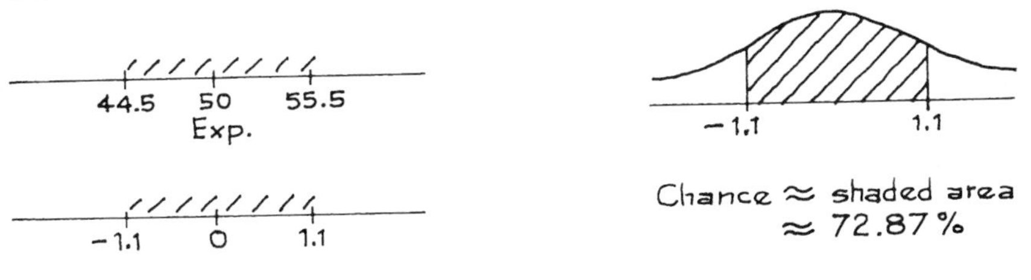

(The exact chance is 72.87%, to two decimals.)

Part (c). The chance of getting 45 to 55 heads exclusive equals the total area of the nine rectangles over the values 46 through 54. That is the area under the histogram between 45.5 and 54.5 on the number-of-heads scale, which correspond to −0.9 and 0.9 on the standard-units scale.

(The exact chance is 63.18%, to two decimals.)

Often, the problem will only ask for the chance that (for instance) the number of heads is between 45 and 55, without specifying whether endpoints are included or excluded. Then, you can use the compromise procedure:

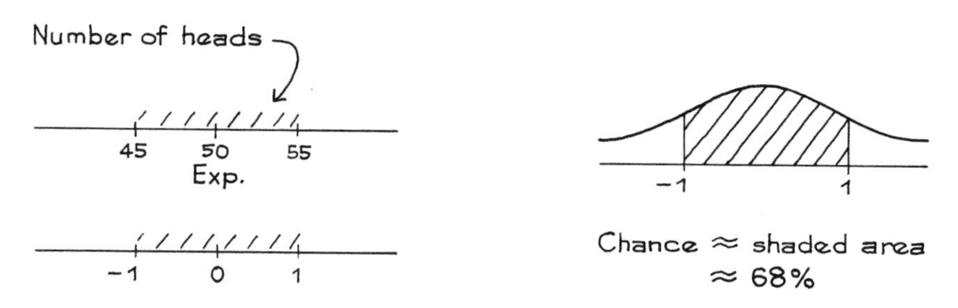

This amounts to replacing the area under the histogram between 45 and 55 by the area under the normal curve between the corresponding values (in standard units). It splits the two end rectangles in half, and does not give quite as much precision as the method used in example 1. Keeping track of the endpoints has an official name—"the continuity correction." The correction is worthwhile if the rectangles are big, or if a lot of precision is needed. Usually, the exercises in this book can be worked without the correction.

The normal approximation consists in replacing the actual probability histogram by the normal curve before computing areas. This is legitimate when the probability histogram follows the normal curve. Probability histograms are often hard to work out, while areas under the normal curve are easy to look up in the table.[5]

Exercise Set B

1. A coin is tossed 10 times. The probability histogram for the number of heads is shown at the top of the next page, with three different shaded areas. One corresponds to the chance of getting 3 to 7 heads inclusive. One corresponds to the chance of getting 3 to 7 heads exclusive. And one corresponds to the chance of getting exactly 6 heads. Which is which, and why?

(i)

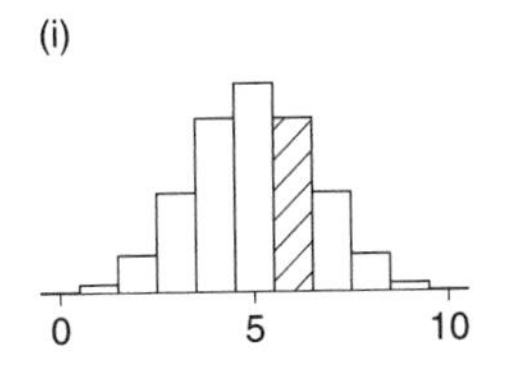

(ii)

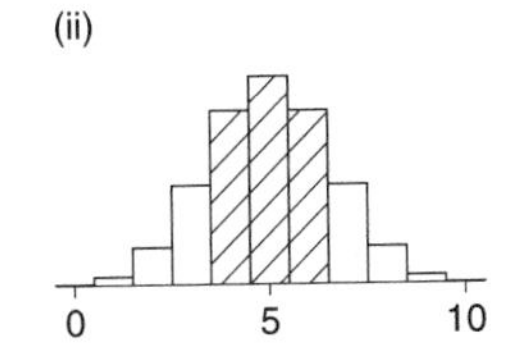

(iii)

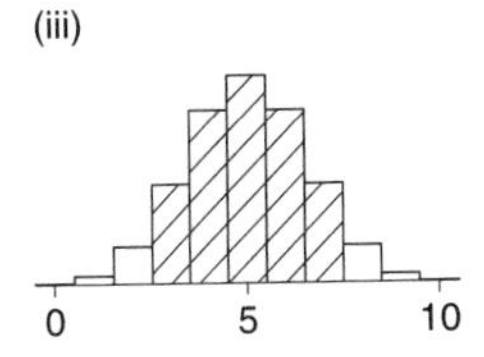

2. In figure 3 on p. 315, the chance of getting 52 heads is exactly equal to the area between ______ and ______ under the ______. Fill in the blanks. For the last one, your options are: normal curve, probability histogram. Explain your answers.

3. A coin is tossed 100 times. Estimate the chance of getting 60 heads.

4. Kerrich's data on 10,000 tosses of a coin can be taken in successive groups of 100 tosses (table 3 on p. 302). About how many groups should show exactly 60 heads? How many actually do?

5. A coin is tossed 10,000 times. Estimate the chance of getting—
   (a) 4,900 to 5,050 heads
   (b) 4,900 heads or fewer
   (c) 5,050 heads or more

6. (a) Suppose you were going to estimate the chance of getting 50 heads or fewer in 100 tosses of a coin. Should you keep track of the edges of the rectangles?
   (b) Same, for the chance of getting 450 heads or fewer in 900 tosses.

   No calculations are needed, just look at figure 4 on p. 316.

The answers to these exercises are on pp. A76–77.

5. THE SCOPE OF THE NORMAL APPROXIMATION

In the preceding section, the discussion has been about a coin, which lands heads or tails with chance 50%. What about drawing from a box? Again, the normal approximation works perfectly well, so long as you remember one thing. The more the histogram of the numbers in the box differs from the normal curve, the more draws are needed before the approximation takes hold. Figure 5 shows the histogram for the tickets in the lopsided box | 9 [0]'s [1] |.

Figure 5. Histogram for the lopsided box | 9 [0]'s [1] |.

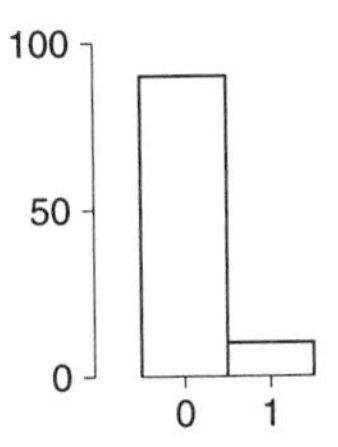

The probability histogram for the sum will be lopsided too, until the number of draws gets fairly large. The computer was programmed to work out the probability histogram for the sum of 25, 100, or 400 draws from the box. The histograms are shown in figure 6 below. With 25 draws, the histogram is a lot higher than the curve on the left, lower on the right. The normal approximation does not apply.

Figure 6. The normal approximation for the sum of draws from the box | 9 [0]'s [1] |. The top panel shows the probability histogram for the sum of 25 draws, the middle panel for 100 draws, the bottom panel for 400 draws. A normal curve is shown for comparison. The histograms are higher than the normal curve on the left and lower on the right, because the box is lopsided.[6] As the number of draws goes up, the histograms follow the curve more and more closely.

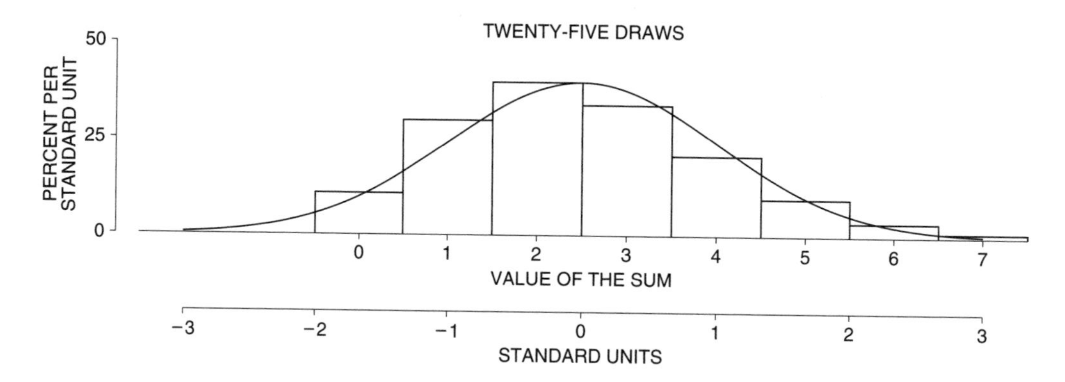

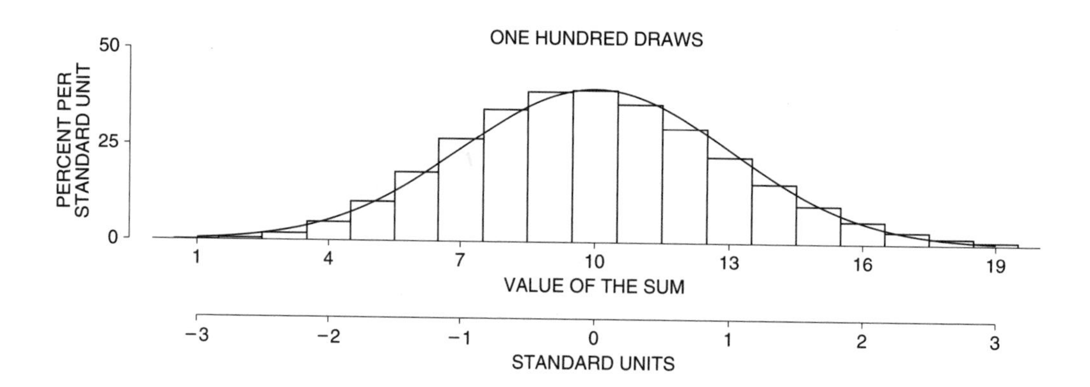

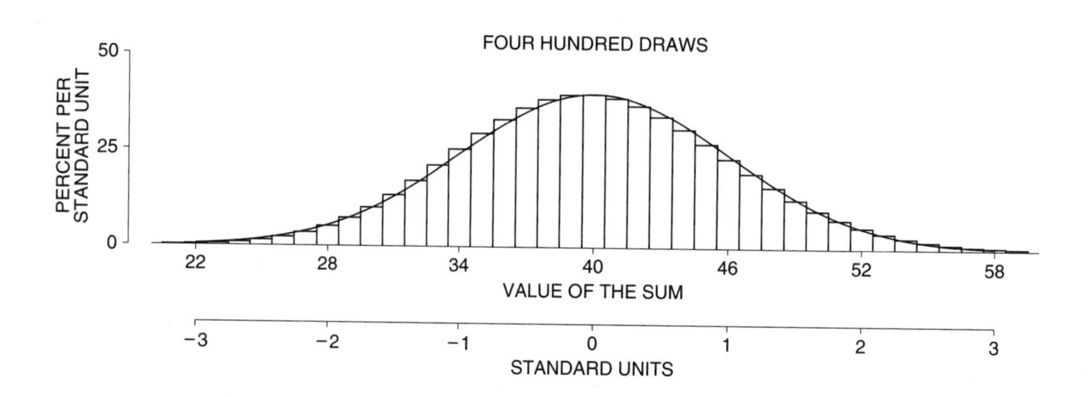

With 100 draws, the histogram follows the curve much better. At 400 draws, you have to look closely to see the difference.

So far, there have only been 0's and 1's in the box. What about other numbers? Our next example is | [1] [2] [3] |. The probability histogram for the sum of 25 draws from this box is already close to the curve; with 50 draws, the histogram follows the curve very closely indeed (figure 7).

Figure 7. Probability histograms for the sum of 25 or 50 draws from the box | [1] [2] [3] |. These histograms follow the normal curve very well.

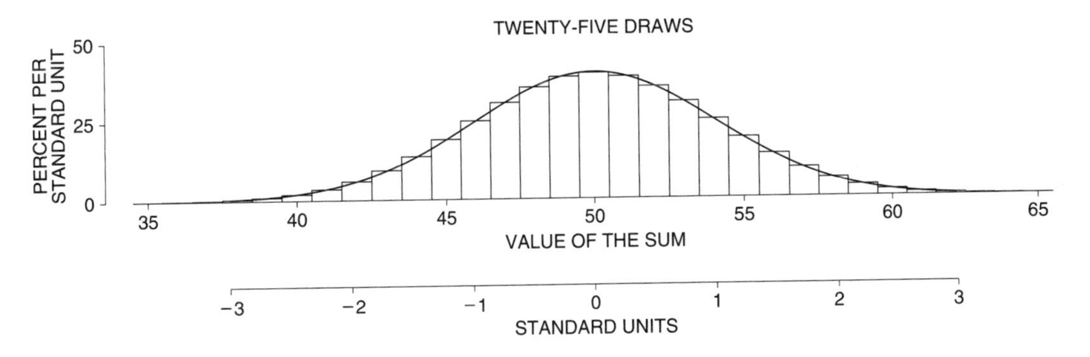

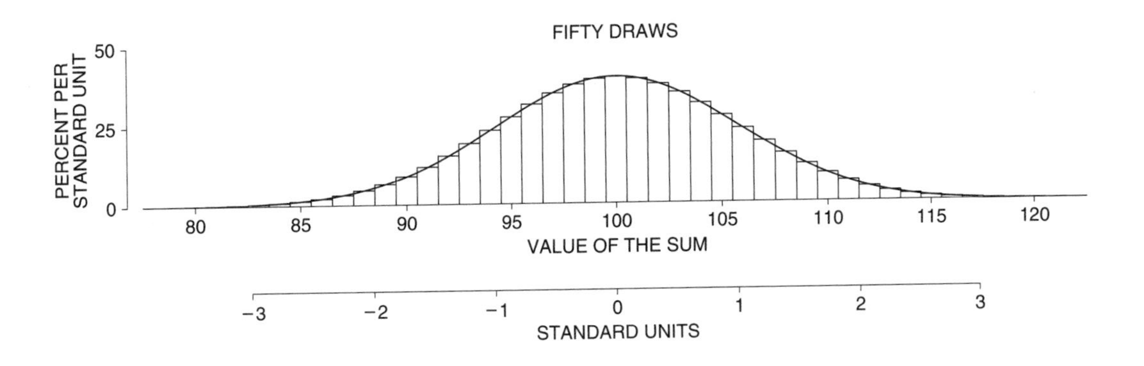

Our last example is the box | [1] [2] [9] |. A histogram for the numbers in the box is shown in figure 8. This histogram looks nothing like the normal curve.

Figure 8. Histogram for the box | [1] [2] [9] |. The histogram is nothing like the normal curve.

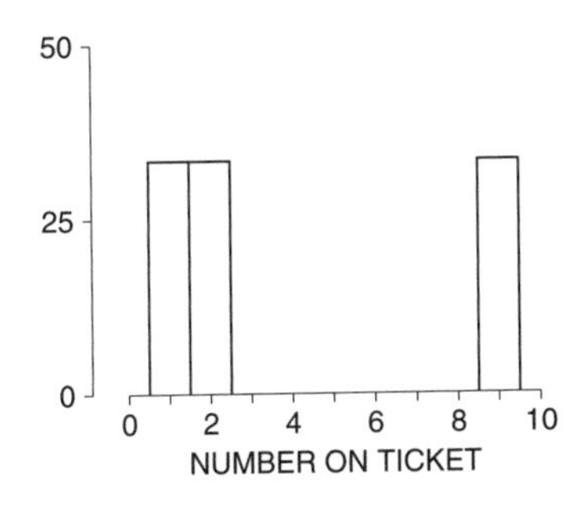

With 25 draws, the probability histogram for the sum is still quite different from the curve—it shows waves (figure 9). With 50 draws, the waves are still there, but much smaller. And by 100 draws, the probability histogram is indistinguishable from the curve.

Figure 9. The normal approximation for a sum. Probability histograms are shown for the sum of draws from the box | 1 2 9 |. The top panel is for 25 draws, and does not follow the normal curve especially well. (Note the waves.[7]) The middle panel is for 50 draws. The bottom panel is for 100 draws. It follows the normal curve very well.

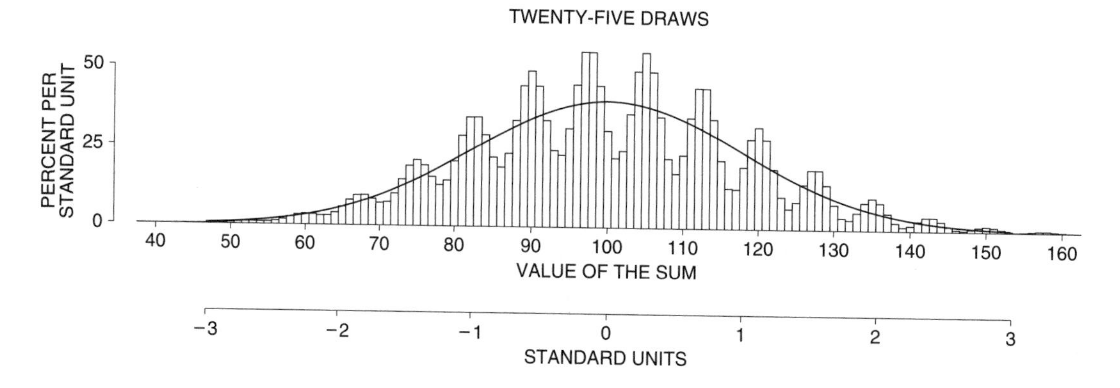

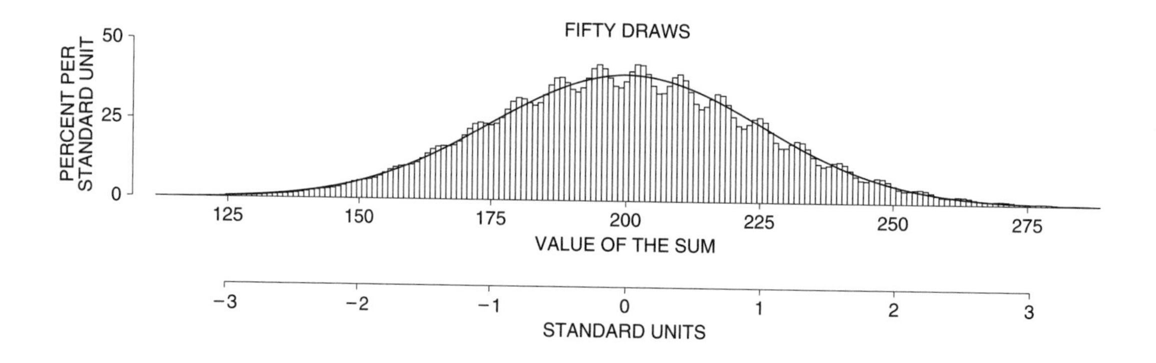

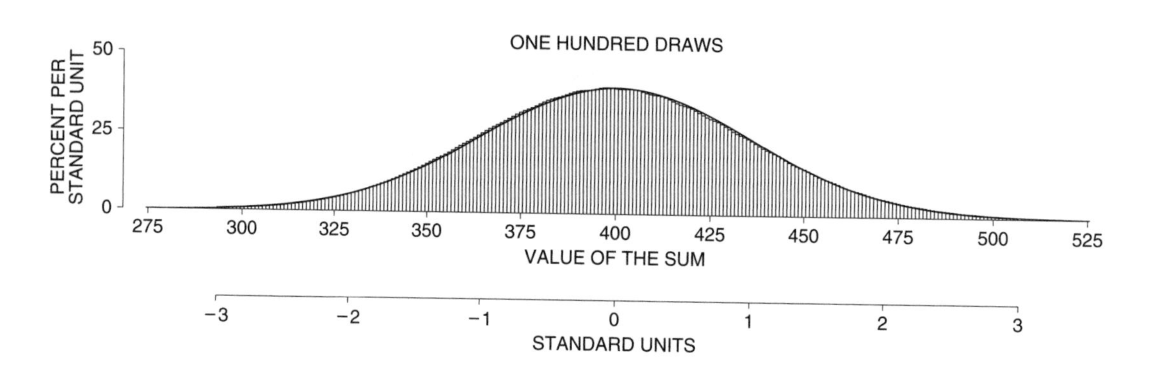

The normal curve is tied to sums. For instance, the probability histogram for a product will usually be quite different from normal. The top panel of figure 10 shows the probability histogram for the product of 10 rolls of a die. This is nothing like the normal curve. Making the number of rolls larger does not make the histogram more normal: the probability histogram for the product of 25 rolls is shown in the bottom panel, and is even worse.[8] Multiplication is different from addition. The normal approximation works for the sum of draws made at random from a box—not for the product.

With 10 rolls, the histogram for the product is shown out to a million; 6% of the area lies beyond that point and is not shown. A million looks like a big number, but products build up fast. The largest value for the product is 6 multiplied by itself 10 times: $6^{10} = 60{,}466{,}176$. On this scale, a million is not so big after all.

Figure 10. Probability histograms for the product of 10 and 25 rolls of a die. The histograms look nothing like the normal curve. The base of each rectangle covers a range of values of the product, and the area of the rectangle equals the chance of the product taking a value in that range. With 10 rolls, about 6% of the area is not shown; with 25 rolls, about 20% is not shown. In the top panel, the vertical scale is percent per 10,000; in the bottom panel, percent per 10^{11}.

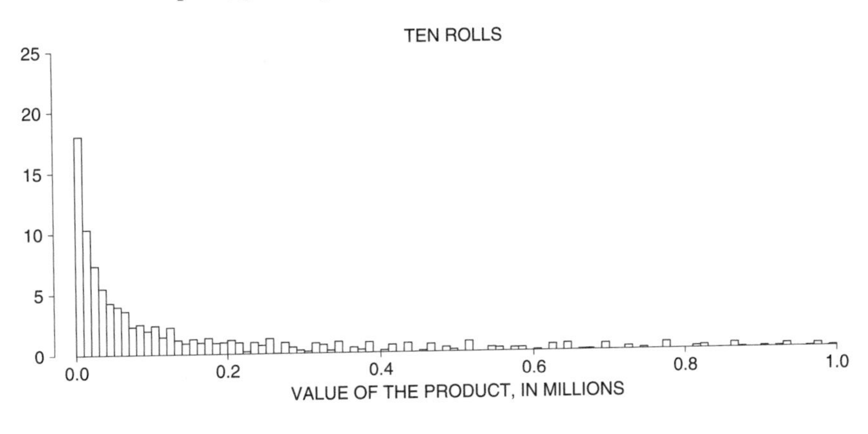

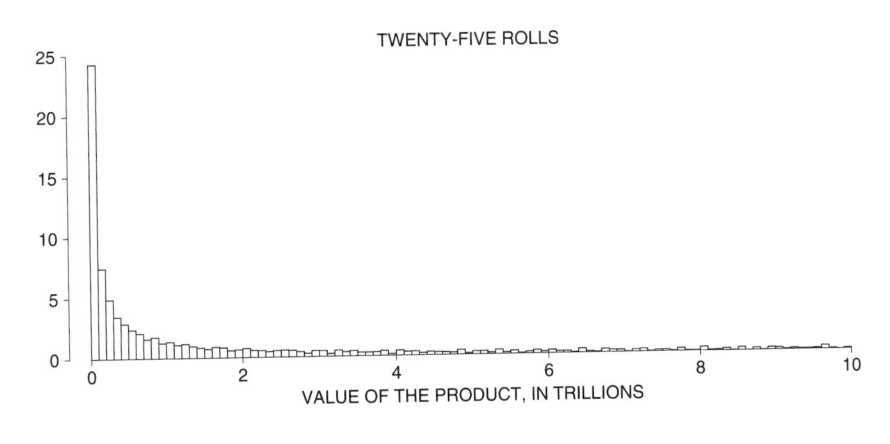

With 25 rolls, the largest possible value for the product really is a big number: $6^{25} \approx 3 \times 10^{19}$, or 3 followed by 19 zeros. (The U.S. federal debt was "only" \$8 trillion in 2006, that is, \$8 followed by 12 zeros.)

Exercise Set C

1. Shown below is the probability histogram for the sum of 15 draws from the box | 0 0 1 |.

   (a) What numbers go into the blanks?
   (b) Which is a more likely value for the sum, 3 or 8? Explain.

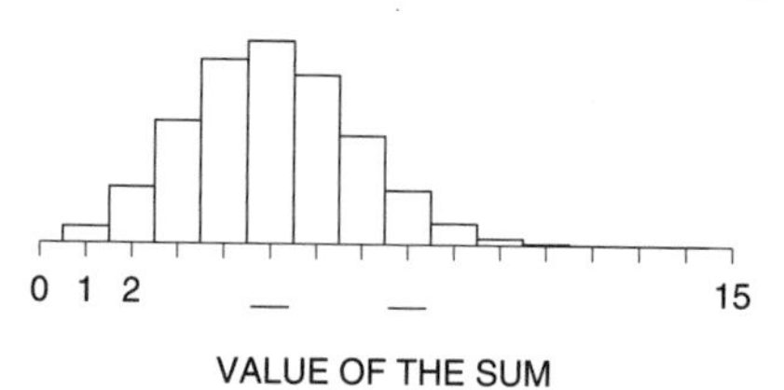

2. A biased coin has one chance in ten of landing heads. It is tossed 400 times. Estimate the chance of getting exactly 40 heads.

3. The coin in exercise 2 is tossed 25 times. Suppose the normal approximation is used to estimate the chance of getting exactly one head. Would the estimate be just about right? too high? too low? No calculations are needed; look at figure 6 on p. 320.

4. The same coin is tossed 100 times. If you were asked to estimate the chance of getting 10 heads or fewer, should you keep track of the edges of the rectangles? No calculations are needed; look at figure 6 on p. 320.

5. Twenty-five draws are made at random with replacement from each of the boxes below.

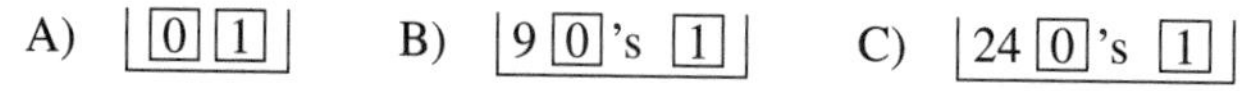

   The probability histograms for the sums are shown below, in scrambled order. Match the histograms with the boxes.

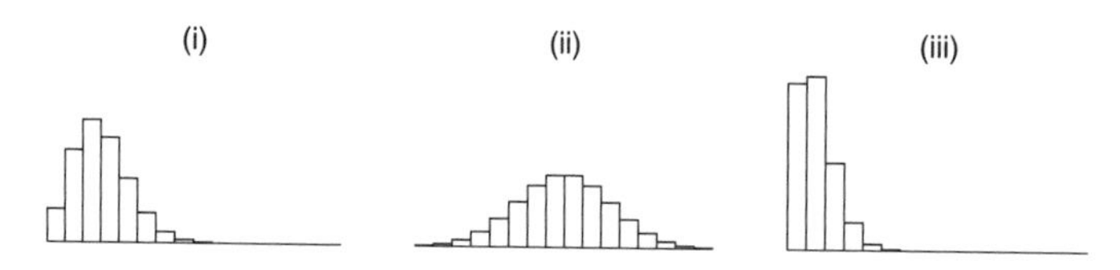

6. Shown below are probability histograms for the sum of 100, 400, and 900 draws from the box | 99 0's 1 |. Which histogram is which?

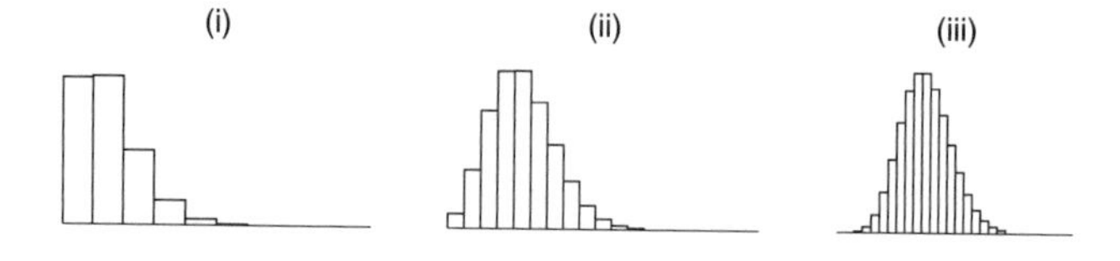

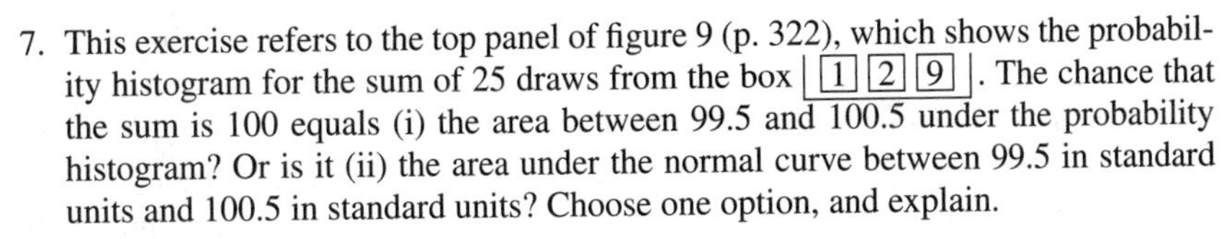

7. This exercise refers to the top panel of figure 9 (p. 322), which shows the probability histogram for the sum of 25 draws from the box | [1] [2] [9] |. The chance that the sum is 100 equals (i) the area between 99.5 and 100.5 under the probability histogram? Or is it (ii) the area under the normal curve between 99.5 in standard units and 100.5 in standard units? Choose one option, and explain.

8. This exercise, like the previous one, can be worked using the top panel of figure 9. Among the options listed below, the sum of 25 draws from the box | [1] [2] [9] | is most likely to equal ______ and least likely to equal ______ even though its expected value is ______. Options:

   100   101   102   103   104   105

9. This exercise refers to the top panel of figure 10 on p. 323.
   (a) The expected value for the product is nearly 276,000. The chance that the product will exceed this number is—

   just about 50%   much bigger than 50%   much smaller than 50%

   Choose one option, and explain.
   (b) There are 100 rectangles in the histogram. The width of each one is

   1   10   100   1,000   10,000   100,000   1,000,000

   (c) Which is a more likely range for the product?

   390,000–400,000   400,000–410,000

The answers to these exercises are on pp. A77–78.

6. CONCLUSION

We have looked at the sum of the draws from four different boxes:

| [0] [1] | | 9 [0]'s [1] | | [1] [2] [3] | | [1] [2] [9] |

There are plenty more where those came from. But the pattern is always the same. With enough draws, the probability histogram for the sum will be close to the normal curve. Mathematicians have a name for this fact. They call it "the central limit theorem," because it plays a central role in statistical theory.[9]

> *The Central Limit Theorem.* When drawing at random with replacement from a box, the probability histogram for the sum will follow the normal curve, even if the contents of the box do not. The histogram must be put into standard units, and the number of draws must be reasonably large.

The central limit theorem applies to sums but not to other operations like products (figure 10). The theorem is the basis for many of the statistical procedures discussed in the rest of the book.

How many draws do you need? There is no set answer. Much depends on the contents of the box—remember the waves in figure 9. However, for many boxes, the probability histogram for the sum of 100 draws will be close enough to the normal curve.

When the probability histogram does follow the normal curve, it can be summarized by the expected value and standard error. For instance, suppose you had to plot such a histogram without any further information. In standard units you can do it, at least to a first approximation:

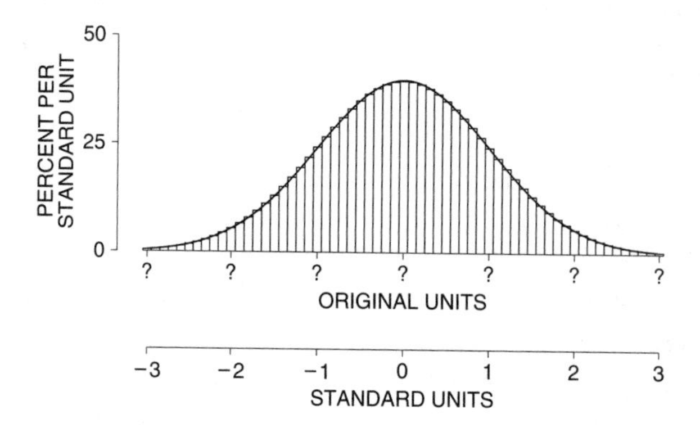

To finish the picture, you have to translate the standard units back into original units by filling in the question marks. This is what the expected value and standard error do. They tell you almost all there is to know about this histogram, because it follows the normal curve.

> The expected value pins the center of the probability histogram to the horizontal axis, and the standard error fixes its spread.

According to the square root law, the expected value and standard error for a sum can be computed from

- the number of draws,
- the average of the box,
- the SD of the box.

These three quantities just about determine the behavior of the sum. That is why the SD of the box is such an important measure of its spread.[10]

This chapter discussed two sorts of convergence for histograms, and it is important to separate them. In figure 1, the number of draws from the box | 1 2 3 4 5 6 | was fixed. It was 2. The basic chance process was drawing from the box and taking the sum. This process was repeated a larger and larger number of times—100, 1,000, 10,000. The empirical histogram for the observed values of the sum (a histogram for data) converged to the probability histogram (a histogram for chances). In section 5, on the other hand, the number of draws from the box got larger and larger. Then the probability histogram for the sum got smoother and smoother, and in the limit became the normal curve. Empirical histograms are one thing; probability histograms quite another.

In part II of the book, the normal curve was used for data. In some cases, this can be justified by a mathematical argument which uses the two types of convergence discussed in this chapter. When the number of repetitions is large, the empirical histogram will be close to the probability histogram. When the number of draws is large, the probability histogram for the sum will be close to the normal curve. Consequently, when the number of repetitions and the number of draws are both large, the empirical histogram for the sums will be close to the curve.[11] This is all a matter of pure logic: a mathematician can prove every step.

But there is still something missing. It has to be shown that the process generating the data is like drawing numbers from a box and taking the sum. This sort of argument will be discussed in part VII. More than mathematics is involved—there will be questions of fact to settle.

7. REVIEW EXERCISES

1. The figure below shows the probability histogram for the total number of spots when a die is rolled eight times. The shaded area represents the chance that the total will be between ______ and ______ (inclusive).

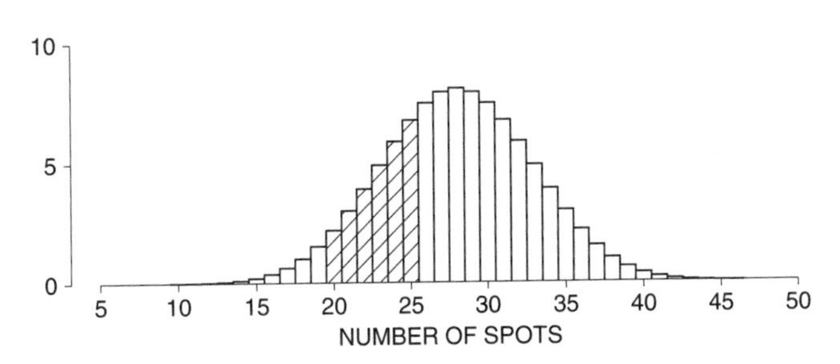

2. Four hundred draws will be made at random with replacement from the box [1] [3] [5] [7].
   (a) Estimate the chance that the sum of the draws will be more than 1,500.
   (b) Estimate the chance that there will be fewer than 90 [3]'s.

3. Ten draws are going to be made at random with replacement from the box [0] [1] [2] [3]. The chance that the sum will be in the interval from 10 to 20 inclusive equals the area under ______ between ______ and ______. Fill in the blanks. For the first one, your options are: the normal curve, the probability histogram for the sum. Explain your answers.

4. A coin is tossed 25 times. Estimate the chance of getting 12 heads and 13 tails.

5. Twenty-five draws are made at random with replacement from the box [1] [1] [2] [2] [3]. One of the graphs below is a histogram for the numbers drawn. One is the probability histogram for the sum. And one is the probability histogram for the product. Which is which? Why?

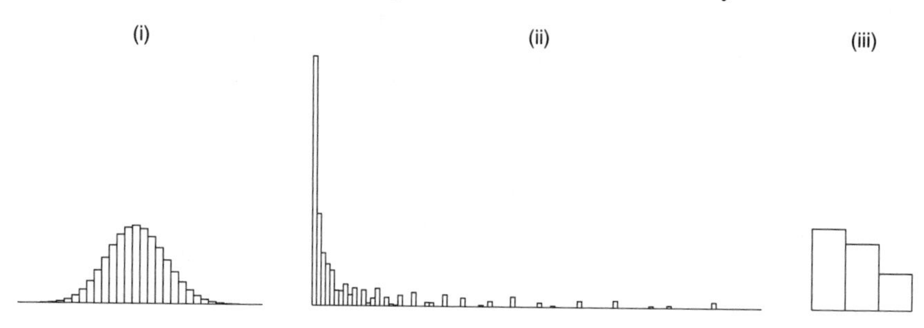

6. A programmer is working on a new program, COIN, to simulate tossing a coin. As a preliminary test, he sets up the program to do one million tosses. The program returns with a count of 502,015 heads. The programmer looks at this and thinks:

> Hmmm. Two thousand and fifteen off. That's a lot. No, wait. Compare it to the million. Two thousand—forget the fifteen—out of a million is two out of a thousand. That's one in five hundred. One fifth of a percent. Very small. Good. COIN passes.

Do you agree? Answer yes or no, and explain.

7. A pair of dice are thrown. The total number of spots is like
   (i) one draw from the box

   [2] [3] [4] [5] [6] [7] [8] [9] [10] [11] [12]

   (ii) the sum of two draws from the box

   [1] [2] [3] [4] [5] [6]

   Explain.

8. A coin is tossed 100 times. True or false, and explain:
   (a) The expected value for the number of heads is 50.
   (b) The expected value for the number of heads is 50, give or take 5 or so.
   (c) The number of heads will be 50.
   (d) The number of heads will be around 50, give or take 5 or so.

9. One hundred draws are made at random with replacement from a box with ninety-nine tickets marked "0" and one ticket marked "1." True or false, and explain:
   (a) The sum will be around 1, give or take 1 or so.
   (b) There is about a 68% chance that the sum will be in the range 0 to 2.

10. Ten thousand draws are made at random with replacement from a box with ninety-nine tickets marked "0" and one ticket marked "1." True or false, and explain:
    (a) The sum will be around 100, give or take 10 or so.
    (b) There is about a 68% chance that the sum will be in the range 90 to 110.

11. One hundred draws are made at random with replacement from the box | [1] [2] [2] [5] |. The draws come out as follows: 17 [1]'s, 54 [2]'s, 29 [5]'s. Fill in the blanks, using the options below; show work.
    (a) For the ______, the observed value is 0.8 SEs above the expected value.
    (b) For the ______, the observed value is 1.33 SEs above the expected value.

    Options (one will be left over):

    sum of the draws   number of 1's   number of 2's

12. A box contains ten tickets, four marked with a positive number and six with a negative number. All the numbers are between −10 and 10. One thousand draws will be made at random with replacement from the box. You are asked to estimate the chance that the sum will be positive.
    (a) Can you do it on the basis of the information already given?
    (b) Can you do it if you are also told the average and SD of the numbers in the box, but are not told the numbers themselves?

    Explain briefly.

13. Repeat exercise 12, if you are asked to estimate the chance of getting 100 or more [3]'s.

14. Repeat exercise 12, if you are asked to estimate the chance of getting 425 or more positive numbers.

15. A box contained 1,500 marbles; 600 were red and the others, blue. The following procedure was repeated many times.

    > One hundred draws were made at random with replacement from the box; the number of red marbles among the draws was counted.

    The first 10 counts were 38, 35, 37, 31, 36, 39, 36, 33, 30, 34. Is anything fishy? Answer yes or no, and explain.

8. SUMMARY

1. If the chance process for getting a sum is repeated many times, the empirical histogram for the observed values converges to the *probability histogram.*

2. A probability histogram represents chance by area.

3. When drawing at random with replacement from a box, the probability histogram for the sum will follow the normal curve, even if the contents of the box do not—the "central limit theorem." The histogram must be put into standard units, and the number of draws must be reasonably large.

4. The normal approximation consists in replacing the actual probability histogram by the normal curve, before computing areas. Often, the accuracy of the approximation can be improved by keeping track of the edges of the rectangles—the "continuity correction."

5. Probability histograms which follow the normal curve can be summarized quite well by the expected value and SE. The expected value locates the center of the probability histogram, and the SE measures the spread.

6. Chapter 16 developed box models for gambling. These models are basic to statistical inference (parts VI–VIII).

7. Chapter 17 introduced the SE for sums of draws from a box. The SE for counts, percents (chapter 20), or averages (chapter 23) are then easily computed. Confidence intervals are derived in chapter 21.

8. Chapter 18 showed that probability histograms for sums converge to the normal curve. That justifies "large-sample" statistical theory—reading confidence levels and P-values off the curve (chapters 21 and 26).

9. "Small-sample" statistical theory includes the t-test (section 26.6) and the sign test (section 27.5). In such cases, distributions other than the normal are used.

PART VI

Sampling

19

Sample Surveys

"Data! data! data!" he cried impatiently. "I can't make bricks without clay."
—Sherlock Holmes[1]

1. INTRODUCTION

An investigator usually wants to generalize about a class of individuals. This class is called the *population*. For example, in forecasting the results of a presidential election in the U.S., one relevant population consists of all eligible voters. Studying the whole population is usually impractical. Only part of it can be examined, and this part is called the *sample*. Investigators will make generalizations from the part to the whole. In more technical language, they make *inferences* from the sample to the population.[2]

Usually, there are some numerical facts about the population which the investigators want to know. Such numerical facts are called *parameters*. In forecasting a presidential election in the U.S., two relevant parameters are

- the average age of all eligible voters,
- the percentage of all eligible voters who are currently registered to vote.

Ordinarily, parameters like these cannot be determined exactly, but can only be estimated from a sample. Then a major issue is accuracy. How close are the estimates going to be?

Parameters are estimated by *statistics*, or numbers which can be computed from a sample. For instance, with a sample of 10,000 Americans, an investigator could calculate the following two statistics:

- the average age of the eligible voters in the sample,
- the percentage of the eligible voters in the sample who are currently registered to vote.

Statistics are what investigators know; parameters are what they want to know.

Estimating parameters from the sample is justified when the sample represents the population. This is impossible to check just by looking at the sample. The reason: to see whether the sample is like the population in the ways that matter, investigators would have to know the facts about the population that they are trying to estimate—a vicious circle. Instead, one has to look at how the sample was chosen. Some methods tend to do badly. Others are likely to give representative samples.

The two main lessons of this chapter:

- the method of choosing the sample matters a lot;
- the best methods involve the planned introduction of chance.

Similar issues come up when assigning subjects to treatment or control in experiments: see part I.

2. THE *LITERARY DIGEST* POLL

In 1936, Franklin Delano Roosevelt was completing his first term of office as president of the U.S. It was an election year, and the Republican candidate was Governor Alfred Landon of Kansas. The country was struggling to recover from the Great Depression. There were still nine million unemployed: real income had dropped by one-third in the period 1929–1933 and was just beginning to turn upward. But Landon was campaigning on a program of economy in government, and Roosevelt was defensive about his deficit financing.[3]

Landon. The spenders must go.

Roosevelt. We had to balance the budget of the American people before we could balance the budget of the national government. That makes common sense, doesn't it?

The Nazis were rearming Germany, and the Civil War in Spain was moving to its hopeless climax. These issues dominated the headlines in the *New York Times*, but were ignored by both candidates.

Landon. We must mind our own business.

Most observers thought Roosevelt would be an easy winner. Not so the *Literary Digest* magazine, which predicted an overwhelming victory for Landon, with Roosevelt getting only 43% of the popular vote. This prediction was based on the largest number of people ever replying to a poll—about 2.4 million individuals. It was backed by the enormous prestige of the *Digest*, which had called the winner in every presidential election since 1916. However, Roosevelt won the 1936 election by a landslide—62% to 38%. (The *Digest* went bankrupt soon after.)

The magnitude of the *Digest*'s error is staggering. It is the largest ever made by a major poll. Where did it come from? The number of replies was more than big enough. In fact, George Gallup was just setting up his survey organization.[4] Using his own methods, he drew a sample of 3,000 people and predicted what the *Digest* predictions were going to be—well in advance of their publication—with an error of only one percentage point. Using another sample of about 50,000 people, he correctly forecast the Roosevelt victory, although his prediction of Roosevelt's share of the vote was off by quite a bit. Gallup forecast 56% for Roosevelt; the actual percentage was 62%, so the error was 62% − 56% = 6 percentage points. (Survey organizations use "percentage points" as the units for the difference between actual and predicted percents.) The results are summarized in table 1.

Table 1. The election of 1936.

	Roosevelt's percentage
The election result	62
The *Digest* prediction of the election result	43
Gallup's prediction of the *Digest* prediction	44
Gallup's prediction of the election result	56

Note: Percentages are of the major-party vote. In the election, about 2% of the ballots went to minor-party candidates.
Source: George Gallup, *The Sophisticated Poll-Watcher's Guide* (1972).

To find out where the *Digest* went wrong, you have to ask how they picked their sample. A sampling procedure should be fair, selecting people for inclusion in the sample in an impartial way, so as to get a representative cross section of the public. A systematic tendency on the part of the sampling procedure to exclude one kind of person or another from the sample is called *selection bias*. The *Digest*'s procedure was to mail questionnaires to 10 million people. The names and addresses of these 10 million people came from sources like telephone books and club membership lists. That tended to screen out the poor, who were unlikely to belong to clubs or have telephones. (At the time, for example, only one household in four had a telephone.) So there was a very strong bias against the poor in the *Digest*'s sampling procedure. Prior to 1936, this bias may not have affected the predictions very much, because rich and poor voted along similar lines. But in 1936, the political split followed economic lines more closely. The poor voted overwhelmingly for Roosevelt, the rich were for Landon. One reason for the magnitude of the *Digest*'s error was selection bias.

> When a selection procedure is biased, taking a large sample does not help. This just repeats the basic mistake on a larger scale.

The *Digest* did very badly at the first step in sampling. But there is also a second step. After deciding which people ought to be in the sample, a survey organization still has to get their opinions. This is harder than it looks. If a large number of those selected for the sample do not in fact respond to the questionnaire or the interview, *non-response bias* is likely.

The non-respondents differ from the respondents in one obvious way: they did not respond. Experience shows they tend to differ in other important ways as well.[5] For example, the *Digest* made a special survey in 1936, with questionnaires mailed to every third registered voter in Chicago. About 20% responded, and of those who responded over half favored Landon. But in the election Chicago went for Roosevelt, by a two-to-one margin.

> Non-respondents can be very different from respondents. When there is a high non-response rate, look out for non-response bias.

In the main *Digest* poll, only 2.4 million people bothered to reply, out of the 10 million who got the questionnaire. These 2.4 million respondents do not even represent the 10 million people who were polled, let alone the population of all voters. The *Digest* poll was spoiled both by selection bias and non-response bias.[6]

Special surveys have been carried out to measure the difference between respondents and non-respondents. It turns out that lower-income and upper-income people tend not to respond to questionnaires, so the middle class is over-represented among respondents. For these reasons, modern survey organizations prefer to use personal interviews rather than mailed questionnaires. A typical response rate for personal interviews is 65%, compared to 25% for mailed questionnaires.[7] However, the problem of non-response bias still remains, even with personal interviews. Those who are not at home when the interviewer calls may be quite different from those who are at home, with respect to working hours, family ties, social background, and therefore with respect to attitudes. Good survey organizations keep this problem in mind, and have ingenious methods for dealing with it (section 6).

> Some samples are really bad. To find out whether a sample is any good, ask how it was chosen. Was there selection bias? non-response bias? You may not be able to answer these questions just by looking at the data.

In the 1936 election, how did Gallup predict the *Digest* predictions? He just chose 3,000 people at random from the same lists the *Digest* was going to use, and mailed them all a postcard asking how they planned to vote. He knew that a random sample was likely to be quite representative, as will be explained in the next two chapters.

3. THE YEAR THE POLLS ELECTED DEWEY

Thomas Dewey rose to fame as a crusading D.A. in New York City, and went on to capture the governor's mansion in Albany. In 1948 he was the Republican candidate for president, challenging the incumbent Harry Truman. Truman began political life as a protégé of Boss Pendergast in Kansas City. After being elected to the Senate, Truman became FDR's vice president, succeeding to the presidency when Roosevelt died. Truman was one of the most effective presidents of the 20th century, as well as one of the most colorful. He kept a sign on his desk, "The buck stops here." Another of his favorite aphorisms became part of America's political vocabulary: "If you can't stand the heat, stay out of the kitchen." But Truman was the underdog in 1948, for it was a troubled time. World War II had barely ended, and the uneasy half-peace of the Cold War had just begun. There was disquiet at home, and complicated involvement abroad.

Three major polls covered the election campaign: Crossley, for the Hearst newspapers; Gallup, syndicated in about 100 independent newspapers across the country; and Roper, for *Fortune* magazine. By fall, all three had declared Dewey the winner, with a lead of around 5 percentage points. Gallup's prediction was based on 50,000 interviews; and Roper's on 15,000. As the *Scranton Tribune* put it,

DEWEY AS GOOD AS ELECTED,
STATISTICS CONVINCE ROPER

The statistics didn't convince the American public. On Election Day, Truman scored an upset victory with just under 50% of the popular vote. Dewey got just over 45% (table 2).

Table 2. The election of 1948.

	The predictions			
The candidates	*Crossley*	*Gallup*	*Roper*	*The results*
Truman	45	44	38	50
Dewey	50	50	53	45
Thurmond	2	2	5	3
Wallace	3	4	4	2

Source: F. Mosteller and others, *The Pre-Election Polls of 1948* (New York: Social Science Research Council, 1949).

To find out what went wrong for the polls, it is necessary to find out how they chose their samples.[8] The method they all used is called *quota sampling*. With this procedure, each interviewer was assigned a fixed quota of subjects to interview. The numbers falling into certain categories (like residence, sex, age, race, and economic status) were also fixed. In other respects, the interviewers were free to select anybody they liked. For instance, a Gallup Poll interviewer in St. Louis was required to interview 13 subjects, of whom:[9]

- exactly 6 were to live in the suburbs, and 7 in the central city,
- exactly 7 were to be men, and 6 women.

Of the 7 men (and there were similar quotas for the women):

- exactly 3 were to be under forty years old, and 4 over forty,
- exactly 1 was to be black, and 6 white.

The monthly rentals to be paid by the 6 white men were specified also:

- 1 was to pay $44.01 or more;
- 3 were to pay $18.01 to $44.00;
- 2 were to pay $18.00 or less.

Remember, these are 1948 prices!

From a common-sense point of view, quota sampling looks good. It seems to guarantee that the sample will be like the voting population with respect to all the important characteristics that affect voting behavior. (Distributions of residence, sex, age, race, and rent can be estimated quite closely from Census data.) But the 1948 experience shows this procedure worked very badly. We are now going to see why.

The survey organizations want a sample which faithfully represents the nation's political opinions. However, no quotas can be set on Republican or Democratic votes. The distribution of political opinion is precisely what the survey organizations do not know and are trying to find out. The quotas for the other variables are an indirect effort to make the sample reflect the nation's politics. Fortunately or unfortunately, there are many factors which influence voting behavior besides the ones the survey organizations control for. There are rich white men in the suburbs who vote Democratic, and poor black women in the central cities who vote Republican. As a result, survey organizations may hand-pick a sample which is a perfect cross section of the nation on all the demographic variables, but find the sample voting one way while the nation goes the other. This possibility must have seemed quite theoretical—before 1948.

The next argument against quota sampling is the most important. It involves a crucial feature of the method, which is easy to miss the first time through. Within the assigned quotas, the interviewers are free to choose anybody they like. That leaves a lot of room for human choice. And human choice is always subject to bias. In 1948, the interviewers chose too many Republicans. On the whole, Republicans are wealthier and better educated than Democrats. They are more likely to own telephones, have permanent addresses, and live on nicer blocks. Within each demographic group, Republicans are marginally easier to interview. If you were an interviewer, you would probably end up with too many Republicans.

The interviewers chose too many Republicans in every presidential election from 1936 through 1948, as shown by the Gallup Poll results in table 3. Prior to 1948, the Democratic lead was so great that it swamped the Republican bias in the polls. The Democratic lead was much slimmer in 1948, and the Republican bias in quota sampling had real impact.

Table 3. The Republican bias in the Gallup Poll, 1936–1948.

Year	*Gallup's prediction of Republican vote*	*Actual Republican vote*	*Error in favor of the Republicans*
1936	44	38	6
1940	48	45	3
1944	48	46	2
1948	50	45	5

Note: Percentages are of the majority-party vote, except in 1948.
Source: F. Mosteller and others, *The Pre-Election Polls of 1948* (New York: Social Science Research Council, 1949).

In quota sampling, the sample is hand-picked to resemble the population with respect to some key characteristics. The method seems reasonable, but does not work very well. The reason is unintentional bias on the part of the interviewers.

The quotas in quota sampling are sensible enough, although they do not guarantee success—far from it. But the method of filling the quotas, free choice by the interviewers, is disastrous.[10] The alternative is to use objective and impartial chance mechanisms to select the sample. That will be the topic of the next section.

4. USING CHANCE IN SURVEY WORK

Even in 1948, a few survey organizations used *probability methods* to draw their samples. Now, many organizations do. What is a probability method for drawing a sample? To get started, imagine carrying out a survey of 100 voters in a small town with a population of 1,000 eligible voters. Then, it is feasible to list all the eligible voters, write the name of each one on a ticket, put all 1,000 tickets in a box, and draw 100 tickets at random. Since there is no point interviewing the same person twice, the draws are made without replacement. In other words, the box is shaken to mix up the tickets. One is drawn out at random and set aside. That leaves 999 in the box. The box is shaken again, a second ticket is drawn out and set aside. The process is repeated until 100 tickets have been drawn. The people whose tickets have been drawn form the sample.

This process is called *simple random sampling*: tickets have simply been drawn at random without replacement. At each draw, every ticket in the box has an equal chance to be chosen. The interviewers have no discretion at all in whom they interview, and the procedure is impartial—everybody has the same chance to get into the sample. Consequently, the law of averages guarantees that the percentage of Democrats in the sample is likely to be close to the percentage in the population.

> *Simple random sampling* means drawing at random without replacement.

What happens in a more realistic setting, when the Gallup Poll tries to predict a presidential election? A natural idea is to take a nationwide simple random sample of a few thousand eligible voters. However, this isn't as easy to do as it sounds. Drawing names at random, in the statistical sense, is hard work. It is not at all the same as choosing people haphazardly.

To begin drawing eligible voters at random, you would need a list of all of them—well over 200 million names. There is no such list.[11] Even if there were, drawing a few thousand names at random from 200 million is a job in itself. (Remember, on each draw every name in the box has to have an equal chance of being selected.) And even if you could draw a simple random sample, the people would be scattered all over the map. It would be prohibitively expensive to send interviewers around to find them all.

It just is not practical to take a simple random sample. Consequently, most survey organizations use a probability method called *multistage cluster sampling*. The name is complicated, and so are the details. But the idea is straightforward. It will be described in the context of the Gallup pre-election surveys during the period from 1952 through 1984; these surveys were all done using just about the same procedure. The Gallup Poll makes a separate study in each of the four geographical regions of the United States—Northeast, South, Midwest, and West (figure 1). Within each region, they group together all the population centers of similar sizes. One such grouping might be all towns in the Northeast with a population between 50 and 250 thousand. Then, a random sample of these towns is selected. Interviewers are stationed in the selected towns, and no interviews are conducted in the other towns of that group. Other groupings are handled the same way. This completes the first stage of sampling.[12]

For election purposes, each town is divided up into *wards*, and the wards are subdivided into *precincts*. At the second stage of sampling, some wards are selected—at random—from each town chosen in the stage before. At the third stage, some precincts are drawn at random from each of the previously selected wards. At the fourth stage, households are drawn at random from each selected precinct.[13] Finally, some members of the selected households are interviewed. Even here, no discretion is allowed. For instance, Gallup Poll interviewers are instructed to "speak to the youngest man 18 or older at home, or if no man is at home, the oldest woman 18 or older."[14]

This design offers many of the advantages of quota sampling. For instance, it is set up so the distribution of the sample by residence is the same as the distribution for the nation. But each stage in the selection procedure uses an objective and impartial chance mechanism to select the sample units. This completely eliminates the worst feature of quota sampling: selection bias on the part of the interviewer.

Figure 1. Multistage cluster sampling.

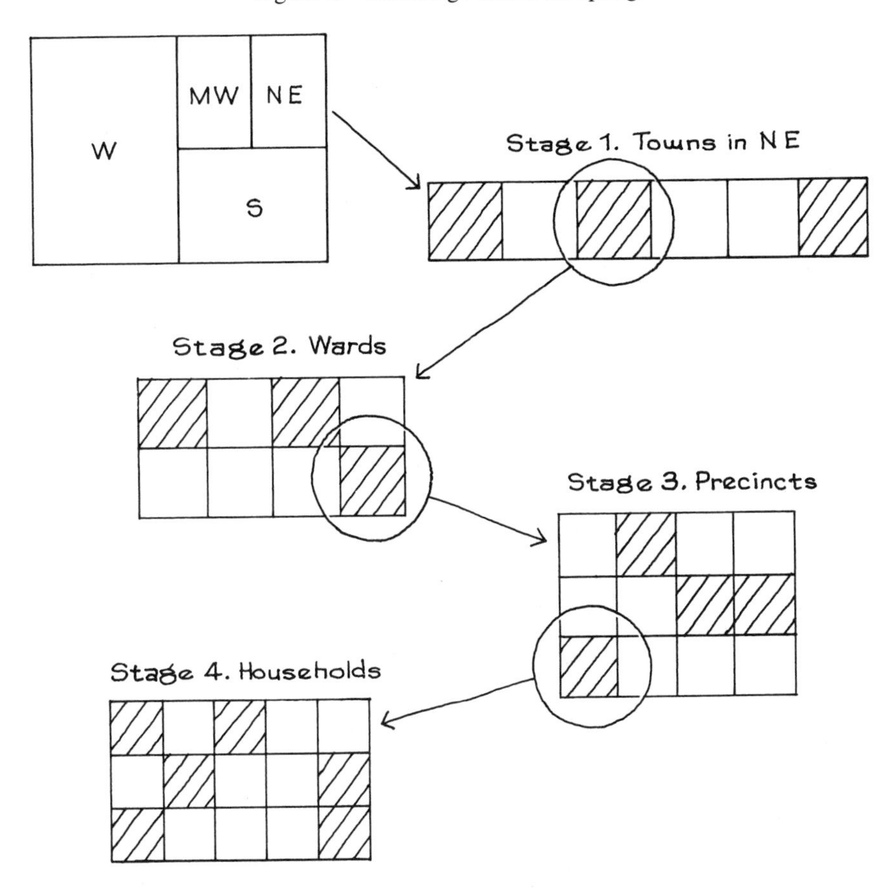

Simple random sampling is the basic probability method. Other methods can be quite complicated. But all probability methods for sampling have two important features:

- the interviewers have no discretion at all as to whom they interview;
- there is a definite procedure for selecting the sample, and it involves the planned use of chance.

As a result, with a probability method it is possible to compute the chance that any particular individuals in the population will get into the sample.[15]

Quota sampling is not a probability method. It fails both tests. The interviewers have a lot of discretion in choosing subjects. And chance only enters in the most unplanned and haphazard way. What kinds of people does the interviewer like to approach? Who is going to be walking down a particular street at a particular time of day? No survey organization can put numbers on these kinds of chances.

5. HOW WELL DO PROBABILITY METHODS WORK?

Since 1948, the Gallup Poll and many other major polls have used probability methods to choose their samples. The Gallup Poll record in post-1948 presidential elections is shown in table 4. There are three points to notice. (i) The sample size has gone down sharply. The Gallup Poll used a sample of size about 50,000 in 1948; they now use samples less than a tenth of that size. (ii) There is no longer any consistent trend favoring either Republicans or Democrats. (iii) The accuracy has gone up appreciably.

From 1936 to 1948, the errors were around 5%. Since then, they are quite a bit smaller. (In 1992, the error went back up to 6%; the reason will be discussed on p. 346.) Using probability methods to select the sample, the Gallup Poll has been able to predict the elections with startling accuracy, sampling less than 5 persons in 100,000—which proves the value of probability methods in sampling.

Table 4. The Gallup Poll record in presidential elections after 1948.

Year	*Sample size*	*Winning candidate*	*Gallup Poll prediction*	*Election result*	*Error*
1952	5,385	Eisenhower	51%	55.1%	4.1%
1956	8,144	Eisenhower	59.5%	57.4%	2.1%
1960	8,015	Kennedy	51%	49.7%	1.3%
1964	6,625	Johnson	64%	61.1%	2.9%
1968	4,414	Nixon	43%	43.4%	0.4 of 1%
1972	3,689	Nixon	62%	60.7%	1.3%
1976	3,439	Carter	48%	50.1%	2.1%
1980	3,500	Reagan	47%	50.7%	3.7%
1984	3,456	Reagan	59%	58.8%	0.2 of 1%
1988	4,089	Bush	56%	53.4%	2.6%
1992	2,019	Clinton	49%	43.0%	6.0%
1996	2,895	Clinton	52%	49.2%	2.8%
2000	3,571	Bush	48%	47.9%	0.1 of 1%
2004	2,014	Bush	49%	50.6%	1.6%

Note: The percentages are of the popular vote. The error is the absolute difference "predicted – actual."
Source: The Gallup Poll (American Institute of Public Opinion) for predictions; *Statistical Abstract*, 2006, Table 384 for actuals.

Why do probability methods work so well? At first, it may seem that judgment is needed to choose the sample. For instance, quota sampling guarantees that the percentage of men in the sample will be equal to the percentage of men in the population. With probability sampling, we can only say that the percentage of men in the sample is likely to be close to the percentage in the population: certainty is reduced to likelihood. But judgment and choice usually show bias, while chance is impartial. That is why probability methods work better than judgment.

> To minimize bias, an impartial and objective probability method should be used to choose the sample.

6. A CLOSER LOOK AT THE GALLUP POLL

Some degree of bias is almost inevitable even when probability methods are used to select the sample, due to the many practical difficulties that survey organizations must overcome. The discussion here is organized around the questionnaire used by the Gallup Poll in the presidential election of 1984. See figures 2 and 3.

Figure 2. The Gallup Poll ballot, 1984. The interviewers use secret ballots, to minimize the number of undecided respondents.

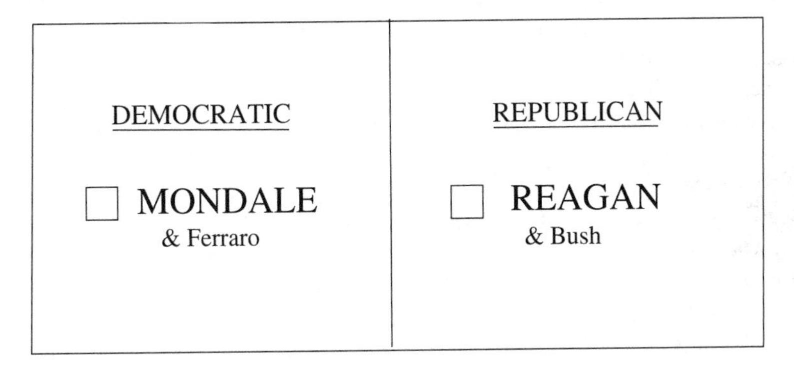

DEMOCRATIC	REPUBLICAN
☐ MONDALE & Ferraro	☐ REAGAN & Bush

"I'D SAY I'M ABOUT FORTY-TWO PERCENT FOR REAGAN, THIRTY-NINE PERCENT FOR MONDALE, AND NINETEEN PERCENT UNDECIDED."

The nonvoters. In a typical presidential election, between one-third and one-half of the eligible voters fail to vote. The job of the Gallup Poll is to predict what the voters will do; the non-voters are irrelevant and should be screened out of the sample as far as possible. That is not so easy. There is a stigma to non-voting, and many respondents say they will vote even if they know better. The problem of screening out non-voters is handled by questions 1–6, and some later questions too: this is a hot topic for pollsters. Question 3, for instance, asks where the respondent would go to vote (figure 3, p. 345). Respondents who know the answer are more likely to vote. Question 13 (p. 347) asks whether the respondent voted in the last election, and is phrased to make a negative answer easy to give—compensating for the stigma attached to non-voting. Respondents who voted last time are more likely to vote this time.

This battery of questions is used to decide whether the respondent is likely to vote; the election forecasts are based only on that part of the sample judged likely to vote. It is a matter of record who actually votes in each election. Post-election studies by the Gallup organization show that their judgments as to who will vote are reasonably accurate. The studies also show that screening out likely non-voters increases the accuracy of the election forecasts, because preferences of likely voters differ from preferences of likely non-voters.[16]

The undecided. Some percentage of the subjects being interviewed are undecided how they will vote. Question 7, which asks for the preferences, is designed to keep the percentage as small as possible. To begin with, it asks how the respondent would vote the day of the interview, rather than Election Day. Subjects who cannot decide are asked to indicate "the candidates toward whom you lean as of today." A final device is the paper ballot (figure 2, p. 343). Instead of naming their preferences out loud, the respondents just mark the ballot and drop it into a box carried by the interviewer.[17]

These techniques have been found to minimize the percentage of undecided. But there are still some left, and if they are thought likely to vote, the Gallup Poll has to guess how. Some information about political attitudes is available from questions 12–14 (p. 347). This information might be used to predict how the undecided respondents are going to vote, but it is difficult to say how well the predictions work.

Response bias. The answers given by respondents are influenced to some extent by the phrasing of the questions, and even the tone or attitude of the interviewer. This kind of distortion is called *response bias.* There was a striking example in the 1948 election survey: changing the order of the candidates' names was found to change the response by 5%, the advantage being with the candidate who was named first. To control response bias, all interviewers use the same questionnaire, and the interview procedure is standardized as far as possible. The ballot technique was found to reduce the effect of the political attitudes of the interviewer on the responses of the subjects.

Non-response bias. Even with personal interviews, many subjects are missed. Since they tend to be different from the subjects available for the interview, a non-response bias is created. To some extent, this bias can be adjusted out, by giving more weight to the subjects who were available but hard to get. This information is obtained by question 20 (p. 347), which asks whether the subject was at home on the previous days. This is done quite subtly, as you can tell by reading the question.

Figure 3. The Gallup Poll questionnaire for the 1984 election. Courtesy of the Gallup Poll News Service.

SURVEY: A1813

DATE: October 25, 1984

No publication, reproduction, dissemination or other use of this questionnaire or any replies thereto, written or oral, is authorized by The Gallup Organization, Inc. Violators of this notice will be prosecuted to the fullest extent of the law.

U.S. & Worldwide

The Gallup Survey

Sponsored by leading Newspapers, Corporations and Agencies.

Copyright 1977 The Gallup Organization, Inc. Princeton, New Jersey 08540

SUGGESTED INTRODUCTION: I'm taking a GALLUP SURVEY. I'd like YOUR opinion on some topics of interest.

Time started: ________

Time ended: ________

Length: ________

1. How much thought have you given to the coming November elections—quite a lot, or only a little?
   1 ☐ Quite a lot
   2 ☐ Some — (volunteered)
   3 ☐ Little
   y ☐ None

2. Have you ever voted in this precinct or district?
   1 ☐ Yes
   2 ☐ No
   y ☐ Don't know

3. Where do people who live in this neighborhood go to vote?
   1 ☐ Specify: ________
   y ☐ Don't know

4a. Are you NOW registered so that you can vote in the election this November?
   1 ☐ Yes — (GO TO Q. 5)
   2 ☐ No
   3 ☐ Don't have to register — (GO TO Q. 5)
   y ☐ Don't know

4b. Do you plan to register so that you can vote in the November election?
   1 ☐ Yes
   2 ☐ No
   3 ☐ Other: ________

5. Generally speaking, how much interest would you say you have in politics — a great deal, a fair amount, only a little, or no interest at all?
   1 ☐ Great deal
   2 ☐ Fair amount
   3 ☐ Little
   y ☐ None

6. How often would you say you vote — always, nearly always, part of the time, or seldom?
   1 ☐ Always
   2 ☐ Nearly always
   3 ☐ Part of the time
   4 ☐ Seldom
   5 ☐ Other: ________
   y ☐ Never vote

7. Suppose you were voting TODAY for president and vice president of the United States. Here is a Gallup Poll secret ballot listing the candidates for these offices. (TEAR OFF ATTACHED BALLOT AND HAND TO RESPONDENT.) Will you please MARK that secret ballot for the candidates you favor today — and then drop the folded ballot into the box.

   INTERVIEWER: IF RESPONDENT HANDS BACK BALLOT AND SAYS HE HASN'T MADE UP HIS MIND OR REFUSES TO MARK IT SAY:

   Well, would you please mark the ballot for the candidates toward whom you lean as of today?

   IF RESPONDENT STILL CAN'T DECIDE OR REFUSES TO MARK THE BALLOT, PLEASE WRITE THAT ON THE BALLOT AND BE SURE TO DROP IT IN THE BOX.

8. Right now, how strongly do you feel about your choice—very strongly, fairly strongly, or not strongly at all?
   1 ☐ Very strongly
   2 ☐ Fairly strongly
   3 ☐ Not strongly at all
   4 ☐ Didn't make choice
   y ☐ Don't know

9a. Do you, yourself, plan to vote in the election this November, or not?
   1 ☐ Yes
   2 ☐ No — (GO TO Q. 10a)
   y ☐ Don't know — (GO TO Q. 10a)

9b. How certain are you that you will vote—ABSOLUTELY certain, FAIRLY certain, or NOT certain?
   1 ☐ Absolutely
   2 ☐ Fairly
   y ☐ Not certain

10a. If the elections for Congress were being held TODAY, which party would you like to see win in this Congressional district, the Democratic Party or the Republican Party?
   1 ☐ Democratic — (GO TO Q. 11)
   2 ☐ Republican — (GO TO Q. 11)
   3 ☐ Other — (GO TO Q. 11)
   y ☐ Undecided, refused

10b. As of today, do you lean more to the Democratic Party or more to the Republican Party?
   1 ☐ Democratic
   2 ☐ Republican
   3 ☐ Other:
   4 ☐ Undecided
   y ☐ Refused

11. Here is a picture of a ladder. (HAND RESPONDENT CARD 1.) Suppose we say the top of the ladder (POINT) marked 10 represents a person who definitely will vote in the election this November, and the bottom of the ladder (POINT) marked zero represents a person who definitely will not vote in the election. How far up or down the ladder would you place yourself? (INTERVIEWER: CIRCLE NUMBER.)

   10 9 8 7 6 5 4 3 2 1 0

   y ☐ Don't know

Check data. The Gallup sample usually includes proportionately too many people with higher education. In a detailed analysis, less weight is put on the responses of those subjects (question 16). Other demographic data can be used in a similar way. This weighting technique is called "ratio estimation." Do not confuse ratio estimation with quota sampling. Ratio estimation is an objective, arithmetic technique applied to the sample after it is chosen, to compensate for various small biases in the sampling procedure. Quota sampling is a method for choosing the sample. It has a large, subjective component—when the interviewer chooses the subjects—and introduces large biases.

Interviewer control. In large-scale survey work, there is always the problem of making sure interviewers follow instructions. Some redundancy is built into the questionnaire, so the answers can be checked for consistency: inconsistencies suggest the interviewer may not be doing the job properly. A small percentage of the subjects are reinterviewed by administrative staff, as a further check on the quality of the work.

Talk is cheap. It is a little risky to predict what people will do on Election Day from what they tell the interviewer they are going to do. People may be unwilling to reveal their true preferences. Even if they do, they may change their minds later. Words and deeds are often different.

The 1992 election. In 1992, there was a fairly large percentage of undecided respondents, and Gallup allocated all of them to Clinton. That turned out to be a bad idea. Many of the undecided seem in the end to have voted for Perot, explaining Gallup's large error for the 1992 election (table 4, p. 342). Predicted and actual votes for Clinton, Bush, and Perot are shown below.

	Clinton	*Bush*	*Perot*
Gallup	49%	37%	14%
Actual	43.0%	37.4%	18.9%

7. TELEPHONE SURVEYS

Many surveys are now conducted by telephone. The savings in costs are dramatic, and—if the work is up to scratch—the results are good. The Gallup Poll changed over to the telephone in 1988, with 200 interviewers covering the whole country in a few days, from offices in Atlanta, Austin, Lincoln, Minneapolis, and Omaha.

How do they pick the sample? In 1988, the Gallup Poll used a multistage cluster sample based on area codes, "exchanges," and "banks:"

Area code	*Exchange*	*Bank*	*Digits*
415	767	26	76

In 1992, they switched to a simpler design. There are 4 time zones in the U.S. The Gallup Poll divided each zone into 3 types of areas, according to population density (heavy, medium, light). That gives $4 \times 3 = 12$ strata. For example, one stratum consisted of heavily populated areas in the Eastern time zone; another consisted of lightly populated areas on Pacific time. Within each stratum, the Gallup Poll just drew a simple random sample of telephone numbers, using the computer to

Figure 3. The Gallup Poll questionnaire for the 1984 election, continued. Courtesy of the Gallup Poll News Service.

NOW, HERE ARE A FEW QUESTIONS SO THAT MY OFFICE CAN KEEP TRACK OF THE CROSS-SECTION OF PEOPLE I'VE TALKED TO:

12. In politics, as of TODAY, do you consider yourself a Republican, Democrat, or Independent?
 1 ☐ Republican
 2 ☐ Democrat
 3 ☐ Independent
 4 ☐ Other: ______________

13. In the election in November 1980 — when Carter ran against Reagan and Anderson — did things come up which kept you from voting, or did you happen to vote? For whom?
 1 ☐ Carter
 2 ☐ Reagan
 3 ☐ Anderson
 4 ☐ Other
 5 ☐ Voted, don't remember for whom
 6 ☐ No, didn't vote
 y ☐ Don't remember if voted

14. Are you, or is your (husband/wife) a member of a labor union?
 1 ☐ Yes, respondent is
 2 ☐ Yes, spouse is
 3 ☐ Yes, both are
 y ☐ No, neither is

15. (HAND RESPONDENT CARD 2) Please tell me which of the categories on this card MOST NEARLY describes the kind of work the chief wage earner in your immediate family does. Just call off the number, please. (INTERVIEWER: IF THE CHIEF WAGE EARNER IS UNEMPLOYED, ASK WHAT TYPE OF WORK HE/SHE WOULD DO IF EMPLOYED.)

1 ☐	11 ☐
2 ☐	12 ☐
3 ☐	13 ☐
4 ☐	14 ☐
5 ☐	15 ☐
6 ☐	16 ☐ Other: ______________
7 ☐	17 ☐ Can't say
8 ☐	
9 ☐	
10 ☐	

16. What was the last grade or class you COMPLETED in school?
 1 ☐ None or Grades 1–4
 2 ☐ Grades 5, 6, 7
 3 ☐ Grade 8
 4 ☐ High school, incomplete (Grades 9–11)
 5 ☐ High school, graduated (Grade 12)
 6 ☐ Technical, trade, or business
 7 ☐ College, university, incomplete
 8 ☐ College, university, graduated

17. What is your religious preference — Protestant, Roman Catholic, Jewish, or an Orthodox church such as the Greek or Russian Orthodox Church?
 1 ☐ Protestant
 2 ☐ Roman Catholic
 3 ☐ Jewish
 4 ☐ Orthodox Church
 5 ☐ Other: ______________
 y ☐ None

18. How many persons 18 years and over are there now living in this household, including yourself? Include lodgers, servants, or other employees living in the household. (CIRCLE NUMBER)

 1 2 3 4 5 6 7 8 9 or more

19. (HAND RESPONDENT CARD 3) From what nationality group or groups are you mainly descended? Just call off the number please.

1 ☐	11 ☐
2 ☐	12 ☐
3 ☐	13 ☐
4 ☐	14 ☐ Don't know (VOLUNTEERED) or refused
5 ☐	
6 ☐	
7 ☐	
8 ☐	
9 ☐	
10 ☐	

20a. We are interested in finding out how often people are at home to watch TV or listen to the radio. Would you mind telling me whether or not you happened to be at home yesterday (last night/last Saturday) at this particular time?

(INTERVIEWER: SEE INTERVIEWER'S BULLETIN FOR HANDLING THIS QUESTION.)
 1 ☐ Yes, at home
 2 ☐ No, not at home

20b. How about the day (night/Saturday) before at this time?
 1 ☐ Yes, at home
 2 ☐ No, not at home

20c. And how about the day (night/Saturday) before at this time? That was ________.
 1 ☐ Yes, at home
 2 ☐ No, not at home

21. And what is your age?
 RECORD AGE: ________

22. CHECK WHETHER:
 1 ☐ White man
 2 ☐ White woman
 3 ☐ Black man
 4 ☐ Black woman
 5 ☐ Other man (SPECIFY) ______________
 6 ☐ Other woman (SPECIFY) ______________

So that my office can check my work in this interview if it wants to, may I have your name, address, and telephone number please?

NAME: ______________________________

ADDRESS: ______________________________
House No. or RFD Route, St. or Rd., Apt. No.)

CITY: ______________ STATE: ________ ZIP ________

TELEPHONE: Area Code ______ Phone No. ________ y ☐ No tel.

PLACE INTERVIEWER BADGE NUMBER HERE

I hereby attest that this is a true and honest interview.

____
(Interviewer's Signature)

Date of interview: ____________

Time interview ended: ____________

exclude businesses by checking the yellow pages. Choosing telephone numbers at random is called RDD, for *random digit dialing*.[18]

People who do not have phones must be different from the rest of us, and that does cause a bias in telephone surveys. The effect is small, because these days nearly everybody has a phone. On the other hand, about one-third of residential telephones are unlisted. Rich people and poor people are more likely to have unlisted numbers, so the telephone book tilts toward the middle class. Sampling from directories would create a real bias, but random digit dialing gets around this difficulty. In 2005, survey organizations were just beginning to work on the questions raised by cell phones. What about dropped calls? Who pays for air time? What to do with people who have land lines and cell phones?

Non-respondents create problems, as usual. So the Gallup Poll does most of its interviewing on evenings and weekends, when people are more likely to be at home. If there is no answer, the interviewer will call back up to 3 times.[19] (Some designs have up to 15 call-backs; that is better, but more expensive.) For many purposes, results are comparable to those from face-to-face interviews, and the cost is about one-third as much. That is why survey organizations are using the telephone.

8. CHANCE ERROR AND BIAS

The previous sections indicated the practical difficulties faced by real survey organizations. People are not at home, or they do not reveal their true preferences, or they change their minds. However, even if all these difficulties are assumed away, the sample is still likely to be off—due to chance error.

To focus the issue, imagine a box with a very large number of tickets, some marked 1 and the others marked 0. That is the population. A survey organization is hired to estimate the percentage of 1's in the box. That is the parameter. The organization draws 1,000 tickets at random without replacement. That is the sample. There is no problem about response—the tickets are all there in the box. Drawing them at random eliminates selection bias. And the tickets do not change back and forth between 0 and 1. As a result, the percentage of 1's in the sample is going to be a good estimate for the percentage of 1's in the box. But the estimate is still likely to be a bit off, because the sample is only part of the population. Since the sample is chosen at random, the amount off is governed by chance:

$$\text{percentage of 1's in sample} = \text{percentage of 1's in box} + \text{chance error.}$$

Now there are some questions to ask about chance errors—

- How big are they likely to be?
- How do they depend on the size of the sample? the size of the population?
- How big does the sample have to be in order to keep the chance errors under control?

These questions will be answered in the next two chapters.

In more complicated situations, the equation has to take bias into account:

$$\text{estimate} = \text{parameter} + \text{bias} + \text{chance error}.$$

Chance error is often called "sampling error:" the "error" comes from the fact that the sample is only part of the whole. Similarly, bias is called "non-sampling error"—the error from other sources, like non-response. Bias is often a more serious problem than chance error, but methods for assessing bias are not well developed. Usually, "bias" means prejudice. However, statistics is a dry subject. For a statistician, bias just means any kind of systematic error in an estimate. "Non-sampling error" is a more neutral term, and may be better for that reason.

Exercise Set A

1. A survey is carried out at a university to estimate the percentage of undergraduates living at home during the current term. What is the population? the parameter?

2. The registrar keeps an alphabetical list of all undergraduates, with their current addresses. Suppose there are 10,000 undergraduates in the current term. Someone proposes to choose a number at random from 1 to 100, count that far down the list, taking that name and every 100th name after it for the sample.

   (a) Is this a probability method?
   (b) Is it the same as simple random sampling?
   (c) Is there selection bias in this method of drawing a sample?

3. The monthly Gallup Poll opinion survey is based on a sample of about 1,500 persons, "scientifically chosen as a representative cross section of the American public." The Gallup Poll thinks the sample is representative mainly because—

   (i) it resembles the population with respect to such characteristics as race, sex, age, income, and education

   or

   (ii) it was chosen using a probability method.

4. In the Netherlands, all men take a military pre-induction exam at age 18. The exam includes an intelligence test known as "Raven's progressive matrices," and includes questions about demographic variables like family size. A study was done in 1968, relating the test scores of 18-year-old men to the number of their brothers and sisters.[20] The records of all the exams taken in 1968 were used.

   (a) What is the population? the sample?
   (b) Is there any sampling error? Explain briefly.

5. Polls often conduct pre-election surveys by telephone. Could this bias the results? How? What if the sample is drawn from the telephone book?

6. About 1930, a survey was conducted in New York on the attitude of former black slaves towards their owners and conditions of servitude.[21] Some of the interviewers were black, some white. Would you expect the two groups of interviewers to get similar results? Give your reasons.

7. One study on slavery estimated that "11.9% of slaves were skilled craftsmen." This estimate turns out to be based on the records of thirty plantations in Plaquemines Parish, Louisiana.[22] Is it trustworthy? Explain briefly.

8. In one study, the Educational Testing Service needed a representative sample of college students.[23] To draw the sample, they first divided up the population of all colleges and universities into relatively homogeneous groups. (One group consisted of all public universities with 25,000 or more students; another group consisted of all private four-year colleges with 1,000 or fewer students; and so on.) Then they used their judgment to choose one representative school from each group. That created a sample of schools. Each school in the sample was then asked to pick a sample of students. Was this a good way to get a representative sample of students? Answer yes or no, and explain briefly.

9. A study was done on the prevalence of chest diseases in a Welsh coal mining town; 600 volunteers had chest X-rays done.[24] At the time, the two main chest diseases in the town were pneumoconiosis (scarring of the lung tissue due to inhalation of dust) and tuberculosis. The data were analyzed by the order in which the volunteers presented themselves. The percentage with tuberculosis among the first 200 subjects to appear for the examination was probably ______ the percentage among the last 200. Fill in the blank, using one of the phrases

   (i) about the same as (ii) quite a bit different from

   Explain your reasoning.

10. Television advertising sales are strongly influenced by the Nielsen ratings. In its annual report, the Nielsen organization does not describe how it takes samples. The report does say:[25]

    > Nielsen, today as in the past, is dedicated to using the newest, most reliable, and thoroughly tested research technologies. This is a commitment to those we serve through the television, cable, and advertising communities . . . .
    >
    > The Nielsen data in this booklet are estimates of the audiences and other characteristics of television usage as derived from Nielsen Television Index and Nielsen Station Index measurements. The use of mathematical terms herein should not be regarded as a representation by Nielsen that such measurements are exact to precise mathematical values . . . .

    Comment briefly.

11. The *San Francisco Examiner* ran a story headlined—

    > 3 IN 10 BIOLOGY TEACHERS BACK BIBLICAL CREATIONISM
    >
    > *Arlington, Texas.* Thirty percent of high school biology teachers polled believe in the biblical creation and 19 percent incorrectly think that humans and dinosaurs lived at the same time, according to a nationwide survey published Saturday.
    >
    > "We're doing something very, very, very wrong in biology education," said Dana Dunn, one of two sociologists at the University of Texas, Arlington.
    >
    > Dunn and Raymond Eve sent questionnaires to 20,000 high school biology teachers selected at random from a list provided by the National Science Teachers Association and received 200 responses . . . .

    The newspaper got it wrong. Dunn and Eve did not send out 20,000 questionnaires: they chose 400 teachers at random from the National Science Teachers association list, sent questionnaires to these 400 people, and received 200 replies.[26] Why do these corrections matter?

12. In any survey, a fair number of people who are in the original sample cannot be contacted by the survey organization, or are contacted but refuse to answer questions. A high non-response rate is a serious problem for survey organizations. True or false, and explain: this problem is serious because the investigators have to spend more time and money getting additional people to bring the sample back up to its planned size.

The answers to these exercises are on pp. A78–79.

9. REVIEW EXERCISES

Review exercises may cover material from previous chapters.

1. A survey organization is planning to do an opinion survey of 2,500 people of voting age in the U.S. True or false, and explain: the organization will choose people to interview by taking a simple random sample.

2. Two surveys are conducted to measure the effect of an advertising campaign for a certain brand of detergent.[27] In the first survey, interviewers ask housewives whether they use that brand of detergent. In the second, the interviewers ask to see what detergent is being used. Would you expect the two surveys to reach similar conclusions? Give your reasons.

3. One study on slavery estimated that a slave had only a 2% chance of being sold into the interstate trade each year. This estimate turns out to be based on auction records in Anne Arundel County, Maryland.[28] Is it trustworthy? Explain briefly.

4. In one study, it was necessary to draw a representative sample of Japanese-Americans resident in San Francisco.[29] The procedure was as follows. After consultation with representative figures in the Japanese community, the four most representative blocks in the Japanese area of the city were chosen. All persons resident in those four blocks were taken for the sample. However, a comparison with Census data shows that the sample did not include a high-enough proportion of Japanese with college degrees. How can this be explained?

5. (Hypothetical.) A survey is carried out by the finance department to determine the distribution of household size in a certain city. They draw a simple random sample of 1,000 households. After several visits, the interviewers find people at home in only 653 of the sample households. Rather than face such a high non-response rate, the department draws a second batch of households, and uses the first 347 completed interviews in the second batch to bring the sample up to its planned strength of 1,000 households. The department counts 3,087 people in these 1,000 households, and estimates the average household size in the city to be about 3.1 persons. Is this estimate likely to be too low, too high, or about right? Why?

6. "Ecstasy" was a popular drug in the 1990s. It produced a sense of euphoria derisively called the "yuppie high." One investigator made a careful sample survey to estimate the prevalence of drug use at Stanford University. Two assistants were stationed on the main campus plaza and instructed to interview all students who passed through at specified times. As it turned out, 39% of 369 students interviewed said they had used Ecstasy at least once.[30] Does the investigator's procedure give a probability sample of Stanford students? Answer yes or no, and explain.

7. A coin is tossed 1,000 times. There are two options:
   (i) To win $1 if the number of heads is between 490 and 510.
   (ii) To win $1 if the percentage of heads is between 48% and 52%.

   Which option is better? Or are they the same? Explain.

8. Can you tell whether the figure below is a probability histogram or a histogram for data? If so, which is it and why? If you can't tell, why not?

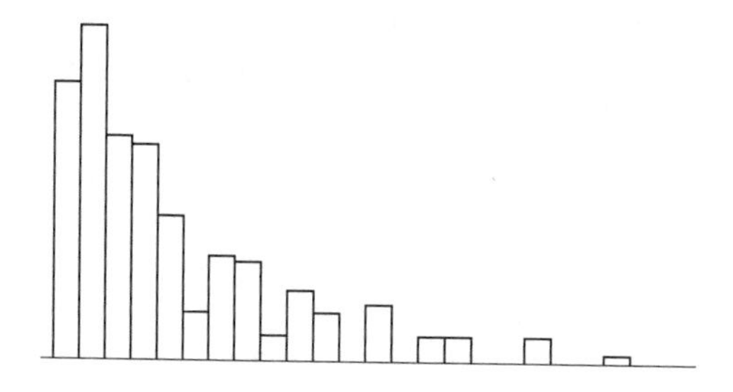

9. One hospital has 218 live births during the month of January.[31] Another has 536. Which is likelier to have 55% or more male births? Or is it equally likely? Explain. (There is about a 52% chance for a live-born infant to be male.)

10. A coin will be tossed 100 times. You get to pick 11 numbers. If the number of heads turns out to equal one of your 11 numbers, you win a dollar. Which 11 numbers should you pick, and what is your chance (approximately) of winning? Explain.

11. A sorcerer has hidden a Porsche in one of an infinite row of boxes

The sorcerer will let you drive away with the car if you can find it. But you are only allowed to look in 11 boxes. He agrees to give you a hint, by tossing a coin 100 times and counting the number of heads. He will not tell you this number, or the number of the box in which he hid the car. But he will tell you the sum of the two numbers.

(a) If the sum is 65, which 11 boxes would you look in?
(b) As in (a), except replace 65 by 95.
(c) What is the general rule?
(d) Following this rule, how likely are you to get the Porsche?

12. The *San Francisco Chronicle* reported on a survey of top high-school students in the U.S. According to the survey,

> Cheating is pervasive. Nearly 80 percent admitted some dishonesty, such as copying someone's homework or cheating on an exam. The survey was sent last spring to 5,000 of the nearly 700,000 high achievers included in the 1993 edition of *Who's Who Among American High School Students*. The results were based on the 1,957 completed surveys that were returned. "The survey does not pretend to be representative of all teenagers," said *Who's Who* spokesman Andrew Weinstein. "Students are listed in *Who's Who* if they are nominated by their teachers or guidance counselors. Ninety-eight percent of them go on to college."

(a) Why isn't the survey "representative of all teenagers"?
(b) Is the survey representative "of the nearly 700,000 high achievers included in the 1993 edition of *Who's Who Among American High School Students*"? Answer yes or no, and explain briefly.

10. SUMMARY

1. A *sample* is part of a *population*.

2. A *parameter* is a numerical fact about a population. Usually a parameter cannot be determined exactly, but can only be estimated.

3. A *statistic* can be computed from a sample, and used to estimate a parameter. A statistic is what the investigator knows. A parameter is what the investigator wants to know.

4. When estimating a parameter, one major issue is accuracy: how close is the estimate going to be?

5. Some methods for choosing samples are likely to produce accurate estimates. Others are spoiled by *selection bias* or *non-response bias*. When thinking about a sample survey, ask yourself:

- What is the population? the parameter?
- How was the sample chosen?
- What was the response rate?

6. Large samples offer no protection against bias.

7. In *quota sampling*, the sample is hand picked by the interviewers to resemble the population in some key ways. This method seems logical, but often

gives bad results. The reason: unintentional bias on the part of the interviewers, when they choose subjects to interview.

8. *Probability methods* for sampling use an objective chance process to pick the sample, and leave no discretion to the interviewer. The hallmark of a probability method: the investigator can compute the chance that any particular individuals in the population will be selected for the sample. Probability methods guard against bias, because blind chance is impartial.

9. One probability method is *simple random sampling*. This means drawing subjects at random without replacement.

10. Even when using probability methods, bias may come in. Then the estimate differs from the parameter, due to bias and chance error:

$$\text{estimate} = \text{parameter} + \text{bias} + \text{chance error}.$$

Chance error is also called "sampling error," and bias is "non-sampling error."

20

Chance Errors in Sampling

To all the ladies present and some of those absent.
—THE TOAST USUALLY PROPOSED BY JERZY NEYMAN

1. INTRODUCTION

Sample surveys involve chance error. This chapter will explain how to find the likely size of the chance error in a percentage, for simple random samples from a population whose composition is known. That mainly depends on the size of the sample, not the size of the population. First, an example. A health study is based on a representative cross section of 6,672 Americans age 18 to 79. A sociologist now wishes to interview these people. She does not have the resources to do them all, in fact she only has enough money to sample 100 of them. To avoid bias, she is going to draw the sample at random. In the imaginary dialogue which follows, she is discussing the problem with her statistician.[1]

Soc. I guess I have to write all the 6,672 names on separate tickets, put them in a box, and draw out 100 tickets at random. It sounds like a lot of work.

Stat. We have the files on the computer, code-numbered from 1 to 6,672. So you could just draw 100 numbers at random in that range. Your sample would be the people with those code numbers.

Soc. Yes, but then I still have to write the numbers from 1 to 6,672 on the tickets. You haven't saved me much time.

Stat. That isn't what I had in mind. With a large box, it's hard to mix the tickets properly. If you don't, most of the draws probably come from the tickets you put in last. That could be a serious bias.

Soc. What do you suggest?

Stat. The computer has a random number generator. It picks a number at random from 1 to 6,672. The person with that code number goes into the sample. Then it picks a second code number at random, different from the first. That's the second person to go into the sample. The computer keeps going until it gets 100 people. Instead of trying to mix the tickets yourself, let the random numbers do the mixing. Besides, the computer saves all that writing.

Soc. OK. But if we use the computer, will my sample be representative?

Stat. What do you have in mind?

Soc. Well, there were 3,091 men and 3,581 women in the original survey: 46% were men. I want my sample to have 46% men. Besides that, I want them to have the right age distribution. Then there's income and education to think about. Of course, what I really want is a group whose attitudes to health care are typical.

Stat. Let's not get into attitudes right now. First things first. I drew a sample to show you. Look at table 1. The first person chosen by the computer was female, so was the second. But the third was male. And so on. Altogether, you got 51 men. That's pretty close.

Table 1. One hundred people were chosen at random and classified by sex. Fifty-one were men (M), and 49 were women (F). In the population, the percentages were 46% and 54%.

F	F	M	F	M	M	F	M	M	M	M	F	M	M	M	M	F	M	F	F
F	M	M	F	M	F	F	M	F	F	M	M	F	F	F	M	F	M	F	M
F	M	F	F	M	M	F	M	M	F	M	F	M	F	M	M	M	F	F	F
F	M	M	M	F	M	F	M	M	F	M	M	M	M	F	F	F	M	F	M
F	M	F	M	M	M	F	F	F	F	M	M	F	M	M	F	F	F	F	F

Soc. But there should only be 46 men. There must be something wrong with the computer.

Stat. No, not really. Remember, the people in the sample are drawn at random. Just by the luck of the draw, you could get too many men—or too few. I had the computer take a lot of samples for you, 250 in all (table 2). The number of men ranged from a low of 34 to a high of 58. Only 17 samples out of the lot had exactly 46 men. There's a histogram (figure 1).

Soc. What stops the numbers from being 46?

Stat. Chance variability. Remember the Kerrich experiment I told you about the other day?

Soc. Yes, but that was about coin tossing, not sampling.

Figure 1. Histogram for the number of men in samples of size 100.

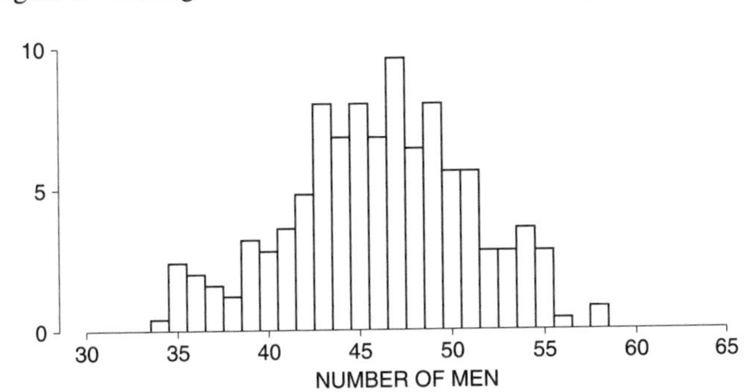

Stat. Well, there isn't much difference between coin tossing and sampling. Each time you toss the coin, you either get a head or a tail, and the number of heads either goes up by one or stays the same. The chances are 50–50 each time. It's the same with sampling. Each time the computer chooses a person for the sample, it either gets a man or a woman, so the number of men either goes up by one or stays the same. The chances are just about 46 to 54 each time—taking 100 tickets out of the box can't change the proportions in the box very much.

Soc. What's the point?

Stat. The chance variability in sampling is just like the chance variability in coin tossing.

Soc. Hmmm. What happens if we increase the size of the sample? Won't it come out more like the population?

Stat. Right. For instance, suppose we increase the sample size by a factor of four, to 400. I got the computer to draw another 250 samples, this time with 400 people in each sample. With some of these samples, the percentage of men is below 46%, with others it is above. The low is 39%, the high is 54%.

Table 2. Two hundred fifty random samples were drawn from the respondents to a health study, of whom 46% were men. The sample size was 100. The number of men in each sample is shown below.

51	40	49	34	36	43	42	45	48	47	51	47	50	54	39	42	47	43	46	46	51	43	53	43	51
42	49	46	44	55	36	49	44	43	45	42	42	45	43	55	53	49	46	45	42	48	44	43	41	44
47	54	54	39	39	52	43	36	39	43	43	46	47	44	55	50	53	55	45	43	47	40	47	40	51
43	56	40	40	49	47	45	49	41	43	45	54	49	50	44	46	48	52	45	47	50	53	46	44	47
47	46	54	42	44	47	47	36	52	50	51	48	46	45	54	48	46	41	49	37	49	45	50	43	54
39	55	38	49	44	43	47	51	46	51	49	42	50	48	52	54	47	51	49	44	37	43	41	48	39
50	41	48	47	50	48	46	37	41	55	43	48	44	40	50	58	47	47	48	45	52	35	45	41	35
38	44	50	44	35	48	49	35	41	37	46	49	42	53	47	48	36	51	45	43	52	46	49	51	44
51	51	39	45	44	40	50	50	46	50	49	47	45	49	39	44	48	42	47	38	53	47	48	51	49
45	42	46	49	45	45	42	45	53	54	47	43	41	49	48	35	55	58	35	47	52	43	45	44	46

Figure 2. Histogram for the percentages of men in samples of size 400. There are 250 samples, drawn at random from the respondents to the health study.

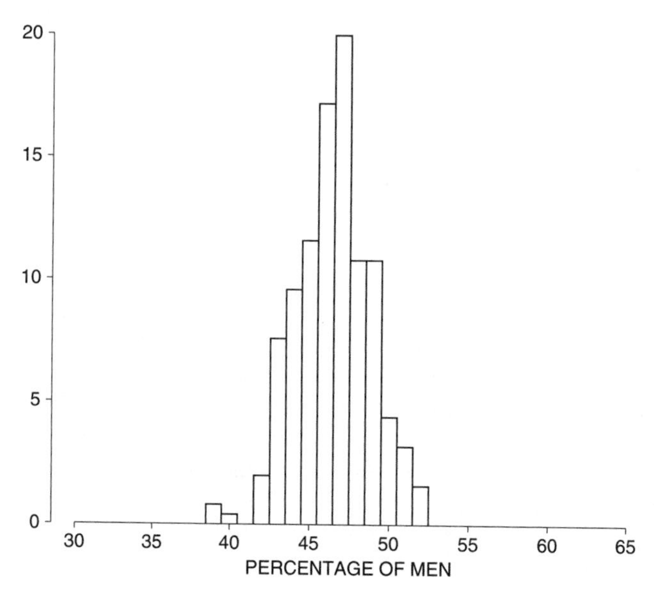

There's a histogram (figure 2). You can compare it with the histogram for samples of size 100. Multiplying the sample size by four cuts the likely size of the chance error in the percentage by a factor of two.

Soc. Can you get more specific about this chance error?

Stat. Let me write an equation:

$$\text{percentage in sample} = \text{percentage in population} + \text{chance error.}$$

Of course, the chance error will be different from sample to sample—remember the variability in table 2.

Soc. So if I let you draw one sample for me, with this random-number business, can you say how big my chance error will be?

Stat. Not exactly, but I can tell you its likely size. If you let me make a box model, I can compute the standard error, and then . . . .

Soc. Wait. There's one point I missed earlier. How can you have 250 different samples with 100 people each? I mean, $250 \times 100 = 25{,}000$, and we only started with 6,672 people.

Stat. The samples are all different, but they have some people in common. Look at the sketch. The inside of the circle is like the 6,672 people, and each shaded strip is like a sample:

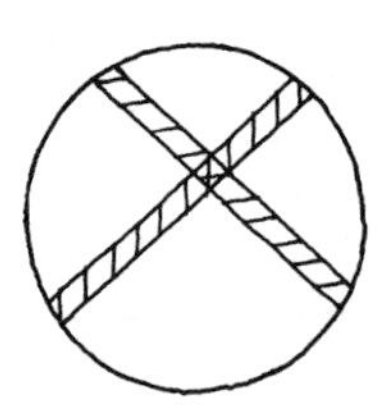

The strips are different, but they overlap. Actually, we only scratched the surface with our sampling. The number of different samples of size 100 is over 10^{200}. That's 1 followed by two hundred 0's. Some physicists don't even think there are that many elementary particles in the whole universe.

2. THE EXPECTED VALUE AND STANDARD ERROR

The sociologist of the previous section was thinking about taking a sample of size 100 from a population of 6,672 subjects in a health study. She knew that the percentage of men in the sample would be somewhere around the percentage of men in the population.

> With a simple random sample, the expected value for the sample percentage equals the population percentage.

However, the sample percentage will not be exactly equal to its expected value—it will be off by a chance error. How big is this error likely to be? The answer is given by the standard error. For the sociologist's problem, the standard error is 5 percentage points. In other words, the sociologist should expect the percentage of men in her sample to be off the percentage in the population by 5 percentage points or so. The method for calculating such standard errors will now be presented. The idea: (i) find the SE for the number of men in the sample; then (ii) convert to percent, relative to the size of the sample. The size of the sample just means the number of sample people—100, in this case.

To compute an SE, you need a box model. The sociologist took a sample of size 100 from a population consisting of 3,091 men and 3,581 women. She classified the people in the sample by sex and counted the men. So there should be only 1's and 0's in the box (section 5 of chapter 17). The number of men in the sample is like the sum of 100 draws from the box

$$\underline{|\ 3{,}091\ \boxed{1}\text{'s}\quad 3{,}581\ \boxed{0}\text{'s}\ |}.$$

She used a simple random sample, so the tickets must be drawn without replacement. This completes the box model.

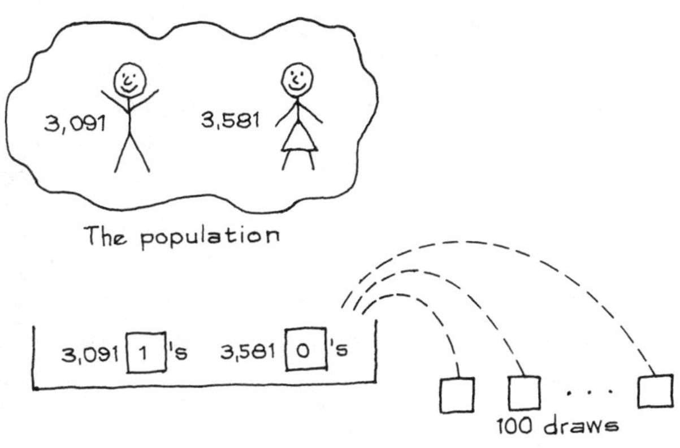

The fraction of 1's in the box is 0.46. Therefore, the SD of the box is $\sqrt{0.46 \times 0.54} \approx 0.50$. The SE for the sum of 100 draws is $\sqrt{100} \times 0.5 = 5$. The sum of 100 draws from the box will be around 46, give or take 5 or so. In other words, the number of men in the sociologist's sample of 100 is likely to be around 46, give or take 5 or so. The SE for the number of men is 5. Now 46 out of 100 is 46%, and 5 out of 100 is 5%. Therefore, the percentage of men in the sample is likely to be around 46%, give or take 5% or so. This 5% is the SE for the percentage of men in the sample.

> To compute the SE for a percentage, first get the SE for the corresponding number; then convert to percent, relative to the size of the sample. As a cold mathematical formula,
>
> $$\text{SE for percentage} = \frac{\text{SE for number}}{\text{size of sample}} \times 100\%.$$

What happens as the sample gets bigger? For instance, if the sociologist took a sample of size 400, the SE for the number of men in the sample would be

$$\sqrt{400} \times 0.5 = 10.$$

Now 10 represents 2.5% of 400, the size of the sample. The SE for the percentage of men in a sample of 400 would be 2.5%. Multiplying the size of the sample by 4 divided the SE for the percentage by $\sqrt{4} = 2$.

> Multiplying the size of a sample by some factor divides the SE for a percentage not by the whole factor—but by its square root.

The formulas are exact when drawing with replacement. And they are good approximations for draws made without replacement, provided the number of draws is small relative to the number of tickets in the box. For example, take the sociologist's SE. No matter which 100 tickets are drawn, among the tickets left in the box, the percentage of 1's will be very close to 46%. So, as far as the chances are concerned, there isn't much difference between drawing with or without replacement. More about this in section 4.

This section showed how the SE for a percentage can be obtained from the SE for the corresponding number. But these two SEs behave quite differently. When the sample size goes up, the SE for the number goes up—and the SE for the percentage goes down. That is because the SE for the number goes up slowly relative to the sample size (pp. 276, 303):

- The SE for the sample number goes up like the square root of the sample size.
- The SE for the sample percentage goes down like the square root of the sample size.

Exercise Set A

1. A town has 30,000 registered voters, of whom 12,000 are Democrats. A survey organization is about to take a simple random sample of 1,000 registered voters. A box model is used to work out the expected value and the SE for the percentage of Democrats in the sample. Match each phrase on list A with a phrase or a number on list B. (Items on list B may be used more than once, or not all.)

List A	*List B*
population	number of 1's among the draws
population percentage	percentage of 1's among the draws
sample	40%
sample size	box
sample number	draws
sample percentage	1,000
denominator for sample percentage	12,000

2. A university has 25,000 students, of whom 10,000 are older than 25. The registrar draws a simple random sample of 400 students.
   (a) Find the expected value and SE for the number of students in the sample who are older than 25.
   (b) Find the expected value and SE for the percentage of students in the sample who are older than 25.
   (c) The percentage of students in the sample who are older than 25 will be around ______, give or take ______ or so.

3. A coin will be tossed 10,000 times. Match the SE with the formula. (One formula will be left over.)

SE for the . . .	*Formula*
percentage of heads	$\sqrt{10{,}000} \times 50\%$
number of heads	$\frac{50}{10{,}000} \times 100\%$
	$\sqrt{10{,}000} \times 0.5$

4. Five hundred draws are made at random with replacement from | 0 0 0 1 |. True or false, and explain:
   (a) The number of 1's among the draws is exactly equal to the sum of the draws.
   (b) The expected value for the percentage of 1's among the draws is exactly equal to 25%.

5. The box | 0 0 0 1 2 | has an average of 0.6, and the SD is 0.8. True or false: the SE for the percentage of 1's in 400 draws can be found as follows—

$$\text{SE for number of 1's} = \sqrt{400} \times 0.8 = 16$$

$$\text{SE for percent of 1's} = \frac{16}{400} \times 100\% = 4\%$$

   Explain briefly.

6. Nine hundred draws are made at random with replacement from a box which has 1 red marble and 9 blue ones. The SE for the percentage of red marbles in the sample is 1%. A sample percentage which is 1 SE above its expected value equals ______.

$$10\% + 1\% \qquad 1.01 \times 10\%$$

Choose one option, and explain briefly.

7. Someone plays a dice game 100 times. On each play, he rolls a pair of dice, and then advances his token along the line by a number of squares equal to the total number of spots thrown. (See the diagram.) About how far does he move? Give or take how much?

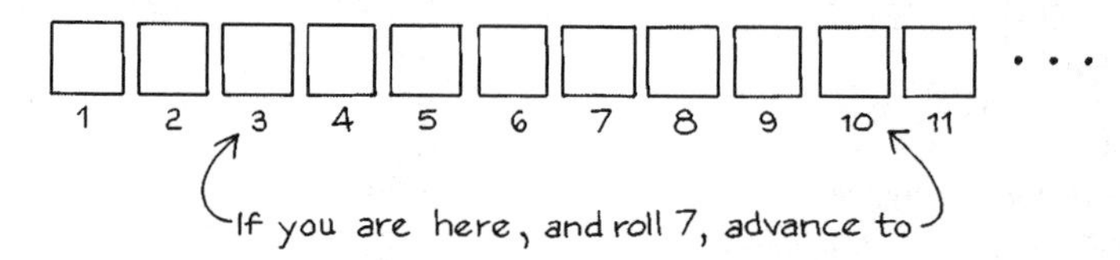

8. According to Sherlock Holmes,

> While the individual man is an insoluble puzzle, in the aggregate he becomes a mathematical certainty. You can, for example, never foretell what any one man will be up to, but you can say with precision what an average number will be up to. Individuals vary, but percentages remain constant. So says the statistician.[2]

The statistician doesn't quite say that. What is Sherlock Holmes forgetting?

The answers to these exercises are on pp. A79–80.

Technical note. When drawing at random with replacement from a 0–1 box, the SE for the number of 1's among the draws is

$$\sqrt{\text{no. of draws}} \times \text{SD of box}.$$

So the SE for the percentage of 1's among the draws is

$$(\sqrt{\text{no. of draws}} \times \text{SD of box}/\text{no. of draws}) \times 100\%.$$

By algebra, this simplifies to (SD of box$/\sqrt{\text{no. of draws}}$) $\times$ 100%. In many books, this would be written $(\sqrt{pq}/\sqrt{n}) \times 100\%$, where p is the fraction of 1's in the box, q is the fraction of 0's, and n is the number of draws.

3. USING THE NORMAL CURVE

This section will review the expected value and SE for a sample percentage, and use the normal curve to compute chances.

Example 1. In a certain town, the telephone company has 100,000 subscribers. It plans to take a simple random sample of 400 of them as part of a market research study. According to Census data, 20% of the company's subscribers earn over $50,000 a year. The percentage of persons in the sample with incomes over $50,000 a year will be around ______, give or take ______ or so.

Solution. The first step is to make a box model. Taking a sample of 400 subscribers is like drawing 400 tickets at random from a box of 100,000 tickets. There is one ticket in the box for each person in the population, and one draw for each person in the sample. The drawing is done at random without replacement.

The problem involves classifying the people in the sample according to whether their incomes are more than \$50,000 a year or not, and then counting the ones whose incomes are above that level. So each ticket in the box should be marked 1 or 0. The people earning more than \$50,000 get 1's and the others get 0's. It is given that 20% of the subscribers earn more than \$50,000 a year, so 20,000 of the tickets in the box are marked 1. The other 80,000 are marked 0. The sample is like 400 draws from the box. And the number of people in the sample who earn more than \$50,000 a year is like the sum of the draws. That completes the first step, setting up the box model.

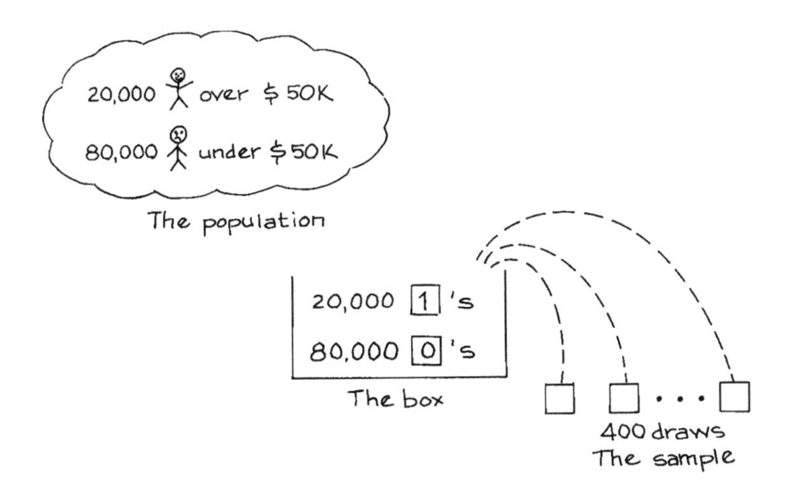

Now you have to work on the sum of the draws from the 0–1 box. The expected value for the sum is $400 \times 0.2 = 80$. To compute the standard error, you need the SD of the box. This is $\sqrt{0.2 \times 0.8} = 0.4$. There are 400 draws, so the SE for the sum is $\sqrt{400} \times 0.4 = 8$. The sum will be around 80, give or take 8 or so. In other words, the number of people in the sample earning more than \$50,000 a year will be around 80, give or take 8 or so.

However, the question is about percent. You convert to percent relative to the size of the sample: 80 out of 400 is 20%, and 8 out of 400 is 2%. The expected value for the sample percentage is 20%, and the SE is 2%. That completes the solution: the percentage of high earners in the sample will be around 20%, give or take 2% or so. (It may be unfortunate, but statisticians use the %-sign as an abbreviation both for "percent" and for "percentage point.")

Of course, the expected value for the sample percent is pretty easy to figure, without the detour through the sample number. When drawing at random from a box of 0's and 1's, the expected value for the percentage of 1's among the draws equals the percentage of 1's in the box (p. 359).

> When drawing at random from a box of 0's and 1's, the percentage of 1's among the draws is likely to be around ______, give or take ______ or so. The expected value for the percentage of 1's among the draws fills in the first blank. The SE for the percentage of 1's among the draws fills in the second blank.

Example 2. (Continues example 1.) Estimate the chance that between 18% and 22% of the persons in the sample earn more than \$50,000 a year.

Solution. The expected value for the sample percentage is 20%, and the SE is 2%. Now convert to standard units:

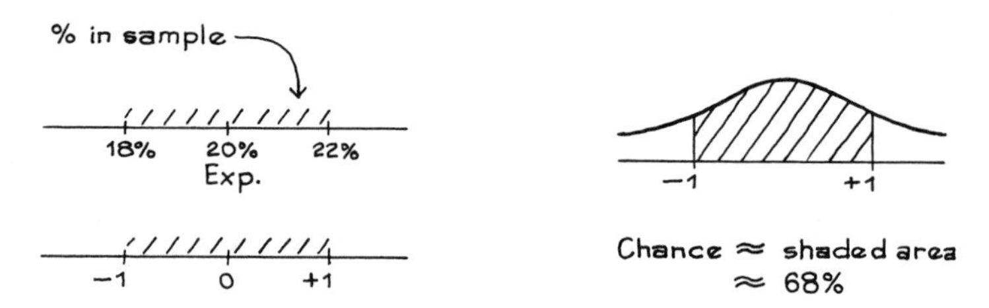

This completes the solution.

Here, the normal curve was used to figure chances. Why is that legitimate? There is a probability histogram for the number of high earners in the sample (figure 3). Areas in this histogram represent chances. For instance, the area between 80 and 90 represents the chance of drawing a sample which has between 80 and 90 high earners. As discussed in chapter 18, this probability histogram follows the normal curve (top panel of figure 3). Conversion to percent is only a change of scale, so the probability histogram for the sample percentage (bottom panel) looks just like the top histogram—and follows the curve too. In example 2, the curve was used on the probability histogram for the sample percentage, not on a histogram for data.

Examples 1 and 2 are about qualitative data. The incomes start out as quantitative data—numbers. However, the problems involve classifying and counting. Each person is classified as earning more than \$50,000 a year, or less. Then the high earners are counted. In other words, the data are treated as qualitative: each income either has or doesn't have the quality of being more than \$50,000 a year.

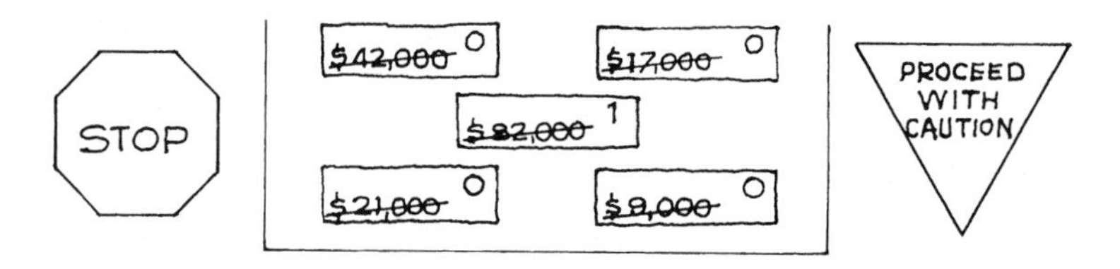

Figure 3. The top panel shows the probability histogram for the number of sample persons with incomes over \$50,000. The bottom panel shows the probability histogram for the percentage of sample persons with incomes over \$50,000. In standard units, the two histograms are exactly the same.[3] (Four hundred persons are chosen at random from a population of 100,000.)

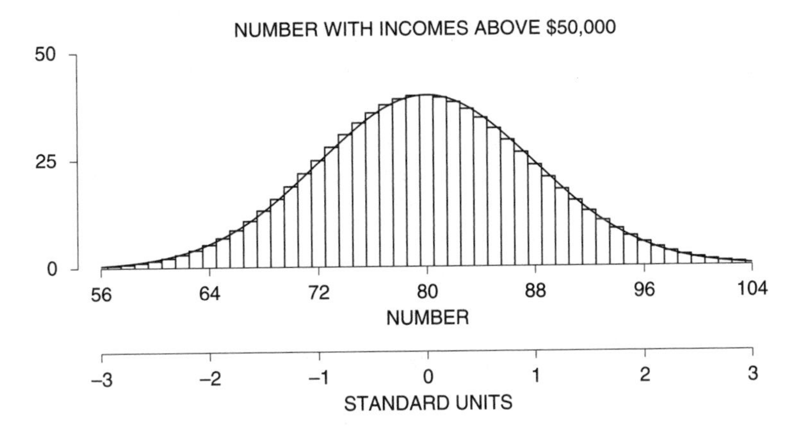

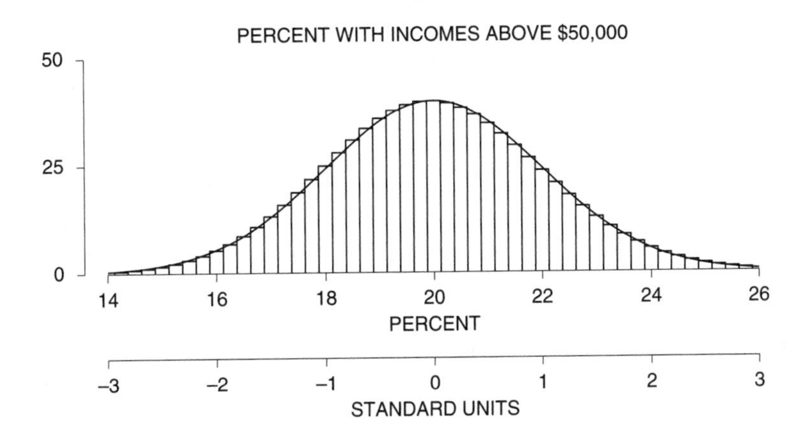

When do you change to a 0–1 box? To answer this question, think about the arithmetic being done on the sample values. The arithmetic might involve:

- adding up the sample values, to get an average;

or

- classifying and counting, to get a percent.

If the problem is about classifying and counting, put 0's and 1's in the box (section 5 of chapter 17).

Exercise Set B

1. You are drawing at random from a large box of red and blue marbles. Fill in the blanks.
   (a) The expected value for the percentage of reds in the ______ equals the percentage of reds in the ______. Options: sample, population
   (b) As the number of draws goes up, the SE for the ______ of reds in the sample goes up but the SE for the ______ of reds goes down. Options: number, percentage

2. In a certain town, there are 30,000 registered voters, of whom 12,000 are Democrats. A survey organization is about to take a simple random sample of 1,000 registered voters.
   (a) The expected value for the percentage of Democrats in the sample is ______. The SE for the percentage of Democrats in the sample is ______.
   (b) The percentage of Democrats in the sample is likely to be around ______, give or take ______ or so.
   (c) Find the chance that between 39% and 41% of the registered voters in the sample are Democrats.

3. According to the Census, a certain town has a population of 100,000 people age 18 and over. Of them, 60% are married, 10% have incomes over $75,000 a year, and 20% have college degrees.[4] As part of a pre-election survey, a simple random sample of 1,600 people will be drawn from this population.
   (a) To find the chance that 58% or less of the people in the sample are married, a box model is needed. Should the number of tickets in the box be 1,600, or 100,000? Explain. Then find the chance.
   (b) To find the chance that 11% or more of the people in the sample have incomes over $75,000 a year, a box model is needed. Should each ticket in the box show the person's income? Explain. Then find the chance.
   (c) Find the chance that between 19% and 21% of the people in the sample have a college degree.

4. The figure below is the probability histogram for the percent of sample persons with incomes above $50,000 (example 1, and bottom panel of figure 3). The shaded area represents ______. Fill in the blank with a phrase.

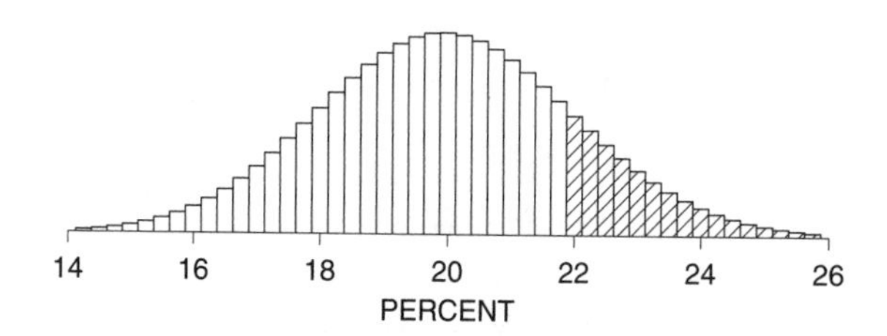

5. (a) In the top panel of figure 3, the area of the rectangle over 88 represents what?
   (b) In the bottom panel of figure 3, the area of the rectangle over 22% represents what?
   (c) The rectangles in parts (a) and (b) have equal areas. Is that a coincidence?

The answers to these exercises are on pp. A80–81.

4. THE CORRECTION FACTOR

It is just after Labor Day, 2004. The presidential campaign (Bush versus Kerry) is in full swing, and the focus is on the Southwest. Pollsters are trying to predict the results. There are about 1.5 million eligible voters in New Mexico, and about 15 million in the state of Texas. Suppose one polling organization takes a simple random sample of 2,500 voters in New Mexico, in order to estimate the percentage of voters in that state who are Democratic. Another polling organization takes a simple random sample of 2,500 voters from Texas. Both polls use exactly the same techniques. Both estimates are likely to be a bit off, by chance error. For which poll is the chance error likely to be smaller?

The New Mexico poll is sampling one voter out of 600, while the Texas poll is sampling one voter out of 6,000. It does seem that the New Mexico poll should be more accurate than the Texas poll. However, this is one of the places where intuition comes into head-on conflict with statistical theory, and it is intuition which has to give way. In fact, the accuracy expected from the New Mexico poll is just about the same as the accuracy to be expected from the Texas poll.

> When estimating percentages, it is the absolute size of the sample which determines accuracy, not the size relative to the population. This is true if the sample is only a small part of the population, which is the usual case.[5]

A box model will help in focusing the issue. We'll need two boxes, NM and TX. Box NM represents New Mexico, box TX represents Texas. Box NM has 1,500,000 tickets, one for each voter. The tickets corresponding to Democrats are marked 1, the others are marked 0. To keep life simple, we make the percentage of 1's in the box equal to 50%. We hire a polling organization to take a simple random sample from box NM, without telling them what is in the box. (Remember, taking a simple random sample means drawing at random without replacement.) The job of the polling organization is to estimate the percentage of 1's in the box. Naturally, they use the percentage of 1's in their sample.

50% [1] 50% [0]		50% [1] 50% [0]
NM		TX

Now for Box TX. This represents Texas, so it has 15,000,000 tickets. Again, we mark 1 on half the tickets in the box, and 0 on the others. Another polling organization is hired to take a simple random sample of 2,500 tickets from box TX, without knowing the composition of the box. This organization too will estimate the percentage of 1's in the box by the percentage in the sample, and will be off by a chance error.

Box NM and box TX have been set up with the same percentage composition, and the two samples are the same size. Intuition would insist that the organization sampling from box NM will have a much smaller chance error, because box NM is so much smaller. But statistical theory shows that the likely size of the chance error is just about the same for the two polls.

The issue has now been stated sharply. How does statistical theory justify itself? To begin with, suppose the samples were drawn with replacement. Then it wouldn't matter at all which box was used. There would be a 50–50 chance to get a 0 or a 1 on each draw, and the size of the box would be completely irrelevant. Box NM and box TX have the same SD of 0.5, so both polling organizations would have the same SE for the number of 1's among the draws:

$$\sqrt{2{,}500} \times 0.5 = 25.$$

As a result, they would both have the same SE for the percentage of 1's among the draws:

$$\frac{25}{2{,}500} \times 100\% = 1\%.$$

If they drew at random with replacement, both organizations would be off by about 1 percentage point or so.

In fact, the draws are made without replacement. However, the number of draws is just a tiny fraction of the number of tickets in the box. Taking the draws without replacement barely changes the composition of the box. On each draw, the chance of getting a 1 must still be very close to 50%, and similarly for 0. As far as the chances are concerned, there is almost no difference between drawing with or without replacement.

In essence, that is why the size of the population has almost nothing to do with the accuracy of estimates. Still, there is a shade of difference between drawing with and without replacement. When drawing without replacement, the box does get a bit smaller, reducing the variability slightly. So the SE for drawing without replacement is a little less than the SE for drawing with replacement. There is a mathematical formula that says how much smaller:

$$\begin{array}{c}\text{SE when drawing}\\ \text{WITHOUT replacement}\end{array} = \begin{array}{c}\text{correction}\\ \text{factor}\end{array} \times \begin{array}{c}\text{SE when drawing}\\ \text{WITH replacement}\end{array}$$

The correction factor itself is somewhat complicated:

$$\sqrt{\frac{\text{number of tickets in box} - \text{number of draws}}{\text{number of tickets in box} - \text{one}}}$$

When the number of tickets in the box is large relative to the number of draws,

Table 3. The correction factor; the number of draws is fixed at 2,500.

Number of tickets in the box	*Correction factor (to five decimals)*
5,000	0.70718
10,000	0.86607
100,000	0.98743
500,000	0.99750
1,500,000	0.99917
15,000,000	0.99992

the correction factor is nearly 1 and can be ignored (table 3, p. 368). Then it is the absolute size of the sample which determines accuracy, through the SE for drawing with replacement. The size of the population does not really matter. On the other hand, if the sample is a substantial fraction of the population, the correction factor must be used.

In our box model, the percentage of 1's was the same for both boxes. In reality, the percentage of Democrats will be different for the two states. However, even quite a large difference will generally not matter very much. In the 2004 presidential election, for example, 50% of the voters in New Mexico chose the Republican candidate (Bush), compared to 61% in Texas.[6] But the SDs for the two states are almost the same:

50% [1] 50% [0]	61% [1] 39% [0]
NM	TX

$$\text{SD} = \sqrt{.50 \times .50} = .50 \qquad\qquad \text{SD} = \sqrt{.61 \times .39} \approx .49$$

A sample of size 2,500 will do as well in Texas as in New Mexico, although Texas is 10 times larger. The Texan in the cartoon is just wrong.

"A BIG STATE NEEDS A BIG SAMPLE, PARDNER."

A non-mathematical analogy may help. Suppose you took a drop of liquid from a bottle, for chemical analysis. If the liquid is well mixed, the chemical composition of the drop should reflect the composition of the whole bottle, and it really wouldn't matter if the bottle was a test tube or a gallon jug. The chemist doesn't care whether the drop is 1% or 1/100 of 1% of the solution.

The analogy is precise. There is one ticket in the box for each molecule in the bottle. If the liquid is well mixed, the drop is like a random sample. The number of molecules in the drop corresponds to the number of tickets drawn. This number—the sample size—is so large that chance error in the percentages is negligible.

Exercise Set C

1. One public opinion poll uses a simple random sample of size 1,500 drawn from a town with a population of 25,000. Another poll uses a simple random sample of size 1,500 from a town with a population of 250,000. The polls are trying to estimate the percentage of voters who favor single-payer health insurance. Other things being equal:

   (i) the first poll is likely to be quite a bit more accurate than the second.
   (ii) the second poll is likely to be quite a bit more accurate than the first.
   (iii) there is not likely to be much difference in accuracy between the two polls.

2. You have hired a polling organization to take a simple random sample from a box of 100,000 tickets, and estimate the percentage of 1's in the box. Unknown to them, the box contains 50% 0's and 50% 1's. How far off should you expect them to be:

   (a) if they draw 2,500 tickets?
   (b) if they draw 25,000 tickets?
   (c) if they draw 100,000 tickets?

3. A survey organization wants to take a simple random sample in order to estimate the percentage of people who have seen a certain television program. To keep the costs down, they want to take as small a sample as possible. But their client will only tolerate chance errors of 1 percentage point or so in the estimate. Should they use a sample of size 100, 2,500, or 10,000? You may assume the population to be very large; past experience suggests the population percentage will be in the range 20%–40%.

4. One hundred draws are made at random with replacement from each of the following boxes. The SE for the percentage of 1's among the draws is smallest for box ______ and largest for box ______. Or is the SE the same for all three boxes?

   (A) | [0] [1] | (B) | 10 [0]'s 10 [1]'s | (C) | 1,000 [0]'s 1,000 [1]'s |

5. A box contains 2 red marbles and 8 blue ones. Four marbles are drawn at random. Find the SE for the percentage of red marbles drawn, when the draws are made

   (a) with replacement.   (b) without replacement.

The answers to these exercises are on p. A81.

5. THE GALLUP POLL

The Gallup Poll predicts the vote with good accuracy, by sampling several thousand eligible voters out of 200 million. How is this possible? The previous section focused on simple random sampling, but the conclusions hold for most probability methods of drawing samples, including the one used by the Gallup

Poll: the likely size of the chance error in sample percentages depends mainly on the absolute size of the sample, and hardly at all on the size of the population. The huge number of eligible voters makes it hard work to draw the sample, but does not affect the standard error.

Is 2,500 a big enough sample? The square root law provides a benchmark. For example, with 2,500 tosses of a coin, the standard error for the percentage of heads is only 1%. Similarly, with a sample of 2,500 voters, the likely size of the chance error is only a percentage point or so. That is good enough unless the election is very close, like Bush versus Gore in 2000. The Electoral College would be a major complication: the Gallup Poll only predicts the popular vote.

6. REVIEW EXERCISES

Review exercises may also cover material from previous chapters.

1. Complete the following table for the coin-tossing game.

	Number of heads		*Percent of heads*	
Number of tosses	*Expected value*	*SE*	*Expected value*	*SE*
100	50	5	50%	5%
2,500				1%
10,000				
1,000,000				

2. A die is rolled one thousand times. The percentage of aces (⚀) should be around ______, give or take ______ or so.
   (a) The first step in solving this problem is
      (i) computing the SD of the box.
      (ii) computing the average of the box.
      (iii) setting up the box model.
   Choose one option and explain.
   (b) Now solve the problem.

3. A group of 50,000 tax forms has an average gross income of \$37,000, with an SD of \$20,000. Furthermore, 20% of the forms have a gross income over \$50,000. A group of 900 forms is chosen at random for audit. To estimate the chance that between 19% and 21% of the forms chosen for audit have gross incomes over \$50,000, a box model is needed.
   (a) Should the number of tickets in the box be 900 or 50,000?
   (b) Each ticket in the box shows

   a zero or a one   a gross income

   (c) True or false: the SD of the box is \$20,000.
   (d) True or false: the number of draws is 900.
   (e) Find the chance (approximately) that between 19% and 21% of the forms chosen for audit have gross incomes over \$50,000.

(f) With the information given, can you find the chance (approximately) that between 9% and 11% of the forms chosen for audit have gross incomes over $75,000? Either find the chance, or explain why you need more information.

4. As in exercise 3, except it is desired to find the chance (approximately) that the total gross income of the audited forms is over $33,000,000. Work parts (a) through (d); then find the chance or explain why you need more information.

5. (Hypothetical.) On the average, hotel guests who take elevators weigh about 150 pounds with an SD of about 35 pounds. An engineer is designing a large elevator for a convention hotel, to lift 50 such people. If she designs it to lift 4 tons, the chance it will be overloaded by a random group of 50 people is about ______. Explain briefly.

6. The Census Bureau is planning to take a sample amounting to 1/10 of 1% of the population in each state in order to estimate the percentage of the population in that state earning over $100,000 a year. Other things being equal:

   (i) The accuracy to be expected in California (population 35 million) is about the same as the accuracy to be expected in Nevada (population 2 million).
   (ii) The accuracy to be expected in California is quite a bit higher than in Nevada.
   (iii) The accuracy to be expected in California is quite a bit lower than in Nevada.

   Explain.

7. Five hundred draws are made at random from the box

   | 60,000 [0]'s 20,000 [1]'s |

   True or false, and explain:

   (a) The expected value for the percentage of 1's among the draws is exactly 25%.
   (b) The expected value for the percentage of 1's among the draws is around 25%, give or take 2% or so.
   (c) The percentage of 1's among the draws will be around 25%, give or take 2% or so.
   (d) The percentage of 1's among the draws will be exactly 25%.
   (e) The percentage of 1's in the box is exactly 25%.
   (f) The percentage of 1's in the box is around 25%, give or take 2% or so.

8. In a certain town, there are 30,000 registered voters, of whom 12,000 are Democrats. A survey organization is about to take a simple random sample of 1,000 registered voters. There is about a 50–50 chance that the percentage of Democrats in the sample will be bigger than ______. Fill in the blank, and explain.

9. Six hundred draws will be made at random with replacement from the box 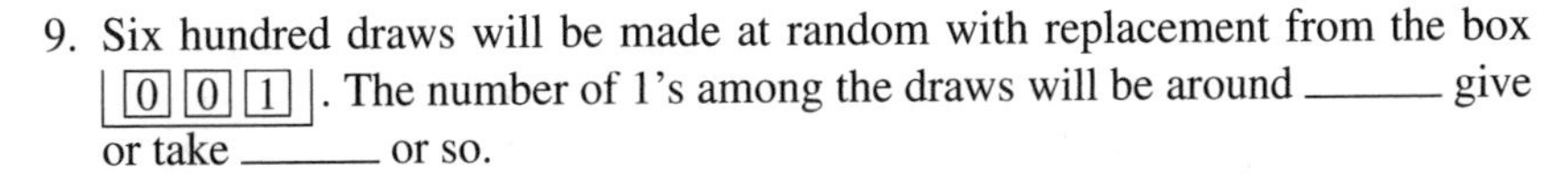 0 0 1 . The number of 1's among the draws will be around ______ give or take ______ or so.

10. A coin is tossed 2,000 times. Someone wishes to compute the SE for the number of heads among the tosses as $\sqrt{2{,}000} \times 0.5 \approx 22$. Is this the right SE? Answer yes or no, and explain briefly.

11. A university has 25,000 students, of whom 17,000 are undergraduates. The housing office takes a simple random sample of 500 students; 357 out of the 500 are undergraduates. Fill in the blanks.
    (a) For the number of undergraduates in the sample, the observed value is ______ but the expected value is ______.
    (b) For the percentage of undergraduates in the sample, the observed value is ______ but the expected value is ______.

12. There are 50,000 households in a certain city. The average number of persons age 16 and over living in each household is known to be 2.38; the SD is 1.87. A survey organization plans to take a simple random sample of 400 households, and interview all persons age 16 and over living in the sample households. The total number of interviews will be around ______, give or take ______ or so. Explain briefly.

7. SUMMARY

1. The sample is only part of the population, so the percentage composition of the sample usually differs by some amount from the percentage composition of the whole population.

2. For probability samples, the likely size of the chance error (the amount off) is given by the standard error.

3. To figure the SE, a box model is needed. When the problem involves classifying and counting, or taking percents, there should only be 0's and 1's in the box. Change the box, if necessary.

4. When drawing at random from a 0–1 box, the expected value for the percentage of 1's in the sample equals the percentage of 1's in the box. To find the SE for the percentage, first get the SE for the corresponding number, then convert to percent. The formula:

$$\text{SE for percentage} = \frac{\text{SE for number}}{\text{size of sample}} \times 100\%.$$

5. When the sample is only a small part of the population, the number of individuals in the population has almost no influence on the accuracy of the sample percentage. It is the absolute size of the sample (that is, the number of individuals in the sample) which matters, not the size relative to the population.

6. The square root law is exact when draws are made with replacement. When the draws are made without replacement, the formula gives a good approximation—provided the number of tickets in the box is large relative to the number of draws.

7. When drawing without replacement, to get the exact SE you have to multiply by the correction factor:

$$\sqrt{\frac{\text{number of tickets in box} - \text{number of draws}}{\text{number of tickets in box} - \text{one}}}$$

When the number of tickets in the box is large relative to the number of draws, the correction factor is nearly one.

21

The Accuracy of Percentages

In solving a problem of this sort, the grand thing is to be able to reason backward. That is a very useful accomplishment, and a very easy one, but people do not practise it much . . . . Most people, if you describe a train of events to them, will tell you what the result would be. They can put those events together in their minds, and argue from them that something will come to pass. There are few people, however, who, if you told them a result, would be able to evolve from their own inner consciousness what the steps were which led up to that result. This power is what I mean when I talk of reasoning backward . . . .

—*Sherlock Holmes*[1]

1. INTRODUCTION

The previous chapter reasoned from the box to the draws. Draws were made at random from a box whose composition was known, and a typical problem was finding the chance that the percentage of 1's among the draws would be in a given interval. As Sherlock Holmes points out, it is often very useful to turn this reasoning around, going instead from the draws to the box. A statistician would call this *inference* from the sample to the population. Inference is the topic of this chapter.

For example, suppose a survey organization wants to know the percentage of Democrats in a certain district. They might estimate it by taking a simple random sample. Naturally, the percentage of Democrats in the sample would be used to estimate the percentage of Democrats in the district—an example of reasoning backward from the draws to the box. Because the sample was chosen at random, it is possible to say how accurate the estimate is likely to be, just from the size and composition of the sample. This chapter will explain how.

The technique is one of the key ideas in statistical theory. It will be presented in the polling context. A political candidate wants to enter a primary in a district with 100,000 eligible voters, but only if he has a good chance of winning. He hires a survey organization, which takes a simple random sample of 2,500 voters. In the sample, 1,328 favor the candidate, so the percentage is

$$\frac{1{,}328}{2{,}500} \times 100\% \approx 53\%.$$

The candidate is discussing this result with his pollster.

Politician. I win.

Pollster. Not so fast. You want to know the percentage you'd get among all the voters in the district. We only have it in the sample.

Politician. But with a good sample, it's bound to be the same.

Pollster. Not true. It's what I said before. The percentage you get in the sample is different from what you'd get in the whole district. The difference is what we call chance error.

Politician. Could the sample be off by as much as three percentage points? If so, I lose.

Pollster. Actually, we can be about 95% confident that we're right to within two percentage points. It looks good.

"I'M BEHIND YOU 100 PERCENT, PLUS OR MINUS 3 PERCENT OR SO."

Politician. What gives you the size of the chance error?

Pollster. The standard error. Remember, we talked about that the other day. As I was telling you . . . .

Politician. Sorry, I'm expecting a phone call now.

The politician has arrived at the crucial question to ask when considering survey data: how far wrong is the estimate likely to be? As the pollster wanted to say, the likely size of the chance error is given by the standard error. To figure that, a box model is needed. There should be one ticket in the box for each voter, making 100,000 tickets in all. Each ticket should be marked 1 or 0, where 1 means a vote for the candidate, 0 a vote against him. There are 2,500 draws made at random from the box. The data are like the draws, and the number of voters in the sample who favor the candidate is like the sum of the draws. This completes the model.

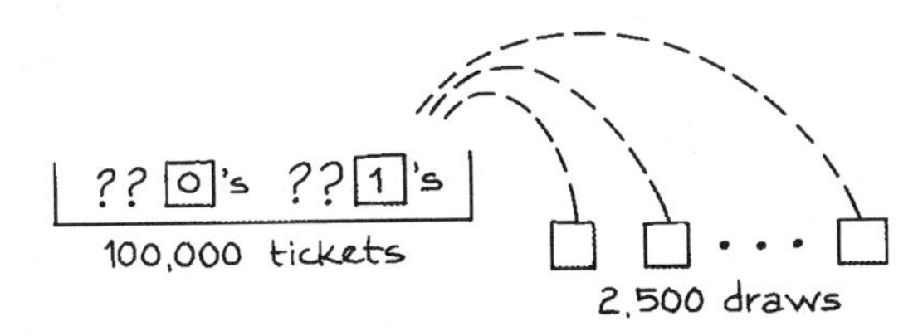

To get the SE for the sum, the survey organization needs the SD of the box. This is

$$\sqrt{(\text{fraction of 1's}) \times (\text{fraction of 0's})}.$$

At this point, the pollsters seem to be stuck. They don't know how each ticket in the box should be marked. They don't even know the fraction of 1's in the box. That parameter represents the fraction of voters in the district who favor their candidate, which is exactly what they were hired to find out. (Hence the question marks in the box.)

Survey organizations lift themselves over this sort of obstacle by their own bootstraps.[2] They substitute the fractions observed in the sample for the unknown fractions in the box. In the example, 1,328 people out of the sample of 2,500 favored the candidate. So $1{,}328/2{,}500 \approx 0.53$ of the sample favored him, and the other 0.47 were opposed. The estimate is that about 0.53 of the 100,000 tickets in the box are marked 1, the other 0.47 being marked 0.

On this basis, the SD of the box is estimated as $\sqrt{0.53 \times 0.47} \approx 0.50$. The SE for the number of voters in the sample who favor the candidate is estimated as $\sqrt{2{,}500} \times 0.50 = 25$. The 25 measures the likely size of the chance error in the 1,328. Now 25 people out of 2,500 (the size of the sample) is 1%. The SE for the percentage of voters in the sample favoring the candidate is estimated as 1 percentage point. This completes the bootstrap procedure for estimating the standard error.

As far as the candidate is concerned, this calculation shows that his pollster's estimate of 53% is only likely to be off by 1 percentage point or so. It is very unlikely to be off by as much as 3 percentage points—that's 3 SEs. He is well on the safe side of 50%, and he should enter the primary.

> *The bootstrap.* When sampling from a 0–1 box whose composition is unknown, the SD of the box can be estimated by substituting the fractions of 0's and 1's in the sample for the unknown fractions in the box. The estimate is good when the sample is reasonably large.

The bootstrap procedure may seem crude. But even with moderate-sized samples, the fraction of 1's among the draws is likely to be quite close to the fraction in the box. Similarly for the 0's. If survey organizations use their sample fractions in the formula for the SD of the box, they are not likely to be far wrong in estimating the SE.

One point is worth more discussion. The expected value for the number of 1's among the draws (translation—the expected number of sample voters who favor the candidate) is

$$2{,}500 \times \text{fraction of 1's in the box.}$$

This is unknown, because the fractions of 1's in the box is unknown. The SE of 25 says about how far the 1,328 is from its expected value. In statistical terminology, the 1,328 is an observed value; the contrast is with the unknown expected value. (Observed values are discussed on p. 292.)

Example 1. In fall 2005, a city university had 25,000 registered students. To estimate the percentage who were living at home, a simple random sample of 400 students was drawn. It turned out that 317 of them were living at home. Estimate the percentage of students at the university who were living at home in fall 2005. Attach a standard error to the estimate.

Solution. The sample percentage is

$$\frac{317}{400} \times 100\% \approx 79\%$$

That is the estimate for the population percentage.

For the standard error, a box model is needed. There are 25,000 tickets in the box, one for each student in the population. There are 400 draws from the box, one for each student in the sample. This problem involves classifying and counting, so each ticket in the box should be marked 1 or 0. We are counting students who were living at home. The tickets corresponding to these students should be marked 1; the others, 0. There are 400 draws made at random from the box. The data are like the draws, and the number of students in the sample who were living at home is like the sum of the draws. That completes the model. (See the sketch at the top of the next page.)

The fraction of 1's in the box is a parameter. It represents the fraction of all the students at this university who were living at home in fall 2005. It is unknown, but can be estimated as 0.79—the fraction observed in the sample. Similarly, the

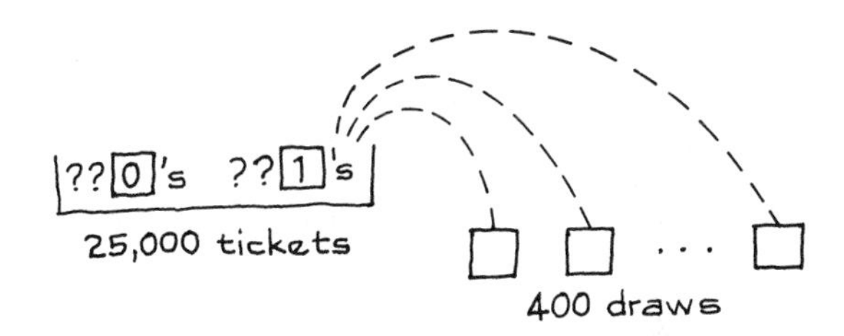

fraction of 0's in the box is estimated as 0.21. So the SD of the box is estimated by the bootstrap method as $\sqrt{0.79 \times 0.21} \approx 0.41$. The SE for the number of students in the sample who were living at home is estimated as $\sqrt{400} \times 0.41 \approx 8$. The 8 gives the likely size of the chance error in the 317. Now convert to percent, relative to the size of the sample:

$$\frac{8}{400} \times 100\% = 2\%$$

The SE for the sample percentage is estimated as 2%. Let's summarize. In the sample, 79% of the students were living at home. The 79% is off the mark by 2 percentage points or so. That is what the SE tells us.

The discussion in this section focused on simple random sampling, where the mathematics is easiest. In practice, survey organizations use much more complicated designs. Even so, with probability methods it is generally possible to say how big the chance errors are likely to be—one of the great advantages of probability methods for drawing samples.

Exercise Set A

1. Fill in the blanks, and explain.
   (a) In example 1 on p. 378, the 317 is the ______ value for the number of students in the sample who were living at home. Options:
   (i) expected (ii) observed
   (b) The SD of the box is ______ 0.41. Options:
   (i) exactly equal to (ii) estimated from the data as
   (c) The SE for the number of students in the sample who were living at home is ______ 8. Options: (i) exactly equal to (ii) estimated from the data as

2. In a certain city, there are 100,000 persons age 18 to 24. A simple random sample of 500 such persons is drawn, of whom 194 turn out to be currently enrolled in college. Estimate the percentage of all persons age 18 to 24 in that city who are currently enrolled in college.[3] Put a give-or-take number on the estimate.
   (a) The first step in solving this problem is:
   (i) finding the SD of the box.
   (ii) finding the average of the box.
   (iii) writing down the box model.
   Choose one option, and explain.
   (b) Now solve the problem.

3. In a simple random sample of 100 graduates from a certain college, 48 were earning $50,000 a year or more. Estimate the percentage of all graduates of that college earning $50,000 a year or more.[4] Put a give-or-take number on the estimate.

4. A simple random sample of size 400 was taken from the population of all manufacturing establishments in a certain state: 11 establishments in the sample had 100 employees or more. Estimate the percentage of manufacturing establishments with 100 employees or more.[5] Attach a standard error to the estimate.

5. In the same state, a simple random sample of size 400 was taken from the population of all persons employed by manufacturing establishments: 187 people in the sample worked for establishments with 100 employees or more. Estimate the percentage of people who worked for establishments with 100 employees or more. Attach a standard error to the estimate.

6. Is the difference between the percentages in exercises 4 and 5 due to chance error?

The next two exercises are designed to illustrate the bootstrap method for estimating the SD of the box.

7. Suppose there is a box of 100,000 tickets, each marked 0 or 1. Suppose that in fact, 20% of the tickets in the box are 1's. Calculate the standard error for the percentage of 1's in 400 draws from the box.

8. Three different people take simple random samples of size 400 from the box in exercise 7, without knowing its contents. The number of 1's in the first sample is 72. In the second, it is 84. In the third, it is 98. Each person estimates the SE by the bootstrap method.
   (a) The first person estimates the percentage of 1's in the box as ______, and figures this estimate is likely to be off by ______ or so.
   (b) The second person estimates the percentage of 1's in the box as ______, and figures this estimate is likely to be off by ______ or so.
   (c) The third person estimates the percentage of 1's in the box as ______, and figures this estimate is likely to be off by ______ or so.

9. In a certain town, there are 25,000 people aged 18 and over. To estimate the percentage of them who watched a certain TV show, a statistician chooses a simple random sample of size 1,000. As it turns out, 308 of the sample people did see the show. Complete the following table; the first 3 lines refer to the sample percentage who saw the show. (N/A = not applicable.)

	Known to be	*Estimated from the data as*
Observed value	30.8%	N/A
Expected value	N/A	30.8%
SE		
SD of box		
Number of draws		

The answers to these exercises are on pp. A81–82.

2. CONFIDENCE INTERVALS

In the example of the previous section, 79% of the students in the sample were living at home: the sample percentage was 79%. How far can the population percentage be from 79%? (Remember, "population percentage" means the percentage of all students at the university who were living at home.) The standard error was estimated as 2%, suggesting a chance error of around 2% in size. So the population percentage could easily be 77%. This would mean a chance error of 2%:

sample percentage	=	population percentage	+	chance error
79%	=	77%	+	2%

The population percentage could also be 76%, corresponding to a chance error of 3%. This is getting unlikely, because 3% represents 1.5 SEs. The population percentage could even be as small as 75%, but this is still more unlikely; 4% represents 2 SEs. Of course, the population percentage could be on the other side of the sample percentage, corresponding to negative chance errors. For instance, the population percentage could be 83%. Then the estimate is low by 4%: the chance error is −4%, which is −2 SEs.

With chance errors, there is no sharp dividing line between the possible and the impossible. Errors larger in size than 2 SEs do occur—infrequently. What happens with a cutoff at 2 SEs? Take the interval from 2 SEs below the sample percentage to 2 SEs above:

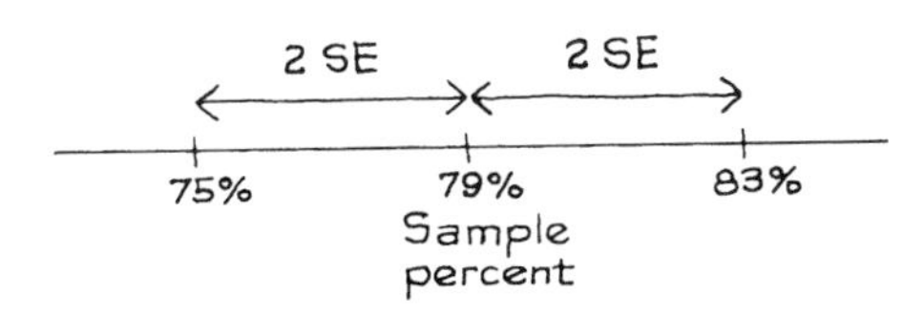

This is a *confidence interval* for the population percentage, with a *confidence level* of about 95%. You can be about 95% confident that the population percentage is caught inside the interval from 75% to 83%.

What if you want a different confidence level? Anything except 100% is possible, by going the right number of SEs in either direction from the sample percentage. For instance:

- The interval "sample percentage ± 1 SE" is a 68%-confidence interval for the population percentage.
- The interval "sample percentage ± 2 SEs" is a 95%-confidence interval for the population percentage.
- The interval "sample percentage ± 3 SEs" is a 99.7%-confidence interval for the population percentage.

However, even 10 SEs may not give 100% confidence, because there is the remote possibility of very large chance errors. There are no definite limits to the normal curve: no matter how large a finite interval you choose, the normal curve has some area outside that interval.[6]

Example 2. A simple random sample of 1,600 persons is taken to estimate the percentage of Democrats among the 25,000 eligible voters in a certain town. It turns out that 917 people in the sample are Democrats. Find a 95%-confidence interval for the percentage of Democrats among all 25,000 eligible voters.

Solution. The percentage of Democrats in the sample is

$$\frac{917}{1{,}600} \times 100\% \approx 57.3\%.$$

The estimate: about 57.3% of the eligible voters in the town are Democrats. For the standard error, a box model is needed. There is one ticket in the box for each eligible voter in the town, making 25,000 tickets in all. There are 1,600 draws, corresponding to the sample size of 1,600. This problem involves classifying (Democrat or not) and counting, so each ticket is marked 1 or 0. It is Democrats that are being counted. So the tickets corresponding to Democrats are marked 1, the others are marked 0. There are 1,600 draws made at random from the box. The data are like the draws, and the number of Democrats in the sample is like the sum of the draws. That completes the model.

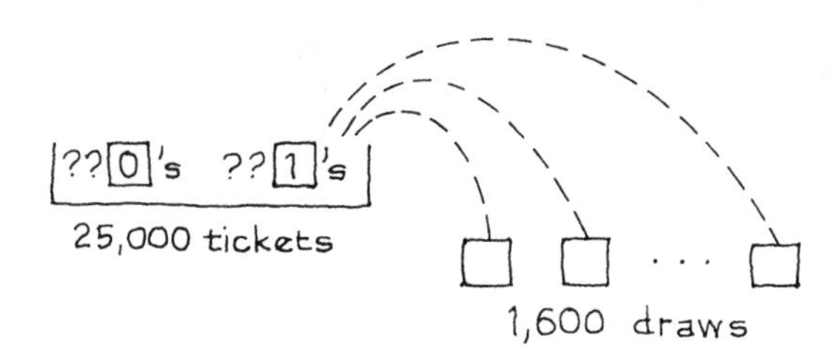

The fraction of 1's in the box (translation—the fraction of Democrats among the 25,000 eligible voters) is unknown, but can be estimated by 0.573, the fraction of Democrats in the sample. Similarly, the fraction of 0's in the box is estimated as 0.427. So the SD of the box is estimated by the bootstrap method as $\sqrt{0.573 \times 0.427} \approx 0.5$. The SE for the number of Democrats in the sample is estimated as $\sqrt{1{,}600} \times 0.5 = 20$. The 20 gives the likely size of the chance error in the 917. Now convert to percent, relative to the size of the sample:

$$\frac{20}{1{,}600} \times 100\% = 1.25\%.$$

The SE for the percentage of Democrats in the sample is 1.25%. The percentage of Democrats in the sample is likely to be off the percentage of Democrats in the population, by 1.25 percentage points or so. A 95%-confidence interval for the percentage of Democrats among all 25,000 eligible voters is

$$57.3\% \ \pm\ 2 \times 1.25\%.$$

That is the answer. We can be about 95% confident that between 54.8% and 59.8% of the eligible voters in this town are Democrats.

Confidence levels are often quoted as being "about" so much. There are two reasons. (i) The standard errors have been estimated from the data. (ii) The normal approximation has been used. If the normal approximation does not apply, neither do the methods of this chapter. There is no hard-and-fast rule for deciding. The best way to proceed is to imagine that the population has the same percentage composition as the sample. Then try to decide whether the normal approximation would work for the sum of the draws from the box. For instance, a sample percentage near 0% or 100% suggests that the box is lopsided, so a large number of draws will be needed before the normal approximation takes over (section 5 of chapter 18). On the other hand, if the sample percentage is near 50%, the normal approximation should be satisfactory when there are only a hundred draws or so.

Exercise Set B

1. Fill in the blanks, and explain.
   (a) In example 2 on p. 382, the 917 is the ______ value for the number of Democrats in the sample. Options: (i) expected (ii) observed
   (b) The SD of the box is ______ $\sqrt{0.573 \times 0.427}$. Options:
   (i) exactly equal to (ii) estimated from the data as
   (c) The SE for the number of Democrats in the sample is ______ 20. Options:
   (i) exactly equal to (ii) estimated from the data as

2. Refer back to exercise 2 on p. 379.
   (a) Find a 95%-confidence interval for the percentage of persons age 18 to 24 in the city who are currently enrolled in college.
   (b) Repeat, for a confidence level of 99.7%.
   (c) Repeat, for a confidence level of 99.7%, supposing the size of the sample was 2,000, of whom 776 were currently enrolled in college.

3. A box contains 1 red marble and 99 blues; 100 marbles are drawn at random with replacement.
   (a) Find the expected number of red marbles among the draws, and the SE.
   (b) What is the chance of drawing fewer than 0 red marbles?
   (c) Use the normal curve to estimate this chance.
   (d) Does the probability histogram for the number of red marbles among the draws look like the normal curve?

4. A box contains 10,000 marbles, of which some are red and the others blue. To estimate the percentage of red marbles in the box, 100 are drawn at random without replacement. Among the draws, 1 turns out to be red. The percentage of red marbles in the box is estimated as 1%, with an SE of 1%. True or false: a 95%-confidence interval for the percentage of red marbles in the box is 1% ± 2%. Explain.

The answers to these exercises are on pp. A82–83.

3. INTERPRETING A CONFIDENCE INTERVAL

In example 1 on p. 378, a simple random sample was taken to estimate the percentage of students registered at a university in fall 2005 who were living at home. An approximate 95%-confidence interval for this percentage ran from 75%

to 83%, because

$$\text{sample percentage} \pm 2\,\text{SE} = 75\% \text{ to } 83\%.$$

It seems more natural to say "There is a 95% chance that the population percentage is between 75% and 83%." But there is a problem here. In the frequency theory, a chance represents the percentage of the time that something will happen. No matter how many times you take stock of all the students registered at that university in the fall of 2005, the percentage who were living at home back then will not change. Either this percentage was between 75% and 83%, or not. There really is no way to define the chance that the parameter will be in the interval from 75% to 83%. That is why statisticians have to turn the problem around slightly.[7] They realize that the chances are in the sampling procedure, not in the parameter. And they use the new word "confidence" to remind you of this.

The chances are in the sampling procedure, not in the parameter.

The confidence level of 95% says something about the sampling procedure, and we are going to see what that is. The first point to notice: the confidence interval depends on the sample. If the sample had come out differently, the confidence interval would have been different. With some samples, the interval "sample percentage ± 2 SE" traps the population percentage. (The word statisticians use is *cover.*) But with other samples, the interval fails to cover. It's like buying a used car. Sometimes you get a lemon—a confidence interval which doesn't cover the parameter.

Three confidence intervals

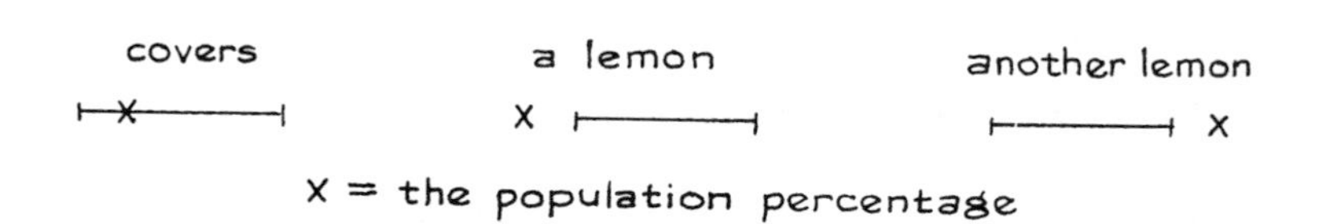

X = the population percentage

The confidence level of 95% can now be interpreted. For about 95% of all samples, the interval

$$\text{sample percentage} \pm 2\,\text{SE}$$

covers the population percentage, and for the other 5% it does not. Of course, investigators usually cannot tell whether their particular interval covers the population percentage, because they do not know that parameter. But they are using a procedure that works 95% of the time: take a simple random sample, and go 2 SEs either way from the sample percentage. It is as if their interval was drawn at random from a box of intervals, where 95% cover the parameter and only 5% are lemons. This beats second-hand cars.

> A confidence interval is used when estimating an unknown parameter from sample data. The interval gives a range for the parameter, and a confidence level that the range covers the true value.

Confidence levels are a bit difficult, because they involve thinking not only about the actual sample but about other samples that could have been drawn. The interpretation is illustrated in figure 1. A hundred survey organizations are hired to estimate the percentage of red marbles in a large box. Unknown to the pollsters,

Figure 1. Interpreting confidence intervals. The 95%-confidence interval is shown for 100 different samples. The interval changes from sample to sample. For about 95% of the samples, the interval covers the population percentage, marked by a vertical line.[8]

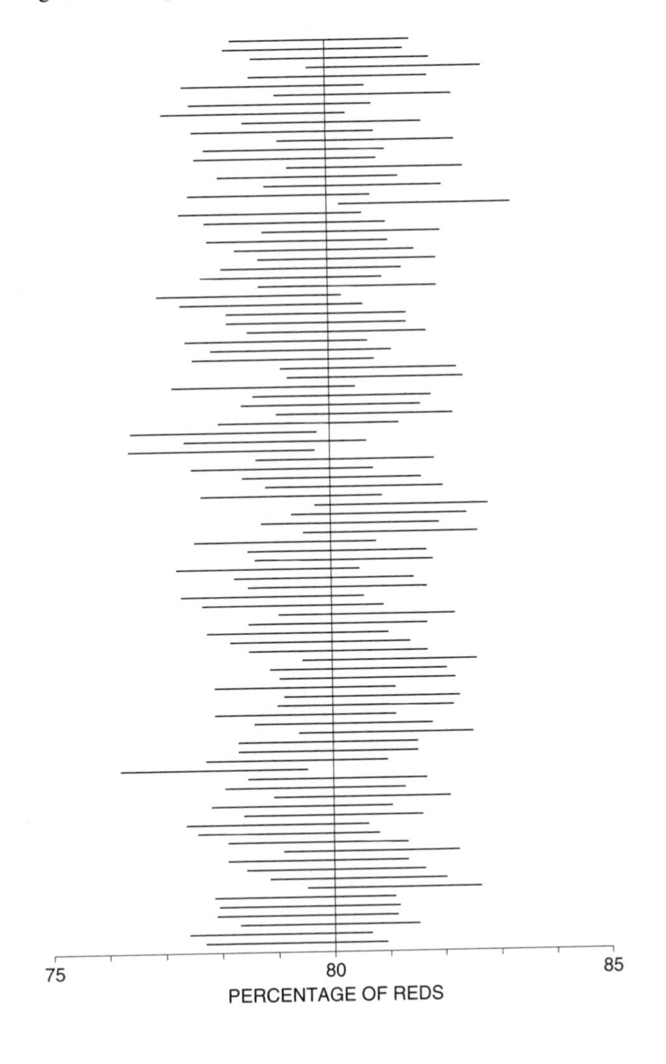

this percentage is 80%. Each organization takes a simple random sample of 2,500 marbles, and computes a 95%-confidence interval for the percentage of reds in the box, using the formula

$$\text{percentage of reds in sample} \pm 2\,\text{SE}.$$

The percentage of reds is different from sample to sample, and so is the estimated standard error. As a result, the intervals have different centers and lengths. Some of the intervals cover the percentage of red marbles in the box, others fail. About 95% of them should cover the percentage, which is marked by a vertical line. In fact, 96 out of 100 do. Of course, this is only a computer simulation, designed to illustrate the theory. In practice, an investigator would have only one sample, and would not know the parameter.

Probabilities are used when you reason forward, from the box to the draws; confidence levels are used when reasoning backward, from the draws to the box (see the chapter opening quote). There is a lot to think about here, but keep the main idea of the chapter in mind.

A sample percentage will be off the population percentage, due to chance error. The SE tells you the likely size of the amount off.

Confidence levels were introduced to make this idea more quantitative.

Exercise Set C

1. Probabilities are used when reasoning from the ______ to the ______; confidence levels are used when reasoning from the ______ to the ______. Options:
   box draws

2. (a) The chance error is in the ______ value. Options: observed, expected
   (b) The confidence interval is for the ______ percentage. Options:
   sample population

3. Refer to exercises 7 and 8 on p. 380. Compute a 95%-confidence interval for the percentage of 1's in the box, using the data obtained by the person in exercise 8(a). Repeat for the other two people. Which of the three intervals cover the population percentage, that is, the percentage of 1's in the box? Which do not? (Remember, the three people in exercise 8 do not know the contents of the box; but you do, from exercise 7.)

4. A box contains a large number of red and blue marbles; the proportion of red marbles is known to be 50%. A simple random sample of 100 marbles is drawn from the box. Say whether each of the following statements is true or false, and explain briefly.
   (a) The percentage of red marbles in the sample has an expected value of 50%, and an SE of 5%.
   (b) The 5% measures the likely size of the chance error in the 50%.

(c) The percentage of reds in the sample will be around 50%, give or take 5% or so.
(d) An approximate 95%-confidence interval for the percentage of reds in the sample is 40% to 60%.
(e) There is about a 95% chance that the percentage of reds in the sample will be in the range from 40% to 60%.

5. A box contains a large number of red and blue marbles, but the proportions are unknown; 100 marbles are drawn at random, and 53 turn out to be red. Say whether each of the following statements is true or false, and explain briefly.
   (a) The percentage of red marbles in the box can be estimated as 53%; the SE is 5%.
   (b) The 5% measures the likely size of the chance error in the 53%.
   (c) The 53% is likely to be off the percentage of red marbles in the box, by 5% or so.
   (d) A 95%-confidence interval for the percentage of red marbles in the box is 43% to 63%.
   (e) A 95%-confidence interval for the percentage of red marbles in the sample is 43% to 63%.

6. A simple random sample of 1,000 persons is taken to estimate the percentage of Democrats in a large population. It turns out that 543 of the people in the sample are Democrats. True or false, and explain:
   (a) The sample percentage is $(543/1{,}000) \times 100\% = 54.3\%$; the SE for the sample percentage is 1.6%.
   (b) $54.3\% \pm 3.2\%$ is a 95%-confidence interval for the population percentage.
   (c) $54.3\% \pm 3.2\%$ is a 95%-confidence interval for the sample percentage.
   (d) There is about a 95% chance for the percentage of Democrats in the population to be in the range $54.3\% \pm 3.2\%$.

7. (Continues exercise 6; hard.) True or false, and explain: If another survey organization takes a simple random sample of 1,000 persons, there is about a 95% chance that the percentage of Democrats in their sample will be in the range $54.3\% \pm 3.2\%$.

8. At a large university, 54.3% of the students are female and 45.7% are male. A simple random sample of 1,000 persons is drawn from this population. The SE for the sample percentage of females is figured as 1.6%. True or false: There is about a 95% chance for the percentage of females in the sample to be in the range $54.3\% \pm 3.2\%$. Explain.

The answers to these exercises are on pp. A83–84.

4. *CAVEAT EMPTOR*

The methods of this chapter were developed for simple random samples. They may not apply to other kinds of samples. Many survey organizations use fairly complicated probability methods to draw their samples (section 4 of chapter 19). As a result, they have to use more complicated methods for estimating their standard errors. Some survey organizations do not bother to use probability methods at all. Watch out for them.

> Warning. The formulas for simple random samples may not apply to other kinds of samples.

Here is the reason. Logically, the procedures in this chapter all come out of the square root law (section 2 of chapter 17). When the size of the sample is small relative to the size of the population, taking a simple random sample is just about the same as drawing at random with replacement from a box—the basic situation to which the square root law applies. The phrase "at random" is used here in its technical sense: at each stage, every ticket in the box has to have an equal chance to be chosen. If the sample is not taken at random, the square root law does not apply, and may give silly answers.[9]

People often think that a statistical formula will somehow check itself while it is being used, to make sure that it applies. Nothing could be further from the truth. In statistics, as in old-fashioned capitalism, the responsibility is on the consumer.

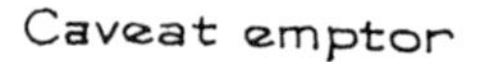

Let the buyer beware

$$\bar{x} \pm z_\alpha \times s/\sqrt{n}$$

Exercise Set D

1. A psychologist is teaching a class with an enrollment of 100. He administers a test of passivity to these students and finds that 20 of them score over 50. The conclusion: approximately 20% of all students would score over 50 on this test. Recognizing that this estimate may be off a bit, he estimates the likely size of the error as follows:

$$\text{SE for number} = \sqrt{100} \times \sqrt{0.2 \times 0.8} = 4$$

$$\text{SE for percent} = (4/100) \times 100\% = 4\%$$

   What does statistical theory say?

2. A small undergraduate college has 1,000 students, evenly distributed among the four classes: freshman, sophomore, junior, and senior. In order to estimate the percentage of students who have ever smoked marijuana, a sample is taken by the following procedure: 25 students are selected at random without replacement from each of the four classes. As it turns out, 35 out of the 100 sample students admit to having smoked. So, it is estimated that 35% out of the 1,000 students at the college would admit to having smoked. A standard error is attached to this estimate, by the following procedure:

$$\text{SE for number} = \sqrt{100} \times \sqrt{0.35 \times 0.65} \approx 5$$

$$\text{SE for percent} = (5/100) \times 100\% = 5\%$$

What does statistical theory say?

The answers to these exercises are on p. A84.

5. THE GALLUP POLL

The Gallup Poll does not use a simple random sample (section 4 of chapter 19). As a result, they do not estimate their standard errors using the method of this chapter. However, it is interesting to compare their samples to simple random samples of the same size. For instance, in 1952 they predicted a 51% vote for Eisenhower, based on a sample of 5,385 people. With a simple random sample,

$$\text{SE for number} = \sqrt{5{,}385} \times \sqrt{0.51 \times 0.49} \approx 37$$

$$\text{SE for percent} = \frac{37}{5{,}385} \times 100\% \approx 0.7 \text{ of } 1\%.$$

In fact, Eisenhower got 54.9% of the vote in that election. The Gallup Poll estimate was off by 3.9 percentage points. This is nearly 6 times the SE for a simple random sample. Table 1 shows the comparison for every presidential election from 1952 to 2004.

Table 1. Comparing the Gallup Poll with a simple random sample. The errors of prediction are on the whole quite a bit bigger than those to be expected from a simple random sample of the same size.

Year	*Sample size*	*SE for simple random sample*	*Actual error*
1952	5,385	0.7 of 1%	3.9%
1956	8,144	0.5 of 1%	2.1%
1960	8,015	0.6 of 1%	1.3%
1964	6,625	0.6 of 1%	2.9%
1968	4,414	0.7 of 1%	0.4 of 1%
1972	3,689	0.8 of 1%	1.8%
1976	3,439	0.9 of 1%	2.0%
1980	3,500	0.8 of 1%	3.5%
1984	3,456	0.8 of 1%	0.5 of 1%
1988	4,089	0.8 of 1%	2.9%
1992	2,019	1.1%	6.1%
1996	2,895	0.9%	2.8%
2000	3,571	0.8 of 1%	0.2%
2004	2,014	1.1%	1.6%

Source: See table 4 in chapter 19.

In 11 elections out of 14, the error was considerably larger than the SE for a simple random sample. One reason is that predictions are based only on part of the sample, namely, those people judged likely to vote (section 6 of chapter 19).

This eliminates about half the sample. Table 2 compares the errors made by the Gallup Poll with SEs computed for simple random samples whose size equals the number of likely voters. The simple random sample formula is still not doing a good job at predicting the size of the errors.

Why not? Well, the Gallup Poll is not drawing tickets at random from a box—although the telephone samples used from 1992 onwards come closer to simple random sampling than designs used before that (pp. 340–341, 346). Three other issues should be mentioned: (i) the process used to screen out the non-voters may break down at times; (ii) some voters may still not have decided how to vote when they are interviewed; (iii) voters may change their minds between the last pre-election poll and election day, especially in close contests. In a volatile, three-way contest like the 1992 election, such problems take their toll (p. 346).

Table 2. The accuracy of the Gallup Poll compared to that of a simple random sample whose size equals the number of likely voters in the Gallup Poll sample.

Year	*Number of likely voters*	*SE for simple random sample*	*Actual error*
1952	3,350	0.9 of 1%	3.9%
1956	4,950	0.7 of 1%	2.1%
1960	5,100	0.7 of 1%	1.3%
1964	4,100	0.8 of 1%	2.9%
1968	2,700	1.0%	0.4 of 1%
1972	2,100	1.1%	1.8%
1976	2,000	1.1%	2.0%
1980	2,000	1.1%	3.5%
1984	2,000	1.1%	0.5 of 1%
1988	2,600	1.0%	2.9%
1992	1,600	1.2%	6.1%
1996	1,100	1.5%	2.8%
2000	2,400	1.0%	0.2%
2004	1,600	1.2%	1.6%

Note: The number of likely voters is rounded.
Source: The Gallup Poll (American Institute of Public Opinion).

Exercise Set E

1. A Gallup Poll pre-election survey based on a sample of 1,000 people estimates a 65% vote for the Democratic candidate in a certain election. True or false, and explain: the likely size of the chance error in this estimate can be figured as follows—

$$\sqrt{1{,}000} \times \sqrt{0.65 \times 0.35} \approx 15, \qquad \frac{15}{1{,}000} \times 100\% = 1.5\%$$

2. One thousand tickets are drawn at random without replacement from a large box, and 651 of the draws show a 1. The fraction of 1's in the box is estimated as 65%. True or false, and explain: the likely size of the chance error in this estimate can be figured as follows—

$$\sqrt{1{,}000} \times \sqrt{0.65 \times 0.35} \approx 15, \qquad \frac{15}{1{,}000} \times 100\% = 1.5\%$$

3. The following article appeared on the *New York Times* Op Ed page of August 27, 1988, headlined MAYBE BUSH HAS ALREADY WON.

> The presidential campaign, only now formally set to begin, is in fact virtually finished. Despite the Niagara of news stories about how the candidates are touting their running mates, haggling over debates and sniping at each other, the die is just about cast.
>
> A significant indicator is the Gallup Poll, which this week shows Vice President Bush ahead of Gov. Michael S. Dukakis by 4 percentage points. In the half century since George Gallup began his electoral opinion surveys in Presidential years, his "trial heats" in the last week or so of September have foretold with notable accuracy the outcome on election day.
>
> The late James A. Farley, the Democrats' peerless tactician of 50 years ago, always argued that voters made up their minds by Labor Day. . . . It is now established, moreover, that when traditional nonvoters—the object of get-out-the-vote efforts—are persuaded to vote, they too cast their ballots in the same proportion as the rest of the electorate . . . . Significant changes in the percentages from September to November are due only to altered voter enthusiasm . . . .

(a) How does the article explain differences in voter opinion between September and November?
(b) What else could explain a difference between Gallup Poll results in late September and election results in early November?
(c) A difference of several percentage points between Gallup Poll results in late September and election results in early November is: very unlikely, unlikely but possible, quite possible. Choose one option, and explain.

The answers to these exercises are on p. A84.

6. REVIEW EXERCISES

Review exercises may cover material from previous chapters.

1. A survey organization draws a simple random sample of 1,000 registered voters in a certain town. In the sample, 32% approve of the Mayor. The organization estimates that 32% of all 50,000 registered voters in the town approve of the Mayor. How to figure the SE? The organization realizes that the number in the sample who approve ______ 1,000 draws ______ box ______. Fill in each blank (33 words or less). Then work out the SE.

2. The Residential Energy Consumption Survey found in 2001 that 47% of American households had internet access.[10] A market survey organization repeated this study in a certain town with 25,000 households, using a simple random sample of 500 households: 239 of the sample households had internet access.

   (a) The percentage of households in the town with internet access is estimated as ______; this estimate is likely to be off by ______ or so.
   (b) If possible, find a 95%-confidence interval for the percentage of all 25,000 households with internet access. If this is not possible, explain why not.

3. Of the 500 sample households in the previous exercise, 7 had three or more large-screen TVs.

   (a) The percentage of households in the town with three or more large-screen TVs is estimated as ______; this estimate is likely to be off by ______ or so.
   (b) If possible, find a 95%-confidence interval for the percentage of all 25,000 households with three or more large-screen TVs. If this is not possible, explain why not.

4. (This continues exercise 3.) Among the sample households, 121 had no car, 172 had one car, and 207 had two or more cars. Estimate the percentage of households in the town with one or more cars; attach a standard error to the estimate. If this is not possible, explain why not.

5. The National Assessment of Educational Progress administers standardized achievement tests to nationwide samples of 17-year-olds in school. One year, the tests covered history and literature. You may assume that a simple random sample of size 6,000 was taken. Only 36.1% of the students in the sample knew that Chaucer wrote *The Canterbury Tales*, but 95.2% knew that Edison invented the light bulb.[11]

   (a) If possible, find a 95%-confidence interval for the percentage of all 17-year-olds in school who knew that Chaucer wrote *The Canterbury Tales*. If this is not possible, why not?
   (b) If possible, find a 95%-confidence interval for the percentage of all 17-year-olds in school who knew that Edison invented the light bulb. If this is not possible, why not?

6. True or false: with a well-designed sample survey, the sample percentage is very likely to equal the population percentage. Explain.

7. (Hypothetical.) One year, there were 252 trading days on the New York Stock Exchange, and IBM common stock went up on 131 of them: $131/252 \approx 52\%$. A statistician attaches a standard error to this percentage as follows:

$$\text{SE for number} = \sqrt{252} \times \sqrt{0.52 \times 0.48} \approx 8$$

$$\text{SE for percent} = \frac{8}{252} \times 100\% \approx 3\%$$

   Is this the right SE? Answer yes or no, and explain.

8. A simple random sample of 3,500 people age 18 or over is taken in a large town to estimate the percentage of people (age 18 and over in that town) who read newspapers. It turns out that 2,487 people in the sample are newspaper readers.[12] The population percentage is estimated as

$$\frac{2{,}487}{3{,}500} \times 100\% \approx 71\%$$

   The standard error is estimated as 0.8 of 1%, because

$$\sqrt{3{,}500} \times \sqrt{0.71 \times 0.29} \approx 27, \qquad \frac{27}{3{,}500} \times 100\% \approx 0.8 \text{ of } 1\%$$

(a) Is 0.8 of 1% the right SE? Answer yes or no, and explain.
(b) 71% ± 1.6% is a ______for the ______. Fill in the blanks and explain.

9. (Hypothetical.) A bank wants to estimate the amount of change people carry. They take a simple random sample of 100 people, and find that on the average, people in the sample carry 73¢ in change. They figure the standard error is 4¢, because

$$\sqrt{100} \times \sqrt{0.73 \times 0.27} \approx 4, \quad 4/100 = .04$$

Are they right? Answer yes or no, and explain.

10. In Keno, there are 80 balls numbered from 1 to 80, and 20 are drawn at random. If you play a double-number, you win if both numbers are chosen. This bet pays 11 to 1, and you have very close to a 6% chance of winning.[13] If you play 100 times and stake $1 on a double-number each time, your net gain will be around ______, give or take ______ or so.

11. One hundred draws will be made at random without replacement from a large box of numbered tickets. There are two options:

(i) To win $1 if the sum of the draws is bigger than 710.
(ii) To win $1 if the average of the draws is bigger than 7.1.

Which is better? Or are they the same? Explain.

12. A monthly opinion survey is based on a sample of 1,500 persons, "scientifically chosen as a representative cross section of the American public." The press release warns that the estimates are subject to chance error, but guarantees that they are "reliable to within two percentage points." The word "reliable" is ambiguous. According to statistical theory, the guarantee should be interpreted as follows:

(i) In virtually all these surveys, the estimates will be within two percentage points of the parameters.
(ii) In most such surveys, the estimates will be within two percentage points of the parameters, but in some definite percentage of the time larger errors are expected.

Explain.

13. One hundred draws are made at random with replacement from the box | [1] [2] [2] [5] |. One of the graphs below is a histogram for the numbers drawn. Another is the probability histogram for the sum. And the third is irrelevant. Which is which? Why?

14. A coin is tossed 1,000 times.

(a) Suppose it lands heads 529 times. Find the expected value for the number of heads, the chance error, and the standard error.

(b) Suppose it lands heads 484 times. Find the expected value for the number of heads, the chance error, and the standard error.
(c) Suppose it lands heads 514 times. Find the expected value for the number of heads, the chance error, and the standard error.

15. A survey organization takes a simple random sample of 1,500 persons from the residents of a large city. Among these sample persons, 1,035 were renters.
(a) The expected value for the percentage of sample persons who rent is ______ 69%.
(b) The SE for the percentage of sample persons who rent is ______ 1.2%.

Fill in the blanks, and explain. Options:

(i) exactly equal to (ii) estimated from the data as

7. SUMMARY

1. With a simple random sample, the sample percentage is used to estimate the population percentage.

2. The sample percentage will be off the population percentage, due to chance error. The SE for the sample percentage tells you the likely size of the amount off.

3. When sampling from a 0–1 box whose composition is unknown, the SD of the box can be estimated by substituting the fractions of 0's and 1's in the sample for the unknown fractions in the box. This *bootstrap estimate* is good when the sample is large.

4. A *confidence interval* for the population percentage is obtained by going the right number of SEs either way from the sample percentage. The confidence level is read off the normal curve. This method should only be used with large samples.

5. In the frequency theory of probability, parameters are not subject to chance variation. That is why confidence statements are made instead of probability statements.

6. The formulas for simple random samples may not apply to other kinds of samples. If the sample was not chosen by a probability method, watch out: SEs computed from the formulas may not mean very much.

22

Measuring Employment and Unemployment

The country is hungry for information; everything of a statistical character, or even a statistical appearance, is taken up with an eagerness that is almost pathetic; the community have not yet learned to be half skeptical and critical enough in respect to such statements.

—GENERAL FRANCIS A. WALKER, SUPERINTENDENT OF THE 1870 CENSUS

1. INTRODUCTION

The unemployment rate is one of the most important numbers published by the government. Unemployment was only 3% in 1929, before the stock market crash (figure 1 on the next page). It reached 25% in the depths of the Depression, and remained fairly high until the U.S. entered World War II. More recently, as a result of anti-inflationary practices adopted by the Federal Reserve Board in 1981, the economy went into a deep recession in 1982–83, and the unemployment rate nearly reached 10%. By the late 1980s, the rate had dropped below 6%; in many metropolitan areas, there were shortages of skilled workers. In 2003, after the Internet bubble collapsed, the rate returned to 6%. Unemployment declined from there to the end of 2005.

The government agency in charge of the employment numbers is the Bureau of Labor Statistics. But how do they know who is employed or unemployed? Employment statistics are estimated from a sample survey—the Current Population Survey. This massive and beautifully organized sample survey is conducted

Figure 1. The unemployment rate from 1929 to 2005.

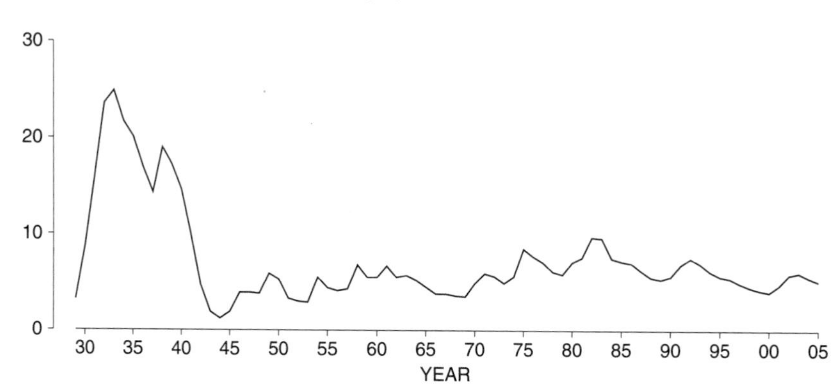

Source: *Employment and Earnings*, January 1976, table A-1; July 1989, table A-3; December 2005, table A-1.

monthly for the Bureau of Labor Statistics by the Census Bureau.[1] During the week containing the 19th day of the month, a field staff of 1,700 interviewers canvasses a nationwide probability sample of about 110,000 people. The size of the labor force, the unemployment rate, and a host of other economic and demographic statistics (like the distribution of income and educational level) are estimated from survey results, at a cost which in 2005 was about $60 million a year. The results are published in:

- *Monthly Labor Review*,
- *Employment and Earnings* (monthly),
- *The Employment Situation* (monthly),
- *Current Population Reports* (irregular),
- *Statistical Abstract of the United States* (annual),
- *Economic Report of the President* (annual).

The object of this chapter is to present the Current Population Survey in detail, from the ground up. This will illustrate and consolidate the ideas introduced in previous chapters. It should also make other large-scale surveys easier to understand. The main conclusions from this case study:

- In practice, fairly complicated probability methods must be used to draw samples. Simple random sampling is only a building-block in these designs.
- The standard-error formulas for simple random samples do not apply to these complicated designs, and other methods must be used for estimating the standard errors.

2. THE DESIGN OF THE CURRENT POPULATION SURVEY

The Current Population Survey is redesigned periodically by the Census Bureau, to take advantage of new information and to accomplish new objectives. There was a major redesign in the early 2000s, using data from the 2000 census.

There were 3,142 counties and independent cities in the U.S. As the first step in the redesign process, the Bureau put these together into groups to form 2,025 *Primary Sampling Units* (or PSUs, for short). Each PSU consisted either of a city, or a county, or a group of contiguous counties.[2] These PSUs were sorted into 824 *strata*, chosen so the PSUs in each stratum would resemble each other on certain demographic and economic characteristics (like unemployment at the time of stratification, the number of large households, and the number of workers in retail trade). The strata do not cross state lines. Many of the larger PSUs, like New York or Los Angeles, were put into strata by themselves.

The sample was chosen in two stages. To begin with, one PSU was chosen from each stratum, using a probability method which ensured that within the stratum, the chance of a PSU getting into the sample was proportional to its population. Since there were 824 strata, the first stage resulted in a sample of 824 PSUs. Until the next redesign (after the 2010 census), all interviewing for the Survey takes place in these 824 PSUs and in no others. The PSUs for an earlier design are shown in figure 2.

Figure 2. Primary Sampling Units for the Current Population Survey: the 1995 sample design with 792 PSUs.

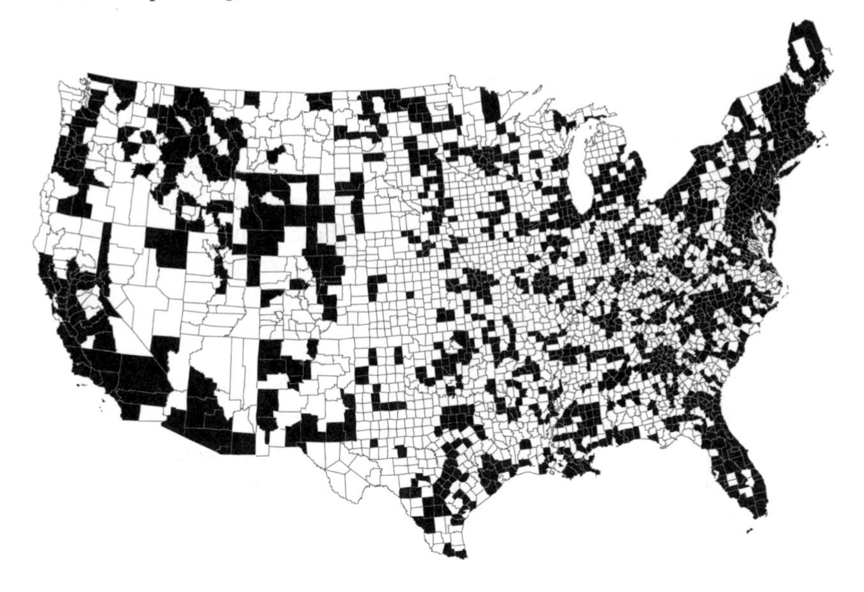

Note: Alaska and Hawaii not shown.
Source: Bureau of the Census, Statistical Methods Division.

Each PSU was divided up into *Ultimate Sampling Units* (or USUs), consisting of about 4 housing units each. At the second stage, some USUs were picked at random for the sample. In the end, every person age 16 and over living in a selected USU in a selected PSU gets into the Current Population Survey. For the U.S. as a whole, the sampling rate is about 1 in 2,000. But the rate varies from about 1 in 300 for D.C. or Wyoming to 1 in 3,000 for large states like California, New York, and Texas.[3] The objective is to estimate unemployment rates in each of the 50 states and the District of Columbia with about the same precision.[4] This meant equalizing, at least roughly, the absolute sizes of the 51 subsamples (section 4 of chapter 20). So the ratio of sample size to population size has to be different from state to state.

The Bureau's choices for the sample to be used from 2005 to 2015 were all made well before 2005. The design even provided for people who were going to live in housing yet to be constructed. And in fact, the Bureau chose not just one sample but 16 different ones, in order to rotate part of the sample every month. After it gets into the sample, a housing unit is kept there for 4 months, dropped out for 8 months, and then brought back for a final 4 months. Why rotate the sample? For one reason, the interviewers may wear out their welcome after a while. Besides that, people's responses probably change as a result of being interviewed, progressively biasing the sample (this is called *panel bias*). For instance, there is some evidence to show that people are more likely to say they are looking for a job the first time they are interviewed than the second time. Then why not change the sample completely every month? Keeping part of it the same saves a lot of money. Besides that, having some overlap in the sample makes it easier to estimate the monthly changes in employment and unemployment.

3. CARRYING OUT THE SURVEY

The Survey design for 2005 produces 72,000 housing units to be canvassed each month. Of these, about 12,000 are ineligible for the sample (being vacant, or even demolished since the sample was designed). Another 4,500 or so are unavailable, because no one is at home, or because those at home will not cooperate. That leaves about 55,500 housing units in the Survey. All persons age 16 or over in these housing units are asked about their work experience in the previous week. On the basis of their answers, they are classified as:

- employed (those who did any paid work in the previous week, or were temporarily absent from a regular job);
- or unemployed (those who were not employed the previous week but were available for work and looking for work during the past four weeks);
- or outside the labor force (defying Aristotle, the Bureau defines this as the state of being neither employed nor unemployed).[5]

The employed are asked about the hours they work and the kind of job they have. The unemployed are asked about their last job, when and why they left it, and how they are looking for work. Those outside the labor force are asked whether they are keeping house, or going to school, or unable to work, or do not work for some other reason (in which case, they are asked to specify what). Results for November 2005 are shown in table 1.

Table 1. The civilian non-institutional population age 16 and over.[6] Bureau of Labor Statistics estimates, November 2005. In millions.

Employed	142.97		
Unemployed	7.27		
Labor force		150.24	
Outside the labor force		76.96	
Total			227.20

Source: *Employment and Earnings*, December 2005, table A-13.

By definition, the *civilian labor force* consists of the civilians who are either employed or unemployed. In November 2005, that amounted to $142.97 + 7.27 = 150.24$ million people.[7] The *unemployment rate* is the percentage of the civilian labor force which is unemployed, and that came to

$$\frac{7.27}{150.24} \times 100\% \approx 4.8\%.$$

This 4.8% is an average rate of unemployment, over all the subgroups of the population. Like many averages, it conceals some striking differences. These differences are brought out by a process of cross-tabulation. Unemployment falls more heavily on teenagers and blacks, as shown by by table 2.

Table 2. Unemployment rates by race, age, and sex. Bureau of Labor Statistics estimates, November 2005. In percent.

		Age group		
Race	*Sex*	*16–19*	*20–64*	*65 and over*
White	Male	15.1	3.5	3.2
White	Female	12.4	3.7	2.4
Black	Male	41.6	9.6	6.8
Black	Female	31.7	9.0	4.5

Source: *Employment and Earnings*, December 2005, table A-13.

The overall unemployment rate is quite variable, as shown in figure 1 on p. 396. But the pattern of rates in table 2 is quite stable in certain respects. For instance, the unemployment rate for blacks has been roughly double the unemployment rate for whites over the period 1961–2005. One development is worth noting. From the 1990s onwards, unemployment rates for men have become higher than those for women—a change from the past.

Unemployment numbers are published for much finer classifications than the ones shown in table 2, including marital status, race, age, sex, type of last job, reason for unemployment (for instance, fired or quit), and duration of unemployment. The Bureau starts with a huge sample. But by the time it comes down to the white men age 35–44 who quit managerial jobs, have been out of work for 5 to 14 weeks, and are looking for work by reading the newspapers, there might not be

Figure 3. Table A-32, *Employment and Earnings*, December 2005.

HOUSEHOLD DATA
NOT SEASONALLY ADJUSTED

A-32. Unemployed persons by reason for unemployment, sex, and age

(Numbers in thousands)

Reason	Total, 16 years and over		Men, 20 years and over		Women, 20 years and over		Both sexes, 16 to 19 years	
	Dec. 2004	Dec. 2005	Dec. 2004	Dec. 2005	Dec. 2004	Dec. 2005	Dec. 2004	Dec. 2005
NUMBER OF UNEMPLOYED								
Total unemployed	7,599	6,956	3,727	3,355	2,802	2,707	1,070	894
Job losers and persons who completed temporary jobs	4,166	3,622	2,573	2,212	1,433	1,281	160	129
On temporary layoff	1,040	1,013	709	679	255	287	77	47
Not on temporary layoff	3,126	2,609	1,864	1,534	1,178	993	84	82
Permanent job losers	2,272	1,866	1,302	1,072	908	743	61	51
Persons who completed temporary jobs	854	743	562	461	270	250	23	31
Job leavers	845	752	398	339	369	335	78	78
Reentrants	2,040	2,083	683	722	894	1,000	462	361
New entrants	548	499	74	82	105	91	369	325
PERCENT DISTRIBUTION								
Total unemployed	100.0	100.0	100.0	100.0	100.0	100.0	100.0	100.0
Job losers and persons who completed temporary jobs	54.8	52.1	69.0	65.9	51.2	47.3	15.0	14.4
On temporary layoff	13.7	14.6	19.0	20.2	9.1	10.6	7.2	5.3
Not on temporary layoff	41.1	37.5	50.0	45.7	42.1	36.7	7.8	9.2
Job leavers	11.1	10.8	10.7	10.1	13.2	12.4	7.3	8.8
Reentrants	26.8	30.0	18.3	21.5	31.9	36.9	43.2	40.4
New entrants	7.2	7.2	2.0	2.4	3.8	3.4	34.5	36.4
UNEMPLOYED AS A PERCENT OF THE CIVILIAN LABOR FORCE								
Job losers and persons who completed temporary jobs	2.8	2.4	3.4	2.9	2.2	1.9	2.3	1.9
Job leavers	.6	.5	.5	.4	.6	.5	1.1	1.1
Reentrants	1.4	1.4	.9	.9	1.4	1.5	6.7	5.3
New entrants	.4	.3	.1	.1	.2	.1	5.4	4.8

NOTE: Beginning in January 2005, data reflect revised population controls used in the household survey.

too many cases left. Figure 3 shows estimates by reason for unemployment, sex, and age. (*Employment and Earnings* goes into much more detail.)

In general, by the time a large sample is cross-tabulated, there will be only very small subsamples in some classifications. Inferences about the corresponding subpopulations would be quite uncertain. Now, suppose that each estimate is within 1% of its true value with probability 95%, say. With a thousand estimates (which is about the number in *Employment and Earnings*), it would not be surprising if a few of them are quite a bit more than 1% off. The Bureau takes a big sample because it has to make many estimates about many subpopulations, and it wants to be reasonably confident that they are all fairly accurate. In fact, the Bureau will not make estimates when a subsample drops below a threshold size of about 50 cases.

"The dip in sales seems to coincide with the decision to eliminate the sales staff."

4. WEIGHTING THE SAMPLE

Suppose that one month, in the Bureau's sample of 110,000 people, there are 3,836 who are unemployed. The Bureau is sampling 1 person in 2,000 from the civilian non-institutional population age 16 and over. So it is natural to think that each person in the sample represents 2,000 people in the country. Then the way to estimate the total number of unemployed in the population is to weight up the sample number of 3,836 by the factor of 2,000:

$$2{,}000 \times 3{,}836 = 7{,}672{,}000$$

However, the Bureau does not do anything that simple. Not everybody in the sample gets the same weight. Instead, the Bureau divides the sample up into groups (by age, sex, race, and area of residence) and weights each group up separately.

There is a good reason for all the complexity. The sampling rate is different from one stratum to another, and the weights have to compensate; otherwise, the estimates could be quite biased. Moreover, the weights are used to control the impact of chance variation. For example, suppose there are too many white males age 16–19 in the sample, relative to their share in the total population. Unemployment is high in this group, which would make the overall unemployment rate in the sample too high. The Bureau has a fix: any group which is over-represented in the sample gets proportionately smaller weights, bringing the sample back into line with the population. On the other hand, if a group is under-represented, the weights are increased. Adjusting the weights this way helps to correct imbalances caused by chance variation. That reduces sampling error.[8]

5. STANDARD ERRORS

In estimating the unemployment rate, precision counts. For example, a definite picture of the economy is given by saying that the unemployment rate is $7.0\% \pm 0.1$ of 1%. However, $7\% \pm 3\%$ covers everything from boom to bust. So, it is important to know how good the estimates really are. Procedures we have discussed in previous chapters do not apply, because the Bureau is not using a simple random sample. In particular, at the second stage of its sampling procedure the Bureau chooses some ultimate sampling units (USUs). A USU is a *cluster* of about four adjacent housing units. Every person age 16 and over living in one of these USUs gets into the sample (section 2). A cluster is all or nothing: either everybody in the cluster gets into the sample, or nobody does. People living in the same cluster tend to be similar to one another in many ways. Information about each one says something about all the others, in terms of family background, educational history, and employment status.

With simple random sampling, by comparison, if one person in the cluster gets into the sample, the other people still have only a small chance of getting in. As a result, each person drawn into a simple random sample provides additional information, independent of the persons drawn previously. The Bureau's cluster sample of 110,000 persons contains less information than a simple random sample of the same size: cluster samples involve a lot of redundancy. Thus, clustering tends to reduce the precision of the Bureau's estimates. On the other hand, the weights improve precision. All in all, computing SEs for the Bureau's estimates is a delicate business.

As it turns out, with a cluster sample the standard errors can themselves be estimated very closely from the data, using the *half-sample method*. Although the details are complicated and take a lot of computer power, the idea is simple. If the Bureau wanted to see how accurate the Current Population Survey was, one thing to do would be make another independent survey following exactly the same procedures. The difference between the two surveys would give some idea of how reliable each set of results was.

Nobody would seriously propose to replicate the Current Population Survey, at a cost of another $60 million a year, just to see how reliable it is. But the Bureau can get almost the same effect by splitting the Survey into two independent pieces which have the same chance behavior (hence the name, "half-sample method"). Suppose for instance that one piece of the survey estimates the civilian labor force at 150.5 million, and the other comes in at 150.7 million. This difference is due to chance error. The pooled estimate of the civilian labor force is

$$\frac{150.5 + 150.7}{2} = 150.6 \text{ million}$$

The two individual estimates are 0.1 million away from their average, and the standard error is estimated by this difference of 0.1 million.

Of course, an estimated standard error based on only one split may not be too reliable. But there are many different ways to split the sample. The Bureau looks at a number of them and combines the standard errors by taking the root-mean-square. This completes the outline of the half-sample method.[9] Some of the estimated standard errors for November 2005 are shown in table 3.

Table 3. Estimated standard errors, November 2005.

	Estimate	*Standard error*
Civilian labor force	150.24 million	300,000
Employment	142.97 million	323,000
Unemployment	7.27 million	155,000
Unemployment rate	4.8%	0.1 of 1%

Source: *Employment and Earnings*, December 2005, tables A13, 1B, and 1C.

How do the estimated standard errors in table 3 compare to those for a simple random sample of the same size and composition? Calculations show that for estimating the size of the labor force, the Bureau's standard error is about 8% smaller than that for a simple random sample: the weights are doing a good job. For estimating the number of unemployed, however, the Bureau's sample is about 30% worse than a simple random sample: the clustering hurts.[10]

So why doesn't the Bureau use simple random sampling? For one thing, there is no list showing all the people age 16 and over in the U.S., with current addresses. Even if there were such a list, taking a simple random sample from it would produce people spread thinly throughout the country, and the cost of interviewing them would be enormous. With the Bureau's procedure, the sample is bound to come out in clumps in relatively small and well-defined areas, so the interviewing cost is quite manageable. In 2005, this was about $100 per interview. The Bureau's sample design turns out to be amazingly cost effective.

The comparison between the Bureau's design and a simple random sample points to a real issue. To compute a standard error properly, you need more than the sample data. You need to know how the sample was picked. With a simple random sample, there is one SE. With a cluster sample, there is another. The formulas which apply to simple random samples will usually underestimate the standard errors in cluster samples. (These issues came up before, in the context of the Gallup Poll: sections 4 and 5 of chapter 21.)

> Cluster samples are less informative than simple random samples of the same size. So the simple random sample formulas for the standard error do not apply.

Exercise Set A

1. One month, the Current Population Survey sample amounted to 100,000 people. Of them, 62,000 were employed, and 3,000 were unemployed. True or false, and explain:
   (a) 65% of the sample was in the labor force.
   (b) The Bureau would estimate that 65% of the population was in the labor force.

2. The Current Population Survey sample is split into two independent halves. From one half, the number of employed persons is estimated as 151.5 million; from the other, it is estimated as 151.3 million. Combine these two estimates, and attach a standard error to the result.

3. (Hypothetical.) The Health Department in a certain city takes a simple random sample of 100 households. In 80 of these households, all the occupants have been vaccinated against polio. So the Department estimates that for 80% of the households in that city, all the occupants have been vaccinated against polio. With this information, can you put a standard error on the 80%? Find the SE, or explain what other information is needed.
4. (Continues exercise 3.) The Department interviews every person age 25 and over in the sample households. They find 144 such persons, of whom 29 have college degrees. They estimate that 20.1% of the people age 25 and over in the city have college degrees. With this information, can you put a standard error on the 20.1%? Find the SE, or explain what other information is needed.
5. In election years, the Bureau makes a special report on voting, using the Current Population Survey sample. In 2000, about 55% of all the people of voting age in the sample said they voted; but only 52% of the total population of voting age did in fact vote.[11] Can the difference be explained as a chance error? If not, how else can it be explained? (You may assume that the Bureau's sample is the equivalent of a simple random sample of 75,000 people.)
6. In table 2 on p. 399, which estimate is more trustworthy: for white males age 20–64, or black males age 20–64? Explain briefly.

The answers to these exercises are on pp. A84–85.

6. THE QUALITY OF THE DATA

The data collected by the Survey are of very high quality: for many purposes, Survey data are considered to be more accurate than Census data. In any large-scale field operation, mistakes are inevitable. Since the Survey operates on a much smaller scale than the Census, it can afford better quality control. The key is careful selection, training, and supervision of the field staff. Interviewers are given about four days of training in survey procedures before they start work, and several hours a month of training while they are on the job. At least once a year, their work is observed by their supervisors. In addition, about 3% of the monthly sample (chosen by a separate probability sampling procedure) is reinterviewed by supervisors. All discrepancies are discussed with the interviewers. The interviewers' reports are *edited*, that is, checked for incomplete or inconsistent entries. For most items, error rates are low; and each error is reviewed with the person who made it.

7. BIAS

Bias is more insidious than chance error, especially if it operates more or less evenly across the sample. SEs computed by the half-sample method—or any other method—will not pick up that kind of bias. Measuring bias, even roughly, is hard work and involves going beyond the sample data.

> When bias operates more or less evenly across the sample, it cannot be detected just by looking at the data.

The Bureau has made unusually careful studies of the biases in the Current Population Survey. On the whole, these seem to be minor, although their exact sizes are not known. To begin with, the Survey design is based on Census data (section 2), and the Census misses a small percentage of the population. This percentage is not easy to pin down. Even if the Bureau knew it, they would still have a hard time adjusting the estimated number of unemployed (say) to compensate for the undercount, because the people missed by the Census are likely to be different from the ones the Census finds. A similar difficulty crops up in another place. The Survey misses about 10% of the people counted by the Census. To some extent, the weights bring these missing people back into the estimates. But non-response bias is not so easy to fix. The people missed by the Survey are probably different from the ones it finds, while the weights pretend they are the same.[12]

Next, the distinction between "employed" and "unemployed" is a little fuzzy around the edges. For example, people who have a part-time job but would like full-time work are classified as employed, but they really are partially unemployed. Moreover, people who want to work but have given up looking are classified as outside the labor force, although they probably should be classified as unemployed. The Bureau's criterion for unemployment, namely being without work, available for work, and looking for work, is necessarily subjective. In practice, it is a bit slippery. Results from the reinterview program (section 6) suggest the number of unemployed is higher than the Bureau's estimate, by several hundred thousand people. In this case, the bias is larger than the sampling error.[13] Over the period from 1980 to 2005, the number of unemployed has ranged from 5 to 10 million. Relatively speaking, both sampling error and non-sampling error are small.

8. REVIEW EXERCISES

Review exercises may cover previous chapters as well.

1. One month, there are 100,000 people in the Current Population Survey sample, of whom 63,000 are employed and 4,000 are unemployed.

    (a) True or false, and explain: the Bureau would estimate the percentage of the population who are unemployed as

    $$\frac{4{,}000}{63{,}000 + 4{,}000} \times 100\% \approx 6\%$$

    (b) What happened to the other 33,000 people?

2. One month, there are 100,000 people in the Current Population Survey sample, and the Bureau estimates the unemployment rate as 6.0%. True or false, and explain: the standard error for this percentage should be estimated as follows—

$$\text{SE for number} = \sqrt{100{,}000} \times \sqrt{0.06 \times 0.94} \approx 75$$

$$\text{SE for percent} = \frac{75}{100{,}000} \times 100\% \approx 0.08 \text{ of } 1\%$$

3. One month, the Current Population Survey sample is split into two independent replicates. Using one replicate, the number of unemployed people is estimated as 7.1 million. The other replicate produces an estimate of 6.9 million. Using this information, estimate the number of unemployed people, and attach a standard error to the estimate.

4. Using the data in exercise 3, what can you say about the bias in the estimate?

5. A simple random sample is drawn at random ______ replacement. Options: with, without.

6. A box contains 250 tickets. Two people want to estimate the percentage of 1's in the box. They agree to use the percentage of 1's in 100 draws made at random from the box. Person A wants to draw with replacement; person B wants to draw without replacement. Which procedure gives a more accurate estimate? Or does it make any difference?

7. (Hypothetical.) A survey organization draws a sample of 100 households from 10,000 in a certain town, by the following procedure. First, they divide the town into 5 districts, with 2,000 households each. Then they draw 2 districts at random. Within each of the 2 selected districts, they draw 50 households at random.

   (a) Is this a probability sample?
   (b) Is this a simple random sample?

   Answer yes or no, and explain.

8. A supermarket chain has to value its inventory at the end of every year, and this is done on a sample basis. There is a master list of all the types of items sold in the stores. Then, auditors take a sample of the items and go through the shelves, finding the amounts in stock and prices for the sample items. To draw the sample, the auditors start by choosing a number at random from 1 to 100. Suppose this turns out to be 17. The auditors take the 17th, 117th, 217th, ... items in the list for the sample. If the random number is 68, they take the 68th, 168th, 268th, ... items. And so forth.

   (a) Is this a probability sample?
   (b) Is this a simple random sample?

   Answer yes or no, and explain.

9. As part of a study on drinking, the attitudes of a sample of alcoholics are assessed by interview.[14] Cases are assigned to interviewers at random. Some of the interviewers are teetotalers, others drink. Would you expect the two groups of interviewers to reach similar conclusions? Answer yes or no, and give reasons.

10. From "The Grab Bag" by L. M. Boyd in the *San Francisco Chronicle*: "The Law of Averages says that if you throw a pair of dice 100 times, the numbers tossed will add up to just about 683." Is this right? Answer yes or no, and explain.

11. A polling organization takes a simple random sample of 750 voters from a district with 18,000 voters. In the sample, 405 voters are for. Fill in the blanks, using the options below. Explain briefly,

    (a) The observed value of the ______ is 405.
    (b) The observed value of the ______ is 54%.
    (c) The expected value of the ______ is equal to the ______.

    Options:

    (i) number of voters in the sample who are for
    (ii) percentage of voters in the sample who are for
    (iii) percentage of voters in the district who are for

12. In 2004, there were 318,390 applications to buy guns in California.[15] A criminologist takes a simple random sample of 193 out of these applications, and finds that only 2 were rejected. True or false, and explain:

    (a) 2 out of 193 is 1.04%.
    (b) The SE on the 1.04% is 0.73%.
    (c) A 95%-confidence interval for the percentage of all 318,390 applications that were rejected is 1.04% $\pm$ 1.46%.

9. SUMMARY

1. Unemployment rates in the U.S. are estimated using the *Current Population Survey*.

2. This survey is based on a nationwide probability sample of about 110,000 persons, who are interviewed monthly. The design is more complicated than simple random sampling.

3. The Survey reweights the sample so it agrees with Census data on age, sex, race, state of residence, and certain other characteristics influencing employment status.

4. When a sample is taken by a probability method, it is possible not only to estimate parameters, but also to figure the likely size of the chance errors in the estimates.

5. The standard errors for *cluster samples* can be obtained by the *half-sample method*, splitting the sample into two halves and seeing how well they agree.

6. The formulas for the standard error have to take into account the details of the probability method used to draw the sample. The formulas which apply to simple random samples will usually underestimate the standard errors in cluster samples.

7. When bias operates more or less evenly across the sample, it cannot be detected just by looking at the sample data. Standard errors ignore that kind of bias.

8. The Current Population Survey, like all surveys, is subject to a number of small biases. The bias in the estimate of the unemployment rate is thought to be larger than the standard error.

23

The Accuracy of Averages

Ranges are for cattle.
—L.B.J.

1. INTRODUCTION

The object of this chapter is to estimate the accuracy of an average computed from a simple random sample. This section deals with a preliminary question: How much chance variability is there in the average of numbers drawn from a box? For instance, take the box

| 1 | 2 | 3 | 4 | 5 | 6 | 7 |

The computer was programmed to make 25 draws at random with replacement from this box:

2 4 3 2 5   7 5 6 4 5   4 4 1 2 4   4 6 4 7 2   7 2 5 7 3

The sum of these numbers is 105, so their average is $105/25 = 4.2$. The computer did the experiment again, and the results came out differently:

5 1 4 3 4   5 2 1 7 7   1 2 3 2 4   7 1 6 5 3   6 6 3 3 4

Now the sum is 95, so the average is $95/25 = 3.8$. The sum of the draws is subject to chance variability, therefore the average is too. The new problem is to calculate the expected value and standard error for the average of the draws. The method will be indicated by example.

Example 1. Twenty-five draws will be made at random with replacement from the box

1	2	3	4	5	6	7

The average of the draws will be around ______, give or take ______ or so.

Solution. The average of the box is 4, so the average of the draws will be around 4. The give-or-take number is the SE. To get the SE for the average, we go back to the sum. The expected value for the sum is

$$\text{number of draws} \times \text{average of box} = 25 \times 4 = 100$$

The SD of the box is 2, and the SE for the sum is

$$\sqrt{\text{number of draws}} \times \text{SD of box} = \sqrt{25} \times 2 = 10$$

The sum will be around 100, give or take 10 or so.

What does this say about the average of the draws? If the sum is one SE above expected value, or $100 + 10$, the average of the 25 draws is

$$\frac{100 + 10}{25} = \frac{100}{25} + \frac{10}{25} = 4 + 0.4$$

On the other hand, if the sum is one SE below expected value, or $100 - 10$, the average is

$$\frac{100 - 10}{25} = \frac{100}{25} - \frac{10}{25} = 4 - 0.4$$

The average of the draws will be about 4, give or take 0.4 or so. The 4 is the expected value for the average of the draws. The 0.4 is the standard error, completing the solution.

The idea in brief:

$$\text{sum of 25 draws} = 100 \pm 10 \text{ or so}$$

$$\text{average of 25 draws} = \frac{100}{25} \pm \frac{10}{25} \text{ or so}$$

In other words, to find the SE for the average of the draws, just go back and get the SE for the sum; then divide by the number of draws.

When drawing at random from a box:

$$\text{EV for average of draws} = \text{average of box}.$$

$$\text{SE for average of draws} = \frac{\text{SE for sum}}{\text{number of draws}}.$$

The SE for the average says how far the average of the draws is likely to be from the average of the box.

Figure 1. The top panel shows a probability histogram for the sum of 25 draws from the box | 1 2 3 4 5 6 7 |. The bottom panel shows the probability histogram for the average of the draws. In standard units, the two histograms are exactly the same.

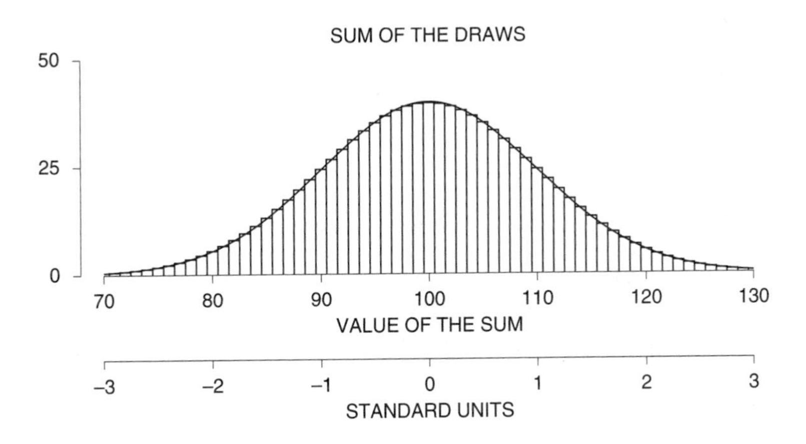

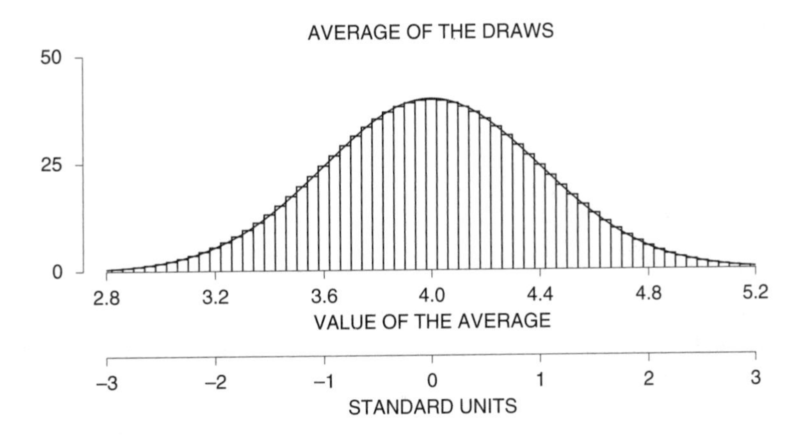

If the number of draws is large enough, the normal curve can be used to figure chances for the average. Figure 1 (bottom panel) shows the probability histogram for the average of 25 draws from the box

| 1 2 3 4 5 6 7 |

The histogram follows the curve, so areas under the histogram can be approximated by areas under the curve.

Why does the probability histogram for the average look like the normal curve? This is a corollary of the mathematics of chapter 18. The probability histogram for the sum of the 25 draws is close to the normal curve (top panel of figure 1). The average of the draws equals their sum, divided by 25. This division is just a change of scale, and washes out in standard units. The two histograms in figure 1 have exactly the same shape, and both follow the curve.

> When drawing at random from a box, the probability histogram for the average of the draws follows the normal curve, even if the contents of the box do not. The histogram must be put into standard units, and the number of draws must be reasonably large.[1]

Example 2. One hundred draws will be made at random with replacement from the box in example 1.

(a) The average of the draws will be around ______, give or take ______ or so.

(b) Estimate the chance that the average of the draws will be more than 4.2.

Solution. As in example 1, the sum of the draws will be around $100 \times 4 = 400$. The give-or-take number is $\sqrt{100} \times 2 = 20$. The sum of the draws will be around 400, give or take 20 or so. The average of the draws will be around $400/100 = 4$, give or take $20/100 = 0.2$ or so. The SE for the average of 100 draws is 0.2.

Part (b) is handled by the normal approximation.

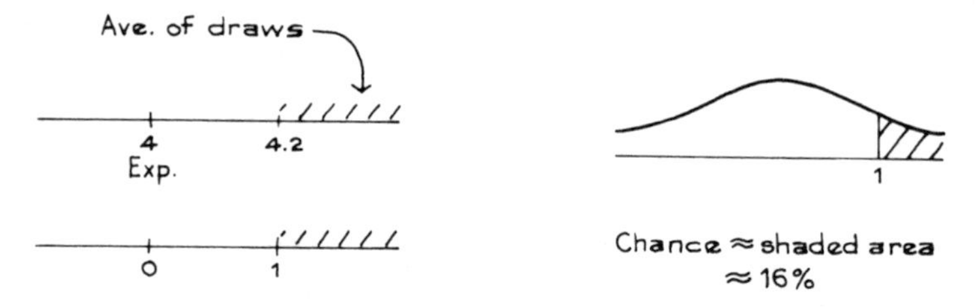

The chance is around 16%. This completes the solution.

In examples 1 and 2, when the number of draws went up by a factor of 4, from 25 to 100, the SE for the average of the draws went down by a factor of $\sqrt{4} = 2$, from 0.4 to 0.2. This is so in general.

> When drawing at random with replacement from a box of tickets, multiplying the number of draws by a factor (like 4) divides the SE for the average of the draws by the square root of that factor ($\sqrt{4} = 2$).

As the number of draws goes up, the SE for the sum gets bigger—and the SE for the average gets smaller. Here is the reason. The SE for the sum goes up, but only by the square root of the number of draws. As a result, while the SE for the sum gets bigger in absolute terms, compared to the number of draws it gets smaller. The division by the number of draws makes the SE for the average go down. Keep this difference between the two SEs in mind.

When drawing without replacement, the exact SE for the average of the draws can be found using the correction factor (section 4 of chapter 20)—

$$\text{SE without} = (\text{correction factor}) \times (\text{SE with}).$$

Usually, the number of draws is small by comparison with the number of tickets in the box, and the correction factor will be so close to 1 that it can be ignored.

Exercise Set A

1. One hundred draws are made at random with replacement from a box.
   (a) If the sum of the draws is 7,611, what is their average?
   (b) If the average of the draws is 73.94, what is their sum?

2. A box of tickets averages out to 75, and the SD is 10. One hundred draws are made at random with replacement from this box.
   (a) Find the chance (approximately) that the average of the draws will be in the range 65 to 85.
   (b) Repeat, for the range 74 to 76.

3. One hundred draws will be made at random with replacement from a box of tickets. The average of the numbers in the box is 200. The SE for the average of the draws is computed, and turns out to be 10. True or false:
   (a) About 68% of the tickets in the box are in the range 190 to 210.
   (b) There is about a 68% chance for the average of the hundred draws to be in the range 190 to 210.

4. You are drawing at random with replacement from a box of numbered tickets.
   (a) The expected value for the average of the ______ equals the average of the ______. Options: box, draws.
   (b) As the number of draws goes up, the SE for the ______ of the draws goes up but the SE for the ______ of the draws goes down. Options:

   sum　　　average

5. A box contains 10,000 tickets. The numbers on these tickets average out to 50, and the SD is 20.
   (a) One hundred tickets are drawn at random with replacement. The average of these draws will be around ______, give or take ______ or so.
   (b) What if 100 draws are made without replacement?
   (c) What if 100 draws are made without replacement, and there are only 100 tickets in the box?

6. The figure below shows the probability histogram for the average of 50 draws from the box [1] [2] [3] [4]. What does the shaded area represent?

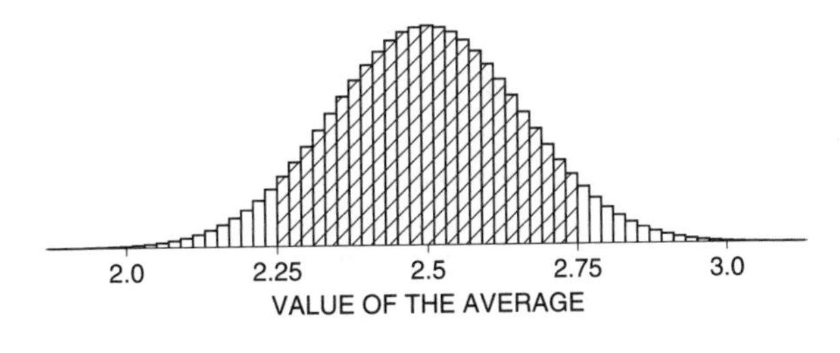

7. The figure below shows a histogram for data generated by drawing 50 times from the box in exercise 6. What does the shaded area represent?

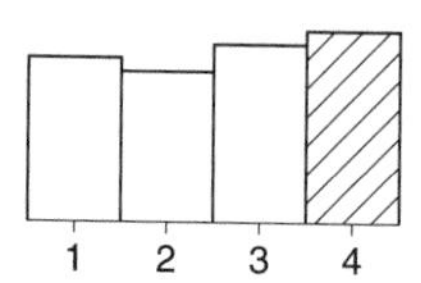

8. (a) In the top panel of figure 1, the area of the rectangle over 90 represents what?
   (b) In the bottom panel of figure 1, the area of the rectangle over 3.6 represents what?
   (c) The rectangles in parts (a) and (b) have exactly the same area. Is that a coincidence? Discuss briefly.

9. Two hundred draws are made at random with replacement from | 1 2 2 3 |. True or false, and explain:
   (a) The expected value for the average of the draws is exactly 2.
   (b) The expected value for the average of the draws is around 2, give or take 0.05 or so.
   (c) The average of the draws will be around 2, give or take 0.05 or so.
   (d) The average of the draws will be exactly 2.
   (e) The average of the box is exactly 2.
   (f) The average of the box is around 2, give or take 0.05 or so.

10. The figure below is a probability histogram for the sum of 25 draws from the box | 1 2 3 |. However, an investigator needs the probability histogram for the average of these draws, by midnight. A research assistant says, "There's nothing to it. All we have to do is change the numbers on the horizontal axis." Is that right? If so, the assistant should change 25 to ______, 50 to ______, and 55 to ______. If the assistant is wrong, what needs to be done? Explain your answers. (No vertical scale is needed.)

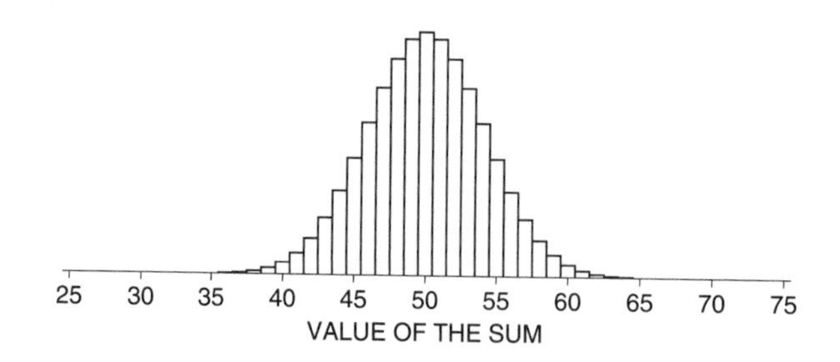

The answers to these exercises are on p. A85.

Technical notes. (i) The bottom panel in figure 1 represents what is called a *sampling distribution*. The histogram shows how the sample averages vary over the set of all possible samples. In more detail, imagine making a list of all possible samples, and computing the sample average for each one. (You would get quite a long list of averages.) Some averages come up more frequently than others. The area of the rectangle over 4.0 shows what percentage of these sample averages are 4.0, and so forth.

(ii) When drawing at random with replacement from a box, the SE for the sum of the draws is

$$\sqrt{\text{no. of draws}} \times \text{SD of box}.$$

So the SE for the average of the draws is

$$(\sqrt{\text{no. of draws}} \times \text{SD of box})/\text{no. of draws}.$$

This simplifies to (SD of box)/$\sqrt{\text{no. of draws}}$, which in most books is written $\sigma/\sqrt{n}$, where σ is the SD and n is the number of draws. The Greek letter σ is read as "sigma."

2. THE SAMPLE AVERAGE

In section 1, the numbers in the box were known, and the problem was to say something about the average of the draws. This section reasons in the opposite—and more practical—direction. A random sample is taken from a box of unknown composition, and the problem is to estimate the average of the box. Naturally, the average of the draws is used as the estimate. And the SE for the sample average can be used with the normal curve to gauge the accuracy of the estimate. (Chapter 21 used the same technique for percentages.)

The method will be presented by example. Along the way, there will be two questions to answer:

- What's the difference between the SD of the sample and the SE for the sample average?
- Why is it OK to use the normal curve in figuring confidence levels?

Now, the example. Suppose that a city manager wants to know the average income of the 25,000 families living in his town. He hires a survey organization to take a simple random sample of 1,000 families. The total income of the 1,000 sample families turns out to be \$62,396,714. Their average income is $\$62{,}396{,}714/1{,}000 \approx \$62{,}400$. The average income for all 25,000 families is estimated as \$62,400. Of course, this estimate is off by a chance error. The problem is to put a give-or-take number on the estimate:

$$\$62{,}400 \pm \$\_\_\_\_\_?$$

The SE is needed, and for that, a box model. There should be one ticket in the box for each family in the town, showing that family's income. The data are like 1,000 draws from the box.

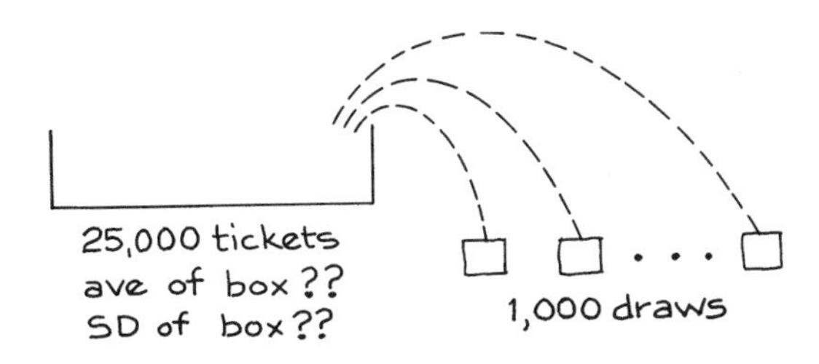

The average income of the sample families is like the average of the draws. The SE for the average of the draws can now be found by the method of section 1. The first step is to find the SE for the sum of the draws. Since 1,000 is such a small fraction of 25,000, there is no real difference between drawing with and without replacement. The SE for the sum is

$$\sqrt{1{,}000} \times \text{SD of box.}$$

Of course, the survey organization does not know the SD of the box, but they can estimate it by the SD of the sample. (This is another example of the bootstrap method discussed in section 1 of chapter 21.)

> With a simple random sample, the SD of the sample can be used to estimate the SD of the box. The estimate is good when the sample is large.

There are 1,000 families in the sample, and the SD of their incomes turns out to be \$53,000. The SD of the box is estimated as \$53,000. The SE for the sum is estimated as

$$\sqrt{1{,}000} \times \$53{,}000 \approx \$1{,}700{,}000.$$

To get the SE for the average, we divide by the number of families in the sample: $\$1{,}700{,}000/1{,}000 = \$1{,}700$. That is the answer. The average of the draws is something like \$1,700 off the average of the box. So the average of the incomes of all 25,000 families in the town can be estimated as

$$\$62{,}400 \pm \$1{,}700.$$

Keep the interpretation of the \$1,700 in mind: it is the margin of error for the estimate. This completes the example.

One point is worth more discussion. The expected value for the sum of the draws—the total income of the sample families—is

$$1{,}000 \times \text{average of the box.}$$

This is unknown because the average of the box is unknown. The total income of the 1,000 sample families turned out to be \$62,396,714. This is the observed value for the sum of the draws. The SE for the sum—\$1,700,000—measures the likely size of the difference between \$62,396,714 and the expected value. In general,

$$\text{observed value} = \text{expected value} + \text{chance error.}$$

The SE measures the likely size of the chance error.

Confidence intervals for percentages (qualitative data) were discussed in section 2 of chapter 21. The same idea can be used to get confidence intervals for the average of the box (quantitative data). For example, a 95%-confidence interval for the average of the incomes of all 25,000 families in the town is obtained by going

2 SEs either way from the sample average:

$$\$62{,}400 \pm 2 \times \$1{,}700 = \$59{,}000 \text{ to } \$65{,}800.$$

("Sample average" is statistical shorthand for the average of the numbers in the sample.)

Two different numbers came up in the calculations: the SD of the sample was \$53,000, and the SE for the sample average was \$1,700. These two numbers do different things.

- The SD says how far family incomes are from average—for typical families.
- The SE says how far sample averages are from the population average—for typical samples.

People who confuse the SD with the SE might think that somehow, 95% of the families in the town had incomes in the range \$62,400 ± \$3,400. That would be ridiculous. The range \$62,400 ± \$3,400 covers only a tiny part of the income distribution: the SD is about \$53,000. The confidence interval is for something else. In about 95% of all samples, if you go 2 SEs either way from the sample average, your confidence interval will cover the average for the whole town; in the other 5%, your interval will miss. The word "confidence" is to remind you that the chances are in the sampling procedure; the average of the box is not moving around. (These issues were discussed before, in section 3 of chapter 21.)

Example 3. As part of an opinion survey, a simple random sample of 400 persons age 25 and over is taken in a certain town in Appalachia. The total years of schooling completed by the sample persons is 4,635. So their average educational level is 4,635/400 ≈ 11.6 years. The SD of the sample is 4.1 years. Find a 95%-confidence interval for the average educational level of all persons age 25 and over in this town.

Solution. First, a box model. There should be one ticket in the box for each person age 25 and over in the town, showing the number of years of schooling completed by that person; 400 draws are made at random from the box. The data are like the draws, and the sample average is like the average of the draws. That completes the model.

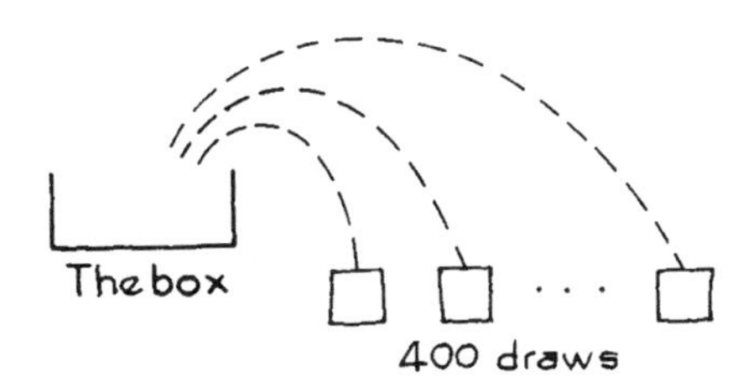

We need to compute the SE for the average of the draws. The SE for the sum is $\sqrt{400} \times$ SD of the box. The SD of the box is unknown, but can be estimated by the SD of the sample, as 4.1 years. So the SE for the sum of the draws is estimated as $\sqrt{400} \times 4.1 = 82$ years. (The 82 measures the likely size of the chance error in the sum, which was 4,635.) The SE for the average is $82/400 \approx 0.2$ years. The average educational level of the persons in the sample will be off the average for the town by 0.2 years or so. An approximate 95%-confidence interval for the average educational level for all persons age 25 and over in the town is

$$11.6 \pm 0.4 \text{ years.}$$

That is the answer.

The confidence level of 95% is the area under the normal curve between -2 and 2. Why is the curve relevant? After all, the histogram for educational levels (p. 39) looks nothing like the curve. However, the curve is not used to approximate the histogram for the data; it is used to approximate the probability histogram for the sample average.

A computer simulation will help. The computer has one ticket in the box for each person age 25 or over in the town, showing his or her educational level. A histogram for the contents of the box is shown at the top of figure 2. This histogram represents the educational level of all people age 25 or over in the town. Its shape is nothing like the normal curve. (Remember, this is just a simulation; in reality, you would not know the contents of the box—but the mathematical theory can still be used.)

Now 400 draws must be made at random without replacement from the box, to get the sample. The computer was programmed to do this. A histogram for the 400 draws is shown in the second panel. This represents the distribution of educational level for the 400 sample people. It is very similar to the first histogram, although there are a few too many people with 8–9 years of education. That is a chance variation. Figure 2 indicates why the SD of the sample is a good estimate for the SD of the box. The two histograms show just about the same amount of spread.

So far, we have seen two histograms, both for data. Now a probability histogram comes in, for the average of the draws. This histogram is shown in the bottom panel. This third histogram does not represent data. Instead, it represents chances for the sample average. For instance, take the area under the probability histogram between 11.6 and 12.4 years. This area represents the chance that the average of 400 draws from the box will be between 11.6 and 12.4 years. The area works out to about 95%. For 95% of samples, the average educational level of the sample families will be in the range 11.6 to 12.4 years. For the other 5%, the sample average will be outside this range. Any area under the probability histogram can be interpreted in a similar way.

Now you can see why the normal approximation is legitimate. As the figure shows, the normal curve is a good approximation to the probability histogram for the average of the draws—even though the data do not follow the curve. That is why the curve can be used to figure confidence levels. Even with large samples, confidence levels read off the normal curve are only approximate, because they depend on the normal approximation; with a small sample, the normal curve should not be used (section 6 of chapter 26).

Figure 2. Computer simulation. The top panel shows the distribution of educational level among people age 25 or over in the whole town. The middle panel shows the distribution of educational level in the sample. These are histograms for data. The bottom panel shows the probability histogram for the average of 400 draws from the box; it is close to the normal curve. The average educational level in the town is 12.0 years, and the SD is 4.0 years; in the sample, the corresponding numbers are 11.6 and 4.1. (The endpoint convention for the data histograms: the class interval 12–13, for example, includes all the people who finished 12 years of schooling but not 13—high school graduates who did not finish a year of college.)

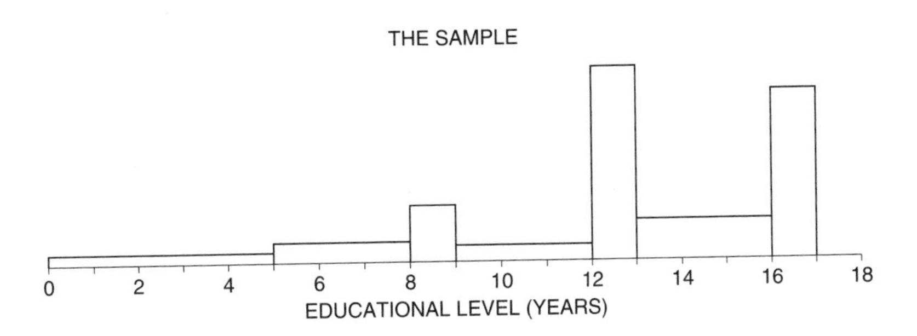

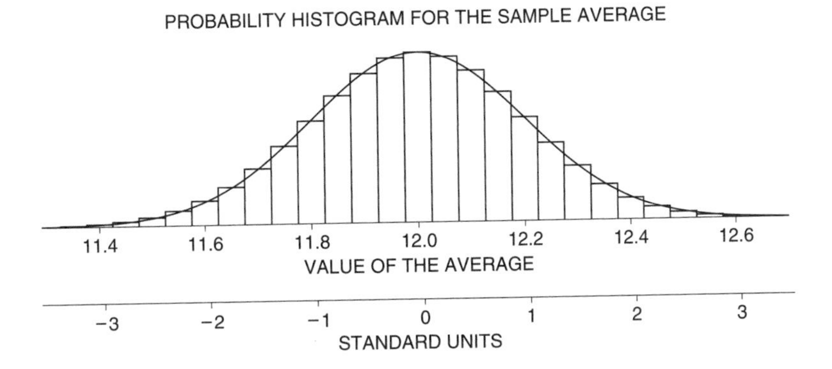

Exercise Set B

1. Match each phrase on list A with one on list B.

List A	*List B*
population	draws
population average	average of the box
sample	box
sample average	number of draws
sample size	average of the draws

2. In each pair of phrases, one makes sense and one does not. Which is which? Explain briefly.

   (a) SE for box, SD of box.
   (b) SE for average of box, SE for average of draws.

3. For the income example on pp. 415–417:

   (a) The SD of the box is ______ \$53,000.
   (b) The SE for the sample average is ______ \$1,700.
   (c) The ______ value for the sample average is \$62,400.

   Fill in the blanks, using the options below, and explain. (At least one option will be left over.)

known to be	estimated from the sample as
expected	observed

4. In example 3 on p. 417, suppose 50 different survey organizations take simple random samples of 400 persons age 25 and over in the town. Each organization gets a 95%-confidence interval "sample average $\pm$ 2 SE." How many of these intervals should cover the population average?

5. The figure below is a computer simulation of the study described in exercise 4. The confidence intervals are plotted at different heights so they can be seen.

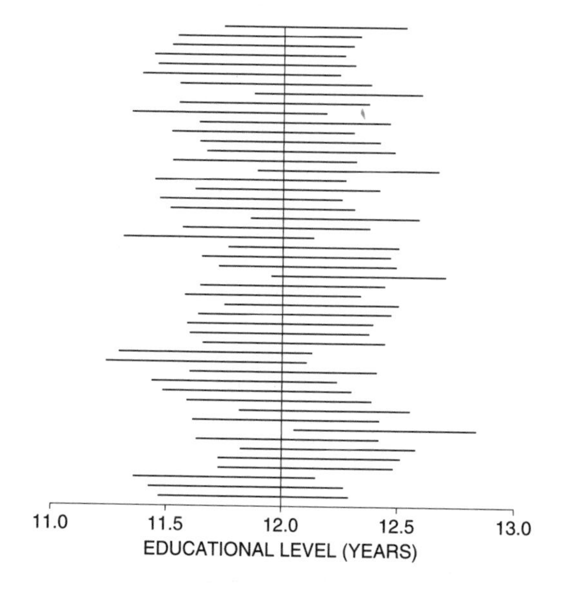

(a) Why do the intervals have different centers?

(b) Why do they have different lengths?

(c) How many of them cover the population average, marked by a vertical line at 12 years?

6. A university has 30,000 registered students. As part of a survey, 900 of these students are chosen at random. The average age of the sample students turns out to be 22.3 years, and the SD is 4.5 years.[2]

   (a) The average age of all 30,000 students is estimated as ______. This estimate is likely to be off by ______ or so.

   (b) Find a 95%-confidence interval for the average age of all 30,000 registered students.

7. The Census Bureau collects information on the housing stock as part of the decennial census. In 2000, for instance, the Bureau found that there were about 105 million occupied housing units in the U.S., one third being rental units. Typical rents varied from about $400 in Wyoming to about $800 in Hawaii.[3] (In some urban markets like east-side Manhattan or San Francisco's Nob Hill, of course, rents are much higher—if you can find an apartment in the first place.)

   A certain town has 10,000 occupied rental units. A local real estate office does a survey of these units: 250 are chosen at random, and the occupants are interviewed. Among other things, the rent paid in the previous month is determined. The 250 sample rents average out to $568, and the SD is $385. A histogram is plotted for the sample rents, and does not follow the normal curve.

   (a) If possible, find a 68%-confidence interval for the average rent paid in the previous month on all 10,000 occupied rental units in this town. If this is not possible, explain why not.

   (b) True or false, and explain: for about 68% of all the occupied rental units in this town, the rent paid in the previous month was between $544 and $592.

8. (Continues exercise 7; hard.) True or false, and explain: if another 250 occupied rental units were taken at random, there would be about a 68% chance for the new sample average to be in the range from $544 to $592.

9. (Hard.) Census data are available on 25,000 families in a certain town. For all 25,000 families, the average income is $61,700 and the SD is $50,000. A market research firm takes a simple random sample of 625 out of the 25,000 families. The figure below is a probability histogram for the average income of the sample families; the histogram is drawn in standard units. The average income of the 625 sample families turned out to be $58,700 and the SD was $49,000.

   (a) On the histogram below, +1 in standard units is ______. Options:

   $60,660 $63,700 $107,700 $111,700

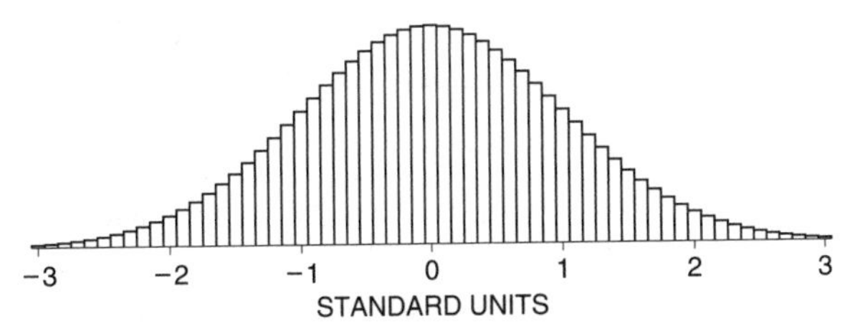

9. Continued.
   (b) In standard units, $58,700 is

   0   − 1.0   − 1.5   other

   Explain your answers.

The answers to these exercises are on pp. A85–87.

3. WHICH SE?

The SE always has the same interpretation: it is the likely size of a chance error. However, there seem to be many SEs. Which to use when? The best thing to do is to write down a box model, and decide what is being done to the draws. That will tell you which formula to use. There are four operations to think about: adding the draws, taking their average, classifying and counting, or taking percents. The corresponding formulas:

$$\text{SE for sum} = \sqrt{\text{number of draws}} \times \text{SD of box}$$

$$\text{SE for average} = \frac{\text{SE for sum}}{\text{number of draws}}$$

$$\text{SE for count} = \text{SE for sum, from a 0–1 box}$$

$$\text{SE for percent} = \frac{\text{SE for count}}{\text{number of draws}} \times 100\%$$

The SE for the sum is basic. The other formulas all come from that one. These formulas are exact for draws made at random with replacement from a box.

Reasoning forward or backward. When reasoning forward from the box to the draws, as in part V, the standard error can be computed exactly from the composition of the box. A chance quantity like the sum of the draws will be around its expected value—but will be off by an SE or so.

When reasoning backward from the draws to the box, you often have to estimate the SD of the box from the sample. So the SE itself is only approximate. However, the interpretation of the SE is almost the same. For instance, suppose the average of the sample is used to estimate the average of the box. This estimate will be off by a little, and the SE says by about how much. When the sample is reasonably large, the error in the SE itself is usually too small to matter.

> The SE shows the likely size of the amount off. It is a give-or-take number.

The terminology may be a bit confusing. Statisticians speak of the standard error for the *sample average*. The idea is that the sample average estimates the population average, but is off by a little. The SE gauges the likely size of the amount off. Statisticians also talk about confidence intervals for the *population average*. The confidence interval is a range computed from the sample. This range covers the population average with some specified degree of confidence.

Exercise Set C

This exercise set also covers material from previous chapters.

1. Fill in the table below, for draws made at random with replacement from the box | 0 2 3 4 6 |.

Number of draws	*EV for sum of draws*	*SE for sum of draws*	*EV for average of draws*	*SE for average of draws*
25				
100				
400				

2. One hundred draws are made at random with replacement from a box. The average of the box is 3.1.
   (a) True or false: the expected value for the average of the draws is exactly equal to 3.1. If this cannot be determined from the information given, what else do you need to know, and why?
   (b) What is the SE for the average of the draws? If this cannot be determined from the information given, what else do you need to know, and why?

3. One hundred draws are made at random with replacement from a box. The average of the draws is 3.1.
   (a) The expected value for the average of the draws is ______ 3.1. Fill in the blank, using one of the options below, and explain.
      (i) exactly equal to
      (ii) estimated from the data as
   (b) What is the SE for the average of the draws? If this cannot be determined from the information given, what else do you need to know, and why?

4. Forty draws are made at random with replacement from the box

| 1 2 3 4 |

   (a) Fill in the blanks with a word or phrase: the SE for the ______ is 7.1, and the SE for the ______ is 0.18. Explain your answers.
   (b) The figure below is a probability histogram for the sum of the draws. What numbers go into the three blanks?

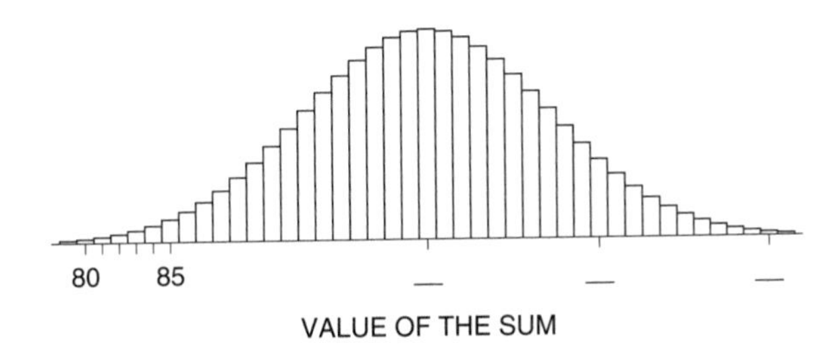

5. What's the worst thing about a sample of size one?

6. There are three boxes of numbered tickets. The average of the numbers in each box is 200. However, the SD of box A is 10, the SD of box B is 20, and the SD of box C is 40. Now

   - 100 draws are made from box A,
   - 200 draws are made from box B,
   - 400 draws are made from box C.

   (The draws are made with replacement.) The average of each set of draws is computed. Here they are, in scrambled order:

   203.6 198.1 200.4

   (a) Which average comes from which box?
   (b) Could it possibly be otherwise?

   Explain briefly.

The answers to these exercises are on p. A87.

4. A REMINDER

This chapter explained how to evaluate the accuracy of an average computed from a simple random sample. Because the draws were made at random, it was possible to gauge the accuracy just from the spread in the data and the size of the sample. This is one of the major achievements of statistical theory.

The arithmetic can be carried out on any list: find the SD, multiply by the square root of the number of entries, then divide by the number of entries. However, the method gives sensible results only when the draws are made at random. If the data do not come from the right kind of sample, the result of the calculation may be nonsense (pp. 387–390, pp. 402–403).

A *sample of convenience* is a sample that is not chosen by a probability method. (An example would be some instructor's first-year psychology class.) Some people use the simple random sample formulas on samples of convenience. That could be a real blunder. With samples of convenience, the chances are hard to define; so are parameters and standard errors.

> The formulas in this chapter are for draws from a box, and should not be applied mechanically to other kinds of samples.

Exercise Set D

This exercise set also covers material from previous chapters.

1. A utility company serves 50,000 households. As part of a survey of customer attitudes, they take a simple random sample of 750 of these households. The average number of television sets in the sample households turns out to be 1.86, and the

SD is 0.80. If possible, find a 95%-confidence interval for the average number of television sets in all 50,000 households.[4] If this isn't possible, explain why not.

2. Out of the 750 households in the survey of the previous exercise, 451 have computers. If possible, find a 99.7%-confidence interval for the percentage of all the 50,000 households with computers. If this isn't possible, explain why not.

3. (Continues exercises 1 and 2.) Out of the 750 households in the survey, 749 have at least one television set. If possible, find a 95%-confidence interval for the percentage of all the 50,000 households with at least one television set. If this isn't possible, explain why not.

4. As part of the survey described in exercise 1, all persons age 16 and over in the 750 sample households are interviewed. This makes 1,528 people. On the average, the sample people watched 5.20 hours of television the Sunday before the survey, and the SD was 4.50 hours. If possible, find a 95%-confidence interval for the average number of hours spent watching television on that Sunday by all persons age 16 and over in the 50,000 households. If this isn't possible, explain why not.

5. (a) As his sample, a psychology instructor takes all the students in his class. Is this a probability sample? a cluster sample?
   (b) A sociologist interviews the first 100 subjects who walk through a shopping mall one day. Does she have a probability sample? a cluster sample?

6. One hundred draws are made at random with replacement from a box. The sum of the draws is 297. Can you estimate the average of the box? Can you attach a standard error to your estimate, on the basis of the information given so far? Explain briefly.

7. A box contains 250 tickets. Two people want to estimate the average of the numbers in the box. They agree to take a sample of 100 tickets, and use the sample average as their estimate. Person A wants to draw the tickets at random without replacement; person B wants to take a simple random sample. Which procedure gives a more accurate estimate? Or does it make any difference?

The answers to these exercises are on p. A88.

5. REVIEW EXERCISES

Review exercises may cover material from previous chapters.

1. A box of tickets has an average of 100, and an SD of 20. Four hundred draws will be made at random with replacement from this box.
   (a) Estimate the chance that the average of the draws will be in the range 80 to 120.
   (b) Estimate the chance that the average of the draws will be in the range 99 to 101.

2. Five hundred draws are made at random with replacement from a box with 10,000 tickets. The average of the box is unknown. However, the average of the draws was 71.3, and their SD was about 2.3. True or false, and explain:

(a) The 71.3 estimates the average of the box, but is likely to be off by 0.1 or so.
(b) A 68%-confidence interval for the average of the box is 71.3 ± 0.1.
(c) About 68% of the tickets in the box are in the range 71.3 ± 0.1.

3. A real estate office wants to make a survey in a certain town, which has 50,000 households, to determine how far the head of household has to commute to work.[5] A simple random sample of 1,000 households is chosen, the occupants are interviewed, and it is found that on average, the heads of the sample households commuted 8.7 miles to work; the SD of the distances was 9.0 miles. (All distances are one-way; if someone isn't working, the commute distance is defined to be 0.)

   (a) The average commute distance of all 50,000 heads of households in the town is estimated as ______, and this estimate is likely to be off by ______ or so.
   (b) If possible, find a 95%-confidence interval for the average commute distance of all heads of households in the town. If this isn't possible, explain why not.

4. (Continues exercise 3.) The real estate office interviewed all persons age 16 and over in the sample households; there were 2,500 such persons. On the average, these 2,500 people commuted 7.1 miles to work, and the SD of the distances was 10.2 miles. (Again, if someone isn't working, the commute distance is defined to be 0; and all distances are one-way.) If possible, find a 95%-confidence interval for the average commute distance for all people age 16 and over in this town. If this isn't possible, explain why not.

5. (Continues exercise 4.) In 721 of the sample households, the head of the household commuted by car. If possible, find a 95%-confidence interval for the percentage of all households in the town where the head of the household commutes by car. If this isn't possible, explain why not.

6. The National Assessment of Educational Progress (NAEP) periodally administers tests on different subjects to high school students.[6] In 2000, the grade 12 students in the sample averaged 301 on the mathematics test; the SD was 30. The likely size of the chance error in the 301 is about ______.

   (a) Can you fill in the blank if a cluster sample of 1,000 students was tested? If so, what is the answer? If not, why not?
   (b) Can you fill in the blank if a simple random sample of 1,000 students was tested? If so, what is the answer? If not, why not?

7. A city government did a survey of working women, to see how they felt about juggling jobs and family responsibilities. Businesses, unions, and community service organizations helped distribute the survey questionnaire to locations where the women could pick up copies. 1,678 out of 2,800 respondents, or 59.9%, checked the item "stress is a serious problem" on the questionnaire. Choose one option, and explain briefly.

(i) The standard error on the 59.9% is 0.9 of 1%.
(ii) The standard error on the 59.9% is some other number.
(iii) Neither of the above.

8. One year, there were about 3,000 institutions of higher learning in the U.S. (including junior colleges and community colleges). As part of a continuing study of higher education, the Carnegie Commission took a simple random sample of 400 of these institutions.[7] The average enrollment in the 400 sample schools was 3,700, and the SD was 6,500. The Commission estimates the average enrollment at all 3,000 institutions to be around 3,700; they put a give-or-take number of 325 on this estimate. Say whether each of the following statements is true or false, and explain. If you need more information to decide, say what you need and why.

   (a) An approximate 68%-confidence interval for the average enrollment of all 3,000 institutions runs from 3,375 to 4,025.
   (b) If a statistician takes a simple random sample of 400 institutions out of 3,000, and goes one SE either way from the average enrollment of the 400 sample schools, there is about a 68% chance that his interval will cover the average enrollment of all 3,000 schools.
   (c) About 68% of the schools in the sample had enrollments in the range $3{,}700 \pm 6{,}500$.
   (d) It is estimated that 68% of the 3,000 institutions of higher learning in the U.S. enrolled between $3{,}700 - 325 = 3{,}375$ and $3{,}700 + 325 = 4{,}025$ students.
   (e) The normal curve can't be used to figure confidence levels here at all, because the data don't follow the normal curve.

9. (Continues exercise 8.) There were about 600,000 faculty members at institutions of higher learning in the U.S. As part of its study, the Carnegie Commission took a simple random sample of 2,500 of these faculty persons.[8] On the average, these 2,500 sample persons had published 1.7 research papers in the two years prior to the survey, and the SD was 2.3 papers. If possible, find an approximate 95%-confidence interval for the average number of research papers published by all 600,000 faculty members in the two years prior to the survey. If this isn't possible, explain why not.

10. A survey organization takes a simple random sample of 625 households from a city of 80,000 households. On the average, there are 2.30 persons per sample household, and the SD is 1.75. Say whether each of the following statements is true or false, and explain.

   (a) The SE for the sample average is 0.07.
   (b) A 95%-confidence interval for the average household size in the sample is 2.16 to 2.44.
   (c) A 95%-confidence interval for the average household size in the city is 2.16 to 2.44.
   (d) 95% of the households in the city contain between 2.16 and 2.44 persons.

(e) The 95%-confidence level is about right because household size follows the normal curve.

(f) The 95%-confidence level is about right because, with 625 draws from the box, the probability histogram for the average of the draws follows the normal curve.

11. The figure below is a probability histogram for the average of 25 draws made at random with replacement from the box | [1] [2] [3] [4] [5] |. Or is something wrong? Explain.

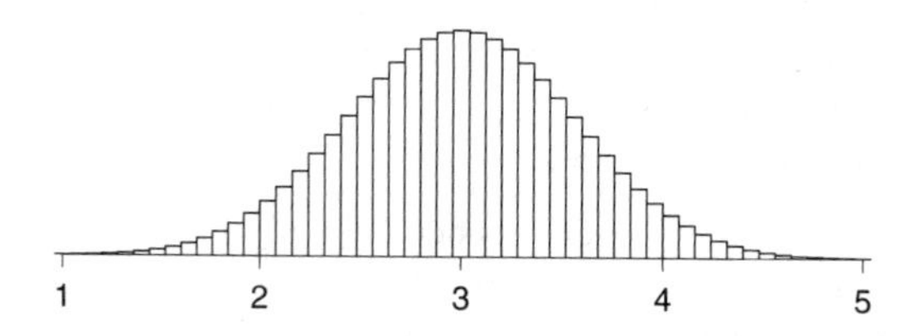

12. One term at the University of California, Berkeley, 400 students took the final in Statistics 2. Their scores averaged 65.3 out of 100, and the SD was 25. Now

$$\sqrt{400} \times 25 = 500, \qquad 500/400 = 1.25$$

Is 65.3 ± 2.5 a 95%-confidence interval? If so, for what? If not, why not?

6. SPECIAL REVIEW EXERCISES

These exercises cover all of parts I–VI.

1. An experiment was carried out to determine the effect of providing free milk to school children in a certain district (Lanarkshire, Scotland).[9] Some children in each school were chosen for the treatment group and got free milk; others were chosen for controls and got no milk. Assignment to treatment or control was done at random, to make the two groups comparable in terms of family background and health.

   After randomization, teachers were allowed to use their judgment in switching children between treatment and control, to equalize the two groups. Was it wise to let the teachers use their judgment this way? Answer yes or no, and explain briefly.

2. For the portacaval shunt (section 2 of chapter 1), survival among the controls in the poorly-designed trials was worse than survival among the controls in the randomized controlled experiments. Is it dangerous to be a control in a poorly-designed study? Answer yes or no, and explain. If your answer is no, what accounts for the difference in survival rates?

3. (a) Epidemiologists find a higher rate of oral cancer among drinkers than non-drinkers. If alcohol causes oral cancer, would that tend to create an asso-

ciation between drinking and oral cancer? Answer yes or no, and discuss briefly.

(b) Epidemiologists find an association between high levels of cholesterol in the blood and heart disease. They conclude that cholesterol causes heart disease. However, a statistician argues that smoking confounds the association, meaning that—

(i) Smoking causes heart disease.
(ii) Smoking causes heart disease, and smokers have high levels of cholesterol in their blood.
(iii) Smokers tend to eat a less healthful diet than non-smokers. Thus, smokers have high levels of cholesterol in the blood, which in turn causes heart disease.
(iv) The percentage of smokers is about the same among persons with high or low levels of cholesterol in the blood.

Choose one option, and discuss briefly.

4. A follow-back study on a large sample of death certificates shows the average age at death is smaller for left-handed people than for right-handers. (In this kind of study, surviving relatives are interviewed.)

(a) Suppose that, other things being equal (age, sex, race, income, etc.), left-handed people are more at risk from accident and disease than right handers. Could that explain a difference in average age at death?
(b) During the twentieth century, there were big changes in child-rearing practices. In the early part of the century, parents insisted on raising children to be right-handed. By mid-century, parents were much more tolerant of left-handedness. Could that explain a difference in average age at death of left-handed and right-handed people in 2005?
(c) What do you conclude from the death certificate data?

5. Before a strike in 1994, the median salary of the 746 major league baseball players was about $500,000. The lowest salary was about $100,000 and the highest was over $5,000,000. Choose one option and explain:

(i) The owners were paying out around $746 \times \$500{,}000 = \$373$ million per year in salaries to the players.
(ii) The owners were paying out substantially less than $373 million per year to the players.
(iii) The owners were paying out substantially more than $373 million per year to the players.

6. In HANES3, the Public Health Service interviewed a representative sample of Americans. Among other things, respondents age 25 and over were asked about their geographic mobility—how often did they move? About 20% of them had moved in the last year. At the other extreme, about 25% of them had been living at the same address for 15 years or more; 5% had been at the same address for 35 years or more! The average time since the last move was

10 years, and the SD was ______. Fill in the blank using one of the options below, and explain briefly.

1 year   2 years   10 years   25 years

7. To measure water clarity in a lake, a glass plate with ruled lines is pushed down into the water until the lines cannot be seen any more. The distance below the surface of the water is called "Secchi depth." To measure pollution by algae, scientists determine the total concentration of chlorophyll in the water. In a certain lake, Secchi depth and total chlorophyll concentration are measured every Thursday at noon, from April through September. Will the correlation between these variables be positive or negative? Explain briefly.

8. An instructor standardizes her midterm and final each semester so the class average is 50 and the SD is 10 on both tests. The correlation between the tests is around 0.50. One semester, she took all the students who scored around 30 at the midterm, and gave them special tutoring. On average, they gained 10 points on the final. Can this be explained by the regression effect? Answer yes or no, and explain briefly.

9. For entering freshmen at a certain university, scores on the Math SAT and Verbal SAT can be summarized as follows:

$$\text{average M-SAT} = 555, \quad \text{SD} = 125$$
$$\text{average V-SAT} = 543, \quad \text{SD} = 115, \quad r = 0.66$$

The scatter diagram is football-shaped. One student is chosen at random and has an M-SAT of 600. You would guess his V-SAT is ______ points, and would have about a 68% chance to be right within ______ points. Fill in the blanks; explain briefly.

10. Pearson and Lee obtained the following results in a study of about 1,000 families:

$$\text{average height of husband} \approx 68 \text{ inches}, \quad \text{SD} \approx 2.7 \text{ inches}$$
$$\text{average height of wife} \approx 63 \text{ inches}, \quad \text{SD} \approx 2.5 \text{ inches}, \quad r \approx 0.25$$

Among the men who were about 5 feet 4 inches tall, estimate the percentage who were shorter than their wives.

11. In a large study of the relationship between incomes of husbands and wives, the following results were obtained:

$$\text{average income of husband} \approx \$50{,}000, \quad \text{SD} \approx \$40{,}000$$
$$\text{average income of wife} \approx \$40{,}000, \quad \text{SD} \approx \$30{,}000, \quad r \approx 0.33$$

(a) The couples were divided into groups according to the income of the husbands ($0–$4,999, $5,000–$9,999, $10,000–$14,999, etc.). The average income for wives in each group was calculated and then plotted above the midpoint of the corresponding range ($2,500, $7,500, $12,500, etc.). It was found that the points on this graph followed a straight line very closely. The slope of this line would be about

0.25   0.75   0.83   1   1.33

Explain briefly. If more information is needed, say what you need and why.

(b) For one couple in the study, the wife's income was $37,500, but the information about her husband's income was lost. At $40,000, the height of the line plotted in part (a) equals $37,500. Is $40,000 a good estimate for the husband's income? Or is the estimate likely to be too high? too low? Why?

12. The figure below shows a scatter diagram, with two lines. One estimates the average value of y for each x. The other estimates the average value of x for each y. Or is something wrong? Explain briefly. (The average of x is 50, and the SD is 17; the statistics for y are just about the same.)

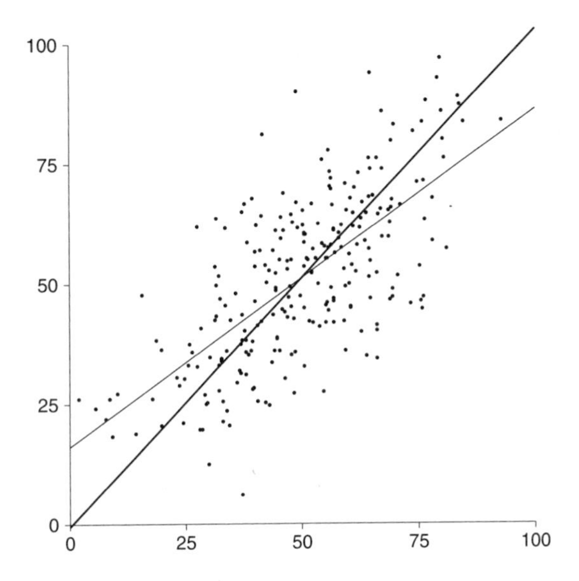

13. Five cards will be dealt from a well-shuffled deck. Find the chance of getting an ace or a king among the 5 cards. (A deck has 52 cards, of which 4 are aces and 4 are kings.)

14. Out of the 300 people enrolled in a large course, 6 got a perfect score on the first midterm and 9 got a perfect score on the second midterm. One person will be chosen at random from the class. If possible with the information given, find the chance that person has a perfect score on both midterms. Otherwise, say what information is needed, and why.

15. A die is rolled 6 times. Find the chance that the first number rolled comes up 3 more times—

(a) If the first roll is an ace.
(b) If the first roll is a six.
(c) If you don't know what happens on the first roll.

(A die has 6 faces, showing 1 through 6 spots; an ace is ⚀; each face is equally likely to come up.)

16. A Nevada roulette wheel has 38 pockets. One is marked "0," another is marked "00," and the rest are numbered from 1 through 36. The wheel is spun and a ball is dropped. The ball is equally likely to end up in any one of the 38 pockets (figure 3 on p. 282). Here are two possibilities:

    (i) You win $1 if any 7's turn up in 15 spins of the wheel.
    (ii) You win $1 if any 7's turn up in 30 spins of the wheel.

    True or false, and explain: the second possibility gives you twice as much of a chance to win as the first.

17. A die will be rolled 20 times. The sum

    number of ones rolled + number of sixes rolled

    will be around ______, give or take ______ or so.

18. A multiple-choice quiz has 50 questions. Each question has 3 possible answers, one of which is correct. Two points are given for each correct answer, but a point is taken off for a wrong answer.

    (a) The passing score is 50. If a student answers all the questions at random, what is the chance of passing?
    (b) Repeat part (a), if the passing score is 10.

19. "Toss a hundred pennies in the air and record the number of heads that come up when they fall. Do this several thousand times and plot a histogram for the numbers that you get. You will have a histogram that closely approximates the normal curve, and the more times you toss the hundred pennies the closer your histogram will get to the curve."[10] If you keep on tossing this group of a hundred pennies, will your histogram get closer and closer to the normal curve? Or will it converge to the probability histogram for the number of heads in 100 tosses of a coin? Choose one option, and explain briefly.

20. Twenty-five draws will be made at random with replacement from the box | 1 2 9 |.

    (a) A statistician uses the normal curve to compute the chance that the sum of the draws will equal 90. The result is

        too low   too high   about right

        Choose one option, and explain.
    (b) Repeat, for the chance that the sum is between 90 and 110.

    No calculations are necessary, just look at figure 9 on p. 322.

21. Imagine making a scatter diagram from table 3 on p. 302 as follows. Plot the point whose x-coordinate is the number of heads in tosses #1–100, and whose y-coordinate is the number of heads in tosses #101–200. This gives (44, 54). Then plot the point whose x-coordinate is the number of heads on tosses #201–300, and whose y-coordinate is the number of heads in tosses #301–400. This gives (48, 53). And so on. One of the scatter diagrams on the next page plots the data. Which one? Explain briefly.

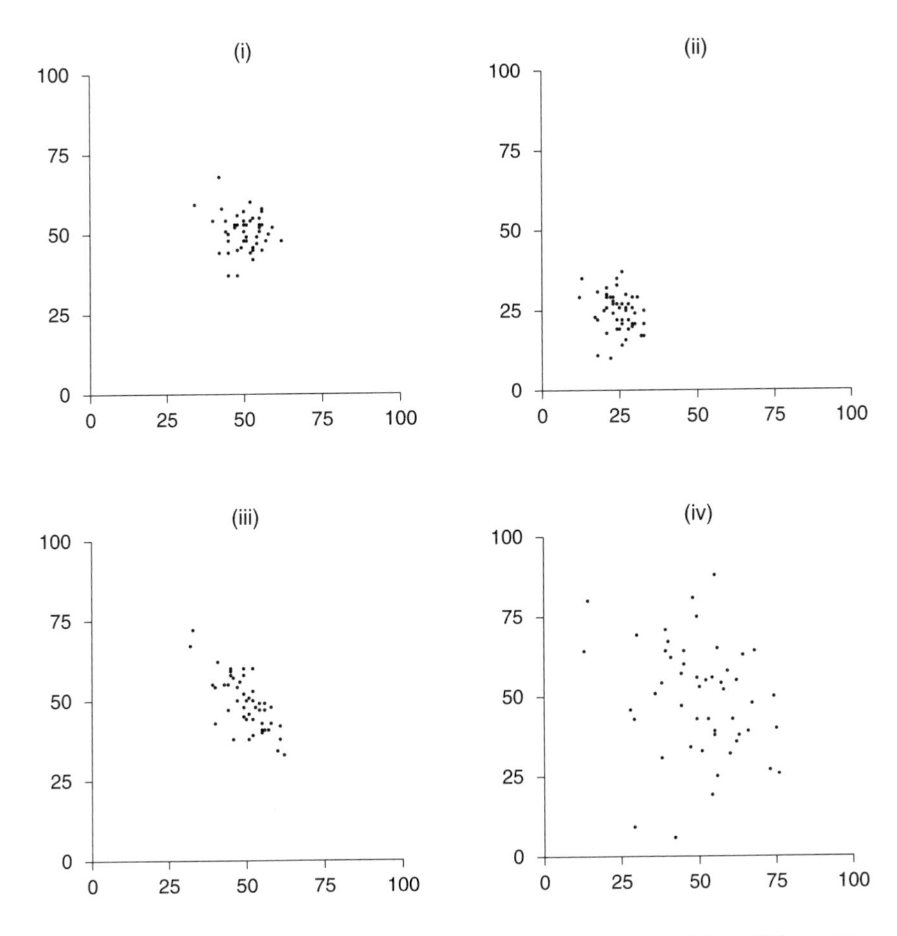

22. A box contains 10,000 marbles: 6,000 are red and 4,000 are blue; 500 marbles are drawn at random without replacement.
    (a) Suppose there are 218 blue marbles in the sample. Find the expected value for the percentage of blues in the sample, the observed value, the chance error, and the standard error.
    (b) Suppose there are 191 blue marbles in the sample. Find the expected value for the percentage of blues in the sample, the observed value, the chance error, and the standard error.

23. The top panel in the figure on the next page shows the probability histogram for the sum of 25 draws made at random with replacement from box A. The bottom panel shows the probability histogram for the average of 25 draws made at random with replacement from box B. Choose one option and explain briefly; if you choose (iii), say what additional information is needed.
    (i) Box A and Box B are the same.
    (ii) Box A and Box B are different.
    (iii) Can't tell without more information.

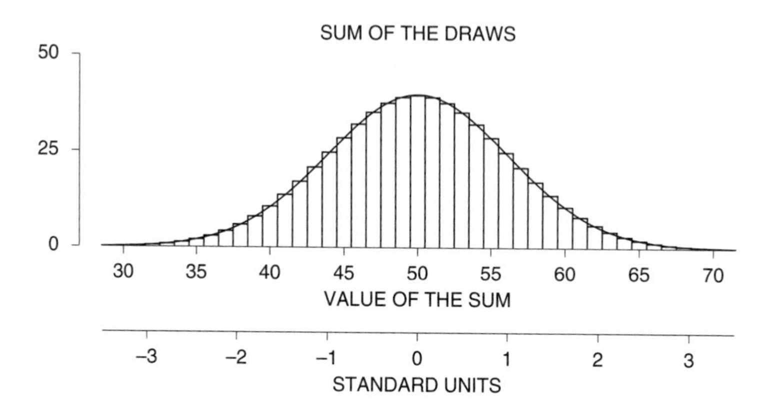

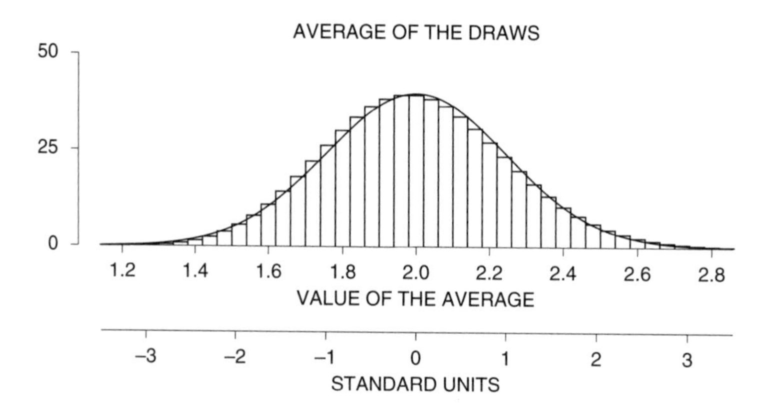

24. Draws are being made at random with replacement from a box. The number of draws is getting larger and larger. Say whether each of the following statements is true or false, and explain. ("Converges" means "gets closer and closer.")

    (a) The probability histogram for the sum of the draws (when put in standard units) converges to the normal curve.
    (b) The histogram for the numbers in the box (when put in standard units) converges to the normal curve.
    (c) The histogram for the numbers drawn (when put in standard units) converges to the normal curve.
    (d) The probability histogram for the product of the draws (when put in standard units) converges to the normal curve.
    (e) The histogram for the numbers drawn converges to the histogram for the numbers in the box.

25. (Hypothetical) A retailer has 1,000 stores nationwide. Each store has 10 to 15

employees, for a national total of 12,000. The personnel department has done a study of these employees, to assess morale. The report begins:

> Findings are based on interviews with 250 employees. We took a simple random sample of 50 stores, and interviewed 5 employees at each of the sample stores. Interviews were done by a team of occupational psychologists provided under contract by an independent survey organization. Since the interviews were anonymous, we do not know the names of the interviewees . . . .

At this point, there should be a question you want answered. What is your question, and why does it matter?

26. In 1965, the U.S. Supreme Court decided the case of *Swain v. Alabama*.[11] Swain, a black man, was convicted in Talladega County, Alabama, of raping a white woman. He was sentenced to death. The case was appealed to the Supreme Court on the grounds that there were no blacks on the jury; even more, no black "within the memory of persons now living has ever served on any petit jury in any civil or criminal case tried in Talladega County, Alabama."

    The Supreme Court denied the appeal, on the following grounds. As provided by Alabama law, the jury was selected from a panel of about 100 persons. There were 8 blacks on the panel. (They did not serve on the jury because they were "struck," through peremptory challenges by the prosecution; such challenges were constitutionally protected until 1986.) The presence of 8 blacks on the panel showed "the overall percentage disparity has been small and reflects no studied attempt to include or exclude a specified number of Negroes."

    At that time in Alabama, only men over the age of 21 were eligible for jury duty. There were 16,000 men over the age of 21 in Talladega County, of whom about 26% were black. If 100 people were chosen at random from this population, what is the chance that 8 or fewer would be black? What do you conclude?

27. The town of Hayward (California) has about 50,000 registered voters. A political scientist takes a simple random sample of 500 of these voters. In the sample, the breakdown by party affiliation is

Republican	115
Democrat	331
Independent	54

    (a) Among all registered voters in Hayward, the percentage of independents is estimated as ______.
    (b) This estimate is likely to be off by ______ or so.
    (c) The range from ______ to ______ is a 95%-confidence interval for the percentage of independents ______.

    Fill in the blanks; explain briefly. (The first four blanks are filled in with numbers; the last blank takes a phrase—25 words or less.)

28. NAEP (National Assessment of Educational Progress) periodically tests scientific knowledge in U.S. schools.[12] Here is one question on the test, administered to students in grade 12.

    The diagram below shows a thermometer. On the diagram, fill in the thermometer so that it reads 37.5 degrees Celsius.

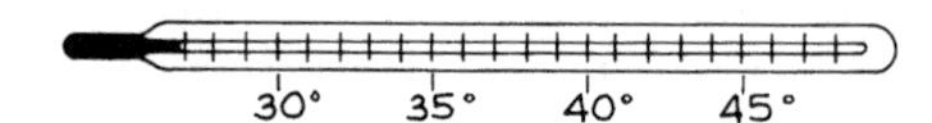

    Only 64% of the students who were tested could answer this question correctly.

    The superintendent of education in a certain state cannot believe these data. To check, he takes a simple random sample of 100 high schools in the state, and tests 10 randomly selected students from Grade 12 in each school. 661 out of the 1,000 students who take the test, or 66.1%, can do the problem.

    With the information given above, can you put a standard error on the 66.1%? Find the SE, or say why this can't be done.

29. Twenty draws are made at random with replacement from the box | 1 1 2 4 |. One of the graphs below is the probability histogram for the average of the draws. Another is the histogram for the numbers drawn. And the third is the histogram for the contents of the box. Which is which? Explain.

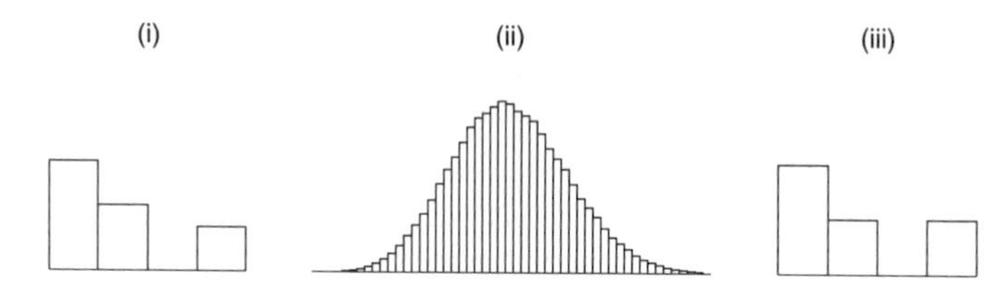

30. A survey research company uses random digit dialing. They have a contract, to estimate the percentage of people watching Spanish-language television in several Southwestern cities. They took a sample of size 1,000 in Austin, Texas—which has about 600,000 residents. They are satisfied with the accuracy of the estimates for Austin.

    Dallas has about twice the population of Austin, but similar demographics. True or false, and explain: to get about the same accuracy in Dallas as in Austin, the company should use a sample size of 2,000.

7. SUMMARY AND OVERVIEW

1. When drawing at random from a box, the expected value for the average of the draws equals the average of the box. The SE for the average of the draws equals the SE for their sum, divided by the number of draws.

2. The average of the draws can be used to estimate the average of the box. The estimate will be off by some amount, due to chance error. The SE for the average tells you the likely size of the amount off.

3. Multiplying the number of draws by some factor divides the SE for their average by the square root of that factor.

4. The probability histogram for the average of the draws will follow the normal curve, even if the contents of the box do not. The histogram must be put into standard units, and the number of draws must be large.

5. With a simple random sample, the SD of the sample can be used to estimate the SD of the box. A confidence interval for the average of the box can be found by going the right number of SEs either way from the average of the draws. The confidence level is read off the normal curve. This method should only be used with large samples.

6. The formulas for simple random samples should not be applied mechanically to other kinds of samples.

7. With *samples of convenience*, standard errors usually do not make sense.

8. This part of the book makes the transition from probability calculations to inference. Chapter 19 distinguishes sampling error from non-sampling error, and shows how important it is to use probability methods when drawing samples. Non-sampling error is often a more subtle and important problem than sampling error. Chapter 20 develops the theory behind simple random sampling. Chapter 21 shows how to estimate population percentages from sample percentages, introducing SEs and confidence intervals based on sample data. Chapter 23 makes the extension to averages.

9. Chapters 20, 21, and 23 build on the probability theory developed in chapters 16–18. These ideas will be applied again in part VII to the study of measurement error; they will be used in part VIII to make tests of significance.

10. The Current Population Survey is discussed in chapter 22, illustrating the concepts in a real survey of some complexity.

PART VII

Chance Models

24

A Model for Measurement Error

Upon the whole of which it appears, that the taking of the Mean of a number of observations, greatly diminishes the chance for all the smaller errors, and cuts off almost all possibility of any great ones: which last consideration, alone, seems sufficient to recommend the use of the method, not only to astronomers, but to all others concerned in making experiments of any kind (to which the above reasoning is equally applicable). And the more observations or experiments there are made, the less will the conclusions be liable to error, provided they admit of being repeated under the same circumstances.

—THOMAS SIMPSON (ENGLISH MATHEMATICIAN, 1710–1761)

1. ESTIMATING THE ACCURACY OF AN AVERAGE

In this part of the book, the frequency theory of chance will be used to study measurement error and genetics. Historically, the frequency theory was developed to handle problems of a very special kind—figuring the odds in games of chance. Some effort is needed to apply the theory to situations outside the gambling context. In each case, it is necessary to show that the situation being studied resembles a process—like drawing at random from a box—to which the theory applies. These box models are sometimes called *chance models* or *stochastic models*. The first example will be a chance model for measurement error.

To review briefly (chapter 6), any measurement is subject to chance error, and if repeated would come out a bit differently. To get at the size of the chance error, the best thing to do is to repeat the measurement several times. The spread in the measurements, as shown by the SD, estimates the likely size of the chance error in a single measurement. Chapter 6 stopped there. This chapter continues the discussion: the focus is on the average of the measurements in the series rather than a single measurement. The problem is to estimate the likely size of the chance error in the average. If the measurements are like draws from a box, the methods of parts V and VI can be used.

Table 1 on p. 99 shows 100 measurements on NB 10. These all fell short of 10 grams, by different amounts. The table gives the amounts, in micrograms. (A microgram is one millionth of a gram, roughly the weight of a speck of dust.) The SD of the 100 numbers in the table is about 6 micrograms: a single measurement is only accurate up to 6 micrograms or so. The best guess for the weight of NB 10 is the average of all 100 measurements, which is 404.6 micrograms short of 10 grams. Since each measurement is thrown off by error, the average cannot be exactly right either. But the average is going to be more accurate than any single measurement, so it is going to be off by less than 6 micrograms.

What is the right give-or-take number to put on the average?

$$\text{average} \pm \underline{\qquad}.$$

The answer is given by the SE for the average, which can be calculated just as in chapter 23. (The calculation rides on a box model, to be discussed in sections 2 and 3 below.) The SE for the sum of 100 measurements can be estimated as

$$\sqrt{100} \times 6 \text{ micrograms} = 60 \text{ micrograms}.$$

Then the SE for the average of the 100 measurements is

$$\frac{60 \text{ micrograms}}{100} = 0.6 \text{ micrograms}.$$

This completes the calculation. The average of all the numbers in the table is 404.6 micrograms. The likely size of the chance error in the average is estimated to be 0.6 micrograms. So NB 10 really weighs about 404.6 micrograms below 10 grams, plus or minus 0.6 micrograms or so.

Two numbers come up in the calculation: 6 micrograms and 0.6 micrograms. The first is the SD of the 100 measurements, the second is the SE for the average. What is the difference between the two?

- The SD says that a single measurement is accurate up to 6 micrograms or so.
- The SE says that the average of all 100 measurements is accurate up to 0.6 micrograms or so.

Example 1. One hundred measurements are made on a certain weight. The average of these measurements is 715 micrograms above one kilogram, and the SD is 80 micrograms.

(a) Is a single measurement likely to be off the exact weight by around 8 micrograms, or 80 micrograms?

(b) Is the average of all 100 measurements likely to be off the exact weight by around 8 micrograms, or 80 micrograms?

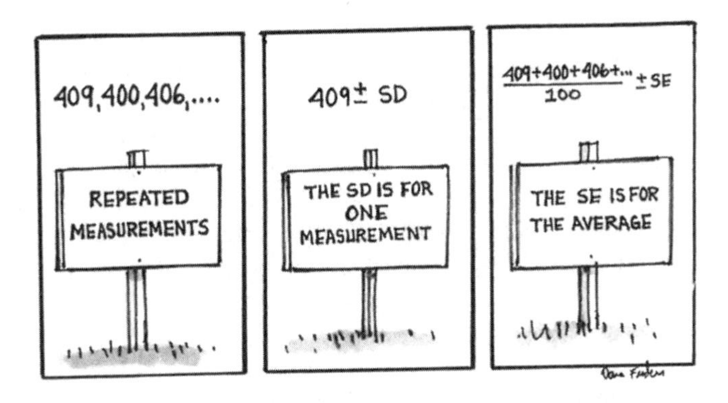

Solution. A single measurement is off by a chance error similar in size to the SD of the measurements. This is 80 micrograms. The answer to (a) is 80 micrograms. For (b), the SE for the sum of the measurements is estimated as $\sqrt{100} \times 80 = 800$ micrograms. So the SE for the average is $800/100 = 8$ micrograms. That is the answer to (b).

In example 1, the give-or-take number for the average of the measurements is 8 micrograms. To make this more precise, statisticians use confidence intervals, just as in sampling. A 95%-confidence interval for the exact weight can be obtained by going 2 SEs in either direction from the average. The average is 715 micrograms above one kilogram, and 2 SEs is $2 \times 8 = 16$ micrograms. So the exact weight is somewhere between 699 and 731 micrograms above one kilogram, with confidence about 95%. The arithmetic:

$$715 - 16 = 699, \quad 715 + 16 = 731.$$

Again, the word "confidence" is there as a reminder that the chances are in the measurement process and not in the thing being measured: the exact weight is not subject to chance variability. (For a similar discussion in the sampling context, see section 3 of chapter 21.)

> The chances are in the measuring procedure, not the thing being measured.

The normal curve should be used to get confidence intervals only when there is a fairly large number of measurements. With fewer than 25 measurements, most statisticians would use a slightly different procedure, based on what is called the t-distribution (section 6 of chapter 26).

Historical note. There is a connection between the theory of measurement error and neon signs. In 1890, the atmosphere was believed to consist of nitrogen (about 80%), oxygen (a little under 20%), carbon dioxide, water vapor—and nothing else. Chemists were able to remove the oxygen, carbon dioxide, and water vapor. The residual gas should have been pure nitrogen.

Lord Rayleigh undertook to compare the weight of the residual gas with the weight of an equal volume of chemically pure nitrogen. One measurement on the weight of the residual gas gave 2.31001 grams. And one measurement of the pure nitrogen gave a bit less, 2.29849 grams. However, the difference of 0.01152 grams was rather small, and in fact was comparable to the chance errors made by the weighing procedure.

Could the difference have resulted from chance error? If not, the residual gas had to contain something heavier than nitrogen. What Rayleigh did was to replicate the experiment, until he had enough measurements to prove that the residual gas from the atmosphere was heavier than pure nitrogen.

He went on to isolate the rare gas called *argon*, which is heavier than pure nitrogen and present in the atmosphere in small quantities. Other researchers later discovered the similar gases neon, krypton, and xenon, all occurring naturally (in trace amounts) in the atmosphere. These gases are what make "neon" signs glow in different colors.[1]

Exercise Set A

1. The total of the 100 measurements on NB 10 was 40,459 micrograms. What is the likely size of the chance error in this total?

2. Some scales use electrical *load cells*. The weight is distributed over a number of cells. Each cell converts the weight it carries to an electrical current, which is fed to a central scanner. This scanner adds up all the currents, and computes the corresponding total weight, which it prints out. This process is repeated several dozen times a second. As a result, a loaded boxcar (weighing about 100,000 pounds) can be weighed as it crosses a special track, with chance errors of only several hundred pounds in size.[2]

   Suppose 25 readings on the weight of a boxcar show an average of 82,670 pounds, and the SD is 500 pounds. The weight of the boxcar is estimated as ______; this estimate is likely to be off by ______ or so.

3. (Hypothetical.) The British Imperial Yard is sent to Paris for calibration against The Meter. Its length is determined 100 times. This sequence of measurements averages out to 91.4402 cm, and the SD is 800 microns. (A *micron* is the millionth part of a meter.)

   (a) Is a single reading off by around 80 microns, or 800 microns?
   (b) Is the average of all 100 readings off by around 80 microns, or 800 microns?
   (c) Find a 95%-confidence interval for the exact length of the Imperial Yard.

4. The 95%-confidence interval for the exact weight of NB 10 is the range from 403.4 to 405.8 micrograms below 10 grams. Say whether each of the following statements is true or false, and explain why.

   (a) About 95% of the measurements are in this range.
   (b) There is about a 95% chance that the next measurement will be in this range.
   (c) About 95% of the time that the Bureau takes 100 measurements and goes 2 SEs either way from the average, they succeed in covering the exact weight.
   (d) If the Bureau took another 100 measurements on NB 10, there is about a 95% chance that the new average would fall in the interval from 403.4 to 405.8 micrograms below 10 grams.

5. Would taking the average of 25 measurements divide the likely size of the chance error by a factor of 5, 10, or 25?

The answers to these exercises are on p. A88.

2. CHANCE MODELS

Section 1 explained how to put a standard error on the average of repeated measurements. The arithmetic is easily carried out on any list of numbers, but the method is legitimate only when the variability in the data is like the variability in repeated draws from a box.

> If the data show a trend or pattern over time, a box model does not apply.

The reason: draws from a box do not show a trend or pattern over time. The following examples illustrate this idea.

Example 2. Table 1 gives the population of the U.S. from 1790 to 2000. Do these numbers look like draws at random from a box?

Table 1. Population of the U.S., 1790 to 2000.

1790	3,929,214
1800	5,308,483
1810	7,239,881
1820	9,638,453
1830	12,866,020
1840	17,069,453
1850	23,191,876
1860	31,443,321
1870	39,818,449
1880	50,189,209
1890	62,979,766
1900	76,212,168
1910	92,228,496
1920	106,021,537
1930	123,202,624
1940	132,164,569
1950	151,325,798
1960	179,323,175
1970	203,302,031
1980	226,542,199
1990	248,718,302
2000	281,422,602

Notes: Resident population. From 1950 onwards, includes Alaska and Hawaii. Revised figures for 1870–1940. Source: *Statistical Abstract*, 2006, Table 1.

Solution. No. The population of the U.S. has been going up steadily. Numbers drawn at random from a box don't do that: sometimes they go up and other times they go down.

Example 3. The 22 numbers in table 1 average out to 94.7 million, and the SD is 89.3 million. An investigator attaches a standard error to the average, by the following procedure:

$$\text{SE for the sum} \approx \sqrt{22} \times 89.3 \text{ million} \approx 419 \text{ million}$$
$$\text{SE for average} \approx 419/22 \approx 19.0 \text{ million.}$$

Is this sensible?

Solution. The average and SD make sense, as descriptive statistics. They summarize part of the information in table 1, although they miss quite a bit—for instance, the fact that the numbers increase steadily. The SE of 19 million, however, is silly. If the investigator wants to know the average of the 22 numbers in the table, that has been computed, and there is no need to worry about chance error. Of course, something else may be involved, like the average of a list showing the population of the U.S. in every year from 1790 to 2000. (Every tenth number on that list is shown in table 1; the numbers in between are known with less precision, because the Census is only taken every ten years.) The investigator would then be making an inference, using the average from table 1 to estimate that other average. And the estimate would be off by some amount. But the square root law cannot help much with the margin of error. The reason is that the numbers in table 1 are not like draws from a box.

The square root law only applies to draws from a box.

Example 4. A list is made, showing the daily maximum temperature at San Francisco airport. Are these data like draws from a box?

Solution. No, there is a definite seasonal pattern to these data—warmer in the summer, colder in the winter. There even are local patterns to the data. The temperature on one day tends to be like the temperature on the day before.

The temperature data are graphed in the top panel of figure 1. There is a dot above each day of the year for 2005, showing the maximum temperature on that day. The seasonal pattern is clear. On the whole, the dots are higher in the summer than in the winter. Also, there is an irregular wavy pattern within each season. The crest of a wave represents a stretch of warm days—a warm spell. The cold spells are in the troughs.

By comparison, the second panel in figure 1 is for a mythical airport where the climate is on average like San Francisco, but the daily maximum temperatures are like draws from a box. These data are random: they show no trend or pattern over the year. With this sort of climate, weather forecasting would be hopeless.

Figure 1. Temperature and box models. The first panel shows the daily maximum temperature at San Francisco airport in 2005.[3] There is a seasonal pattern to the data, warmer in summer than winter. Also, there are local patterns: warm spells and cold spells. A box model would not apply. The second panel shows what the temperatures would look like if they were generated by drawing from a box.

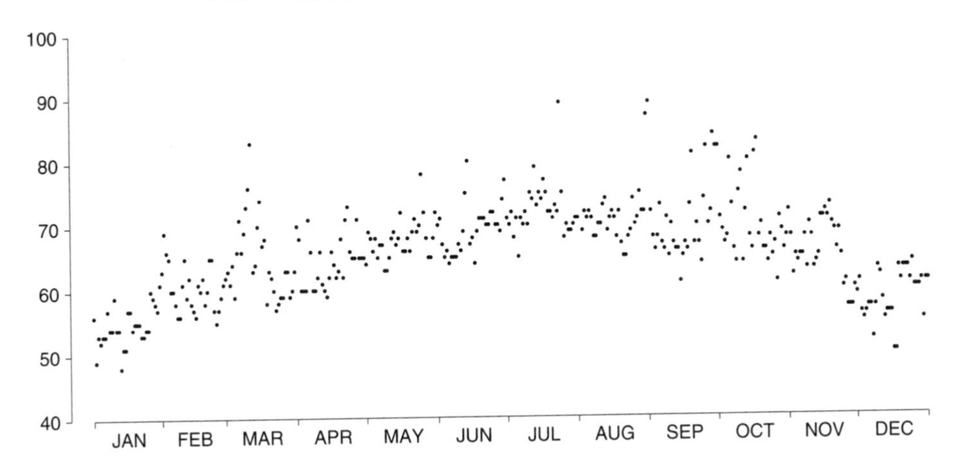

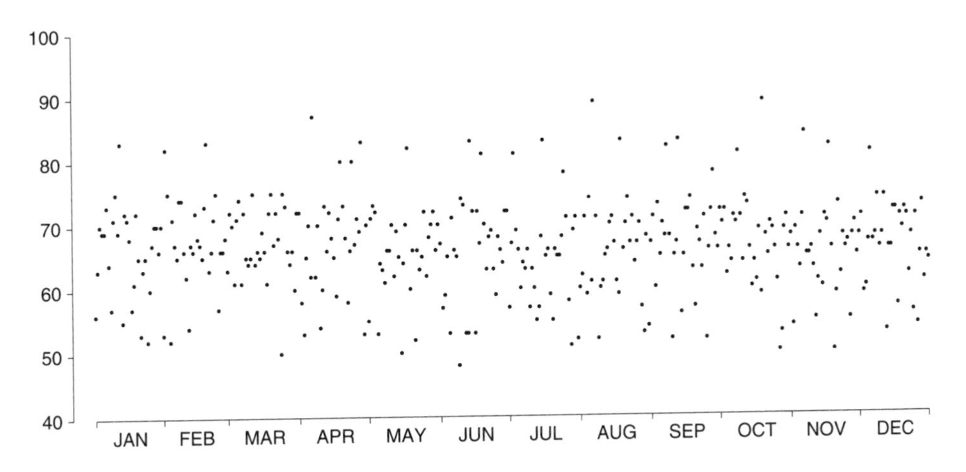

In section 1, we acted as if the measurements on NB 10 were like draws from a box. Was this sensible? The top panel in figure 2 (next page) is a graph of the data. There is one point for each measurement. The x-coordinate says which measurement it was: first, or second, or third, and so on. The y-coordinate says how many micrograms below 10 grams the measurement was. The points do not show any trend or pattern over time; they look as random as draws from a box. In

Figure 2. The top panel graphs the repeated measurements on NB 10 (p. 99). The middle panel graphs hypothetical data, generated by computer simulation of a box model. These two panels are very similar, showing how well the box model represents the real data. The bottom panel graphs data showing a strong pattern: a box model would not apply.

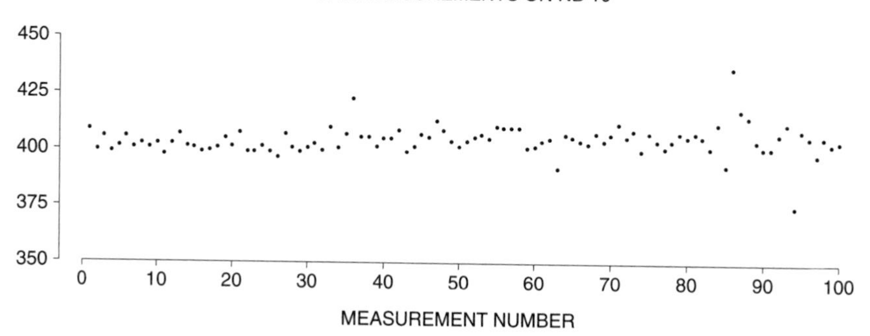

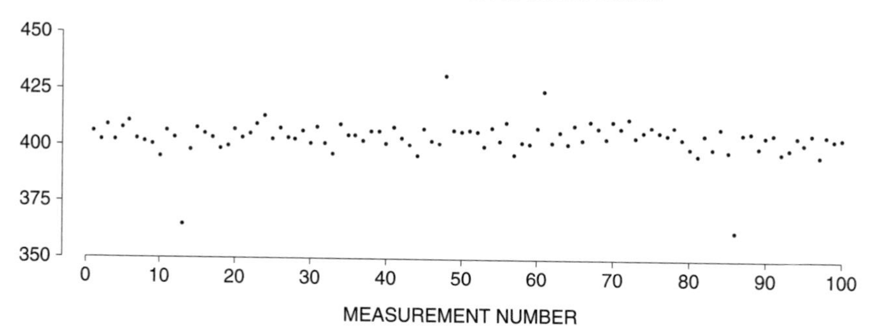

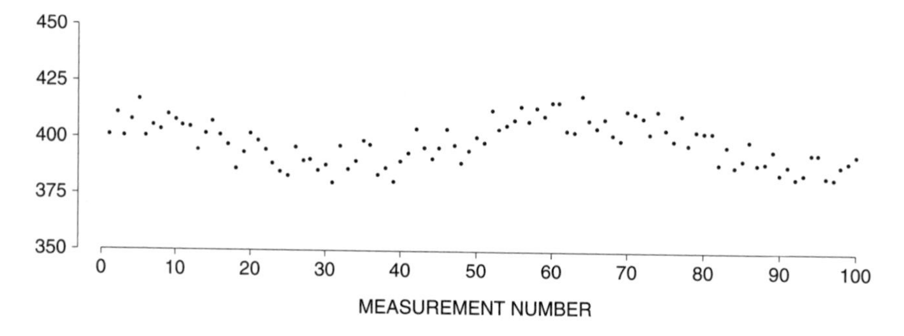

fact, the second panel shows hypothetical data generated on the computer using a box model.[4] If you did not know which was which, it would be hard to tell the difference between these two panels. By comparison the third panel (also for computer-generated data) shows a strong pattern: a box model would not apply.

It is no accident that the data on NB 10 look like draws from a box. Investigators at the Bureau use pictures of the data, like the top panel in figure 2, to check their work. A trend or pattern is a signal that something is wrong and needs to be fixed. This idea is basic to precision measurement work—and to quality control in manufacturing, where the number of defective units is plotted against time.

Exercise Set B

1. A thumbtack is thrown in the air. It lands either point up or point down.

   Someone proposes the following box model: drawing with replacement from the box [U] [D], where U means "point up" and D means "point down." Someone else suggests the box [U] [D] [D]. How could you decide which box was better?

2. In San Francisco, it rains on about 17% of the days in an average year. Someone proposes the following chance model for the sequence of dry and rainy days: draw with replacement from a box containing one card marked "rainy" and five cards marked "dry." Is this a good model?

3. Someone goes through the phone book, and makes a list showing the last digit of each phone number. Can this be modeled by a sequence of draws (with replacement) from the box

   [0] [1] [2] [3] [4] [5] [6] [7] [8] [9]

   What about a list of first digits?

4. Someone makes a list showing the first letter of each family name in the phone book, going name by name through the book in order. Is it sensible to model this sequence of letters by drawing at random with replacement from a box? (There would be 26 tickets in the box, each ticket marked with one letter of the alphabet.) Explain.

5. "The smart professional gambler, when heads comes up four times in a row, will bet that it comes up again. A team that's won six in a row will win seven. *He believes in the percentages.* The amateur bettor will figure that heads can't come up again, that tails is 'due.' He'll bet that a team on a losing streak is 'due' to win. *The amateur believes in the law of averages.*"

   —Jimmy the Greek, *San Francisco Chronicle*, July 2, 1975

   Kerrich's coin (chapter 16) will be tossed until it lands heads four times in a row. Suppose Jimmy the Greek offers 5 to 4 that the coin will land heads on the next toss. (On heads, he pays you $5; on tails, you pay him $4.) Do you take the bet?

The answers to these exercises are on p. A89.

3. THE GAUSS MODEL

The box model for measurement error will now be described in more detail. The basic situation is that a series of repeated measurements are made on some quantity. According to the model, each measurement differs from the exact value by a chance error; this error is like a draw made at random from a box of tickets—the *error box*. Successive measurements are done independently and under the same conditions, so the draws from the error box are made with replacement. To capture the idea that the chance errors aren't systematically positive or systematically negative, it is assumed that the average of the numbers in the error box equals 0. This model is named after Carl Friedrich Gauss (Germany, 1777–1855), who worked on measurement error in astronomical data.

> In the Gauss model, each time a measurement is made, a ticket is drawn at random with replacement from the error box. The number on the ticket is the chance error. It is added to the exact value to give the actual measurement. The average of the error box is equal to 0.

In the model, it is the SD of the box which gives the likely size of the chance errors. Usually, this SD is unknown and must be estimated from the data. Take the 100 measurements on NB 10, for example. According to the model, each measurement is around the exact weight, but it is off by a draw from the error box:

1st measurement = exact weight + 1st draw from error box
2nd measurement = exact weight + 2nd draw from error box
.
.
.
100th measurement = exact weight + 100th draw from error box

Carl Friedrich Gauss (Germany, 1777–1855)

Wolff-Leavenworth collection, courtesy of the Syracuse University Art Collection.

With the NB 10 data, the SD of the 100 draws would be a fine estimate for the SD of the error box.[5] The catch is that the draws cannot be recovered from the data, because the exact weight is unknown. However, the variability in the measurements equals the variability in the draws, because the exact weight does not change from measurement to measurement. More technically, adding the exact value to all the errors does not change the SD (pp. 92–93). That is why statisticians use the SD of the measurements when computing the SE. And that completes the reasoning behind the calculation in section 1.[6]

> When the Gauss model applies, the SD of a series of repeated measurements can be used to estimate the SD of the error box. The estimate is good when there are enough measurements.

There may be another way to get at the SD of the error box. When there is a lot of experience with the measurement process, it is better to estimate the SD from all the past data rather than a few current measurements. The reason: the error box belongs to the measurement process, not the thing being measured.

Example 5. (Hypothetical.) After making several hundred measurements on NB 10 and finding the SD to be about 6 micrograms, the Bureau's investigators misplace this checkweight. They go out and buy a new one. They measure its weight by exactly the same procedure as for NB 10, and on the same scale. After a week, they accumulate 25 measurements. These average out to 605 micrograms above 10 grams, and the SD is 7 micrograms. Assuming the Gauss model, the new weight is 605 micrograms above 10 grams, give or take about

6 micrograms 7 micrograms 1.2 micrograms 1.4 micrograms.

Solution. According to the model, the chance error in each measurement is like a draw from the error box. The error box belongs to the scales, not the weight. The SD of the error box should be estimated by the SD of the large amount of past data on NB 10, not the small amount of current data on the new weight. The SD of the error box is estimated as 6 micrograms. This tells the likely size of the chance error in a single measurement. But the likely size of the chance error in the average of 25 measurements is smaller. The SE for the average is 1.2 micrograms. That is the answer.

In the model, the error box belongs to the scales, not the weight. This seems reasonable for chunks of metal which are similar in size. However, if we change from a 10-gram weight to a 100-gram weight, the error box could change too. And for weights that wiggle around more actively—like babies—the separation between "true value" and "chance error" might not be so convincing.

A final point. The version of the Gauss model presented here makes the assumption that there is no bias in the measuring procedure. When bias is present, each measurement is the sum of three terms:

$$\text{exact value} + \text{bias} + \text{chance error.}$$

Then the SE for the average no longer says how far the average of the measurements is from the exact value, but only how far it is from

$$\text{exact value} + \text{bias.}$$

The methods of this chapter are no help in judging bias. We did not take it into account for the measurements on NB 10, because other lines of reasoning suggest that the bias in precision weighing at the National Bureau of Standards is negligible. In other situations, bias can be more serious than chance errors—and harder to detect.[7]

Exercise Set C

1. (a) A 10-gram checkweight is being weighed. Assume the Gauss model with no bias. If the exact weight is 501 micrograms above 10 grams, and the number drawn from the error box is 3 micrograms, what would the measurement be?
   (b) Repeat, if the exact weight is 510 micrograms above 10 grams, and the number drawn from the error box is −6 micrograms.

2. The first measurement on NB 10 was 409 micrograms below 10 grams. According to the Gauss model (with no bias),

   $$409 = \text{exact value} + \text{chance error.}$$

   Can you figure out the numerical value for each of the two terms? Explain briefly.

3. In the Gauss model for the measurements on NB 10, the SD of the error box is ______ 6 micrograms. Fill in the blank using one of the two phrases below, and explain briefly.

   known to be   estimated from the data as

4. The figure below shows the result of a computer simulation: 50 imaginary investigators set out to weigh NB 10, following the procedure used by the Bureau. Each investigator takes 100 measurements and computes the average, the SD, and the SE for the average. The 50 confidence intervals "average ± 2 SE" are plotted at different heights in the figure so they can be seen. In the simulation, the exact weight is taken as 405 micrograms below 10 grams.
   (a) Why do the intervals have different centers?
   (b) Why do they have different lengths?
   (c) How many should cover the exact weight?
   (d) How many do?

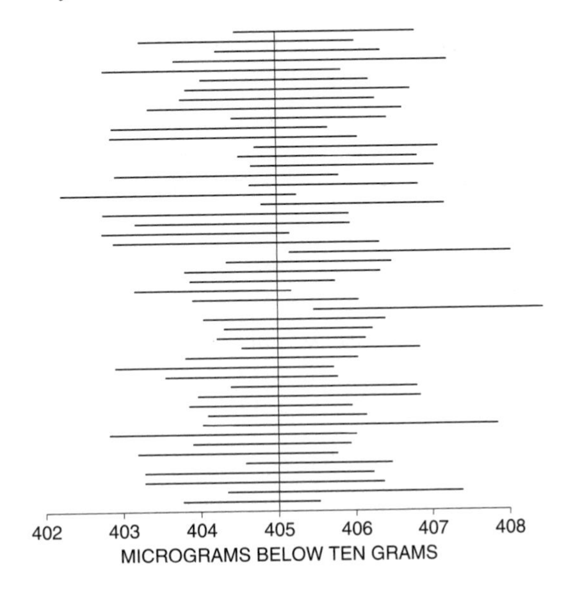

5. The Bureau is about to weigh a one-kilogram checkweight 100 times, and take the average of the measurements. They are willing to assume the Gauss model, with no bias, and on the basis of past experience they estimate the SD of the error box to be 50 micrograms.
   (a) The average of all 100 measurements is likely to be off the exact weight by ______ or so.
   (b) The SD of all 100 measurements is likely to be around ______.
   (c) Estimate the probability that the average of all 100 measurements will be within 10 micrograms of the exact weight.

6. Suppose you sent a nominal 10-gram weight off to the Bureau, asking them to weigh it 25 times and tell you the average. They will use the same procedure as on NB 10, where the SD of several hundred measurements was about 6 micrograms. The 25 measurements average out to 307 micrograms above 10 grams, and the SD is about 5 micrograms. Your weight is 307 micrograms above 10 grams, give or take around

   5 micrograms   6 micrograms   1 microgram   1.2 micrograms

   (You may assume the Gauss model, with no bias.)

7. Twenty-five measurements are made on the speed of light. These average out to 300,007 and the SD is 10, the units being kilometers per second. Fill in the blanks in part (a), then say whether each of (b–f) is true or false. Explain your answers briefly. (You may assume the Gauss model, with no bias.)
   (a) The speed of light is estimated as ______. This estimate is likely to be off by ______ or so.
   (b) The average of all 25 measurements is off 300,007 by 2 or so.
   (c) Each measurement is off 300,007 by 10 or so.
   (d) A 95%-confidence interval for the speed of light is $300{,}007 \pm 4$.
   (e) A 95%-confidence interval for the average of the 25 measurements is $300{,}007 \pm 4$.
   (f) If a 26th measurement were made, there is a 95% chance that it would be off the exact value for the speed of light by less than 4.

8. A surveyor is measuring the distance between five points A, B, C, D, E. They are all on a straight line. She finds that each of the four distances AB, BC, CD, and DE measures one mile, give or take an inch or so. These four measurements are made independently, by the same procedure.

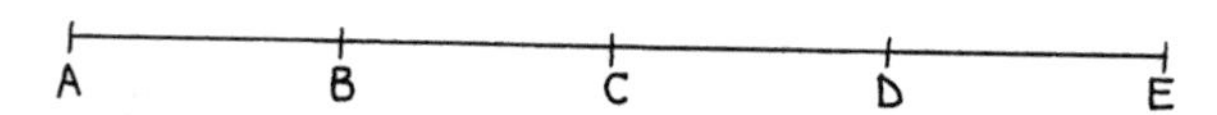

   The distance from A to E is about four miles; but this estimate is likely to be off by around

   4 inches 2 inches 1 inch 1/2 inch 1/4 inch.

   Explain briefly. (You may assume the Gauss model, with no bias.)

9. The concept of measurement error is often applied to the results of psychological tests. The equation is

   $$\text{actual test score} = \text{true test score} + \text{chance error}.$$

   The chance error term reflects accidental factors, like the mood of the subject, or luck. Do you think that the Gauss model applies?

The answers to these exercises are on pp. A89–90.

4. CONCLUSION

NB 10 is just a chunk of metal. It is weighed on a contraption of platforms, gears, and levers. The results of these weighings have been subjected to a statistical analysis involving the standard error, the normal curve, and confidence intervals. It is the Gauss model which connects the mathematics to NB 10. The chance errors are like draws from a box; their average is like the average of the draws. The number of draws is so large that the probability histogram for the average will follow the normal curve very closely. Without the model there would be no box, no standard error, and no confidence levels.

Statistical inference uses chance methods to draw conclusions from data. Attaching a standard error to an average is an example. Now it is always possible to go through the SE procedure mechanically. Many computer programs will do the work for you. It is even possible to label the output as a "standard error."

Do not get hypnotized by the arithmetic or the terminology. The procedure only makes sense because of the square root law. The implicit assumption is that the data are like the results of drawing from a box (an old point, but worth repeating). Many investigators don't pay attention to assumptions. The resulting "standard errors" are often meaningless.[8]

> Statistical inference can be justified by putting up an explicit chance model for the data. No box, no inference.

Parts II and III focused on *descriptive statistics*—drawing diagrams or calculating numbers which summarize data and bring out the salient features. Such techniques can be used very generally, because they do not involve any hidden assumptions about where the data came from. For statistical inference, however, models are basic.

5. REVIEW EXERCISES

1. Laser altimeters can measure elevation to within a few inches, without bias, and with no trend or pattern to the measurements. As part of an experiment, 25 readings were made on the elevation of a mountain peak. These averaged out to 81,411 inches, and their SD was 30 inches. Fill in the blanks in part (a), then say whether each of (b–f) is true or false. Explain your answers briefly.
   (a) The elevation of the mountain peak is estimated as ______; this estimate is likely to be off by ______ or so.
   (b) 81,411 ± 12 inches is a 95%-confidence interval for the elevation of the mountain peak.
   (c) 81,411 ± 12 inches is a 95%-confidence interval for the average of the 25 readings.
   (d) There is about a 95% chance that the next reading will be in the range 81,411 ± 12 inches.
   (e) About 95% of the readings were in the range 81,411 ± 12 inches.
   (f) If another 25 readings are made, there is about a 95% chance that their average will be in the range 81,411 ± 12 inches.

2. The first measurement on NB 10 was 409 micrograms below 10 grams. The average of all 100 measurements was 404.6 micrograms below 10 grams, with an SD of 6.4 micrograms; the data are shown in figure 2 on p. 102. Fill in the blanks with a word or phrase. Explain briefly. You may assume the Gauss model, with no bias.
   (a) $404.6 \pm 2 \times 0.64$ is a 95%-confidence interval for the weight of NB 10 because, with 100 draws from the box, the ______ follows the normal curve.
   (b) $409 \pm 2 \times 6.4$ isn't a 95%-confidence interval for the weight of NB 10 because the ______ doesn't follow the normal curve.

3. The speed of light was measured 2,500 times. The average was 299,774 kilometers per second, and the SD was 14 kilometers per second.[9] Assume the Gauss model, with no bias. Find a 95%-confidence interval for the speed of light.

4. In exercise 3, light was timed as it covered a certain distance. The distance was measured 57 times, and the average of these measurements was 1.594265 kilometers. What else do you need to know to decide how accurate this value is?

5. Exercise 4 points to one possible source of bias in the measurements described in exercise 3. What is it?

6. In 2005, the average of the daily maximum temperature at San Francisco airport was 65.8 degrees, and the SD was 7.0 degrees (figure 1, p. 447). Now

$$\sqrt{365} \times 7.0 \approx 134 \text{ degrees}, \quad 134/365 \approx 0.4 \text{ degrees}.$$

True or false: a 95%-confidence interval for the average daily maximum temperature at San Francisco airport is 65.8 ± 0.8 degrees. Explain briefly.

7. A calibration laboratory has been measuring a one-kilogram checkweight by the same procedure for several years. They have accumulated several hundred measurements, and the SD of these measurements is 18 micrograms. Someone now sends in a one-kilogram weight to be calibrated by the same procedure. The lab makes 50 measurements on the new weight, which average 78.1 micrograms above a kilogram, and their SD is 20 micrograms. If possible, find a 95%-confidence interval for the value of this new weight. (You may assume the Gauss model, with no bias.)

8. In a long series of trials, a computer program is found to take on average 58 seconds of CPU time to execute, and the SD is 2 seconds. There is no trend or pattern in the data. It will take about ______ seconds of CPU time to execute the program 100 times, give or take ______ seconds or so. (The CPU is the "central processing unit," where the machine does logic and arithmetic.)

9. A machine makes sticks of butter whose average weight is 4.0 ounces; the SD of the weights is 0.05 ounces. There is no trend or pattern in the data. There are 4 sticks to a package.
   (a) A package weighs ______, give or take ______ or so.
   (b) A store buys 100 packages. Estimate the chance that they get 100 pounds of butter, to within 2 ounces.

10. True or false, and explain: "If the data don't follow the normal curve, you can't use the curve to get confidence levels."

11. "All measurements were made twice. If two staff members were present, the duplicate measurements were made by different people. In order to minimize gross errors, discrepancies greater than certain arbitrary limits were measured a third time, and if necessary a fourth, until two measurements were obtained which agreed within the set limits. In cases of discrepancy, the mea-

surers decided which of the three or four results was most 'representative' and designated it for inclusion in the statistical record. In cases of satisfactory agreement, the statistical record was based routinely on the first measurement recorded." Comment briefly.[10]

6. SUMMARY

1. According to the *Gauss model* for measurement error, each time a measurement is made, a ticket is drawn at random with replacement from the *error box*. The number on the ticket is the chance error. It is added to the exact value of the thing being measured, to give the actual measurement. The average of the error box is equal to 0. Here, bias is assumed to be negligible.

2. When the Gauss model applies, the SD of many repeated measurements is an estimate for the SD of the error box. This tells the likely size of the chance error in an individual measurement.

3. The average of the series is more precise than any individual measurement, by a factor equal to the square root of the number of measurements. The calculation assumes that the data follow the Gauss model.

4. An approximate confidence interval for the exact value of the thing being measured can be found by going the right number of SEs either way from the average of the measurements; the confidence level is taken from the normal curve. The approximation is good provided the Gauss model applies, with no bias, and there are enough measurements.

5. With the Gauss model, the chance variability is in the measuring process, not the thing being measured. The word "confidence" is to remind you of this.

6. If the model does not apply, neither does the procedure for getting confidence intervals. In particular, if there is any trend or pattern in the data, the formulas may give silly answers.

7. *Statistical inference* is justified in terms of an explicit *chance model* for the data.

25

Chance Models in Genetics

I shall never believe that God plays dice with the world.
—ALBERT EINSTEIN (1879–1955)

1. HOW MENDEL DISCOVERED GENES

This chapter is hard, and it can be skipped without losing the thread of the argument in the book. It is included for two reasons:

- Mendel's theory of genetics is great science.
- The theory shows the power of simple chance models in action.

In 1865, Gregor Mendel published an article which provided a scientific explanation for heredity, and eventually caused a revolution in biology.[1] By a curious twist of fortune, this paper was ignored for about thirty years, until the theory was simultaneously rediscovered by three men, Correns in Germany, de Vries in Holland, and Tschermak in Austria. De Vries and Tschermak are now thought to have seen Mendel's paper before they published, but Correns apparently found the idea by himself.

Mendel's experiments were all carried out on garden peas; here is a brief account of one of these experiments. Pea seeds are either yellow or green. (As the phrase suggests, seed color is a property of the seed itself,[2] and not of the parental plant: indeed, one parent often has seeds of both colors.) Mendel bred a pure yellow strain, that is, a strain in which every plant in every generation had

Gregor Mendel (Austria, 1822–1884).
From the collection of the Moravian Museum, Brno.

only yellow seeds. Separately, he bred a pure green strain. He then crossed plants of the pure yellow strain with plants of the pure green strain. For instance, he used pollen from the yellows to fertilize ovules on plants of the green strain. (The alternative method, using pollen from the greens to fertilize plants of the yellow strain, gave exactly the same results.) The seeds resulting from a yellow-green cross, and the plants into which they grow, are called *first-generation hybrids.* First-generation hybrid seeds are all yellow, indistinguishable from seeds of the pure yellow strain. The green seems to have disappeared completely.

These first-generation hybrid seeds grew into first-generation hybrid plants which Mendel crossed with themselves, producing *second-generation hybrid* seeds. Some of these second-generation seeds were yellow, but some were green. So the green disappeared for one generation, but reappeared in the second. Even more surprising, the green reappeared in a definite, simple proportion. Of the second-generation hybrid seeds, about 75% were yellow and 25% were green.

What is behind this regularity? To explain it, Mendel postulated the existence of the entities now called *genes.*[3] According to Mendel's theory, there were two different variants of a gene which paired up to control seed color. They will be denoted here by *y* (for yellow) and *g* (for green). It is the gene-pair in the seed—not the parent—which determines what color the seed will be, and all the cells making up a seed contain the same gene-pair.

There are four different gene-pairs: y/y, y/g, g/y, and g/g. Gene-pairs control seed color by the rule

- y/y, y/g, and g/y make yellow,
- g/g makes green.

As geneticists say, *y* is *dominant* and *g* is *recessive*. This completes the first part of the model.

Now the seed grows up and becomes a plant. All the cells in this plant will also carry the seed's color gene-pair—with one exception. Sex cells, either sperm or eggs, contain only one gene of the pair.[4] For instance, a plant whose ordinary cells contain the gene-pair y/y will produce sperm cells containing the gene y. Similarly, the plant will produce egg cells containing the gene y. On the other hand, a plant whose ordinary cells contain the gene-pair y/g will produce some sperm cells containing the gene y, and some sperm cells containing the gene g. In fact, half its sperm cells will contain y, and the other half will contain g; half its eggs will contain y, the other half will contain g.

This model accounts for the experimental results. Plants of the pure yellow strain have the color gene-pair y/y, so the sperm and eggs all just contain the gene y. Similarly, plants of the pure green strain have the gene-pair g/g, so their pollen and ovules just contain the gene g. Crossing a pure yellow with a pure green amounts for instance to fertilizing a g-egg by a y-sperm, producing a fertilized cell having the gene-pair y/g. This cell reproduces itself and eventually becomes a seed, in which all the cells have the gene-pair y/g and are yellow in color. The model has explained why all first-generation hybrid seeds are yellow, and none are green.

What about the second generation? A first-generation hybrid seed grows into a first-generation hybrid plant, with the gene-pair y/g. This plant produces sperm cells, of which half will contain the gene y and the other half will contain the gene g. The plant also produces eggs, of which half will contain y and the other half will contain g. When two first-generation hybrids are crossed, a resulting second-generation hybrid seed gets one gene at random from each parent—because the seed is formed by the random combination of a sperm cell and an egg. From the point of view of the seed, it's as if one ticket was chosen at random from each of two boxes. In each box, half the tickets are marked y and the other half are marked g. The tickets are the genes, and there is one box for each parent (figure 1).

Figure 1. Mendel's chance model for the genetic determination of seed-color: one gene is chosen at random from each parent. The chance of each combination is shown. (The sperm gene is listed first; in terms of seed color, the combinations y/g and g/y are not distinguishable after fertilization.[5])

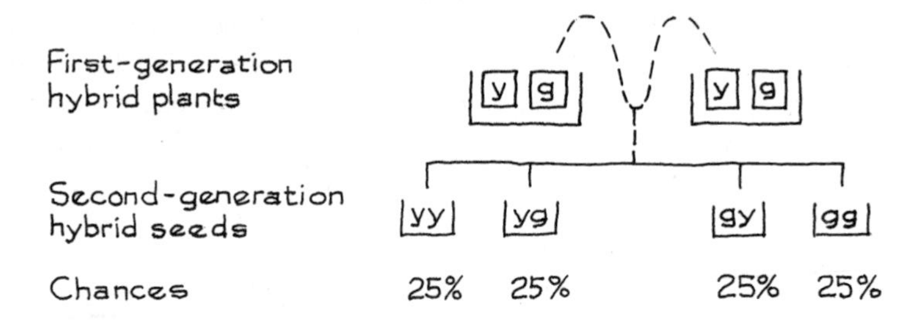

As shown in Figure 1, the seed has a 25% chance to get a gene-pair with two g's and be green. The seed has a 75% chance to get a gene-pair with one or two y's and be yellow. The number of seeds is small by comparison with the number of pollen grains, so the selections for the various seeds are essentially independent. The conclusion: the color of second-generation hybrid seeds will be determined as if by a sequence of draws with replacement from the box

yellow	yellow	yellow	green

And that is how the model accounts for the reappearance of green in the second generation, for about 25% of the seeds.

Mendel made a bold leap from his experimental evidence to his theoretical conclusions. His reconstruction of the chain of heredity was based entirely on statistical evidence of the kind discussed here. And he was right. Modern research in genetics and molecular biology is uncovering the chemical basis of heredity, and has provided ample direct proof for the existence of Mendel's hypothetical entities. As we know today, genes are segments of DNA on chromosomes—the dark patches in Figure 2 on the next page.

Essentially the same mechanism of heredity operates in all forms of life, from dolphins to fruit flies. So the genetic model proposed by Mendel unlocks one of the great mysteries of life. How is it that a pea-seed always produces a pea, and never a tomato or a whale? Furthermore, the answer turns out to involve chance in a crucial way, despite Einstein's quote at the opening of the chapter.

Exercise Set A

1. In some experiments, a first-generation hybrid pea is "back-crossed" with one parent. If a y/g plant is crossed with a g/g, about what percentage of the seeds will be yellow? Of 1,600 such seeds, what is the chance that over 850 will be yellow?

2. Flower color in snapdragons is controlled by one gene-pair. There are two variants of the gene, *r* (for red) and *w* (for white). The rules are:

   r/r makes red flowers,
   r/w and w/r make pink flowers,
   w/w makes white flowers.

   So neither *r* nor *w* is dominant. Their effects are *additive*, like mixing red paint with white paint.

   (a) Work out the expected percentages of red-, pink-, and white-flowered plants resulting from the following crosses: white × red, white × pink, pink × pink.
   (b) With 400 plants from pink × pink crosses, what is the chance that between 190 and 210 will be pink-flowered?

3. Snapdragon leaves come in three widths: wide, medium, and narrow. In breeding trials, the following results are obtained:

   wide × wide → 100% wide
   wide × medium → 50% wide, 50% medium
   wide × narrow → 100% medium
   medium × medium → 25% narrow, 50% medium, 25% wide.

   [Exercise continues on p. 463.]

Figure 2. Photomicrograph. These cells are from the root tip of a pea plant, and are magnified about 2,000 times. The cell shown in the center is about to divide. At this stage, each individual chromosome consists of two identical pieces, lying side by side. There are fourteen chromosomes arranged in seven homologous pairs, indicated by the Roman numerals from I to VII. The gene-pair controlling seed-color is located on chromosome pair I, one of the genes being on each chromosome.[6]

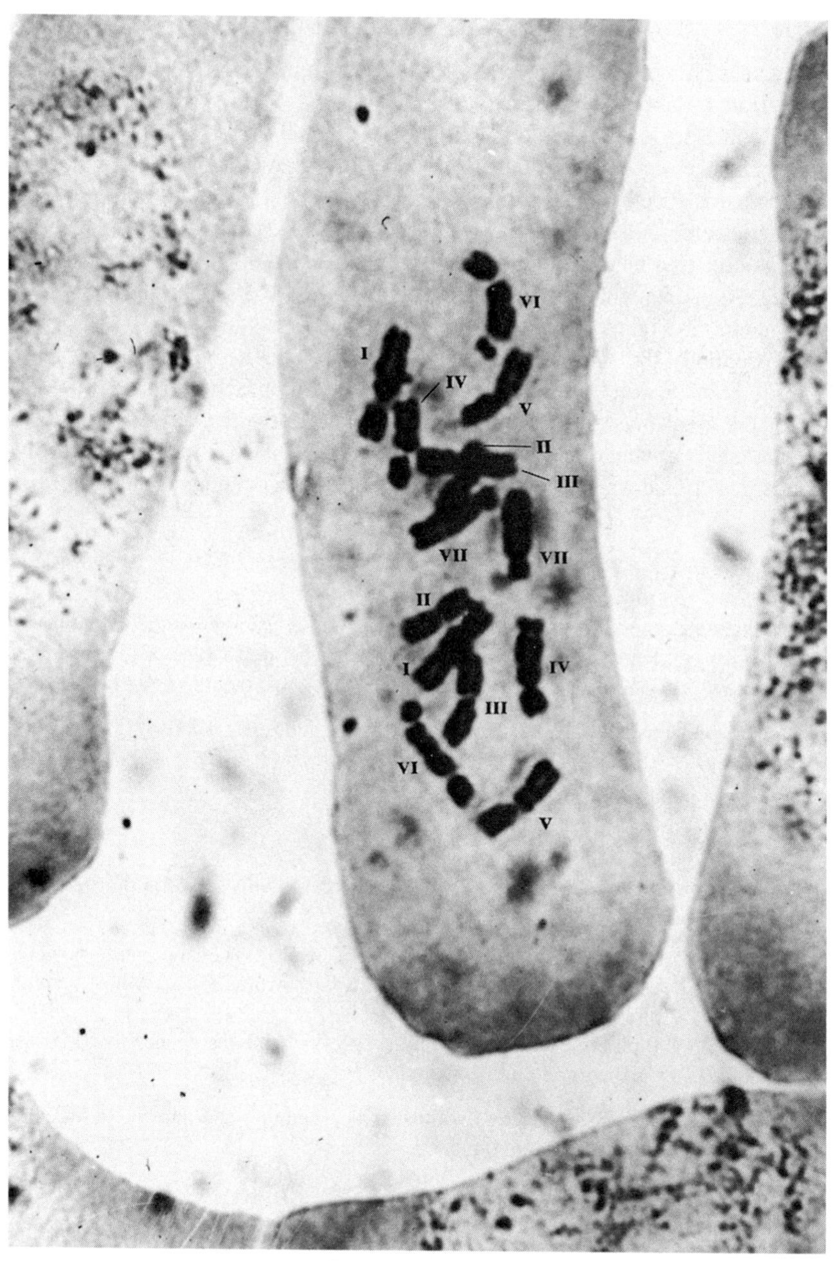

Source: New York State Agriculture Experiment Station, Geneva, N.Y.

(a) Can you work out a genetic model to explain these results?
(b) What results would you expect from each of the following crosses: narrow × narrow, narrow × medium?

4. Eye color in humans is determined by one gene-pair, with brown dominant and blue recessive. In a certain family, the husband had a blue-eyed father; he himself has brown eyes. The wife has blue eyes. They plan on having three children. What is the chance that all three will have brown eyes? (It is better to work this out exactly rather than using the normal approximation.)

The answers to these exercises are on p. A90.

2. DID MENDEL'S FACTS FIT HIS MODEL?

Mendel's discovery ranks as one of the greatest in science. Today, his theory is amply proved and extremely powerful. But how good was his own experimental proof? Did Mendel's data prove his theory? Only too well, answered R. A. Fisher:

> ...the general level of agreement between Mendel's expectations and his reported results shows that it is closer than would be expected in the best of several thousand repetitions. The data have evidently been sophisticated systematically, and after examining various possibilities, I have no doubt that Mendel was deceived by a gardening assistant, who knew only too well what his principal expected from each trial made.[7]

Leave the gardener aside for now. Fisher is saying that Mendel's data were fudged. The reason: Mendel's observed frequencies were uncomfortably close to his expected frequencies, much closer than ordinary chance variability would permit.

In one experiment, for instance, Mendel obtained 8,023 second-generation hybrid seeds. He expected $1/4 \times 8{,}023 \approx 2{,}006$ of them to be green, and observed 2,001, for a discrepancy of 5. According to his own chance model, the data on seed color are like the results of drawing 8,023 times with replacement from the box

| yellow | yellow | yellow | green |

In this model, what is the chance of observing a discrepancy of 5 or less between the number of greens and the expected number? In other words, what is the probability that the number of greens will be

$$\text{between } 1/4 \times 8{,}023 - 5 \approx 2{,}001 \text{ and } 1/4 \times 8{,}023 + 5 \approx 2{,}011?$$

That is like drawing 8,023 times with replacement from the box

| 0 | 0 | 0 | 1 |

and asking for the chance that the sum will be between 2,001 and 2,011 inclusive. This chance can be estimated using the normal approximation, keeping track of the edges of the rectangles, as on p. 317.

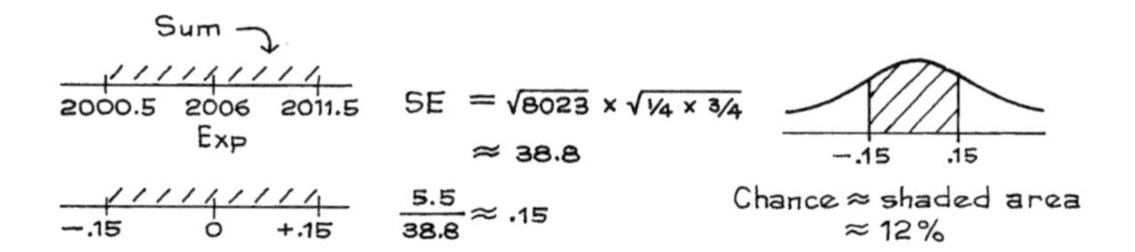

About 88% of the time, chance variation would cause a discrepancy between Mendel's expectations and his observations greater than the one he reported.

By itself, this evidence is not very strong. The trouble is, every one of Mendel's experiments (with an exception to be discussed below) shows this kind of unusually close agreement between expectations and observations. Using the χ^2-test to pool the results (chapter 28), Fisher showed that the chance of agreement as close as that reported by Mendel is about four in a hundred thousand. To put this another way, suppose millions of scientists were busily repeating Mendel's experiments. For each scientist, imagine measuring the discrepancy between his observed frequencies and the expected frequencies by the χ^2-statistic. Then by the laws of chance, about 99,996 out of every 100,000 of these imaginary scientists would report a discrepancy between observations and expectations greater than the one reported by Mendel. That leaves two possibilities:

- either Mendel's data were massaged
- or he was pretty lucky.

The first is easier to believe.

One aspect of Fisher's argument deserves more attention. However, the discussion is technical, and readers can skip to the beginning of the next section. Mendel worked with six characteristics other than seed color. One of them, for instance, was the shape of the pod, which was either inflated (the dominant form) or constricted (the recessive form). The hereditary mechanism is very similar to that for seed color. Pod shape is controlled by one gene-pair. There are two variants of the shape-gene, denoted by i (inflated) and c (constricted). The gene i is dominant, so i/i or i/c or c/i make inflated pods, and c/c makes constricted pods. (The gene-pair controlling seed color acts independently of the pair controlling pod shape.)

There is one difference between seed color and pod shape. Pod shape is a characteristic of the parent plant, and is utterly unaffected by the fertilizing pollen. Thus, if a plant of a pure strain showing the recessive constricted form of seed pods is fertilized with pollen from a plant of pure strain showing the dominant inflated form, all the resulting seed pods will have the recessive constricted form. But when the seeds of this cross grow up into mature first-generation hybrid plants and make their own seed pods, they will all exhibit the dominant inflated form.

If first-generation hybrids are crossed with each other, of the second-generation hybrid plants about 3/4 will exhibit the dominant form and 1/4 the recessive form. As Figure 1 shows, of the second-generation hybrid plants with the dominant inflated form, about

$$\frac{25\%}{25\% + 25\% + 25\%} = \frac{1}{3}$$

should be i/i's and the other 2/3 should be i/c or c/i. Mendel checked this out on 600 plants, finding 201 i/i's, a result too close to the expected 200 for comfort.[8] (The chance of such close agreement is only 10%.)

But worse is yet to come. You can't tell the i/i's from the i/c's or c/i's just by looking, the appearances are identical. So how did Mendel classify them? Well, if undisturbed by naturalists, a pea plant will pollinate itself. So Mendel took his second-generation hybrid plants showing the dominant inflated form, and selected 600 at random. He then raised 10 offsprings from each of his selected plants. If the plant bred true and all 10 offsprings showed the dominant inflated form, he classified it as i/i. If the plant produced any offspring showing the recessive constricted form, he classified it as i/c or c/i.

There is one difficulty with this scheme, which Mendel seems to have overlooked. As Figure 1 shows, the chance that the offspring of a self-fertilized i/c will contain at least one dominant gene i, and hence show the dominant inflated form, is 3/4. So the chance that 10 offsprings of an i/c crossed with itself will all show the dominant form is $(3/4)^{10} \approx 6\%$. Similarly for c/i's. The expected frequency of plants classified as i/i is therefore a bit higher than 200, because about 6% of the 400 i/c's and c/i's will be incorrectly classified as i/i. Indeed, the expected frequency of plants classified as i/i—correctly or incorrectly—is

$$200 + 0.06 \times 400 = 224.$$

Mendel's observed frequency (201 classified as i/i) is rather too far from expectation: the chance of such a large discrepancy is only about 5%. As Fisher concludes, "There is no easy way out of the difficulty."

3. THE LAW OF REGRESSION

This section is difficult, and readers can skip to the next section. Part III discussed Galton's work on heredity, and presented his finding that on the average a child is halfway between the parent and the average. In 1918, Fisher proposed a chance model[9] based on Mendel's ideas, which explained Galton's finding on regression as well as the approximate normality of many biometric characteristics like height (chapter 5). The model can be made quite realistic at the expense of introducing complications. This section begins with a stripped-down version which is easier to understand; later, some refinements will be mentioned. The model will focus on heights, although exactly the same argument could be made for other characteristics. The first assumption in the model is

(1) height is controlled by one gene-pair.

The second assumption is

(2) the genes controlling height act in a *purely additive* way.

The symbols h^*, h^{**}, h', h'' will be used to denote four typical variants of the height-gene. (Variants of a gene are called "alleles.") Assumption (2) means, for instance, that h^* always contributes a fixed amount to an individual's height, whether it is combined with another h^*, or with an h', or with any other variant of the height gene. These genes act very differently from the y's and g's controlling seed color in Mendel's peas: g contributes green to the seed color when it is combined with another g, but when it is combined with a y it has no effect. The height genes are more like the snapdragon genes in exercises 2 and 3 on p. 461 above.

With assumption (2), each gene contributes a fixed amount to an individual's height. This contribution (say in inches) will be denoted by the same letter as used to denote the gene, but in capitals. Thus, an individual with the gene-pair h^*/h' will have height equal to the sum $H^* + H'$. In the first instance, the letters refer to the genes; in the second, to the contributions to height.

Fisher assumed with Mendel that a child gets one gene of the pair controlling height at random from each parent (figure 3). To be more precise, the father has a gene-pair controlling height, and so does the mother. Then one gene is drawn at random from the father's pair and one from the mother's pair to make up the child's pair.

Figure 3. The simplified Mendel-Fisher model for the genetic determination of height. Height is controlled by one gene-pair, with purely additive genetic effects. One gene is drawn at random from each parent's gene-pair to make up the child's gene-pair.

For the sake of argument, suppose the father has the gene-pair h^*/h^{**}, and the mother has the gene-pair h'/h''. The child has chance $1/2$ to get h^* and chance $1/2$ to get h^{**} from the father. Therefore, the father's expected contribution to the child's height is $1/2\,H^* + 1/2\,H^{**} = 1/2(H^* + H^{**})$, namely one-half the father's height. Similarly, the mother's expected contribution equals one-half her height. If you take a large number of children of parents whose father's height is fixed at one level, and mother's height is fixed at another level, the average height of these children must be about equal to

(3) $1/2$(father's height + mother's height).

The expression (3) is called the *mid-parent height*. For instance, with many families where the father is 72 inches tall and the mother is 68 inches tall, the mid-parent height is $^1/_2(72 + 68) = 70$, and on the average the children will be about 70 inches tall at maturity, give or take a small chance error. This is the biological explanation for Galton's law of regression to mediocrity (pp. 169–173).

The assumption (1), that height is controlled by one gene-pair, isn't really needed in the argument; it was made to avoid complicated sums. If three gene-pairs are involved, you only have to assume additivity of the genetic effects and randomness in drawing one gene from each pair for the child (figure 4).

Figure 4. The simplified Mendel-Fisher model for the genetic determination of height, assuming three gene-pairs with purely additive effects. One gene is drawn at random from each gene-pair of each parent to make up the corresponding gene-pair of the child.

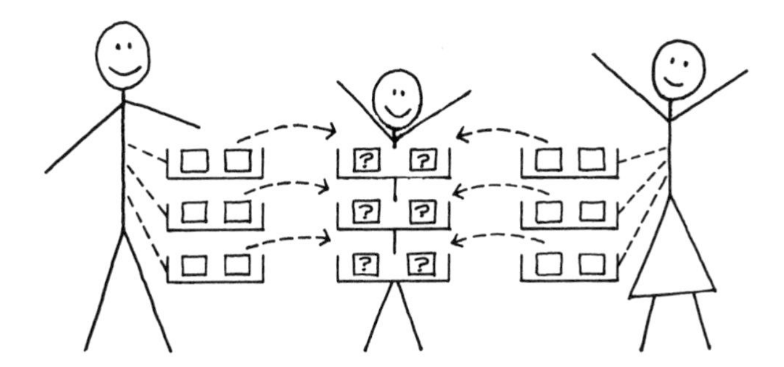

So far, the model has not taken into account sex differences in height. One way to get around this is by "adjusting" women's heights, increasing them by around 8% so that women are just as tall as men—at least in the equations of the model. More elegant (and more complicated) methods are available too.

How well does the model fit? For the Pearson–Lee study (p. 119), the regression of son's height on parents' heights was approximately[10]

$$\text{(4)} \qquad \text{estimated son's ht.} = 15'' + 0.8 \times \frac{\text{father's ht.} + 1.08 \times \text{mother's ht.}}{2}$$

The regression coefficient of 0.8 is noticeably lower than the 1.0 predicted by a purely additive genetic model. Some of the discrepancy may be due to environmental effects, and some to nonadditive genetic effects. Furthermore, the sons averaged 1 inch taller than the fathers. This too cannot be explained by a purely additive genetic model.[11]

The regression of son's height on father's height was very nearly

$$\text{(5)} \qquad \text{estimated son's height} = 35'' + 0.5 \times \text{father's height.}$$

Equation (5) can be derived from equation (3) in the additive model, by assuming that there is no correlation between the heights of the parents.[12] Basically, however, this is a case of two mistakes cancelling. The additive model is a bit off, and the heights of parents are somewhat correlated; but these two facts work in opposite directions, and balance out in equation (5).

Technical note. To derive equation (3) from the model, no assumptions are necessary about the independence of draws from different gene-pairs; all that mattered was each gene having a 50% chance to get drawn. No assumptions are necessary about statistical relationships between the genes in the different parents (such as independence). And no assumptions are necessary about the distribution of the genes in the population (like equilibrium).

4. AN APPRECIATION OF THE MODEL

Genetics represents one of the most satisfying applications of statistical methods. To review the development, Mendel found some striking empirical regularities—like the reappearance of a recessive trait in one-fourth of the second-generation hybrids. He made up a chance model involving what are now called genes to explain his results. He discovered these entities by pure reasoning—he never saw any. Independently, Galton and Pearson found another striking empirical regularity: on the average, a son is halfway between his father and the overall average for sons.

At first sight, the Galton-Pearson results look very different from Mendel's, and it is hard to see how they can both be explained by the same biological mechanism. But Fisher was able to do it. He explained why the average height of children equaled mid-parent height, and even why there are deviations from average. These deviations are caused by chance variation, when genes are chosen at random to pass from the parents to the children.

Chance models are now used in many fields. Usually, the models only assert that certain entities behave as if they were determined by drawing tickets at random from a box, and little effort is spent establishing a physical basis for the claim of randomness. Indeed, the models seldom say explicitly what is like the box, or what is like the tickets.

The genetic model is quite unusual, in that it answers such questions. There are two main sources of randomness in the model:

(i) the random allotment of chromosomes (one from each pair) to sex cells;

(ii) the random pairing of sex cells to produce the fertilized egg.

The two sources of randomness will now be discussed in more detail.

Chromosomes naturally come in *homologous* pairs. The one matching C is denoted C'; the chromosomes C and C' are similar, but not identical. A gene-pair has one gene located on each chromosome of a homologous chromosome-pair. A body cell can divide to form other cells. As a preliminary step, each chromosome in the parent cell doubles itself, as shown in figure 2 and (schematically) in figure 5. When chromosome C is in this doubled condition, it will be denoted C-C. The two pieces are chemically identical and loosely joined together.

The production of ordinary body cells is shown in figure 5a. The parent cell splits in two. Each fragment becomes a separate cell with one-half of each doubled chromosome, winding up with exactly the same complement of chromo-

Figure 5. Production of sex-cells and body-cells by splitting. Chromosomes are denoted here by capital letters, like C. Chromosomes come in homologous pairs, as indicated by primes. Thus, C and C' form a homologous pair; they are chemically similar but not identical. As a preliminary to splitting, a cell doubles all its chromosomes. When doubled, C will be denoted by C-C. The two pieces are chemically identical, and loosely attached. Similarly, C' doubles to C'-C'.[13]

(a) Splitting to make body cells

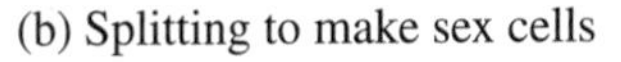
(b) Splitting to make sex cells

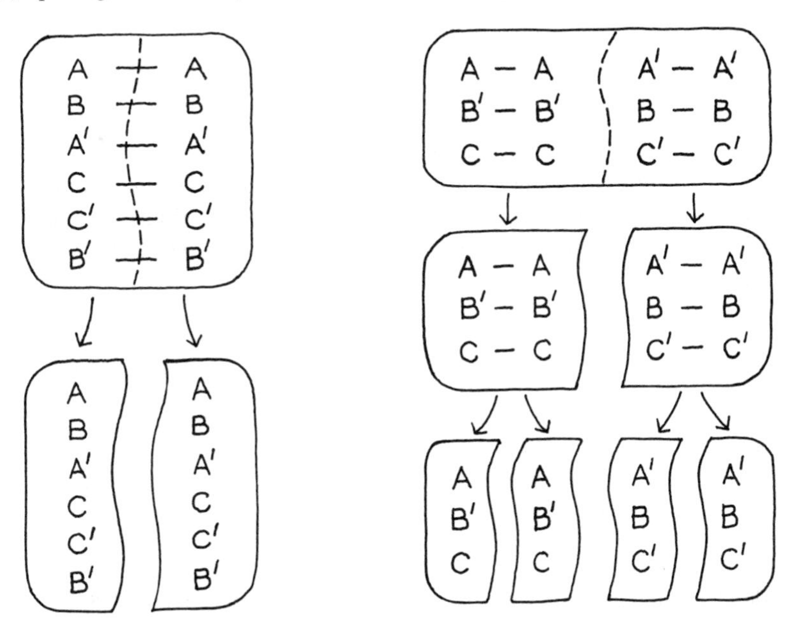

somes as the parent cell (before doubling). There is nothing random about the resulting chromosomes—it is a matter of copying the whole set. Homologous chromosomes are not treated in any special way.

The production of sex cells is shown in figure 5b. The doubled chromosomes move into position, with one doubled chromosome from each homologous pair on opposite sides of the line along which the cell will split (top of figure 5b). Which side of the line? This seems to be random, like coin-tossing. Sometimes one side, sometimes the other, just as a coin sometimes lands heads, sometimes tails. In the model, the choice of side is assumed to be random.

The cell then splits as shown in the middle of figure 5b. Each fragment contains doubled chromosomes—but only one chromosome of each homologous pair is represented. Finally, each of these fragments splits again, as shown at the bottom of figure 5b, and the results of the second split are the sex cells.[14] The lining-up of the homologous pairs (top of figure 5b) is a critical step. The sex cell contains ordinary, undoubled chromosomes—but only one chromosome out of each homologous pair. Which one? One chosen at random. This is one physical source of randomness in Mendelian genetics.

A fertilized egg results from the union of one male sex cell and one female, out of the many which are produced. Which ones? This seems to be random, like drawing tickets at random from a box. In the model, the pairing is assumed to be random. This is the second main physical source of randomness in Mendelian genetics.

When thinking about any other chance model, it is good to ask two questions:

- What are the physical entities which are supposed to act like the tickets and the box?
- Do they really act like that?

5. REVIEW EXERCISES

1. Mendel discovered that for peas, the unripe pods are green or yellow. Their color is controlled by one gene-pair, with variants g for green and y for yellow, g being dominant. In a set of breeding trials, plants with known pod color but unknown genetic makeup were crossed. The results are tabulated below.[15] For each line of the table, guess the genetic makeup of the parents:
   (i) g/g (ii) y/g or g/y (iii) y/y

Pod color of parents	*Number of progeny with*	
	green pods	*yellow pods*
green × yellow	82	78
green × green	118	39
yellow × yellow	0	50
green × yellow	74	0
green × green	90	0

2. Mendel found that pea seeds were either smooth or wrinkled. He bred a pure smooth strain and a pure wrinkled strain. Interbreeding these two strains gave first-generation hybrids, which all turned out to be smooth. Mendel crossed the first-generation hybrids with themselves to get second-generation hybrids; of 7,324 second-generation hybrid plants, 5,474 turned out to be smooth, and 1,850 were wrinkled. Make up a genetic model to account for these results. In the model, what is the chance of agreement between the expected frequency of smoothies and the observed frequency as close as that reported by Mendel?

3. Peas flower at three different times: early, intermediate, and late.[16] Breeding trials gave the following results:

   early × early → early
   early × late → intermediate
   late × late → late.

   Suppose you have 2,500 plants resulting from the cross

   intermediate × intermediate.

   What is the chance that 1,300 or more are intermediate-flowering?

4. In humans, there is a special chromosome-pair which determines sex. Males

have the pair X-Y, while females have the pair X-X. A child gets one X-chromosome automatically, from the mother; from the father, it has half a chance to get an X-chromosome and be female, half a chance to get Y and be male. Some genes are carried only on the X-chromosome: these are said to be *sex-linked*. An example is the gene for male-pattern baldness. (Color blindness and hemophilia are other sex-linked characteristics; the model for baldness is simplified.)

(a) If a man has a bald father, is he more likely to go bald?
(b) If a man's maternal grandfather was bald, is he more likely to go bald?

Explain briefly.

5. Sickle-cell anemia is a genetic disease. In the U.S., it is especially prevalent among blacks: one person in four hundred suffers from it. The disease is controlled by one gene-pair, with variants A and a, where a causes the disease but is recessive:

$$A/A,\ A/a,\ a/A\text{—healthy person}$$
$$a/a\text{—sickle-cell anemia.}$$

(a) Suppose one parent has the gene-pair A/A. Can the child have sickle-cell anemia? How?
(b) Suppose neither parent has sickle-cell anemia. Can the child have it? How?
(c) Suppose both parents have sickle-cell anemia. Can the child avoid having it? How?

6. SUMMARY AND OVERVIEW

1. Whenever reproduction is sexual, the mechanism of heredity is based on gene-pairs. The offspring gets one gene of each pair drawn at random from the corresponding pair in the maternal organism, and one at random from the corresponding pair in the paternal organism. The two genes in a pair are very similar, but not identical.

2. Gene-pairs can control biological characteristics in several ways. One way is *dominance*. In this case, there may be only two varieties (alleles) of the gene, say d and r. The gene-pairs d/d, d/r, r/d all produce the dominant characteristic, while r/r produces the recessive characteristic. (Seed color in peas is an example.) Another way is *additivity*. In this case, each variety of the gene has an effect, and the effect of the gene-pair is the sum of the individual effects of the two genes in the pair. (Flower color in snapdragons is an example.)

3. Fisher showed that Galton's law of regression was a mathematical consequence of Mendel's rules, assuming additive genetic effects.

4. The genetic model explains (at least part of the reason) why children resemble their parents, and also why they differ.

5. This part of the book discussed two chance models: the Gauss model for measurement error and Mendel's model for genetics. These models show how complicated phenomena can be analyzed using the techniques built up in parts II and IV–VI.

6. Chance models are now used in many fields. Usually, the models only assert that some things behave like tickets drawn at random from a box. The genetic model is unusual, because it establishes a physical basis for the claim of randomness.

7. In the next part of the book, we will look at some of the procedures statisticians use for testing models.

PART VIII

Tests of Significance

26

Tests of Significance

Who would not say that the glosses [commentaries on the law] increase doubt and ignorance? It is more of a business to interpret the interpretations than to interpret the things.

—MICHEL DE MONTAIGNE (FRANCE, 1533–1592)[1]

1. INTRODUCTION

Was it due to chance, or something else? Statisticians have invented *tests of significance* to deal with this sort of question. Nowadays, it is almost impossible to read a research article without running across tests and significance levels. Therefore, it is a good idea to find out what they mean. The object in chapters 26 through 28 is to explain the ideas behind tests of significance, and the language. Some of the limitations will be pointed out in chapter 29. This section presents a hypothetical example, where the arguments are easier to follow.

Suppose two investigators are arguing about a large box of numbered tickets. Dr. Nullsheimer says the average is 50. Dr. Altshuler says the average is different from 50. Eventually, they get tired of arguing, and decide to look at some data. There are many, many tickets in the box, so they agree to take a sample—they'll draw 500 tickets at random. (The box is so large that it makes no difference whether the draws are made with or without replacement.) The average of the draws turns out to be 48, and the SD is 15.3.

Dr. Null The average of the draws is nearly 50, just like I thought it would be.

Dr. Alt The average is really below 50.

Dr. Null Oh, come on, the difference is only 2, and the SD is 15.3. The difference is tiny relative to the SD. It's just chance.

Dr. Alt Hmmm. Dr. Nullsheimer, I think we need to look at the SE not the SD.

Dr. Null Why?

Dr. Alt Because the SE tells us how far the average of the sample is likely to be from its expected value—the average of the box.

Dr. Null So, what's the SE?

Dr. Alt Can we agree to estimate the SD of the box as 15.3, the SD of the data?

Dr. Null I'll go along with you there.

Dr. Alt OK, then the SE for the sum of the draws is about $\sqrt{500} \times 15.3 \approx 342$. Remember the square root law.

Dr. Null But we're looking at the average of the draws.

Dr. Alt Fine. The SE for the average is $342/500 \approx 0.7$.

Dr. Null So?

Dr. Alt The average of the draws is 48. You say it ought to be 50. If your theory is right, the average is about 3 SEs below its expected value.

Dr. Null Where did you get the 3?

Dr. Alt Well,

$$\frac{48 - 50}{0.7} \approx -3.$$

Dr. Null You're going to tell me that 3 SEs is too many SEs to explain by chance.

Dr. Alt That's my point. You can't explain the difference by chance. The difference is real. In other words, the average of tickets in the box isn't 50, it's some other number.

Dr. Null I thought the SE was about the difference between the sample average and its expected value.

Dr. Alt Yes, yes. But the expected value of the sample average *is* the average of the tickets in the box.

Our first pass at testing is now complete. The issue in the dialog comes up over and over again: one side thinks a difference is real but the other side might say it's only chance. The "it's only chance" attack can fended off by a calculation, as in the dialog. This calculation is called a *test of significance*. The key idea: if an observed value is too many SEs away from its expected value, that is hard to explain by chance. Statisticians use rather technical language when making this sort of argument, and the next couple of sections will introduce the main terms: *null hypothesis, alternative hypothesis, test statistic*, and *P-value*.[2]

Exercise Set A

1. Fill in the blanks. In the dialog—

(a) The SD of the box was ______ 15.3. Options: known to be, estimated from the data as
(b) The 48 is an ______ value. Options: observed, expected

2. In the dialog, suppose the 500 tickets in the sample average 48 but the SD is 33.6. Who wins now, Dr. Null or Dr. Alt?

3. In the dialog, suppose 100 tickets are drawn, not 500. The sample average is 48 and the SD is 15.3. Who wins now, Dr. Null or Dr. Alt?

4. A die is rolled 100 times. The total number of spots is 368 instead of the expected 350. Can this be explained as a chance variation, or is the die loaded?

5. A die is rolled 1,000 times. The total number of spots is 3,680 instead of the expected 3,500. Can this be explained as a chance variation, or is the die loaded?

The answers to these exercises are on p. A91.

2. THE NULL AND THE ALTERNATIVE

In the example of the previous section, there was sample data for 500 tickets. Both sides saw the sample average of 48. In statistical shorthand, the 48 was "observed." The argument was about the interpretation: what does the sample tell us about the other tickets in the box? Dr. Altshuler claimed that the observed difference was "real." That may sound odd. Of course 48 is different from 50. But the question was whether the difference just reflected chance variation—as Dr. Nullsheimer thought—or whether the average for all the tickets in the box was different from 50, as Dr. Altshuler showed.

In statistical jargon, the null hypothesis and the alternative hypothesis are statements about the box, not just the sample. Each hypothesis represents one side of the argument.

- Null hypothesis—the average of the box equals 50.
- Alternative hypothesis—the average of the box is less than 50.

In the dialog, Dr. Nullsheimer is defending the null hypothesis. According to him, the average of the box was 50. The sample average turned out to be lower than 50 just by the luck of the draw. Dr. Altshuler was arguing for the alternative hypothesis. She thinks the average of the box is lower than 50. Her argument in a nutshell: the sample average is so many SEs below 50 that Dr. Nullsheimer almost has to be wrong. Both sides agreed about the data. They disagreed about the box.

> The null hypothesis corresponds to the idea that an observed difference is due to chance. To make a test of significance, the null hypothesis has to be set up as a box model for the data. The alternative hypothesis is another statement about the box, corresponding to the idea that the observed difference is real.

The terminology may be unsettling. The "alternative hypothesis" is often what someone sets out to prove. The "null hypothesis" is then an alternative (and dull) explanation for the findings, in terms of chance variation. However, there is no way out: the names are completely standard.

Every legitimate test of significance involves a box model. The test gets at the question of whether an observed difference is real, or just chance variation. A real difference is one that says something about the box, and isn't just a fluke of sampling. In the dialog, the argument was about all the numbers in the box, not the 500 numbers in the sample. A test of significance only makes sense in a debate about the box. This point will be discussed again, in section 4 of chapter 29.

Exercise Set B

1. In order to test a null hypothesis, you need
   (i) data
   (ii) a box model for the data
   (iii) both of the above
   (iv) none of the above

2. The ______ hypothesis says that the difference is due to chance but the ______ hypothesis says that the difference is real. Fill in the blanks. Options: null, alternative.

3. In the dialog of section 1, Dr. Alt needed to make a test of significance because
   (i) she knew what was in the box but didn't know how the data were going to turn out, or
   (ii) she knew how the data had turned out but didn't know what was in the box.

   Choose one option, and explain briefly.

4. In the dialog, the null hypothesis says that the average of the ______ is 50. Options: sample, box.

5. One hundred draws are made at random with replacement from a box. The average of the draws is 22.7, and the SD is 10. Someone claims that the average of the box equals 20. Is this plausible?

The answers to these exercises are on p. A91.

3. TEST STATISTICS AND SIGNIFICANCE LEVELS

In the dialog of section 1, Dr. Altshuler temporarily assumed the null hypothesis to be right (the average of the box is 50). On this basis, she calculated how many SEs away the observed value of the sample average was from its expected value:

$$\frac{48 - 50}{0.7} \approx -3.$$

This is an example of a *test statistic*.

> A test statistic is used to measure the difference between the data and what is expected on the null hypothesis.

Dr. Altshuler's test statistic is usually called z:

$$z = \frac{\text{observed} - \text{expected}}{\text{SE}}$$

Tests using the z-statistic are called *z-tests*. Keep the interpretation in mind.

> z says how many SEs away an observed value is from its expected value, where the expected value is calculated using the null hypothesis.

It is the null hypothesis which told Dr. Altshuler to use 50 as the benchmark, and not some other number, in the numerator of z. That is the exact point where the null hypothesis comes into the procedure. Other null hypotheses will give different benchmarks in the numerator of z. The null hypothesis did not tell us the SD of the box. That had to be estimated from the data, in order to compute the SE in the denominator of z.

The z-statistic of -3 stopped Dr. Nullsheimer cold. Why was it so intimidating? After all, 3 is not a very big number. The answer, of course, is that the area to the left of -3 under the normal curve is ridiculously small. The chance of getting a sample average 3 SEs or more below its expected value is about 1 in 1,000.

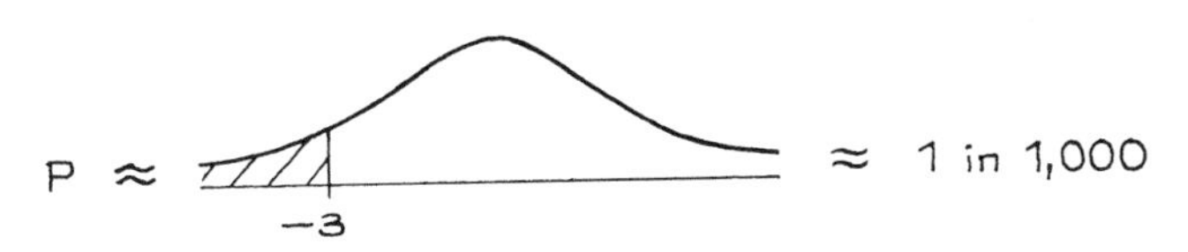

(From the normal table on p. A104, the area is 0.135 of 1%; rounding off, we get 0.1 of 1%; this is 0.1 of $0.01 = 0.001 = 1/1{,}000$.)

The chance of 1 in 1,000 forced Dr. Nullsheimer to concede that the average of the box—not just the average of the sample—was below 50. This chance of 1 in 1,000 is called an *observed significance level*. The observed significance level is often denoted P, for probability, and referred to as a *P-value*. In the example, the P-value of the test was about 1 in 1,000.

Why look at the area to the left of -3? The first point to notice: the data could have turned out differently, and then z would have been different too. For instance, if the sample average is 47.2 and the SD is 14.1,

$$z = \frac{47.2 - 50}{0.63} \approx -4.4$$

This is stronger evidence against the null hypothesis: 4.4 SEs below 50 is even worse for "it's just chance" than 3 SEs. On the other hand, if the sample average is 46.9 and the SD is 37,

$$z = \frac{46.9 - 50}{1.65} \approx -1.9$$

This is weaker evidence. The area to the left of -3 represents the samples which give even more extreme z-values than the observed one, and stronger evidence against the null hypothesis.

> The observed significance level is the chance of getting a test statistic as extreme as, or more extreme than, the observed one. The chance is computed on the basis that the null hypothesis is right. The smaller this chance is, the stronger the evidence against the null.

The z-test can be summarized as follows:

$$z = \frac{\text{observed} - \text{expected}}{\text{SE}}, \quad P \approx$$ [area to the left of z under the normal curve]

Since the test statistic z depends on the data, so does P. That is why P is called an "observed" significance level.

At this point, the logic of the z-test can be seen more clearly. It is an argument by contradiction, designed to show that the null hypothesis will lead to an absurd conclusion and must therefore be rejected. You look at the data, compute the test statistic, and get the observed significance level. Take, for instance, a P of 1 in 1,000. To interpret this number, you start by assuming that the null hypothesis is right. Next, you imagine many other investigators repeating the experiment.

What the 1 in 1,000 says is that your test statistic is really far out. Only one investigator in a thousand would get a test statistic as extreme as, or more extreme than, the one you got. The null hypothesis is creating absurdities, and should be rejected. In general, the smaller the observed significance level, the more you want to reject the null. The phrase "reject the null" emphasizes the point that with a test of significance, the argument is by contradiction.

Our interpretation of P may seem convoluted. It is convoluted. Unfortunately, simpler interpretations turn out to be wrong. If there were any justice in the world, P would be the probability of the null hypothesis given the data. However, P is computed using the null. Even worse, according to the frequency theory, there is no way to define the probability of the null hypothesis being right.

The null is a statement about the box. No matter how often you do the draws, the null hypothesis is either always right or always wrong, because the box does not change.[3] (A similar point for confidence intervals is discussed in section 3 of chapter 21.) What the observed significance level gives is the chance of getting evidence against the null as strong as the evidence at hand—or stronger—if the null is true.

> The *P*-value of a test is the chance of getting a big test statistic—assuming the null hypothesis to be right. *P* is not the chance of the null hypothesis being right.

The z-test is used for reasonably large samples, when the normal approximation can be used on the probability histogram for the average of the draws. (The average has already been converted to standard units, by z.) With small samples, other techniques must be used, as discussed in section 6 below.

Exercise Set C

1. (a) Other things being equal, which of the following *P*-values is best for the null hypothesis? Explain briefly.

   0.1 of 1% 3% 17% 32%

   (b) Repeat, for the alternative hypothesis.

2. According to one investigator's model, the data are like 50 draws made at random from a large box. The null hypothesis says that the average of the box equals 100. The alternative says that the average of the box is more than 100. The average of the draws is 107.3 and the SD is 22.1. The SE for the sample average is 3.1. Now

   $$z = (107.3 - 100)/3.1 = 2.35 \text{ and } P = 1\%.$$

   True or false, and explain:

   (a) If the null hypothesis is right, there is only a 1% chance of getting a z bigger than 2.35.
   (b) The probability of the null hypothesis given the data is 1%.

3. True or false, and explain:

   (a) The observed significance level depends on the data.
   (b) If the observed significance level is 5%, there are 95 chances in 100 for the alternative hypothesis to be right.

4. According to one investigator's model, the data are like 400 draws made at random from a large box. The null hypothesis says that the average of the box equals 50; the alternative says that the average of the box is more than 50. In fact, the data averaged out to 52.7, and the SD was 25. Compute z and *P*. What do you conclude?

5. In the previous exercise, the null hypothesis says that the average of the ______ is 50. Fill in the blank, and explain briefly. Options: box, sample

6. In the dialog of section 1, suppose the two investigators had only taken a sample of 10 tickets. Should the normal curve be used to compute *P*? Answer yes or no, and explain briefly.

7. Many companies are experimenting with "flex-time," allowing employees to choose their schedules within broad limits set by management.[4] Among other things, flex-time is supposed to reduce absenteeism. One firm knows that in the past few years, employees have averaged 6.3 days off from work (apart from vacations). This year, the firm introduces flex-time. Management chooses a simple random sample of 100 employees to follow in detail, and at the end of the year, these employees average 5.5 days off from work, and the SD is 2.9 days. Did absenteeism really go down, or is this just chance variation? Formulate the null and alternative hypotheses in terms of a box model, then answer the question.
8. Repeat exercise 7 for a sample average of 5.9 days and an SD of 2.9 days.

The answers to these exercises are on pp. A91–92.

4. MAKING A TEST OF SIGNIFICANCE

Making a test of significance is a complicated job. You have to

- set up the null hypothesis, in terms of a box model for the data;
- pick a test statistic, to measure the difference between the data and what is expected on the null hypothesis;
- compute the observed significance level P.

The choice of test statistic depends on the model and the hypothesis being considered. So far, we've discussed the "one-sample z-test." Two-sample z-tests will be covered in chapter 27. There are also "t-tests" based on the t-statistic (section 6), "χ^2-tests" based on the χ^2-statistic (chapter 28), and many other tests not even mentioned in this book. However, all tests follow the steps outlined above, and their P-values can be interpreted in the same way.

It is natural to ask how small the observed significance level has to be before you reject the null hypothesis. Many investigators draw the line at 5%.

- If P is less than 5%, the result is called *statistically significant* (often shortened to *significant*).

There is another line at 1%.

- If P is less than 1%, the result is called *highly significant*.

These somewhat arbitrary lines will be discussed again in section 1 of chapter 29.

Do not let the jargon distract you from the main idea. When the data are too far from the predictions of a theory, that is bad for the theory. In statistics, the null hypothesis is rejected when the observed value is too many SEs away from the expected value.

Exercise Set D

1. True or false:
   (a) A "highly significant" result cannot possibly be due to chance.
   (b) If a difference is "highly significant," there is less than a 1% chance for the null hypothesis to be right.

(c) If a difference is "highly significant," there is better than a 99% chance for the alternative hypothesis to be right.

2. True or false:

   (a) If P is 43%, the null hypothesis looks plausible.
   (b) If P is 0.43 of 1%, the null hypothesis looks implausible.

3. True or false:

   (a) If the observed significance level is 4%, the result is "statistically significant."
   (b) If the P-value of a test is 1.1%, the result is "highly significant."
   (c) If a difference is "highly significant," then P is less than 1%.
   (d) If the observed significance level is 3.6%, then $P = 3.6\%$.
   (e) If $z = 2.3$, then the observed value is 2.3 SEs above what is expected on the null hypothesis.

4. An investigator draws 250 tickets at random with replacement from a box. What is the chance that the average of the draws will be more than 2 SEs above the average of the box?

5. One hundred investigators set out to test the null hypothesis that the average of the numbers in a certain box equals 50. Each investigator takes 250 tickets at random with replacement, computes the average of the draws, and does a z-test. The results are plotted in the diagram. Investigator #1 got a z-statistic of 1.9, which is plotted as the point (1, 1.9). Investigator #2 got a z-statistic of 0.8, which is plotted as (2, 0.8), and so forth. Unknown to the investigators, the null hypothesis is true.

   (a) True or false, and explain: the z-statistic is positive when the average of the draws is more than 50.
   (b) How many investigators should get a positive z-statistic?
   (c) How many of them should get a z-statistic bigger than 2? How many of them actually do?
   (d) If $z = 2$, what is P?

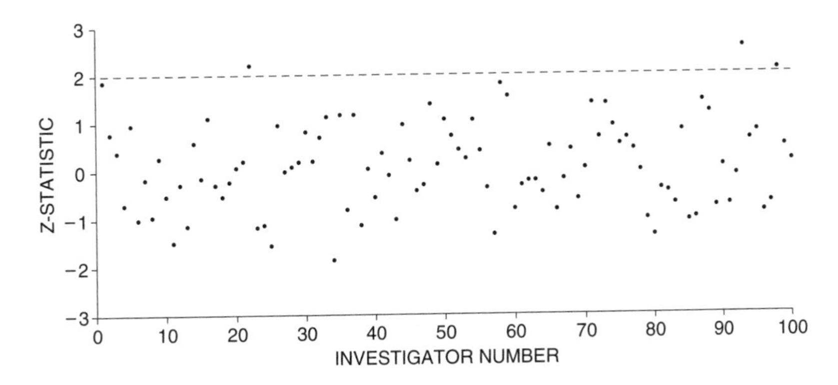

The answers to these exercises are on p. A92.

5. ZERO-ONE BOXES

The z-test can also be used when the situation involves classifying and counting. It is a matter of putting 0's and 1's in the box (section 5 of chapter 17). This section will give an example. Charles Tart ran an experiment at the University of California, Davis, to demonstrate ESP.[5] Tart used a machine called the "Aquarius." The Aquarius has an electronic random number generator and 4 "targets." Using its random number generator, the machine picks one of the 4 targets at random. It does not indicate which. Then, the subject guesses which target was chosen, by pushing a button. Finally, the machine lights up the target it picked, ringing a bell if the subject guessed right. The machine keeps track of the number of trials and the number of correct guesses.

Tart selected 15 subjects who were thought to be clairvoyant. Each of the subjects made 500 guesses on the Aquarius, for a total of $15 \times 500 = 7,500$ guesses. Out of this total, 2,006 were right. Of course, even if the subjects had no clairvoyant abilities whatsoever, they would still be right about $1/4$ of the time. In other words, about $1/4 \times 7,500 = 1,875$ correct guesses are expected, just by chance. True, there is a surplus of $2,006 - 1,875 = 131$ correct guesses, but can't this be explained as a chance variation?

Tart could—and did—fend off the "it's only chance" explanation by making a test of significance. To set up a box model, he assumed that the Aquarius generates numbers at random, so each of the 4 targets has 1 chance in 4 to be chosen. He assumed (temporarily) that there is no ESP. Now, a guess has 1 chance in 4 to be right.

The data consist of a record of the 7,500 guesses, showing whether each one is right or wrong. The null hypothesis says that the data are like 7,500 draws from the box

$$\boxed{\boxed{1}\ \boxed{0}\ \boxed{0}\ \boxed{0}} \qquad 1 = \text{right}, \quad 0 = \text{wrong}$$

The number of correct guesses is like the sum of 7,500 draws from the box. This completes the box model for the null hypothesis.

The machine is classifying each guess as right or wrong, and counting the number of correct guesses. That is why a zero-one box is needed. Once the null hypothesis has been translated into a box model, the z-test can be used:

$$z = \frac{\text{observed} - \text{expected}}{\text{SE}}$$

The "observed" is 2,006, the number of correct guesses. The expected number of correct guesses comes from the null hypothesis, and is 1,875. The numerator of the z-statistic is $2,006 - 1,875 = 131$, the surplus number of correct guesses.

Now for the denominator. You need the SE for the number of correct guesses. Look at the box model. In this example, the null hypothesis tells you exactly what is in the box: a 1 and three 0's. The SD of the box is $\sqrt{0.25 \times 0.75} \approx 0.43$. The SE is $\sqrt{7,500} \times 0.43 \approx 37$. So

$$z = 131/37 \approx 3.5$$

The observed value of 2,006 is 3.5 SEs above the expected value. And P is tiny:

The surplus of correct guesses is hard to dismiss as a chance variation. This looks like strong evidence for ESP. However, there are other possibilities to consider. For example, the Aquarius random number generator may not be very good (section 5 of chapter 29). Or the machine may be giving the subject some subtle clues as to which target it picked. There may be many reasonable explanations for the results, besides ESP. But chance variation isn't one of them. That is what the test of significance shows, finishing the ESP example.

The same z-statistic is used here as in section 1:

$$z = \frac{\text{observed} - \text{expected}}{\text{SE}}$$

Although the formula is the same, there are some differences between the z-test in this section and the z-test in section 1.

1) In section 1, the SE was for an average. Here, the SE is for the number of correct guesses. To work out z, first decide what is "observed" in the numerator. Are you dealing with a sum, an average, a number, or a percent? That will tell you which SE to use in the denominator. In the ESP example, the number of correct guesses was observed. That is why the SE for the number goes into the denominator, as indicated by the sketch.

$$z = \frac{\overset{\text{number}}{\text{observed}} - \overset{\text{number}}{\text{expected}}}{\underset{\text{for number}}{\text{SE}}}$$

2) In section 1, the SD of the box was unknown. The investigators had to estimate it from the data. Here, the SD of the box is given by the null hypothesis. You do not have to estimate it. The diagram summarizes points 1) and 2).

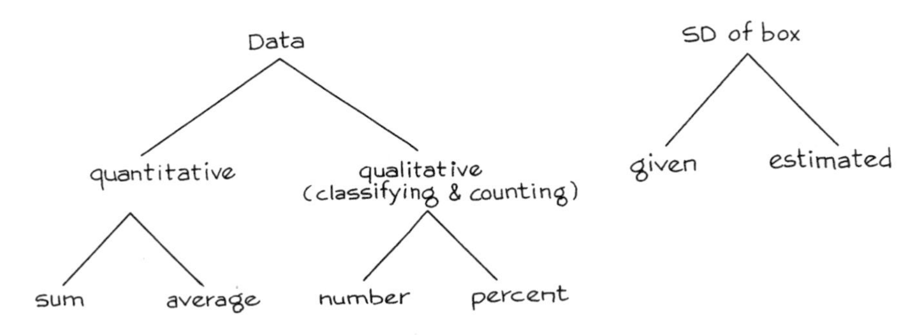

3) In section 1, there was an alternative hypothesis about the box: its average was below 50. With ESP, there is no sensible way to set up the alternative hypothesis as a box model. The reason: if the subjects do have ESP, the chance for each guess to be right may well depend on the previous trials, and may change from trial to trial. Then the data will not be like draws from a box.[6]

4) In section 1, the data were like draws from a box, because the investigators agreed to take a simple random sample of tickets. The argument was only about the average of the box. Here, part of the question is *whether* the data are like draws from a box—any box.

Chapters 19–24 were about estimating parameters from data, and getting margins of error. *Testing*, the topic of this chapter, is about another kind of question. For example, is a parameter equal to some prespecified value, or isn't it? Estimation and testing are related, but the goals are different.

Exercise Set E

This exercise set also covers material from previous sections.

1. In Tart's experiment, the null hypothesis says that ______. Fill in the blank, using one of the options below.

   (i) The data are like 7,500 draws from the box | [0] [0] [0] [1] |.
   (ii) The data are like 7,500 draws from the box | [0] [0] [1] |.
   (iii) The fraction of 1's in the box is 2,006/7,500.
   (iv) The fraction of 1's among the draws is 2,006/7,500.
   (v) ESP is real.

2. As part of a statistics project in the early 1970s, Mr. Frank Alpert approached the first 100 students he saw one day on Sproul Plaza at the University of California, Berkeley, and found out the school or college in which they enrolled. There were 53 men in his sample. From Registrar's data, 25,000 students were registered at Berkeley that term, and 67% were male. Was his sampling procedure like taking a simple random sample?

   Fill in the blanks. That will lead you step by step to the box model for the null hypothesis. (There is no alternative hypothesis about the box.)

   (a) There is one ticket in the box for each ______.

   person in the sample   student registered at Berkeley that term

   (b) The ticket is marked ______ for the men and ______ for the women.
   (c) The number of tickets in the box is ______ and the number of draws is ______. Options: 100, 25,000.
   (d) The null hypothesis says that the sample is like ______ ______ made at random from the box. (The first blank must be filled in with a number; the second, with a word.)
   (e) The percentage of 1's in the box is ______. Options: 53%, 67%.

3. (This continues exercise 2.) Fill in the blanks. That will lead you step by step to z and P.

   (a) The observed number of men is ______.
   (b) The expected number of men is ______.
   (c) If the null hypothesis is right, the number of men in the sample is like the ______ of the draws from the box. Options: sum, average.
   (d) The SE for the number of men is ______.
   (e) $z =$ ______ and $P =$ ______.

4. (This continues exercises 2 and 3.) Was Alpert's sampling procedure like taking a simple random sample? Answer yes or no, and explain briefly.

5. This also continues exercises 2 and 3.

   (a) In 3(b), the expected number was ______.

   computed from the null hypothesis   estimated from the data

   (b) In 3(d), the SE was ______.

   computed from the null hypothesis   estimated from the data

6. Another ESP experiment used the "Ten Choice Trainer." This is like the Aquarius, but with 10 targets instead of 4. Suppose that in 1,000 trials, a subject scores 173 correct guesses.

   (a) Set up the null hypothesis as a box model.
   (b) The SD of the box is ______. Fill in the blank, using one of the options below, and explain briefly.

   $$\sqrt{0.1 \times 0.9} \qquad \sqrt{0.173 \times 0.827}$$

   (c) Make the z-test.
   (d) What do you conclude?

7. A coin is tossed 10,000 times, and it lands heads 5,167 times. Is the chance of heads equal to 50%? Or are there too many heads for that?

   (a) Formulate the null and alternative hypotheses in terms of a box model.
   (b) Compute z and P.
   (c) What do you conclude?

8. Repeat exercise 7 if the coin lands heads 5,067 times, as it did for Kerrich (section 1 of chapter 16).

9. One hundred draws are made at random with replacement from a box of tickets; each ticket has a number written on it. The average of the draws is 29 and the SD of the draws is 40. You see a statistician make the following calculation:

   $$z = \frac{29 - 20}{4} = 2.25, \quad P \approx 1\%$$

   (a) She seems to be testing the null hypothesis that the average of the ______ is 20. Options: box, sample.
   (b) True or false: there is about a 1% chance for the null hypothesis to be right.

   Explain briefly.

10. A colony of laboratory mice consisted of several hundred animals. Their average weight was about 30 grams, and the SD was about 5 grams. As part of an experiment, graduate students were instructed to choose 25 animals haphazardly, without any definite method.[7] The average weight of these animals turned out to be around 33 grams, and the SD was about 7 grams. Is choosing animals haphazardly the same as drawing them at random? Or is 33 grams too far above average for that? Discuss briefly; formulate the null hypothesis as a box model; compute z and P. (There is no need to formulate an alternative hypothesis about the box; you must decide whether the null hypothesis tells you the SD of the box: if not, you have to estimate the SD from the data.)

11. (Hard.) Discount stores often introduce new merchandise at a special low price in order to induce people to try it. However, a psychologist predicted that this practice would actually reduce sales. With the cooperation of a discount chain, an experiment was performed to test the prediction.[8] Twenty-five pairs of stores were selected, matched according to such characteristics as location and sales volume. These stores did not advertise, and displayed their merchandise in similar ways.

    A new kind of cookie was introduced in all 50 stores. For each pair of stores, one was chosen at random to introduce the cookies at a special low price, the price increasing to its regular level after two weeks. The other store in the pair introduced the cookies at the regular price. Total sales of the cookies were computed for each store for six weeks from the time they were introduced.

    In 18 of the 25 pairs, the store which introduced the cookies at the regular price turned out to have sold more of them than the other store. Can this result be explained as a chance variation? Or does it support the prediction that introducing merchandise at a low price reduces long-run sales? (Formulate the null hypothesis as a box model; there is no alternative hypothesis about the box.)

The answers to these exercises are on pp. A92–93.

6. THE *t*-TEST

With small samples, the *z*-test has to be modified. Statisticians use the *t*-test, which was invented by W. S. Gosset (England, 1876–1936). Gosset worked as an executive in the Guinness Brewery, where he went after taking his degree at Oxford. He published under the pen name "Student" because his employers didn't want the competition to realize how useful the results could be.[9]

This section will show how to do the *t*-test, by example. However, the discussion is a bit technical, and can be skipped. In Los Angeles, many studies have been conducted to determine the concentration of CO (carbon monoxide) near freeways with various conditions of traffic flow. The basic technique involves capturing air samples in special bags, and then determining the CO concentrations in the bag samples by using a machine called a *spectrophotometer*. These machines can measure concentrations up to about 100 ppm (parts per million by volume) with errors on the order of 10 ppm. Spectrophotometers are quite delicate and have to be calibrated every day. This involves measuring CO concentration in a manufactured gas sample, called *span gas*, where the concentration is precisely controlled at 70 ppm. If the machine reads close to 70 ppm on the span gas, it's ready for use; if not, it has to be adjusted. A complicating factor is that the size of the measurement errors varies from day to day. On any particular day, however, we assume that the errors are independent and follow the normal curve; the SD is unknown and changes from day to day.[10]

One day, a technician makes five readings on span gas, and gets

78 83 68 72 88

Four out of five of these numbers are higher than 70, and some of them by quite a

bit. Can this be explained on the basis of chance variation? Or does it show bias, perhaps from improper adjustment of the machine?

A test of significance is called for, and a box model is needed. The one to use is the Gauss model (section 3 of chapter 24). According to this model, each measurement equals the true value of 70 ppm, plus bias, plus a draw with replacement from the error box. The tickets in the error box average out to 0, and the SD is unknown.

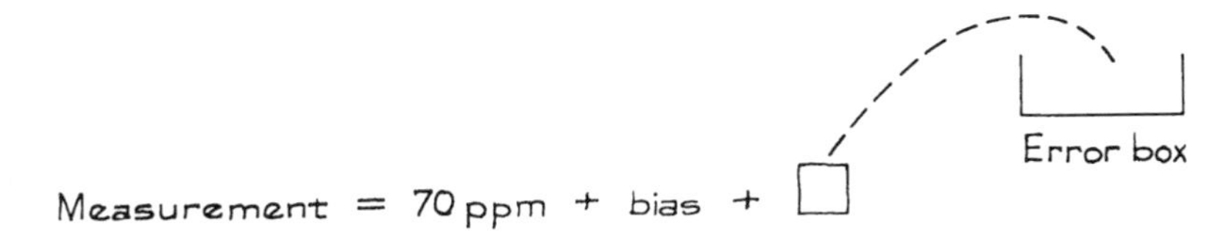

The key parameter is the bias. The null hypothesis says that the bias equals 0. On this hypothesis, the average of the 5 measurements has an expected value of 70 ppm; the difference between the average and 70 ppm is explained as a chance variation. The alternative hypothesis says that the bias differs from 0, so the difference between the average of the measurements and 70 ppm is real.

As before, the appropriate test statistic to use is

$$\frac{\text{observed} - \text{expected}}{\text{SE}}$$

The average of the 5 measurements is 77.8 ppm, and their SD is 7.22 ppm. It seems right to estimate the SD of the error box by 7.22 ppm. The SE for the sum of the draws is $\sqrt{5} \times 7.22 \approx 16.14$ ppm. And the SE for the average is $16.14/5 \approx 3.23$ ppm. The test statistic is

$$\frac{77.8 - 70}{3.23} \approx 2.4$$

In other words, the average of the sample is about 2.4 SEs above the value expected on the null hypothesis. Now the area to the right of 2.4 under the normal curve is less than 1%. This *P*-value looks like strong evidence against the null hypothesis.

There is something missing, however. The SD of the measurements is only an estimate for the SD of the error box. And the number of measurements is so small that the estimate could be way off. There is extra uncertainty that has to be taken into account, which is done in two steps.

Step 1. When the number of measurements is small, the SD of the error box should not be estimated by the SD of the measurements. Instead, SD^+ is used[11]

$$SD^+ = \sqrt{\frac{\text{number of measurements}}{\text{number of measurements} - \text{one}}} \times SD.$$

This estimate is larger. (See p. 74 for the definition of SD^+, and p. 495 for the logic behind the definition.)

In the example, the number of measurements is 5 and their SD is 7.22 ppm. So $SD^+ \approx \sqrt{5/4} \times 7.22 \approx 8.07$ ppm. Then, the SE is figured in the usual way. The SE for the sum is $\sqrt{5} \times 8.07 \approx 18.05$ ppm; the SE for the average is $18.05/5 = 3.61$ ppm. The test statistic becomes

$$\frac{77.8 - 70}{3.61} \approx 2.2$$

Step 2. The next step is to find the *P*-value. With a large number of measurements, this can be done using the normal curve. But with a small number of measurements, a different curve must be used, called *Student's curve*. As it turns out, the *P*-value from Student's curve is about 5%. That is quite a bit more than the 1% from the normal curve.

Using Student's curve takes some work. Actually, there is one of these curves for each number of *degrees of freedom*. In the present context,

$$\text{degrees of freedom} = \text{number of measurements} - \text{one}.$$

Student's curves for 4 and 9 degrees of freedom are shown in figure 1, with the

Figure 1. Student's curves. The dashed line is Student's curve for 4 degrees of freedom (top panel) or 9 degrees of freedom (bottom). The solid line is a normal curve, for comparison.

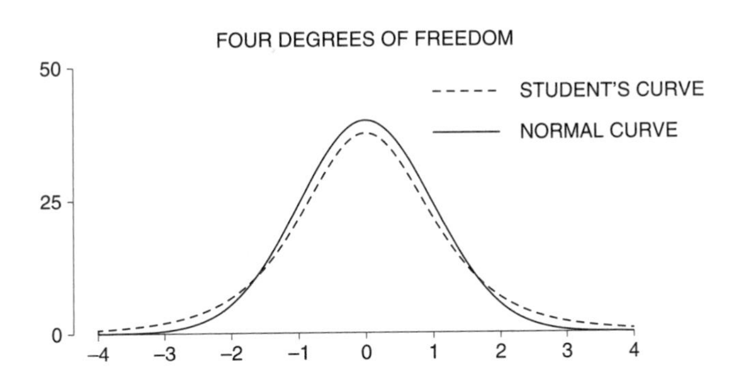

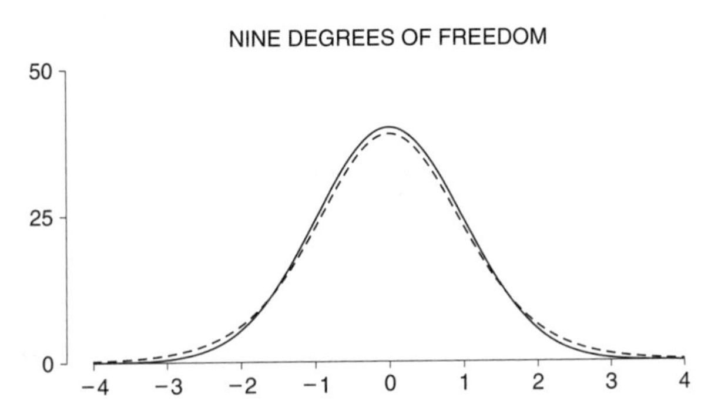

normal curve for comparison. Student's curves look quite a lot like the normal curve, but they are less piled up in the middle and more spread out. As the number of degrees of freedom goes up, the curves get closer and closer to the normal, reflecting the fact that the SD of the measurements is getting closer and closer to the SD of the error box. The curves are all symmetric around 0, and the total area under each one equals 100%.[12]

In the example, with 5 measurements there are $5-1=4$ degrees of freedom. To find the P-value, we need to find the area to the right of 2.2 under Student's curve with 4 degrees of freedom:

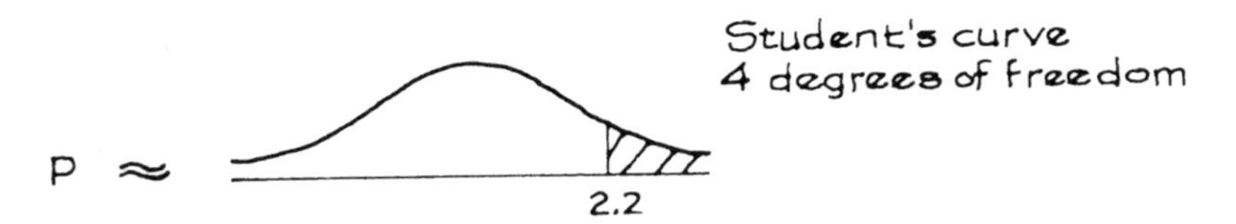

The area can be found with the help of a special table (p. A105), part of which is shown in table 1 (next page). The rows are labeled by degrees of freedom. Look across the row for 4 degrees of freedom. The first entry is 1.53, in the column headed 10%. This means the area to the right of 1.53 under Student's curve with 4 degrees of freedom equals 10%. The other entries can be read the same way.

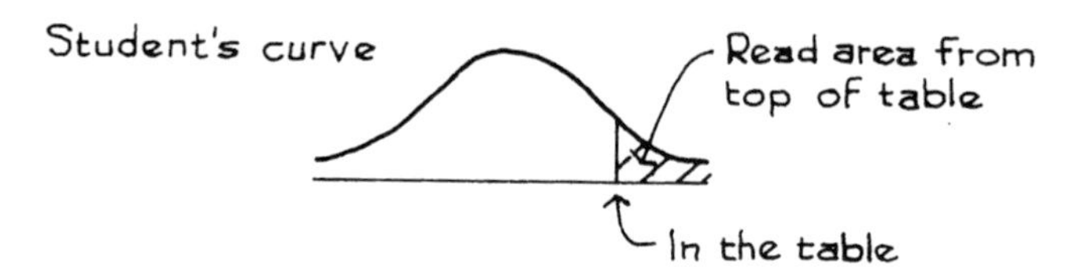

Table 1. A short t-table.

Degrees of freedom	*10%*	*5%*	*1%*
1	3.08	6.31	31.82
2	1.89	2.92	6.96
3	1.64	2.35	4.54
4	1.53	2.13	3.75
5	1.48	2.02	3.36

In the example, there are 4 degrees of freedom, and t is 2.2. From table 1, the area under Student's curve to right of 2.13 is 5%. So the area to the right of 2.2 must be about 5%. The P-value is about 5%.

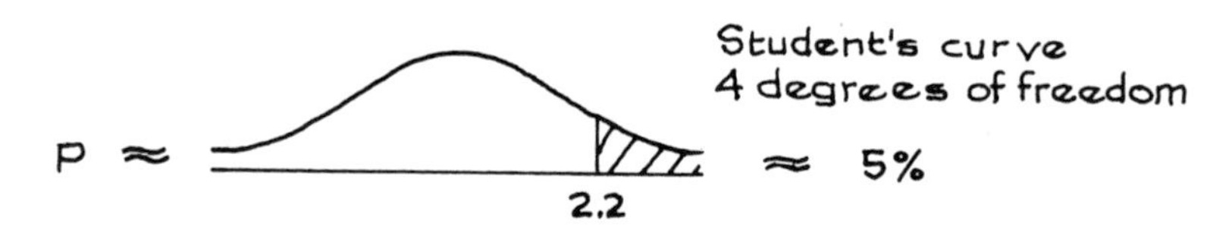

The evidence is running against the null hypothesis, though not very strongly. This completes the example.

Student's curve should be used under the following circumstances.

- The data are like draws from a box.
- The SD of the box is unknown.
- The number of observations is small, so the SD of the box cannot be estimated very accurately.
- The histogram for the contents of the box does not look too different from the normal curve.

With a large number of measurements (say 25 or more), the normal curve would ordinarily be used. If the SD of the box is known, and the contents of the box follow the normal curve, then the normal curve can be used even for small samples.[13]

Example 1. On another day, 6 readings on span gas turn out to be

72 79 65 84 67 77.

Is the machine properly calibrated? Or do the measurements show bias?

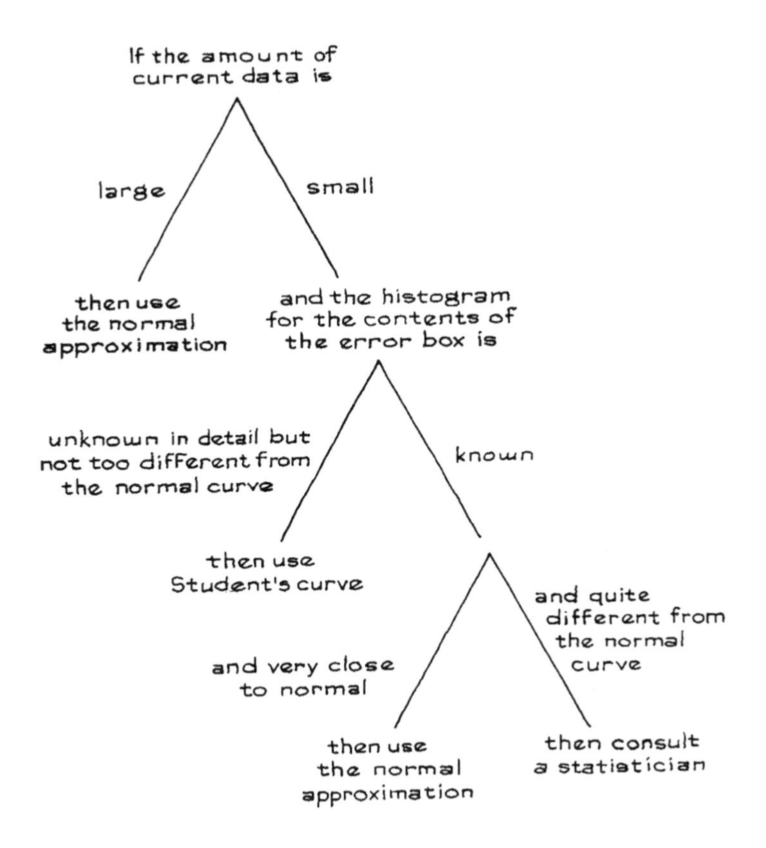

Solution. The model is the same as before. The average of the new measurements is 74 ppm, and their SD is 6.68 ppm. Since there are only 6 observations, the SD of the error box should be estimated by SD^+ of the data, not the SD. The SD^+ is $\sqrt{6/5} \times 6.68 \approx 7.32$ ppm, so the SE for the average is 2.99 ppm. Now

$$t = \frac{74 - 70}{2.99} \approx 1.34$$

To compute the *P*-value, Student's curve is used instead of the normal, with $6 - 1 = 5$ degrees of freedom.

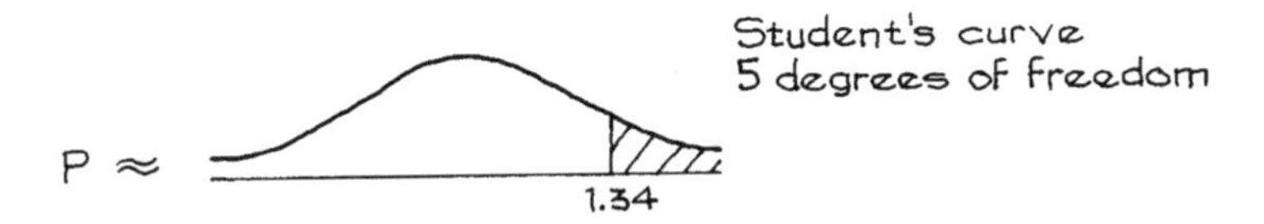

From table 1, the area to the right of 1.34 under Student's curve with 5 degrees of freedom is a little more than 10%. There does not seem to be much evidence of bias. The machine is ready to use. The reasoning on the 10%: from the table, the

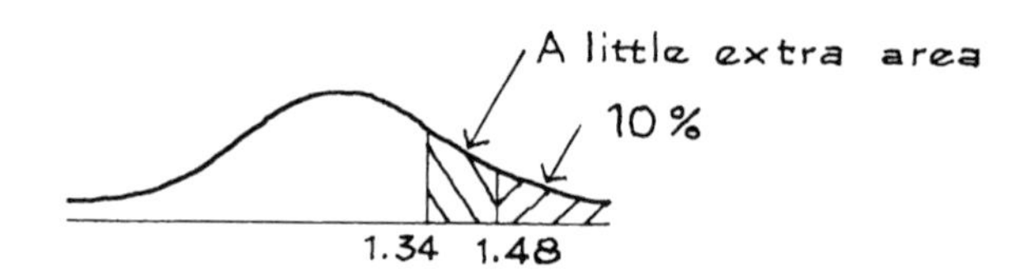

area to the right of 1.48 is 10%. And 1.34 is just to the left of 1.48. So the area to the right of 1.34 is a little more than 10%.

Exercise Set F

1. Find the area under Student's curve with 5 degrees of freedom:

   (a) to the right of 2.02
   (b) to the left of −2.02
   (c) between −2.02 and 2.02
   (d) to the left of 2.02.

2. The area to the right of 4.02 under Student's curve with 2 degrees of freedom is

   less than 1% between 1% and 5% more than 5%

   Choose one option, and explain.

3. True or false, and explain: to make a *t*-test with 4 measurements, use Student's curve with 4 degrees of freedom.

4. Each (hypothetical) data set below represents some readings on span gas. Assume the Gauss model, with errors following the normal curve. However, bias may be present. In each case, make a *t*-test to see whether the instrument is properly calibrated or not. In one case, this is impossible. Which one, and why?

   (a) 71, 68, 79
   (b) 71, 68, 79, 84, 78, 85, 69
   (c) 71
   (d) 71, 84

5. A new spectrophotometer is being calibrated. It is not clear whether the errors follow the normal curve, or even whether the Gauss model applies. In two cases, these assumptions should be rejected. Which two, and why? The numbers are replicate measurements on span gas.

   (a) 71, 70, 72, 69, 71, 68, 93, 75, 68, 61, 74, 67
   (b) 71, 73, 69, 74, 65, 67, 71, 69, 70, 75, 71, 68
   (c) 71, 69, 71, 69, 71, 69, 71, 69, 71, 69, 71, 69

6. A long series of measurements on a checkweight averages out to 253 micrograms above ten grams, and the SD is 7 micrograms. The Gauss model is believed to apply, with negligible bias. At this point, the balance has to be rebuilt, which may introduce bias as well as changing the SD of the error box. Ten measurements on the checkweight, using the rebuilt scale, show an average of 245 micrograms above ten grams, and the SD is 9 micrograms. Has bias been introduced? Or is this chance variation? (You may assume that the errors follow the normal curve.)

7. Several thousand measurements on a checkweight average out to 512 micrograms above a kilogram; the SD is 50 micrograms. Then, the weight is cleaned. The next 100 measurements average out to 508 micrograms above one kilogram; the SD is 52 micrograms. Apparently, the weight got 4 micrograms lighter. Or is this chance variation? (You may assume the Gauss model with no bias.)
   (a) Formulate the null and alternative hypotheses as statements about a box model.
   (b) Would you estimate the SD of the box as 50 or 52 micrograms?
   (c) Would you make a z-test or a t-test?
   (d) Did the weight get lighter? If so, by how much?

The answers to these exercises are on p. A94.

Technical notes. (i) The term "degrees of freedom" is slightly baroque; here is the idea behind the phrase. The SE for the average depends on the SD of the measurements, and that in turn depends on the deviations from the average. But the sum of the deviations has to be 0, so they cannot all vary freely. The constraint that the sum equals 0 eliminates one degree of freedom. For example, with 5 measurements, the sum of the 5 deviations is 0. If you know 4 of them, you can compute the 5th—so there are only 4 degrees of freedom.

(ii) Why use SD^+? Suppose we have some draws made at random with replacement from a box whose SD is unknown. If we knew the average of the box, the r.m.s. difference between the sample numbers and the average of the box could be used to estimate the SD of the box. However, we usually do not know the average of the box and must estimate that too, using the average of the draws. Now there is a little problem. The average of the draws follows the draws around; deviations from the average of the *draws* tend to be smaller than deviations from the average of the *box*. SD^+ corrects this problem.

7. REVIEW EXERCISES

Review exercises may cover material from previous chapters.

1. True or false, and explain:
   (a) The P-value of a test equals its observed significance level.
   (b) The alternative hypothesis is another way of explaining the results; it says the difference is due to chance.

2. With a perfectly balanced roulette wheel, in the long run, red numbers should turn up 18 times in 38. To test its wheel, one casino records the results of 3,800 plays, finding 1,890 red numbers. Is that too many reds? Or chance variation?
   (a) Formulate the null and alternative hypotheses as statements about a box model.
   (b) The null says that the percentage of reds in the box is ______. The alternative says that the percentage of reds in the box is ______. Fill in the blanks.

(c) Compute z and P.
(d) Were there too many reds?

3. One kind of plant has only blue flowers and white flowers. According to a genetic model, the offsprings of a certain cross have a 75% chance to be blue-flowering, and a 25% chance to be white-flowering, independently of one another. Two hundred seeds of such a cross are raised, and 142 turn out to be blue-flowering. Are the data consistent with the model? Answer yes or no, and explain briefly.

4. One large course has 900 students, broken down into section meetings with 30 students each. The section meetings are led by teaching assistants. On the final, the class average is 63, and the SD is 20. However, in one section the average is only 55. The TA argues this way:

   > If you took 30 students at random from the class, there is a pretty good chance they would average below 55 on the final. That's what happened to me—chance variation.

   Is this a good defense? Answer yes or no, and explain briefly.

5. A newspaper article says that on the average, college freshmen spend 7.5 hours a week going to parties.[14] One administrator does not believe that these figures apply at her college, which has nearly 3,000 freshmen. She takes a simple random sample of 100 freshmen, and interviews them. On average, they report 6.6 hours a week going to parties, and the SD is 9 hours. Is the difference between 6.6 and 7.5 real?

   (a) Formulate the null and alternative hypotheses in terms of a box model.
   (b) Fill in the blanks. The null says that the average of the box is ______. The alternative says that average of the box is ______.
   (c) Now answer the question: is the difference real?

6. In 1969, Dr. Spock came to trial before Judge Ford, in Boston's federal court house. The charge was conspiracy to violate the Military Service Act. "Of all defendants, Dr. Spock, who had given wise and welcome advice on child-rearing to millions of mothers, would have liked women on his jury."[15] The jury was drawn from a "venire," or panel, of 350 persons selected by the clerk. This venire included only 102 women, although a majority of the eligible jurors in the district were female. At the next stage in selecting the jury to hear the case, Judge Ford chose 100 potential jurors out of these 350 persons. His choices included 9 women.

   (a) 350 people are chosen at random from a large population, which is over 50% female. How likely is it that the sample includes 102 women or fewer?
   (b) 100 people are chosen at random (without replacement) from a group consisting of 102 women and 248 men. How likely is it that the sample includes 9 women or fewer?
   (c) What do you conclude?

"YOUR HONOR, THE PROSECUTION OBJECTS TO THE COMPOSITION OF THIS JURY."

7. I. S. Wright and associates did a clinical trial on the effect of anticoagulant therapy for coronary heart disease.[16] Eligible patients who were admitted to participating hospitals on odd days of the month were given the therapy; eligible patients admitted on even days were the controls. In total, there were 580 patients in the therapy group and 442 controls. An observer says,

   > Since the odd-even assignment to treatment or control is objective and impartial, it is just as good as tossing a coin.

   Do you agree or disagree? Explain briefly. Assume the trial was done in a month with 30 days.

8. Bookstores like education, one reason being that educated people are more likely to spend money on books. National data show the nationwide average educational level to be 13 years of schooling completed, with an SD of about 3 years, for persons age 18 and over.[17]

   A bookstore is doing a market survey in a certain county, and takes a simple random sample of 1,000 people age 18 and over. They find the average educational level to be 14 years, and the SD is 5 years. Can the difference in average educational level between the sample and the nation be explained by chance variation? If not, what other explanation can you give?

9. A computer is programmed to make 100 draws at random with replacement from the box | [0] [0] [0] [0] [1] |, and take their sum. It does this 144 times; the average of the 144 sums is 21.13. The program is working fine. Or is it?

   Working fine Something is wrong

   Choose one option, and explain your reason.

10. On November 9, 1965, the power went out in New York City, and stayed out for a day—the Great Blackout. Nine months later, the newspapers suggested that New York was experiencing a baby boom. The table below shows the number of babies born every day during a 25 day period, centered nine months and ten days after the Great Blackout.[18] These numbers average out to 436. This turns out not to be unusually high for New York. But there is an interesting twist to the data: the 3 Sundays only average 357. How likely is it that the average of 3 days chosen at random from the table will be 357 or less? Is chance a good explanation for the difference between Sundays and weekdays? If not, how would you explain the difference?

Number of births in New York, August 1–25, 1966

Date	*Day*	*Number*	*Date*	*Day*	*Number*
1	Mon.	451	15	Mon.	451
2	Tues.	468	16	Tues.	497
3	Wed.	429	17	Wed.	458
4	Thur.	448	18	Thur.	429
5	Fri.	466	19	Fri.	434
6	Sat.	377	20	Sat.	410
7	Sun.	344	21	Sun.	351
8	Mon.	448	22	Mon.	467
9	Tue.	438	23	Tues.	508
10	Wed.	455	24	Wed.	432
11	Thur.	468	25	Thur.	426
12	Fri.	462			
13	Sat.	405			
14	Sun.	377			

11. According to the census, the median household income in Atlanta (1.5 million households) was $52,000 in 1999.[19] In June 2003, a market research organization takes a simple random sample of 750 households in Atlanta; 56% of the sample households had incomes over $52,000. Did median household income in Atlanta increase over the period 1999 to 2003?

    (a) Formulate null and alternative hypotheses in terms of a box model.
    (b) Calculate the appropriate test statistic and P.
    (c) Did median family income go up?

12. (Hard.) Does the psychological environment affect the anatomy of the brain? This question was studied experimentally by Mark Rosenzweig and his associates.[20] The subjects for the study came from a genetically pure strain of rats. From each litter, one rat was selected at random for the treatment group, and one for the control group. Both groups got exactly the same kind of food and drink—as much as they wanted. But each animal in the treatment group lived with 11 others in a large cage, furnished with playthings which were changed daily. Animals in the control group lived in isolation, with no toys. After a month, the experimental animals were killed and dissected.

Cortex weights (in milligrams) for experimental animals. The treatment group (T) had an enriched environment. The control group (C) had a deprived environment.

Expt. #1		*Expt. #2*		*Expt. #3*		*Expt. #4*		*Expt. #5*	
T	*C*	*T*	*C*	*T*	*C*	*T*	*C*	*T*	*C*
689	657	707	669	690	668	700	662	640	641
656	623	740	650	701	667	718	705	655	589
668	652	745	651	685	647	679	656	624	603
660	654	652	627	751	693	742	652	682	642
679	658	649	656	647	635	728	578	687	612
663	646	676	642	647	644	677	678	653	603
664	600	699	698	720	665	696	670	653	593
647	640	696	648	718	689	711	647	660	672
694	605	712	676	718	642	670	632	668	612
633	635	708	657	696	673	651	661	679	678
653	642	749	692	658	675	711	670	638	593
		690	621	680	641	710	694	649	602

On the average, the control animals were heavier and had heavier brains, perhaps because they ate more and got less exercise. However, the treatment group had consistently heavier cortexes (the "grey matter," or thinking part of the brain). This experiment was repeated many times; results from the first 5 trials are shown in the table: "T" means treatment, and "C" is for control. Each line refers to one pair of animals. In the first pair, the animal in treatment had a cortex weighing 689 milligrams; the one in control had a lighter cortex, weighing only 657 milligrams. And so on.

Two methods of analyzing the data will be presented in the form of exercises. Both methods take into account the pairing, which is a crucial feature of the data. (The pairing comes from randomization within litter.)

(a) *First analysis.* How many pairs were there in all? In how many of these pairs did the treatment animal have a heavier cortex? Suppose treatment had no effect, so each animal of the pair had a 50–50 chance to have the heavier cortex, independently from pair to pair. Under this assumption, how likely is it that an investigator would get as many pairs as Rosenzweig did, or more, with the treatment animal having the heavier cortex? What do you infer?

(b) *Second analysis.* For each pair of animals, compute the difference in cortex weights "treatment − control." Find the average and SD of all these differences. The null hypothesis says that these differences are like draws made at random with replacement from a box whose average is 0—the treatment has no effect. Make a z-test of this hypothesis. What do you infer?

(c) To ensure the validity of the analysis, the following precaution was taken. "The brain dissection and analysis of each set of littermates was done in immediate succession but in a random order and identified only by code number so that the person doing the dissection does not know which cage the rat comes from." Comment briefly on the following: What was the point of this precaution? Was it a good idea?

8. SUMMARY

1. A *test of significance* gets at the question of whether an observed difference is real (the *alternative hypothesis*) or just a chance variation (the *null hypothesis*).

2. To make a test of significance, the null hypothesis has to be set up as a box model for the data. The alternative hypothesis is another statement about the box.

3. A *test statistic* measures the difference between the data and what is expected on the null hypothesis. The *z-test* uses the statistic

$$z = \frac{\text{observed} - \text{expected}}{\text{SE}}$$

The expected value in the numerator is computed on the basis of the null hypothesis. If the null hypothesis determines the SD of the box, use this information when computing the SE in the denominator. Otherwise, you have to estimate the SD from the data.

4. The *observed significance level* (also called P, or the P-value) is the chance of getting a test statistic as extreme as or more extreme than the observed one. The chance is computed on the basis that the null hypothesis is right. Therefore, P does not give the chance of the null hypothesis being right.

5. Small values of P are evidence against the null hypothesis: they indicate something besides chance was operating to make the difference.

6. Suppose that a small number of tickets are drawn at random with replacement from a box whose contents follow the normal curve, with an average of 0 and an unknown SD. Each draw is added to an unknown constant to give a measurement. The null hypothesis says that this unknown constant equals some given value c. An alternative hypothesis says that the unknown constant is bigger than c. The SD of the box is estimated by the SD^+ of the data. Then the SE for the average of the draws is computed. The test statistic is

$$t = \frac{\text{average of draws} - c}{\text{SE}}$$

The observed significance level is obtained not from the normal curve but from one of the Student's curves, with

$$\text{degrees of freedom} = \text{number of measurements} - \text{one}.$$

This procedure is a *t-test*.

27

More Tests for Averages

Vive la différence!

1. THE STANDARD ERROR FOR A DIFFERENCE

This chapter is about comparing two samples. The SE for the difference between their averages is needed. We begin with an example to illustrate the mathematics. (Real examples come later.) Suppose two boxes A and B have the averages and SDs shown below.

Box A	Box B
Average = 110	Average = 90
SD = 60	SD = 40

Four hundred draws are made at random with replacement from box A, and independently 100 draws are made at random with replacement from box B.

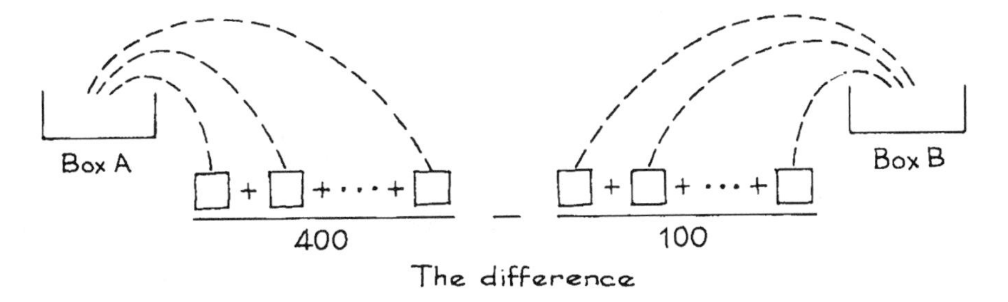

The problem is to find the expected value and standard error for the difference between the two sample averages. The first step is to compute the expected value and SE for each average separately (section 1 of chapter 23):

$$\text{average of 400 draws from box A} = 110 \pm 3 \text{ or so}$$
$$\text{average of 100 draws from box B} = 90 \pm 4 \text{ or so}$$

The expected value for the difference is just $110 - 90 = 20$. The next problem is how to put the SEs together:

$$(110 \pm 3) - (90 \pm 4) = 20 \pm \underline{\qquad}?$$

A natural guess is to add the SEs: $3 + 4 = 7$. This ignores the possibility of cancellation in the two chance errors. The right SE, which is noticeably less than 7, can be found using a square root law.[1]

> The standard error for the difference of two independent quantities is $\sqrt{a^2 + b^2}$, where
> - a is the SE for the first quantity;
> - b is the SE for the second quantity.

In the example, the draws from the two boxes are made independently, so the two averages are independent, and the square root law applies. Now a is 3 and b is 4. So the SE for the difference between the two averages is

$$\sqrt{3^2 + 4^2} = \sqrt{25} = 5.$$

Example 1. One hundred draws are made at random with replacement from box C, shown below. Independently, 100 draws are made at random with replacement from box D. Find the expected value and SE for the difference between the number of 1's drawn from box C and the number of 4's drawn from box D.

(C) [1] [2] (D) [3] [4]

Solution. The number of 1's will be around 50, give or take 5 or so. The number of 4's will also be around 50, give or take 5 or so. The expected value for the difference is $50 - 50 = 0$. The draws are made independently, so the two numbers are independent, and the square root law applies. The SE for the difference is $\sqrt{5^2 + 5^2} \approx 7$.

Example 2. One hundred draws are made at random with replacement from the box

[1] [2] [3] [4]

The expected number of 1's is 25, with an SE of 4.3. The expected number of 4's is also 25 with an SE of 4.3. True or false: the SE for the difference between the number of 1's and the number of 4's is $\sqrt{4.3^2 + 4.3^2}$.

Solution. This is false. The two numbers are dependent: if one is large, the other is likely to be small. The square root law does not apply.

Exercise Set A

1. Two hundred draws are made at random with replacement from the box in example 2. Someone is thinking about the difference

   "number of 1's in draws 1–100" – "number of 5's in draws 101–200"

   True or false, and explain: the SE for the difference is $\sqrt{4^2 + 4^2}$.

2. Box A has an average of 100 and an SD of 10. Box B has an average of 50 and an SD of 18. Now 25 draws are made at random with replacement from box A, and independently 36 draws are made at random with replacement from box B. Find the expected value and standard error for the difference between the average of the draws from box A and the average of the draws from box B.

3. A coin is tossed 500 times. Find the expected value and SE for the difference between the percentage of heads in the first 400 tosses and the percentage of heads in the last 100 tosses.

4. A coin is tossed 500 times. True or false, and explain.
   (a) The SE for the percentage of heads among the 500 tosses is 2.2 percentage points.
   (b) The SE for the percentage of tails among the 500 tosses is 2.2 percentage points.
   (c) The SE for the difference

   percentage of heads – percentage of tails

   is $\sqrt{2.2^2 + 2.2^2} \approx 3.1$ percentage points.

5. A box contains 5,000 numbered tickets, which average out to 50; the SD is 30. Two hundred tickets are drawn at random without replacement. True or false, and explain: the SE for the difference between the average of the first 100 draws and the average of the second 100 draws is approximately $\sqrt{3^2 + 3^2}$. (Hint: What if the draws were made with replacement?)

6. One hundred draws are made at random with replacement from box F: the average of these draws is 51 and their SD is 3. Independently, 400 draws are made at random with replacement from box G: the average of these draws is 48 and their SD is 8. Someone claims that both boxes have the same average. What do you think?

The answers to these exercises are on pp. A94–95.

2. COMPARING TWO SAMPLE AVERAGES

The National Assessment of Educational Progress (NAEP) monitors trends in school performance. Each year, NAEP administers tests on several subjects to a nationwide sample of 17-year-olds who are in school.[2] The reading test was given in 1990 and again in 2004. The average score went down from 290 to 285. The difference is 5 points. Is this real, or just a chance variation?

A z-test can be used, but the calculation is more complicated than it was in chapter 26. There, a sample average was compared to an external standard. Here, there are two samples, and the difference of their averages is the issue:

average score in 2004 sample − average score in 1990 sample.

Both averages are subject to chance variability, and the SE for the difference must take that into account. The method of section 1 can be used.

To compute standard errors, you need a box model, and the model depends on the design of the sample. In fact, the NAEP design was quite complicated, but a simplified version can be presented here. Suppose that in 2004 and in 1990, the test was administered to a nationwide simple random sample of one thousand 17-year-olds currently enrolled in school.

With this design, the model is straightforward. There have to be two boxes, one for each of the two test years. The 2004 box has millions of tickets—one for each person who was 17 years old back then, and enrolled in school. The number on the ticket shows what that person would have scored, if he or she had taken the NAEP reading test. The 2004 data are like 1,000 draws at random from the box. The 1990 box is set up the same way. That completes the model.

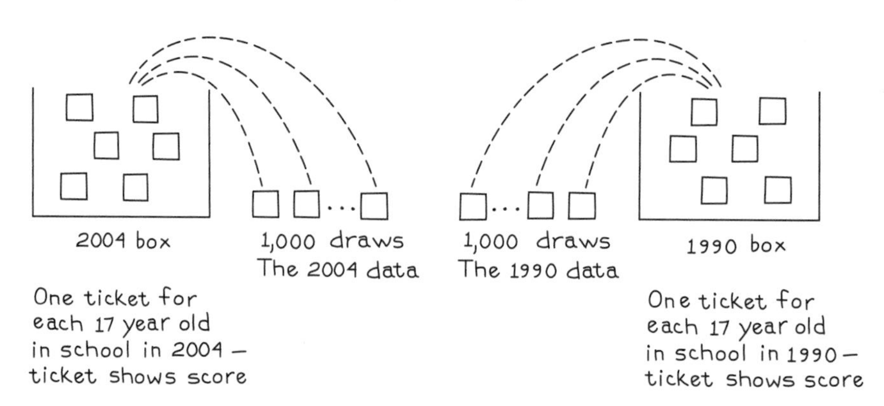

The null hypothesis says that the averages of the two boxes are equal. On that basis, the difference between the sample averages is expected to be 0, and the observed difference just reflects the luck of the draw. Schools are not getting worse. The alternative hypothesis says that the average of the 2004 box is smaller than the average of the 1990 box—reading scores really did go down, and that is why the two sample averages are different. The two-sample z-statistic will help in choosing between these hypotheses.

$$z = \frac{\text{observed difference} - \text{expected difference}}{\text{SE for difference}}$$

We begin with the numerator of the z-statistic. It is the difference between the sample averages that is observed: $285 - 290 = -5$ points. Therefore, the relevant benchmark in the numerator of z is the expected value of the difference. The expected value is computed using the null hypothesis. On that basis, the difference between the two sample averages is expected to be 0. So the numerator of the z-statistic is

$$-5 - 0 = -5$$

Now the denominator. The SE for the difference between the sample averages is needed. Take the samples one at a time. In 2004, the SD of the 1,000 test scores turned out to be 37. So the SD of the 2004 box is estimated as 37. The SE for the sum of the 1,000 test scores in 2004 is estimated as $\sqrt{1{,}000} \times 37 \approx 1{,}170$. The SE for the average is $1{,}170/1{,}000 \approx 1.2$. In 1990, the SD was 40 and the SE for the 1990 average is 1.3. The SE for the difference can be computed using the method of the previous section, because the samples are independent:

$$\sqrt{1.2^2 + 1.3^2} \approx 1.8$$

Finally,

$$z \approx -5/1.8 \approx -2.8$$

In other words, the difference between 2004 and 1990 was about 2.8 SEs below the value expected on the null hypothesis—pushing the envelope of chance variation. We reject the null hypothesis, and are left with the alternative hypothesis that the difference is real. On the other hand, the difference is small, and other measures of school performance give more optimistic results. Chapter 29 continues the discussion.

The two-sample z-statistic is computed from—
- the sizes of the two samples,
- the averages of the two samples,
- the SDs of the two samples.

The test assumes two independent simple random samples.

With NAEP, the samples are big enough so that the probability histogram for each sample average follows the normal curve. Then z follows the normal curve. The two-sample z-test can also be used for percents, as the next example shows.[3]

Example 3. In 1999, NAEP found that 13% of the 17-year-old students had taken calculus, compared to 17% in 2004. Is the difference real, or a chance variation?

Solution. Again, let's assume that in each of the two years, NAEP took a nationwide simple random sample of one thousand 17-year-olds currently enrolled in school. There are two samples here, so the two-sample z-test is needed rather than the one-sample test (chapter 26). As in example 2, there are two boxes, one for 2004 and one for 1999. The data are qualitative rather than quantitative, so 0's and 1's go on the tickets. A ticket is marked 1 if the student took calculus and 0 otherwise. The 2004 data are like 1,000 draws from 2004 box. The 1999 box can be set up the same way. The null hypothesis says that the percentage of 1's in the two boxes is the same. The alternative hypothesis says that the percentage for the 2004 box is bigger than the percentage for the 1999 box.

To make the z-test, we need to put an SE on the difference between the sample percentages. Take the samples one at a time. The SE for the number of 1's in the 2004 sample is estimated as

$$\sqrt{1{,}000} \times \sqrt{0.17 \times 0.83} \approx 11$$

The SE for the percentage is

$$\frac{11}{1{,}000} \times 100\% = 1.2\%$$

Similarly, the SE for the 1999 percentage is 1.1%. The SE for the difference is

$$\sqrt{1.2^2 + 1.1^2} \approx 1.6\%$$

On the null hypothesis, the expected difference is 0%. The observed difference is $17 - 13 = 4\%$. So the test statistic is

$$z = \frac{\text{observed difference} - \text{expected difference}}{\text{SE for difference}} \approx \frac{4\% - 0\%}{1.6\%} \approx 2.5$$

Again, we reject the null: $P \approx 1\%$.

Exercise Set B

1. "Is the difference between two sample averages just due to chance?" To help answer this question, statisticians use a ______ ______ z-test. Fill in the blanks, and explain briefly.

2. In 1990 and 2004, NAEP tested the 17-year-olds on mathematics as well as reading. The average score went up from 305 to 307. You may assume the NAEP took simple random samples of size 1,000 in each of the two years; the SD for the 1990 data was 34, and the SD for the 2004 data was 27. (In fact, NAEP used a more complicated sample design.[4]) Can the difference between the 305 and 307 be explained as a chance variation?
   (a) Should you make a one-sample z-test or a two-sample z-test? Why?
   (b) Formulate the null and alternative hypotheses in terms of a box model. Do you need one box or two? Why? How many tickets go into each box? How many draws? Do the tickets show test scores, or 0's and 1's. Why?
   (c) Now answer the main question: is the difference real, or can it be explained by chance?

3. In 1970, 59% of college freshmen thought that capital punishment should be abolished; by 2005, the percentage had dropped to 35%.[5] Is the difference real, or can it be explained by chance? You may assume that the percentages are based on two independent simple random samples, each of size 1,000.

4. A study reports that freshmen at public universities work 10.2 hours a week for pay, on average, and the SD is 8.5 hours; at private universities, the average is 8.1

hours and the SD is 6.9 hours. Assume these data are based on two independent simple random samples, each of size 1,000.[6] Is the difference between the averages due to chance? If not, what else might explain it?

5. A university takes a simple random sample of 132 male students and 279 females; 41% of the men and 17% of the women report working more than 10 hours during the survey week. To find out whether the difference in percentages is statistically significant, the investigator starts by computing $z = (41 - 17)/.048$. Is anything wrong?

6. Cycle III of the Health Examination Survey used a nationwide probability sample of youths age 12 to 17. One object of the survey was to estimate the percentage of youths who were illiterate.[7] A test was developed to measure literacy. It consisted of seven brief passages, with three questions about each, like the following:

   There were footsteps and a knock at the door. Everyone inside stood up quickly. The only sound was that of the pot boiling on the stove. There was another knock. No one moved. The footsteps on the other side of the door could be heard moving away.

   - The people inside the room
     (a) Hid behind the stove
     (b) Stood up quickly
     (c) Ran to the door
     (d) Laughed out loud
     (e) Began to cry
   - What was the only sound in the room?
     (a) People talking
     (b) Birds singing
     (c) A pot boiling
     (d) A dog barking
     (e) A man shouting
   - The person who knocked at the door finally
     (a) Walked into the room
     (b) Sat down outside the door
     (c) Shouted for help
     (d) Walked away
     (e) Broke down the door.

   This test was designed to be at the fourth-grade level of reading, and subjects were defined to be literate if they could answer more than half the questions correctly.

   There turned out to be some difference between the performance of males and females on this test: 7% of the males were illiterate, compared to 3% of the females. Is this difference real, or the result of chance variation? You may assume that the investigators took a simple random sample of 1,600 male youths, and an independent simple random sample of 1,600 female youths.

7. Cycle II of the Health Examination Survey used a nationwide probability sample of children age 6 to 11. One object of the survey was to study the relationship between the children's scores on intelligence tests and the family backgrounds.[8] The WISC vocabulary scale was used. This consists of 40 words which the child has to define; 2 points are given for a correct answer, and 1 point for a partially

correct answer. There was some relationship between test scores and the type of community in which the parents lived. For example, big-city children averaged 26 points on the test, and their SD was 10 points. But rural children only averaged 25 points with the same SD of 10 points. Can this difference be explained as a chance variation?

You may assume that the investigators took a simple random sample of 400 big-city children, and an independent simple random sample of 400 rural children.

8. Repeat the previous exercise, if both samples were of size 1,000 instead of 400.

9. Review exercise 12 in chapter 26 described an experiment in which 59 animals were put in treatment (enriched environment), and 59 were in control. The cortex weights for the treatment group averaged 683 milligrams, and the SD was 31 milligrams. The cortex weights for the control group averaged 647 milligrams, and the SD was 29 milligrams. Someone proposes to make a two-sample z-test:

$$\text{SE for sum of treatment weights} \approx \sqrt{59} \times 31 \approx 238 \text{ milligrams}$$
$$\text{SE for average of treatment weights} \approx 238/59 \approx 4.0 \text{ milligrams}$$
$$\text{SE for sum of control weights} \approx \sqrt{59} \times 29 \approx 223 \text{ milligrams}$$
$$\text{SE for average of control weights} \approx 223/59 \approx 3.8 \text{ milligrams}$$
$$\text{SE for difference} \approx \sqrt{4.0^2 + 3.8^2} \approx 5.5 \text{ milligrams}$$
$$z = 36/5.5 \approx 6.5, \quad P \approx 0$$

What does statistical theory say?

The answers to these exercises are on pp. A95–96.

3. EXPERIMENTS

The method of section 2 can also be used to analyze certain kinds of experimental data, where the investigators choose some subjects at random to get treatment "A" and others to get "B." In the Salk vaccine field trial, for instance, treatment A would be the vaccine; treatment B, the placebo given to the control group (chapter 1). We begin with an example to illustrate the mechanics, and then say why the method works.

Example 4. There are 200 subjects in a small clinical trial on vitamin C. Half the subjects are assigned at random to treatment (2,000 mg of vitamin C daily) and half to control (2,000 mg of placebo). Over the period of the experiment, the treatment group averaged 2.3 colds, and the SD was 3.1. The controls did a little worse: they averaged 2.6 colds and the SD was 2.9. Is the difference in averages statistically significant?

Solution. The difference between the two averages is -0.3, and you need to put a standard error on this number. Just pretend that you have two independent samples drawn at random with replacement. The SE for the treatment sum is $\sqrt{100} \times 3.1 = 31$; the SE for the treatment average is $31/100 = 0.31$. Similarly, the SE for the control average is 0.29. The SE for the difference is

$$\sqrt{0.31^2 + 0.29^2} \approx 0.42$$

Suppose the null hypothesis is right: vitamin C has no effect. On this basis, the expected value for the difference is 0.0. The observed difference was −0.3. So

$$z = \frac{\text{observed difference} - \text{expected difference}}{\text{SE for difference}} = \frac{-0.3 - 0.0}{0.42} \approx -0.7$$

The difference could easily be due to chance: a few too many susceptible people were assigned to the control group.[9]

Now, a look behind the scenes. In working the example, you were asked to pretend that the treatment and control samples were drawn independently, at random with replacement, from two boxes. However, the experiment wasn't done that way. There were 200 subjects; 100 were chosen at random—without replacement—to get the vitamin C; the other 100 got the placebo. So the draws are made without replacement. Furthermore, the samples are dependent. For instance, one subject might be quite susceptible to colds. If this subject is in the vitamin C group, he cannot be in the placebo group. The assignment therefore influences both averages.

Why does the SE come out right, despite these problems? The reasoning depends on the box model. The investigators are running an experiment. They choose one group of subjects at random to get treatment A and another group to get treatment B. As usual, the model has a ticket for each subject. But now the ticket has two numbers. One shows what the response would be to treatment A; the other, to treatment B. See figure 1. Only one of the two numbers can be observed, because the subject can be given only one of the two treatments.

Figure 1. A randomized controlled experiment comparing treatments A and B. There is a ticket for each subject. The ticket has two numbers: one shows the subject's response to treatment A; the other, to treatment B. Only one of the two numbers can be observed.

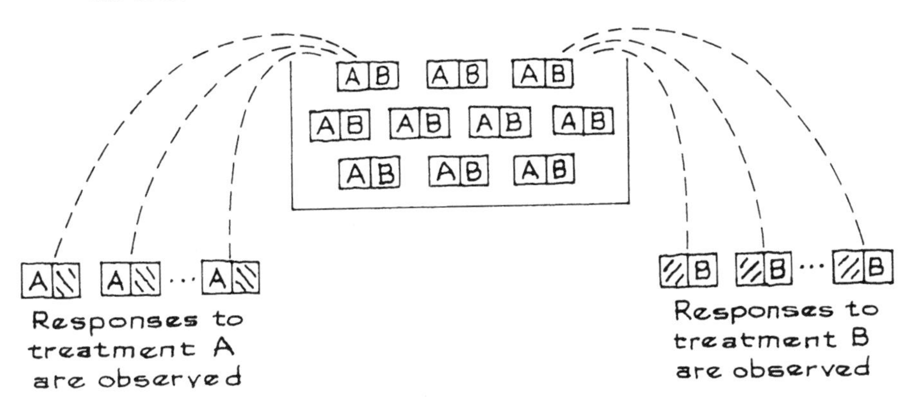

In the model, some tickets are drawn at random without replacement from the box and the responses to treatment A are observed. The data on treatment A are like this first batch of responses. Then, more draws are made at random without replacement from the box and the responses to treatment B are observed. The data on treatment B are like this second batch of responses. In example 4, every one of the 200 subjects was assigned either to vitamin C or to the placebo. In such a case, the second sample just amounts to the tickets left behind in the box after the first sample has been drawn.

The null hypothesis says that the response is the same for both treatments.[10] To test this hypothesis, investigators usually compare averages (or percents):

average response in group A − average response in group B.

What is the SE for this difference? The solution to example 4 seems to involve the two mistakes mentioned earlier—

- The draws are made without replacement, but the SEs are computed as if drawing with replacement.
- The two averages are dependent, but the SEs are combined as if the averages were independent.

When the number of draws is small relative to the number of tickets in the box, neither mistake is serious. There is little difference between drawing with or without replacement, and the dependence between the averages is small too. There almost are two separate boxes, one for the treatment group and one for the controls. However, the "two-box" model is unrealistic for a randomized controlled experiment—unless the subjects really are chosen as a random sample from a large population. That is unusual, although exercise 8 (p. 520) gives one example.

If the number of draws is large relative to the size of the box—and this is the usual case—then the impact of each mistake by itself can be substantial. For instance, when half the subjects are assigned to each treatment group, as in example 4, the correction factor will be noticeably less than 1 (section 4 of chapter 20). Dependence can also be strong. It is a lucky break that when applied to randomized experiments, the procedure of section 2 is conservative, tending to overestimate the SE by a small amount. That is because the two mistakes offset each other.

- The first mistake inflates the SE.
- The second mistake cuts the SE back down.

> There is a box of tickets. Each ticket has two numbers. One shows what the response would be to treatment A; the other, to treatment B. Only one of the numbers can be observed. Some tickets are drawn at random without replacement from the box. In this sample, the responses to treatment A are observed. Then, a second sample is drawn at random without replacement from the remaining tickets. In the second sample, the responses to treatment B are observed.
>
> The SE for the difference between the two sample averages can be conservatively estimated as follows:
>
> (i) compute the SEs for the averages as if the draws were made with replacement;
>
> (ii) combine the SEs as if the samples were independent.

To make the mathematics work, the SEs for the two sample averages must be computed on the basis of drawing WITH replacement—even though the draws are made WITHOUT replacement. That is what compensates for the dependence between the two samples.[11] In summary: when the data come from a randomized experiment (like example 4), the procedure of section 2 can be used even though there is dependence.

Exercise Set C

1. (Hypothetical.) Does coaching for the Math SATs work? A group of 200 high-school seniors volunteer as subjects for an experiment; 100 are selected at random for coaching, and the remaining 100 are controls. After six months, all 200 subjects take the Math SAT. A box model is set up for this experiment, as shown.

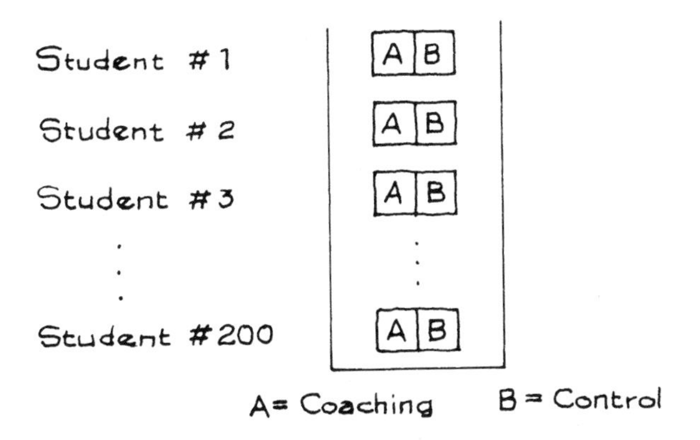

   (a) John Doe participated in the experiment—he was student #17. He got assigned to the coaching group. There was a ticket in the box for him. Did this ticket have one number on it, or two numbers?
   (b) His sister Jane Doe participated also (she was student #18). She was assigned to the control group—no coaching. Did her ticket have an A-number on it? If so, what does this number mean? Do the investigators know what this number was?
   (c) The coaching group averaged 486 on the Math SAT; their SD was 98. The control group averaged 477, and had an SD of 103. Did the coaching work? Or was it chance?

2. (Hypothetical.) Is Wheaties a power breakfast? A study is done in an elementary statistics class; 499 students agree to participate. After the midterm, 250 are randomized to the treatment group, and 249 to the control group. The treatment group is fed Wheaties for breakfast 7 days a week. The control group gets Sugar Pops.
   (a) Final scores averaged 66 for the treatment group; the SD was 21. For the control group, the figures were 59 and 20. What do you conclude?
   (b) What aspects of the study could have been done "blind?"

3. This continues exercise 2.
   (a) Midterm scores averaged 61 for the treatment group; the SD was 20. For the control group, the figures were 60 and 19. What do you conclude?

(b) Repeat, if the average midterm score for the treatment group is 68, and the SD is 21; for the control group, the figures are 59 and 18.

4. Suppose the study in example 4 is repeated on 2,000 subjects, with 1,000 assigned to the vitamin C group, and 1,000 to control. Suppose the average number of colds in the vitamin C group is 2.4 and the SD is 2.9; the average in the control group is 2.5 and the SD is 3.0.

   (a) Is the difference in averages statistically significant? What do you conclude?
   (b) Why would the averages change from one study to the next?

5. In the box below, each ticket has a left-hand number and a right-hand number:

   | 0 4 | | 2 0 | | 3 6 | | 4 12 | | 6 8 |

   (For instance, the left-hand number on | 0 4 | is 0 and the right-hand number is 4.) One hundred draws are made at random with replacement from this box. One investigator computes the average of the left-hand numbers. A second investigator computes the average of the right-hand numbers. True or false, and explain—

   (a) The SE for the first average is 0.2.
   (b) The SE for the second average is 0.4.
   (c) The SE for the difference of the two averages is $\sqrt{0.2^2 + 0.4^2}$.

The answers to these exercises are on pp. A96–97.

4. MORE ON EXPERIMENTS

The technique described in the previous section can also be used for experiments where the response is qualitative rather than quantitative, so the tickets must show 0's and 1's. This section will give an example; but first, some background material. The standard theory of economic behavior assumes "rational" decision making, according to certain formal (and perhaps unrealistic) rules. In particular, the theory says that decision makers respond to facts, not to the way the facts are presented. Psychologists, on the other hand, tend to think that "framing"—the manner of presentation—counts. Empirical work favors the psychological view.[12]

One study, by Amos Tversky and others, involved presenting information on the effectiveness of surgery or radiation as alternative therapies for lung cancer. The subjects were a group of 167 doctors in a summer course at Harvard.[13] The information was presented in two different ways. Some of the doctors got form A, which reports death rates.

> *Form A)* Of 100 people having surgery, 10 will die during treatment, 32 will have died by one year, and 66 will have died by five years. Of 100 people having radiation therapy, none will die during treatment, 23 will die by one year, and 78 will die by five years.

Other doctors got form B, which reports survival rates.

> *Form B)* Of 100 people having surgery, 90 will survive the treatment, 68 will survive one year or longer, and 34 will survive five years or longer. Of 100 people having radiation therapy, all will survive the treatment, 77 will survive one year or longer, and 22 will survive five years or longer.

Both forms contain exactly the same information. For example, 10 patients out of 100 will die during surgery (form A), so 90 out of 100 will survive (form B). By the fifth year, the outlook for lung cancer patients is quite bleak.

In the experiment, 80 of the 167 doctors were picked at random and given form A. The remaining 87 got form B. After reading the form, each doctor wrote down the therapy he or she would recommend for a lung cancer patient. In response to form A, 40 out of 80 doctors chose surgery (table 1). But in response to form B, 73 out of 87 favored surgery: 40/80 is 50%, and 73/87 is 84%. Style of presentation seems to matter.

Table 1. Results from an experiment on the effect of presentation of data.

	Form A	*Form B*
Favored surgery	40	73
Favored radiation	40	14
Total	80	87
Percent favoring surgery	50%	84%

An economist who is defending standard decision theory might argue that the difference is just due to chance. Based on the information, which is the same on both forms, some doctors will recommend surgery while others will choose radiation. But the decision can't depend on how the information is presented. By the luck of the draw, the economist might say, too many of the doctors who favor surgery were picked to get form B, and too few of them got form A. After all, there were only 80 people in group A, and 87 in group B. There seems to be a lot of room for chance variation.

To evaluate this argument, a significance test is needed. The difference between the two percentages in table 1 is 34%, and you need to put a standard error on this difference. The method of example 4 can be used. Pretend that you have two independent samples, drawn at random with replacement. The first sample is the doctors who got form A. There were 80 of them, and 40 favored surgery. The SE for the number favoring surgery is

$$\sqrt{80} \times \sqrt{0.50 \times 0.50} \approx 4.5$$

The SE for the percentage favoring surgery is $4.5/80 \times 100\% \approx 5.6\%$. The second sample is the doctors who got form B, and the SE for the percentage favoring surgery in response to form B is 3.9%. The SE for the difference is

$$\sqrt{5.6^2 + 3.9^2} \approx 6.8\%$$

On the null hypothesis, the expected difference between the percentages in the two samples is 0.0%. The observed difference is 34%. The test statistic is

$$z = \frac{\text{observed difference} - \text{expected difference}}{\text{SE for difference}} = \frac{34\% - 0.0\%}{6.8\%} = 5.0$$

Chance is not a good explanation for the results in table 1.

Another look behind the scenes: A box model for this experiment needs a ticket for each doctor. As in the previous section, each ticket shows a pair of numbers [A|B]. The first number on the ticket codes the response to form A. It is 1 if the doctor would favor surgery when presented with form A, and 0 if she would prefer radiation. Similarly, the second number on the ticket codes the response to form B.

Eighty draws are made at random without replacement from the box, and the responses to form A are observed [A|▧]. The responses to form A in the experiment are like this first batch of 80 draws. The 50% in table 1 is like the percentage of 1's in this batch of draws. The 87 tickets left in the box are the second sample. With this second sample, the responses to form B are observed [▨|B]. The responses to form B are like this second batch of 0's and 1's. The 84% in table 1 is like the percentage of 1's in the second sample.

Now the null hypothesis can be set up in terms of the model. Because both forms convey the same information, the economist thinks that a doctor's response to the two forms must be the same, so both numbers on the ticket are the same (figure 2). The box model can be used to show that our method gives a conservative estimate for the SE.[14]

Figure 2. The null hypothesis for the experiment: deciding between radiation and surgery based on form A or form B with the same information. The first number on the ticket codes the response to form A; the second, to form B. Responses favoring surgery are coded "1."

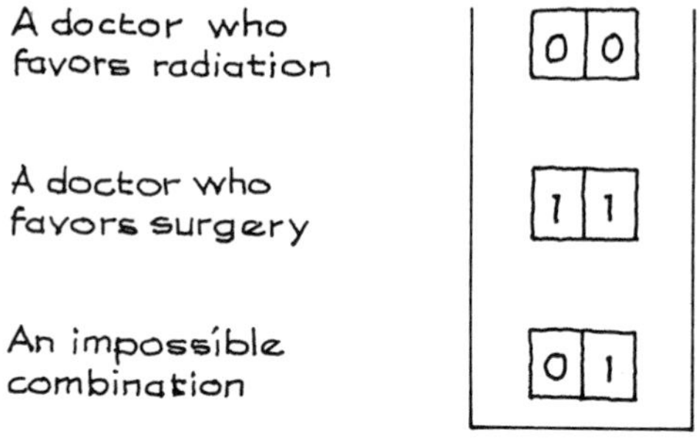

The experimental design used in this study may seem a bit indirect. Why not give both forms of the questionnaire, one after the other, to all the doctors? The reason is simple. Asking both questions at the same time pushes the subjects to be more consistent: perhaps they see that both forms describe the same data.[15]

Exercise Set D

1. The study described in the text was replicated on another group of subjects: MBA students at Harvard and Stanford.
   (a) One MBA student would prefer radiation therapy if presented with form A, but surgery if given form B. Fill in his ticket [A|B].
   (b) Another MBA student has the ticket [1|0]. How would she respond to form A? form B?

(c) Which of the three tickets is consistent with the null hypothesis?

(i) [1 0] (ii) [0 0] (iii) [0 1]

(d) The results came out as follows.

	Form A	*Form B*
Favored surgery	112	84
Favored radiation	84	17

Can the difference in response to the forms be explained by chance? (Hint: to get started, find how many students got form A; of them, what percentage favored radiation? Then do the same for form B.)

2. In the Salk vaccine field trial, 400,000 children were part of a randomized controlled double-blind experiment. Just about half of them were assigned at random to the vaccine group, and the other half to the placebo.[16] In the vaccine group, there were 57 cases of polio, compared to 142 in the placebo group. Is this difference due to chance? If not, what explains it?

3. (a) In the HIP trial (pp. 22–23), there were 39 deaths from breast cancer in the treatment group, and 63 deaths in the control group. Is the difference statistically significant?
   (b) In the treatment group, there were 837 deaths from all causes, compared to 879 in the control group. Is the difference statistically significant?

4. Many observational studies conclude that low-fat diets protect against cancer and cardiovascular "events" (heart attacks, stroke, and so forth). Experimental results, however, are generally negative. In 2006, the Women's Health Initiative (WHI) published its results.[17] This was a large-scale randomized trial on women who had reached menopause. As one part of the study, 48,835 women were randomized: 19,541 were assigned to the treatment group and put on a low-fat diet. The other 29,294 women were assigned to the control group and ate as they normally would. Subjects were followed for 8 years.

   Among other things, the investigators found that 1,357 women on the low-fat diet experienced at least one cardiovascular event, compared to 2,088 in the control group. Can the difference between the two groups be explained by chance? What do you conclude about the effect of the low-fat diet?

5. A geography test was given to a simple random sample of 250 high-school students in a certain large school district. One question involved an outline map of Europe, with the countries identified only by number. The students were asked to pick out Great Britain and France. As it turned out, 65.6% could find France, compared to 70.4% for Great Britain.[18] Is the difference statistically significant? Or can this be determined from the information given?

6. Some years, the Gallup Poll asks respondents how much confidence they have in various American institutions. You may assume that results are based on a simple random sample of 1,000 persons each year; the samples are independent from year to year.[19]
   (a) In 2005, only 41% of the respondents had "a great deal or quite a lot" of confidence in the Supreme Court, compared to 50% in 2000. Is the difference real? Or can you tell from the information given?

(b) In 2005, only 22% of the respondents had "a great deal or quite a lot" of confidence in Congress, whereas 24% of the respondents had "a great deal or quite a lot" of confidence in organized labor. Is the difference between 24% and 22% real? Or can you tell from the information given?

Discuss briefly.

7. Breast-feeding infants for the first few months after their birth is considered to be better for their health than bottle feeding. According to several observational studies, withholding the bottle in hospital nurseries increases the likelihood that mothers will continue to breast-feed after leaving the hospital. As a result, withholding supplementation has been recommended.

A controlled experiment was done by K. Gray-Donald, M. S. Kramer, and associates at the Royal Victoria Hospital in Montreal.[20] There were two nurseries. In the "traditional" nursery, supplemental bottle-feedings were given as usual—at 2 A.M., and whenever the infant seemed hungry. In the experimental nursery, mothers were awakened at 2 A.M. and asked to breast-feed their babies; bottle-feeding was discouraged.

Over the four-month period of the experiment, 393 mothers and their infants were assigned at random to the traditional nursery, and 388 to the experimental one. The typical stay in the hospital was 4 days, and there was followup for 9 weeks after release from the hospital.

(a) At the end of 9 weeks, 54.7% of the mothers who had been assigned to the traditional nursery were still breast-feeding their infants, compared to 54.1% in the experimental nursery. Is this difference statistically significant? What do you conclude?

(b) It was really up to the mothers whether to breast-feed or bottle-feed. Were their decisions changed by the treatments? To answer that question, the investigators looked at the amounts of bottle-feeding in the two nurseries, expressed as milliliters per day (ml/day). In the traditional nursery, this averaged 36.6 ml/day per infant, and the SD was 44.3. In the experimental nursery, the figures were 15.7 and 43.6. What do you conclude?

(c) Did the different treatments in the two nurseries affect the infants in any way? To answer that question, the investigators looked at the weight lost by each infant during the stay, expressed as a percentage of birth weight. In the traditional nursery, this averaged 5.1% and the SD was 2.0%. In the experimental nursery, the average was 6.0% and the SD was 2.0%. What do you conclude? (It may be surprising, but most newborns lose a bit of weight during the first few days of life.)

(d) Was the randomization successful? To find out, the investigators looked at the birth weights themselves (among other variables). In the traditional nursery, these averaged 3,486 grams and the SD was 438 grams. In the experimental nursery, the average was 3,459 grams and the SD was 434 grams. What do you conclude?

The answers to these exercises are on pp. A97–98.

5. WHEN DOES THE z-TEST APPLY?

The square root law in section 1 was designed for use with two independent simple random samples. Example 1 in section 1 illustrates this application. So do the NAEP results in section 2. The procedure can also be used with a randomized controlled experiment, where each subject has two possible responses but only one is observed. The investigators see the response to treatment for the subjects who are randomly selected into the treatment group. They see the other response for subjects in the control group. Sections 3 and 4 (vitamin C and rational decision making) illustrate this application, which involves a minor miracle—two mistakes that cancel.

You are not expected to derive the formulas, but you should learn when to use them and when not to. The formulas should not be used when two correlated responses are observed for each subject. Exercise 5 on p. 515 (the geography test) is an example of when not to use the formulas. Each subject makes two responses, by answering (i) the question on Great Britain, and (ii) the question on France. Both responses are observed, because each subject answers both questions. And the responses are correlated, because a geography whiz is likely to be able to answer both questions correctly, while someone who does not pay attention to maps is likely to get both of them wrong. By contrast, if you took two independent samples—asking one group about France and the other about Great Britain—the formula would be fine. (That would be an inefficient way to do the study.)

Exercise 9 on p. 508 is another case when you should not use the formulas. This is a bit subtle, because the data were collected in a randomized controlled experiment—but you get two correlated responses for each of the 59 pairs of animals. By contrast, if 59 of the 118 rats had been selected at random and put into treatment, while the remaining 59 were used as controls, our formulas would be fine. (Again, the design used by the investigators turns out to be more efficient.)

> The z-test (sections 1 and 2) applies to two independent samples. Generally, the formulas give the wrong answer when applied to dependent samples. There is an exception: the z-test can be used to compare the treatment and control groups in a randomized controlled experiment—even though the groups are dependent (sections 3 and 4).

The square root law in section 1 gives the wrong answer with dependent samples because it does not take the dependence into account. Other formulas are beyond our scope. However, it is easy to do the z-test on the differences, as in exercise 12 on pp. 498–499.[21] Also see exercise 6 on pp. 258–259, exercise 11 on pp. 262–263, exercise 15 on p. 329, or exercise 11 on p. 488, which all use a technique called "the sign test."

6. REVIEW EXERCISES

Review exercises may cover material from previous chapters.

1. Five hundred draws are made at random with replacement from a box of numbered tickets; 276 are positive. Someone tells you that 50% of the tickets in the box show positive numbers. Do you believe it? Answer yes or no, and explain.

2. One hundred draws are made at random with replacement from box A, and 250 are made at random with replacement from box B.

   (a) 50 of the draws from box A are positive, compared to 131 from box B: 50.0% versus 52.4%. Is this difference real, or due to chance?
   (b) The draws from box A average 1.4 and their SD is 15.3; the draws from box B average 6.3 and their SD is 16.1. Is the difference between the averages statistically significant?

3. The Gallup poll asks respondents how they would rate the honesty and ethical standards of people in different fields—very high, high, average, low, or very low.[22] The percentage who rated clergy "very high or high" dropped from 60% in 2000 to 54% in 2005. This may have been due to scandals involving sex abuse; or it may have been a chance variation. (You may assume that in each year, the results are based on independent simple random samples of 1,000 persons in each year.)

   (a) Should you make a one-sample z-test or a two-sample z-test? Why?
   (b) Formulate the null and alternative hypotheses in terms of a box model. Do you need one box or two? Why? How many tickets go into each box? How many draws? What do the tickets show? What do the null and alternative hypotheses say about the box(es)?
   (c) Can the difference between 60% and 54% be explained as a chance variation? Or was it the scandals? Or something else?

4. This continues exercise 3. In 2005, 65% of the respondents gave medical doctors a rating of "very high or high," compared to a 67% rating for druggists. Is the difference real, or a chance variation? Or do you need more information to decide? If the difference is real, how would you explain it? Discuss briefly. You may assume that the results are based on a simple random sample of 1,000 persons taken in 2005; each respondent rated clergy, medical doctors, druggists, and many other professions.[23]

5. One experiment involved 383 students at the University of British Columbia. 200 were chosen at random to get item A, and 92 of them answered "yes." The other 183 got item B, and 161 out of the second group answered "yes."[24]

   Item A) Imagine that you have decided to see a play and paid the admission price of $20 per ticket. As you enter the theatre, you discover that you have lost the ticket. The seat was not marked, and the ticket cannot be recovered. Would you pay $20 for another ticket?

Item B) Imagine that you have decided to see a play where admission is $20 per ticket. As you enter the theatre, you discover that you have lost a $20 bill. Would you still pay $20 for a ticket for the play? [In Canada, "theatre" is the right spelling.]

From the standpoint of economic theory, both items present the same facts and call for the same answer; any difference between them must be due to chance. From a psychological point of view, the framing of the question can be expected to influence the answer. What do the data say?

6. An experiment is performed to see whether calculators help students do word problems.[25] The subjects are a group of 500 thirteen-year-olds in a certain school district. All the subjects work the problem below. Half of them are chosen at random and allowed to use calculators; the others do the problem with pencil and paper. In the calculator group, 18 students get the right answer; in the pencil-and-paper group, 59 do. Can this difference be explained by chance? What do you conclude?

   The problem. An army bus holds 36 soldiers. If 1,128 soldiers are being bussed to their training site, how many buses are needed?

   Note. 1,128/36 = 31.33, so 32 buses are needed. However, 31.33 was a common answer, especially in the calculator group; 31 was another common answer.

7. When convicts are released from prison, they have no money, and there is a high rate of "recidivism:" the released prisoners return to crime and are arrested again. Would providing income support to ex-convicts during the first months after their release from prison reduce recidivism? The Department of Labor ran a randomized controlled experiment to find out.[26] The experiment was done on a selected group of convicts being released from certain prisons in Texas and Georgia. Income support was provided, like unemployment

insurance. There was a control group which received no payment, and four different treatment groups (differing slightly in the amounts paid).

The exercise is on the results for Georgia, and combines the four treatment groups into one. Assume that prisoners were randomized to treatment or control.

(a) 592 prisoners were assigned to the treatment group, and of them 48.3% were rearrested within a year of release. 154 were assigned to the control group, and of them 49.4% were rearrested within a year of release. Did income support reduce recidivism? Answer yes or no, and explain briefly.

(b) In the first year after their release from prison, those assigned to the treatment group averaged 16.8 weeks of paid work; the SD was 15.9 weeks. For those assigned to the control group, the average was 24.3 weeks; the SD was 17.3 weeks. Did income support reduce the amount that the ex-convicts worked? Answer yes or no, and explain briefly.

8. One experiment contrasted responses to "prediction-request" and to "request-only" treatments, in order to answer two research questions.[27]

(i) Can people predict how well they will behave?

(ii) Do their predictions influence their behavior?

In the prediction-request group, subjects were first asked to predict whether they would agree to do some volunteer work. Then they were requested to do the work. In the request-only group, the subjects were requested to do the work; they were not asked to make predictions beforehand. In parts (a-b-c), a two-sample z-test may or may not be legitimate. If it is legitimate, make it. If not, why not?

(a) 46 residents of Bloomington, Indiana were chosen at random for the "prediction-request" treatment. They were called and asked to predict "whether they would agree to spend 3 hours collecting for the American Cancer Society if contacted over the telephone with such a request." 22 out of the 46 said that they would. Another 46 residents of that town were chosen at random for the "request-only" treatment. They were requested to spend the 3 hours collecting for the American Cancer Society. Only 2 out of 46 agreed to do it. Can the difference between 22/46 and 2/46 be due to chance? What do the data say about the research questions (i) and (ii)?

(b) Three days later, the prediction-request group was called again, and requested to spend 3 hours collecting for the American Cancer Society: 14 out of 46 agreed to do so. Can the difference between 14/46 and 2/46 be due to chance? What do the data say about the research questions (i) and (ii)?

(c) Can the difference between 22/46 and 14/46 be due to chance? What do the data say about the research questions (i) and (ii)?

9. A researcher wants to see if the editors of journals in the field of social work are biased. He makes up two versions of an article, "in which an asthmatic child was temporarily separated from its parents in an effort to relieve the symptoms of an illness that is often psychosomatic." In one version, the separation has a positive effect; in another, negative.[28] The article is submitted to a group of 107 journals; 53 are chosen at random to get the positive version, and 54 get the negative one. The results are as follows:

	Positive	*Negative*
Accept	28	8
Reject	25	46

The first column of the table says that 28 of the journals getting the positive version accepted it for publication, and 25 rejected it. The second column gives the results for the journals that got the negative version. Is chance a good explanation for the results? If not, what can be concluded about journal publication policy?

10. An investigator wants to show that first-born children score higher on IQ tests than second-borns. He takes a simple random sample of 400 two-child families in a school district, both children being enrolled in elementary school. He gives these children the WISC vocabulary test (described in exercise 7 on pp. 507–508), with the following results.

- The 400 first-borns average 29 and their SD is 10.
- The 400 second-borns average 28 and their SD is 10.

(Scores are corrected for age differences.) He makes a two-sample z-test:

SE for first-born average ≈ 0.5

SE for second-born average ≈ 0.5

SE for difference $= \sqrt{0.5^2 + 0.5^2} \approx 0.7$

$z = 1/0.7 \approx 1.4, \quad P \approx 8\%$

Comment briefly on the use of statistical tests.

11. (Hard.) The logic of the two-sample z-test in section 27.2 relies on two mathematical facts: (i) the expected value of a difference equals the difference of the expected values, and (ii) the expected value of the sample average equals the population average. Explain briefly, with reference to the NAEP reading scores.

7. SUMMARY

1. The expected value for the difference of two quantities equals the difference of the expected values. (Independence is not required here.)

2. The standard error for the difference of two independent quantities is $\sqrt{a^2 + b^2}$, where

- a is the SE for the first quantity;
- b is the SE for the second quantity.

For dependent quantities, this formula is usually wrong.

3. Suppose that two independent and reasonably large simple random samples are taken from two separate boxes. The null hypothesis is about the difference between the averages of the two boxes. The appropriate test statistic is

$$z = \frac{\text{observed difference} - \text{expected difference}}{\text{SE for difference}}$$

In the formula, the "difference" is between the averages of the two samples. (If the null hypothesis says that the two boxes have the same average, the expected difference between the sample averages is 0.)

4. Tests based on this statistic are called *two-sample z-tests*.

5. The two-sample z-test can handle situations which involve classifying and counting, by putting 0's and 1's in the boxes.

6. The two-sample z-test can also be used to compare treatment and control averages or rates in an experiment. Suppose there is a box of tickets. Each ticket has two numbers: one shows what the response would be to treatment A; the other, to treatment B. For each ticket, only one of the two numbers can be observed. Some tickets are drawn at random without replacement from the box, and the responses to treatment A are observed. Then, a second sample is drawn at random without replacement from the remaining tickets. In the second sample, the responses to treatment B are observed. The SE for the difference between the two sample averages can be conservatively estimated as follows:

(i) compute the SEs for the averages as if drawing with replacement;
(ii) combine the SEs as if the two samples were independent.

28

The Chi-Square Test

Don't ask what it means, but rather how it is used.
—L. WITTGENSTEIN (1889–1951)

1. INTRODUCTION

How well does it fit the facts? Sooner or later, this question must be asked about any chance model. And in many cases, it can be settled by the χ^2-*test* (invented in 1900 by Karl Pearson).[1] χ is a Greek letter, often written as "chi," read like the "ki" in kite, so χ^2 is read as "ki-square." Section 5 of chapter 26 explained how to test a chance model for a parapsychology experiment. There, each guess was classified into one of two categories—right or wrong. According to the model, a guess had 1 chance in 4 to be right, so the number of correct guesses was like the sum of draws from the box

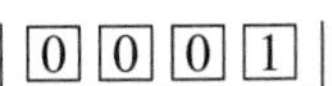

In that case, the z-test was appropriate, but only two categories were involved. If there are more than two categories, statisticians use the χ^2-test rather than the z-test. For instance, you might want to see if a die is fair. Each throw can be classified into one of 6 categories:

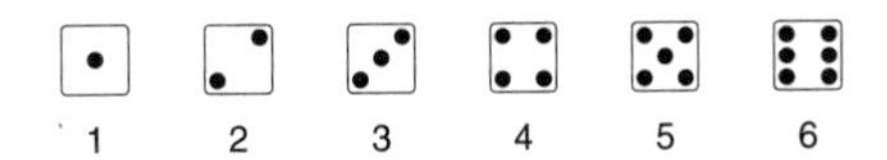

The χ^2-test will help to check whether these categories are equally likely, as in the next example.

Example 1. A gambler is accused of using a loaded die, but he pleads innocent. A record has been kept of the last 60 throws (table 1). There is disagreement about how to interpret the data and a statistician is called in.

Table 1. Sixty rolls of a die, which may be loaded.

4	3	3	1	2	3	4	6	5	6
2	4	1	3	3	5	3	4	3	4
3	3	4	5	4	5	6	4	5	1
6	4	4	2	3	3	2	4	4	5
6	3	6	2	4	6	4	6	3	2
5	4	6	3	3	3	5	3	1	4

Discussion. If the gambler is innocent, the numbers in table 1 are like the results of drawing 60 times (at random with replacement) from the box

| [1] [2] [3] [4] [5] [6] |

According to this model, each number should turn up about 10 times: the *expected frequency* is 10. To find out how the data compare with expectations, you have to count and see how many times each number did in fact turn up. The *observed frequencies* are shown in table 2. A check on the arithmetic: the sum of each frequency column must be 60, the total number of entries in table 1. ("Frequency" is statistical jargon for the number of times something happens.)

Table 2. Observed and expected frequencies for the data in table 1.

Value	*Observed frequency*	*Expected frequency*
1	4	10
2	6	10
3	17	10
4	16	10
5	8	10
6	9	10
sum	60	60

As the table indicates, there are too many 3's. The SE for the number of 3's is $\sqrt{60} \times \sqrt{1/6 \times 5/6} \approx 2.9$, so the observed number is about 2.4 SEs above the expected number. But don't shoot the gambler yet. The statistician won't advise taking the table one line at a time.

- Several lines in the table may look suspicious. For example, in table 2 there are also too many 4's.
- On the other hand, with many lines in the table, there is high probability that at least one of them will look suspicious—even if the die is fair. It's like playing Russian roulette. If you keep on going, sooner or later you're going to lose.

For each line of the table, there is a difference between observed and expected frequencies. The idea is to combine all these differences into one overall measure of the distance between the observed and expected values. What χ^2 does is to square each difference, divide by the corresponding expected frequency, and take the sum:

$$\chi^2 = \text{sum of } \frac{(\text{observed frequency} - \text{expected frequency})^2}{\text{expected frequency}}$$

There is one term for each line in the table. At first sight, the formula may seem quite arbitrary. However, every statistician uses it because of one very convenient feature, which will be pointed out later.

With the data in table 2, the χ^2-statistic is

$$\frac{(4-10)^2}{10} + \frac{(6-10)^2}{10} + \frac{(17-10)^2}{10} + \frac{(16-10)^2}{10} + \frac{(8-10)^2}{10} + \frac{(9-10)^2}{10} = \frac{142}{10} = 14.2$$

When the observed frequency is far from the expected frequency, the corresponding term in the sum is large; when the two are close, this term is small. Large values of χ^2 indicate that observed and expected frequencies are far apart. Small values of χ^2 mean the opposite: observeds are close to expecteds. So χ^2 does give a measure of the distance between observed and expected frequencies.[2]

Of course, even if the data in table 1 had been generated by rolling a fair die 60 times, χ^2 could have turned out to be 14.2, or more—the chance variation defense. Is this plausible? To find out, we need to know the chance that when a fair die is rolled 60 times and χ^2 is computed from the observed frequencies, its value turns out to be 14.2 or more.

Why "or more"? The observed value 14.2 may be evidence against the model because it is too big, meaning that the observed frequencies are too far from the expected frequencies. If so, values larger than 14.2 would be even stronger evidence against the model. What is the chance that the model will produce such strong evidence against itself? To find out, we calculate the chance of getting a χ^2-statistic of 14.2 or more.

Calculating this chance looks like a big job, but the computer does it in a flash, and the answer is 1.4%. If the die is fair, there is only a 1.4% chance for it to produce a χ^2-statistic as big as (or bigger than) the observed one. At this point, the statistician has finished. Things do not look good for the gambler.

The 1.4% is called "the observed significance level" and denoted by P, as in chapter 26. In Pearson's time, there were no computers to find the chances. So he developed a method for approximating P by hand. This method involved a new curve, called the χ^2-*curve*. More precisely, there is one curve for each number of *degrees of freedom*.[3] The curves for 5 and 10 degrees of freedom are shown in figure 1.

Figure 1. The χ^2-curves for 5 and 10 degrees of freedom. The curves have long right-hand tails. As the degrees of freedom go up, the curves flatten out and move off to the right. (The solid curve is for 5 degrees of freedom; dashed, for 10.)

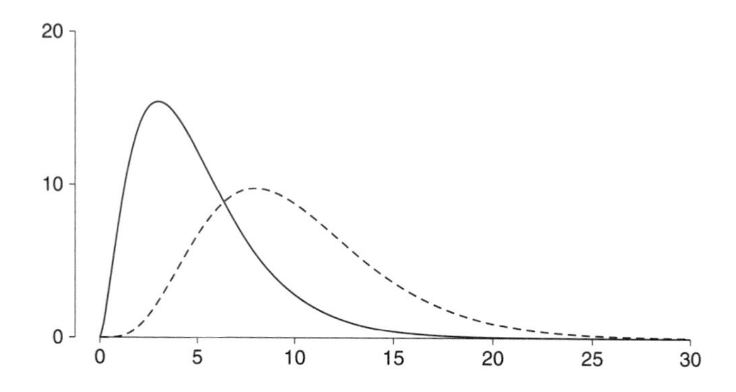

Sometimes, it is hard to work out the degrees of freedom. However, in example 1, the model was *fully specified*. There were no parameters to estimate from the data, because the model told you what was in the box. When the model is fully specified, computing the degrees of freedom is easy:

$$\text{degrees of freedom} = \text{number of terms in } \chi^2 - \text{one}.$$

In example 1, there are $6 - 1 = 5$ degrees of freedom. Why? In table 2, the six observed frequencies have to add up to 60. If you know any five of them, you can compute the sixth. Only five of the frequencies can vary freely. (Compare section 6 of chapter 26.)

> For the χ^2-test, P is approximately equal to the area to the right of the observed value for the χ^2-statistic, under the χ^2-curve with the appropriate number of degrees of freedom. When the model is fully specified (no parameters to estimate),
>
> degrees of freedom = number of terms in χ^2 − one.

For example 1,

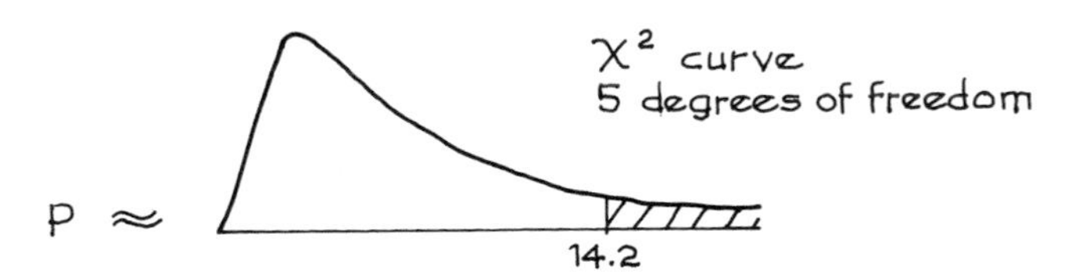

This area can be found using tables or a statistical calculator. In principle, there is one table for each curve but this would be so awkward that a different arrangement is used, as shown in table 3 (extracted from a bigger one on p. A106). Areas, in percent, are listed across the top of the table; degrees of freedom are listed down the left side. For instance, look at the column for 5% and the row for 5 degrees of freedom. In the body of the table there is the entry 11.07, meaning that the area to the right of 11.07 under the curve for 5 degrees of freedom is 5%. The area to the right of 14.2 under the curve for 5 degrees of freedom cannot be

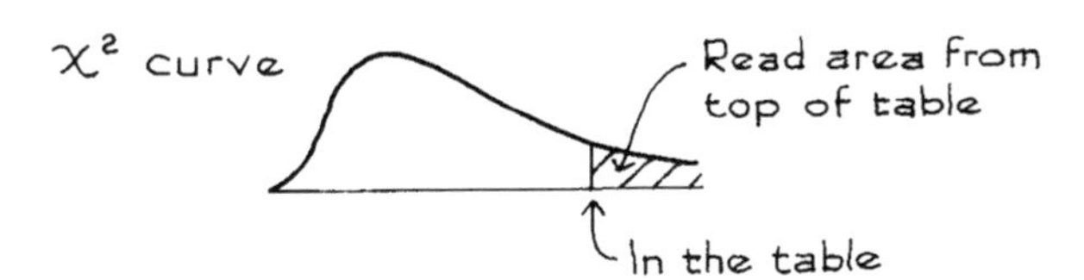

Table 3. A short χ^2 table extracted from the bigger one on p. A106.

Degrees of freedom	*90%*	*50%*	*10%*	*5%*	*1%*
1	0.016	0.46	2.71	3.84	6.64
2	0.21	1.39	4.60	5.99	9.21
3	0.58	2.37	6.25	7.82	11.34
4	1.06	3.36	7.78	9.49	13.28
5	1.61	4.35	9.24	11.07	15.09
6	2.20	5.35	10.65	12.59	16.81
7	2.83	6.35	12.02	14.07	18.48
8	3.49	7.34	13.36	15.51	20.09
9	4.17	8.34	14.68	16.92	21.67
10	4.86	9.34	15.99	18.31	23.21

read from the table, but it is between 5% (the area to the right of 11.07) and 1% (the area to the right of 15.09). It is reasonable to guess that the area under the curve to the right of 14.2 is just a bit more than 1%.

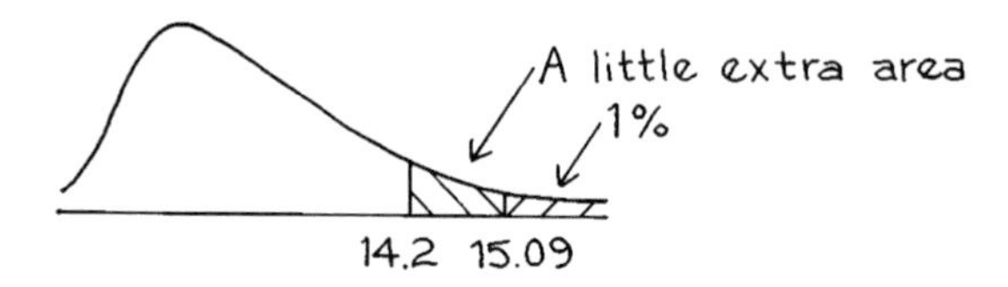

Pearson developed the formulas for the χ^2-statistic and the χ^2-curves in tandem. His objective was to approximate the P-values without having to do a computation that was—by the standards of his time—quite formidable. How good is his approximation? Figure 2 shows the probability histogram for the χ^2-statistic with 60 rolls of a fair die. A χ^2-curve with 5 degrees of freedom is plotted too.

Figure 2. Pearson's approximation. The top panel shows the probability histogram for the χ^2-statistic with 60 rolls of a fair die, compared with a χ^2-curve (5 degrees of freedom). The bottom panel shows the ratio of tail areas. For example, take 14.2 on the horizontal axis. The area under the histogram to the right of 14.2 is 1.4382%. The area under the curve is 1.4388%. The ratio $1.4382/1.4388 \approx 0.9996$ is plotted above 14.2. Other ratios are plotted the same way.

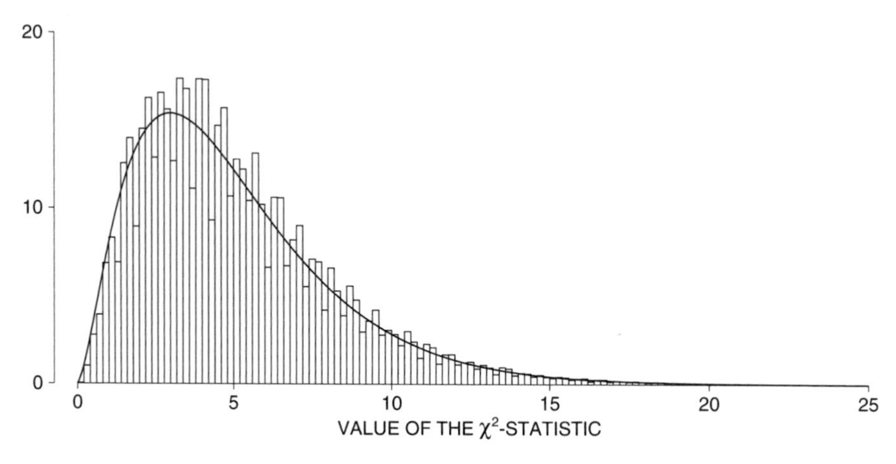

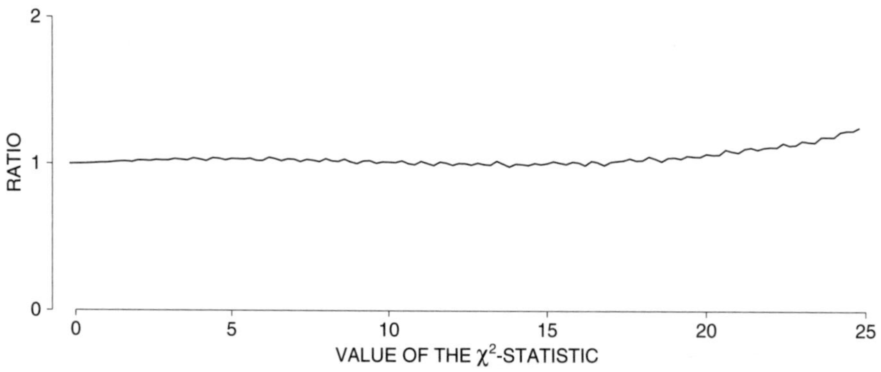

The histogram is quite a bit bumpier than the curve, but follows it rather well. The area under the histogram to the right of any particular value is going to be close to the corresponding area under the curve. The ratio of these tail areas is graphed in the bottom panel.

In example 1, the area to the right of 14.2 under the histogram gives the exact value of P. This is 1.4382%. The area to the right of 14.2 under the curve gives Pearson's approximate value for P. This is 1.4388%. Not bad. When the number of rolls goes up, the approximation gets better, and the histogram gets less bumpy.[4]

As a rule of thumb, the approximation will be good when the expected frequency in each line of the table is 5 or more. In table 2, each expected frequency was 10, and the approximation was excellent. On the other hand, the approximation would not be so good for 100 draws from the box

$$\boxed{1}\ \boxed{2}\ \boxed{3}\ \ 96\ \boxed{4}\text{'s}$$

In this case, the expected number of $\boxed{1}$'s is only 1; similarly for $\boxed{2}$ and $\boxed{3}$. The expected numbers are too small for the approximation to be reliable.

When should the χ^2-test be used, as opposed to the z-test? If it matters how many tickets of each kind are in the box, use the χ^2-test. If it is only the average of the box that matters, use the z-test. For instance, suppose you are drawing with replacement from a box of tickets numbered 1 through 6; the percentages of the different kinds of tickets are unknown. To test the hypothesis that each value appears on $16\frac{2}{3}\%$ of the tickets, use the χ^2-test. Basically, there is only one box which satisfies this hypothesis:

$$\boxed{1}\ \boxed{2}\ \boxed{3}\ \boxed{4}\ \boxed{5}\ \boxed{6}$$

On the other hand, to test the hypothesis that the average of the box is 3.5, use the z-test. Of course, there are many boxes besides $\boxed{1}\ \boxed{2}\ \boxed{3}\ \boxed{4}\ \boxed{5}\ \boxed{6}$ where the average is 3.5: for instance,

$$\boxed{1}\ \boxed{2}\ \boxed{3}\ \boxed{3}\ \boxed{4}\ \boxed{4}\ \boxed{5}\ \boxed{6} \quad \text{or} \quad \boxed{1}\ \boxed{1}\ \boxed{2}\ \boxed{3}\ \boxed{4}\ \boxed{5}\ \boxed{6}\ \boxed{6}$$

To sum up:

- The χ^2-test says whether the data are like the result of drawing at random from a box whose contents are given.
- The z-test says whether the data are like the result of drawing at random from a box whose average is given.[5]

The balance of this section tells how χ^2 was used on a wheel of fortune.[6] Some winners in the California State Lottery are chosen to appear on a television game show called "The Big Spin." Each contestant spins a heavy cast aluminum wheel, with 100 slots numbered from 1 through 100. A hard rubber ball bounces around inside the wheel and then settles down into one slot or another, determining the prize given to the contestant.

Millions of dollars are at stake, so the wheel has to be tested quite carefully. The State Lottery Commission's statistical consultant Don Ylvisaker had the operators spin the wheel 800 times and count the number of times the ball landed in each slot. Then he made a χ^2-test of the observed frequencies against the expected frequencies. The χ^2-statistic turned out to be 119. There were $100 - 1 = 99$ degrees of freedom, and $P \approx 8\%$. This seemed marginal.

Slot number 69 came up most often and 19 least often. These two numbers were opposite each other. The wheel was then examined more carefully. A metal weight was found on the back, attached to the rim near slot number 69. Apparently, this had been done to balance the wheel, just as you would balance an automobile tire. The weight was removed, the wheel was rebalanced, and the tests were run again. The first 400 numbers did not look especially random, but things improved from there. As it turned out, the operators had oiled the wheel around spin 400 because it squeaked. The wheel was accepted and works well. (It is oiled regularly.)

"NO! YOU MAY NOT OIL THE WHEEL."

2. THE STRUCTURE OF THE χ^2-TEST

Section 1 described the χ^2-test. What are the ingredients?

(a) The basic data. This consists of some number of observations, usually denoted N. For the die, N was 60 and the basic data were in table 1. For the wheel of fortune, N was 800. The basic data were the 800 numbers generated on the trial spins.

(b) The chance model. Only one kind of chance model has been considered so far in this chapter. There is a box of tickets, whose contents are given. Draws are made at random with replacement from the box. According to the model, the data are like the draws. For the die, the box was

| [1] [2] [3] [4] [5] [6] |

For the wheel of fortune, the box had 100 tickets, numbered from 1 through 100.

(c) The frequency table. For each value, the observed frequency is obtained from the basic data by counting.[7] The expected frequency is obtained from N and the chance model. Table 2 reported the observed and expected frequencies for the die. A frequency table for the wheel would have 100 rows; it is omitted.

(d) The χ^2-statistic. This is computed from the formula. For the die, the χ^2-statistic was 14.2; for the wheel, the χ^2-statistic was 119.

(e) The degrees of freedom. This is one less than the number of terms in the sum for χ^2 (when the contents of the box are specified by the model). For the die, there were 5 degrees of freedom; for the wheel of fortune, there were 99. The degrees of freedom are computed from the model, not from the data.

(f) The observed significance level. This is approximated by the area to the right of the χ^2-statistic, under the χ^2-curve with the appropriate number of degrees of freedom. For the die, $P \approx 1.4\%$; for the wheel, $P \approx 8\%$.

The terminology is complicated because everything starts with "χ^2."

- The χ^2-test involves steps (a–f).
- The χ^2-statistic is calculated from the data each time you make the test.
- Two χ^2-curves are shown in figure 1.
- The χ^2-table is based on the curves and is used to look up P-values.

Whatever is in the box, the same χ^2-curves and tables can be used to approximate P, provided N is large enough. That is what motivated the formula. With other test statistics, a new curve would be needed for every box.

Exercise Set A

1. Find the area under the χ^2-curve with 5 degrees of freedom to the right of
   (a) 1.61 (b) 9.24 (c) 15.09
2. Find the area to the right of 15.09 under the χ^2-curve with 10 degrees of freedom.
3. Suppose the observed frequencies in table 2 had come out as shown in table 4A below. Compute the value of χ^2, the degrees of freedom, and P. What can be inferred?

Table 4A

Value	*Observed frequency*
1	5
2	7
3	17
4	16
5	8
6	7

Table 4B

Value	*Observed frequency*
1	9
2	11
3	10
4	8
5	12
6	10

Table 4C

Value	*Observed frequency*
1	90
2	110
3	100
4	80
5	120
6	100

Table 4D

Value	*Observed frequency*
1	10,287
2	10,056
3	9,708
4	10,080
5	9,935
6	9,934

4. Suppose the observed frequencies in table 1 had come out as shown in table 4B. Make a χ^2-test of the null hypothesis that the die is fair.

5. Suppose that table 1 had 600 entries instead of 60, with observed frequencies as shown in table 4C. Make a χ^2-test of the null hypothesis that the die is fair.

6. Suppose that table 1 had 60,000 entries, with the observed frequencies as shown in table 4D.

   (a) Compute the percentage of times each value showed up.
   (b) Does the die look fair?
   (c) Make a χ^2-test of the null hypothesis that the die is fair.

7. One study of grand juries in Alameda County, California, compared the demographic characteristics of jurors with the general population, to see if the jury panels were representative.[8] The results for age are shown below. The investigators wanted to know whether these 66 jurors were selected at random from the population of Alameda County. (Only persons 21 and over are considered; the county age distribution is known from Public Health Department data.)

   (a) True or false: to answer the investigators' question, you should make a z-test on each line in the table.
   (b) Fill in the blank: the ______-test combines information from all the lines in the table into an overall measure of the difference between the observed frequencies and expected frequencies. Options: z, χ^2.
   (c) True or false: the right-hand column in the table gives the observed frequencies.
   (d) Fill in the blank: to make the χ^2-test, you need to compute the ______ frequency in each age group. Options: expected, observed.
   (e) Now answer the investigators' question.

Age	*County-wide percentage*	*Number of jurors*
21 to 40	42	5
41 to 50	23	9
51 to 60	16	19
61 and over	19	33
Total	100	66

8. Someone tells you to work exercise 7 as follows. (i) Convert each number to a percent: for instance, 5 out of 66 is 7.6%. (ii) Take the difference between the observed and expected percent. (iii) Square the difference. (iv) Divide by the expected percent. (v) Add up to get χ^2. Is this right?

9. Another device tested by the California State Lottery has a set of 10 Ping-Pong balls, numbered from 0 through 9. These balls are mixed in a glass bowl by an air jet, and one is forced out at random. In the trial runs described below, the mixing machine seemed to be working well, but some of the ball sets may not have been behaving themselves. On each run, the machine made 120 draws from the bowl, with replacement.

   (a) Suppose everything is going as it should. In 120 draws from the bowl, each ball is expected to be drawn ______ times.

(b) The table below shows the results of testing 4 sets of balls. Sets A and D seemed marginal and were retested. Set B was rejected outright. Set C was accepted. How do these decisions follow from the data? (The table is read as follows: with ball set A, ball no. 0 was drawn 13 times; ball no. 1 was drawn 11 times; and so forth.)

(c) After retesting, what would you do with sets A and D? Explain briefly.

FREQUENCIES

Ball no.	*Ball set A* test	*Ball set A* retest	*Ball set B* test	*Ball set C* test	*Ball set D* test	*Ball set D* retest
0	13	19	22	12	16	8
1	11	9	8	10	7	15
2	16	10	7	14	12	22
3	11	12	8	10	14	11
4	5	7	19	11	15	15
5	12	15	20	10	5	8
6	12	19	10	20	10	17
7	19	10	11	12	21	9
8	5	12	6	12	11	8
9	16	7	9	9	9	7

10. (a) A statistician wants to test the null hypothesis that his data are like 100 draws made at random with replacement from the box | [1] [2] [3] [4] [5] [6] |. The alternative hypothesis: the data are like 100 draws made at random with replacement from the box | [1] [1] [2] [3] [4] [5] [6] [6] |. Can the χ^2-test do the job?

(b) As in (a), but the boxes are

| [1] [2] [3] [4] [5] [6] |

| [1] [2] [3] [4] [5] [6] [1] [2] [3] [4] [5] [6] |

The answers to these exercises are on pp. A99–100.

3. HOW FISHER USED THE χ^2-TEST

Fisher used the χ^2-statistic to show that Mendel's data (chapter 25) were fudged.[9] For each of Mendel's experiments, Fisher computed the χ^2-statistic. These experiments were all independent, for they involved different sets of plants. Fisher *pooled* the results.

> With independent experiments, the results can be pooled by adding up the separate χ^2-statistics; the degrees of freedom add up too.

For instance, if one experiment gives $\chi^2 = 5.8$ with 5 degrees of freedom, and another independent experiment gives $\chi^2 = 3.1$ with 2 degrees of freedom, the two together have a pooled χ^2 of $5.8 + 3.1 = 8.9$, with $5 + 2 = 7$ degrees of freedom. For Mendel's data, Fisher got a pooled χ^2-value under 42, with 84 degrees of freedom. The area to the left of 42 under the χ^2-curve with 84 degrees of freedom is about 4 in 100,000. The agreement between observed and expected is too good to be true.

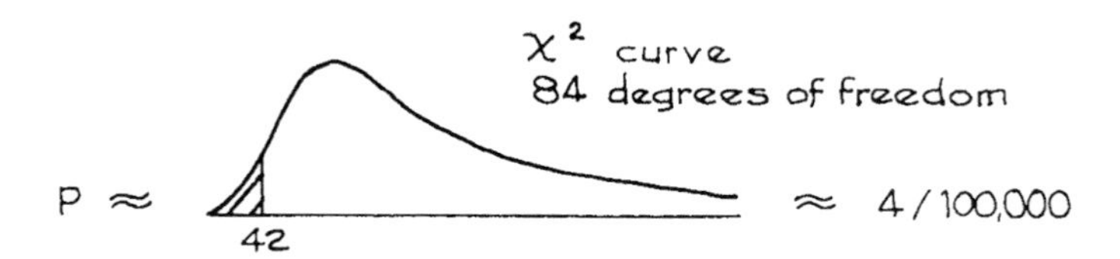

At this point, a new principle seems to be involved: P was computed as a left-hand tail area, not a right-hand one. Why?

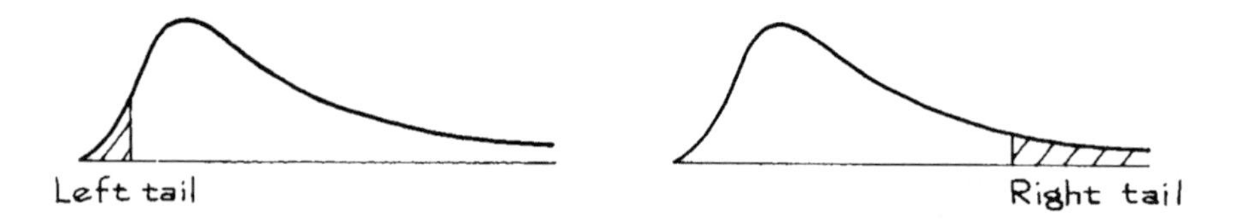

Here is the reason. Fisher was not testing Mendel's chance model; he took that for granted. Instead, he was comparing two hypotheses—

- The null hypothesis: Mendel's data were gathered honestly.
- The alternative hypothesis: Mendel's data were fudged to make the reported frequencies closer to the expected ones.

Small values of χ^2 say the observed frequencies are closer to the expected ones than chance variation would allow, and argue for the alternative hypothesis. Since it is small values of χ^2 that argue against the null hypothesis, P must be computed as a left-hand tail area. It is straightforward to set up the null hypothesis as a box model (chapter 25). The alternative hypothesis would be more complicated.

Exercise Set B

1. Suppose the same die had been used to generate the data in tables 4A and 4C (p. 531), rolling it first 60 times for table 4A, and then 600 times for table 4C. Can you pool the results of the two tests? If so, how?
2. Suppose the same die had been used to generate the data in tables 4A and 4C (p. 531), rolling it 600 times in all. The first 60 rolls were used for table 4A; but table 4C reports the results on all 600 rolls. Can you pool the results of the two tests? If so, how?

3. One of Mendel's breeding trials came out as follows.[9] Make a χ^2-test to see whether these data were fudged. Which way does the evidence point? Is it decisive?

Type of pea	*Observed number*	*Expected number*
Smooth yellow	315	313
Wrinkled yellow	101	104
Smooth green	108	104
Wrinkled green	32	35

The answers to these exercises are on p. A100.

4. TESTING INDEPENDENCE

The χ^2-statistic is also used to test for independence, as will be explained in this section. The method will be indicated by example: Are handedness and sex independent? More precisely, take people age 25–34 in the U.S. The question is whether the distribution of "handedness" (right-handed, left-handed, ambidextrous) among the men in this population differs from the distribution among the women.

If data were available, showing for each man and woman in the population whether they were right-handed, left-handed, or ambidextrous, it would be possible to settle the issue directly, just by computing percentages. Such information is not available. However, HANES (p. 58) took a probability sample of 2,237 Americans 25–34. One of the things they determined for each sample person was handedness. Results are shown in table 5.

Table 5. Handedness by sex.

	Men	*Women*
Right-handed	934	1,070
Left-handed	113	92
Ambidextrous	20	8

This is a "3 × 2 table," because it has 3 rows and 2 columns. In general, when studying the relationship between two variables, of which one has m values and the other has n values, an $m \times n$ table is needed. In Table 5, it is hard to compare the distribution of handedness for men and women, because there are more women than men. Table 6 converts the data to percentages.

Table 6. Handedness by sex.

	Men	*Women*
Right-handed	87.5%	91.5%
Left-handed	10.6%	7.9%
Ambidextrous	1.9%	0.7%

As you can see, the distributions are different. The women are a bit likelier than men to be right-handed; they are less likely to be left-handed or ambidextrous. According to some neurophysiologists, right-handedness is associated with left-hemisphere dominance in the brain, the rational faculty ruling the emotional.[10] Does the sample show that women are more rational than men? Another interpretation: right-handedness is socially approved, left-handedness is socially deviant. Are women under greater pressure than men to follow the social norm for handedness?

A less dramatic interpretation: it's just chance. Even if handedness is distributed the same way for men and women in the population, the distributions could be different in the sample. Just by the luck of the draw, there could be too few right-handed men in the HANES sample, or too many right-handed women. To decide whether the observed difference is real or just due to chance, a statistical test is needed. That is where the χ^2-test comes in.

The HANES sampling design is too complicated to analyze by means of the χ^2-test. (This issue came up in sampling, where the formula for the SE depended on the design; pp. 388, 403, 424.) To illustrate technique, we are going to pretend that table 5 is based on a simple random sample, with 2,237 people chosen at random without replacement from the population. A box model for the data can be set up on that basis. There is one ticket in the box for each person in the population (Americans age 25–34). Each of these millions of tickets is marked in one of the following ways:

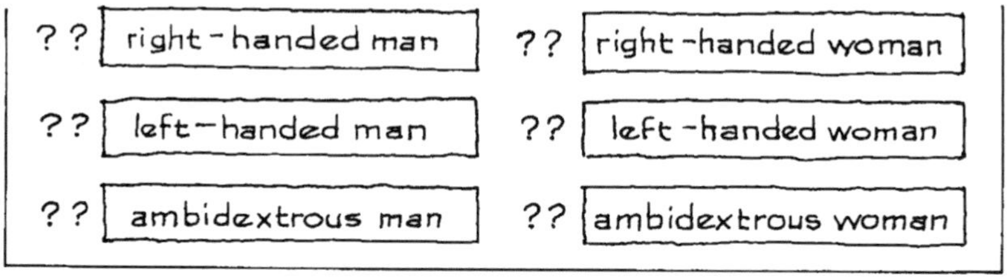

?? right-handed man ?? right-handed woman

?? left-handed man ?? left-handed woman

?? ambidextrous man ?? ambidextrous woman

Millions of tickets

Our model says that the numbers in table 5 were generated by drawing 2,237 tickets at random without replacement from this huge box, and counting to see how many tickets there were for each of the 6 different types. The percentage composition of the box is unknown, so there are 6 parameters in the model.

Now we can formulate the null hypothesis and the alternative in terms of the box. The null hypothesis says that handedness and sex are independent. More explicitly, the percentage of right-handers among all men in the population equals the corresponding percentage among women; similarly for left-handers and the ambidextrous. On the null hypothesis, the differences in the sample percentages (table 6) just reflect chance variation. The alternative hypothesis is dependence—in the population, the distribution of handedness among the men differs from the distribution for women. On the alternative hypothesis, the differences in the sample reflect differences in the population.

To make a χ^2-test of the null hypothesis, we have to compare the observed frequencies (table 5) with the expected frequencies. Getting these expected frequencies is a bit complicated, and the technique is explained later. The expected frequencies themselves are shown in table 7. They are based on the null hypothesis of independence.

Table 7. Observed and expected frequencies.

	Observed		*Expected*	
	Men	*Women*	*Men*	*Women*
Right-handed	934	1,070	956	1,048
Left-handed	113	92	98	107
Ambidextrous	20	8	13	15

The next step is to compute

$$\begin{aligned}\chi^2 &= \text{sum of } \frac{(\text{observed frequency} - \text{expected frequency})^2}{\text{expected frequency}} \\ &= \frac{(934-956)^2}{956} + \frac{(1{,}070-1{,}048)^2}{1{,}048} \\ &\quad + \frac{(113-98)^2}{98} + \frac{(92-107)^2}{107} \\ &\quad + \frac{(20-13)^2}{13} + \frac{(8-15)^2}{15} \\ &\approx 12\end{aligned}$$

How many degrees of freedom are there?

> When testing independence in an $m \times n$ table (with no other constraints on the probabilities), there are $(m-1) \times (n-1)$ degrees of freedom.

There are 6 terms in the sum for χ^2, but there are only $(3-1) \times (2-1) = 2$ degrees of freedom. To see why, look at the differences.

−22	22
15	−15
7	−7

(The arithmetic for the first one: $934 - 956 = -22$, see table 7.) The differences add up to 0, horizontally and vertically. So, if you know the −22 and the 15, say, you can compute all the rest: only 2 of the differences are free to vary.

Now that we have the χ^2-statistic and its degrees of freedom, P can be worked out on the computer (or looked up in a table):

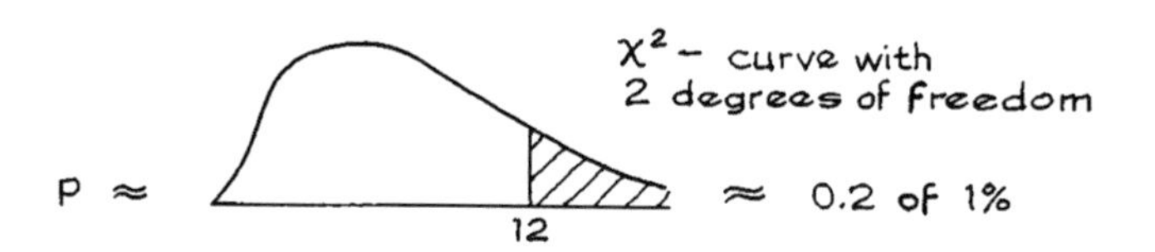

The observed significance level P is the area to the right of 12 under the χ^2-curve with 2 degrees of freedom, and this is about 0.2 of 1%. (The table will only tell you that the area is quite a bit less than 1%, which is good enough for present purposes.) The null hypothesis should be rejected. There is strong evidence to show that the distribution of handedness among the men in the population is different from the distribution for women. The observed difference in the sample seems to reflect a real difference in the population, rather than chance variation. That is what the χ^2-test says. (A more careful analysis would have to take the design of the sample into account, but the conclusion stays the same.[11])

What is left is to compute the expected frequencies in table 7, and this will take some effort. To get started, you compute the row and column totals for table 5, as shown in table 8.

Table 8. Row and column totals.

	Men	*Women*	*Total*
Right-handed	934	1,070	2,004
Left-handed	113	92	205
Ambidextrous	20	8	28
Total	1,067	1,170	2,237

How do you get the 956 in table 7? From table 8, the percentage of right-handers in the sample is

$$\frac{2{,}004}{2{,}237} \times 100\% \approx 89.6\%$$

The number of men is 1,067. If handedness and sex are independent, the number of right-handed men in the sample should be

$$89.6\% \text{ of } 1{,}067 \approx 956.$$

The other expected frequencies in table 7 can be worked out the same way.

Expected values ought to be computed directly from the box model. In table 7, however, the "expected frequencies" are estimated from the sample—and the null hypothesis of independence. "Estimated expected frequencies" would be a better phrase, but "expected frequencies" is what statisticians say.[12]

Exercise Set C

1. The percentage of women in the sample (table 8) is $1,170/2,237 \approx 52.3\%$. Someone wants to work out the expected number of ambidextrous women as 52.3% of 28. Is that OK?

2. (Hypothetical.) In a certain town, there are about one million eligible voters. A simple random sample of size 10,000 was chosen, to study the relationship between sex and participation in the last election. The results:

	Men	*Women*
Voted	2,792	3,591
Didn't vote	1,486	2,131

Make a χ^2-test of the null hypothesis that sex and voting are independent.

The next few exercises will help you learn which test to use when.

3. The table below shows the distribution of marital status by sex for persons age 25–34 in Wyoming.[13]

Question: Are the distributions really different for men and women?

You may assume the data are from a simple random sample of 299 persons, of whom 143 were men and 156 were women. To answer the question, you use—

(i) the one-sample z-test.
(ii) the two-sample z-test.
(iii) the χ^2-test, with a null hypothesis that tells you the contents of the box (section 1).
(iv) the χ^2-test for independence (section 4).

Now answer the question. If the distributions are different, who are the women marrying?

	Men	*Women*
Never married	31.5%	19.2%
Married	60.1%	67.3%
Widowed, divorced, separated	8.4%	13.5%

4. Suppose all the numbers in exercise 3 had come from the Current Population Survey for March 2005, by extracting the data for people age 25–34 in Wyoming. Would that affect your answers? Explain briefly.

5. A study is made of incomes among full-time workers age 25–54 in a certain town. A simple random sample is taken, of 250 people with high school degrees: the sample average income is \$30,000 and the SD is \$25,000. Another simple random sample is taken, of 250 people with college degrees: the sample average income is \$50,000 and the SD is \$40,000.

Question: Is the difference in averages real, or due to chance?

To answer this question, you use—

(i) the one-sample z-test.
(ii) the two-sample z-test.
(iii) the χ^2-test, with a null hypothesis that tells you the contents of the box (section 1).
(iv) the χ^2-test for independence (section 4).

Now answer the question.

6. Demographers think that about 55% of newborns are male. In a certain hospital, 568 out of 1,000 consecutive births are male.

   Question: Are the data consistent with the theory?

   To answer this question, you use—

   (i) the one-sample z-test.
   (ii) the two-sample z-test.
   (iii) the χ^2-test, with a null hypothesis that tells you the contents of the box (section 1).
   (iv) the χ^2-test for independence (section 4).

   Now answer the question.

7. To test whether a die is fair, someone rolls it 600 times. On each roll, he just records whether the result was even or odd, and large (4, 5, 6) or small (1, 2, 3). The observed frequencies turn out as follows:

	Large	*Small*
Even	183	113
Odd	88	216

   Question: Is the die fair?

   To answer this question, you use—

   (i) the one-sample z-test.
   (ii) the two-sample z-test.
   (iii) the χ^2-test, with a null hypothesis that tells you the contents of the box (section 1).
   (iv) the χ^2-test for independence (section 4).

   Now answer the question.

The answers to these exercises are on pp. A100–101.

5. REVIEW EXERCISES

Review exercises may cover material from previous chapters.

1. You are drawing 100 times at random with replacement from a box. Fill in the blanks, using the options below.

   (a) To test the null hypothesis that the average of the box is 2, you would use ______.
   (b) To test the null hypothesis that the box is [1] [2] [3], you would use ______.

Options (some may not be used):

(i) the one-sample z-test.
(ii) the two-sample z-test.
(iii) the χ^2-test, with a null hypothesis that tells you the contents of the box (section 1).
(iv) the χ^2-test for independence (section 4).

2. As part of a study on the selection of grand juries in Alameda county, the educational level of grand jurors was compared with the county distribution:[14]

Educational level	*County*	*Number of jurors*
Elementary	28.4%	1
Secondary	48.5%	10
Some college	11.9%	16
College degree	11.2%	35
Total	100.0%	62

Could a simple random sample of 62 people from the county show a distribution of educational level so different from the county-wide one? Choose one option and explain.

(i) This is absolutely impossible.
(ii) This is possible, but fantastically unlikely.
(iii) This is possible but unlikely—the chance is around 1% or so.
(iv) This is quite possible—the chance is around 10% or so.
(v) This is nearly certain.

3. Each respondent in the Current Population Survey of March 2005 was classified as employed, unemployed, or outside the labor force. The results for men in California age 35–44 can be cross-tabulated by marital status, as follows:[15]

	Married	*Widowed, divorced, or separated*	*Never married*
Employed	790	98	209
Unemployed	56	11	27
Not in labor force	21	7	13

Men of different marital status seem to have different distributions of labor force status. Or is this just chance variation? (You may assume the data come from a simple random sample.)

4. (a) Does the histogram in figure 2 represent data, or chance?
(b) There is a block over the interval from 5 to 5.2. What does the area of this block represent? (Ranges include the left endpoint, but not the right.)
(c) Which chance is larger for 60 throws of a die? Or can this be determined from figure 2?
(i) The chance that the χ^2-statistic is in the range from 4.8 to 5.0.
(ii) The chance that the χ^2-statistic is in the range from 5.0 to 5.2.

(d) If $\chi^2 = 10$, then P is about

1%   10%   25%   50%   cannot be determined from the figure

5. An investigator makes a χ^2-test, to see whether the observed frequencies are too far from the expected frequencies.
   (a) If $\chi^2 = 15$, the P-value will be bigger with ______ degrees of freedom than with ______ degrees of freedom. Options: 5, 10.
   (b) If there are 10 degrees of freedom, the P-value will be bigger with $\chi^2 =$ ______ than with $\chi^2 =$ ______. Options: 15, 20.

   No calculations are needed, just look at figure 1.

6. Someone claims to be rolling a pair of fair dice. To test his claim, you make him roll the dice 360 times, and you count up the number of times each sum appears. The results are shown below. (For your convenience, the chance of throwing each sum with a pair of fair dice is shown too.) Should you play craps with this individual? Or are the observed frequencies too close to the expected frequencies?

Sum	*Chance*	*Frequency*
2	1/36	11
3	2/36	18
4	3/36	33
5	4/36	41
6	5/36	47
7	6/36	61
8	5/36	52
9	4/36	43
10	3/36	29
11	2/36	17
12	1/36	8

7. The International Rice Research Institute in the Philippines develops new lines of rice which combine high yields with resistance to disease and insects. The technique involves crossing different lines to get a new line which has the most advantageous combination of genes. Detailed genetic modeling is required. One project involved breeding new lines to resist the "brown plant hopper" (an insect): 374 lines were raised, with the results shown below.[16]

	Number of lines
All plants resistant	97
Mixed: some plants resistant, some susceptible	184
All plants susceptible	93

   According to the IRRI model, the lines are independent: each line has a 25% chance to be resistant, a 50% chance to be mixed, and a 25% chance to be susceptible. Are the facts consistent with this model?

8. Two people are trying to decide whether a die is fair. They roll it 100 times, with the results shown at the top of the next page. One person wants to make

a z-test, the other wants to make a χ^2-test. Who is right? Explain briefly.

21 ⚀'s 15 ⚁'s 13 ⚂'s 17 ⚃'s 19 ⚄'s 15 ⚅'s

Average of numbers rolled $\approx$ 3.43, SD $\approx$ 1.76

9. Each respondent in the Current Population Survey of March 2005 can be classified by age and marital status. The table below shows results for women age 20–29 in Montana.

   Question A. Women of different ages seem to have different distributions of marital status. Or is this just chance variation?

   Question B. If the difference is real, what accounts for it?

   (a) Can you answer these questions with the information given? If so, answer them. If not, why not?
   (b) Can you answer these questions if the data in the table resulted from a simple random sample of women age 20–29 in Montana? If so, answer them. If not, why not?

	A G E	
	20–24	*25–29*
Never married	46	21
Married	17	32
Widowed, divorced, separated	1	6

10. The U.S. has bilateral extradition treaties with many countries. (A person charged with a crime in his home country may escape to the U.S.; if he is captured in the U.S., authorities in his home country may request that he be "extradited," that is, turned over for prosecution under their laws.)

    The Senate attached a special rider to the treaty governing extradition to Northern Ireland: fugitives cannot be returned if they will be discriminated against on the basis of religion. In a leading case, the defense tried to establish discrimination in Northern Ireland's criminal justice system.

    One argument was based on 1991 acquittal rates for persons charged with terrorist offenses.[17] According to a defense expert, these rates were significantly different for Protestants and Catholics: $\chi^2 \approx 6.2$ on 1 degree of freedom, $P \approx 1\%$. The data are shown below: 8 Protestants out of 15 were acquitted, compared to 27 Catholics out of 65.

    (a) Is the calculation of χ^2 correct? If not, can you guess what the mistake was? (That might be quite difficult.)
    (b) What box model did the defense have in mind? Comment briefly on the model.

	Protestant	*Catholic*
Acquitted	8	27
Convicted	7	38

6. SUMMARY

1. The χ^2-statistic can be used to test the hypothesis that data were generated according to a particular chance model.

2. $$\chi^2 = \text{sum of } \frac{(\text{observed frequency} - \text{expected frequency})^2}{\text{expected frequency}}$$

3. When the model is fully specified (no parameters to estimate from the data),

$$\text{degrees of freedom} = \text{number of terms} - \text{one}.$$

4. The observed significance level P can be approximated as the area under the χ^2-curve to the right of the observed value for χ^2. The significance level gives the chance of the model producing observed frequencies as far from the expected frequencies as those at hand, or even further, distance being measured by χ^2.

5. Sometimes the model can be taken as true, and the problem is to decide whether the data have been fudged to make the observed frequencies closer to the expected ones. Then P would be computed as the area to the left of the observed value for χ^2.

6. If experiments are independent, the χ^2-statistics can be pooled by addition. The degrees of freedom are just added too.

7. The χ^2-statistic can also be used to test for independence. This is legitimate when the data have been obtained from a simple random sample, and an inference about the population is wanted. With an $m \times n$ table (and no extra constraints on the probabilities) there are $(m - 1) \times (n - 1)$ degrees of freedom.

29

A Closer Look at Tests of Significance

One of the misfortunes of the law [is that] ideas become encysted in phrases and thereafter for a long time cease to provoke further analysis.

—OLIVER WENDELL HOLMES, JR. (UNITED STATES, 1841–1935)[1]

1. WAS THE RESULT SIGNIFICANT?

How small does P have to get before you reject the null hypothesis? As reported in section 4 of chapter 26, many investigators draw lines at 5% and 1%. If P is less than 5%, the result is "statistically significant," and the "null hypothesis is rejected at the 5% level." If P is less than 1%, the result is "highly significant." However, the question is almost like asking how cold it has to get before you are entitled to say, "It's cold." A temperature of 70°F is balmy, −20°F is cold indeed, and there is no sharp dividing line. Logically, it is the same with testing. There is no sharp dividing line between probable and improbable results.

A P-value of 5.1% means just about the same thing as 4.9%. However, these two P-values can be treated quite differently, because many journals will only publish results which are "statistically significant"—the 5% line. Some of the more prestigious journals will only publish results which are "highly significant"—the 1% line.[2] These arbitrary lines are taken so seriously that many investigators only report their results as "statistically significant" or "highly significant." They don't even bother telling you the value of P, let alone what test they used.

Investigators should summarize the data, say what test was used, and report the P-value instead of just comparing P to 5% or 1%.

Historical note. Where do the 5% and 1% lines come from? To find out, we have to look at the way statistical tables are laid out. The t-table is a good example (section 6 of chapter 26). Part of it is reproduced below as table 1.

Table 1. A short t-table.

Degrees of freedom	*10%*	*5%*	*1%*
1	3.08	6.31	31.82
2	1.89	2.92	6.96
3	1.64	2.35	4.54
4	1.53	2.13	3.75
5	1.48	2.02	3.36

How is this table used in testing? Suppose investigators are making a t-test with 3 degrees of freedom. They are using the 5% line, and want to know how big the t-statistic has to be in order to achieve "statistical significance"—a P-value below 5%. The table is laid out to make this easy. They look across the row for 3 degrees of freedom and down the column for 5%, finding the entry 2.35 in the body of the table. The area to the right of 2.35 under the curve for 3 degrees of freedom is 5%. So the result is "statistically significant" as soon as t is more than 2.35. In other words, the table gives the cutoff for "statistical significance." Similarly, it gives the cutoff for the 1% line, or for any other significance level listed across the top.

R. A. Fisher was one of the first to publish such tables, and it seems to have been his idea to lay them out that way. There is a limited amount of room on the page. Once the number of levels was limited, 5% and 1% stood out as nice round numbers, and they soon acquired a magical life of their own. With computers everywhere, this kind of table is almost obsolete. So are the 5% and 1% levels.[3]

Exercise Set A

1. True or false, and explain:

   (a) If $P = 1.1\%$, the result is "significant" but not "highly significant."
   (b) If $P = 0.9$ of 1%, the result is "highly significant."

2. True or false, and explain:

   (a) The P-value of a test is the chance that the null hypothesis is true.

(b) If a result is statistically significant, there are only 5 chances in 100 for it to be due to chance, and 95 chances in 100 for it to be real.

The answers to these exercises are on p. A101.

2. DATA SNOOPING

The point of testing is to help distinguish between real differences and chance variation. People sometimes jump to the conclusion that a result which is statistically significant cannot be explained as chance variation. This is false. Once in a while, the average of the draws will be 2 SEs above the average of the box, just by chance. More specifically, even if the null hypothesis is right, there is a 5% chance of getting a difference which the test will call "statistically significant." This 5% chance could happen to you—an unlikely event, but not impossible. Similarly, on the null hypothesis, there is 1% chance to get a difference which is highly significant but just a fluke.

Put another way, an investigator who makes 100 tests can expect to get five results which are "statistically significant" and one which is "highly significant" even if the null hypothesis is right in every case—so that each difference is just due to chance. (See exercise 5 on p. 483.) You cannot determine, for sure, whether a difference is real or just coincidence.

To make bad enough worse, investigators often decide which hypotheses to test only after they have seen the data. Statisticians call this *data snooping*. Investigators really ought to say how many tests they ran before statistically significant differences turned up. And to cut down the chance of being fooled by "statistically significant" flukes, they ought to test their conclusions on an independent batch of data—for instance by replicating the experiment.[4] This good advice is seldom followed.

Data-snooping makes P-values hard to interpret.

Example 1. Clusters. Liver cancer is a rare disease, often thought to be caused by environmental agents. The chance of having 2 or more cases in a given year in a town with 10,000 inhabitants is small—perhaps 1/2 of 1%. A cluster of liver cancer cases (several cases in a small community) prompts a search for causes, like the contamination of the water supply by synthetic chemicals.[5]

Discussion. With (say) 100 towns of this size and a 10-year time period, it is likely that several clusters will turn up, just by chance. There are $100 \times 10 = 1{,}000$ combinations of towns and years; and $0.005 \times 1{,}000 = 5$. If you keep on testing null hypotheses, sooner or later you will get significant differences.

One form of data snooping is looking to see whether your sample average is too big or too small—before you make the statistical test. To guard against this kind of data snooping, many statisticians recommend using *two-tailed* rather than *one-tailed* tests. The point is easiest to see in a hypothetical example. Someone wants to test whether a coin is fair: does it land heads with probability 50%? The coin is tossed 100 times, and it lands heads on 61 of the tosses. If the coin is fair, the expected number of heads is 50, so the difference between 61 and 50 just represents chance variation. To test this null hypothesis, a box model is needed. The model consists of 100 draws from the box

$$\boxed{?? \ \boxed{0}\text{'s} \ \ ?? \ \boxed{1}\text{'s}} \qquad 0 = \text{tails}, \quad 1 = \text{heads}.$$

The fraction of 1's in this box is an unknown parameter, representing the probability of heads. The null hypothesis says that the fraction of 1's in the box is 1/2. The test statistic is

$$z = \frac{\text{observed} - \text{expected}}{\text{SE}} = \frac{61 - 50}{5} = 2.2$$

One investigator might formulate the alternative hypothesis that the coin is biased toward heads: in other words, that the fraction of 1's in the box is bigger than 1/2. On this basis, large positive values of z favor the alternative hypothesis, but negative values of z do not. Therefore, values of z bigger than 2.2 favor the alternative hypothesis even more than the observed value does.

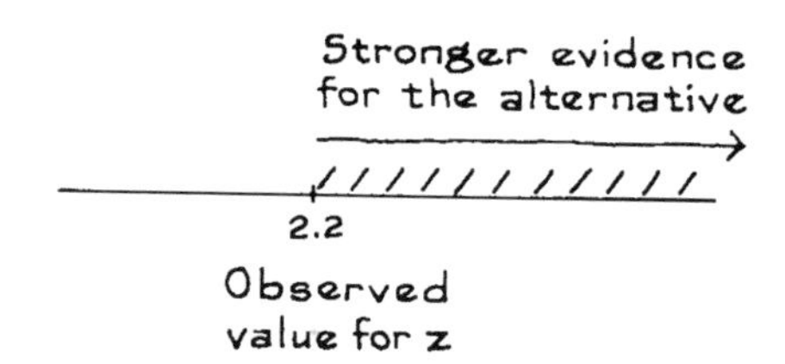

So P is figured as the area to the right of 2.2 under the normal curve:

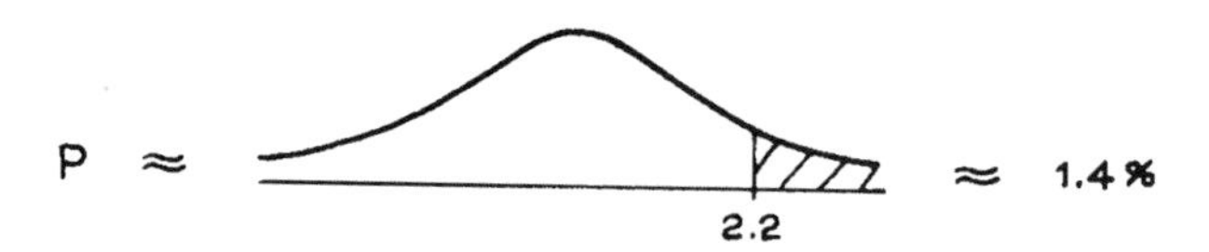

Another investigator might formulate a different alternative hypothesis: that the probability of heads differs from 50%, in either direction. In other words, the fraction of 1's in the box differs from 1/2, and may be bigger or smaller. On this basis, large positive values of z favor the alternative, but so do large negative values. If the number of heads is 2.2 SEs above the expected value of 50, that is bad for the null hypothesis. And if the number of heads is 2.2 SEs below the expected value, that is just as bad. The z-values more extreme than the observed 2.2 are:

- 2.2 or more

or

- −2.2 or less.

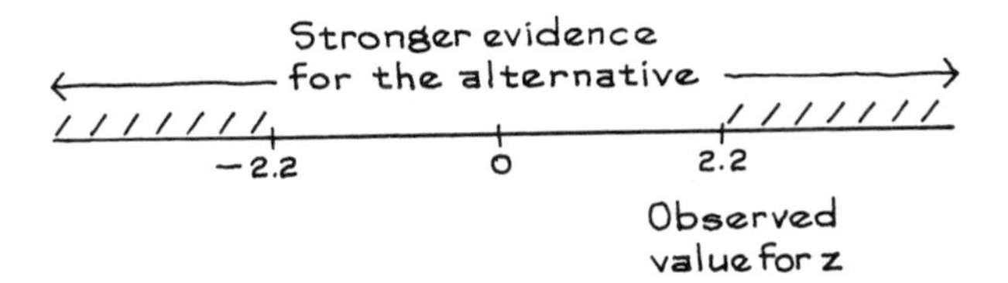

Now P is figured differently:

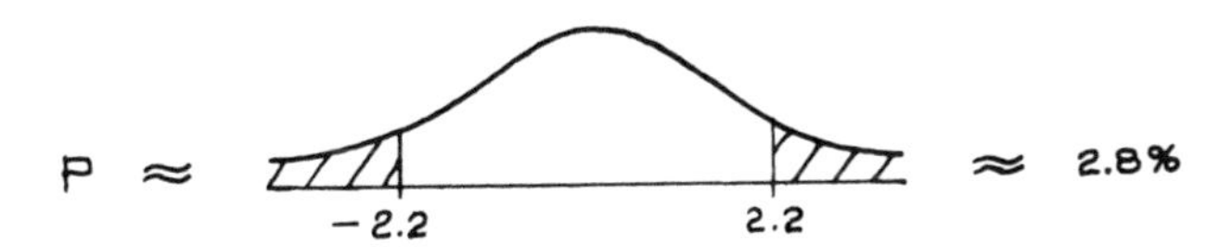

The first way of figuring P is the *one-tailed z-test*; the second is *two-tailed.* Which should be used? That depends on the precise form of the alternative hypothesis. It is a matter of seeing which z-values argue more strongly for the alternative hypothesis than the one computed from the data. The one-tailed test is appropriate when the alternative hypothesis says that the average of the box is bigger than a given value. The two-tailed test is appropriate when the alternative hypothesis says that the average of the box differs from the given value—bigger or smaller.

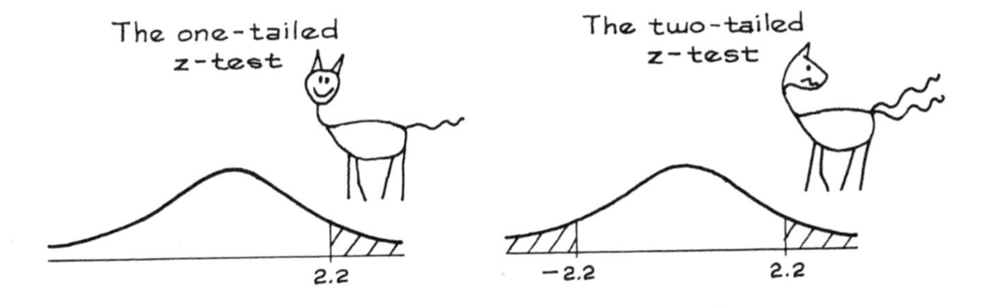

In principle, it doesn't matter very much whether investigators make one-tailed or two-tailed tests, as long as they say what they did. For instance, if they made a one-tailed test, and you think it should have been two-tailed, just double the P-value.[6] To see why such a fuss is made over this issue, suppose a group of investigators makes a two-tailed z-test. They get $z = 1.85$, so $P \approx 6\%$.

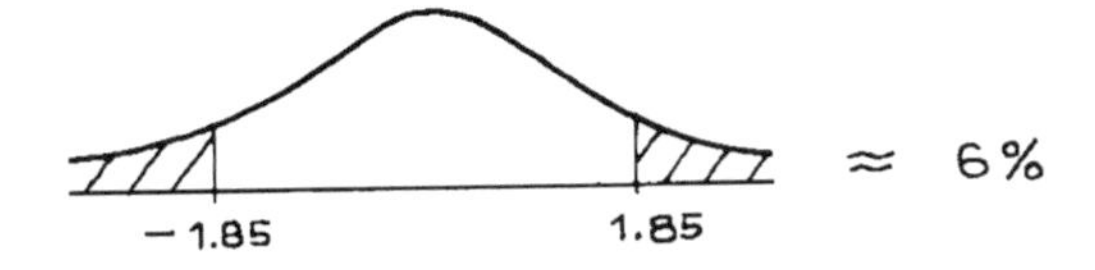

Naturally, they want to publish. But as it stands, most journals won't touch the report—the result is not "statistically significant."

What can they do? They could refine the experimental technique, gather more data, use sharper analytical methods. This is hard. The other possibility is simpler: do a one-tailed test. It is the arbitrary lines at 5% and 1% which make the distinction between two-tailed and one-tailed tests loom so large.

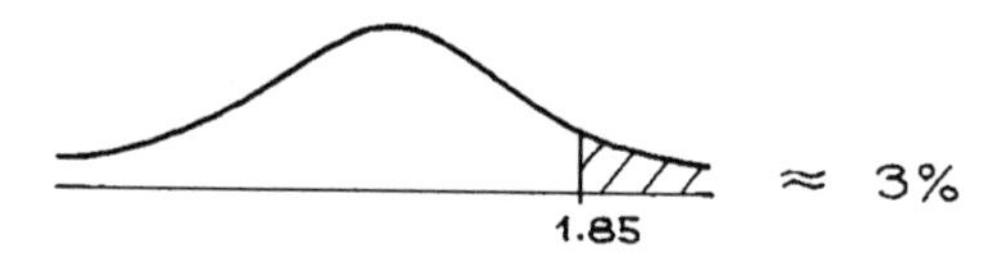

Example 2. Cholesterol. A randomized controlled double-blind experiment was performed to demonstrate the efficacy of a drug called "cholestyramine" in reducing blood cholesterol levels and preventing heart attacks. There were 3,806 subjects, who were all middle-aged men at high risk of heart attack; 1,906 were chosen at random for the treatment group and the remaining 1,900 were assigned to the control group. The subjects were followed for 7 years. The drug did reduce the cholesterol level in the treatment group (by about 8%). Furthermore, there were 155 heart attacks in the treatment group, and 187 in the control group: 8.1% versus 9.8%, $z \approx -1.8$, $P \approx 3.5\%$ (one-tailed). This was called "strong evidence" that cholesterol helps cause heart attacks.[7]

Discussion. With a two-tailed test, $P \approx 7\%$ and the difference is not significant. (The article was published in the *Journal of the American Medical Association*, whose editors are quite strict about the 5% line.) The investigators are overstating their results, and the emphasis on "statistical significance" encourages them to do so.

Exercise Set B

1. One hundred investigators each set out to test a different null hypothesis. Unknown to them, all the null hypotheses happen to be true. Investigator #1 gets a P-value of 58%, plotted in the graph below as the point (1, 58). Investigator #2 gets a P-value of 42%, plotted as (2, 42). And so forth. The 5%-line is shown.

   (a) How many investigators should get a statistically significant result?
   (b) How many do?
   (c) How many should get a result which is highly significant?

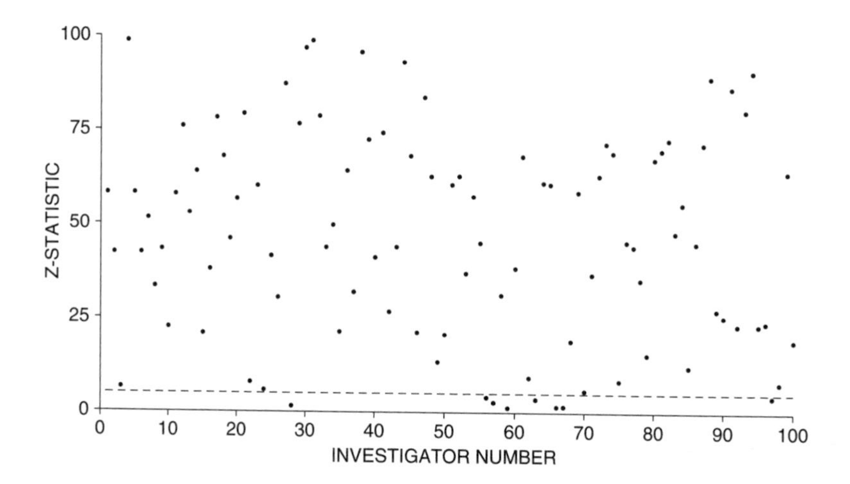

2. In "Ganzfeld" experiments on ESP, there are two subjects, a sender and a receiver, located in separate rooms.[8] There is a standard set of patterns, arranged in 25 sets of 4. The experimenter goes through the 25 sets in order. From each set, one pattern is chosen at random, and shown to the sender (but not to the receiver). The sender tries to convey a mental image of the pattern to the receiver. The receiver is shown the 4 patterns, and ranks them from 1 = most likely to 4 = least likely. After going through all 25 sets of patterns, the experimenter makes a statistical analysis to see if the receiver did better than the chance level. Three test statistics are used.

   - The number of "hits." A receiver scores a hit by assigning rank 1 to the pattern that was in fact chosen. The number of hits ranges from 0 to 25. (If the number of hits is large, that is evidence for ESP.)
   - The number of "high ranks." A receiver scores a high rank by assigning rank 1 or rank 2 to the pattern that was chosen. The number of high ranks ranges from 0 to 25. (If the number of high ranks is large, that is evidence for ESP.)
   - The sum of the ranks assigned to the 25 chosen patterns. This sum ranges from 25 to 100. (If the sum is small, that is evidence for ESP.)

   Suppose there is no ESP, no cheating, and the choice of patterns is totally random.

   (a) The number of hits is like the sum of ______ draws from [1] [0] [0] [0]. Fill in the blank and explain.

   (b) The number of high ranks is like the sum of 25 draws from the box ______. Fill in the blank, and explain.

   (c) Make a box model for the sum of the ranks.

For use in exercise 3, you are given the following information. Suppose 25 tickets are drawn at random with replacement from the box [1] [2] [3] [4].

- There is about a 3% chance of getting 11 or more tickets marked 1.
- There is about a 5% chance of getting 17 or more tickets marked 1 or 2.
- There is about a 5% chance that the sum of the draws will be 53 or less.

3. (This continues exercise 2.) Suppose there is no ESP, no cheating, and the choice of patterns is totally random.

   (a) One hundred investigators do Ganzfeld experiments. They will publish "significant" evidence for ESP if the number of hits is 11 or more. About how many of them will get significant evidence?

   (b) Repeat, if the definition of significant evidence is changed to "the number of high ranks is 17 or more."

   (c) Repeat, if the definition of significant evidence is changed to "the sum of the ranks is 53 or less."

4. (This continues exercises 2 and 3.) Suppose there is no ESP, no cheating, and the choice of patterns is totally random. One hundred investigators do Ganzfeld experiments. They will decide on a statistical test after seeing the data.

   - If the number of hits is 11 or more, they will base the test on the number of hits.
   - If not, but the number of high ranks is 17 or more, they will base the test on the number of high ranks.
   - If not, but the sum of the ranks is 53 or less, they will base the test on the sum of the ranks.

The number of these investigators who get "significant" evidence of ESP will be ______ 5. Fill in the blank, using one of the options below, and explain briefly.

just about   somewhat more than   somewhat less than

5. New chemicals are screened to see if they cause cancer in lab mice. A "bioassay" can be done with 500 mice: 250 are chosen at random and given the test chemical in their food, the other 250 get a normal lab diet. After 33 months, cancer rates in the two groups are compared, using the two-sample z-test.[9]

   Investigators look at cancer rates in about 25 organs and organ systems—lungs, liver, circulatory system, etc. With one chemical, $z \approx -1.8$ for the lungs, $z \approx 2.4$ for the liver, $z \approx -2.1$ for leukemia, and there are another 22 values of z that range from -1.6 to $+1.5$. The investigators conclude that the chemical causes liver cancer ($z \approx 2.4$, $P \approx 1\%$, one-tailed). Comment briefly.

6. One hundred draws are made at random from box X. The average of the draws is 51.8, and their SD is 9. The null hypothesis says that the average of the box equals 50, while the alternative hypothesis says that the average of the box differs from 50. Is a one-tailed or a two-tailed z-test more appropriate?

7. One hundred draws are made at random from box Y. The average of the draws is 51.8, and their SD is 9. The null hypothesis says that the average of the box equals 50, while the alternative hypothesis says that the average of the box is bigger than 50. Is a one-tailed or a two-tailed z-test more appropriate?

8. An investigator has independent samples from box A and from box B. Her null hypothesis says that the two boxes have the same average. She looks at the difference

   average of sample from A − average of sample from B.

   The two-sample z-test gives $z \approx 1.79$. Is the difference statistically significant—

   (a) if the alternative hypothesis says that the average of box A is bigger than the average of box B?
   (b) if the alternative hypothesis says that the average of box A is smaller than the average of box B?
   (c) if the alternative hypothesis says that the average of box A is different from the average of box B?

9. (Hard.) Transfusion of contaminated blood creates a risk of infection. (AIDS is a case in point.) A physician must balance the gain from the transfusion against the risk, and accurate data are important. In a survey of the published medical literature on serum hepatitis resulting from transfusions, Chalmers and associates found that the larger studies had lower fatality rates.[10] How can this be explained?

The answers to these exercises are on pp. A101–102.

3. WAS THE RESULT IMPORTANT?

If a difference is statistically significant, then it is hard to explain away as a chance variation. But in this technical phrase, "significant" does not mean "important." Statistical significance and practical significance are two different ideas.[11]

The point is easiest to understand in the context of a hypothetical example (based on exercise 7, pp. 507–508). Suppose that investigators want to compare WISC vocabulary scores for big-city and rural children, age 6 to 9. They take a simple random sample of 2,500 big-city children, and an independent simple random sample of 2,500 rural children. The big-city children average 26 on the test, and their SD is 10 points. The rural children only average 25, with the same SD of 10 points. What does this one-point difference mean? To find out, the investigators make a two-sample z-test. The SE for the difference can be estimated as 0.3, so

$$z \approx 1/0.3 \approx 3.3, \qquad P \approx 5/10{,}000.$$

The difference between big-city children and rural children is highly significant, rural children are lagging behind in the development of language skills, and the investigators launch a crusade to pour money into rural schools.

The commonsense reaction must be, slow down. The z-test is only telling us that the one-point difference between the sample averages is almost impossible to explain as a chance variation. To focus the issue, suppose that the samples are a perfect image of the population, so that all the big-city children in the U.S. (not just the ones in the sample) would average 26 points on the WISC vocabulary scale, while the average for all the rural children in the U.S. would be 25 points. Then what? There is no more chance variation to worry about, so a test of significance cannot help. All the facts are in, and the problem is to find out what the difference means.

To do that, it is necessary to look at the WISC vocabulary scale itself. There are forty words which the child has to define. Two points are given for a correct definition, and one point for a partially correct definition. So the one-point difference between big-city and rural children only amounts to a partial understanding of one word out of forty. This is not a solid basis for a crusade. Quite the opposite: the investigators have proved there is almost no difference between big-city and rural children on the WISC vocabulary scale.[12]

Of course, the sample does not reflect the population perfectly, so a standard error should be attached to the estimate for the difference. Based on the two samples of 2,500 children, the difference in average scores between all the big-city and rural children in the U.S. would be estimated as 1 point, give or take 0.3 points or so. The z-statistic is impressive because 1 is a lot, relative to 0.3.

A big sample is good because it enables the investigator to measure a difference quite accurately—with a small SE. But the z-test compares the difference to the SE. Therefore, with a large sample even a small difference can lead to an impressive value for z. The z-test can be too sensitive for its own good.

> The P-value of a test depends on the sample size. With a large sample, even a small difference can be "statistically significant," that is, hard to explain by the luck of the draw. This doesn't necessarily make it important. Conversely, an important difference may not be statistically significant if the sample is too small.

Example 3. As reported in section 2 of chapter 27, reading test scores declined from 290 in 1990 to 285 in 2004. These were averages based on nationwide samples; $z \approx -2.8$ and $P \approx 1/4$ of 1% (one-tailed). Does the increase matter?

Solution. The P-value says the increase is hard to explain away as chance error. The P-value does not say whether the increase matters. More detailed analysis of the data suggests that each extra year of schooling is associated with about a 6-point increase in average test scores. On this basis, a 5-point decline is worrisome. Other measures of school performance, however, are more reassuring.[13]

Exercise Set C

1. True or false, and explain:
   (a) A difference which is highly significant must be very important.
   (b) Big samples are bad because small differences will look significant.
2. A large university wants to compare the performance of male and female undergraduates on a standardized reading test, but can only afford to do this on a sample basis. An investigator chooses 100 male undergraduates at random, and independently 100 females. The men average 49 on the test, and their SD is 10 points. The women average 51 on the test, with the same SD of 10 points. Is the difference in the average scores real, or a chance variation? Or does the question make sense?
3. Repeat exercise 2, keeping the averages and SDs the same, but increasing the sample sizes from 100 to 400.
4. Someone explains the point of a test of significance as follows.[14] "If the null hypothesis is rejected, the difference isn't trivial. It is bigger than what would occur just by chance." Comment briefly.
5. Other things being equal, which is stronger evidence for the null hypothesis: $P = 3\%$ or $P = 27\%$?
6. Before publication in a scholarly journal, papers are reviewed. Is this process fair? To find out, a psychologist makes up two versions of a paper.[15] Both versions describe a study on the effect of rewarding children for classroom performance. The versions are identical, except for the data. One data set shows that rewards help motivate learning; the other, that rewards don't help. Some reviewers were chosen at random to get each version. All the reviewers were associated with a journal whose position was "behaviorist:" rewards for learning should work. As it turned out, both versions of the paper contained a minor inconsistency in the description of the study. The investigator did a two-sample z-test, concluding that—

   > Of the individuals who got the positive version, only 25% found the mistake. Of those who got the negative version, 71.5% found the mistake. By the two-sample z-test, this difference must be considered substantial, $P \approx 2\%$, one-tailed.

   (a) Why is the two-sample z-test legitimate? Or is it?
   (b) The standard error for the difference was about ______ percentage points.
   (c) Is the difference between 71.5% and 25% substantial? Answer yes or no, and discuss briefly.

(d) What do the results of the z-test add to the argument?
(e) What do the data say about the fairness of the review process?

7. An economist makes a study of how CALTRANS chose freeway routes in San Francisco and Los Angeles.[16] He develops a chance model in order to assess the effect of different variables on the decisions. "External political and public variables" include the views of other state agencies, school boards, businesses, large property owners, and property owners' associations. To find out how much influence these variables have on the freeway decisions, the economist makes a test of significance. The null hypothesis says that the external political and public variables make no difference in the decisions. The observed significance level is about 3%.

   Since the result is statistically significant but not highly significant, the economist concludes "these factors do influence the freeway decisions, but their impact is relatively weak." Does the conclusion follow from the statistical test?

8. An economist estimates the price elasticity for refined oil products as −6. (An elasticity of −6 means that a 1% increase in prices leads to a 6% drop in sales; an elasticity of −4.3 means that a 1% increase in prices leads to a 4.3% drop in sales, and so forth.)

   The standard error on the estimated elasticity is 2.5. The economist tests the null hypothesis that the elasticity is 0, and gets $z = -6/2.5 = -2.4$, so $P \approx 1\%$ (one-tailed). The conclusion: he is "99% confident of the estimate."[17]

   The economist seems to have assumed something like the Gauss model—

$$\text{estimated elasticity} = \text{true elasticity} + \text{error.}$$

   The error has an expected value of 0, an SE of 2.5, and follows the normal curve. You can use these assumptions, but comment on them when you answer part (d).

   (a) Find a 99% confidence interval for the "true" elasticity.
   (b) What does the P-value of 1% mean?
   (c) Was the economist using statistical tests in an appropriate way?
   (d) What do you think of the "Gauss model for elasticity?"

The answers to these exercises are on pp. A102–103.

4. THE ROLE OF THE MODEL

To review briefly, a test of significance answers the question, "Is the difference due to chance?" But the test can't do its job until the word "chance" has been given a precise definition. That is where the box model comes in.[18]

To make sense out of a test of significance, a box model is needed.

This idea may be a little surprising, because the arithmetic of the test does not use the box model. Instead, the test seems to generate the chances directly from the data. That is an illusion. It is the box model which defines the chances. The formulas for the expected values and standard errors make a tacit assumption:

that the data are like draws from a box. So do the statistical tables—normal, t, and χ^2. If the box model is wrong, the formulas and the tables do not apply, and may give silly results. This section discusses some examples.

Example 4. Census data show that in 1980, there were 227 million people in the U.S., of whom 11.3% were 65 or older. In 2000, there were 281 million people, of whom 12.3% were 65 or older.[19] Is the difference in the percentages statistically significant?

Discussion. The arithmetic of a two-sample z-test is easy enough to do, but the result is close to meaningless. We have Census data on the whole population. There is no sampling variability to worry about. Census data are subject to many small errors, but these are not like draws from a box. The aging of the population is real. It makes a difference to the health care and social security systems. However, the concept of statistical significance does not apply. The P-value would not help us to interpret the data.

> If a test of significance is based on data for the whole population, watch out.

Example 5. The Graduate Division at the University of California, Berkeley compares admission rates for men and women. For one year and one graduate major, this came out as follows: 825 men applied, and 61.7% were admitted; 108 women applied, and 82.4% were admitted. Is the difference between admission rates for men and women statistically significant?

Discussion. Again, there is nothing to stop you from doing a two-sample z-test. However, to make sense out of the results, a box model would be needed, and there doesn't seem to be one in the neighborhood. It is almost impossible to identify the pool of potential applicants. Even if you could, the actual applicants were not drawn from this pool by any probability method. Nor do departments admit candidates by drawing names from a hat (although that might not be such a bad idea). The concept of statistical significance does not apply.

Statisticians distinguish between samples drawn by probability methods and samples of convenience (section 4 of chapter 23). A sample of convenience consists of whoever is handy—students in a psychology class, the first hundred people you bump into, or all the applicants to a given department in a given year. With a sample of convenience, the concept of chance becomes quite slippery, the phrase "the difference is due to chance" is hard to interpret, and so are P-values. Example 5 was based on a sample of convenience.[20]

> If a test of significance is based on a sample of convenience, watch out.

Example 6. Academic gains were made by minority children in the Headstart preschool program, but tended to evaporate when the children went on to regular schools. As a result, Congress established Project Follow Through to provide continued support for minority children in regular schools. Seven sponsors were given contracts to run project classrooms according to different educational philosophies, and certain other classrooms were used as controls. SRI (a consulting firm based in Stanford) was hired to evaluate the project, for the Department of Health, Education, and Welfare.[21] One important question was whether the project classrooms really were different from the control classrooms.

To see whether or not there were real differences, SRI devised an implementation score to compare project classrooms with control classrooms. This score involved observing the classrooms to determine, for instance, the amount of time children spent playing, working independently, asking questions of the teacher, and so on. The results for one sponsor, Far West Laboratory, are shown in table 2.

Table 2. SRI Implementation Scores for 20 Far West Laboratory classrooms. Scores are between 0 and 100.

Site	*Classroom scores*			
Berkeley	73	79	76	72
Duluth	76	84	81	80
Lebanon	82	76	84	81
Salt Lake City	81	86	76	80
Tacoma	78	72	78	71

The average of these 20 scores is about 78; their SD is about 4.2. The average score for the control classrooms was about 60, so the difference is 18 points. As far as the SRI implementation score is concerned, the Far West classrooms are very different from the control classrooms. So far, so good. However, SRI was not satisfied. They wished to make a z-test,

> to test whether the average implementation score for Follow Through was significantly greater than the average for Non-Follow Through.

The computation is as follows.[22] The SE for the sum of the scores is estimated as $\sqrt{20} \times 4.2 \approx 19$. The SE for their average is $19/20 \approx 1$ and $z \approx (78-60)/1 = 18$. Now

The inference is:

> the overall Far West classroom average is significantly different from the Non-Follow Through classroom average of 60.

Discussion. The arithmetic is all in order, and the procedure may seem reasonable at first. But there is a real problem, because SRI did not have a chance model for the data. It is hard to invent a plausible one. SRI might be thinking of the 20 treatment classrooms as a sample from the population of all classrooms. But they didn't choose their 20 classrooms by simple random sampling, or even by some more complicated probability method. In fact, no clear procedure for choosing the classrooms was described in the report. This was a sample of convenience, pure and simple.

SRI might be thinking of measurement error. Is there some "exact value" for Far West, which may or may not be different from the one for controls? If so, is this a single number? Or does it depend on the site? on the classroom? the teacher? the students? the year? Or are these part of the error box? If so, isn't the error box different from classroom to classroom, or site to site? Why are the errors independent?

The report covers 500 pages, and there isn't a single one which touches on these problems. It is taken as self-evident that a test of significance can be used to compare the average of any sample, no matter where it comes from, with an external standard. The whole argument to show that the project classrooms differ from the controls rests on these tests, and the tests rest on nothing. SRI does not have a simple random sample of size 20, or 20 repeated measurements on the same quantity. It has 20 numbers. These numbers have chance components, but almost nothing is understood about the mechanism which generated them. Under these conditions, a test of significance is an act of intellectual desperation.

We went down to SRI to discuss these issues with the investigators. They insisted that they had taken very good statistical advice when designing their study, and were only doing what everybody else did. We pressed our arguments. The discussion went on for several hours. Eventually, the senior investigator said:

> Look. When we designed this study, one of our consultants explained that some day, someone would arrive out of the blue and say that none of our statistics made any sense. So you see, everything was very carefully considered.

Exercise Set D

1. One term, there were 600 students who took the final in Statistics 2 at the University of California, Berkeley. The average score was 65, and the SD was 20 points. At the beginning of the next academic year, the 25 teaching assistants assigned to the course took exactly the same test. The TAs averaged 72, and their SD was 20 points too.[23] Did the TAs do significantly better than the students? If appropriate, make a two-sample z-test. If this isn't appropriate, explain why not.

2. The five planets known to the ancient world may be divided into two groups: the *inner planets* (Mercury and Venus), which are closer to the Sun than the Earth; and the *outer planets* (Mars, Jupiter, and Saturn), which are farther from the Sun. The densities of these planets are shown below; the density of the Earth is taken as 1.

Mercury	Venus	Mars	Jupiter	Saturn
0.68	0.94	0.71	0.24	0.12

The two inner planets have an average density of 0.81, while the average density for the three outer planets is 0.36. Is this difference statistically significant?[24] Or does the question make sense?

3. Two researchers studied the relationship between infant mortality and environmental conditions in Dauphin County, Pennsylvania. As a part of the study, the researchers recorded, for each baby born in Dauphin County during a six-month period, in what season the baby was born, and whether or not the baby died before reaching one year of age.[25] If appropriate, test to see whether infant mortality depends on season of birth. If a test is not appropriate, explain why not.

	Season of birth	
	July–Aug.–Sept.	*Oct.–Nov.–Dec.*
Died before one year	35	7
Lived one year	958	990

4. In the WISC block design test, subjects are given colored blocks and asked to assemble them to make different patterns shown in pictures. As part of Cycle II of the Health Examination Survey, this test was given to a nationwide sample of children age 6 to 9, drawn by probability methods. Basically, this was a multistage cluster sample of the kind used by the Current Population Survey (chapter 22). There were 1,652 children in the sample with family incomes in the range $5,000 to $7,000 a year: these children averaged 14 points on the test, and the SD was 8 points. There were 813 children in the sample with family incomes in the range $10,000 to $15,000 a year: these children averaged 17 points on the test, and the SD was 12 points. (The study was done in 1963–65, which explains the dollars.[26]) Someone asks whether the difference between the averages can be explained as chance variation.

    (a) Does this question make sense?
    (b) Can it be answered on the basis of the information given?

    Explain briefly.

5. Political analysts think that states matter: different states have different political cultures, which shape voters' attitudes.[27] After controlling for certain demographic variables, investigators estimate the effect of state of residence on party affiliation (Republican or Democratic). The data base consists of 55,145 persons surveyed by CBS/*New York Times* over a six-year period in the U.S. The null hypothesis—no difference among states—is rejected ($P \approx 0$, adjusted for multiple comparisons across states). True or false, and explain briefly: since P is tiny, there are big differences in state political cultures.

6. An investigator asked whether political repression of left-wing views during the McCarthy era was due to "mass opinion or elite opinion."[28] He measured the effect of mass and elite opinion on the passage of repressive laws. (Effects were measured on a standardized scale going from -1 to $+1$.) Opinions were measured by surveys of—

    > .... a sample of the mass public and the political elites .... The elites selected were in no sense a random sample of the state elites .... Instead, the elite samples represent only themselves .... The [effect of] mass opinion is -0.06; for elite opinion it is -0.35 (significant beyond .01). Thus political repression

occurred in states with relatively intolerant elites. Beyond the intolerance of elites, the preferences of the mass public seemed to matter little.

Comment briefly on the use of statistical tests.

The answers to these exercises are on p. A103.

5. DOES THE DIFFERENCE PROVE THE POINT?

Usually, an investigator collects data to prove a point. If the results can be explained by chance, the data may not prove anything. So the investigator makes a test of significance to show that the difference was real. However, the test has to be told what "chance" means. That is what the box model does, and if the investigator gets the box model wrong, the results of the test may be quite misleading. Section 4 made this point, and the discussion continues here.

For example, take an ESP experiment in which a die is rolled, and the subject tries to make it land showing six spots.[29] This is repeated 720 times, and the die lands six in 143 of these trials. If the die is fair, and the subject's efforts have no effect, the die has 1 chance in 6 to land six. So in 720 trials, the expected number of sixes is 120. There is a surplus of $143 - 120 = 23$ sixes.

Is the difference real, or a chance variation? That is where a test of significance comes in. The null hypothesis can be set up as a box model: the number of sixes is like the sum of 720 draws from the box | [0] [0] [0] [0] [0] [1] |. The SE for the sum of the draws is

$$\sqrt{720} \times \sqrt{1/6 \times 5/6} = 10.$$

So $z = (143 - 120)/10 = 2.3$, and $P \approx 1\%$. The difference looks real.

Does the difference prove that ESP exists? As it turned out, in another part of the experiment the subject tried to make the die land showing aces, and got too many sixes. In fact, whatever number the subject tried for, there were too many sixes. This is not good evidence for ESP.

Did the z-test lead us astray? It did not. The test was asked whether there were too many sixes to explain by chance. It answered, correctly, that there were. But the test was told what "chance" meant: rolling a fair die. This assumption was used to compute the expected value and SE in the formula for z. The test proves that the die was biased, not that ESP exists.

A test of significance can only tell you that a difference is there. It cannot tell you the cause of the difference. The difference could be there because the investigator got the box model wrong, or made some other mistake in designing the study.

> A test of significance does not check the design of the study.

Tests of significance have to be told what chances to use. If the investigator gets the box model wrong, as in the ESP example, do not blame the test.

Example 7. Tart's experiment on ESP was discussed in section 5 of chapter 26. A machine called the "Aquarius" picked one of 4 targets at random, and subjects tried to guess which one. The subjects scored 2,006 correct guesses in 7,500 tries, compared to the chance level of $1/4 \times 7{,}500 = 1{,}875$. The difference was $2{,}006 - 1{,}875 = 131$, $z \approx 3.5$, and $P \approx 2/10{,}000$ (one-tailed). What does this prove?

Discussion. The difference is hard to explain as a chance variation. That is what the z-test shows. But was it ESP? To rule out other explanations, we have to look at the design of the study. Eventually, statisticians got around to checking Tart's random number generators. These generators had a flaw: they seldom picked the same target twice in a row. In the experiment, the Aquarius lit up the target after each guess. Subjects who noticed the pattern, or picked new targets each time for some other reason, may have improved their chances due to the flaw in the random number generator. Tart's box model—which defined the chances for the test—did not correspond to what the random number generator was really doing.

Tart began by denying that the non-randomness in the numbers made any difference. Eventually, he replicated the experiment. He used better random number generators, and tightened up the design in other ways too. In the replication, subjects guessed at about the chance level. There was no ESP. The subjects in both experiments were students at the University of California, Davis. Tart's main explanation for the failure to replicate—"a dramatic change" in student attitudes between experiments.[30]

> In the last year or two, students have become more serious, competitive and achievement-oriented than they were at the time of the first experiment. Such "uptight" attitudes are less compatible with strong interest and motivation to explore and develop a "useless" talent such as ESP. Indeed, we noticed that quite a few of our participants in the present experiment did not seem to really "get into" the experiment and were anxious to "get it over with."

Exercise Set E

1. Exercise 7 on p. 482 discussed an experiment where flex-time was introduced at a plant, for a sample of 100 employees. For these employees, on average, absenteeism dropped from 6.3 to 5.5 days off work. A test indicated that this difference was real. Is it fair to conclude that flex-time made the difference? If not, what are some other possible explanations for the drop in absenteeism?

2. Chapter 1 discussed the Salk vaccine field trial, where there were many fewer polio cases in the vaccine group than in the control group. A test of significance showed that the difference was real (exercise 2 on p. 515). Is it fair to conclude that the vaccine protected the children against polio? If not, what are some other possible explanations?

3. Saccharin is used as an artificial low-calorie sweetener in diet soft drinks. There is some concern that it may cause cancer. Investigators did a bioassay on rats. (Bioassays are discussed in exercise 5 on p. 552.) In the treatment group, the animals got

2% of their daily food intake in the form of saccharin. The treatment group had a higher rate of bladder cancer than the control group, and the difference was highly significant. The investigators concluded that saccharin probably causes cancer in humans. Is this a good way to interpret the P-value?

4. A company has 7 male employees and 16 female. However, the men earn more than the women, and the company is charged with sex discrimination in setting salaries. One expert reasons as follows:

   There are $7 \times 16 = 112$ pairs of employees, where one is male and the second female. In 68 of these pairs, the man earns more. If there was no sex discrimination, the man would have only a 50–50 chance to earn more. That's like coin tossing. In 112 tosses of a coin, the expected number of heads is 56, with an SE of about 5.3. So

   $$z = \frac{\text{obs} - \text{exp}}{\text{SE}} \approx \frac{68 - 56}{5.3} \approx 2.3$$

   And $P \approx 1\%$. That's sex discrimination if I ever saw it.

   Do you agree? Answer yes or no, and explain.

The answers to these exercises are on p. A103.

6. CONCLUSION

When a client is going to be cross-examined, lawyers often give the following advice:

> Listen to the question, and answer the question. Don't answer the question they should have asked, or the one you wanted them to ask. Just answer the question they really asked.

Tests of significance follow a completely different strategy. Whatever you ask, they answer one and only one question:

> How easy is it to explain the difference between the data and what is expected on the null hypothesis, on the basis of chance variation alone?

Chance variation is defined by a box model. This model is specified (explicitly or implicitly) by the investigator. The test will not check to see whether this model is relevant or plausible. The test will not measure the size of a difference, or its importance. And it will not identify the cause of the difference.

Often, tests of significance turn out to answer the wrong question. Therefore, many problems should be addressed not by testing but by estimation. That involves making a chance model for the data, defining the parameter you want to estimate in terms of the model, estimating the parameter from the data, and attaching a standard error to the estimate.

Nowadays, tests of significance are extremely popular. One reason is that the tests are part of an impressive and well-developed mathematical theory. Another reason is that many investigators just cannot be bothered to set up chance models. The language of testing makes it easy to bypass the model, and talk about "statistically significant" results. This sounds so impressive, and there is so much mathematical machinery clanking around in the background, that tests seem truly scientific—even when they are complete nonsense. St. Exupéry understood this kind of problem very well:

> When a mystery is too overpowering, one dare not disobey.
>
> —*The Little Prince*[31]

7. REVIEW EXERCISES

Review exercises may cover material from previous chapters.

1. True or false and explain briefly.
   (a) A difference which is highly significant can still be due to chance.
   (b) A statistically significant number is big and important.
   (c) A P-value of 4.7% means something quite different from a P-value of 5.2%.

2. Which of the following questions does a test of significance deal with?
   (i) Is the difference due to chance?
   (ii) Is the difference important?
   (iii) What does the difference prove?
   (iv) Was the experiment properly designed?

   Explain briefly.

3. Two investigators are testing the same null hypothesis about box X, that its average equals 50. They agree on the alternative hypothesis, that the average differs from 50. They also agree to use a two-tailed z-test. The first investigator takes 100 tickets at random from the box, with replacement. The second investigator takes 900 tickets at random, also with replacement. Both investigators get the same SD of 10. True or false: the investigator whose average is further from 50 will get the smaller P-value. Explain briefly.[32]

4. In employment discrimination cases, courts have held that there is proof of discrimination when the percentage of blacks among a firm's employees is lower than the percentage of blacks in the surrounding geographical region, provided the difference is "statistically significant" by the z-test. Suppose that in one city, 10% of the people are black. Suppose too that every firm in the city hires employees by a process which, as far as race is concerned, is equivalent to simple random sampling. Would any of these firms ever be found guilty of discrimination by the z-test? Explain briefly.

5. The inner planets (Mercury, Venus) are the ones closer to the sun than the Earth. The outer planets are farther away. The masses of the planets are shown below, with the mass of the Earth taken as 1.

Mercury	Venus	Mars	Jupiter	Saturn	Uranus	Neptune	Pluto
0.05	0.81	0.11	318	95	15	17	0.8

   The masses of the inner planets average 0.43, while the masses of the outer planets average 74. Is this difference statistically significant?[33] Or does the question make sense? Explain briefly.

6. Using election data, investigators make a study of the various factors influencing voting behavior. They estimate that the issue of inflation contributed about 7 percentage points to the Republican vote in a certain election. However, the standard error for this estimate is about 5 percentage points. Therefore, the increase is not statistically significant. The investigators conclude that "in fact, and contrary to widely held views, inflation has no impact on voting behavior."[34] Does the conclusion follow from the statistical test? Answer yes or no, and explain briefly.

7. According to Census data, in 1950 the population of the U.S amounted to 151.3 million persons, and 13.4% of them were living in the West. In 2000, the population was 281.4 million, and 22.5% of them were living in the West.[35] Is the difference in percentages practically significant? statistically significant? Or do these questions make sense? Explain briefly.

8. According to Current Population Survey data for 1985, 50% of the women age 16 and over in the United States were employed. By 2005, the percentage had increased to 59%.[36] Is the difference in percentages statistically significant?
   (a) Does the question make sense?
   (b) Can you answer it based on the information given?
   (c) Can you answer it if you assume the Current Population Survey was based on independent simple random samples in each year of 50,000 women age 16 and over?

9. In 1970, 36% of first-year college students thought that "being very well off financially is very important or essential." By 2000, the percentage had increased to 74%.[37] These percentages are based on nationwide multistage cluster samples.
   (a) Is the difference important? Or does the question make sense?
   (b) Does it make sense to ask if the difference is statistically significant? Can you answer on the basis of the information given?
   (c) Repeat (b), assuming the percentages are based on independent simple random samples of 1,000 first-year college students drawn each year.

10. R. E. Just and W. S. Chern claimed that the buyers of California canning tomatoes exercised market power to fix prices. As proof, the investigators estimated the price elasticity of demand for tomatoes in two periods—before and after the introduction of mechanical harvesters. (An elasticity of -5, for instance, means that a 1% increase in prices causes a 5% drop in demand.) They put standard errors on the estimates.

    In a competitive market, the harvester should make no difference in demand elasticity. However, the difference between the two estimated elasticities—pre-harvester and post-harvester—was almost statistically significant ($z \approx 1.56$, $P \approx 5.9\%$, one-tailed). The investigators tried several ways of estimating the price elasticity before settling on the final version.[38] Comment briefly on the use of statistical tests.

11. A market research company interviews a simple random sample of 3,600 persons in a certain town, and asks what they did with their leisure time last

year: 39.8% of the respondents read at least one book, whereas 39.3% of them entertained friends or relatives at home.[39] A reporter wants to know whether the difference between the two percentages is statistically significant. Does the question make sense? Can you answer it with the information given?

12. There have been arguments about the validity of identification based on DNA matching in criminal cases. One problem is that different subgroups may have different frequencies of "alleles," that is, variants of a gene. What is rare in one group may be common in another. Some empirical work has been done, to measure differences among subgroups. According to one geneticist,[40]

> Statistical significance is an objective, unambiguous, universally accepted standard of scientific proof. When differences in allele frequencies among ethnic groups are statistically significant, it means that they are real—the hypothesis that genetic differences among ethnic groups are negligible cannot be supported.

Comment briefly on this interpretation of statistical significance.

8. SPECIAL REVIEW EXERCISES

1. Oregon has an experimental boot camp program to rehabilitate prisoners before their release. The object is to reduce the "recidivism rate"—the percentage who will be back in prison within three years. Prisoners volunteer for the program, which lasts several months. Some prisoners drop out before completing the program.

   To evaluate the program, investigators compared prisoners who completed the program with prisoners who dropped out. The recidivism rate for those who completed the program was 29%. For the dropouts, the recidivism rate was 74%. On this basis, the investigators argued that the program worked. However, a second group of investigators was skeptical.

   (a) Was this an observational study or a controlled experiment?
   (b) What was the treatment group? the control group?
   (c) Why might the second group of investigators have been skeptical?
   (d) The second group of investigators combined the graduates of the program with the dropouts. For the combined group, the recidivism rate was 36%. By comparison, among prisoners who did not volunteer for treatment, the recidivism rate was 37%.
      (i) What were the treatment and control groups used by the second group of investigators?
      (ii) Do their data support the idea that treatment works? or that the program has no effect? Explain your answer.
   (e) The description of the studies and the data is paraphrased from the *New York Times*.[41] Assume the numbers are correct. Did most of the prisoners who volunteered for the program complete it, or did most drop out? Explain briefly.

2. A study of baseball players shows that left-handed players have a higher death rate than right-handers. One observer explained this as "due to confounding: baseball players are more likely to be left-handed than the general population, and the players have higher death rates too." Is that a good explanation for the data? Answer yes or no, and explain briefly.

3. Schools in Northern Ireland are run on the English system. "Grammar Schools" and "Secondary Intermediate Schools" are both roughly equivalent to U.S. high schools, but students who plan on attending college generally go to Grammar Schools. Before graduation, students in both types of schools take standardized proficiency examinations.

   At Grammar Schools, Catholic students do a little better on the proficiency exams than Protestant students. At the Secondary Intermediate Schools too, Catholic students do a little better.[42] True or false, and explain: if you combine the results from both kinds of schools, the Catholic students must do a little better on the proficiency exams than the Protestants.

4. The City University of New York has about 200,000 students on 21 campuses. The figure below (adapted from the *New York Times*) shows the distribution of these students by age. For example, 21.1% of them were age 19 and under. The percentages start high, rise a little, then drop, climb, and finally drop again. How can this pattern be explained?

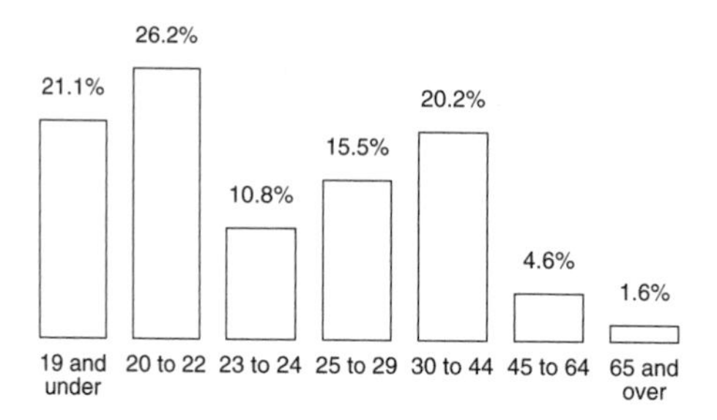

Note: Figure redrawn from original, copyright 1991 by the *New York Times*; reproduced by permission.

5. Data from one month of the National Health Interview Survey are shown below. (The survey is done monthly by the Census Bureau for the National Center for Health Statistics.) For example, 70% of the people age 18–64 ate breakfast every day, compared to 90% of the people age 65 and over. True or false: the data show that as people age, they adopt more healthful life-styles. Explain your answer. If false, how do you account for the pattern in the data?

Age	*Eats breakfast*	*Current drinker*	*Current smoker*
18–64	70%	40%	35%
65 and over	90%	10%	15%

Note: Percents are rounded. Source: *Statistical Abstract*, 1988, Table 178.

6. The U.S. Department of Justice made a study of 12,000 civil jury cases that were decided one year in state courts in the nation's 75 largest counties.[43] Juries gave money damages to plaintiffs in 55% of the cases. The median amount was $30,000, and the average was $600,000. Percentiles were computed for this distribution. Investigator A looks at the difference between the 90th percentile and the 50th percentile. Investigator B looks at the difference between the 50th percentile and the 10th percentile. Which difference is bigger? Or are they about the same? Explain briefly.

7. The scatter diagram below shows ages of husbands and wives in Ohio. Data were extracted from the March Current Population Survey. Or did something go wrong? Explain your answer.

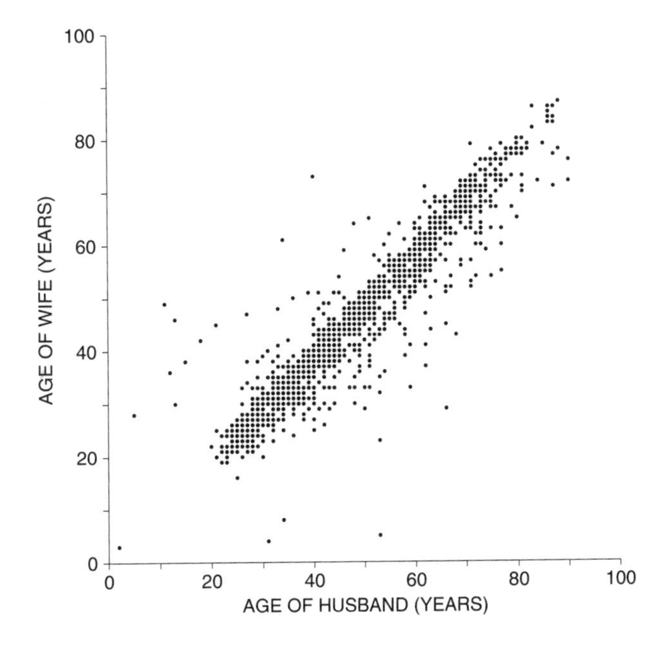

8. For the data set below, find the regression equation for predicting y from x.

x	y
1	1
8	4
10	6
10	12
14	12
17	7

9. Investigators are studying the relationship between income and education, for women age 25–54 who are working.

(a) Investigator A computes the correlation between income and education for all these women. Investigator B computes the correlation only for women who have professional, technical, or managerial jobs. Who gets the higher correlation? Or should the correlations be about the same? Explain.

(b) Investigator C computes the correlation between income and education for all the women. Investigator D looks at each state separately, computes the average income and average education for that state—and then computes the correlation coefficient for the 50 pairs of state averages. Which investigator gets the higher correlation? Or should the correlations be about the same? Explain.

10. Data on the heights of fathers and sons can be summarized as follows:

average height of fathers $\approx$ 68 inches, SD $\approx$ 2.7 inches
average height of sons $\approx$ 69 inches, SD $\approx$ 2.7 inches, $r \approx 0.5$

The scatter diagram is football-shaped. On average, the sons of the 72-inch fathers are ______ 6 inches taller than the sons of the 66-inch fathers. Fill in the blank, using one of the options below, and explain your reasoning.

(i) just about
(ii) somewhat more than
(iii) somewhat less than

11. A university made a study of all students who completed the first two years of undergraduate work. The average first-year GPA was 3.1, and the SD was 0.4. The correlation between first-year and second-year GPA was 0.4. The scatter diagram was football shaped.

Sally Davis was in the study. She had a first-year GPA of 3.5, and her second-year GPA was just about average—among those who had a first-year GPA of 3.5. What was her percentile rank on the second-year GPA, relative to all the students in the study? If this cannot be determined from the information given, say what else you need to know, and why.

12. A report by the Environmental Defense Association is discussing the relationship between air pollution and annual death rates for a sample of 47 major cities in the U.S. The average death rate is reported as 9/1,000 and the SD is 3/1,000. The r.m.s. error of the regression line for predicting death rates from air pollution is reported as 4/1,000. Is there anything wrong with the numbers? Or do you need more information to decide? Explain briefly.

13. For women age 25–54 working full time in the U.S. in 2005, the relationship between income and education (years of schooling completed) can be summarized as follows:

average education $\approx$ 14.0 years, SD $\approx$ 2.5 years
average income $\approx$ \$38,000, SD $\approx$ \$35,000, $r \approx 0.34$

If you take the women with 8 years of education, the SD of their incomes will be ______ $\sqrt{1 - .34^2} \times$ \$35,000. Fill in the blank, and explain briefly. Options: less than, just about, more than.

14. About 1.5 million high-school students took the SATs in 2005. The regression equation for predicting the Math SAT score from the Verbal SAT score is

$$\text{predicted M-SAT} = 0.6 \times \text{V-SAT} + 220$$

The r.m.s. error of the regression line is 80 points. (The scatter diagram is football-shaped; numbers have been simplified a little.) About 50,000 students scored 500 points on the V-SAT. Of these students, about how many scored better than 500 on the M-SAT? Or do you need more information?

15. Three cards are dealt off the top of a well-shuffled deck. Find the chance that—

    (a) You only get kings.
    (b) You get no kings.
    (c) You get no face cards.
    (d) You get at least one face card.

Reminder. A deck has 52 cards. There are 4 suits—clubs, diamonds, hearts, and spades. In each suit, there are 4 face cards—jack, queen, king, ace—and 9 cards numbered 2 through 10.

16. A die is rolled 6 times. Find the chance of getting 3 aces and 3 sixes.
Reminder. A die has 6 faces, showing 1 through 6 spots. An ace is ⚀. Each face is equally likely to come up.

17. According to *Esquire Magazine*,

> If you want to play roulette, do it in Atlantic City, where the house lets you "surrender" on the results of 0 and 00—that is, it returns half your wager.

A gambler in Atlantic City plays roulette 100 times, staking $1 on red each time. Find the chance that he comes out ahead of the game.

Reminder. The roulette wheel has 38 pockets, numbered 0, 00, and 1 through 36 (figure 3 on p. 282). The green numbers are 0 and 00. Of the other numbers, half are red and half are black. If you bet $1 on red and a red number comes up, you win $1. If a black number comes up, you lose $1. But if 0 or 00 comes up, you only lose $0.50—because of the "surrender."

18. A nationwide telephone survey used random digit dialing. Out of 1,507 respondents, 3% said they had been homeless at some point in the last five years. Is there selection bias in this 3% estimate? Which way does the bias go? Discuss briefly.

19. R. C. Lewontin wrote a critical review of *The Social Organization of Sexuality* by E. O. Laumann and others. Laumann was using data from a sample survey, in which respondents answered questions about their sexual behavior, including the number of partners in the previous five-year period. On average, among heterosexuals, men reported having about twice as many partners as women. Lewontin thought this was a serious inconsistency, showing that respondents "are telling themselves and others enormous lies." Laumann

replied that you should not use averages to summarize such skewed and long-tailed distributions.[44]

(a) Why is it inconsistent for men to report having twice as many partners as women?
(b) Evaluate Laumann's response.
(c) One objective of Laumann's study was to get baseline data on the epidemiology of AIDS. However, about 3% of the population (including homeless people and people in jail) were deliberately excluded from the sample. Lewontin considered this to be a serious flaw in the design of the study. Do you agree or disagree? Why?
(d) The non-response rate was about 20%. Does this matter? Explain your answer.

20. A certain town has 25,000 families. These families own 1.6 cars, on the average; the SD is 0.90. And 10% of them have no cars at all. As part of an opinion survey, a simple random sample of 1,500 families is chosen. What is the chance that between 9% and 11% of the sample families will not own cars? Show work.

21. The Census Bureau is planning to take samples in several cities, in order to estimate the percentage of the population in those areas with incomes below the poverty level. They will interview 1,000 people in each city that they study. Other things being equal:

(i) The accuracy in New York (population 8,000,000) will be about the same as the accuracy in Buffalo (population 300,000).
(ii) The accuracy in New York will be quite a bit higher than in Buffalo.
(iii) The accuracy in New York will be quite a bit lower than in Buffalo.

Choose one option, and explain briefly.

22. A market research company knows that out of all car owners in a certain large town, 80% have cell phones. The company takes a simple random sample of 500 car owners. What is the chance that exactly 400 of the car owners in the sample will have cell phones?

23. (a) What's wrong with quota samples?
(b) What's the difference between a cluster sample and a sample of convenience?
(c) What are the advantages and disadvantages of a cluster sample compared to a simple random sample?

24. (Hypothetical.) The Plaintiff's Bar Association estimates that 10% of its members favor no-fault auto insurance. This estimate is based on 2,500 questionnaires filled out by members attending a convention. True or false, and explain: the SE for this estimate is 0.6 of 1%, because

$$\sqrt{2{,}500} \times \sqrt{0.1 \times 0.9} = 15, \quad \frac{15}{2{,}500} = 0.6 \text{ of } 1\%.$$

25. A cable company takes a simple random sample of 350 households from a city with 37,000 households. In all, the 350 sample households had 637 TV sets. Fill in the blanks, using the options below.

    (a) The observed value of the ______ is 637.
    (b) The observed value of the ______ is 1.82.
    (c) The expected value of the ______ is equal to the ______.

    Options:

    (i) total number of TV sets in the sample households
    (ii) average number of TV sets per household in the sample
    (iii) average number of TV sets per household in the city

26. An airline does a market research survey on travel patterns. It takes a simple random sample of 225 people aged 18 and over in a certain city, and works out the 95%-confidence interval for the average distance they travelled on vacations in the previous year. This was 488 to 592 miles. Say whether each statement below is true or false; give reasons. If there is not enough information to decide, explain what else you need to know.

    (a) The average of the 225 distances is about 540 miles.
    (b) The SD of the 225 distances is about 390 miles.
    (c) The histogram for the 225 distances follows the normal curve.
    (d) The probability histogram for the sample average is close to the normal curve.
    (e) The probability histogram for the population average is close to the normal curve.
    (f) A 95%-confidence interval based on a sample of 450 people will be about half as wide as one based on a sample of 225 people.

27. The National Assessment of Educational Progress (NAEP) tests nationwide samples of students in school.[45] Here is an item from one of the mathematics tests.

    One plan for a state income tax requires those persons with income of $10,000 or less to pay no tax and those persons with income greater than $10,000 to pay a tax of 6 percent only on the part of their income that exceeds $10,000. A person's effective tax rate is defined as the percent of total income that is paid in tax. Based on this definition, could any person's effective tax rate be 5 percent? Could it be 6 percent?
    [Answer: People with incomes of $60,000 pay 5%, nobody pays 6%.]

    Of the grade 12 students in the sample, only 3% could answer this question correctly. The likely size of the chance error in the 3% is about ______.

    (a) Can you fill in the blank if a cluster sample of 1,000 students was tested? If so, what is the answer? If not, why not?
    (b) Can you fill in the blank if a simple random sample of 1,000 students was tested? If so, what is the answer? If not, why not?

28. Courts have ruled that standard errors and confidence intervals take bias into account.[46] Do you agree? Answer yes or no, and explain briefly.

29. One month, the Current Population Survey interviewed 54,000 households, and estimated that 94.2% of all households in the U.S. had telephones. Choose one option, and explain.

    (i) The standard error on the 94.2% can be computed as follows:

    $$\sqrt{54{,}000} \times \sqrt{0.942 \times 0.058} \approx 54, \quad \frac{54}{54{,}000} \times 100\% \approx 0.1 \text{ of } 1\%$$

    (ii) The standard error on the 94.2% can be computed some other way.

    (iii) Neither of the above

30. You may assume the Gauss model with no bias. Say whether each assertion is true or false, and why. If (c) is true, say how to do the calculations.

    (a) If all you have is one measurement, you can't estimate the likely size of the chance error in it—you'd have to take another measurement, and see how much it changes.

    (b) If all you have is one hundred measurements, you can't estimate the likely size of the chance error in their average—you'd have to take another hundred measurements, and see how much the average changes.

    (c) If all you have is one hundred measurements, you can estimate (i) the likely size of the chance error in a single measurement, and (ii) the likely size of the chance error in the average of all one hundred measurements.

31. A laboratory makes 25 repeated measurements on the molecular weight of a protein (in "kilo-Daltons"). The average is 119, and the SD is 15. The lab now wants to estimate the likely size of certain chance errors. Fill in the blanks, using the options below; some options will be left over. You may assume the Gauss model, with no bias. Explain your answers.

    (a) The chance error in one measurement is about ______.

    (b) The chance error in the average of the measurements is about ______.

    Options:

    $$15/25 \qquad 15/\sqrt{25} \qquad 15 \qquad 15 \times \sqrt{25} \qquad 15 \times 25$$

32. Feather color in Leghorn chickens is controlled by one gene pair with variants C and c. The variant C is dominant and makes colored feathers; c is recessive and makes white feathers. A geneticist mates a C/c rooster with some C/c hens and gets 24 chicks. Find the chance that half the chicks have colored feathers.[47]

33. In the U.S., there are two sources of national statistics on crime rates: (i) the FBI's Uniform Crime Reporting Program, which publishes summaries on all crimes reported to police agencies in jurisdictions covering virtually 100%

of the population; (ii) the National Crime Survey, based on interviews with a nationwide probability sample of households.[48]

In 2001, 3% of the households in the sample told the interviewers they had experienced at least one burglary within the past 12 months. The same year, the FBI reported a burglary rate of 20 per 1,000 households, or 2%. Can this difference be explained as chance error? If not, how would you explain it? You may assume that the Survey is based on a simple random sample of 50,000 households out of 100 million households.

34. A statistician tosses a coin 100 times and gets 60 heads. His null hypothesis says that the coin is fair; the alternative, that the coin is biased—the probability of landing heads is more than 50%. True or false, and explain:
    (a) If the coin is fair, the chance of getting 60 or more heads is about 3%.
    (b) Given that it lands heads 60 times, there is only about a 3% chance for the coin to be fair.
    (c) Given that it lands heads 60 times, there is about a 97% chance for the coin to be biased.

35. The Multiple Risk Factor Intervention Trial tested the effect of an intervention to reduce three risk factors for coronary heart disease—serum cholesterol, blood pressure, and smoking. The subjects were 12,866 men age 35–57, at high risk for heart disease. 6,428 were randomized to the intervention group and 6,438 to control. The intervention included counseling on diet and smoking, and in some cases therapy to reduce blood pressure. Subjects were followed for a minimum of 6 years.[49]
    (a) On entry to the study, the diastolic blood pressure of the intervention group averaged 91.0 mm Hg; their SD was 7.6 mm Hg. For the control group, the figures were 90.9 and 7.7. What do you conclude? (Blood pressure is measured in millimeters of mercury, or mm Hg.)
    (b) After 6 years, the diastolic blood pressure of the intervention group averaged 80.5 mm Hg; their SD was 7.9 mm Hg. For the control group, the figures were 83.6 and 9.2. What do you conclude?
    (c) On entry to the study, the serum cholesterol level of the intervention group averaged 253.8 mg/dl; their SD was 36.4 mg/dl. For the control group, the figures were 253.5 and 36.8. What do you conclude? (mg/dl is milligrams per deciliter.)
    (d) After 6 years, the serum cholesterol level of the intervention group averaged 235.5 mg/dl; their SD was 38.3 mg/dl. For the control group, the figures were 240.3 and 39.9. What do you conclude?
    (e) On entry to the study, 59.3% of the intervention group were smoking, compared to 59.0% for the control group. What do you conclude?
    (f) After 6 years, the percentage of smokers was 32.3% in the intervention group and 45.6% in the control group. What do you conclude?
    (g) In the treatment group, 211 men had died after 6 years, compared to 219 in the control group. What do you conclude?

36. The Gallup Poll asks respondents how they would rate the honesty and ethical standards of people in different fields—very high, high, average, low, or very low. In 2005, only 8% of the respondents gave car salesmen a rating of "very high or high," while 7% rated telemarketers as "very high or high." Is the difference between 8% and 7% real, or a chance variation? Or do you need more information? Discuss briefly. You may assume that the results are based on a simple random sample of 1,000 persons taken in 2005; each respondent rated car salesmen, telemarketers, and many other professions.[50]

37. Each respondent in the Current Population Survey of March 2005 can be classified by education and occupation. The table below shows the observed frequencies for civilian women age 25–29 in Virginia.

    (i) Women with different educational levels seem to have different occupations. Or is this just chance variation?
    (ii) If the difference is real, what accounts for it?

    (a) Can you answer these questions with the information given? If so, answer them. If not, why not?
    (b) Can you answer these questions if the data in the table resulted from a simple random sample of women age 25–29 in Virginia? If so, answer them. If not, why not?

	Educational level	
	High school or less	*More than high school*
Professional, managerial, technical	12	34
Other white collar	15	17
Blue collar	5	2
Not in labor force	31	14

Notes: "Other white collar" includes sales and clerical. "Blue collar" includes hotel and restaurant service, factory work, and so forth, as well as unemployed workers with no civilian experience.
Source: March 2005 Current Population Survey; CD-ROM supplied by the Bureau of the Census.

38. Defining statistical significance, a court writes: "Social scientists consider a finding of two SEs significant, meaning there is about one chance in 20 that the explanation for the difference could be random." Do you agree or disagree with this interpretation of significance? Explain briefly.[51]

39. M. S. Kanarek and associates studied the relationship between cancer rates and levels of asbestos in the drinking water, in 722 Census tracts around San Francisco Bay.[52] After adjusting for age and various demographic variables, but not smoking, they found a "strong relationship" between the rate of lung cancer among white males and the concentration of asbestos fibers in the drinking water: $P < 1/1{,}000$.

    Multiplying the concentration of asbestos by a factor of 100 was associated with an increase in the level of lung cancer by a factor of about 1.05, on average. (If tract B has 100 times the concentration of asbestos fibers in the water as tract A, and the lung cancer rate for white males in tract A is 1 per 1,000 persons per year, a rate of 1.05 per 1,000 persons per year is predicted in tract B.)

The investigators tested over 200 relationships—different types of cancer, different demographic groups, different ways of adjusting for possible confounding variables. The P-value for lung cancer in white males was by far the smallest one they got.

Does asbestos in the drinking water cause lung cancer? Is the effect a strong one? Discuss briefly.

40. Belmont and Marolla conducted a study on the relationship between birth order, family size, and intelligence.[53] The subjects consisted of all Dutch men who reached the age of 19 between 1963 and 1966. These men were required by law to take the Dutch army induction tests, including Raven's intelligence test. The results showed that for any particular birth order, intelligence decreased with family size. For example, first-borns in two-child families did better than first-borns in three-child families. Results remained true even after controlling for the social class of the parents. Moreover, for each family size, measured intelligence decreased with birth order: first-borns did better than second-borns, second-borns did better than third-borns, and so on. For instance, with two-child families:

    - the first-borns averaged 2.575 on the test;
    - the second-borns averaged 2.678 on the test.

    (Raven test scores range from 1 to 6, with 1 being best and 6 worst.) The difference is small, but it could have interesting implications.

    To show that the difference was real, Belmont and Marolla made a two-sample z-test. The SD for the test scores was around 1 point, both for the first-borns and the second-borns, and there were 30,000 of each, so

$$\text{SE for sum} \approx \sqrt{30{,}000} \times 1 \text{ point} \approx 173 \text{ points}$$

$$\text{SE for average} \approx 173/30{,}000 \approx 0.006 \text{ points}$$

$$\text{SE for difference} \approx \sqrt{(0.006)^2 + (0.006)^2} \approx 0.008 \text{ points.}$$

    Therefore, $z \approx (2.575 - 2.678)/0.008 \approx -13$, and P is astonishingly small. Belmont and Marolla concluded:

    > Thus the observed difference was highly significant . . . . a high level of statistical confidence can be placed in each average because of the large number of cases.

    (a) What was the population? the sample? What parameters were estimated from the sample?
    (b) Was the two-sample z-test appropriate? Answer yes or no, and explain.

9. SUMMARY AND OVERVIEW

1. A result is "statistically significant" if P is less than 5%, and "highly significant" if P is less than 1%. However, these lines are somewhat arbitrary because there is no sharp dividing line between probable and improbable results.

2. Investigators should summarize the data, say what test was used, and report the P-value instead of just comparing P to 5% or 1%.

3. Even if the result is statistically significant, it can still be due to chance. Data-snooping makes P-values hard to interpret.

4. A z-test can be done either one-tailed or two-tailed, depending on the form of the alternative hypothesis.

5. The P-value of a test depends on the sample size. With a large sample, even a small difference can be "statistically significant," that is, hard to explain as a chance variation. That doesn't necessarily make it important. Conversely, an important difference may be statistically insignificant if the sample is too small.

6. To decide whether a difference observed in the sample is important, pretend it applies to the whole population and see what it means in practical terms. This is a "test of real significance."

7. A test of significance deals with the question of whether a difference is real or due to chance variation. It does not say how important the difference is or what caused it. Nor does the test check the design of the study.

8. Usually, a test of significance does not make sense when data are available for the whole population, because there is no chance variation to screen out.

9. To test whether a difference is due to chance, you have to define the chances. That is what the box model does. A test of significance makes little sense unless there is a chance model for the data. In particular, if the test is applied to a sample of convenience, the P-value may be hard to interpret.

10. Part VIII of the book introduced two tests, the z-test and the χ^2-test. The z-test can be used to compare the average of a sample with an external standard (chapter 26). The test can also be used to compare the averages of two samples (chapter 27). The χ^2-test compares observed and expected frequencies (chapter 28).

11. Tests ask whether a difference is too large to explain by chance. Procedures are based on the descriptive statistics of part II, and the mathematical theory of part V—including the square root law (chapter 17) and the central limit theorem (chapter 18).

12. There are many pitfalls in testing, as chapter 29 indicates.

NOTES

ANSWERS TO EXERCISES

TABLES

INDEX

Notes

Part I. Design of Experiments

Chapter 1. Controlled Experiments

1. The method of comparison was used in the early nineteenth century, to show that bleeding was not such an effective treatment for pneumonia. See Pierre Charles-Alexandre Louis, *Recherches sur les effets de la saignée dans quelques maladies inflammatoires: et sur l'action de l'émetique et des vésicatoires dans la pneumonie* (J. B. Baillière, Paris, 1835; English translation, 1836; reprinted by The Classics of Medicine Library, Birmingham, Alabama, 1986). For discussion, see R. H. Shryock, *The Development of Modern Medicine* (University of Pennsylvania Press, 1936, p. 163). Lind's trial on vitamin C for scurvy should also be mentioned: see K. J. Carpenter, *The History of Scurvy and Vitamin C* (Cambridge University Press, 1986).
2. Thomas Francis, Jr. et al., "An evaluation of the 1954 poliomyelitis vaccine trials—summary report," *American Journal of Public Health* vol. 45 (1955) pp. 1–63. Also see the article by P. Meier, "The biggest public health experiment ever: The 1954 field trial of the Salk poliomyelitis vaccine," in J. M. Tanur et al., *Statistics: A Guide to the Unknown*, 3rd ed. (Wadsworth, 1989). There is a less-formal account in Jane S. Smith, *Patenting the Sun* (Anchor, 1990).
3. One example: anti-arrhythmic drugs probably killed substantial numbers of people. See Thomas J. Moore, *Deadly Medicine* (Simon & Schuster, 1995). For a survey of drug trials, see N. Freemantle et al., "Composite outcomes in randomized trials," *Journal of the American Medical Association* vol. 289 (2003) pp. 2554–59.
4. "Control what you can and randomize the rest" is the advice often given by statisticians. Matching or blocking will reduce variance, at the expense of complicating the analysis. Also see note 12 to chapter 19, and note 16 to chapter 27.
5. H. K. Beecher, *Measurement of Subjective Responses* (Oxford University Press, 1959, pp. 66–67). Also see Berton Roueché, *The Medical Detectives* (Washington Square Press, New York, 1984, vol. II, chapter 9). More recent references include K. B. Thomas, "General practice consultations: Is there any point in being positive?" *British Medical Journal* vol. 294 (1987) pp. 1200–2 and J. A. Turner et al., "The importance of placebo effects in pain treatment and research," *Journal of the American Medical Association* vol. 271 (1994) pp. 1609–14.
6. N. D. Grace, H. Muench and T. C. Chalmers, "The present status of shunts for portal hypertension in cirrhosis," *Gastroenterology* vol. 50 (1966) pp. 684–91. We found this example in J. P. Gilbert, R. J. Light and F. Mosteller, "Assessing social innovations: An empirical guide for policy," *Benefit Cost and Policy Analysis Annual* (1974). For a review of more recent therapies, see A. J. Stanley and P. C. Hayes, "Portal hypertension and variceal haemorrhage," *Lancet* vol. 350 (1997) pp. 1235–39; there does not seem to be any survival advantage for current surgical therapies (including "TIPS," see p. 1238). But see A. J. Sanyal et al., "The North American study for the treatment of refractory ascites," *Gastroenterology* vol. 124 (2003) pp. 634–41.
7. The definition of "randomized controlled trial" is not strict. The original table included data on anticoagulants after myocardial infarct. Even in the 1980s, there was some controversy about the interpretation of clinical trials on anticoagulants. Since then, thrombolytic therapies have changed considerably, and there are many new experiments. For reviews, see—

   Coronary Artery Disease vol. 5 no. 4 (1994).

   "ACC/AHA guidelines for the management of patients with ST-elevation myocardial infarction: A report of the American College of Cardiology/American Heart Association Task Force on Practice Guidelines," *Circulation* vol. 110 (2004) pp. 588–636.

   J. D. Talley, "Review of thrombolytic intervention for acute myocardial infarction—is it valuable?" *Journal of the Arkansas Medical Society* 91 (1994) pp.70–79.

   C. H. Hennekens, "Thrombolytic therapy: Pre- and post-GISSI-2, ISIS-3, and GUSTO-1," *Clinical Cardiology* vol. 17 suppl. I (1994) pp. I15–7.

   R. Collins, R. Peto, S. Parish and P. Sleight, "ISIS-3 and GISSI-2: No survival advantage with tissue plasminogen activator over streptokinase, but a significant excess of strokes with tissue plasminogen activator in both trials," *American Journal of Cardiology* vol. 71 (1993) pp. 1127–30.

   M. J. Stampfer et al., "Effect of intravenous streptokinase on acute myocardial infarction: Pooled results from randomized trials," *New England Journal of Medicine* vol. 307 (1982) pp. 1180–82.

8. T. C. Chalmers, "The impact of controlled trials on the practice of medicine," *Mount Sinai Journal of Medicine* vol. 41 (1974) pp. 753–59.

Chapter 2. Observational Studies

1. "Epidemiology" is the statistical study of disease. An excellent introductory text is Leon Gordis, *Epidemiology*, 3rd ed. (Elsevier Saunders, Philadelphia, 2004). Some references on the epidemiology of smoking—

   J. Berkson, "The statistical study of association between smoking and lung cancer," *Proceedings of the Mayo Clinic* vol. 30 (1955) pp. 319–48.

   R. A. Fisher, *Smoking: The Cancer Controversy* (Oliver and Boyd, 1959)

   J. Cornfield, W. Haenszel, E. C. Hammond, A. M. Lilienfeld, M. B. Shimkin and E. L. Wynder, "Smoking and lung cancer: Recent evidence and a discussion of some questions," *Journal of the National Cancer Institute* vol. 22 (1959) pp. 173–203.

   U.S. Public Health Service, *Smoking and Health: Report of the Advisory Committee to the Surgeon General* (Washington, D.C., 1964).

   International Agency for Research on Cancer, *Tobacco Smoking*. Monograph 38 (Lyon, France, 1986).

   U.S. Public Health Service, *The Health Benefits of Smoking Cessation. A Report of the Surgeon General* (Washington, D.C., 1990).

   For evidence on mechanism, see—

   M. F. Denissenko et al., "Preferential formation of Benzo[a]pyrene adducts at lung cancer mutational hotspots in p53," *Science* vol. 274 (1996) pp. 430–2.

2. The Coronary Drug Project Research Group, "Influence of adherence to treatment and response of cholesterol on mortality in the Coronary Drug Project," *New England Journal of Medicine* vol. 303 (1980) pp. 1038–41. The other drugs were estrogens (at two dose levels), dextrothyroxine, and nicotinic acid. Enrollment ran from March 1966 to October 1969. The estrogens and dextrothyroxine were discontinued, due to adverse side effects. Clofibrate and nicotinic acid lowered the cholesterol level, but did not reduce mortality. See "Clofibrate and niacin in coronary heart disease," *Journal of the American Medical Association* vol. 231 (1975) pp. 360–81.

   The Lancet (March 4, 1989) pp. 473–74 suggests a positive effect from nicotinic acid in late followup. However, there is evidence of harm from such drugs. See Toronto Working Group on Cholesterol Policy, *Journal of Clinical Epidemiology* vol. 43 (1990) no. 10 pp. 1021ff. Other papers are cited in note 7, chapter 29.

   An interesting sidelight: about 40 risk factors were measured at baseline. As a group, the non-adherers did seem to be in worse shape when the study began. But it was impossible to adjust out the difference between adherers and non-adherers by regression using the measured risk factors; adherence to protocol was not well related to the covariates. This suggests caution in using regression models to control for confounding in observational studies. Also see R. T. Tsuyuki et al., "A meta-analysis of the association between adherence to drug therapy and mortality," *British Medical Journal* vol. 333 (2006) pp. 15–19.

3. The quote is from D. A. Roe, *A Plague of Corn* (Cornell University Press, 1973). Another excellent reference, which reprints many of the original papers with commentary, is K. J. Carpenter, *Pellagra* (Academic Press, 1981). Also see M. Terris, editor, *Goldberger on Pellagra* (Louisiana State University Press, 1964)

   Corn was brought to Europe from America. The Indians treated corn with alkali before cooking it, which released the niacin; pellagra was not a problem for them. The pellagra epidemic in the U.S. seems to have begun when millers started extracting the germ from corn; the germ contains much of the available niacin or tryptophan. (Tryptophan is an amino acid which can be converted to niacin in the body.) In the U.S., pellagra deaths peaked around 1930, and declined fairly steadily from then on, presumably because of a general improvement in economic conditions and diet; enriching the flour came too late to have much impact. Pellagra is still endemic in parts of Africa and India. In the late 1980s, research in South Africa suggested a possible etiologic role for mycotoxins; so the infection theory may have a little truth to it after all. We would like to thank K. J. Carpenter (U.C. Berkeley) for his help with this example.

4. References—

   E. L. Wynder, J. Cornfield, P. D. Schroff and K. R. Doraiswami, "A study of environmental factors in carcinoma of the cervix," *American Journal of Obstetrics and Gynecology* vol. 68 (1954) pp. 1016–52.

   C. Buck et al., editors, *The Challenge of Epidemiology: Issues and Selected Readings*, Scientific Publication No. 505 (World Health Organization, Geneva, 1989).

   N. Muñoz, F. X. Bosch, K. V. Shah and A. Meheus, editors, *The Epidemiology of Cervical*

Cancer and Human Papillomavirus. International Agency for Research on Cancer, Scientific Publication no. 119 (1992).

A. S. Evans, *Causation and Disease: A Chronological Journey* (Plenum, 1993).

S. A. Cannistra and J. M. Niloff, "Cancer of the uterine cervix," *New England Journal of Medicine* vol. 334 (1996) pp. 1030–38.

A. Storey et al., "Role of a p53 polymorphism in the development of human papillomavirus-associated cancer," *Nature* vol. 393 (1998) pp. 229–34.

X. Castellsague et al., "Male circumcision, penile human papillomavirus infection, and cervical cancer in female partners," *New England Journal of Medicine* vol. 346 (2002) pp. 1105–12.

Wynder et al. found that circumcision was protective. The history is discussed by Evans, and some of the key papers are reprinted in Buck et al. Castellsague et al. conclude that circumcision is protective if the man is highly active sexually. The death rate from cervical cancer has been declining for some time. Smoking is a risk factor for this disease, so the decline in smoking may explain the decline in death rates, and screening is protective. A vaccine against papilloma virus is now available. The example was suggested by Michael Kramer (Montreal).

5. R. M. Moore et al., "The relationship of birthweight and intrauterine diagnostic ultrasound exposure," *Journal of Obstetrics and Gynecology* vol. 71 (1988) pp. 513–17. The confounding variables: race, registration status (public or private), smoking status, delivery status (full-term or premature), spontaneous abortion history, alcohol history, amniocentesis status, fetal monitoring, method of delivery, education, weeks pregnant at registration, number of prenatal visits, maternal weight and weight gain, gestational age at delivery.

   The clinical trial is U. Waldenstrom et al., "Effects of routine one-stage ultrasound screening in pregnancy: A randomized controlled trial," *Lancet* (Sept. 10, 1988) pp. 585–88. Babies exposed to ultrasound had higher weights, on average, than the controls. In the treatment group, women watched ultrasound images of the babies they were carrying. Many of them gave up smoking as a result, and smoking does cause low birthweight. The change in smoking behavior may account for the protective effect.
6. "Suicide and the Samaritans," *Lancet* (Oct. 7, 1978) pp. 772–73 (editorial). The original investigator was C. Bagley (*Social Science and Medicine*, 1968). He did not match the towns by type of gas used, and these data do not seem to be available now. The replication was by B. Barraclough et al. (*Lancet*, 1977; *Psychological Medicine*, 1978). We found this example in D. C. Hoaglin, R. J. Light, B. McPeek, F. Mosteller and M. A. Stoto, *Data for Decisions* (University Press of America, 1982, p. 133).
7. The paradox in the Berkeley data was noticed by Eugene Hammel, then associate dean of the graduate division. He resolved it with the help of two colleagues, P. Bickel and J. W. O'Connell. We are following their report, "Is there a sex bias in graduate admissions?" *Science* vol. 187 (1975) pp. 398–404. The admissions data are from fall, 1973.
8. For a review, see Myra Samuels, "Simpson's Paradox and related phenomena," *Journal of the American Statistical Association* vol. 88 (1993) pp. 81–88.
9. Some typical examples:

   (i) The confounder may be a common cause of exposure and disease.
   (ii) The confounder may be associated with exposure and cause disease.
   (iii) The confounder may be associated with disease and cause exposure.

   A common effect of exposure and disease will generally not explain the association. Paradoxically, selecting on a common effect may create a negative correlation: see the Berkson paper cited in note 1.
10. *Statistical Abstract*, 2003, table 108.
11. See note 2, chapter 1.
12. References—

    L. M. Friedman, C. D. Furberg and D. L. DeMets, *Fundamentals of Clinical Trials*, 3rd corr. ed. (Springer, 2006, p. 83).

    T. L. Lewis, T. R. Karlowski, A. Z. Kapikian, J. M. Lynch, G. W. Shaffer, D. A. George and T. C. Chalmers, "A controlled clinical trial of ascorbic acid for the common cold," *Annals of the New York Academy of Science* vol. 258 (1975) pp. 505–12.

    T. R. Karlowski, T. C. Chalmers, L. D. Frenkel, A. Z. Kapikian, T. L. Lewis and J. M. Lynch, "Ascorbic acid for the common cold," *Journal of the American Medical Association* vol. 231 (1975) pp. 1038–42.

    K. J. Carpenter, *The History of Scurvy and Vitamin C* (Cambridge University Press, 1986).
13. "Nicotinic acid" is the technical term for niacin, the pellagra-preventive factor. Apparently, the term "niacin" was introduced because "nicotinic acid" looked too ominous on flour labels. Nicotinic acid was tried in the Coronary Drug Project and had no effect.

14. The savings in lives persist over many years, and other trials give quite similar results. Screening speeds up detection by a year or so, and that seems to be enough to matter. Unpublished data were kindly provided by the late Sam Shapiro, professor of epidemiology, Johns Hopkins. In the HIP trial, there was an initial screening examination and three annual rescreenings, each including breast examination by a doctor and mammography.

    The risk of breast cancer is modulated by hormone balance, and pregnancy is protective; early first pregnancy has a marked effect. Presumably, that accounts for the gradient with income. On social gradients in disease risk, with further references, see

    J. N. Morris et al., "Levels of mortality, education, and social conditions in the 107 local education authority areas of England," *Journal of Epidemiology and Community Health* vol. 50 (1996) pp. 15–17.

    J. Pekkanen et al., "Social class, health behavior, and mortality among men and women in eastern Finland," *British Medical Journal* vol. 311 (1995) pp. 589–93.

    M. G. Marmot et al., "Contribution of job control and other risk factors to social variations in coronary heart disease," *Lancet* vol. 350 (1997) pp. 235–9.

    The key reference on the HIP trial is S. Shapiro, W. Venet, P. Strax, and L. Venet, *Periodic Screening for Breast Cancer: The Health Insurance Plan Project and its Sequelae, 1963–1986* (Hopkins, 1988). In 2000, questions were raised again about the value of screening, but the critics seem to have misinterpreted much of the evidence. For a review and further references, see D. A. Freedman, D. B. Petitti, and J. M. Robins, "On the efficacy of screening for breast cancer," *International Journal of Epidemiology* vol. 33 (2004) pp. 43–73, 1404–6.
15. For references, see note 4.
16. This example was suggested by Shanna Swan (Rochester), based on data from an observational study done at Kaiser Permanente in Walnut Creek, California.
17. *Statistical Abstract*, 2003, tables 17, 307.
18. *Federal Register*, vol. 69, no. 169, Sept. 1, 2004, pp. 53354–59. Technically, it is not sales figures that are reported, but vehicles "manufactured for [model year] 2002, as reported to the Environmental Protection Agency."
19. *Statistical Abstract*, 1971, table 118. The study was done in 1964; the same effect turns up in many other studies. If you quit smoking and survive more than a few years, your risk will drop relative to continuing smokers. See U.S. Public Health Service, *The Health Benefits of Smoking Cessation. A Report of the Surgeon General* (Washington, D.C., 1990).
20. We found the example in Friedman et al., cited in note 12 above. References—

    P. J. Schechter, W. T. Friedewald, D. A. Bronzert, M. S. Raff and R. I. Henkin, "Idiopathic hypoguesia: a description of the syndrome and a single-blind study with zinc sulfate," *International Review of Neurobiology* (1972) Supplement 1 pp. 125–39.

    R. I. Henkin, P. J. Schechter, W. T. Friedewald, D. L. DeMets and M. S. Raff, "A double blind study of the effects of zinc sulfate on taste and smell dysfunction," *American Journal of the Medical Sciences* vol. 272 (1976) pp. 285–99.
21. This example was suggested by Shanna Swan. See E. Peritz et al., "The incidence of cervical cancer and duration of oral contraceptive use," *American Journal of Epidemiology* vol. 106 (1977) pp. 462–69. Adjustments were also made for religion, smoking (a risk factor for cervical cancer), number of Pap smears before entry, and "selected infections." For additional references, see note 4.
22. Quoted by Herb Caen in the *San Francisco Chronicle*, Wednesday, August 9, 1995.
23. References—

    E. R. Greenberg et al., "A clinical trial of antioxidant vitamins to prevent colorectal adenoma," *New England Journal of Medicine* vol. 331 (1994) pp. 141–47.

    O. P. Heinonen et al., "Effect of vitamin E and beta carotene on the incidence of lung cancer and other cancers in male smokers," *New England Journal of Medicine* vol. 330 (1994) pp. 1029–35.

    For other trials and additional discussion, see—

    C. H. Hennekens et al., "Lack of effect of long-term supplementation with beta carotene on the incidence of malignant neoplasms and cardiovascular disease," *New England Journal of Medicine* vol. 334 (1996) pp. 1145–9.

    J. Virtamo, P. Pietinen, J. K. Huttunen et al., "Incidence of cancer and mortality following alpha-tocopherol and beta-carotene supplementation: A postintervention follow-up," *Journal of the American Medical Association* 290 (2003) pp. 476–85.

    D. A. Lawlor, G. D. Smith, K. R. Bruckdorfer et al., "Those confounded vitamins: What can we learn from the differences between observational vs randomised trial evidence," *Lancet* 363 (2004) pp. 1724–27.

G. S. Omenn et al., "Effects of a combination of beta carotene and vitamin A on lung cancer and cardiovascular disease," *New England Journal of Medicine* vol. 334 (1996) pp. 1150–5.

24. The story ran November 9, 1994. The source was S. L. Johnson and L. L. Birch, "Parents' and children's adiposity and eating style," *Pediatrics* vol. 94 (1994) pp. 653–61. "Mothers who were more controlling of their children's food intake had children who showed less ability to self-regulate energy intake ($r = -.67$, $P < .0001$)."
25. This exercise is based on a story in the *San Francisco Chronicle*, January 19, 1993. The quote is edited to simplify the study design. Generally, prisoners are offered early parole as an inducement to volunteer.

Part II. Descriptive Statistics

Chapter 3. The Histogram

1. By Antoine de St. Exupéry. Reproduced by permission of the publisher, Harcourt Brace Jovanovich.
2. *Money Income in 1973 of Families and Persons in the United States*, Current Population Reports, Series P-60, No. 97 (January, 1975). U.S. Department of Commerce.
3. For the 1973 data, see note 2. The 2004 data are from the March 2005 Current Population Survey; a CD-ROM was supplied by the Bureau of the Census. Generally, "family income" means the total income of primary family members in a household, and includes the income of related sub-family members. Income from self-employment is net, and can be quite negative if the business lost money for a year.

   The March survey over-samples large families to address questions about availability of health insurance. We use sample weights for issues that involve the number of children in the family; see, e.g., the histogram for family size shown in figure 6, or exercise 2 in set D. We do not use weights for other questions, e.g., the correlation between education and income: the weights make surprisingly little difference.

   In many cases, unequal class intervals are forced by the design of the questionnaire. For instance, a respondent may only be asked to specify which of several ranges best describes his income; the ranges—class intervals—will be of unequal length.

   Adjusting for inflation by price indices may be problematic, because quality improvements are hard to account for. See W. D. Nordhaus, "Do real output and real wage measures capture reality? The history of lighting suggests not," in T. F. Bresnahan and R. J. Gordon, editors, *The Economics of New Goods* (Chicago University Press, 1997).
4. See note 2.
5. Among other things, data histograms in chapter 3 prepare the way for probability histograms in chapter 18. The latter can be approximated by probability densities; in particular, a histogram can "follow" the normal curve, in the sense that the curve is a good approximation to the histogram. If f is a probability density on the line, areas under f represent probability; the height of f at x represents probability per unit length, rather than probability: indeed, the probability of x is 0, not $f(x)$.

   Generally, mathematicians do not bother with physical units—inches, pounds, or dollars. For statistical purposes, such units matter. The units for f are inverse to the units for x. Thus, we present the density scale for an income histogram as "percent per thousand dollars"; the density scale for a weight histogram would be "percent per pound." For non-mathematicians, "2 percent per pound" may be more palatable than .02 lb^{-1}.
6. This is exact for class intervals, approximate for other intervals.
7. *Statistical Abstract*, 1971, table 118.
8. Many variables can be classified either way, depending on how you view them. Incomes, for instance, can never differ by less than a penny. Nevertheless, it is convenient to treat income as continuous—because the range is so much larger than the minimum difference.
9. With narrow class intervals, the histogram may be so ragged that its shape is impossible to make out. With wider class intervals, the shape of the histogram may be easier to see, even though some information is lost. For discussion, see P. Diaconis and D. Freedman, "On the histogram as a density estimator: L_2 theory," *Z. Wahrscheinlichkeitstheorie* vol. 57 (1981) pp. 453–76.
10. See I. R. Fisch, S. H. Freedman and A. V. Myatt, "Oral contraceptives, pregnancy, and blood pressure," *Journal of the American Medical Association* vol. 222 (1972) pp. 1507–10. Our discussion follows this paper. We are grateful to Shanna Swan and Michael Grossman for technical advice.

    Blood pressure is taken in two phases, *systolic* and *diastolic*. We are looking at the systolic phase. Results on the diastolic phase are quite similar. In the Contraceptive Drug Study, blood pressures were measured by a machine. The study excluded about 3,500 women who were pregnant, post-partum, or taking hormonal medication other than the pill; these factors affect blood pressure. The Drug Study found that four age groups were enough: 17–24, 25–34, 35–44, and 45–58. The age distributions of users or non-users within each of these age groups were quite similar.

11. R. C. Tryon, "Genetic differences in maze-learning techniques in rats," 39th yearbook, *National Society for the Study of Education* part I (1940) pp. 111–19. This article is reprinted in a very nice book of readings: Anne Anastasi, *Individual Differences* (John Wiley & Sons, 1965). Tryon uses a non-linear scale for his histograms, so they look quite different from our sketches.
12. *1970 Census of Population.* See vol. 1, part 1, section 2, appendix, p. 14. U.S. Department of Commerce. Only persons age 23–99 are counted in the column for 1880; only persons age 23–82 are counted in the column for 1970.
13. K. Bemesderfer and J. May, *Social and Political Inquiry* (Belmont, California: Duxbury Press, 1972, p. 6).
14. References—

    R. A. Baron and V. M. Ransberger, "Ambient temperature and the occurrence of collective violence: The 'long, hot summer' revisited," *Journal of Personality and Social Psychology* vol. 36 (1978) pp. 351–60. The quote is edited slightly.

    J. M. Carlsmith and C. A. Anderson, "Ambient temperature and the occurrence of collective violence: A new analysis," *Journal of Personality and Social Psychology* vol. 37 (1979) pp. 337–44.

    The figure is redrawn from Baron and Ransberger, by permission of the authors and copyright holder (the American Psychological Association).

Chapter 4. The Average and the Standard Deviation

1. *Natural Inheritance* (Macmillan, London, 1889; reprinted by the American Mathematical Society Press, 1973).
2. The point where the histogram is highest, the *mode*, is sometimes used to indicate the center. This is not recommended, as minor changes in the data can cause major shifts in the mode.
3. Tom Alexander, "A revolution called plate tectonics," *Smithsonian Magazine* vol. 5, no. 10 (1975). A. Hallam, "Alfred Wegener and the hypothesis of continental drift," *Scientific American* vol. 232, no. 2 (1975). Ursula Marvin, *Continental Drift* (Smithsonian Press, 1973).
4. The Public Health Service and the National Center for Health Statistics (NCHS) are in the Department of Health and and Human Services. Data on HANES2 are from series 11 of the Vital and Health Statistics publications, and from tapes supplied by the National Center for Health Statistics and by the Inter-University Consortium for Political and Social Research. Data on HANES3 were kindly supplied on a CD-ROM by NCHS. These data, and data for HANES5, are now available on the internet, at

    http://www.cdc.gov/nchs/about/major/nhanes/datalink.htm

    (URLs cited here were alive in February 2006, but time will doubtless take its toll.) We are responsible for the interpretation of the data. For help with earlier editions of the book, we thank Dale Hitchcock, Arthur McDowell, and Bob Murphy at NCHS, as well as Dorothy Rice (UCSF). For the fourth edition, we thank Wim van Veen (Health Council of the Netherlands).

    The histograms in figures 4, 8, 9 are based on sample counts, unweighted, ages 18+; likewise for the scatter diagrams discussed in part III. Summary statistics are heavily rounded. Sample weights made little difference in HANES2, but have more noticeable effects in HANES5. The table below compares HANES5 to HANES2 (ages 18–74).

HANES2: 1976–80

	Men 18–74 unweighted	*Men 18–74 weighted*	*Women 18–74 unweighted*	*Women 18–74 weighted*
Height	68.78 ± 2.83	69.11 ± 2.82	63.46 ± 2.62	63.71 ± 2.60
Weight	170.92 ± 30.13	172.19 ± 29.75	145.71 ± 32.65	144.18 ± 32.27

HANES5: 2003–04

	Men 18–74 unweighted	*Men 18–74 weighted*	*Women 18–74 unweighted*	*Women 18–74 weighted*
Height	69.11 ± 3.10	69.61 ± 2.97	63.67 ± 2.76	64.09 ± 2.65
Weight	188.92 ± 42.95	193.94 ± 41.95	165.84 ± 43.76	165.32 ± 44.19

5. The groups in figure 3: 18–24, 25–34, 35–44, 45–54, 55–64, 65–74. On HANES2, see *Anthropometric Reference Data and Prevalence of Overweight: United States, 1976–80*; data are from

the National Health Survey, series 11, no. 238, U.S. Department of Health and Human Services, Washington, D.C. In the 1970s, the secular trend was estimated at about 0.4 inch per decade; and, over the 20-year period 1960–80, Americans did become 0.8 inches taller, on average. Furthermore, people seem to lose 0.5–1.5 inch of height as they age from 50 to 75. (One possible explanation: about 2 inches of height is made up of air spaces between the bones in the body; the body settles in on itself with age, so these air spaces get smaller and smaller.) The secular trend and the shrinking would suggest a total drop of 2.5–3.5 inches from age 20 to age 70. The observed drop in HANES2 was 2.3 inches for the men and 2.1 inches for the women, so there may have been other factors at work. We would like to thank Reubin Andres (NIH) and Stanley Garn (University of Michigan) for their help. See R. Floud, K. Wachter and A. Gregory, *Height, Health, and History* (Cambridge University Press, 1991) for a discussion of trends in height as indicators of social change. Also see Gina Kolata, *The New York Times*, July 30, 2006, p. 1.

6. See note 4 above for the data source. Cases with missing or implausible values (for instance, diastolic pressure below 30 mm) were excluded. The good news is that blood pressures have dropped by 5–10 mm since HANES2. Some of the decline may be due to increased use of anti-hypertensive medications.
7. This is exact for integer data and class intervals centered at the integers; more generally, if the mean over each class interval is the midpoint of the interval. Otherwise, it is only an approximation.
8. Data from the Current Population Survey, March 2005 (note 3 to chapter 3). See section 5.4 for discussion.
9. The basic reason is called *orthogonality* by statisticians. When errors in some situation arise from several independent sources, there is a simple and exact formula for getting the r.m.s. size of the total error: the r.m.s. errors combine like the sides of a right-angled triangle. With two orthogonal sources of error,

$$c = \sqrt{a^2 + b^2}$$

where a is the r.m.s. size of the errors coming from one source, b is the r.m.s. size of the errors coming from another source, and c is the r.m.s. size of the total error. This fact will be used several times in the book: in regression (part III), in computing the standard error for a sum (part V), and in computing the standard error for a difference (part VIII). No such formulas are possible for the average absolute value.

10. The 68%–95% rule works quite well even for many data sets which do not follow the normal curve. Take, for example, the lengths of the reigns of the 61 English monarchs through George VI. These average 18.1 years, with an SD of 15.5 years. Their histogram is shown below, and it is nothing like the normal curve. Still, 42 out of 61, or 69%, were within 1 SD of average. And 57 out of 61, or 93%, were within 2 SDs of average. (By definition, the length of a reign is the difference between its first and last years, as reported on pp. 274–75 of the 1988 *Information Please Almanac*; this example was contributed by David Lane, Modena, Italy.)

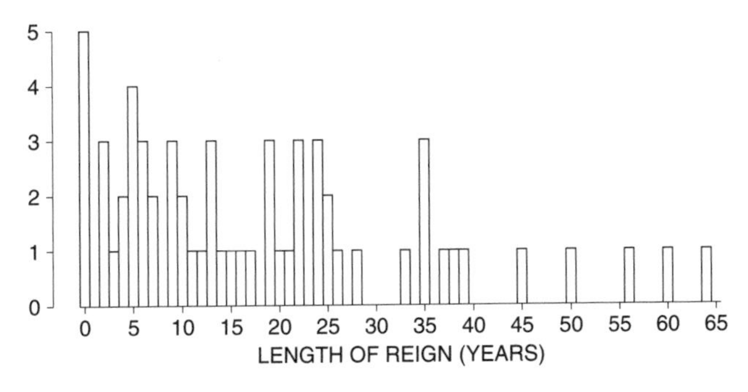

11. The square of the SD is called the *variance*. This is often used as a measure of spread, but we do not recommend it as a descriptive statistic. For instance, the SD of weight for American men is about 40 pounds: individual men are roughly 40 pounds away from average weight. The variance of weight is

$$(40 \text{ pounds})^2 = 1600 \text{ square pounds.}$$

12. However, this formula may be vulnerable to roundoff error.
13. See note 4 above for the data source. Cases with missing or implausible values (for instance, diastolic pressure below 30 mm) were excluded.
14. See note 4 above for the data source.
15. Patricia Ruggles, *Drawing the Line* (Urban Institute Press, Washington, D.C., 1990). The description of the underclass is paraphrased from p. 105. The book discusses the impact of definitions; see also chapter 5 on the time dimension. More recent data are available from SIPP (Survey of

Income and Program Participation). See *Dynamics of Economic Well-Being: Poverty 1996–1999* (P70-91), http://www.sipp.census.gov/sipp/. Also see *Statistical Abstract*, 2003, table 700. But see Ann Huff Stevens, "Climbing out of poverty, falling back in: Measuring the persistence of poverty over multiple spells," *Journal of Human Resources* (Summer 1999).

Chapter 5. The Normal Approximation for Data

1. Also called standard scores, *z*-scores, sigma-scores.
2. Here, we use "estimate" in its ordinary sense, "to compute approximately." "Estimate" also has a technical meaning in statistics, to be discussed in part VI.
3. See note 3 to chapter 3. The histogram, the SD, and the percentiles are computed from the CD-ROM, without weights, for primary families. The mean and SD are about right when incomes above $200,000 are censored. The percentiles are for the full distribution of primary-family incomes, the corresponding mean and SD are both about $70,000. (Note that high incomes have been top-coded by the Bureau to protect respondent confidentiality.)
4. The data are from a College Board press release, September 20, 1988; *1994 Profile of SAT and Achievement Test Takers* (Educational Testing Service, Princeton, N.J.); *Statistical Abstract*, 1993, table 265. Data for the fourth edition are from *Statistical Abstract*, 2003, table 264 and from *2005 College-Bound Seniors: Total Group Profile Report*, available at http://www.collegeboard.com. "College bound seniors" are students in a high-school graduating class who took the SAT at some point in their high-school years.

   The SD seems to drift upwards over time; it was 109 in 1975 and 113 in 1994 on the Verbal SAT. In 1994, the SAT was re-normed to a mean of 500 and an SD of 100; changes took effect in 1995. By 2005, the SD was 115. On the Math SAT (exercise 4) and many other such tests, the men have higher SDs than the women—120 vs 110; that is one reason why there are more men at the extremes of the distributions. See L. V. Hedges and A. Nowell, "Sex differences in mental test scores, variability, and numbers of high-scoring individuals," *Science* vol. 269 (1995) pp. 41–45.

   For a discussion of the decline in SAT scores, see Willard Wirtz et al., *On Further Examination. Report of the Advisory Panel on the Scholastic Aptitude Test Score Decline* (College Entrance Examination Board, New York, 1977). We thank Susan Bryce (ETS), Paul Holland (ETS), and Howard Wainer (National Board of Medical Examiners) for their help with the third edition.
5. March 2005 Current Population Survey; data from a CD-ROM supplied by the Census Bureau.
6. *2005 College-Bound Seniors: Total Group Profile Report.*

Chapter 6. Measurement Error

1. NBS started out as the Bureau of Weights of Measures, under the leadership of Ferdinand Hassler (born in 1770 in Switzerland, Director of the Bureau from 1830 to 1843). Hassler's great achievement was to bring the weights and measures used all over the U.S. into a high degree of consistency. The Bureau is now called NIST—The National Institute of Standards and Technology. We would like to thank H. H. Ku of the Bureau for his help with earlier editions.
2. Weight is used here instead of the more technical word *mass*. In 1983, the General Conference on Weights and Measures superseded the Treaty of the Meter, but weight was still defined by reference to a standard object. For an update on The Kilogram, see *Science* (May 12, 1995) p. 804.
3. Two major sources of chance error in the precision weighing at the Bureau are thought to be:
   - minute amount of play in the balance mechanism, especially at the knife edge;
   - slight variations in the position of the weights on the balance pans.
4. P. E. Pontius, "Measurement philosophy of the pilot program for mass calibration," *NBS Technical Note No. 288* (1966). The Bureau rejects outliers only "for cause, such as door-slam or equipment malfunction." Also see H. H. Ku, editor, *Precision Measurement and Calibration*, NBS Special Publication no. 300, vol. 1 (Washington, D.C., 1969).
5. Data were extracted from tapes supplied by the National Center for Health Statistics and by the Inter-University Consortium for Political and Social Research. Also see note 4 to chapter 4.
6. The data are from the March 2005 Current Population Survey; a CD-ROM was supplied by the Bureau of the Census. Primary families only; includes the income of related sub-family members.
7. J. N. Morris and J. A. Hardy, "Physique of London busmen," *Lancet* (1956) pp. 569–570. This reference was supplied by Eric Peritz (Jerusalem).
8. See note 14 in chapter 2.
9. The observational data are discussed in H. Zeisel, H. Kalven, Jr., and B. Buchholz, *Delay in the Court*, 4th ed. (Little, Brown & Co., 1959). The experiment was reported in Maurice Rosenberg, *The Pretrial Conference and Effective Justice* (Columbia University Press, 1964). The definitions are on p. 19, the numbers on p. 20, and the data in tables 8 and 9, pp. 48 and 52. For another discussion, see H. Zeisel. *Say It with Figures*, 6th ed. (Harper & Row, 1985, p. 141).

Part III. Correlation and Regression

Chapter 8. Correlation

1. There are methods for dealing with more than two variables, but these are quite complicated. Some matrix algebra is needed to follow the discussion. References—

   M. L. Eaton, *Multivariate Statistics: A Vector Space Approach* (John Wiley & Sons, 1983).
   D. A. Freedman, *Statistical Models: Theory and Practice* (Cambridge University Press, 2005).
   C. R. Rao, *Linear Statistical Inference and Its Applications*, 2nd ed. (John Wiley & Sons, 1973).
   J. A. Rice, *Mathematical Statistics and Data Analysis*, 3d ed. (Duxbury Press, 2005).
   H. Scheffé, *The Analysis of Variance* (John Wiley & Sons, 1961).
   G. A. F. Seber and Alan J. Lee, *Linear Regression Analysis* (John Wiley & Sons, 2003).

   There is a brief discussion of multiple regression in section 3 of chapter 12.
2. K. Pearson and A. Lee, "On the laws of inheritance in man," *Biometrika* vol. II (1903) pp. 357–462. Their table xxii gives the joint distribution, with heights rounded to the nearest inch. We added uniform noise to get continuous data. Due to independent randomization, the data set here differs slightly from the one in our first edition.
3. "Point of averages" is not a standard term.
4. H. N. Newman, F. N. Freeman, and K. J. Holzinger, *Twins: A Study of Heredity and Environment* (University of Chicago Press, 1937). In twin studies, the convention is to plot each twin pair twice; once as (x, y), and once as (y, x).
5. Data from the March 2005 Current Population Survey; a CD-ROM was supplied by the Bureau of the Census. Income is from 2004, and was censored at $150,000: this reduces the mean and the SD, but increases the correlation by a little. Starting in 1992, the Current Population Survey (CPS) reports educational attainment in categories, for instance, "1st–4th grade" or "some college, no degree." See *Monthly Labor Review*, September 1993, pp. 34–38. Single years of education were imputed from grouped data. The correlation between the imputed educational level and the ordered categorical variable used in the CPS is about 0.97.
6. See note 5. "Number of children" is number of own, never-married children under the age of 18. (The data are for women age 25–39, and the correlation depends to some extent on the age range that is used.) Weights must be used here, because the March CPS over-samples large families (note 3 to chapter 3).
7. When the correlation is 0, either slope can be used. "SD line" is not a standard term.
8. However, this formula can be vulnerable to roundoff error.
9. *Consumption Patterns of Household Vehicles 1985*. Residential Transportation Energy Consumption Survey. Energy Information Administration, Washington, D.C.
10. Marjorie Honzik (Institute of Human Development, Berkeley) was kind enough to supply the data.

Chapter 9. More about Correlation

1. The New York temperatures are measured at JFK; Boston, at Logan Airport. Data are from the Weather Undergound,

   http://www.wunderground.com
2. R. Doll, "Etiology of lung cancer," *Advances in Cancer Research* vol. 3 (1955) pp. 1–50. Report of the U.S. Surgeon General, *Smoking and Health* (Washington, D.C., 1964).
3. The idea goes back to W. S. Robinson, "Ecological correlations and the behavior of individuals," *American Sociological Review* vol. 15 (1950) pp. 351–57. Robinson gives the example of literacy and race, based on 1930 Census data. Our example is a replication; see note 5 to chapter 8 on the data source.

   If each cluster is bivariate normal, with a common regression line, then the slope and intercept can be estimated from the averages. Also see L. Goodman, "Ecological regression and the behavior of individuals," *American Sociological Review* vol. 18 (1953) pp. 663–64. For more discussion, see S. P. Klein and D. A. Freedman, "Ecological regression in voting rights cases," *Chance* vol. 6 (1993) pp. 38–43. Also see D. A. Freedman, "Ecological inference and the ecological fallacy," in N. J. Smelser and P. B. Baltes, editors, *International Encyclopedia of the Social & Behavioral Sciences* (Elsevier, 2001, vol. 6, pp. 4027–30).
4. E. Durkheim, *Suicide* (Macmillan, 1951, p. 164). We computed the correlation. Durkheim looked at averages of clusters of provinces, for which the correlation was 0.9. His conclusion was "Public instruction and suicide are identically distributed."
5. Multiple regression is some help, but may raise more questions than it answers (section 3, chapter 12). Also see note 11 to chapter 12.

6. For more discussion, see H. Zeisel, *Say It With Figures*, 6th ed. (Harper & Row, 1985, pp. 152ff.)
7. Data supplied by M. Russell from table 1 in D. Jablonski, "Larval ecology and macroevolution in marine invertebrates," *Bulletin of Marine Science* vol. 39 part 2 (1986) pp. 565–87. Also see *Science* vol. 240 (1988) p. 969.
8. References—

   R. Doll and R. Peto, *The Causes of Cancer* (Oxford University Press, 1981).
   B. E. Henderson, R. K. Ross and M. C. Pike, "Toward the primary prevention of cancer," *Science* vol. 254 (1991) pp. 1131–38.
   B. N. Ames, L. S. Gold and W. C. Willett, "The causes and prevention of cancer," *Proceedings of the National Academy of Science U.S.A.* vol. 92 (1995) pp. 5258–65.
   B. S. Hulka and A. T. Stark, "Breast cancer: Cause and prevention," *Lancet* vol. 346 (September 30, 1995) pp. 883–887.

   Figure 8 controls for age, but number of children would seem to be an important confounder (note 14 to chapter 2). Diet in the 1950s and 1960s would be at issue in the figure. There is strong evidence from epidemiology—and animal experiments—to show that over-eating is carcinogenic. The impact of fat (in isocaloric diets) is less clear. Two prospective studies support the ecological analysis: A. Schatzkin et al., "Serum cholesterol and cancer in the NHANES I epidemiologic followup study," *Lancet* ii (1987) pp. 298–301; W. C. Willett et al., "Relation of meat, fat, and fiber intake to the risk of colon cancer in a prospective study among women," *New England Journal of Medicine*, December 13, 1990, pp. 1664–71. But see D. Hunter et al., "Cohort studies of fat intake and the risk of breast cancer—a pooled analysis," *New England Journal of Medicine* vol. 334 (1996) pp. 356–61. Recent experimental evidence contradicts the hypothesis that low-fat diets are protective against cancer.

   A Schatzkin et al., "Lack of effect of a low-fat, high-fiber diet on the recurrence of colorectal adenomas," *New England Journal of Medicine* vol. 342 (2000) pp. 1149–55.
   R. L. Prentice et al., "Low-fat dietary pattern and risk of invasive breast cancer: The Women's Health Initiative randomized controlled dietary modification trial," *Journal of the American Medical Association* vol. 295 (2006) 629–642.
   S. A. Beresford et al., "Low-fat dietary pattern and risk of colorectal cancer: The Women's Health Initiative randomized controlled dietary modification trial," *Journal of the American Medical Association* vol. 295 (2006) 643–54.
9. National Assessment of Educational Progress, *The Reading Report Card* (ETS/NAEP, Princeton, N.J., 1985, p. 53). There is also a negative correlation with scores on standardized knowledge tests. See Lee R. Jones et al., *The 1990 Science Report Card: NAEP's Assessment of 4th, 8th, and 12th Graders* (U.S. Department of Education, Office of Educational Research and Improvement, Washington, D.C., 1992).
10. T. R. Dawber et al., "Coffee and cardiovascular disease: Observations from the Framingham study," *New England Journal of Medicine* vol. 291 (1974) pp. 871–74.
11. M. P. Rogin and J. L. Shover, *Political Change in California* (Greenwood Press, Westport, Connecticut, 1970, p. xvii).
12. See note 5 to chapter 8.
13. This replicates a study by M. and B. Rodin, "Student evaluations of teachers," *Science* vol. 177 (1972) pp. 1164–66. At the individual level, the correlations would be weaker; however, it is the sign which is interesting. More recent papers include the following—

   L. D. Barnett, "Are teaching questionnaires valid?" *Journal of Collective Negotiations in the Public Sector* vol. 25 (1996) pp. 335–49.
   A. G. Greenwald and J. M. Gillmore, "No pain, no gain? The importance of measuring course workload in student ratings of instruction," *Journal of Educational Psychology* vol. 89 (1997) pp. 743–51.
   M. Scriven, "A unified theory approach to teacher evaluation," *Studies in Educational Evaluation* vol. 21 (1995) pp. 111–29.
14. http://www.collegeboard.com/about/news_info/cbsenior/yr2005/links.html, table 3. In Connecticut, 86% of the seniors took the test. In Iowa, only 5% took the test. The reason: in Iowa and neighboring states, most seniors take the ACT—only those planning to attend elite schools take the SAT. The data are quite non-linear, but Connecticut and Iowa seem close to average, after adjustment for participation rate.

Chapter 10. Regression

1. These figures are rounded. The exact figures (unweighted):

   average height = 69.6 inches SD = 3.19 inches
   average weight = 177 pounds SD = 46.8 pounds $r = 0.414$

The data source is http://www.cdc.gov/nchs/about/major/nhanes/datalink.htm

2. See note 5 to chapter 8.
3. The term "graph of averages" is not standard. In principle, the graph depends on how finely the x-values are subdivided.
4. Surprisingly, the "graph of medians" is no more regular.
5. The average height of the fathers was 67.7 inches, with an SD of 2.74 inches; the average height of the sons was 68.7 inches, with an SD of 2.81 inches; r was 0.501. From the original cross-tab, the SDs were 2.72 and 2.75 inches, $r \approx 0.514$; the averages were barely affected by unrounding. See note 2 to chapter 8.
6. D. Kahneman and A. Tversky, "On the psychology of prediction," *Psychological Review* vol. 80 (1973) pp. 237–51.
7. Marjorie Honzik supplied the data. For comparison, in the March 2005 Current Population Survey, the correlation between educational level of husbands and wives was 0.63.

Chapter 11. The R.M.S. Error for Regression

1. *Statistical Methods for Research Workers* (Oliver and Boyd, 1958, p.182).
2. In multiple regression, the residuals can be plotted against the dependent variable, each independent variable, the fitted values, and omitted variables.
3. There were 60 families where the father was 64 inches tall (to the nearest inch); the sons averaged 66.7 inches in height, with an SD of 2.29 inches. There were 50 families where the father was 72 inches tall (to the nearest inch); the sons averaged 70.7 inches in height, with an SD of 2.30 inches.
4. A "homoscedastic" scatter diagram has more or less the same SD in any narrow vertical strip, but may have a non-linear trend. A "football-shaped" scatter diagram is homoscedastic, and the trend is linear; these diagrams look like sample data from a bivariate normal distribution.
5. In this part of the book, the focus is mainly descriptive. There is a finite data set, and each point is given equal weight. The regression line is viewed as smoothing $E\{Y \mid X\}$ in this finite population. Likewise, Y is predicted from X for a random element in the given population. The $\sqrt{1 - r^2}$ formula for prediction error is correct in this setting.

   If a regression line is fitted to a training sample and then used to make forecasts, there is a component of variance around the regression line as discussed above. The sample regression line also has a component of variance around the population line. This latter depends on X, of course, and is beyond the scope of this book. With a large training sample, however, the second component of variance may be rather small. Take example 1 in section 5, with a training sample of size 100. The first component of variance (around the regression line) is about 64. For a student who is z SDs away from average on the LSAT, the second component of variance—due to sampling error in the regression line—is about $0.64(1 + z^2)$. A fairly extreme case is $z = 3$. The total r.m.s. prediction error is about 8.4; ignoring the second component of variance gives 8.0.

   Of course, there is no free lunch. If you get far enough away from the center of the scatter diagram, the regression estimates become quite untrustworthy. For one thing, linearity may break down. For another thing, sampling error takes its toll. Here is one way to visualize the latter issue. The regression line has to go through the point of averages, and that point is quite solidly anchored—at least with a large-enough sample. However, there is some uncertainty about the slope. The further you go from the point of averages, the more impact the slope has: its uncertainty gets magnified.
6. See note 5 to chapter 8. Summary statistics are unweighted and simplified. Each dot may represent several women. Incomes were censored above at \$250,000.
7. Ages censored above 79. In public-use files, to protect respondent confidentiality, the Bureau reports ages 80–84 as 80, and ages 85+ as 85.
8. Station 3 in Lake Mead; 53 monthly geometric averages of available data; year-round; sampling period 1976–86; data provided by the late Jerome Horowitz in connection with a hearing on water quality standards.
9. The data were supplied by Marjorie Honzik.
10. Pitchers are excluded; however, sophomore slump can also be observed on the earned run average for pitchers. In some years, there are two "Rookies of the Year" in the same league. The problem was originally suggested by David Lane. We thank Sam Buttrey and Oren Tversky for expert advice on baseball statistics. Data for the third edition were provided by STATS, Skokie, Illinois. For the fourth edition, we used

    http://www.baseball-almanac.com

    In present format, awards for Rookies of the Year go back to 1949. Summary statistics depend on number of times at bat, and therefore whether averages are weighted by number of at-bats. (The good batters go to bat more frequently, but the effect diminishes as number of at-bats increases.)

There are some minor differences between the two leagues and between years; but on the whole, the results are fairly stable over time.

The following data are for the 1992–1993 seasons. Both leagues are pooled and simple averages are used. There were 588 men who played in both seasons; 438 had at least 25 at-bats in both seasons. The summary statistics for the 438 pairs of batting averages—

1992 average = 241 SD = 55
1993 average = 250 SD = 55
year-to-year correlation = 0.52.

There were 298 players who had at least 100 at-bats in both seasons. The summary statistics—

1992 average = 260 SD = 30
1993 average = 269 SD = 35
year-to-year correlation = 0.26.

The correlation may be attenuated due to restriction of range: many players with 25 to 100 at-bats had batting averages below 200, few players with over 100 at-bats do that poorly. Measurement error plays some role, too. There were 186 players who had at least 250 at-bats in both seasons. The summary statistics—

1992 average = 268 SD = 27
1993 average = 276 SD = 31
year-to-year correlation = 0.40.

11. HANES5 only has categorical data on education (less than high school, high school, more than high school); years are imputed from the Current Population Survey.

Chapter 12. The Regression Line

1. *Abhandlungen zur Methode der kleinsten Quadrate* (Berlin, 1887, p. 6). We follow the translation by L. Le Cam and J. Neyman, *Bayes-Bernoulli-Laplace Seminar* (Springer-Verlag, 1965, p. viii).
2. See note 5 to chapter 8. Each dot may represent several men. Incomes were censored above at $250,000, and summary statistics were rounded. The exact figures (unweighted):

   average educational level = 12.62 years SD = 3.31 years
   average income = $30,161 SD = $24,007 $r = 0.2713$

3. Data are from the Current Population Survey of March 2005. See note 5 to chapter 8.
4. Returns to education are discussed in—

   Orley Ashenfelter, Colm Harmon, and Hessel Oosterbeek,"A review of estimates of the schooling/earnings relationship, with tests for publication bias," *Journal of Labor Economics* vol. 6 (1999) pp. 453–70.

   David Card, "The causal effect of education on earnings," in Orley Ashenfelter and David Card, editors, *The Handbook of Labor Economics* (Elsevier, Amsterdam, 1999).

   David Card, "Estimating the return to schooling: Progress on some persistent econometric problems," *Econometrica* vol. 69 (2001) pp. 1127–60.

5. Suppose, for instance, that $y = a + bx + cx^2$. If subjects are randomized to different levels of x, and a regression line is fitted to the data, the slope does not predict the response of y to changes in x, except in some overall, average sense—averaged across subjects and their values of x. Of course, it may be possible to estimate the correct functional form from the data. In example 1, the relationship between education and income is non-linear. The value of a college degree relative to a high school degree—measured by the difference in average earnings—is substantially more than the slope would suggest; the relative value of a post-graduate degree is substantially less, probably because the women with advanced degrees are just starting out on their careers. For comparing high school and middle school, the slope is fine.
6. This equation is based on rounded values of the summary statistics supplied by IRRI.
7. Paraphrase of testimony by Franklin Fisher (MIT) in *Cuomo v. Baldrige* (80 civ. 4550 S.D.N.Y. 1987) on regression models for adjusting the Census, transcript pp. 2149ff.
8. Carried out by the late William Fretter, former professor of physics, University of California, as a demonstration in his elementary course; details are simplified.
9. Regression is appropriate here, in Berkson's case of the errors-in-variable model. The nominal values of the weights are fixed by the investigator, the actual value is subject to error; it is the nominal value which goes into the regression. When the value of the weight is measured subject to error, and the measured value goes into the regression, then the usual regression estimates are biased. A reference is G. W. Snedecor and W. G. Cochran, *Statistical Methods*, 6th ed. (Iowa State University Press, 1973). In effect, we are summarizing the data in table 1 by the two averages,

the two SDs, and r. A more natural summary would be based on

(i) the slope, intercept, and r.m.s. error of the regression line, and
(ii) the average and SD of the weights.

The statistics in (i) would be more or less invariant (up to sampling error) across experiments with different weights. Perhaps that is why many statisticians find it more natural to begin with the regression line and do correlation later. From our perspective, however, there are not so many examples with the kind of invariance shown by Hooke's law. With examples of a different texture—say, educational levels of husbands and wives—the correlation coefficient seems to offer the more natural summary. Furthermore, starting from the line makes it very difficult (at least in our experience) to explain r. That is why we start with r—and hope that devotees of the other approach will bear with us.

10. The sample size is 1,036, so the slope is real, not a fluke of sampling. (Cases with missing data on income or education are excluded; summary statistics are rounded.) Also see T. W. Teasdale et al., "Fall in association of height with intelligence and educational level," *British Medical Journal* vol. 298 (1989) pp. 1292–93. In data from HANES3 (1988–91), the slope of height on education for men age 25–34 was about 0.20 inches per year of schooling completed; there were 763 sample persons. See note 4 to chapter 4 for data sources.
11. For more discussion of regression models in the social sciences, see the Summer 1987 issue of *Journal of Educational Statistics*; *Sociological Methodology 1991*; *Foundations of Science* vol. 1, no. 1 (1995). Also see D. A. Freedman, *Statistical Models: Theory and Practice* (Cambridge University Press, 2005).
12. See note 4 to chapter 4. HANES5 reports incomes by category only; we used the correlation and summary statistics on heights from HANES5, summary statistics on income from the Current Population Survey of March 2005.
13. HANES3 had detailed questions on diet, including items on pizza and beer. The statistics in the exercise were about right for the U.S. population age 18–24. See note 4 to chapter 4 for data sources.
14. For a review of the literature, see S. J. Pocock, M. Smith and P. Baghurst, "Environmental lead and children's intelligence: a systematic review of the epidemiological evidence," *British Medical Journal* vol. 309 (1994) pp. 1189–97.
15. "Intersalt: an international study of electrolyte excretion and blood pressure. Results for 24 hour urinary sodium and potassium excretion," *British Medical Journal* vol. 297 (1988) pp. 319–28. The first analysis was a multiple regression, with 52×200 subjects pooled across centers; the second had separate regressions within each center. For commentary, see—

    D. A. Freedman and D. B. Petitti, "Salt and blood pressure: Conventional wisdom reconsidered," *Evaluation Review* vol. 25 (2001) pp. 267–87.
    N. Graudal and A. Galløe, "Should dietary salt restriction be a basic component of antihypertensive therapy? *Cardiovascular Drugs and Therapy* vol. 14 (2000) pp. 381–6.
    G. Taubes, "The (political) science of salt," *Science* vol. 281 (1998) pp. 898–907.

Part IV. Probability

Chapter 13. What Are the Chances?

1. For other views of chance, see—

    R. A. Fisher, *Statistical Methods and Scientific Inference*, 13th ed., reprinted by Oxford University Press, 1993, in J. H. Bennett, editor, *Statistical Methods, Experimental Design and Scientific Inference*.
    L. J. Savage, *Foundations of Statistics*, 2nd ed. (Dover, 1972).

    For discussion, see *Foundations of Science*, vol. 1 (1995) no. 1.
2. The 3rd edition was published in 1756, after de Moivre's death. It has been reprinted by Chelsea Publishing, New York, 1967.
3. *Statistical Abstract*, 2003, table 11.
4. W. Fairley and F. Mosteller, "A conversation about Collins," *University of Chicago Law Review* (1974).
5. The prosecutor calculated two "chances" for two "events," slipping back and forth between them. The first event was that the accused were guilty. The second event was that no other couple in Los Angeles matched the description. For a frequentist, the concept of chance does not apply so well. Even a Bayesian might find some difficulty here, because there is no reasonable chance model to connect the data with the hypothesis of guilt or innocence. (Also see note 6.)

    Were there other couples in Los Angeles matching the description? In principle, this might seem like a statistical issue, which could be settled by taking a sample. However, a calculation will show that sampling the couples in the city does not settle the issue with any reasonable level of confidence: a complete census is needed.

6. The "characteristics" of DNA used in matching are the variable number of tandem repeats (VNTRs) between loci on non-coding segments of DNA. References—

   Jurimetrics, vol. 34, no. 1 (1993).

   National Academy of Sciences/National Research Council, *DNA Technology in Forensic Science* (Washington, D.C., 1992).

   National Academy of Sciences/National Research Council, *DNA Forensic Science: An Update* (Washington, D.C., 1996).

   Federal Judicial Center, *Reference Manual on Scientific Evidence*, 2nd ed. (Washington, D.C., 2000).

   The "prosecutor's fallacy" consists in confusing the rate at which defendant's DNA occurs in the population (however well or poorly that may be estimated) with the probability that defendant is innocent; more generally—at least from a Bayesian perspective—of confusing

   $$P\{\text{evidence} \mid \text{innocence}\} \quad \text{with} \quad P\{\text{innocence} \mid \text{evidence}\}.$$

   See W. C. Thompson and E. L. Schumann, "Interpretation of statistical evidence in criminal trials: the prosecutor's fallacy and the defense attorney's fallacy," *Law and Human Behavior* vol. 11 (1987) pp. 167–87.

7. This exercise was suggested by D. Kahneman and A. Tversky, "Judgment under uncertainty: heuristics and bias," *Science* vol. 185 (1974) pp. 1124–31. Also see D. Kahneman, P. Slovic, and A. Tversky, editors, *Judgment under Uncertainty: Heuristics and Biases* (Cambridge University Press, 1982).

Chapter 14. More about Chance

1. From the dedication to the *Doctrine of Chances*.
2. Of course, P(A or B) = P(A) + P(B) if P(A and B) = 0. More generally,

   $$P(\text{A or B}) = P(\text{A}) + P(\text{B but not A}),$$

   which is analogous to the multiplication rule for dependent events.
3. For a more historical account of the correspondence between Pascal and Fermat, see pp. 88–89 of F. N. David, *Games, Gods and Gambling* (Buckinghamshire, England: Charles Griffin & Co., 1962). Sandrine Dudoit (U.C. Berkeley) gave assistance exceptionelle on *franglais*.

Chapter 15. The Binomial Coefficients

1. Apparently, Jia Xian discovered the binomial formula a little earlier—in the eleventh century. See Li Yan and Du Shiran, *Chinese Mathematics* (Oxford University Press, 1987, pp. 121 ff). Pascal's triangle is also called, perhaps more justly, Yang Hui's triangle.
2. The model is only approximate: there is a slightly better than even chance for a newborn to be male, and successive births in the family are slightly dependent.
3. This exercise was suggested by D. Kahneman and A. Tversky, "Judgment under uncertainty: heuristics and bias," *Science* vol. 185 (1974) pp. 1124–31.
4. On the Finnish twin study, see J. Kaprio and M. Koskenvuo, "Twins, smoking and mortality: A 12-year prospective study of smoking-discordant twin pairs," *Social Science and Medicine* vol. 29 (1989) pp. 1083–89. For references on the health effects of smoking, see note 1 to chapter 2. Some researchers in artificial intelligence remain skeptical about the evidence: P. Spirtes, C. Glymour, R. Scheines, *Causation, Prediction, and Search*, 2nd ed. (MIT Press, 2000).

   Data in the table are for current smokers; males and females are pooled. Kaprio and Koskenvuo considered a number of potential confounders. On the following, they saw little difference between smokers and non-smokers: alcohol, blood pressure, cholesterol, diabetes, coffee consumption, education, occupation, marital status, Eysenck personality inventory. Smokers exercised less, which may increase the risk of coronary heart disease; however, they were also thinner, which may reduce their risk. For U.S. data, see D. Carmelli and W. F. Page, "Twenty-four year mortality in World War II U.S. male veteran twins discordant for cigarette smoking," *International Journal of Epidemiology* vol. 25 (1996) pp. 554–59.
5. *Statistical Abstract*, 1988, table 268; 1994, table 305; 2003, table 309.
6. *San Francisco Chronicle*, December 9, 1975. The research report itself is much more sober: D. J. Ullyot et al., "Improved survival after coronary artery surgery in patients with extensive coronary artery disease," *Journal of Thoracic and Cardiovascular Surgery* vol. 70 (1975) pp. 405–13.

7. *Bouman v. Block*, 940 F.2d 1211 (9th Cir. 1991).
8. For national data, see J. H. Pryor et al., *The American Freshman: National Norms for Fall 2005* (Higher Education Research Institute, UCLA, 2006).
9. This exercise was suggested by *Economic Report of the President 1974*, pp. 147 ff.
10. T. W. Teasdale et al., "Degree of myopia in relation to intelligence and educational level," *Lancet* (December 10, 1988) pp. 1351–54.

Part V. Chance Variability

Chapter 16. The Law of Averages

1. *An Experimental Introduction to the Theory of Probability* (University of Witwatersrand Press, 1964). Kerrich went to teach in South Africa after World War II.
2. We distinguish between the difference as a number (in "absolute terms") and the difference as a percent. Absolute values also come in at a more technical level. Let X_n be the chance error after n tosses, that is, the number of heads minus half the number of tosses. Then X_n is a martingale, so $E\{X_{n+m}|X_n\} = X_n$; but $E\{|X_{n+m}|\big|X_n\} > |X_n|$.
3. "Chance process" is used in a non-technical sense. A "number generated by a chance process" is the observed value of a random variable.
4. Computer programs are deterministic, and therefore cannot generate numbers in a truly random way. However, a program can generate a sequence of numbers which look quite random. One method involves a multiplier M, which is a very big number. A "seed" x is chosen by the programmer: x is between 0 and 1. The computer works out M times x, which has an integer part and a decimal part:

   $$\ldots.\ aaaaaaaaaaaa\,.\,bbbbbbbbbbbb\ \ldots.$$

   Digits to the left of the decimal point are printed out as the first random number, and the decimal part is used as the seed for the next random number. For more discussion, see

   Jerry Banks, editor, *Handbook of Simulation* (Wiley, 1998).
   P. L'Ecuyer, "Efficient and portable combined random number generators," *Communications of the ACM* vol. 31 (1988) pp. 742–74.
   J. E. Gentle, W. Haerdle, and Y. Mori, editors, *Handbook of Computational Statistics* (Springer-Verlag, 2004).
   D. Knuth, *The Art of Computer Programming* vol. II (Addison–Wesley, 1998).
   P. A. W. Lewis and E. J. Orav, *Simulation Methodology for Statisticians, Operations Analysts, and Engineers* (Wadsworth & Brooks/Cole, 1988, chapter 5).
   G. Marsaglia, "Random numbers fall mainly in the planes," *Proceedings of the National Academy of Sciences* vol. 60 (1968) pp. 25–28.
   ———, "A current view of random number generators," *Proceedings of the Sixteenth Symposium on the Interface between Computer Science and Statistics* (1985) pp. 3–10.
5. "Sum of draws from a box" is not a standard term but it is lighter than "sum of independent, identically distributed, random variables." "Box model" is not standard either, although it seems to be catching on.
6. Let S_N be binomial with N trials and success probability p. We claim that $P\{S_{2n+2} > n+1\} > P\{S_{2n} > n\}$ for all p with $1/2 \le p < 1$. Indeed, let $f_{Nm}(p) = P\{S_N > m\}$. Then

   $$f'(p) = NP\{S_{N-1} = m\}. \tag{1}$$

   Let $g_n(p) = P\{S_{2n+2} > n+1\} - P\{S_{2n} > n\}$. By (1),

   $$\frac{d}{dp}g_n(p) = \frac{(2n)!}{n!\,(n-1)!}\left[\frac{4n+2}{n}p(1-p) - 1\right]p^n(1-p)^{n-1}. \tag{2}$$

   Now $(4n+2)/n > 4$. There is a p_0 with $1/2 < p < p_0$ such that $g_n'(p)$ is positive for $1/2 \le p < p_0$, negative for $p_0 < p < 1$, and 0 for $p = p_0$. Thus, g_n increases between $1/2$ and p_0, then decreases between p_0 and 1. Suppose $n > 1$. Then $g_n(1) = 0$, and it suffices to check the claim for $p = 1/2$. However,

   $$f_{2n,n}\left(\frac{1}{2}\right) = \frac{1}{2} - P\left\{S_{2n} = n \,\middle|\, p = \frac{1}{2}\right\}$$

   is strictly increasing with n. If $n = 1$, the claim reduces to showing that $3p^2 - 4p + 1 < 0$ for $1/2 \le p < 1$, which is easily verified.

Chapter 17. The Expected Value and Standard Error

1. Keno is the Las Vegas equivalent of bingo. There are 80 balls, numbered 1 through 80. On each play, 20 balls are chosen at random without replacement. If you bet on the single number 17, for example, you are betting that ball number 17 will be among the 20 that are chosen. Your chance of winning is $20/80 = 1/4$.
2. If X_i are independent and identically distributed with mean μ and variance σ^2, then
$$E(X_1 + \cdots + X_n) = n\mu$$
and
$$\text{var}\,(X_1 + \cdots + X_n) = \text{var}\,X_1 + \cdots + \text{var}\,X_n = n\sigma^2.$$
The SE, which is the square root of the variance, is then $\sqrt{n}\,\sigma$. That is the square root law.
3. In this book, we use SD for data and SE for chance quantities (random variables). This distinction is not standard, and the term SD is often used in both situations.
4. In parts II and III, we used standard units for data, centering on the average and scaling by the SD. Here, we make the transition to random variables, centering on the expected value and scaling by the SE.
5. We are using "estimate" in its ordinary sense, of approximation. Statisticians also use "estimate" in a more technical sense, to be taken up in part VI.
6. Consider 11 cells labelled $-5, -4, \ldots, +4, +5,$ and 10 balls. There are altogether
$$\binom{20}{10} = 184{,}756$$
ways to distribute the 10 balls into the 11 cells; each distribution defines a "box" with 10 tickets: for instance, if you put 4 balls into the cell "−1" and another 6 into the cell "3," you get the box $\lfloor 4\ \boxed{-1}\text{'s}\ 6\ \boxed{3}\text{'s} \rfloor$. We wrote a computer program to check all 184,756 boxes; 5,448 of them have mean 0, but one of these consists of 10 $\boxed{0}$'s. The program checked the remaining possibilities to verify the intuitively obvious result. On the combinatorics, see section II.4 of W. Feller, *An Introduction to Probability Theory and its Applications*, vol. I, 3rd ed. (John Wiley & Sons, 1968). Also see note 6 to chapter 16. For an example with non-monotone behavior, consider the sum of 100 or 200 draws from the box
$$\lfloor 5\ \boxed{-16}\text{'s}\ 5\ \boxed{16}\text{'s} \rfloor$$
Let S_n be the sum of n draws from the box, which is necessarily a multiple of 32. Thus, if $-15 \le S_{100} \le 15$ then $S_{100} = 0$; if $-30 \le S_{200} \le 30$ then $S_{200} = 0$. And $P\{S_{100} = 0\} > P\{S_{200} = 0\}$.
7. The idea of using "big" and "small" to label the values is due to the statistics group at Southern Methodist University.
8. E. O. Thorp, *Beat the Dealer* (Random House, 1966). Some side bets at baccarat also have positive expected values. So do some bets on the favorites (curiously) at race tracks. See R. T. Thaler and W. T. Ziemba, "Parimutuel betting markets: Racetracks and lotteries," *Journal of Economic Perspectives* vol. 2 (1988) pp. 161–74.

Chapter 18. The Normal Approximation for Probability Histograms

1. The Hewlett Packard HP 15C.
2. A mathematical discussion can be found in Chapter 7 of W. Feller, *An Introduction to Probability Theory and its Applications* vol. I, 3rd ed. (John Wiley & Sons, 1968).
3. If the number of draws is very large, it may be helpful to group ranges of values together, as in figure 10 (for products). Likewise if the tickets do not show whole numbers. Even if the number of draws is moderate, and there are whole numbers on the tickets, it may be helpful to use wider class intervals: this is so when the differences of the numbers on the tickets have a common divisor bigger than 1. For example, suppose you bet \$1 on the toss of a coin, 100 times. Your net gain is like the sum of 100 draws made at random with replacement from the box $\lfloor \boxed{-1}\boxed{+1} \rfloor$. The possible values for the net gain are even: $0, \pm 2, \pm 4, \ldots$ The histogram can be drawn with rectangles of width 1 centered at these values. However, with the method of section 4, rectangles of width 2 give better results. The problems in this book do not raise this sort of issue.
4. Computed on the HP 15C.
5. The continuity correction can be used when the tickets in the box are integers whose differences have no common divisor (except 1). This is called "aperiodicity." If the numbers have a common divisor bigger than 1, or the tickets have values other than integers, don't use the continuity correction without further thought. Also see notes 3 and 9.

6. A mathematical analysis of the skewness is provided by the Edgeworth expansion. See Chapter 16 in W. Feller, *An Introduction to Probability Theory and its Applications* vol. II, 2nd ed. (John Wiley & Sons, 1970).
7. The waves can be explained as follows. If the box were [1] [1] [9], the possible values for the sum would be 25, 33, 41, ... separated by gaps of 8. If the box were [2] [2] [9], the possible values for the sum would be 50, 57, 64, ... separated by gaps of 7. The box in figure 9 is intermediate between these two, and the peak-to-peak distance alternates between 7 and 8. Another way to look at it: the peaks reflect the distribution of the number of 9's among the 25 draws.
8. The shape of the histograms in figure 10 may be a little surprising. However, if $X_1, X_2, \ldots$ are the successive rolls of the die, then it is

$$(X_1 X_2 \cdots X_n)^{1/\sqrt{n}}$$

which is approximately log normal after centering. A probability histogram for the 5th root of the product of 25 rolls is shown below, and it has the right shape. The probabilities were computed using a combinatorial algorithm, and the wiggles are real. (The product of 25 rolls of a die has the form $2^a 3^b 5^c$ for non-negative integers a, b, c, lending itself to gaps and wiggles.)

The logarithm (base 10) of the product of 25 rolls is the sum of 25 logarithms. Each has mean 0.4762 and SD 0.2627, so the sum of 25 logs has expected value $25 \times 0.4762 \approx 11.91$ and standard error $\sqrt{25} \times 0.2627 \approx 1.31$. The sum of 25 logs is already quite close to normally distributed. Take the bottom panel in figure 10, for the product of 25 rolls. The axis cuts off at 10^{13}, which is 13 on the log scale, or 0.83 in standard units. About 20% of the probability is to the right of this value. The width of each rectangle in the histogram is 10^{11}. The first rectangle covers the interval from $-\infty$ to 11 on the log scale, which in standard units is $(-\infty, -0.69)$. This interval contains about 25% of the probability!

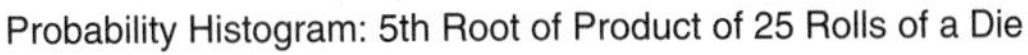

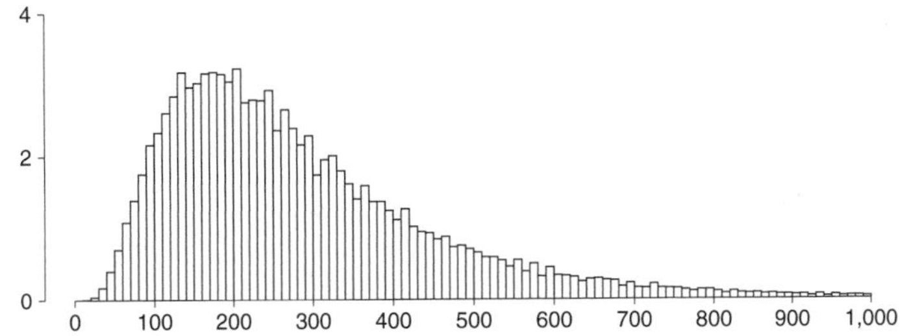

9. The tacit assumptions: nonzero SD, and a finite number of tickets in the box with integer values. Suppose for simplicity that the numbers in the box are aperiodic; let μ be their mean and σ their SD. Let $h_n(x)$ be the histogram for the sum of n draws, plotted by our convention: each rectangle has width 1, centered on a possible value. Let $\phi(z)$ be the standard normal density. Then

$$\sigma\sqrt{n}\, h_n(n\mu + \sigma\sqrt{n}z) \to \phi(z).$$

The "histogram in standard units" euphemizes this change of scale. See W. Feller, *An Introduction to Probability Theory and its Applications*, vol. II, 2nd ed. (John Wiley & Sons, 1971, pp. 517, 540).

10. Suppose the tickets in two boxes have the same average, and average absolute deviation from average. If they also have the same SD, the asymptotic behavior of the sums will be the same. If not, not. An example would be

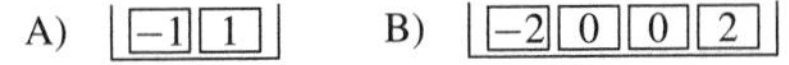

In both boxes, the tickets average out to 0, and the average absolute deviation from average is 1. But the SD for box A is 1, while the SD for box B is about 1.4. Consequently, the sum of 100 draws from box B is about 1.4 times as spread out (by any reasonable measure of spread) than the sum of 100 draws from box A. It is the average and SD of the numbers in the box which control the asymptotic distribution of the sum: other measures of location and spread do not.

11. Let n denote the number of draws, and k the number of repetitions. The implicit condition is that $k/\sqrt{n} \log n \to \infty$. See D. A. Freedman, "A central limit theorem for empirical histograms," *Zeitschrift für Wahrscheinlichkeitstheorie* vol. 41 (1977) pp. 1–11.

Part VI. Sampling

Chapter 19. Sample Surveys

1. The chapter opening quote is from *The Adventure of the Copper Beeches*. We found it in Don McNeil, *Interactive Data Analysis* (John Wiley & Sons, 1977).
2. References on sampling

   LESS TECHNICAL

   N. M. Bradburn and S. Sudman, *Polls and Surveys* (Jossey-Bass Inc., 1988).
   A. Campbell, G. Gurin and W. Miller, *The Voter Decides* (Row-Peterson, Evanston, 1954).
   Jean M. Converse, *Survey Research in the United States: Roots and Emergence, 1890–1960* (University of California Press, 1987).
   Shari Seidman Diamond, Reference Guide on Survey Research, in *Reference Manual on Scientific Evidence*, 2nd ed. (Federal Judicial Center, Washington, D.C., 2000).
   George Gallup, *The Sophisticated Poll Watcher's Guide* (Princeton Opinion Press, 1972).
   Herbert Hyman et al., *Interviewing in Social Research* (University of Chicago Press, 1954).
   Frederick Mosteller et al., *The Pre-Election Polls of 1948* (Social Science Research Council, New York, 1949).
   Mildred Parten, *Surveys, Polls and Samples* (Harper & Row, 1950).
   F. F. Stephan and P. J. McCarthy, *Sampling Opinions* (John Wiley & Sons, 1958).
   Hans Zeisel and David Kaye, *Prove It with Figures* (Springer, 1997).

   MORE TECHNICAL

   W. G. Cochran, *Sampling Techniques*, 3rd ed. (John Wiley & Sons, 1977).
   Robert M. Groves et al., *Survey Methodology* (Wiley-Interscience, 2004).
   M. H. Hansen, W. N. Hurwitz and W. G. Madow, *Sample Survey Methods and Theory* (John Wiley & Sons, 1953).
   Leslie Kish, *Statistical Design for Research* (John Wiley & Sons, 1987).
   Seymour Sudman, *Applied Sampling* (Academic Press, 1976).
3. All quotes are from the *New York Times* (Oct. 1–15, 1936).
4. For help with the first and second editions, we thank Diane Colasanto, Laura Kalb, Jack Ludwig, Coleen McMurray, and Paul Perry of the Gallup Poll. Figure 3 was typeset from copy provided by the Gallup organization, and reproduced with their kind permission. For the third edition, we thank David Moore, Kim Neighbor, and Lydia Saad; the description of the 1992 survey is based on conversations with them. For the fourth edition, we again thank Lydia Saad.
5. See Parten, p. 393, and Stephan and McCarthy, pp. 241–70 (note 2).
6. For another discussion, see M. C. Bryson, "The *Literary Digest* poll: Making of a statistical myth," *American Statistician* vol. 30 (1976) pp. 184–85. Bryson agrees that the *Digest* poll was spoiled by non-response bias. However, he discounts selection bias as a problem, and questions whether the *Digest* really drew on phone books for its mailing list (that is, the list of people to be polled). Our primary source of information about the *Digest* poll was George Gallup—a shrewd, interested, and first-hand observer. He maintained that the *Digest* used phone books, lists of automobile owners, and its own subscription list as the source for the mailing list. This account is confirmed, at least in essentials, by others like Parten, or Stephan and McCarthy (see note 2).

   The *Digest* did not publish any very full account of its procedures that we could find. However, something can be learned by reviewing the issues of the *Digest* for the period August 22 through November 14, 1936. For instance,

   > The Poll represents thirty years' constant evolution and perfection. Based on "commercial sampling" methods used for more than a century by publishing houses to push book sales, the present mailing list is drawn from *every telephone book in the United States*, from the rosters of clubs and associations, from city directories, lists of registered voters, classified mail order and occupational data. [Aug. 22, p. 3, our italics.]

   The article goes on to explain that the list was put together for the 1924 election, but was subsequently revised by "trained experts." Most of the names on the list were held over from year to year, and the list was used for polls between elections (Aug. 29, p. 6). Drawing on lists of registered voters seems to have been an innovation for 1936, and such lists were used only for certain "big cities" (Oct. 17, p. 7). Clearly, the *Digest* expected to get the percentages right (Aug. 22, p. 3):

   > Once again, THE DIGEST was asking more than ten million voters—one out of four, representing every county in the United States—to settle November's election in October. Next week, the first answers from these ten million will begin the incoming tide of marked ballots, to be *triple-checked*, verified, *five times* cross-classified and totaled. When the last figure has

> been totted and checked, if past experience is a criterion, the country will know *to within a fraction of 1 percent* the actual popular vote of forty millions. [Their italics.]

The *Digest* was off by 19 percentage points. Why? By modern standards, the *Digest*'s mailing list was put together in a somewhat arbitrary way, and it was biased: it excluded substantial, identifiable portions of the community. Bryson suggests that if the *Digest* had somehow managed to get 100% response from its list of 10 million names, it would have been able to predict the election results. This seems unlikely. As we say in the text, there were two main reasons for the *Digest*'s error: selection bias and non-response bias.

7. This 65% is typical of four-call probability samples in the late 1980s. The response rate declined from about 75% in 1975, and 85% in 1960. This decline is a major worry for polling organizations. In 2005, the best face-to-face research surveys in the U.S., interviewing a randomly-selected adult in a household, get response rates over 80%. Response rates for the Current Population Survey—around 95%—are discussed in chapter 22.
8. This section draws on the book by Mosteller et al. (note 2).
9. Stephan and McCarthy, p. 286 (note 2).
10. It is tempting to confuse quota sampling with stratified sampling, but the two are different. Suppose, for instance, that it is desired to draw a sample of size 200 from a certain town, controlling for sex; in fact, making the number of men equal to the number of women. A quota sampler could in principle hire two interviewers, one to interview 100 men, the other to interview 100 women. In other respects, the two interviewers would pick whomever they wanted. This is not such a good design. By contrast, a stratified sample would be drawn as follows:

    - Take a simple random sample of 100 men.
    - Independently, take a simple random sample of 100 women.

    This is a better design, because human bias is ruled out.
11. The list of units to be sampled is the "sampling frame," and the first step in taking a probability sample is drawing up the sampling frame. This can be quite difficult, and there is often some degree of mismatch between the frame and the population. With area samples, the frame is a list of geographic units.
12. Details of such designs are discussed in chapter 22. We suggest that stratification is needed to draw the sample in a way that keeps the costs reasonable, but in many polls the stratification does little to reduce sampling error. To take a hypothetical example, suppose a country consisted of two regions, East and West. In the East, 60% of the voters are Democrats; in the West, only 40% are. East and West are equal in size, so the overall percentage of Democrats is 50%. Now, two survey organizations take samples to estimate the overall percentage of Democrats. The first one uses a simple random sample of size n. The standard error is $50\%/\sqrt{n}$. The second one stratifies, taking a simple random sample of size $n/2$ in the East, and an independent simple random sample of size $n/2$ in the West. The standard error is $\sqrt{0.4 \times 0.6} \times 100\%/\sqrt{n}$. Since $\sqrt{0.4 \times 0.6} \approx 0.49$, the reduction in SE is minimal. Furthermore, in this artificial example, the difference between the regions is much larger than the difference observed in real elections. So the advantage of stratification in predicting real elections is even less. (By contrast, when sampling economic units like companies or establishments, stratification can really help to reduce variance; also see note 5 to chapter 20.)
13. The Gallup Poll uses variants of random-start list sampling. In the first 3 stages, probability is proportional to size; in effect, each unit appears on the list with multiplicity equal to its size. Within each of the four geographic regions, there is a stratum of rural areas, which is handled somewhat differently from the urban areas.
14. The Gallup organization explains "This method of selection within the household has been developed empirically to produce an age distribution by men and women separately which compares closely with the age distribution of the population."
15. Strictly speaking, for the Gallup Poll it is possible to compute sampling probabilities only for households, not for individuals—due to the rule used in selecting individuals within households. Non-response is another complication. We thank Ben King (Florida) for useful discussions on this point. Often, probability methods are designed so that each individual in the population will get into the sample with an equal chance, so the sample is "self-weighting." However, the Gallup poll interviews only one person in each household selected for the survey. This discriminates against people who live in large households; not enough of them are represented in the sample. (See sketch at top of next page.) An adjustment is made to correct for this bias, by giving more weight to the people from large households who do get into the sample. Household size is obtained from question 18, figure 3, p. 347.
16. Paul Perry, "A comparison of the voting preferences of likely voters and likely nonvoters," *Public Opinion Quarterly* vol. 37 (1973) pp. 99–109. Who has voted is a matter of public record; how they voted, of course, is not.

Household bias. Imagine selecting one of the two households below at random: then select a person at random from the selected household. This produces a sample of size one. A person in the small household has a better chance of getting into the sample than a person in the large household.

17. The Gallup Poll "secret ballot" is not secret; ballots are connected to questionnaires.
18. After 1992, the Gallup Poll changed the design. They stratified the sample by four census regions. Within each region, they chose a random sample of residential telephone banks, and dialed random numbers within sampled banks.
19. In 2005, for a good commercial telephone survey, about 1/3 of the telephone numbers dialed do not answer. If someone answers the phone, about 2/3 hang up rather quickly. However, if the interviewer gets through to a person, and engages them for a minute or two, the completion rate is around 95%.
20. L. Belmont and F. Marolla, "Birth-order, family-size, and intelligence," *Science* vol. 182 (1973) pp. 1096–1101. On the average, intelligence decreases with birth order and family size, even after controlling for family background. Also see R. B. Zajonc, "Family configuration and intelligence," *Science* vol. 192 (1976) pp. 227–36. However, the association may be to due to residual confounding by social class. See J. L. Rodgers, H. H. Cleveland, E. van den Oord, and D. C. Rowe, "Resolving the debate over birth order, family size, and intelligence," *American Psychologist* vol. 55 (2000) pp. 599—612. The Belmont-Marolla study is discussed again in exercise 40 on p. 575.
21. Kenneth Stampp, Professor Emeritus of History, University of California, Berkeley. This was a WPA project, and the subjects must have been in their seventies!
22. R. W. Fogel and S. L. Engerman, *Time on the Cross* (New York: W. W. Norton & Company, 1989, p. 39); *Evidence and Methods* (Little, Brown & Company, 1974, p. 37). A careful critique is by Richard Sutch, "The treatment received by American slaves," *Explorations in Economic History* vol. 12 (1975) pp. 335–438.
23. L. L. Bairds, *The Graduates* (ETS, Princeton, N.J., 1973).
24. Discussion by A. L. Cochrane in The Medical Research Council, *The Application of Scientific Methods to Industrial and Service Medicine* (HMSO, London, 1951, pp. 36–39).
25. A. C. Nielsen, *1987 Annual Report on Television*; *New York Times*, March 10, 1997, p. C1.
26. The story was published on September 11, 1988. The source was Raymond A. Eve and Dana Dunn, "Psychic powers, astrology and creationism in the classroom," *American Biology Teacher* vol. 52 (1990) pp. 10–21. The investigators got 190 responses out of their sample of 387 drawn from the list of 20,000 names, which in turn was a systematic sample from the National Register of High School Life Science and Biology Teachers. This is a good study which merits attention. Unfortunately, in the first few printings of the second edition, we relied on the newspaper description, which omitted crucial details about the sample; we drew the wrong conclusion about non-response bias.
27. Based on an example in Parten's book (note 2).
28. From *Time on the Cross* (note 22). Anne Arundel was the wife of the second Lord Proprietary of Maryland, Cecil Calvert. The two main slave auction houses of the time were at Annapolis (Arundel County) and Charleston (South Carolina). We thank Sharon Tucker for the Maryland history.
29. E. K. Strong, *Japanese in California* (Stanford University Press, 1933).
30. *San Francisco Chronicle*, December 10, 1987; letter by Stephen Peroutka to *New England Journal of Medicine* vol. 317 (1987) pp. 1542–43.
31. This example was suggested by D. Kahneman and A. Tversky, "Judgment under uncertainty: heuristics and bias," *Science* vol. 185 (1974) pp. 1124–31.

Chapter 20. Chance Errors in Sampling

1. The example is loosely suggested by followup studies on Cycle I of the Health Examination Survey, a probability sample of persons age 18–79 in the U.S. The study was done in 1960–61. The sample size was 6,672, of whom 3,091 were male.
2. Sir Arthur Conan Doyle, *The Sign of Four* (J. B. Lippincott, 1899; Ballantine Books, 1974, p. 91). Holmes attributes the thought to Winwood Reade.

3. The histograms in figure 3, like the calculations in example 2, are based on sampling with replacement. In this example—with a sample of 400 from a population of 100,000—there is little difference between sampling with or without replacement. Details are in the next section. The vertical axis is drawn in percent per standard unit.
4. Data for the whole U.S. are available from *Statistical Abstract*, 2003: table 63 gives marital status; table 229, educational level for age 25+; table 693, personal income; and tables 490ff, income tax returns.
5. The issues may be different in other contexts. For instance, suppose you are sampling from two different strata, and want to allocate a fixed number of sampling units between the two. If the object is to equalize accuracy of the two estimated percentages, a reasonable first cut is to use equal sample sizes. If the object is to equalize accuracy of estimated numbers, or to estimate a percentage that is pooled across the strata, a larger sample should generally be drawn from the larger stratum. Gains in accuracy from stratification—as opposed to simple random sampling—should not be overestimated (note 12 to chapter 19).
6. Voting-age population by state comes from *Statistical Abstract*, 2006, table 408; election results by state from table 388. The population for NM was closer to 1.4 million; for TX, 16 million.

Chapter 21. The Accuracy of Percentages

1. Sir Arthur Conan Doyle, *A Study in Scarlet* (J. B. Lippincott, 1893; Ballantine Books, 1975, p. 136).
2. The technique described in the text is a special case of what statisticians now call the "bootstrap" method for estimating standard errors. See Brad Efron and Rob Tibshirani, *An Introduction to the Bootstrap* (Chapman & Hall, 1993). For a cautionary note, see L. D. Brown, T. T. Cai, and A. DasGupta, "Confidence intervals for a binomial proportion and asymptotic expansions,' *Annals of Statistics* vol. 30 (2002) pp. 160–201.
3. For college enrollments in the whole U.S., see *Statistical Abstract*, 2003, tables 278ff.
4. Family income data by educational level of head of household, for the whole U.S., is reported in *Statistical Abstract*, 2003, table 689.
5. *Statistical Abstract*, 2003, table 744. Exercises 4 and 5 more or less match these data.
6. Suppose we draw at random with replacement. As the sample size $n \to \infty$,

$$P\{\hat{p} - k\,\text{SE} < p < \hat{p} + k\,\text{SE}\}$$

tends to the area under the normal curve between $-k$ and k; this is less than 1.
7. This book takes a strict frequentist line. Other views are cited in note 1 to chapter 13. Many colleagues will feel that we shut our eyes and walked across an intellectual minefield in this section. We hope they will be charitable in their judgment of the results.
8. This picture was proposed by Juan Ludlow, CIMASS/UNAM, Mexico.
9. The standard errors applicable to simple random samples are often computed for samples of convenience. In some contexts, the results may be useful. For example, the objective may be to show that the sampling procedure at issue is quite different from simple random sampling (as in exercise 26 on p. 435).
10. *Statistical Abstract*, 2003, table 977. For national data on automobile ownership, see table 976.
11. D. Ravitch and C. E. Finn, Jr., *What Do Our 17-Year-Olds Know?* (Harper & Row, 1987). The sample was highly designed.
12. For national data, see *Statistical Abstract*, 2003, tables 1125–27.
13. The rules of Keno are explained in note 1 to chapter 17. The chance for a single number is $20/80$, because there are 20 draws. The chance for a double number is $(20 \times 19)/(80 \times 79)$.

Chapter 22. Measuring Employment and Unemployment

1. We are grateful to many people at Census Bureau for their help with previous editions, including Sherry Courtland, Charles Jones (deceased), Donna Kostanich, Marty Riche, and Jay Waite. For the 4th edition, we thank Louis Kincannon, Greg Weyland, and Cindy Taeuber.

   The Bureau of the Census is responsible for the sample design, collection, and production of data, as well as calculation of the estimates and their standard errors. The Bureau of Labor Statistics does the seasonal adjustments, and is responsible for the publication and economic interpretation of the results. Some useful references on the Current Population Survey—

   http://www.bls.gov/cps
   http://www.census.gov/cps
   Bureau of the Census, *The Current Population Survey: Design and Methodology*, Technical Paper No. 66 (2006).

Employment and Earnings vol. 52, no. 12 (December, 2005).
Technical Documentation, March 2005 Current Population Survey.
M. Thompson and G. Shapiro, "The Current Population Survey: An overview," *Annals of Economics and Social Measurement* vol. 2 (1973).

2. There are a few exceptions in the Northeast; Hawaii is also exceptional.
3. In November 2005, the sampling fraction was 1/2160 overall, 1/283 for Wyoming, 1/3286 for Texas. Rates will change over time, especially when the sample is redesigned after the 2010 Census. A number of interesting points are glossed over in the main text. (i) The discussion focused on the "unit frame," for the household population. There is a separate frame for group quarters, and an "area frame" to sample geographical areas which are sparesly population, or where addresses are poorly defined. (ii) Some large USUs are treated differently. (iii) Within a PSU, the Bureau takes a random-start list sample. The list is organized to reduce variance. In effect, this stratifies the USUs. (iv) Fairly detailed information is collected on persons age 15, and on military personnel living off-post; some information is collected on persons age 14 and below. These data are not published.
4. For the 1995 design, the precision of monthly estimates in the 11 largest states was equalized, as well as the precision of the annual averages in the remaining 40 states; the District of Columbia counts as a state, and "precision" means the coefficient of variation of the estimated number of unemployed persons. For 2005, there were separate controls on several large substate areas, including the county of Los Angeles and the city of New York.
5. Persons who worked in their own business or profession, or on their own farm, are counted as employed; so are persons who worked at least 15 hours (even without pay) in a family business. Persons on layoff, but expecting to be recalled, need not be looking for work in order to be counted as unemployed. There is another classification—"discouraged workers"—for persons who want a job, but are no longer actively looking for a job because they believe no jobs are available. The official definitions have remained essentially unchanged since the survey was first conducted in 1940. Some revisions were made in 1994, as part of the redesign of the survey; and the questionnaire changed appreciably. See note 13 below.
6. Inmates of penal and mental institutions, and the military, are excluded from "the civilian noninstitutional population." Data in tables 1–3 are not seasonally adjusted.
7. The total labor force equals the civilian labor force plus the military.
8. The procedure for getting the weights is sometimes called "ratio estimation." The technique actually used by the Bureau is a bit more complicated, since they also cross-classify by other demographic variables. Furthermore, they make an adjustment to correct for non-interviews, and for known demographic differences between the sample PSUs and the country, using Census data. They make another adjustment to the current estimates using information from the previous month's sample. Finally, they adjust the weights in an effort to compensate for differential coverage in the Census (note 12 below).
9. The procedure used in the 1970s involved linearizing the estimates first, and computing some building-block variances by the half-sample method. It is sketched by R. S. Woodruff, "A simple method for approximating the variance of a complicated estimate," *Journal of the American Statistical Association* vol. 66 (1971) pp. 411–14. In the 1980s, a partially balanced replication method was used. See Janice Lent, "Variance estimation for Current Population Survey small area estimates," *Proceedings of the Section on Survey Research Methods* (American Statistical Association, August, 1991). A complete description of procedures for the period 1995–2005 is available in Technical Paper No. 66.
10. The stratification reduces the standard errors, as does the use of ratio estimates. But the clustering pushes the standard errors up.
11. *Statistical Abstract*, 2003, tables 396 and 419.
12. Census undercount in 1980 is discussed in R. E. Fay, J. S. Passel, J. G. Robinson and C. D. Cowan, *The Coverage of the Population in the 1980 Census*, Bureau of the Census, 1988; also see *Survey Methodology* vol. 18, no. 1, June, 1992. For discussions of the undercount in 1990, and proposals for adjustment, see *Jurimetrics* vol. 34, no. 1 (fall 1993), *Statistical Science* (November 1994), and *Evaluation Review* (August 1996).

    On proposed adjustments for Census 2000, see *Society*, vol. 39 no. 1 (November, 2001); D. A. Freedman and K. W. Wachter, "On the likelihood of improving the accuracy of the census through statistical adjustment," in *Science and Statistics: A Festscrift for Terry Speed*, Institute of Mathematical Statistics Monograph 40 (2003).

    On coverage differences between the Current Population Survey (CPS) and the Census, see pp. G5–6 in the technical documentation to the March 2005 CPS.
13. The evidence suggests that the Bureau can find out reasonably well who has a full-time job, and who is outside the labor force. The problem is with a third group, the marginal workers who are

classified either as part-time workers, or with a job but not at work, or unemployed. For example, results from the reinterview program for the last quarter of 1987 can be tabulated as shown below.

Thus, 7,511 people were reinterviewed; 3,015 were classified as working full time in non-agricultural industries at the original interview, but 2,997 were classified that way—presumably correctly—at reinterview. The decrease is 0.6 of 1%. On the other hand, the number of part-time workers went up by 4.5%, and the number of unemployed went up by 3.7%. The overall number of unemployed—based on the original interviews—was estimated as about 7,000,000. Since 3.7% of 7,000,000 = 250,000, the bias in the estimate amounts to several hundred thousand people. The number of unemployed persons in these data is small, so the calculation is only to illustrate the idea. Also see K. W. Clarkson and R. F. Meiners, "Institutional changes, reported unemployment, and induced institutional changes," Supplement to *Journal of Monetary Economics* (1979).

	Labor force status at original interview						
Labor force status at reinterview	Agri-culture	Employed in nonagricultural industry Full-time	Part-time	Not at work	Unem-ployed	Not in labor force	Total
Agriculture	117	1	2	0	0	2	122
Non-agriculture							
full-time	0	2,967	22	2	3	3	2,997
part-time	1	45	1,187	5	4	28	1,270
not at work	0	2	0	137	2	4	145
Unemployed	0	0	2	2	226	21	251
Not in labor force	0	0	2	1	7	2,716	2,726
Total	118	3,015	1,215	147	242	2,774	7,511

Notes: After reconciliation, before weighting; 75% sample.
Source: Bureau of the Census, Statistical Methods Division

In 1994, there was a major revision to the CPS questionnaire; new "probe" questions were added on hours of work and duration of unemployment; the definitions of "discouraged workers" and involuntary part-time workers were changed. See the *Monthly Labor Review* for September 1993, and *Employment and Earnings* for February 1994. Changing the questions made a noticeable impact on the numbers, confirming that biases in the data (although small) are probably larger than sampling error. Also see T. J. Plewes, "Federal agencies introduce redesigned Current Population Survey," *Chance* vol. 7, no. 1 (1994) pp. 35–41.

In theory, ratio estimates can create small biases. In practice, however, with reasonably large samples the bias from this source is negligible. There is one problem the Bureau does not have: household bias (note 15 to chapter 19). The reason is that the sample includes all persons age 16 and over in the selected households, not just one person that the interviewer finds at home.

14. Based on an example in Hyman's book (note 2 to chapter 19).
15. http://ag.ca.gov/newsalerts/2005/05-018.htm

Chapter 23. The Accuracy of Averages

1. The draws can be made with replacement, or without. In the second case, the number of draws—and the number of tickets left in the box—both have to be large; the correction factor may be needed for computing the SE (chapter 20, section 3). See T. Höglund, "Sampling from a finite population: a remainder term estimate," *Scandinavian Journal of Statistics* vol. 5 (1978) pp. 69–71. If the number of draws is small, the distribution of the sum depends strongly on the contents of the box, and may be quite far from normal: see chapter 18, or section 26.6 on the t-test.
2. *Statistical Abstract*, 2003 gives enrollments by age and sex for the whole U.S.: see tables 286–87.
3. *Statistical Abstract*, 2003, tables 953ff reports information about housing stock, based on the Census of 2000 and the American Housing Survey. Table 970 gives median rents for states. Table 971 does large metropolitan areas: San Francisco-Oakland-San Jose is the highest, at $968. "Specified" units are defined in the headnote to table 965.
4. *Statistical Abstract*, 2003, tables 977, 1126 gives data for the whole U.S. (Strange but true: more households have a TV than an oven.) For hours spent watching television, see tables 1125, 1127.
5. *Statistical Abstract*, 2003, tables 40, 1093.

6. *The Nation's Report Card: Mathematics 2000*, http://nces.ed.gov/nationsreportcard.
7. The study was done in 1976. A previous phase of the Carnegie survey is discussed in Martin Trow, editor, *Teachers and Students* (McGraw-Hill, 1975). In fact, a stratified sample was used. In 1992, there were about 3,600 institutions; the average enrollment was about 4,000: *Statistical Abstract*, 1994, table 275. Also see the *Digest of Educational Statistics*, although definitions are not exactly comparable across sources. In 2000, there were about 4,200 institutions, with an average entrollment of about 3,700. Over the period 1975–2000, the professoriate swelled by 60%, while student enrollments went up by about 40%.
8. See note 7, and pp. 6–7 of Trow's book.
9. E. S. Pearson and J. Wishart, editors, *Student's Collected Papers* (Cambridge University Press, 1942).
10. A. R. Jensen, "Environment, heredity and intelligence," *Harvard Educational Review* (1969, p. 20). The quote was edited slightly. For a recent discussion of the substantive issues, see J. P. Rushton and A. R. Jensen, "Thirty years of research on race differences in cognitive ability," *Psychology, Public Policy, and Law* vol. 11 (2005) 235-294.
11. 380 U.S. 202 (1965). In subsequent cases, courts have held that juries must be representative of the community; random sampling gives the permissible deviations from community percentage makeup. The leading case is *Castaneda* 430 U.S. 482 (1977), especially note 17 at 496; also see *Avery* 345 U.S. 559 (1953), where Justice Frankfurter held that "the mind of justice, not merely its eyes, would have to be blind to attribute such an occasion to mere fortuity." In *Batson* 476 U.S. 79 (1986) the Supreme Court narrowed the right to peremptory challenges.

    "Standard deviation analysis" is now used not only in jury selection cases, but in employment discrimination litigation, and even in antitrust matters. *McCleskey* 481 U.S. 279 (1987) suggests that, for better or for worse, in capital punishment cases the courts are reluctant to accept statistical evidence of discrimination. For some discussion of statistical evidence from various points of view, see—

    B. Black, "Evolving legal standards for the admissibility of scientific evidence," *Science* vol. 239 (1988) pp. 1508–12; vol. 241 (1988) pp. 1413–14.

    S. Fienberg, editor, *The Evolving Role of Statistical Assessments as Evidence in the Courts* (Springer-Verlag, 1989).

    M. Finkelstein, "The application of statistical decision theory to the jury discrimination cases," *Harvard Law Review* vol. 338 (1966) pp. 353–56.

    M. Finkelstein and B. Levin, *Statistics for Lawyers*, 2nd ed. (Springer-Verlag, 2001).

    D. Kaye and D. Freedman, *Reference Guide on Statistics*, in *Reference Manual on Scientific Evidence*, 2nd ed. (Federal Judicial Center, Washington, D.C., 2000).

    P. Meier, J. Sacks and S. L. Zabell, "What happened in Hazelwood: Statistics, employment discrimination, and the 80% rule," *American Bar Foundation Research Journal* (1984) pp. 139–86.

    D. W. Peterson, editor, "Statistical inference in litigation," *Law & Contemporary Problems* vol. 46, no. 4 (1983).

    D. L. Rubinfeld, "Econometrics in the courtroom," *Columbia Law Review* vol. 85 (1985) pp. 1048–97.
12. Lee R. Jones et al., *The 1990 Science Report Card* (U.S. Department of Education, Office of Educational Research and Improvement, Washington, D.C., 1992). See pp. 135 and 165.

Part VII. Chance Models

Chapter 24. A Model for Measurement Error

1. W. J. Youden, *Experimentation and Measurement* (National Science Teachers Association, Washington, D.C., 1962).
2. Such equipment is manufactured by Toledo scale, using four load cells. Railway cars can move up to 6 mph as they cross the weigh-bridge.
3. Data from http://www.wunderground.com
4. The error box was a bit complicated: 95% of the tickets followed the normal curve, with an average of 0 and an SD of 4 micrograms; the other 5% followed the normal curve with an average of 0 and an SD of 25 micrograms. Two normal curves were needed, one for the middle and one for the outliers.
5. The root-mean-square might be even better, since the average of the box is assumed to be 0.
6. The empirical distribution of the data on NB 10 is skewed and long-tailed (figure 2 in chapter 6). However, the 100-fold convolution of this distribution with itself is quite close to normal; the minor deviations from normality are described quite well by an Edgeworth expansion to order $1/n$.

7. W. J. Youden, "Enduring values," *Technometrics* vol. 14 (1972) pp. 1–11. Also see M. Henrion and B. Fischhoff, "Assessing uncertainty in physical constants," *American Journal of Physics* vol. 54 (1986) pp. 791–97.
8. Dependence between repeated measurements is often caused by observer bias: the person making the measurements subconsciously wants the second measurement to be close to the first one. The Bureau takes elaborate precautions to eliminate this kind of bias. For instance, the value of NB 10 is obtained by comparing total masses of different sets of weights. These sets are varied according to a design chosen by the Bureau. The person who actually makes the measurements does not know how these sets are related to one another, and so cannot form any opinion about what the scales "should" read.
9. By Michelson, Pease, and Pearson at the Irvine Ranch in 1929–33. The results were rounded off a bit in the exercise. Their average value for the speed of light, converted to miles per second, is about 186,270. The measurements were taken in several groups, and there is some evidence to show that the error SD changed from group to group.

   In essence, the speed of light is now a definition: "In 1983 the General Conference on Weights and Measures officially redefined the meter as the distance that light travels in vacuum in 1/299,792,458 of a second." See E. M. Purcell, *Electricity and Magnetism*, 2nd ed. (McGraw-Hill, 1985, Appendix E).
10. The quote is from R. D. Tuddenham and M. M. Snyder, *Physical Growth of California Boys and Girls from Birth to Eighteen Years* (University of California Press, 1954, p. 191). It was edited slightly. As the authors continue,

    > With the wisdom of hindsight, we recognized in the later years of the study that a more accurate estimate of the theoretical "true value" would have been not the first measurement recorded, nor even the "most representative," but simply the [average] of the set.

Chapter 25. Chance Models in Genetics

1. We are grateful for expert advice (some of which we took) from Everett Dempster and Michael Freeling of the Genetics Department, University of California, Berkeley. G. A. Marx and D. K. Ourecky of the New York State Agricultural Experiment Station were also extremely helpful. Finally, we thank Ann Lane, University of Minnesota. Standard textbooks on molecular genetics include—

   B. Alberts, D. Bray, J. Lewis, M. Raff, K. Roberts, and J. D. Watson, *Molecular Biology of the Cell*, 4th ed. (Garland Publishing, New York, 2002).

   A. J. F. Griffiths, J. H. Miller, D. T. Suzuki, R. C. Lewontin, and W. M. Gelbart, *An Introduction to Genetic Analysis*, 8th ed. (W. H. Freeman & Co., 2004).

   Desmond S. T. Nicholl, *An Introduction to Genetic Engineering*, 2nd ed. (Cambridge, 2002).

   Cancer biology provides fascinating insights into genetics. There are two books by Robert A. Weinberg that provide an excellent introduction to the field: *One Renegade Cell* (Basic Books, 1999) and *Biology of Cancer* (Garland Science, 2006).
2. Strictly speaking, this refers only to one part of the seeds, the *cotyledons* or first leaves.
3. The term "gene" is slightly ambiguous, but refers to a region of DNA coding for a protein or major protein chain, or to a specific variant of DNA in that region. ("Allele" is perhaps the better term for a specific segment of DNA occupying a coding region.) Presumably, several proteins are needed to determine seed color. If so, the pure yellow and pure green strains would have many of the corresponding alleles in common, but would differ on one pair—the y/y and g/g in the text. Mendel himself referred to "entities" which controlled phenotypes.
4. Sperm are carried by the pollen, eggs are in the ovules. Technically, these are nuclei not cells.
5. Maternal and paternal genes are sometimes distinguishable due to "imprinting." See C. Sapienza, "Parental imprinting of genes," *Scientific American* vol. 263 (October, 1990) pp. 52ff. Also see K. Peterson and C. Sapienza, "Imprinting the genome—imprinting, genes, and a hypothesis for their interaction," *Annual Review of Genetics* vol. 27 (1993) pp. 7–31.
6. The location of the genes on the chromosomes was worked out by Lamprecht (*Agric. Hortique Genetica*, 1961). There is a discussion in English by S. Blixt (same journal, 1972). Also see S. Blixt, "The Pea," chapter 9, vol. 2, *Handbook of Genetics (Plants, Viruses and Protista)* edited by R. C. King, Plenum, 1974; S. Blixt, "Why didn't Gregor Mendel find linkage?" *Nature* vol. 256 (1975) p. 206; E. Novitski and S. Blixt, "Mendel, linkage, and synteny," *Biosciences* vol. 28 (1978) pp. 34–35.
7. *Experiments in Plant Hybridisation* (Oliver & Boyd, 1965, p. 53). That book reprints Mendel's original paper, and some commentaries by Fisher, based on an article in the *Annals of Science* vol. 1 (1936) pp. 115–37. Some geneticists are quite critical of Fisher's reasoning; see, for example, F. Weiling, "What about R. A. Fisher's statement of the 'too good' data of J. G.

Mendel's Pisum paper?" *Journal of Heredity* vol. 77 (1986) pp. 281–83. On balance, Fisher's argument seems persuasive.

8. This experiment used five characteristics, not just the one discussed here. One trial was repeated, since Mendel thought the fit was poor. He used 100 plants in each trial, making the total of 600 referred to in the text.
9. "On the correlation between relatives on the assumption of Mendelian inheritance," *Transactions of the Royal Society of Edinburgh* vol. 52 pp. 399–433.
10. *Biometrika* (1903). The factor 1.08 more or less adjusts for the sex difference in heights. The equation is rounded off from the one in the paper.
11. There were 1,078 families in the study, so chance variation on this scale is very unlikely.
12. To get equation (5) from equation (3), take the conditional expectation given father's height; with non-assortative mating, mother's height is replaced by its overall average value. In fact, however, the correlation between parental heights was about 0.25.
13. Chromosomes may not replicate exactly in ordinary cell division. The "telomeres" (chromosome ends) seem to get shorter when the cell does not manufacture the enzyme telomerase. References—

    C. W. Greider and E. H. Blackburn, "Telomeres, telomerase, and cancer," *Scientific American* (February 1996) pp. 92–97,

    M. Barinaga, "Cells count proteins to keep their telomeres in line," *Science* vol. 275 (1997) p. 928.

    D. A. Banks and M. Fossel, "Telomeres, cancer, and aging," *Journal of the American Medical Asociation* vol. 278 (1997) pp. 1345–48.

    A. G. Bodnar et al., "Extension of life-span by introduction of telomerase into normal human cells," *Science* vol. 279 (1998) pp. 349–52.

    C. Bischoff et al., "No association between telomere length and survival among the elderly and oldest old," *Epidemiology* vol. 17 (2006) pp. 190–94.
14. This discussion ignores more-complicated phenomena like mutation and crossover.
15. This exercise is adapted from M. W. Strickberger, *Genetics*, 3rd ed. (Macmillan, 1985). The focus here is the color of the pods, which may be quite different from the color of the seeds.
16. Rasmusson, *Hereditas* vol. 20 (1935). This problem too is from Strickberger.

Part VIII. Tests of Significance

Chapter 26. Tests of Significance

1. From "Of experience," quoted in Jerome Frank, *Courts on Trial* (Princeton University Press, 1949).
2. Additional reading, in order of difficulty—

    J. L. Hodges, Jr. and E. Lehmann, *Basic Concepts of Probability and Statistics*, 2nd ed. (SIAM, 2004).

    L. Breiman, *Statistics with a View towards Applications* (Houghton Mifflin, 1973).

    J. Rice, *Mathematical Statistics and Data Analysis*, 3d ed. (Duxbury Press, 2005).

    P. Bickel and K. Doksum, *Mathematical Statistics*, 2nd ed. (Prentice Hall, 2001).

    E. Lehmann, *Theory of Point Estimation*, 2nd ed. with G. Casella (Springer, 1998).

    ———, *Testing Statistical Hypotheses*, 3rd ed. with J. Romano (Springer, 2005).
3. For a Bayesian, the frequentist P-value of a test can be substantially different from the posterior probability of the null hypothesis; indeed, the latter must depend on (i) the power of the test and (ii) the prior probability of the null. For more discussion, see J. Berger and T. Sellke, "Testing a point null hypothesis: The irreconcilability of P-values and evidence," *Journal of the American Statistical Association* vol. 82 (1987) pp.112–39.
4. In 2001, about 30% of people with full-time jobs were on flexible schedules: *Statistical Abstract*, 2003, table 606.
5. We are using ESP loosely to cover PK and clairvoyance as well. The experiment is described in C. Tart, *Learning to Use Extrasensory Perception* (University of Chicago Press, 1976). One subject did not, in fact, complete all the runs. And there are also results from a "Ten Choice Trainer," see exercise 6 on p. 487.
6. For the reasons given in the text, we do not consider it suitable to formulate the alternative hypothesis as drawing from a 0-1 box where the fraction of 1's is bigger than 1/4. It may be more reasonable to say that the number of correct guesses is stochastically larger than the sum of 7,500 draws from | [1] [0] [0] [0] |.
7. One such experiment was conducted by the former Professor W. Meredith, Psychology Department, University of California, Berkeley.

8. A. N. Doob et al., "Effect of initial selling price on subsequent sales," *Journal of Personality and Social Psychology*, vol. 11 (1969) pp. 345–50.
9. The anecdote about Student is reported in W. J. Youden, *Experimentation and Measurement* (Washington, D.C., 1963).
10. The t-test is one of the most popular statistical techniques, and we regret having to present it in a context which is both dry and partially hypothetical. (The story in the text is true, up to where they make a t-test; in practice, they don't.) We didn't run across any examples which were simultaneously real, interesting, and plausible. Our difficulty was the following. The t-test is used to compute significance levels. With small samples, some departures from normality can throw the computation off by a large factor. By way of illustration, the figure shows a probability histogram for the t-statistic, based on 10 draws made at random with replacement from the box | [−3] [−2] [5] |. The distribution is far from t-like.

Probability histogram for t-statistic based on 10 draws with replacement from | [−3] [−2] [5] |

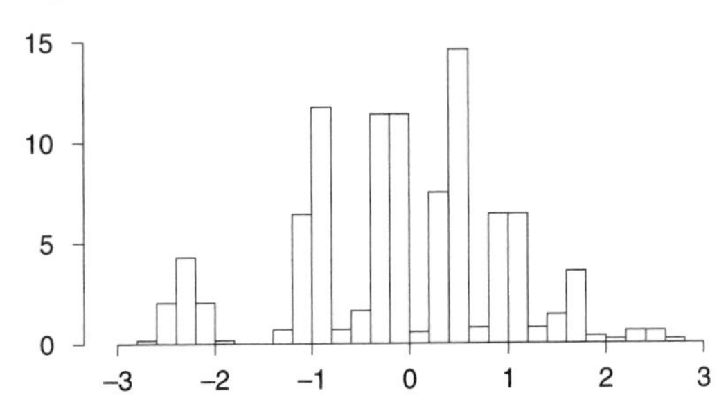

The histogram was computed exactly, by considering all

$$\binom{12}{2} = 66$$

possible divisions of 10 objects into 3 groups; for each group, we computed the probability and the t-statistic. For 3 of the divisions, the t-statistic is undefined, but their mass is only 10^{-5}; another 2% of the mass is outside the range $[-3, 3]$. The example can easily be modified so there is a smooth density with a shift parameter. For the combinatorics, see W. Feller, *An Introduction to Probability Theory and its Applications*, vol. I, 3rd ed. (John Wiley & Sons, 1968, section II.4).

To rely on the t-test, it seems to be necessary to know that the distribution of the errors is close to normal, without having a fair idea about the spread of the errors: if you knew the spread, you wouldn't be using the t-distribution. But how would you know the shape of the distribution without knowing its spread?

With large samples, departures from normality don't matter so much. Student's curves merge with the normal, and the t-statistic follows the normal curve (by the central limit theorem and the consistency of $\hat{\sigma}^2$ as an estimator of σ^2). This is one thing statisticians mean by the "robustness of the t-test." In our terms, this concept of robustness applies to the z-test not the t-test. Two references—

H. D. Posten, "The robustness of the one-sample t-test over the Pearson system," *Journal of Statistical Computation and Simulation* vol. 9 (1979) pp. 133–49.

E. Lehmann and W.-Y. Loh, "Pointwise vs. uniform robustness of some large sample tests and confidence intervals," *Scandinavian Journal of Statistics* vol. 17 (1990) pp. 177–87.

Small departures from independence can have large impacts on both the z-test and the t-test. Also see notes 12–13 below.

11. For present purposes, this is just a convention: the factor $\sqrt{n/(n-1)}$ could be absorbed into the multiplier derived from Student's curve. In some contexts, however, SD^+ is preferred to the SD of the sample as estimator for the SD of the population: $(\text{SD}^+)^2$ is unbiased, and this matters when pooling variances estimated from a large number of small samples.
12. The equation for the curve is

$$y = \text{constant}\left(1+\frac{t^2}{d}\right)^{-(d+1)/2}$$

$$\text{constant} = 100\%\,\frac{\Gamma\left(\frac{d+1}{2}\right)}{\sqrt{\pi d}\,\Gamma\left(\frac{d}{2}\right)}$$

$$d = \text{degrees of freedom}$$

$$\Gamma = \text{Euler's gamma function}$$

The t-test was put on a rigorous mathematical footing by R. A. Fisher, who also showed that the procedure can give good approximations even when the errors did not follow the normal curve exactly: some departures from normality do not matter. This small-sample property is called "robustness" too. (But see note 10.)

13. If the tickets in the box follow the normal curve, then the probability histogram for the sum of the draws does too—even with only a few draws. Technically, the convolution of a normal curve with itself gives another normal curve. If the tickets in the box have a known distribution, which is not normal, statisticians can work out the probability histogram for the sum or average of the draws, using convolutions.
14. For national data, see J. H. Pryor et al., *The American Freshman: National Norms for Fall 2005* (Higher Education Research Institute, UCLA, 2006).
15. After Zeisel published the 1969 article, the next group of jurors chosen by Judge Ford was 24% female. References—

Hans Zeisel, "Dr. Spock and the case of the vanishing women jurors," *University of Chicago Law Review* vol. 37 (1969) pp. 1–18.

———, "Race bias in the administration of the death penalty: the Florida experience," *Harvard Law Review* vol. 95 (1981) pp. 456–68.

16. S. C. Truelove, "Therapeutic trials," in L. J. Witts, editor, *Medical Surveys and Clinical Trials* (Oxford University Press, 1959). Blinding the randomization is discussed in T. C. Chalmers, P. Celano, H. S. Sacks and H. Smith, Jr., "Bias in treatment assignment in controlled clinical trials," *New England Journal of Medicine* vol. 309 (1983) pp. 1358–61.
17. *Statistical Abstract*, 2003, tables 229, 1138, 1244 gives national data on education and reading. Also see *Reading At Risk: A Survey of Literary Reading in America* (National Endowment for the Arts, Washington, D.C., 2004). The latter publication takes a rather alarmist view of the prospects for the book, as the title indicates. By contrast, the data in *Statistical Abstract* suggest that books remain quite popular. For example, more people read books than surf the net.
18. These data originate with the Public Health Department of New York. We got them from Sandy Zabell, Professor of Statistics, Northwestern University. A reference is A. J. Izenman and S. L. Zabell, "Babies and the blackout: The genesis of a misconception," *Social Science Research* vol. 10 (1981) pp. 282–99. Apparently, the *New York Times* sent a reporter around to a few hospitals on Monday, August 8, and Tuesday, August 9, nine months after the blackout. The hospitals reported that their obstetrics wards were busier than usual—probably because of the general pattern that weekends are slow, Mondays and Tuesdays are busy. These "findings" were published in a front-page article on Wednesday, August 10, 1966, under the headline "Births Up 9 Months After the Blackout." That seems to be the origin of the baby-boom myth.

19. *Statistical Abstract*, 2003, table 681. The survey is hypothetical.
20. For a recent survey, see M. R. Rosenzweig and E. L. Bennett, "Psychobiology of plasticity: Effects of training and experience on brain and behavior," *Behavioural Brain Research* vol. 78 (1996) pp. 57–65. In fact, the experiment used not pairs but triplets, assigned at random to enriched, standard, and deprived environments. Data kindly provided by the investigators.

Chapter 27. More Tests for Averages

1. Suppose X and Y are random variables, with variances σ^2 and τ^2 respectively, and correlation ρ. Then $\text{var}(X - Y) = \sigma^2 + \tau^2 - 2\rho\sigma\tau$. If $\rho = 1$, the "chance errors" (i.e., departures from expected values) necessarily have the same sign, and offset each other—cancellation. Then the SE for the difference is $|\sigma - \tau|$. If $\rho = -1$, the errors reinforce each other, and the SE is $\sigma + \tau$. The case of independence corresponds to $\rho = 0$, and the SE is intermediate between the two extremes: $\text{SE} = \sqrt{\sigma^2 + \tau^2}$. If X and Y are independent, then $\text{SE}(X + Y) = \text{SE}(X - Y)$.
2. *NAEP 2004: Trends in Academic Progress*, http://nces.ed.gov/nationsreportcard. Average scores are as reported in the text; the sample was much larger, and highly designed; the SDs were close to 30. However, the SEs reported in the text are about right. Scores on these standardized performance tests seem to have bottomed out around 1970, with minor variations up and down since then. Using the pooled sample variance in the denominator of the test statistic would not be appropriate here, because the population variances may differ—even if the null hypothesis is true.
3. For the data source, see note 2. Three other tests are widely used for this sort of problem, besides the one presented in the text:

(i) Given the null hypothesis, the percentage of 1's in the two boxes is the same, and can be estimated by pooling the two samples:

$$\frac{107 + 132}{200 + 300} = \frac{239}{500} \approx 48\%.$$

On this basis, the common SD of the two boxes is estimated as

$$\sqrt{0.48 \times 0.52} \approx 0.50.$$

This pooled SD can be used to compute the SE in the denominator of the test statistic. (However, the pooled SD should not be used for other purposes, like putting confidence intervals around the difference.)

(ii) Fisher's "exact" test conditions on the total number of 1's in the two samples. The test statistic is the number of 1's in (say) the first sample; the sample sizes are fixed. The null distribution is the hypergeometric.

(iii) The χ^2 statistic may be computed from the 2×2 table, and referred to a χ_1^2-distribution.

Conditional on the total number of 1's in the two samples, test statistic (i) is a monotone function of the number of 1's in the first sample; so tests (i) and (ii) are equivalent. In effect, the hypergeometric has been approximated by the normal, which is fine for reasonably large samples. Likewise, (iii) is equivalent to (i) and (ii); indeed, $\chi^2 \equiv {Z_1}^2$, where Z_1 is the z-statistic computed with the pooled SD. If Z is the z-statistic computed from the separate SDs, as in the text, then $Z^2 > {Z_1}^2$. However, as will be shown below, conditional on the total number of 1's in the two samples, Z is a monotone function of the number of 1's in the first sample. So our test is equivalent to tests (i)-(ii)-(iii).

Some notation will be helpful. Let ξ be the number of heads in n tosses of a p-coin, and $\hat{p} = \xi/n$. Let ζ be the number of heads in m tosses of a q-coin, and $\hat{q} = \zeta/m$. The two coins are independent. The null hypothesis is that $p = q$; the alternative, $p \neq q$. We condition on $s = \xi + \zeta$. Let $r = s/(m + n)$. This r will be held constant.

For (i), the variance is computed as

$$v = r(1 - r)\left(\frac{1}{n} + \frac{1}{m}\right),$$

and the test statistic is

$$Z_1 = (\hat{p} - \hat{q})/\sqrt{v}.$$

For our test statistic, the variance is computed as

$$w = \frac{\hat{p}(1 - \hat{p})}{n} + \frac{\hat{q}(1 - \hat{q})}{m}$$

and the test statistic is

$$Z = (\hat{p} - \hat{q})/\sqrt{w}.$$

We view m, n, and $r = s/(m+n)$ as fixed; tacitly, we have assumed $0 < \hat{p}, \hat{q}, r < 1$. It will be convenient to introduce the new variable $x = \hat{p} - r$, so $\hat{q} = r - (nx/m)$. Clearly, x takes only finitely many values. (Later, however, it will be convenient to think of x as running through the whole interval from its minimum value to its maximum.) Now

$$\hat{p}(1-\hat{p}) = r(1-r) + (1-2r)x - x^2$$

$$\hat{q}(1-\hat{q}) = r(1-r) - \frac{n}{m}(1-2r)x - \frac{n^2}{m^2}x^2.$$

The two variances are related as follows:

$$w = v + bx - cx^2,$$

where

$$b = \left(\frac{1}{n} - \frac{n}{m^2}\right)(1-2r), \qquad c = \frac{1}{n} + \frac{n^2}{m^3}.$$

Finally, our test statistic is

$$Z = \frac{m+n}{m}x \Big/ \sqrt{v + bx - cx^2}$$

Of course, w, v, and c are all positive; b may be positive or negative. The monotonicity of Z as a function of x follows from the lemma below.

Lemma. Let $v, c > 0$ and let b be real. Confine x to the interval where $v + bx - cx^2 > 0$. Let

$$f(x) = x/\sqrt{v + bx - cx^2}$$

Then $f(x)$ is monotone increasing with x.

Proof. If $b \le 0$, this is trivial; so let $b > 0$. Then

$$\frac{df}{dx} = \frac{2v + bx}{2(v + bx - cx^2)^{3/2}} > 0.$$

4. See note 2.
5. *Statistical Abstract*, 2003, table 284. Most of the drop occurred between 1970 and 1980; the percentage bottomed out around 1995, and has been edging back up. Also see A. W. Astin et al., *The American Freshman: Thirty-Five Year Trends, 1966–2001* (Higher Education Research Institute, UCLA, 1991).
6. For national data, see J. H. Pryor et al., *The American Freshman: National Norms for Fall 2005* (Higher Education Research Institute, UCLA, 2006).
7. "Literacy among youths 12–17 years," *Vital and Health Statistics*, series 11, no. 131 (Washington, D.C., 1973). The sample design was like the Current Population Survey, and the investigators estimated the standard errors by the half-sample method. Simple random samples of the size indicated in the exercise will have standard errors about equal to the real ones.
8. "Intellectual development of children by demographic and socioeconomic factors," *Vital and Health Statistics*, series 11, no. 110 (Washington, D.C., 1971). For a discussion of the standard errors, see the previous note. The correlation between the children's test scores and parental education was 0.5, dropping to 0.3 when parental income was held constant. "Big city" means a population of 3 million or more. In fact, children in cities with a population in the range 1 to 3 million did best, averaging around 28 points.
9. For references to real trials on vitamin C, with interesting sidelights on blinding the randomization, see note 12, chapter 2.
10. In the text, we are using a strict null hypothesis: treatment has no effect on any subject. There is really only one number for each ticket, copied into both fields. The strict null has the charm of simplicity, and is appropriate in many cases. A weaker version of the null is sometimes used: the average response is the same for both treatments. This may be more realistic if the new treatment hurts some subjects but helps others. The alternative hypothesis usually does require two numbers for each subject, one for the response to treatment and one for the response to placebo. The discussion continues in note 11.
11. Consider a clinical trial to compare treatments A and B. We consider the weak form of the null, as in note 10; and the alternative, which does not constrain the responses at all. Suppose there are N subjects, indexed by $i = 1, \ldots, N$. Let x_i be the response of subject i to treatment A; likewise,

y_i is the response to B. For each i, either x_i or y_i can be observed, but not both. Let

$$\overline{x} = \frac{1}{N}\sum_{i=1}^{N} x_i \qquad \overline{y} = \frac{1}{N}\sum_{i=1}^{N} y_i$$

$$\sigma^2 = \frac{1}{N}\sum_{i=1}^{N} (x_i - \overline{x})^2 \qquad \tau^2 = \frac{1}{N}\sum_{i=1}^{N} (y_i - \overline{y})^2$$

$$\text{cov}\,(x, y) = \frac{1}{N}\sum_{i=1}^{N} (x_i - \overline{x})(y_i - \overline{y})$$

This model is sufficiently flexible to handle the weak form of the null hypothesis (note 10), as well as subject-to-subject heterogeneity under the alternative hypothesis. Thus, for instance, the average difference between treatments A and B—averaged over all the subjects in the study—is $\overline{x} - \overline{y}$. This "average causal effect" measures the difference between putting all the subjects into regime A, or putting all of them into regime B. The average causal effect is often the key parameter. And it is estimable, although the two responses are not simultaneously observable for any individual subject. Indeed, $\overline{x}$, $\overline{y}$, σ^2, and τ^2 are all estimable; on the other hand, cov (x, y) cannot be estimated by a sample covariance.

Responses in treatment and control are often modeled, for instance, as independent binomial with two different p's, or independent normals with two different μ's. These parametric models seem less realistic. Independence of the two sample averages is generally wrong, and there is no reason to assume subjects are exchangeable within each treatment group. Such assumptions are *not* secured by randomization, which only makes the two groups comparable *as groups*. Thus, theoretical underpinnings are absent for, e.g., the t-test. It is surprising—and reassuring—that the permutation distributions of the conventional test statistics more or less coincide with the model-based distributions, at least in the contexts we are considering.

We now compute the variance of $\overline{X} - \overline{Y}$ under the alternative hypothesis, in our permutation setup. Let S be a random subset of $\{1, \ldots, N\}$, with n elements; this group gets treatment A, so x_i is observed for $i \in S$. Let T be a random subset of $\{1, \ldots, N\}$, with m elements, disjoint from S. This group gets treatment B, so y_j is observed for $j \in T$. We estimate the population means $\overline{x}$ and $\overline{y}$ by the sample means

$$\overline{X} = \frac{1}{n}\sum_{i\in S} x_i \qquad \overline{Y} = \frac{1}{m}\sum_{j\in T} y_j$$

By combinatorial calculations,

$$\text{var}\,\overline{X} = \frac{N-n}{N-1}\frac{\sigma^2}{n} \qquad \text{var}\,\overline{Y} = \frac{N-m}{N-1}\frac{\tau^2}{m}$$

$$\text{cov}\,(\overline{X}, \overline{Y}) = -\frac{1}{N-1}\,\text{cov}\,(x, y)$$

Thus

$$\begin{aligned}\text{var}\,(\overline{X} - \overline{Y}) &= \frac{N-n}{N-1}\frac{\sigma^2}{n} + \frac{N-m}{N-1}\frac{\tau^2}{m} + \frac{2}{N-1}\,\text{cov}\,(x, y)\\ &= \frac{N}{N-1}\left(\frac{\sigma^2}{n} + \frac{\tau^2}{m}\right) + \frac{1}{N-1}\Big[2\,\text{cov}\,(x, y) - \sigma^2 - \tau^2\Big]\\ &\le \frac{N}{N-1}\left(\frac{\sigma^2}{n} + \frac{\tau^2}{m}\right)\end{aligned}$$

because cov $(x, y) \le \sigma\tau$ and $2\sigma\tau - \sigma^2 - \tau^2 \le 0$. The "conservative estimate" in the text is $\sigma^2/n + \tau^2/m$. In practice, σ^2 and τ^2 would be estimated by sample variances.

The signs may be a little perplexing. In general, we expect x and y to be positively correlated over all subjects. If too many subjects with high x-values are assigned to treatment A, then too few with high y-values are left for B. So the sample averages $\overline{X}$ and $\overline{Y}$ are negatively correlated. In principle, cov (x, y) should be near its upper limit $\sigma\tau$, at least when x and y are highly correlated across subjects. Then the "conservative estimate" should be reasonably accurate for large samples. The strict null hypothesis in the text specifies that $x \equiv y$. Then $\sigma = \tau$, and the calculation is exact under the null hypothesis. Also see note 14 below. Of course, if N is large relative to m and n, then $\overline{X}$ and $\overline{Y}$ are nearly independent; again, the "conservative estimate" will be nearly right.

The impact of other variables may be handled as follows. Let η denote treatment status. Let ω denote the state of other variables influencing the response. We assume there is a function f such that the response of subject i to treatment is $f(i, \eta, \omega)$. Let ρ denote the assignment variable: if $\rho(i) = A$ then subject i is assigned to treatment A, and likewise for B. We assume that ρ and ω are independent: given ω, the law of ρ is uniform over all partitions of the subjects into a group S of cardinality n assigned to A and another group of cardinality m assigned to B. The object of randomization, blinding, etc. is to secure this assumption. Then our argument can be done separately for each ω, with

$$x_i = f(i, A, \omega) \quad \text{for } i \in S$$
$$y_j = f(j, B, \omega) \quad \text{for } j \in T$$

Few experiments are done on random samples of subjects. Instead, there is some initial screening process. Only subjects who pass the screen are randomized, and these subjects are best viewed as a sample of convenience. Therefore, some care is needed in setting up the inference problem. In our model, each subject has two potential responses, one to the treatment regime and one to the control regime. The "population" consists of pairs of responses. Both responses cannot be simultaneously observed for any subject. The experiment generates data not for the whole population, but for part of it. We observe responses to the treatment regime for subjects in the treatment group, and responses to the control regime for subjects in the control group. The statistical inference is from these observations to parameters characterizing the set of pairs of responses for the subjects that are randomized. The inference is not to some larger population of subjects—that kind of generalization would not be automatically justified by randomization. This is one aspect of Campbell's distinction between "internal validity" and "external validity:" see W. R. Shadish, T. D. Cook, W. T. Campbell, *Experimental and Quasi-Experimental Designs for Generalized Causal Inference* (Houghton Mifflin, 2002).

We are thinking primarily of experiments where subjects are divided into two random groups. However, similar comments apply if, for instance, subjects are paired by some ad hoc procedure; then a coin is tossed for each pair, choosing one subject for the treatment regime and one for the control regime. Again, the inference is to parameters characterizing the set of possible responses, and is made conditionally on the set of subjects and the pairing.

The model seems to go back to Neyman's early work on agricultural experiments. Some references:

J. Neyman, "Sur les applications de la théorie des probabilités aux experiences agricoles: Essai des principes," *Roczniki Nauk Rolniczki* vol. 10 (1923) pp. 1–51, in Polish; English translation by D. Dabrowska and T. Speed, *Statistical Science*, vol. 5 (1990) pp. 463–80.

H. Scheffé, "Alternative models in the analysis of variance," *Annals of Mathematical Statistics* vol. 27 (1956) pp. 251–71.

J. L. Hodges, Jr. and E. Lehmann, *Basic Concepts of Probability and Statistics* (Holden-Day, 1964, section 9.4; 2nd ed. reprinted by SIAM, 2004).

D. Rubin, "Estimating causal effects of treatments in randomized and nonrandomized studies," *Journal of Educational Psychology* vol. 66 (1974) pp.688–701.

J. Robins, "Confidence interval for causal parameters," *Statistics in Medicine* vol. 7 (1988) pp. 773–85.

P. Holland, "Causal inference, path analysis, and recursive structural equations models," *Sociological Methodology 1988*, C. Clogg, editor (American Sociological Association, Washington, D.C., Chapter 13.)

L. Dümbgen, "Combinatorial stochastic processes," *Stochastic Processes and their Applications* vol. 52 (1994) pp. 75–92.

D. A. Freedman, *Statistical Models: Theory and Practice* (Cambridge University Press, 2005).

Minor technical issues: (i) The relevant central limit theorem is for sampling without replacement (note 1, chapter 23). (ii) For small samples, the t-distribution may not provide a better approximation than the normal: the assumptions underlying the t-test do not hold.

12. A. Tversky and D. Kahneman, "Rational choice and the framing of decisions," *Journal of Business* vol. 59, no. 4, part 2 (1986) pp. S251–78. Also see D. Kahneman and A. Tversky, "On the reality of cognitive illusions," *Psychological Review* vol. 103 (1996) pp. 582–96 (with discussion); D. Kahneman and A. Tversky, editors, *Choices, Values, and Frames* (Cambridge University Press, 2000); A. K. Sen, *Rationality and Freedom* (Harvard University Press, 2002).
13. B. J. McNeil, S. G. Pauker, H. C. Sox, Jr., and A. Tversky, "On the elicitation of preferences for alternative therapies," *New England Journal of Medicine* vol. 306 (1982) pp. 1259–62.
14. There were $80 + 87 = 167$ subjects in all (table 1). Of them, $40 + 73 = 113$ favored surgery; the remaining 54 favored radiation. The strict null hypothesis (note 10) specifies $x \equiv y$, so $\sigma = \tau$ and both are computable from the data. Indeed, on the null hypothesis, the percentage of doctors favoring surgery is $113/167 \times 100\% \approx 68\%$. Then

$$\sigma = \tau \approx \sqrt{0.68 \times 0.32} \approx 0.47$$

Likewise, the covariance between $\overline{X}$ and $\overline{Y}$ can be computed exactly. This term achieves the upper bound $\sigma\tau = \sigma^2$, because the correlation between x and y across subjects is 1. Now

$$\text{var}\,(\overline{X} - \overline{Y}) = \frac{N}{N-1}\left(\frac{1}{n} + \frac{1}{m}\right)\sigma^2$$

The two forms of the test statistic (pooled or separate SDs, see note 3) are virtually identical. For example, if the null hypothesis defines the model, the r.m.s. difference between the values of the two statistics is only 0.013. Furthermore, the normal approximation is quite good: for either statistic, the chance of exceeding 2 in absolute value is about 4.8%, compared to the normal tail probability of 4.6%.

15. D. Kahneman and A. Tversky, "Choices, values, and frames," *American Psychologist* vol. 39 (1984) pp. 341–50.
16. In fact, the randomization was a bit more complicated. Inoculation required 3 separate injections over time, and hence the control group was given 3 injections (of the placebo) too. Vials containing the injection material were packed 6 to a box; 3 contained the vaccine and had a common code number; the other 3 contained the placebo, with another common code number. Each vial had enough fluid for 10 injections.

    When the time came for the 1st round of injections, one vial was taken out of the box, and 10 children got their injections from that vial; the investigator recorded its code number against these 10 children; these 10 children got their 2nd and 3rd injections from the other 2 vials with the same code number in the box. The next 10 children got their 1st round injection from 1 of the 3 vials of the other group in that box (with a code number different from the 1st one used); the code number of the vial was recorded against them; and their subsequent injections were from the remaining 2 vials in the group.

    In effect, then, the children were blocked into pairs of groups of 10; a coin was tossed for each pair; one whole group went into treatment, and the other group into control, with a 50–50 chance. The calculation in the text is exact, on the plausible assumption that no 2 polio cases got injections from the same box. Otherwise, the calculation has to be modified. This particular trial is usually analyzed by the two-sample z-test, without taking account of the blocking (note 2 to chapter 1). We follow suit.
17. Barbara V. Howard et al., "Low-fat dietary pattern and risk of cardiovascular disease: The Women's Health Initiative randomized controlled dietary modification trial," *Journal of the American Medical Association* vol. 295 (2006) pp. 655–66.
18. D. Ravitch and C. E. Finn, Jr., *What Do Our 17-Year-Olds Know?* (Harper & Row, 1987, p. 52). The Soviet Union had the highest recognition factor.
19. http://www.gallup.com
20. References—

    K. Gray-Donald, M. S. Kramer, S. Munday et al., "Effect of formula supplementation in the hospital on duration of breast-feeding: A controlled clinical trial," *Pediatrics* vol. 75 (1985) pp. 514–18.

    K. Gray-Donald and M. S. Kramer, "Causality inference in observational vs. experimental studies: An empirical comparison," *American Journal of Epidemiology* vol. 127 (1988) pp. 885–92.

    Prior to running the controlled experiment, these investigators also ran an observational study, where both nurseries followed standard supplementation practice. There was a strong negative association between supplementation in the nurseries and breast-feeding later, as in the previous studies. Technically, assignment to the nurseries was not random. When a mother presented, she was assigned to the nursery with a bed available; this was done by clerical personnel not involved with the study. Eligibility was determined on objective criteria specified in the protocol. Unpublished data were kindly provided by the investigators.
21. Let (X_i, Y_i) be independent and identically distributed pairs of random variables, with $E\{X_i\} = \alpha$, var $X_i = \sigma^2$, $E\{Y_i\} = \beta$, and var $Y_i = \tau^2$; let ρ be the correlation between X_i and Y_i, so cov $(X_1, Y_i) = \rho\sigma\tau$. Let $\overline{X} = (X_1 + \cdots + X_n)/n$ and $\overline{Y} = (Y_1 + \cdots + Y_n)/n$. The sample means are correlated, and var $(\overline{X} - \overline{Y}) = v/n$ with

$$v = \sigma^2 + \tau^2 - 2\rho\sigma\tau.$$

The variance v would be estimated from sample data as

$$\hat{v} = \hat{\sigma}^2 + \hat{\tau}^2 - 2r\hat{\sigma}\hat{\tau},$$

where

$$\hat{\sigma}^2 = \frac{1}{n}\sum_{i=1}^{n}(X_i - \overline{X})^2, \qquad \hat{\tau}^2 = \frac{1}{n}\sum_{i=1}^{n}(Y_i - \overline{Y})^2$$

and r is the sample correlation coefficient,

$$r = \frac{1}{n}\sum_{i=1}^{n}(X_i - \overline{X})(Y_i - \overline{Y})/\hat{\sigma}\hat{\tau}.$$

The z-test would use the statistic $(\overline{X} - \overline{Y})/\sqrt{\hat{v}/n}$.

We now make the connection with the z-test based on the differences $X_i - Y_i$. Plainly, $\overline{X} - \overline{Y} = \overline{X - Y}$, the latter being $\frac{1}{n}\sum_{i=1}^{n}(X_i - Y_i)$. The differences $X_i - Y_i$ are independent and identically distributed, with $E\{X_i - Y_i\} = \alpha - \beta$ and var $\{X_i - Y_i\} = \sigma^2 + \tau^2 - 2\rho\sigma\tau = v$; of course, var $\{\overline{X - Y}\} = v/n =$ var $\{\overline{X} - \overline{Y}\}$, where v was defined above. The natural estimator for v based on the differences is

$$\frac{1}{n}\sum_{i=1}^{n}\left[(X_i - Y_i) - (\overline{X - Y})\right]^2 = \hat{v},$$

coinciding with the variance estimator based on the paired data. (The equality takes a little algebra.) As a result, the z-statistic computed from the pairs must equal the z-statistic computed from the differences.

22. http://www.gallup.com
23. See note 22 for the source. The question was, "How would you rate the honesty and ethical standards of the people in these different fields—very high, high, average, low, or very low?" The percentage ratings of "very high or high" are shown in the table below, for some of the fields.

Nurses	82%
Druggists	67%
Medical doctors	65%
High school teachers	64%
Clergy	54%
Journalists	28%
Building contractors	20%
Lawyers	18%
Congressmen	14%
Car salesmen	8%
Telemarketers	7%

24. A. Tversky and D. Kahneman, "The framing of decisions and the psychology of choice," *Science* vol. 211 (1981) pp. 453–458. Prices in the exercise were adjusted for inflation.
25. *The Third National Mathematics Assessment: Results, Trends and Issues* (Princeton: ETS/NAEP, 1983). The item is from the assessment, and the results are about as reported; the calculator group really did worse. However, it is not clear from the report whether the study was done observationally or experimentally.
26. P. H. Rossi, R. A. Berk and K. J. Lenihan, *Money, Work and Crime: Experimental Evidence* (San Diego: Academic Press, 1980, especially table 5.1). The study was done in 1976. We have simplified the experimental design, but not in any essential way; likewise, we changed the percents a little to make the testing problem sharper. Rossi et al. argue that income support did reduce recidivism, but the effect was masked by the impact on weeks worked. Their analysis has been criticized by H. Zeisel, "Disagreement over the evaluation of a controlled experiment," *American Journal of Sociology* vol. 88 (1982) pp. 378–96, with discussion.
27. S. J. Sherman, "On the self-erasing nature of errors of prediction," *Journal of Personality and Social Psychology* vol. 19 (1980) pp. 211–21.
28. William Epstein, as reported in the *New York Times*, September 27, 1988.

Chapter 28. The χ^2-Test

1. K. Pearson, "On the criterion that a given system of deviations from the probable in the case of a correlated system of variables is such that it can reasonably be supposed to have arisen from random sampling," *Phil. Mag.*, series V, vol. 1 (1900) pp. 157–75.
2. If the chance model is right, each term is expected to be a bit less than one; the sum of all the terms is expected to be $n - 1$, where n is the number of terms.
3. The equation for the curve is

$$y = \frac{100\%}{\Gamma(d/2)}\left(\frac{1}{2}\right)^{d/2} x^{(d/2)-1}e^{-x/2}$$

d = degrees of freedom
Γ = Euler's gamma function

4. The exact distribution was obtained using a program that stepped through all six-tuples of numbers adding up to 60, arranged in lexicographic order. It computed the χ^2-statistic for each six-tuple, and the corresponding probability (using the multinomial formula). These probabilities were summed to give the answer—and the probability histogram in figure 2. The calculation seemed to be accurate to about 15 decimal places, since the sum of all the probabilities was $1 - 10^{-15}$. The wiggles in figure 2 are real.

   Many books recommend the Yates correction (subtracting 0.5 from the absolute difference before squaring, when this difference exceeds 0.5). With one degree of freedom, this is equivalent to the continuity correction (p. 317) and is a good thing to do. With more than one degree of freedom, numerical calculations show that it is often a bad thing to do. The histogram can be shifted much too far to the left. Numerical computations also show that with 5 observations expected per cell, and only a few degrees of freedom, the χ^2-curve can be trusted out to the 5% point or so. With 10 observations expected per cell, the curve can be trusted well past the 1% point. Even if one or two cells in a moderate-size table have expecteds in the range 1–5, the approximation is often good.
5. When there are only two kinds of tickets in the box, the χ^2-statistic is equal to the square of the z-statistic. Since the square of a normal variable is χ^2 with 1 degree of freedom, the χ^2-test will in this case give exactly the same results as a (two-tailed) z-test. Also see note 3 to chapter 27.
6. The data for this example, and for exercise 9 on p. 532, were kindly supplied by the California State Lottery through their statistical consultant Don Ylvisaker (UCLA).
7. In some cases (e.g., with only a few observations per cell), it is advisable to group the data.
8. *UCLA Law Review*, vol. 20 (1973) p. 615.
9. See note 7 to chapter 25.
10. A. R. Luria, *The Working Brain* (Basic Books, New York, 1973).
11. The HANES design involved a cluster sample, so there is some dependence in the data, which the χ^2-test would not take into account. The half-sample method could be used to generate the null distribution. Women are consistently more right-handed than men, in all age groups. See *Anthropometric Reference Data and Prevalence of Overweight: United States, 1976–80.* Data from the National Health Survey, series 11, no. 238. (U.S. Department of Health and Human Services, Washington, D.C.). The numbers in table 5 are close to the real data, and make the arithmetic easier to follow.
12. Of course, if the test is done conditional on the marginals, the expecteds may be viewed as given. Also see note 3 to chapter 27.
13. Unweighted counts from a CD-ROM supplied by the Census Bureau, for the March 2005 Current Population Survey. The χ^2-test does not take the design of the sample into account, but the difference is real.
14. *UCLA Law Review*, vol. 20 (1973) p. 616.
15. Unweighted counts from a CD-ROM supplied by the Census Bureau. The table is restricted to civilians. The χ^2-test does not take the design of the sample into account. In many such surveys, across all age groups, the never-married men are less successful at work. For women, however, the unemployment rate for never-marrieds is about the same as for the married group. Also see R. M. Kaplan and R. G. Kronick, "Marital status and longevity in the United States population," *Journal of Epidemiology and Community Health* vol. 60 (2006) pp. 760–5.
16. This exercise is adapted from data supplied by IRRI.
17. Paraphrased from evidence presented at an extradition hearing for James Smyth, Federal District Court (N.D. Cal., 1993). See Defense brief of December 10, 1993 (pp. 7–8), Plaintiffs' exhibit 72.15, and Declaration of Robert Koyak. The District Court's decision not to extradite on grounds of probable discrimination was reversed on appeal.

Chapter 29. A Closer Look at Tests of Significance

1. 225 U.S. 391, quoted from Jerome Frank, *Courts on Trial* (Princeton University Press, 1949).
2. When he was editor of the *Journal of Experimental Psychology*, Arthur Melton defended the practice in these words:

   > The next step in the assessment of an article involved a judgment with respect to the confidence to be placed in the findings—confidence that the results of the experiment would be repeatable under the conditions described. In editing the *Journal* there has been a strong reluctance to accept and publish results related to the principal concern of the research when those results were [only] significant at the .05 level, whether by one- or two-tailed test! This has not implied a slavish worship of the .01 level or any other level, as some critics may have implied. Rather, it reflects a belief that it is the responsibility of the investigator in a science to reveal his effect in such a way that no reasonable man would be in a position to discredit the results by saying they were the product of the way the ball bounced.

There is a better way to make sure results are repeatable: namely, to insist that important experiments be replicated. The quote comes from an editorial in the *Journal* vol. 64 (1962) pp. 553–57. We found it in an article by David Bakan, reprinted in J. Steger, editor, *Readings in Statistics* (Holt, Rinehart and Winston, 1971). Also see note 4 below.

3. The history is on the authority of G. A. Barnard, formerly the professor of statistics, Imperial College of Science and Technology.
4. Unfortunately, even a relatively modest amount of data-snooping can produce off-scale P-values. Of course, the problems created for P-values should not stop investigators from looking at their data. One good research strategy is to *cross-validate:* develop the model on half the data, then see how well the fit holds up when the equations are applied to the other half. Real replication is even better. Replication is a crucial idea, and we do not do it justice in the text. References on data snooping and replication include—

    R. Abelson, *Statistics as Principled Argument* (Lawrence Erlbaum Associates, Hillsdale, N.J., 1995).
    T. K. Dijkstra, editor, *On Model Uncertainty and its Statistical Implications*. Springer Lecture Notes No. 307 in *Economics and Mathematical Systems* (1988).
    A. S. C. Ehrenberg and J. A. Bound, "Predictability and prediction," *Journal of the Royal Statistical Society*, series A, vol. 156, part 2 (1993) pp. 167–206.
    D. A. Freedman, *Statistical Models: Theory and Practice* (Cambridge, 2005).
    M. Oakes, *Statistical Inference* (ERI, Chestnut Hill, 1986).
5. The example is stylized, but the problem is real. We are assuming an incidence rate of 1 per 100,000 per year, and using a Poisson model. Despite concerns about environmental pollution, liver cancer rates have been falling steadily in the U.S. since the 1930s. For discussion and other references, see D. Freedman and H. Zeisel, "From mouse to man: The quantitative assessment of cancer risks," *Statistical Science* vol. 3 (1988) pp. 3–56, with discussion. Also see B. N. Ames, L. S. Gold and W. C. Willett, "The causes and prevention of cancer," *Proceedings of the National Academy of Sciences, U.S.A.* vol. 92 (1995) pp. 5258–65. For a controversial example of a cluster, see S. W. Lagakos, B. S. Wessen and M. Zelen, "An analysis of contaminated well water and health effects in Woburn, Massachusetts," *Journal of the American Statistical Association* vol. 81 (1986) pp. 583–614, with discussion. There is a fascinating account of the Woburn litigation by Jonathan Harr, *A Civil Action* (Random House, 1995). Also see R. B. Schinazi, "The probability of a cancer cluster due to chance alone," *Statistics in Medicine* vol. 19 (2000) pp. 2195–98.
6. In other cases, it is harder to correct the P-value for data snooping. See the book by Dijkstra, cited in note 4. For some discussion of the impact on journal publications, see—

    L. J. Chase and R. B. Chase, "A statistical power analysis of applied psychological research," *Journal of Applied Psychology* vol. 61 (1976) pp. 234–37.
    K. Dickersin, S. Chan, T. C. Chalmers, H. S. Sacks and H. R. Smith, Jr., "Publication bias and clinical trials," *Journal of Controlled Clinical Trials* vol. 8 (1987) pp. 343–53.
    A. Tversky and D. Kahneman, "Belief in the law of small numbers," *Psychological Bulletin* vol. 2 (1971) pp. 105–10.
    C. B. Begg and J. A. Berlin, "Publication bias and dissemination of clinical research," *Journal of the National Cancer Institute* vol. 81 (1989) pp. 107–15.
7. "The Lipid Research Clinics Primary Prevention Trial Results," *Journal of the American Medical Association* vol. 251 (1984) pp. 351–64. The investigators quote $z \approx -1.92$, based on lifetable analysis and blocking. The protocol did not state whether one- or two-tailed tests would be used; it noted "significant morbidity and mortality associated with cholesterol-lowering agents"; and declared that a significance level of 1% "was chosen as the standard for showing a convincing difference between treatment groups." There was a strong suggestion that fatal and non-fatal heart attacks would be analyzed separately—in which case the differences are not significant. See *Journal of Chronic Diseases* vol. 32 (1979) pp. 609–31. The investigators do not appear to have followed protocol. Also see *Journal of Clinical Epidemiology* vol. 43 no. 10 (1990) pp. 1021ff. There are less-formal accounts by T. J. Moore, *Heart Failure* (Random House, 1989) and *Lifespan* (Simon & Schuster, 1993).

    Another experiment is reported by H. Buchwald et al., "Effect of partial ileal bypass surgery on mortality and morbidity from coronary heart disease in patients with hypercholesterolemia," *New England Journal of Medicine* vol. 323 (1990) pp. 946–55. But see G. D. Smith and J. Pekkanen, "Should there be a moratorium on the use of cholesterol lowering drugs?" *British Medical Journal* vol. 304 (1992) pp. 431–34: the evidence from several trials suggests that cholesterol-lowering drugs actually increase the death rate. On the other hand, a large Scandinavian study on Simvastatin obtained a 30% reduction in mortality, among subjects with a history of heart disease. See "Randomised trial of cholesterol lowering in 4444 patients with coronary heart disease: the Scandinavian Simvastatin Survival Study," *Lancet* vol. 344 (November 19, 1994) pp. 1383–89.

There is also the Scottish study on pravastatin, see the *New England Journal of Medicine* (November 16, 1995). For a review, see A. M. Garber, W. S. Browner and S. B. Hulley, "Cholesterol screening in asymptomatic adults, revisited," *Annals of Internal Medicine* vol. 124 (1996) pp. 518–31.

8. K. R. Rao, editor, "The Ganzfeld debate," *Journal of Parapsychology* vol. 49, no. 1 (1985) and vol. 50, no. 4 (1986). The discreteness of the distributions matters, and significance probabilities must be computed by convolution.
9. The evaluation of bioassay results is a complicated issue, but the multiple-endpoint problem is a real one. Many chemicals do seem to cause liver cancer but prevent leukemia in mice. See the paper by Freedman and Zeisel referenced in note 5. Also see T. S. Davies and A. Monro, "The rodent carcinogenicity bioassay produces a similar frequency of tumor increases and decreases: Implications for risk assessment," *Regulatory Toxicology and Pharmacology* vol. 20 (1994) pp. 281–301; T. H. Lin et al., "Carcinogenicity tests and inter-species concordance," *Statistical Science* vol. 10 (1995) pp. 337–53.
10. T. C. Chalmers, R. S. Koff and G. F. Grady, "A note on fatality in serum hepatitis," *Journal of Gastroenterology and Hepatology* vol. 69 (1965) pp. 22–26.
11. The confusion between "statistical significance" and importance gets worse with correlation coefficients. Instead of looking at the value of r, some investigators will test whether $r = 0$, and then use P as the measure of association. Regression coefficients often get the same treatment. However, it is the analysis of variance which presents the problem in its most acute form: some investigators will report P-values, F-statistics, everything except the magnitude of their effect. For some discussion, see P. E. Meehl, "Theoretical risks and tabular asterisks: Sir Karl, Sir Ronald, and the slow progress of soft psychology," *Journal of Consulting and Clinical Psychology* vol. 46 (1978) pp. 806–34.
12. On the other hand, there may be noticeable differences in reading abilities between big-city children and rural children, in later ages. See I. S. Kirsch and A. Jungeblut, *Literacy: Profiles of America's Young Adults* (ETS/NAEP, Princeton, N.J., 1986).
13. The 6 points comes from a rough-and-ready regression analysis of auxiliary data, and includes selection effects. Other indicators of school quality are discussed in review exercise on p. 94 and exercise 2 on p. 506.
14. This is a close paraphrase of a comment (taken out of context) by D. T. Campbell, "Reforms as experiments," *American Psychologist* vol. 24 (1969) pp. 409–29. The reference was supplied by the late Merrill Carlsmith, formerly professor of psychology, Stanford University.
15. M. J. Mahoney, "Publication prejudices: An experimental study of confirmatory bias in the peer review system," *Journal of Cognitive Therapy and Research* vol. 1 (1977) pp. 161–75. The experimental design, and the quotes, have been simplified a little.
16. Daniel McFadden, "The revealed preferences of a government bureaucracy: Empirical evidence," *Bell Journal of Economics* vol. 7 (1971) pp. 55–72. The study period was 1958–66. The "effect" of a variable is a coefficient in a model; of course, the model may be open to question. This reference was supplied by Chris Achen, professor of political science, University of Michigan.
17. Paraphrase of testimony by W. Hogan and J. Kalt (Harvard) in a 1987 administrative hearing on violations of oil price controls. Elasticity is a price coefficient in a regression model.
18. To paraphrase Keynes, the significance tester who thinks he doesn't need a box model may just have a naive one. J. M. Keynes, *The General Theory of Employment, Interest, and Money* (Harcourt Brace Jovanovich, 1935, pp. 383–84).

    > Practical men, who believe themselves to be quite exempt from any intellectual influences, are usually the slaves of some defunct economist.

19. *Statistical Abstract*, 2003, table 11.
20. This study was discussed in section 4 of chapter 2; also see note 7 to that chapter, for references. In this example, $z \approx 5$ so P is rather small. We can interpret P as a descriptive statistic. Altogether there were 933 candidates, of whom 825 were men and 108 were women. If you think that sex and admissions were unrelated, comparing admission rates for men and women is like comparing the admission rate for any group of 825 people with the admission rate for the remaining group of 108 people. (After all, there are many irrelevant splits, based on fingerprints and so forth.) There are

$$\binom{933}{825} \approx 7 \times 10^{143}$$

possible ways to split the 933 candidates into two groups, one of size 825 and the other of size 108. For each split, compute z. This population of z-values is close to normally distributed, so the observed z-value of 5 is quite unusual. See D. Freedman and D. Lane, "A nonstochastic interpretation of reported significance levels," *Journal of Business and Economic Statistics* vol. 1

(1983) pp. 292–98. The idea goes back to R. A. Fisher. See E. J. G. Pitman, "Significance tests which may be applied to samples from any population," *Journal of the Royal Statistical Society* Series B vol. 4 (1936) pp. 119–30.

21. *Project Follow Through Classroom Evaluation*, published by SRI at Menlo Park, California. The senior investigator was Jane Stallings. The quotes were edited slightly. The study was done in 1972–73.
22. This assumes the control average of 60 to be known without error. In fact, SRI made a two-sample t-test. However, the SRI scoring procedure was bound to introduce dependence between treatment and control scores—it was based on pooled ranks.
23. These are real numbers, from 1976. About half the TAs had participated in grading the final, and many had graded similar finals in previous years. Over time, the graduate students did learn how to handle Statistics 2 problems.
24. F. Mosteller and R. Rourke, *Sturdy Statistics* (Addison-Wesley, 1973, p. 54).
25. T. A. Ryan, B. L. Joiner and B. F. Ryan, *Minitab Student Handbook* (Duxbury Press, Boston, 1976, p. 228).
26. "Intellectual development of children by demographic and socioeconomic factors," *Vital and Health Statistics* series 11, no. 110 (Washington, D.C., 1971).
27. R. S. Erikson, J. P. McIver and G. C. Wright, Jr., "State political culture and public opinion," *American Political Science Review* vol. 81 (1987) pp. 797–813. The analytic technique was multiple regression on dummy variables for demographic categories (e.g., low income, etc.); then dummies were added for regions and states. Adding in the state dummies increased the adjusted R^2 from 0.0898 to 0.0953, but the F to enter was 8.35, with 40 degrees of freedom in the numerator—and 55,072 in the denominator. The authors say that the state effects are significant in practical terms as well; the R^2's suggest otherwise. The authors acknowledge that state dummies may be proxies for omitted variables, but argue against this interpretation. The papers cited in this note and the next are discussed by D. A. Freedman, "Statistical models and shoe leather," in P. Marsden, editor, *Sociological Methodology 1991* (American Sociological Association, Washington, D.C., chapter 10). Also see D. A. Freedman, *Statistical Models: Theory and Practice* (Cambridge University Press, 2005).
28. J. L. Gibson, "Political intolerance and political repression during the McCarthy era," *American Political Science Review* vol. 82 (1988) pp. 511–39. "Effects" are coefficients in a path model. Presumably, the author would view the randomness in the estimates as generated by the model. On the other hand, the adequacy of the model may be open to question.
29. The experiment is discussed by C. E. M. Hansel, *ESP: A Scientific Evaluation* (Charles Scribner's Sons, 1966, chapter 11). The numbers have been changed to simplify the arithmetic. The point of the experiment was to illustrate the fallacy discussed in the text. The reference was supplied by Charles Yarbrough, Santa Rosa, Calif.
30. The random number generator on the Aquarius itself does not seem to have been tested, but the generator is similar to ones that were tested. In ESP research, nothing is simple, and Tart would not agree with much of what we write: C. Tart et al., "Effects of immediate feedback on ESP performance: A second study," *Journal of the American Society for Psychical Research* vol. 73 (1979) pp. 151–65. For a lively discussion of the issues, see Martin Gardner, *Science: Good, Bad, and Bogus* (Avon Books, 1981, chapters 18 and 31).
31. Reproduced by permission of the publisher, Harcourt Brace Jovanovich, Inc.
32. Based on a question used by A. Tversky and D. Kahneman. Also see p. 298 in Steger's book, referenced in note 2 above.
33. See p. 68 of Mosteller and Rourke, note 24.
34. F. Arcelus and A. H. Meltzer, "The effect of aggregate economic variables on congressional elections," *American Political Science Review* vol. 69 (1965) pp. 1232–69, with discussion. This reference was supplied by Chris Achen. The argument uses a regression model, and is therefore more subtle than indicated in the exercise. (Of course, the validity of the model is open to question.) However, the investigators' position on hypothesis testing is brutal; see the rejoinder by Arcelus and Meltzer to the comments by Goodman and Kramer.
35. *Statistical Abstract*, 1988, table 21; *Statistical Abstract*, 1994, table 26; *Statistical Abstract*, 2003, table 17.
36. *Statistical Abstract*, 1994, tables 616 and 621. *Employment and Earnings* vol. 52, no. 12 (December, 2005), table A-2.
37. *Statistical Abstract*, 2003, table 284. Also see A. W. Astin et al., *The American Freshman: Thirty-Five Year Trends, 1966–2001* (Higher Education Research Institute, UCLA, 1991). Most of the change occurred between 1970 and 1980.
38. R. E. Just and W. S. Chern, "Tomatoes, technology and oligopsony," *Bell Journal of Economics* vol. 11 (1980) pp. 584–602. For discussion, see R. Daggett and D. Freedman, "Econometrics and the law: A case study in the proof of antitrust damages," in L. M. LeCam and R. A. Olshen, editors, *Proceedings of the Berkeley Conference in Honor of Jerzy Neyman and Jack Kiefer* vol. 1,

pp. 123–72 (Wadsworth, Belmont, California, 1985). Just and Chern estimated both linear and log-linear demand functions; the t-test reported in the exercise was applied to the coefficient of price in a linear demand function.

39. For national data, see *Statistical Abstract*, 2003, table 1244. By this measure, dining out was the most popular activity, followed by reading, and entertaining at home.
40. The quote is from D. L. Hartl, Letter, *Nature* vol. 372 (1994) p. 398; we thank David Kaye (Arizona State University) for calling it to our attention. Also see note 6 to chapter 13.
41. June 27, 1993.
42. Paraphrased from evidence presented at an extradition hearing for James Smyth, Federal District Court (N.D. Cal., 1993). Defense Exhibit 31, *Secondary Analysis of the School Leavers Survey* (1989), Standing Advisory Commission on Human Rights, by Cormack et al.
43. Data are from Thomas H. Cohen and Steven K. Smith (2004), *Civil Trial Cases and Verdicts in Large Counties 2001*, Bureau of Justice Statistics, U.S. Department of Justice. Results were simplified a little. Jury awards have declined over the period 1991–2001. Interestingly enough, judges tend to be more generous to plaintiffs than are juries.
44. See R. C. Lewontin, "Sex, lies, and social science," in *New York Review of Books*, April 20, May 25, and August 10, 1995. Lewontin is reviewing R. T. Michael et al., *Sex In America: A Definitive Survey* (Little Brown, 1994), which is a popularized version of E. O. Laumann et al., *The Social Organization of Sexuality: Sexual Practices in the United States* (University of Chicago Press, 1994). Also see Devon D. Brewer et al., "Prostitution and the sex discrepancy in reported number of sexual partners," *Proceedings of the National Academy of Sciences of the U.S.A.*, vol. 97 (2000) pp. 12385–388. Brewer et al. find that female prostitutes—who have very large numbers of male partners—are substantially under-represented in the survey; and "men are reluctant to acknowledge that their reported partners include prostitutes."
45. John A. Dossey et al., *Can Students Do Mathematical Problem Solving?* (U.S. Department of Education, Office of Educational Research and Improvement, Washington, D.C., 1992, pp. 141, 172).
46. *Brock v. Merrell Dow Pharmaceuticals, Inc.*, 874 F.2d 307, 311–12 (5th Cir.), modified, 884 F.2d 166 (5th Cir. 1989), cert. denied, 494 U.S. 1046 (1990); D. H. Kaye and D. A. Freedman, *Reference Guide on Statistics*, 2nd ed. (Federal Judicial Center, Washington, D.C., 2000, p. 121).
47. See W. T. Keeton, J. L. Gould, and C. G. Gould, *Biological Science*, 5th ed. (W. W. Norton & Company, 1993, p. 445).
48. *Statistical Abstract*, 2003: table 66 gives 108 million households, table 305 gives 2.11 million burglaries reported to the police, table 321 gives 3.14 million burglaries reported to the survey. The survey uses a highly designed sample, but a simple random sample of 50,000 gives (roughly) the right standard errors. Also see J. P. Lynch and L. A. Addington, *Understanding Crime Statistics* (Cambridge, 2007).
49. The randomization included blocking, not accounted for here. The averages were published; the SDs were kindly provided by J. D. Neaton (professor of biostatistics, University of Minnesota). An interesting sidelight: logistic regressions fitted to the Framingham data predicted a very substantial reduction in mortality due to the modest-looking decrements in risk factors (3 mm in blood pressure, 5 mg/dl in serum cholesterol, 13% in smoking). There was some concern that smoking was under-reported by the treatment group, and an adjustment was made for this by blood chemistry. References—

    "Multiple Risk Factor Intervention Trial," *Journal of the American Medical Association* vol. 248 (1982) pp. 1465–77.

    "Statistical design considerations in the NHLI Multiple Risk Factor Intervention Trial (MR-FIT)," *Journal of Chronic Diseases* vol. 30 (1972) pp. 261–75.

    "Mortality rates after 10.5 years for participants in the Multiple Risk Factor Intervention Trial," *Journal of the American Medical Association* vol. 263 (1990) pp. 1795–1801.

50. http://www.gallup.com
51. *Waisome v. Port Authority*, 948 F.2d 1370, 1376 (2nd Cir. 1991); D. H. Kaye and D. A. Freedman, *Reference Guide on Statistics*, 2nd ed. (Federal Judicial Center, 2000, Washington, D.C., p. 124). The quote is edited slightly.
52. M. S. Kanarek et al., "Asbestos in drinking water and cancer incidence in the San Francisco Bay Area," *American Journal of Epidemiology* vol. 112 (1980) pp. 54–72. There was no relationship between asbestos in the water and lung cancer for blacks or women. Data in the paper strongly suggest that smoking was a confounder. For more discussion, see D. A. Freedman, "From association to causation: Some remarks on the history of statistics," *Statistical Science*, vol. 14 (1999) pp. 243–58; reprinted in *Journal de la Société Française de Statistique*, vol. 140 (1999) pp. 5–32 and in *Stochastic Musings: Perspectives from the Pioneers of the Late 20th Century* (Lawrence Erlbaum Associates, 2003, pp. 45–71), edited by J. Panaretos.
53. See note 20 to chapter 19. Children with no siblings are an exception, scoring slightly below first-borns in two-child families.

Answers to Exercises

Part I. Design of Experiments

Chapter 2. Observational Studies

Set A, page 20

1. False. The population got bigger too. You need to look at the number of deaths relative to total population size. The population in 2000 was about 281 million, and in 1970 it was about 203 million: 2.4 out of 281 is smaller than 1.9 out of 203, so the death rate was lower in 2000. There was a very considerable increase in life expectancy between 1970 and 2000.

   Comment. Between 1970 and 2000, the population got older, on average, so the reduction in death rates is even more impressive.

2. The basic facts: richer families are more likely to volunteer for the experiment, and their children more vulnerable to polio (section 1 of chapter 1).

   (a) From line 1 of the table, the polio rates in the two vaccine groups were about the same. If (for example) the consent group in the NFIP study had been richer, their rate would have been higher.

   (b) From line 3 of the table, the polio rates in the two no-consent groups were about the same.

   (c) From line 2 of the table, the polio rate in the NFIP control group was quite a bit lower than the rate in the other control group.

   (d) The no-consent group is predominantly lower-income, and the children are more resistant to polio. The NFIP control group has a range of incomes, including the more vulnerable children from the higher-income families.

   (e) The ones who consent are different from the ones who don't consent (p. 4).

   Comment on (c). The NFIP controls had a whole range of family backgrounds. The controls in the randomized experiment were from families who consented to participate. These families were richer, and their children more vulnerable to polio. The NFIP design was biased against the vaccine.

3. Children who were vaccinated might engage in more risky behavior—a bias against the vaccine. On the other hand, the placebo effect goes in favor of the vaccine. (The similarity of rates in line 1 of table 1, p. 6, suggests biases are small.)

4. No, because the experimental areas were selected in those parts of the country most at risk from polio. See section 1 of chapter 1.

5. The people who broke the blind found out whether or not they were getting vitamin C. The ones who knew they were getting vitamin C for prevention tended to get fewer colds. Those on vitamin C for therapy tended to get shorter colds. This is the placebo effect. Blinding is important.

6. $558/1{,}045 \approx 53\%$, and $1{,}813/2{,}695 \approx 67\%$. Adherence is lower in the nicotinic acid group. Something went wrong with the randomization or the blind. (For example, nicotinic acid might have unpleasant side effects, which causes subjects to stop taking it.)

7. In trial (i), something must have gone wrong with the randomization. The difference between 49.3% and 69.0% shows that the treatment group smoked less to begin with, which would bias any further comparisons. The difference cannot be due to the treatment, because baseline data say what the subjects were like before assignment to treatment or control. (More about this in chapter 27.)

8. Option (ii) explains the association, option (i) does not. Choose (ii). See p. 20.

9. (a) Yes: 39 deaths from breast cancer in the treatment group, versus 63 in the control group.
   (b) The death rate in the treatment group (screened and refused together) is about the same as the death rate in the control group because screening has little impact on deaths from causes other than breast cancer.
   (c) Compare A) the control group with B) those who refused screening in the treatment group. Group A includes women who would accept screening as well as those who would refuse. On average, then, group A is richer than group B. Neither group is affected by screening, and group A has a higher death rate from breast cancer.
   (d) Most deaths are from causes other than breast cancer; those rates are not affected by screening. However, the women who refuse screening are poorer and more vulnerable to most diseases. That is why their death rates are higher.

   Comments. (i) In part (a), you should compare the whole treatment group with the whole control group. This is the "intention to treat" principle. It is conservative, that is, it understates the benefit of screening. (If all the women had come in for screening, the benefit would have been higher.) You should not compare the "examined" with the "refused" or with the controls: that is biased against treatment, see exercise 10(a).

   (ii) The Salk vaccine field trial could have been organized like HIP: (1) define a study population of, say, 1,000,000 children; (2) randomize half of them to treatment and half to control, where treatment is the invitation to come in and be vaccinated; (3) compare polio rates for the whole treatment group versus the whole control group. In this setup, it would not be legitimate to compare just the vaccinated children with the controls; you would have to compare the whole treatment group with the whole control group. The design actually used in the Salk field trial was better, because of the blinding (section 1 of chapter 1); however, this seems to have been a relatively minor issue for HIP, and the design they used is substantially easier to manage.

10. (a) This is not a good comparison. There is a bias against screening. The comparison between the "examined" and "refused" groups is observational, even though the context is an experiment: it is the women who decide whether to be examined or not. This is just like adherence to protocol in the clofibrate trial (section 2). There are confounding variables, like income and education, to worry about. These matter. The comparison is biased against screening because the women who come in for examination are richer, and more vulnerable to breast cancer.
    (b) This is not a good theory: the overall death rate in the treatment group from diseases other than breast cancer is about the same as that in the control group, and the reduction in breast cancer death rate is due to screening.
    (c) False. Screening detects breast cancers which are there and would otherwise be detected later. That is the point of screening.

Comments. (i) In the HIP trial, the number of deaths from other causes is large, and subject to moderately large chance effects, so the difference $837-879 = -42$ is not such a reliable statistic. More about this in chapter 27. The comparison of 1.1 and 1.5 in 10(a) is very unreliable, because the number of breast cancers is so small—23 and 16. However, the difference between 39 and 63 in 9(a) is hard to explain as a chance variation.

(ii) In part 10(c), within the treatment group, the screened women had a higher incidence rate of diagnosed breast cancer, compared to the women who refused. The two main reasons: (1) screening detects cancers; (2) breast cancer—like polio and unlike most other diseases—hits the rich harder than it hits the poor, and the rich are more likely to accept screening.

(iii) The benefits of mammography for women age 50–70 are now generally recognized; there remains some question whether the benefits extend to women below the age of 50. For references, see note 14 to chapter 2.

11. The women who have been exposed to herpes are the ones who are more active sexually; this evidence is not convincing. (See example 2 on p. 16.)

    Comment. In the 1970s, herpes (HSV-2) was thought to be causal. In the 1980s, new evidence from molecular biology suggested that HSV was not a primary causal agent, and implicated strains of human papilloma virus (HPV-16,18). For references, see note 4 to chapter 2.

12. If a woman has already aborted in a previous pregnancy—and is therefore more at risk in her current pregnancy—a physician is likely to tell her to cut down on exercise. In this instance, exercise is a marker of good health, not a cause.

13. False. Altogether, 900 out of 2,000 men are admitted, or 45%; while 360 out of 1,100 women are admitted, or 33%. This is because women tend to apply to department B, which is harder to get into. See section 4.

14. (a) 39 out of 398 is like 40 out of 400, or 10 out of 100, or 10%.
    (b) 25% (c) 25% (d) 50%

15. (a) 10%. That's spread over a $10,000 range, so for the next three parts, guess about 1% in each $1,000 range.
    (b) 1% (c) 1% (d) 2%

Part II. Descriptive Statistics

Chapter 3. The Histogram

Set A, page 33

1. (a) 2% (b) 3% (c) 4% (d) 5% (e) 15% (f) 15%
2. More between $10,000 and $11,000.
3. (a) B (b) 20% (c) 70%
4. (a) Well over 50%. (b) Well under 50%. (c) About 50%.
5. Class (b).

6. There were more in the range 90 to 100.

7. A (ii), B (i), C (iii)

8. The figure does not adjust for inflation, so the comparison is not a good one.

Comment. In 1973, a dollar bought roughly 4 times as much as in 2004. The figure below compares the 2004 histogram with the 1973 histogram—corrected for this change in purchasing power. Family income went up by a factor of about 4 in "nominal" dollars, but in "real" dollars—corrected for inflation—there was not that much improvement. (We shifted the 2004 histogram to the right a little; data on the consumer price index are from *Statistical Abstract*, 1993, table 756; 2003, table 713; table 690 in the latter publication suggests about a 15% increase in real family income over the period 1980–2000; prices indices are not the most reliable of statistics, because they may not reflect quality improvements.)

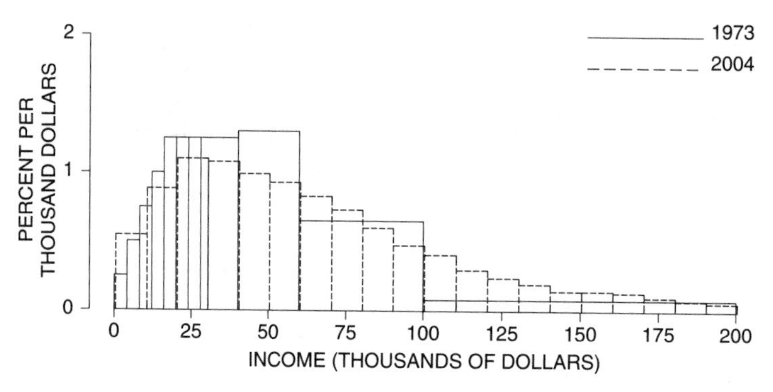

Set B, page 38

1. The 1991 histogram is shown in figure 5 on p. 39, and the reason for the spikes is discussed on that page.
2. Smooths out the graph between 0 and 8.
3. The educational level went up. For example, more people finished high school and went on to college in 1991 than in 1970.

   Comment. In this century, there has been a remarkable and steady increase in the educational level of the population. In 1940, only 25% of the population age 25+ had finished high school. By 1993, this percentage was up to 80%, and still climbing. In that year, about 7% of the population age 25+ had completed a master's degree or better. In 2005, about 85% of the population age 25+ had a high school degree, and 9% had a master's degree or better.
4. Went up.

Set C, page 41

1. 15% per $100.
2. Option (ii) is the answer, because (i) doesn't have units, and (iii) has the wrong units for density.
3. 1,750, 2,000, 1, 0.5. The idea on density: If you spread 10 percent evenly over 1 cm = 10 mm, there is 1 percent in each mm, that is, 1 percent per mm.
4. (a) 1.5% per cigarette × 10 cigarettes = 15%.
   (b) 30% (c) 30% + 20% = 50% (d) 10% (e) 3.5%

Set D, page 44

1. (a) qualitative
   (b) qualitative
   (c) quantitative, continuous
   (d) quantitative, continuous
   (e) quantitative, discrete

2. (a) Number of children is a discrete variable.
   (b)

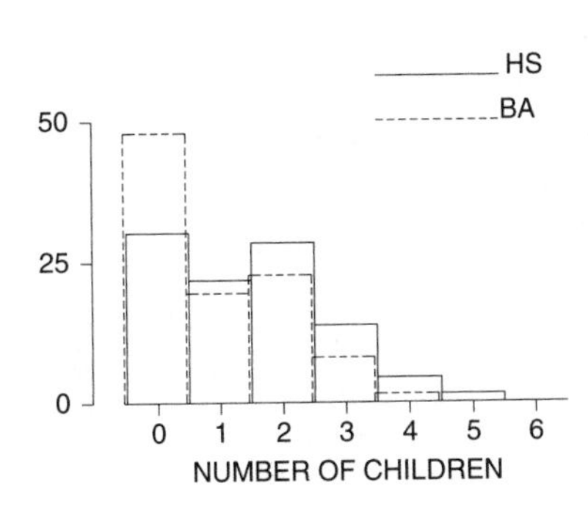

   (c) Better-educated women have fewer children.

Set E, page 46

1. On the whole, the mothers with four children have higher blood pressures. Causality is not proved, there is the confounding factor of age. The mothers with four children are older. (After controlling the age, the Drug Study found there was no association left between number of children and blood pressure.)

2. Left: adds 10 mm Right: adds 10%

Set F, page 48

1. (a) 7% (b) 5%
   (c) The users tend to have higher blood pressures.

2. Use of the pill is associated with an increase in blood pressure of several mm.

3. The younger women have slightly higher blood pressures.

   Comment. This is a definite anomaly. Most U.S. studies show that systolic blood pressure goes up with age. By comparison, the younger women in the Contraceptive Drug Study have blood pressures which are too high, while the older women have blood pressures which are too low. This probably results from bias in the procedure used to measure blood pressures at the multiphasic, which tended to minimize the prevalence of blood pressures above 140 mm.

Chapter 4. The Average and the Standard Deviation

Set A, page 60

1. (a)

3 4 5

(b)

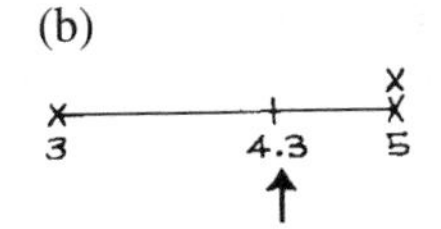

(c)

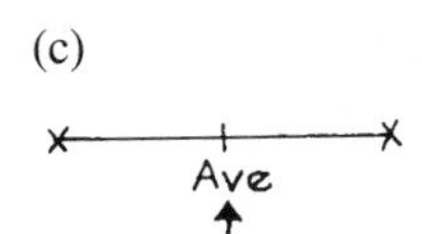

Comment. With two numbers, the average is half way between. If you add bigger numbers to the list, the average moves up. (Smaller numbers move it down.) The average is always somewhere between the smallest and biggest number on the list.

2. If the average is 1, the list consists of ten 1's. If the average is 3, the list consists of ten 3's. The average cannot be 4: it has to between 1 and 3.
3. The average of (ii) is bigger, it has the large entry 11.
4. $(10\times66 \text{ inches}+77 \text{ inches})/11 = 67 \text{ inches} = 5$ feet 7 inches. Or reason this way: the new person is 11 inches taller than the old average. So he adds $11 \text{ inches}/11 =$ 1 inch to the average.
5. 5 feet $6\frac{1}{2}$ inches. As the number of people in the room goes up, each additional person has less of an effect on the average.
6. 5 feet 6 inches + 22 inches = 7 feet 4 inches: it's a giraffe.

7. The Rocky Mountains are at the right end, Kansas is around 0 (sea level), and the Marianas trench is at the left end.
8. The conclusion does not follow, the data are cross-sectional not longitudinal. The men with higher diastolic blood pressures are likely to die earlier; they will not be represented in the graph. Furthermore, men with higher blood pressure are more likely to be put on medications that reduce blood pressure.
9. During the recessions, firms tend to lay off the workers with lowest seniority, who are also the lowest paid. This raises the average wage of those left on the payroll. When the recession ends, these low-paid workers are rehired.

   Comment. It matters who is included in an average—and who is excluded.

Set B, page 65

1. (a) 50 (b) 25 (c) 40
2. (a) median = average (b) median = average
   (c) median is to the left of the average—long right-hand tail at work.

3. 20
4. The average has to be bigger than the median, so guess 25. (The exact answer is 27.)
5. The average: long right-hand tail.
6. (a) 1 (b) 10 (c) 5 (d) 5
   ("Size" means, neglecting signs.)

Set C, page 67

1. (a) average = 0, r.m.s. size = 4
   (b) average = 0, r.m.s. size = 10.
   On the whole, the numbers in list (b) are bigger in size.
2. (a) 10 (to one decimal place, the exact answer is 9.0).
   (b) 20 (to one decimal place, the exact answer is 19.8).
   (c) 1 (to one decimal place, the exact answer is 1.3).
   The average of the lists is 0; the r.m.s. operation wipes out the signs.
3. For both lists, it's 7; all the entries have the same size, 7.
4. The r.m.s. size is 3.2.
5. The r.m.s. size is 3.1.
   Comment. The r.m.s. in exercise 5 is smaller than in exercise 4. There is a reason. Suppose we are going to compare each number on a list to some common value. The r.m.s. size of the amounts off depends on this value. For some values the r.m.s. is larger, for others the r.m.s. is smaller. When is the r.m.s. smallest? It can be proved mathematically that the r.m.s. size of the amounts off is smallest for the average.
6. The errors are way bigger than 3.6, which is supposed to be the r.m.s. size. Something is wrong with the computer.

Set D, page 70

1. (a) 170 cm is 24 cm above average, the SD is 8 cm, so 24 cm represents 3 SDs.
   (b) 2 cm is 0.25 SDs.
   (c) $1.5 \times 8 = 12$ cm, the boy is $146 - 12 = 134$ cm tall.
   (d) shortest, $146 - 18 = 128$ cm; tallest, $146 + 18 = 164$ cm.
2. (a) 150 cm—about average; 4 cm is only 0.5 SDs.
   130 cm—unusually short; 16 cm is 2 SDs.
   165 cm—unusually tall.
   140 cm—about average.
   (b) About 68% were in the range 138 to 154 cm (ave $\pm$ 1 SD), and 95% were in the range 130 to 162 cm (ave $\pm$ 2 SD).
3. biggest, (iii); smallest, (ii).
   Comment. All three lists have the same average of 50 and the same range, 0 to 100. But in list (iii), more of the numbers are further away from 50. In list (ii), more of the numbers are closer to 50. There is more to "spread" than the range.
4. (a) 1, since all deviations from the average of 50 are ± 1.
   (b) 2 (c) 2 (d) 2 (e) 10

Comment. The SD says how far off average the entries are, on the whole. Just ask yourself whether the amounts off are on the whole more like 1, 2, or 10 in size.

5. 25 years. The average is maybe 30 years, so if 5 years were the answer, many people would be 4 SDs away from the average; with 50 years, everybody would be within 1 SD of the average.

6. (a) (i) (b) (ii) (c) (v)

7. In trial (i), something went wrong: the treatment group is much heavier than the control group. (See exercise 7 on p. 22.)

8. The averages and SDs should be about the same, but the investigator with the bigger sample is likely to get the tallest man, as well as the shortest. The bigger the sample, the bigger the range. The SD and the range measure different things.

9. Guess the average, 69 inches. You have about 1/3 of a chance to be off by more one SD, which is 3 inches.

10. 3 inches. The SD is the r.m.s. deviation from average.

Set E, page 72

1. The SD of (ii) is larger; in fact, the SD of (i) is 1, the SD of (ii) is 2.

2. No, the SD is different from the average absolute deviation, so the method is wrong.

3. No, the 0 does count, so the method is wrong.

4. (a) All three classes have the same average, 50.
   (b) Class B has the biggest SD; there are more students far away from average.
   (c) All three classes have the same range. There is more to spread than the range; see exercise 3 on p. 70.

5. (a) (i) average $= 4$; deviations $= -3, -1, 0, 1, 3$; SD $= 2$.
   (ii) average $= 9$; deviations $= -3, -1, 0, 1, 3$; SD $= 2$.
   (b) List (ii) is obtained from list (i) by adding 5 to each entry. This adds 5 to the average, but does not affect the deviations from the average. So, it does not affect the SD. Adding the same number to each entry on a list does not affect the SD.

6. (a) (i) average $= 4$; deviations $= -3, -1, 0, 1, 3$; SD $= 2$.
   (ii) average $= 12$; deviations $= -9, -3, 0, 3, 9$; SD $= 6$.
   (b) List (ii) is obtained from list (i) by multiplying each entry by 3. This multiplies the average by 3. It also multiplies the deviations from the average by a factor of 3, so it multiplies the SD by a factor of 3. Multiplying each entry on a list by the same positive number just multiplies the SD by that number.

7. (a) (i) average $= 2$; deviations $= 3, -6, 1, -3, 5$; SD $= 4$.
   (ii) average $= -2$; deviations $= -3, 6, -1, 3, -5$; SD $= 4$.
   (b) List (ii) is obtained from list (i) by changing the sign of each entry. This changes the sign of the average and all the deviations from the average, but does not affect the SD.

8. (a) This would increase the average by \$250 but leave the SD alone.
   (b) This would increase the average and SD by 5%.

9. The r.m.s. size is 17, and the SD is 0.

10. The SD is much smaller than the r.m.s. size. See p. 72.

11. No.

12. Yes; for instance, the list 1, 1, 16 has an average of 6 and an SD of about 7.

Chapter 5. The Normal Approximation for Data

Set A, page 82

1. (a) 60 is 10 above average; that's 1 SD. So 60 is $+1$ in standard units. Similarly, 45 is -0.5 and 75 is $+2.5$.
   (b) 0 corresponds to the average, 50. The score which is 1.5 in standard units is 1.5 SDs above average; that's $1.5 \times 10 = 15$ points above average, or 65 points. The score 22 is -2.8 in standard units.

2. The average is 10; the SD is 2.
   (a) In standard units, the list is $+1.5$, -0.5, $+0.5$, -1.5, 0.
   (b) The converted list has an average of 0 and an SD of 1. (This is always so: when converted to standard units, any list will average out to 0 and the SD will be 1.)

Set B, page 84

1. (a) 11% (b) 34% (c) 79%
   (d) 25% (e) 43% (f) 13%

2. (a) 1 (b) 1.15

3. (a) 1.65
   (b) 1.30. It's NOT the same z as in (a).

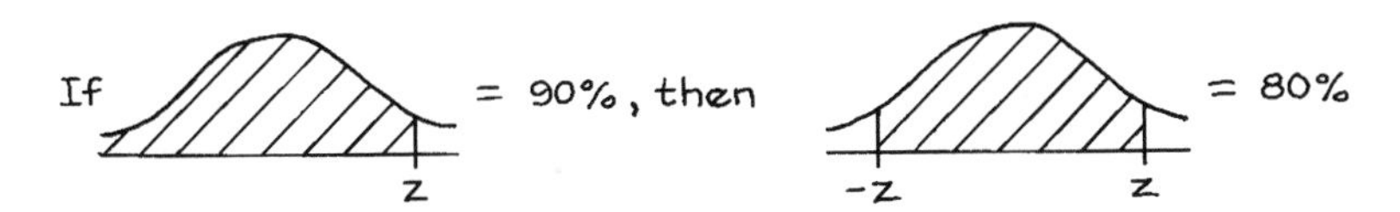

4. (a) $100\% - 39\% = 61\%$.
   (b) impossible without further information

5. (a) $58\% \div 2 = 29\%$ (b) $50\% - 29\% = 21\%$.
   (c) impossible without further information.

Set C, page 88

1. (a)

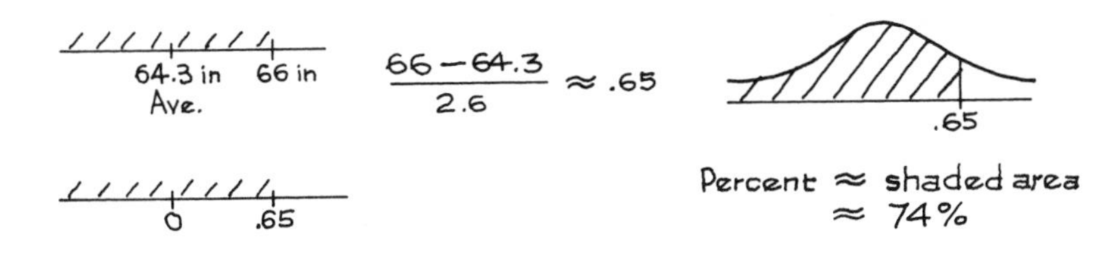

   (b) 69% (c) 0.2 of 1%.

2. (a) 77% (b) 69%

3. In figure 2, the percentage of women with heights between 61 inches and 66 inches is exactly equal to the area under the histogram and approximately equal to the area under the normal curve.

Set D, page 89

1. (a) 75% (b) $29,000
   (c) 75%. Reason: 90% − 10% = 80% are in the range $15,000 to $135,000; and $15,000 to $125,000 is about the same range but a little smaller.

2. 5, 95.

3. $7,000.

4. The area to the left of the 25th percentile has to be 25% of the total area, so the 25th percentile must be quite a bit smaller than 25 mm.

5. (a) It has fatter tails.
   (b) The interquartile range is about 15.

Set E, page 92

1. She was 2.15 SDs above average, at the 98th percentile.

2. The score is 0.85 SDs above average, which is $0.85 \times 100 \approx 85$ points above average. That's $535 + 85 = 620$.

3. 2.75 points—0.50 SDs below average.

Set F, page 93

1. (a) The average is

$$\frac{5}{9} \times (98.6 - 32) = 37.0$$

The SD is

$$\frac{5}{9} \times 0.3 = 0.17$$

   (b) In standard units, the change of scale washes out, so the answer is 1.5.

Chapter 7. Plotting Points and Lines

Set A, page 111

1. A = (1, 2) B = (4, 4) C = (5, 3) D = (5, 1) E = (3, 0).

2. x up by 3, y up by 2.

3. Point D.

Set B, page 112

1. The four points all lie on a line.

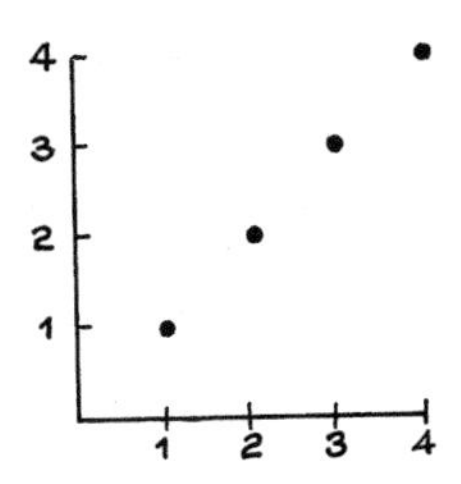

2. The maverick is (1, 2) and it is above the line.

3. The points all lie on a line.

x	y
1	3
2	5
3	7
4	9

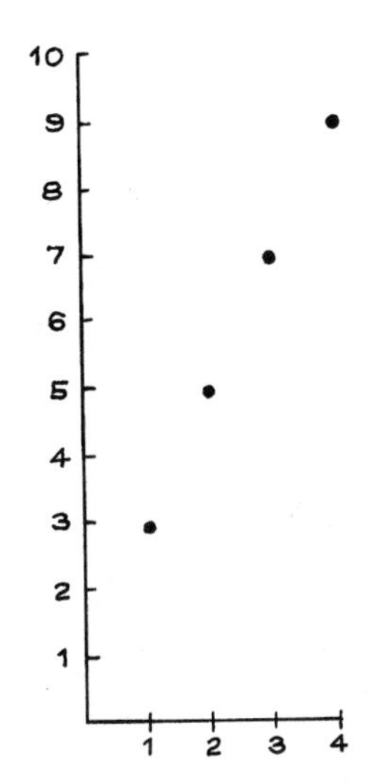

4. (1, 2) is out; (2, 1) is in.

5. (1, 2) is in; (2, 1) is out.

6. (1, 2) is in; (2, 1) is out.

Set C, page 114

1.

	Fig. 16	*Fig. 17*	*Fig. 18*
Slope	−1/4 in per lb	5	1
Intercept	1 in	−10	0

Note: In Figure 18, the axes cross at (2, 2).

Set D, page 115

1.

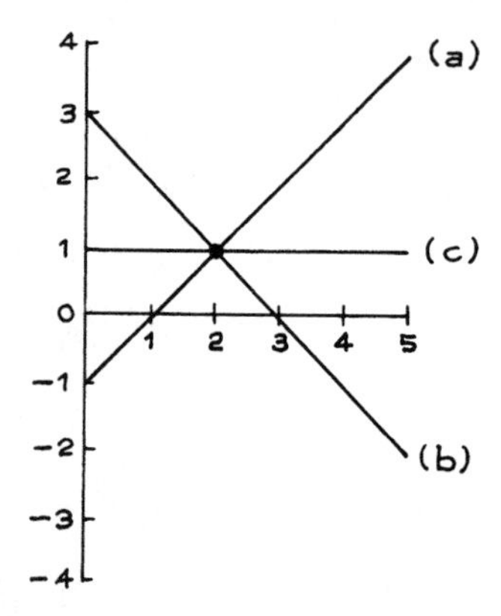

2. On the line.
3. On the line.
4. Above the line.
5.

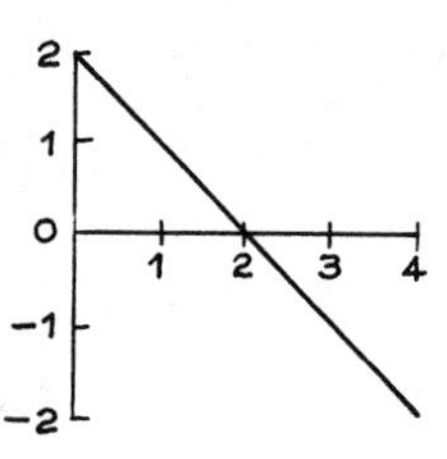

6.

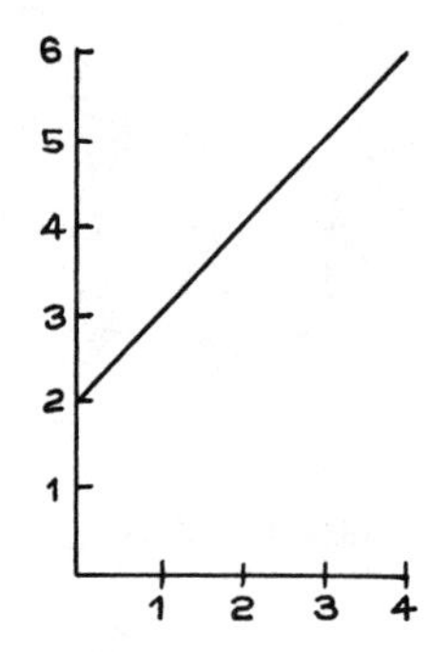

Set E, page 116

1.

	Slope	*Intercept*	*Height at $x = 2$*
(a)	2	1	5
(b)	1/2	2	3

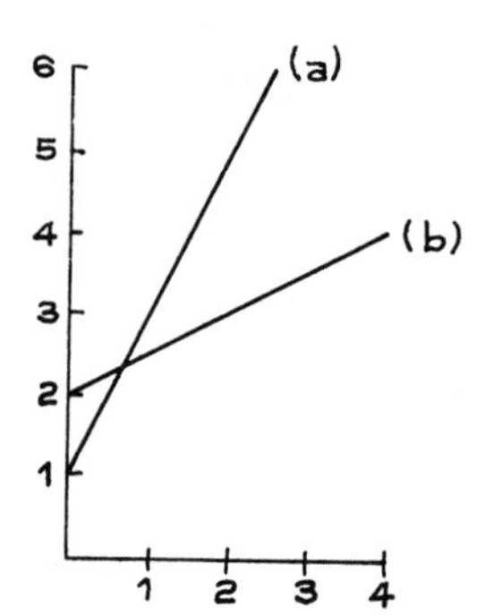

2. (a) $y = \frac{3}{4}x + 1$ (b) $y = -\frac{1}{4}x + 4$ (c) $y = -\frac{1}{2}x + 2$

3. They are all on the line $y = 2x$.

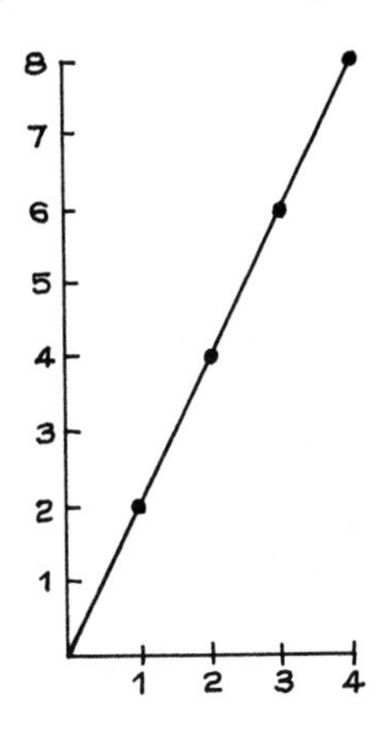

4. They are all on the line $y = x$.

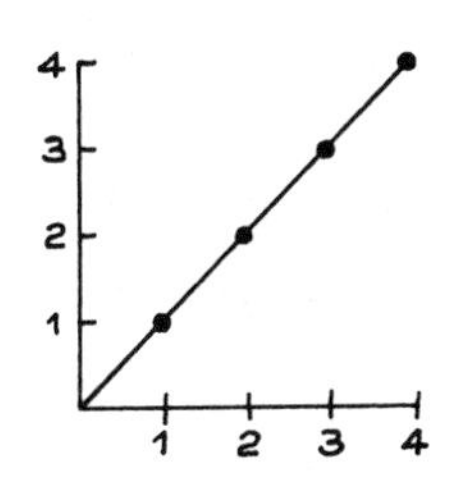

5. (a) on the line. (b) above the line. (c) below the line.

6. All three statements are true. If you understand exercises 4, 5, and 6, you are in good shape for part III.

Part III. Correlation and Regression

Chapter 8. Correlation

Set A, page 122

1. (a) shortest father, 59 inches; his son, 65 inches.
   (b) tallest father, 75 inches; his son, 70 inches.
   (c) 76 inches, 64 inches.
   (d) two: 69 inches, 70 inches.
   (e) ave = 68 inches. (f) SD = 3 inches.

2.

x	y
1	4
2	3
3	1
4	1
4	2

3. (a) ave $x = 1.5$ (b) SD of $x = 0.5$
   (c) ave $y = 2$ (d) SD of $y = 1.5$

4.

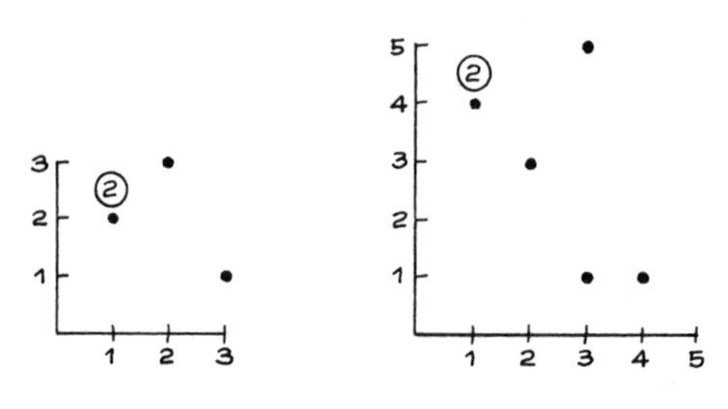

5. (a) A, B, F (b) C, G, H (c) ave ≈ 50
   (d) SD ≈ 25 (e) ave ≈ 30
   (f) False. (g) False, the association is negative.

6. (a) 75 (b) 10 (c) 20
   (d) The final. (e) The final. (f) True.

Set B, page 128

1. (a) Negative. The older the car, the lower the price.
   (b) Negative. The heavier the car, the less efficient.

2. Left: ave $x = 3.0$, SD $x = 1.0$, ave $y = 1.5$, SD $y = 0.5$, positive correlation.
   Right: ave $x = 3.0$, SD $x = 1.0$, ave $y = 1.5$, SD $y = 0.5$, negative correlation.

3. The left hand diagram has correlation closer to 0, it's less like a line.

4. The correlation is about 0.5.

5. The correlation is nearly 0.

   Comment. Psychologists call this "attenuation." If you restrict the range of one variable, that usually cuts the correlation down.

6. (a) All the points on the scatter diagram would lie on a line sloping up, so the correlation would be 1.
   (b) Close to 1; this is like part (a), with some noise thrown into the data.

   Comment. In the March 2005 Current Population Survey, the correlation between the ages of the husbands and wives was about 0.93; the husbands were, on average, 2.3 years older than their wives.

7. (a) Nearly -1: the older you are, the earlier you were born; but there is some fuzz, depending on whether your birthday is before or after the day of the questionnaire.
   (b) Somewhat positive.

8. (a) Somewhat positive. Although wife's income must be less than family income, the two are positively associated.
   (b) Nearly -1. If family income is practically constant, the more the wife makes, the less the husband can make.

   Comment. In the March 2005 Current Population Survey, the correlation between wife's income and total income was about 0.70. Among families with total income in the range \$80,000–\$90,000, the correlation between husband's income and wife's income was about -0.98.

9. False: see p. 126.

Set C, page 131

1. (a) True. (b) False.
2. Dashed.
3. He is one SD above average in height and must weigh $140 + 20 = 160$ pounds.
4. (a) Yes. (b) No. (c) Yes.

Set D, page 134

1. (a) ave of $x = 4$, SD of $x = 2$
   ave of $y = 4$, SD of $y = 2$

Standard units		
x	y	*Product*
−1.5	1.0	−1.50
−1.0	1.5	−1.50
−0.5	0.5	−0.25
0.0	0.0	0.00
0.5	−0.5	−0.25
1.0	−1.5	−1.50
1.5	−1.0	−1.50

r = average of products ≈ -0.93

   (b) $r = 0.82$, by calculation.
   (c) No calculation is necessary: $r = -1$. The points all lie on a line sloping down, $y = 8 - x$.
2. About 50%.
3. About 25%.
4. About 5%.

Chapter 9. More about Correlation

Set A, page 143

1. (a) About the same.
   (b) The maximum has to be bigger than the minimum.
2. No: the correlation between x and y is the same as the correlation between y and x.
3. r stays the same.
4. r stays the same.
5. r changes.
6. (a) Up. (b) Down. (c) Reverses the sign.
7. (a) 1 (b) Goes down.
   (c) r will be less than 1—measurement error.
8. The correlation would go down (to about 0.25, in fact).

9. The correlation for the whole year is bigger; for example, it will be very cold in the winter, very hot in the summer—in both cities.

   Comment. This is another example of "attenuation" (exercise 5 on p. 130). In the scatter diagram below, the crosses show the data for June 2005 ($r = 0.42$); the dots show the data for days in other months; the correlation for all 365 days is 0.92. Focusing on June restricts the range of the temperatures, and attenuates (weakens) the correlation.

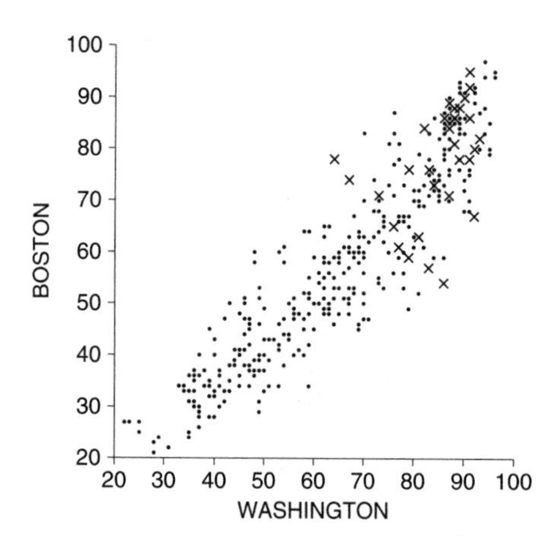

10. Data set (iii) is the same as (ii), with x and y switched; so r is 0.7857. Data set (iv) comes from (i), by adding 1 to each x-value, so r is 0.8571. Data set (v) comes from (i) by doubling each y-value, so r is 0.8571 too. Data set (vi) comes from (ii) by subtracting 1 from each x-value, and multiplying each y-value by 3, so r is 0.7857.

Set B, page 145

1. Each diagram separately has correlation near 0.6. But all together, things look much more like a line, and the correlation is closer to 0.9—this is attenuation in reverse.
2. Somewhat more than 0.67. This is like the previous exercise: when you put all the children together, the data are much more linear. Also see exercise 9 on p. 144.
3. Yes; the only difference is a change of scale.
4. Yes; it's like any of the diagrams in the previous exercise, so $r \approx 0.7$.

Set C, page 148

1. (i) should be summarized using r, (ii) and (iii) should not.
2. False: like diagram (iii) in exercise 1.
3. Nearly 1. There is a strong association, but the relationship is quadratic not linear, so the correlation cannot be 1.
4. Both are false. You need to look at the scatter diagram to check for outliers or non-linearity.

Set D, page 149

1. (a) Diagram is not given. (b) True.
   (c) This cannot be determined from the data (but is true by other studies).
2. No. This correlation might well exaggerate the strength of the relationship—it's based on rates.

Set E, page 152

1. Duration is only measured to the nearest 2 million years; this variable is not easy to determine very accurately.
2. Yes, and this would exaggerate the strength of the association.
3. (a) True. (b) True. (c) True. (d) False.
   Moral: association is not the same as causation.
4. Probably, but this doesn't follow from the data. It could be, for example, that people who have trouble reading watch more television—so causality runs in the other direction. After all, the correlation between x and y equals the correlation between y and x.
5. The best explanation is the association between coffee drinking and cigarette smoking. Coffee drinkers are likelier to smoke, smoking causes heart trouble.
6. This is an observational study, not a controlled experiment, and plotting points from the fifties or seventies on the graph just makes a mud pie.

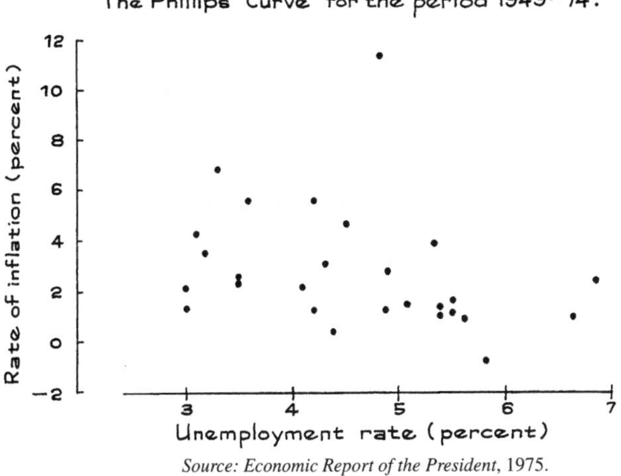

Source: Economic Report of the President, 1975.

Chapter 10. Regression

Set A, page 161

1. (a) 67.5 (b) 45 (c) 60
   Work for (a). A score of 75 is 1 SD above average. However, r is only 0.5. If you take the students who are 1 SD above average on the midterm, their average score on the

final will only be about 0.5 SDs above average on the final, that is, $0.5 \times 15 = 7.5$ points. So, the estimated average score on the final for this group is $60 + 7.5 = 67.5$.

Comment. The regression estimates always lie on a line—the regression line. More about this in chapter 12.

2. (a) 190 pounds (b) 173 pounds
   (c) −68 pounds (d) −206 pounds.

   Comment on (c). This is getting ridiculous, but the Public Health Service didn't run into any little men 2 feet tall, so the regression line doesn't pay much attention to this possibility. The regression line should be trusted less and less the further away it gets from the center of the scatter diagram.

3. False. Think of the scatter diagram for the heights and weights of all the men. Take a vertical strip over 69 inches, representing all the men whose height was just about average. Their average weight should be just about the overall average. But the men aged 45–74 are represented by a different collection of points, some of which are in the strip, and many of which aren't. The regression line says how average weight depends on height, not age. (The older men actually weigh a little more than average—middle-age spread has set in.)

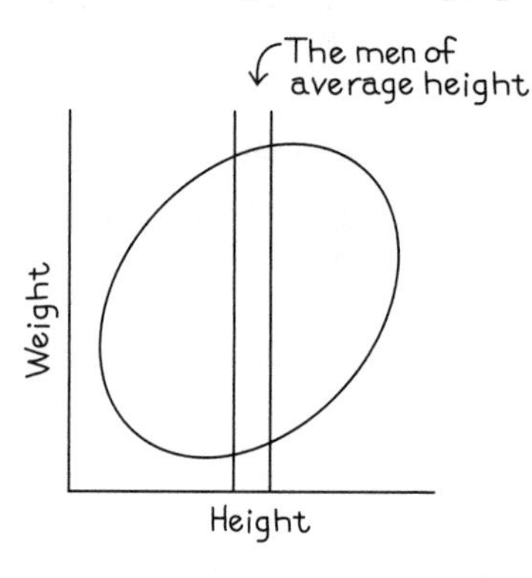

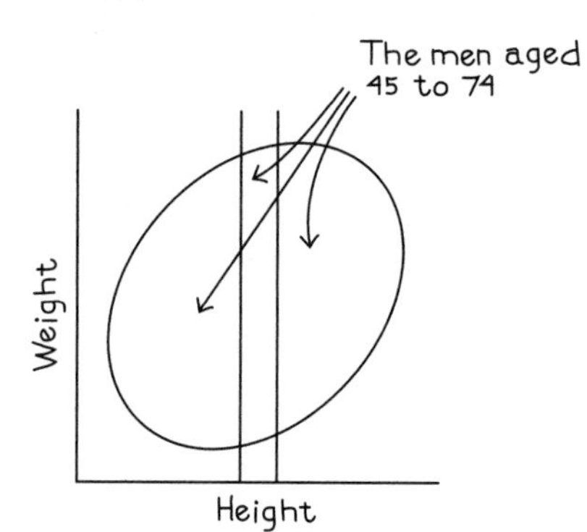

4. These women have completed 12 years of schooling, which is 2 years below average. They are $2/2.4 \approx 0.83$ SDs below average in schooling. The estimate is that they are below average in income, but not by 0.83 SDs—only by $r \times 0.83 \approx 0.28$ SDs of income. In dollars, that's $0.28 \times \$26{,}000 \approx \$7{,}300$. Their average income is estimated as

$$\text{overall average} - \$7{,}300 = \$32{,}000 - \$7{,}300 = \$24{,}700.$$

5. The points must all lie on the SD line, which slopes down; the rate is one SD of y per SD of x.

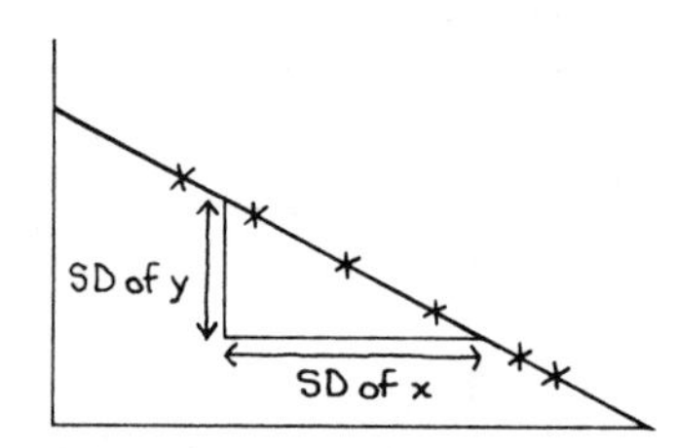

Set B, page 163

1. (a) True: the graph of averages slopes upward. Generally, men with higher incomes have wives with higher incomes. People often choose mates with similar educational levels and family backgrounds, which tends to bring incomes into line as well.
   (b) Chance error. The data are from a sample, and there are only 4 couples behind the dot.
   (c) The regression estimates would be a little too low: the line runs below the dots.

2. 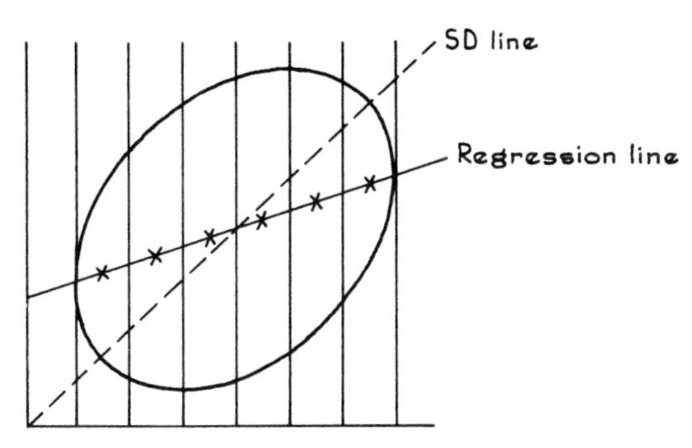

   The crosses fall on the solid regression line, the dashed line is the SD line.

3. For the two diagrams on the left, the SD line is dashed and the regression line is solid. For the two on the right, the SD line is solid and the regression line is dashed. Moral: the regression line isn't as steep as the SD line.

4. 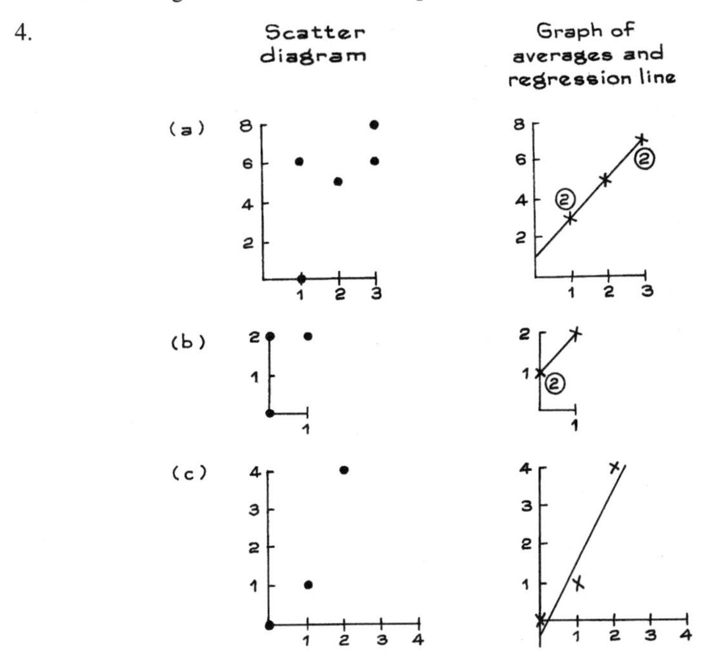

4. 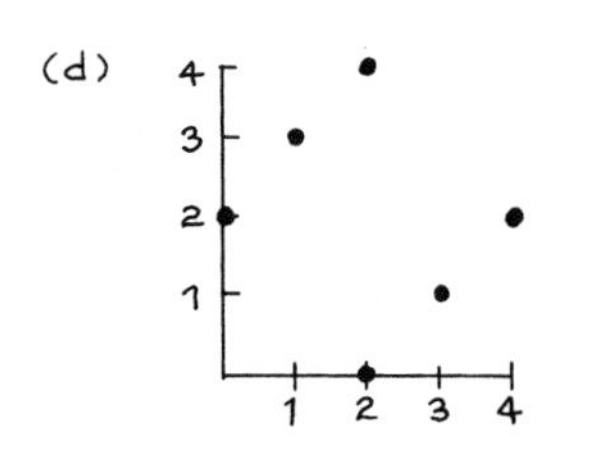

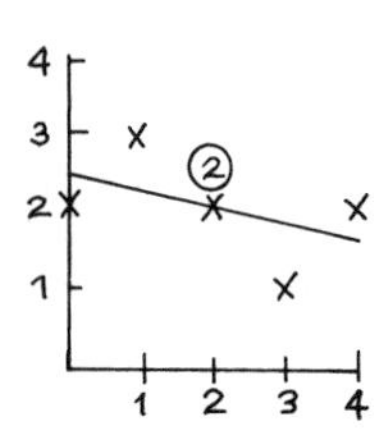

Set C, page 167

1. (a) 67.5 (b) 45 (c) 60 (d) 60
   This exercise is about individuals; exercise 1 on p. 161 was about groups. The arithmetic for parts (a–c) is the same; pp. 165–66.
2. (a) 79% (b) 38% (c) 50% (d) 50%
   Work for (a):

   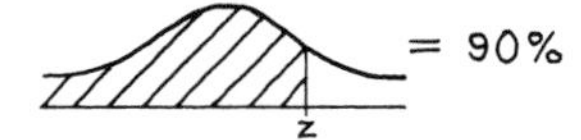

   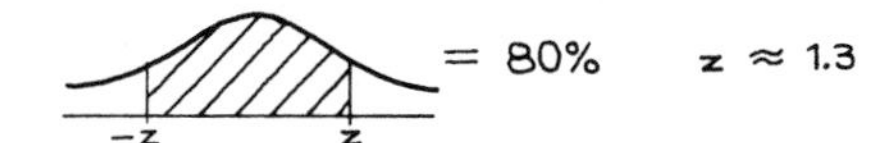

   In standard units, his SAT score was 1.3. The regression prediction for his first-year score is $0.6 \times 1.3 \approx 0.8$ in standard units.

   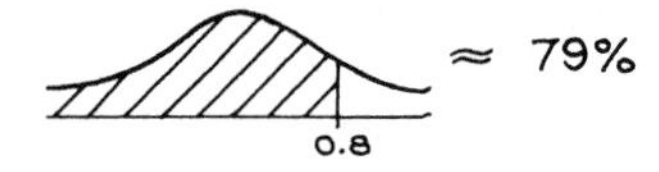

   This corresponds to a percentile rank of 79%. In example 2, the predicted percentile rank was only 69%, which is closer to 50%. That is because the correlation was lower in example 2. There is more regression to the mean in example 2.
3. (a) The SD line—dashed. (b) The regression line—solid.
4. (a) There is a minimum age for marriage.
   (b) Age is reported as a whole year; there are a lot of husbands age 30, but none aged 30.33; likewise for the wives.
5. False. The regression line says how average weight depends on height, not on age. See exercise 3 on p. 161.

Set D, page 174

1. No, this looks like the regression effect. Imagine a controlled experiment. At one airport, the instructors discuss the ratings with the pilots. At another, the instructors keep the ratings to themselves. Even at the second airport, the ratings on the two landings will not be identical—differences come in. So the regression effect appears: on the average, the bottom group improves a bit, and the top group falls back. That is probably all the air force saw in their data.
2. No. It looks like the tutoring had an effect—regression would only take them closer to the average, but they got to the other side.
3. The sons of the 61-inch fathers are taller, on the average, than the sons of the 62-

inch fathers. This is just chance variation. By the luck of the draw, Pearson got too many families where the father was 61 inches tall and the son was extra tall.

Comment. There were only 8 families where the father was about 61 inches tall, and 15 where the father was 62 inches—lots of room for chance error.

Set E, page 175

1. False. There are two completely different groups of men here. (See the diagram below.) The ones who are 63 inches tall are in the vertical strip. They average 138 pounds in weight, as shown by the cross. The ones who weighed 138 pounds are in the horizontal strip. Their average height is shown by a heavy dot, and it's a lot more than 63 inches.

   Remember, there are two regression lines—

   - one for weight on height,
   - one for height on weight.

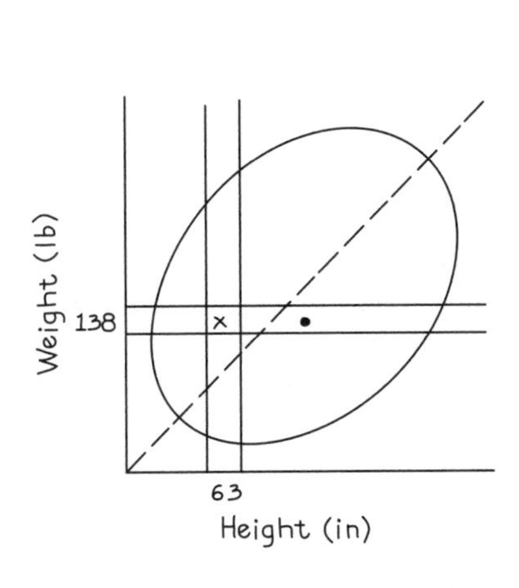

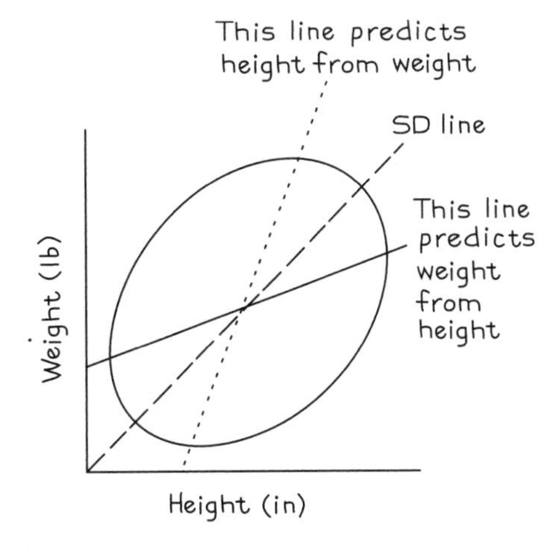

2. False. The fathers only average 69 inches; you have to use the other line.
3. False. This is just like exercises 1 and 2. (A typical student at the 69th percentile of the first-year tests should be at the 58th percentile on the SAT; use the other line.)

Chapter 11. The R.M.S. Error for Regression

Set A, page 184

1. B is tall and chubby, while D is short and skinny.
2. (a) False. (b) True.
3. Prediction errors = −7, 1, 3, −1, 4; r.m.s. error = 3.9.
4. (a) 0.2 (b) 1 (c) 5.
5. A few thousand dollars.
6. The one with the smaller r.m.s. error should be used, as it will be more accurate overall.

7. (a) 8 points—one r.m.s. error. (b) 16 points—two r.m.s. errors.

8. (a) $20,000. (b) The horizontal line. See p. 183.

Set B, page 187

1. $\sqrt{1-0.6^2} \times 10 = 8$ points.

2. (a) Guess the average, 65.
   (b) 10. If you use the regression line, the r.m.s. error is given by the formula (exercise 1). If you use the average, the r.m.s. error is the SD. (See exercises 9–10 on p. 71.)
   (c) Use the regression line, and the r.m.s. error is given by the formula as 8 points (exercise 1).

3. Generally, it helps to have more information. The r.m.s. error will be smaller for person B, by the factor $\sqrt{1-0.6^2} = 0.8$. See p. 186.

Set C, page 189

1. (a) (iii) (b) (ii) (c) (i)

2. (a) (i) (ii) (b) not used (c) (iii)

3. (a) SD of $y \approx 1$
   (b) SD of residuals ≈ 0.6
   (c) SD of y in strip ≈ 0.6, about the same as the SD of the residuals.

   Comment. The vertical scatter in the strip is about the same as the r.m.s. error of the regression line—but the vertical scatter in the whole diagram is a lot more than the vertical scatter in the strip.

Set D, page 193

1. (a) True.
   (b) True; the scatter diagram is homoscedastic, so the subjects are off the regression line by similar amounts in each vertical strip.
   (c) False, because the scatter diagram is heteroscedastic; 9 points is a sort of average amount off, but the prediction errors are going to be bigger with high scores.

2. (a) $\sqrt{1-0.5^2} \times 2.7 \approx 2.3$ inches.
   (b) 71 inches—regression method.
   (c) 2.3 inches. The scatter diagram is homoscedastic, so the sons' heights are off the regression line by similar amounts, for any father's height. The amount off is the r.m.s. error of the line.
   (d) The prediction is 68 inches, and it is likely to be off by 2.3 inches or so.

3. (a) $\sqrt{1-0.37^2} \times \$20{,}000 \approx \$18{,}600$.
   (b) $24,500—regression method.
   (c) This cannot be determined from the information given. The $18,600 is sort of the average amount off the line. But the scatter diagram is heteroscedastic, so the amount off the line changes from strip to strip. The spread in incomes is larger for more highly educated people, so the amount off will be larger than $18,600.
   (d) The prediction is $7,100. The amount off cannot be determined, but will be less than $18,600.

4. The husband is between 20 and 30 years of age.

5. (a) 50, 15 (b) 50, 15 (c) 0.95 (d) 25, 5
   (e) 0.5—attenuation. See exercise 9 on p. 144 and exercises 1–2 on pp. 145–146.

6. (a) The SD for all the wives is much bigger. That is the main point of exercises 4–6. See the comments below.
   (b) The two SDs are about the same.

   Comments. If you just take the families where the husband is 20 to 30 years of age, the wives are going to be much more similar in age, their SD drops from about 15 years to about 5 years. If you take the husbands born in March, that does not cut down the variability in the ages of their wives. Smaller samples do not generally have smaller SDs (exercise 8 on p. 71). But if you restrict the range of x, that will generally reduce the SD of y.

7. (a) 68 inches, the average.
   (b) 3 inches, the SD.
   (c) Regression. If one twin is 6 ft 6 in, guess 6 ft $5\frac{1}{2}$ in for the other one.
   (d) $\sqrt{1 - 0.95^2} \times 3 \approx 0.9$ inches.

   Comments. (i) If $r = 1$, you should guess that the height of the second twin equals the height of the first one. But r is a little less than 1. So you regress the second twin back toward the mean—a little bit.

   (ii) The answer to (d) is quite a bit smaller than the answer to (b). When $r = 0.95$, there is quite a large reduction in r.m.s. error when you use the regression line.

Set E, page 197

1. (a)

63 in 68 in
Ave

0 2

2

Percent ≈ 2%

   (b) new average ≈ 63.9 inches, new SD ≈ 2.4 inches

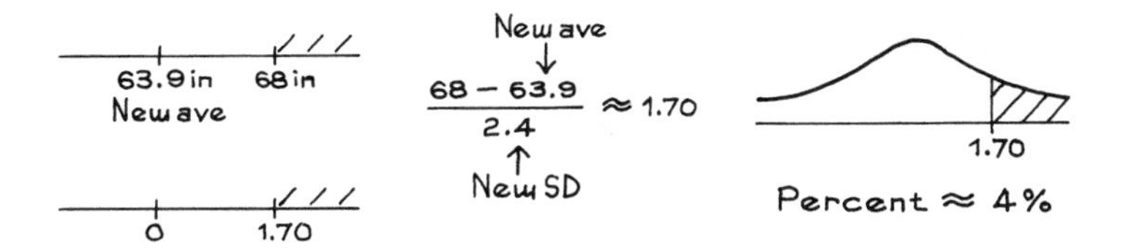

2. (a) 14% (b) 33%

3. (a) 38% (b) 60%

Chapter 12. The Regression Line

Set A, page 207

1. (a) $2,000 × 8 + $5,000 = $21,000
   (b) $2,000 × 12 + $5,000 = $29,000
   (c) $2,000 × 16 + $5,000 = $37,000

2. (a) 240 ounces = 15 pounds (b) 20 ounces.
   (c) 3 ounces of nitrogen yields 18 lb 12 oz of rice, 4 ounces of nitrogen yields 20 pounds of rice.
   (d) Controlled.
   (e) Yes. The line fits quite well ($r = 0.95$), and 3 ounces is close to a value that was used.
   (f) No. That's too far away from the amounts used.

3. (a) Predicted son's height = $0.5 \times$ father's height + 35 inches.
   (b) Predicted father's height = $0.5 \times$ son's height + 33.5 inches.

   Comment. There are two regression lines, one predicts son's height from father's height, the other predicts father's height from son's height (section 5 of chapter 10).

4. This testimony is overstatement. Associations in the data may be due to confounding. Without doing the experiment, or working very hard at the observational data, you can't be sure what the impact of interventions will be.

Set B, page 210

1. With 12 years of education, height is predicted as 69.75 inches; with 16 years, height is predicted as 70.75 inches. Going to college clearly has no effect on height. This observational study picked up a correlation between height and education due to some third factor in family background.

2. 439.16 cm, 439.26 cm. Hanging a bigger weight on the wire makes it stretch more. You can trust the regression line in exercise 2 because it is based on an experiment. In exercise 1, the line was fitted to data from an observational study.

3. (a) $540 + 110 = 650$ (b) 540 (c) Greater than (p. 208).

4. (a) 540 (b) 540 (c) Greater than (p. 208).

   Comment. if you use the average value of y to predict y, the r.m.s. error is the SD of y; see p. 183.

5. The regression line makes the smallest r.m.s. error (p. 208).

Part IV. Probability

Chapter 13. What Are the Chances?

Set A, page 225

1. (a) (vi) (b) (iii) (c) (iv) (d) (i)
   (e) (ii) (f) (v) (g) (vi)

2. About 500.

3. About 1,000.

4. About 14.

5. Box (ii), because [3] pays more than [2], and the other ticket is the same.

Set B, page 227

1. (a) The question is about the second ticket, not the first: see part (a) of example 2. The answer is 1/4.

(b) 1/3; there are 3 tickets left after [2] is drawn.

2. (a) 1/4 (b) 1/4

   With replacement, the box stays the same.

3. (a) 1/2 (b) 1/2

   The chances for the 5th toss of the penny do not depend on the results of the first 4 tosses.

4. (a) 1/52 (b) 1/48

   This is like example 2 on p. 226.

Set C, page 229

1. (a) 12/51 (b) $13/52 \times 12/51 = 1/17 \approx 6\%$.
2. (a) 1/6 (b) $1/6 \times 1/6 \times 1/6 = 1/216 \approx 1/2$ of 1%.
3. (a) 4/52 (b) $4/52 \times 4/51 \times 4/50 \approx 5/10{,}000$.

   Comment. In this exercise, the cards are dependent; in exercise 2, the rolls were independent.

4. "At least one ace" is the better option: you would choose an exam in which you had to get at least one question right out of six, over an exam in which you had to get all six right.
5. This is fine, it's the multiplication rule.
6. The coin has to land "tails, heads"; the chance is 1/4.
7. (a) 1/8
   (b) $1 - 1/8 = 7/8$
   (c) 7/8; you get at least one tail when you don't get three heads: so (b) and (c) are the same.
   (d) 7/8; just switch heads and tails in (c).

Set D, page 232

1. (a) independent: if you get a white ticket, there is 1 chance in 3 to get "1" and 2 chances in 3 to get "2"; if you get the black ticket, the chances for the numbers stay the same.
   (b) independent
   (c) dependent: with the white tickets, there is only 1 chance in 3 to get "2"; with the black tickets, there are 2 chances in 3.
2. (a,b) independent (c) dependent

   Comment. This kind of box will come up again in chapter 27. Here is the argument for (a). Suppose you draw a ticket, and see the first number is 4 but don't see the second number: the chance that the second number will be 3 is 1/2. Likewise if the first number is 1. That is independence.

3. Ten years is 520 weeks, so the chance is $(999{,}999/1{,}000{,}000)^{520} \approx 0.9995$.

   Comment. In the New York State Lotto, your chance of winning something is about 1/12,000,000.

4. This is false. It's like saying someone doesn't have a temperature because you can't find the thermometer. To figure out whether two things are independent or not, you pretend to know how the first one turned out, and then see if the chances for the second change. The emphasis is on the word "pretend."

5. (a) 5% (b) 20%

   To figure (a) out, suppose you have 80 men and 20 women in the class. You also have 15 cards marked "freshman" and 85 cards marked "sophomore." You want to give out a card to each student, so that as few women as possible get "sophomore." The strategy is to give a sophomore card to each man; you are left with 5, which have to go to 5 women. The 15 freshman cards go to the other 15 women.

   Comment. If year and sex are independent, the percentage of sophomore women would be 85% of 20% = 17%, between the two extremes.

6. Same as previous exercise: the chance of getting a sophomore woman equals the percentage of sophomore women in the class.

7. False. The calculation assumes that the percentage of women is the same across all age groups, and it isn't: women live longer than men. (Actually, women age 85 and over accounted for nearly 1.1% of the U.S. population in 2002.)

8. If the subject draws the ace of spades from the small pile, he has 13 chances in 52 to draw a spade from the big deck, and win the prize. Likewise if he draws the deuce of clubs. Or any other card. So the answer is $13/52 = 1/4$.

Chapter 14. More about Chance

Set A, page 240

1. ⚀⚃ ⚁⚂ ⚂⚁ ⚃⚀. The chance is 4/36.

2. There are 25 possible results; for 5 of them, the sum is 6. So the chance is 5/25. (The figure is not shown.)

3. Most often, 7; least often, 2, 12. (Use figure 1 to get the chance of each total, as in exercise 1.)

4. (a) 2/4 (b) 2/6 (c) 3/6

Set B, page 242

1. False. The question is about the number of children who had either cookies or ice cream, including the gluttons who had both. The number depends on the choices made by the children, and two possibilities are shown in the table.

Cookies only	*Ice cream only*	*Both*	*Neither*
12	17	0	21
3	8	9	30

   In the first case, 12 children had cookies only, 17 children had ice cream only, 0 had both, and 21 had neither. So $12 + 17 = 29$ had cookies or ice cream. The second line shows another possibility, where 9 children had both cookies and ice cream. In this situation, the number with cookies or ice cream is $3 + 8 + 9 = 20$. Just as a check: the number with cookies is $3 + 9 = 12$, and the number with ice cream is $8 + 9 = 17$, as given in the problem. But the number with cookies or ice cream is not $12 + 17$, because the addition double counts the 9 gluttons. The number who had cookies or ice cream depends on the number of gluttons who had both.

2. (a) 4/20 (b) 8/20 (c) 12/20 (d) 14/20

   Comment. $(4 + 8 + 12)/20$ gives the wrong answer to (d)—by double-counting some dots and triple-counting others.

3. They are the same.

4. False. Simply adding the two chances double counts the chance of ⚀⚁. See example 5 on p. 242.

5. False. There is 1 chance in 10 of getting [7] on any particular draw, but these events are not mutually exclusive.

6. True. $100\% - (10\% + 20\%) = 70\%$. Use the addition rule, and p. 223 for the subtraction.

Set C, page 246

1. (a) 1/52 of the contestants step forward.
   (b) 1/52 of the contestants step forward; example 2 in chapter 13.
   (c) The ones who got both the ace of hearts on the first card and the king of hearts on the second card step forward twice. (In terms of getting the weekend, that's overkill.) The fraction who step forward twice is $1/52 \times 1/51$.
   (d) False; as (c) shows, the events aren't mutually exclusive, so addition double counts the chance that both occur.

   Comment. The chance in (d) is

$$1/52 + 1/52 - 1/52 \times 1/51.$$

2. (a) 1/52 of the contestants step forward.
   (b) 1/52 of the contestants step forward.
   (c) If you get the ace of hearts on the first card, you can't get it on the second card; nobody steps forward twice.
   (d) True; as (c) shows, the events are mutually exclusive, so addition is legitimate.

   Comment. In exercise 2, the two ways to win are mutually exclusive; not so in exercise 1. Addition is legitimate in exercise 2, not in 1.

3. (a,b) True; see example 2 in chapter 13.
   (c) False. "Top card is the jack of clubs" and "bottom card is the jack of diamonds" aren't mutually exclusive, so you can't add the chances.
   (d) True. "Top card is the jack of clubs" and "bottom card is the jack of clubs" are mutually exclusive.
   (e,f) False; these events aren't independent, you need the conditional chances.

4. (a) False; $1/2 \times 1/3 = 1/6$, but A and B may be dependent: you need the conditional chance of B given A.
   (b) True; see section 4 of chapter 13.
   (c) False. ("Mutually exclusive" implies dependence, and the chance is actually 0.)
   (d) False; $1/2 + 1/3 = 5/6$, but you can't add the chances because A and B may not be mutually exclusive.
   (e) False; if they're independent, they have some chance of happening together, so they can't be mutually exclusive: don't add the chances.
   (f) True.

   Comment. If you have trouble with exercises 3 and 4, look at example 6, p. 244.

5. See example 2 in chapter 13.
   (a) $4/52$ (b) $4/51$ (c) $4/52 \times 4/51$

Set D, page 250

1. (a) (i) (b) (i) (ii)
   (c) (iii) (d) (ii) (iii)
   (e) (i) (ii) (f) (i)
2. Bets (a) and (f) say the same thing in different language. So do (b) and (e). Bet (d) is better than (c).
3. (a) 3/4 (b) 3/4 (c) 9/16 (d) 9/16 (e) $1 - 9/16 = 7/16$
4. (a) Chance of no aces $= (5/6)^3 \approx 58\%$, so chance of at least one ace $\approx 42\%$. Like de Méré, with 3 rolls instead of 4.
   (b) 67% (c) 89%
5. $1 - (35/36)^{36} \approx 64\%$
6. The chance that the point 17 will not come up in 22 throws is $(31/32)^{22} \approx 49.7\%$. The chance that it will come up in 22 throws is therefore $100\% - 49.7\% = 50.3\%$. So this wager (laid at even money) was also favorable to the Master of the Ball. Poor Adventurers.
7. The chance of surviving 50 missions is $(0.98)^{50} \approx 36\%$. Deighton is adding chances for events that are not mutually exclusive.

Chapter 15. The Binomial Coefficients

Set A, page 258

1. The number is 4.
2. The number is 6.
3. (a) $(5/6)^4 = 625/1{,}296 \approx 48\%$
   (b) $4(1/6)(5/6)^3 = 500/1{,}296 \approx 39\%$
   (c) $6(1/6)^2(5/6)^2 = 150/1{,}296 \approx 12\%$
   (d) $4(1/6)^3(5/6) = 20/1{,}296 \approx 1.5\%$
   (e) $(1/6)^4 = 1/1{,}296 \approx 0.08$ of 1%
   (f) Addition rule: $(150 + 20 + 1)/1{,}296 \approx 13\%$.
4. This is the same as exercise 3(a–c). Rolling an ace is like drawing a red marble, while 2 through 6 correspond to green. To see why, imagine two people, A and B, performing different chance experiments:

- A rolls a die four times and counts the number of aces.
- B draws four times at random with replacement from the box
  | R G G G G G | and counts the number of R's.

The equipment is different, but as far as the chance of getting any particular number of reds is concerned, the two experiments are equivalent.

- There are four rolls, just as there are four draws.
- The rolls are independent; so are the draws.
- Each roll has 1 chance in 6 to contribute one to the count (ace); similarly for each draw (red).

5. The chance of getting exactly 5 heads is $\frac{10!}{5!\,5!}\left(\frac{1}{2}\right)^{10} = \frac{252}{1{,}024} \approx 25\%$. The chance of getting exactly 4 heads is $\frac{10!}{4!\,6!}\left(\frac{1}{2}\right)^{10} = \frac{210}{1{,}024} \approx 21\%$. The chance of getting exactly 6 heads is the same. By the addition rule, the chance of getting 4 through 6 heads is $672/1{,}024 \approx 66\%$.

6. You need the chance of getting 7, 8, 9, or 10 heads when a coin is tossed 10 times. Use the binomial formula, and the addition rule:

$$\frac{10!}{7!\,3!}\left(\frac{1}{2}\right)^{10} + \frac{10!}{8!\,2!}\left(\frac{1}{2}\right)^{10} + \frac{10!}{9!\,1!}\left(\frac{1}{2}\right)^{10} + \frac{10!}{10!\,0!}\left(\frac{1}{2}\right)^{10} = \frac{176}{1{,}024} \approx 17\%.$$

Comment. Looks like chance, not vitamins.

Part V. Chance Variability

Chapter 16. The Law of Averages

Set A, page 277

1. The error is 50 in absolute terms, 5% in percentage terms.
2. The error is 1,000 in absolute terms, 1/10 of 1% in percentage terms. Compare this with the previous exercise: the chance error has gone up in absolute terms (from 50 to 1,000) but down in percentage terms (from 5% to 1/10 of 1%).
3. False. The chance stays at 50%. See p. 274.
4. (a) Ten tosses. As the number of tosses goes up, you are more and more likely to be close to 50% heads, less and less likely to be above 60% heads. Here, chance variability in the percentages helps you, a small number of tosses is better than a large number.
   (b) One hundred tosses: now chance variability in the percentages hurts you—because you want to be close to 50%. With more tosses, there is less chance variability in the percentages. More tosses are better.
   (c) One hundred tosses; like (b).
   (d) Ten tosses. As the number of tosses goes up, there is less and less chance for the number of heads to exactly equal the expected number. Let's take a more extreme case: suppose you toss the coin 1,000,000 times. The chance of getting exactly 500,000 heads—rather than 500,001 or 500,043 or 499,997 or some other number close to 500,000—is quite slim.

5. Option (i) is better. This is just like exercise 4(a).
6. Option (ii), the reason is chance error.
7. It's about the same with or without replacement.
8. Same. Both have 50% [−1]'s and 50% [+1]'s.
9. Eventually, the chance error would be large and negative. Then, it would get positive again. In absolute terms, the swings get wilder and wilder.

Set B, page 280

1. $47 \times 1 + 53 \times 2 = 153$.
2. (a) 100, 200 (b) 50, 50 (c) $50 \times 1 + 50 \times 2 = 150$.
3. (a) 100, 900.
   (b) $33 \times 1 + 33 \times 2 + 33 \times 9 \approx 400$.
   Comment. 400 isn't halfway between 100 and 900.
4. Guess 500 in all three cases; (iii) is best, (i) worst.
5. The chance for "1" is 1 in 10; the chance for "3 or less" is 3 in 10; the chance for "4 or more" is 7 in 10—there are 7 numbers from 4 through 10 inclusive. Drawing at random from boxes is discussed in chapters 13–14.
6. Box (i) is better, it has fewer −1's, and the same 2.
7. Options (i) and (ii) do it. Your net gain is the sum of your wins and losses, taking signs into account.

Set C, page 284

1. (i) and (ii) are the same. (iii) means that all ten draws must be "1," which is worse than (i).
2. Option (i) is no good; the sum of the draws is unrelated to the net gain. Option (ii) is no good; it says you win $17 with 2 chances in 36 on a single play, but your chances are 2 in 38. Option (iii) is right. If in doubt, review example 1 on p. 283.
3. Your net gain is like the sum of 10 draws made at random with replacement from the box

   | 1 ticket [$36] 215 tickets [−$1] |

   This is a terrible game.

Chapter 17. The Expected Value and Standard Error

Set A, page 290

1. (a) $100 \times 2 = 200$ (b) −25 (c) 0 (d) $66\frac{2}{3}$
   Comment on (d). The "expected value" need not be one of the possible values. It's like saying that the average family has 2.1 children. This is sensible, even though the "average family" is a statistical fiction.
2. This is the same as the expected value for the sum of two draws from the box | [1] [2] [3] [4] [5] [6] |. So the answer is $2 \times 3.5 = 7$ squares.

3. The model is given on pp. 283–284. The average of the the numbers in the box is
$$(\$35 - \$37)/38 = -\$2/38 \approx -\$0.05$$
(To compute the average, you have to add up the tickets in the box; [+$35] adds $35 to the total, but the 37 [−$1]'s take $37 away; then you have to divide by the number of tickets in the box, which is 38.) The expected net gain is equal to $100 \times (-\$.05) = -\$5$. You can expect to lose around $5.

4. The box is on p. 283. The average of the box is
$$(\$18 - \$20)/38 = -\$2/38 \approx -\$0.05$$
(The average is the total of the numbers in the box, divided by 38; the 18 tickets marked "+$1" contribute $18 to the total, while the 20 tickets marked "−$1" take $20 away.) The expected net gain is $100 \times (-\$0.05) = -\$5$.

   Comment. Exercises 3 and 4 show that with either bet (number or red-or-black), you can expect to lose 1/19 of your stake on each play.

5. −$50. Moral: the more you play, the more you lose.

6. The average of the box is $(18x - \$20)/38$. To be fair, this has to equal 0. The equation is $18x - \$20 = 0$. So $x \approx \$1.11$. They should pay you $1.11.

7. The Master of the Ball should have paid 31 pounds, just as the Adventurers thought. Moral: the Adventurers may have the fun, but it is the Master of the Ball who has the profit.

Set B, page 293

1. (a) The average of the box is 4; the SD is 2. So the expected value for the sum is $100 \times 4 = 400$; the SE for the sum is $\sqrt{100} \times 2 = 20$.
   (b) Around 400, give or take 20 or so.
   (c) Guess 400, off by 20 or so. Parts (b) and (c) interpret the numbers in (a).

2. The net gain is like the sum of 100 draws from the box | [−$1] [$1] |. The average of the box is $0; the SD is $1. The sum of 100 draws has expected value $0; the SE for the sum is $\sqrt{100} \times \$1 = \$10$. So your net gain will be around $0, give or take $10 or so.

3. With option (ii), the numbers are too close to 50; no number is more than 5 away. With option (iii), the numbers alternate much too regularly. Option (i) is it.

4. The expected value is 150, the observed value is 157, the chance error is 7, the standard error is 10.

5. Multiplying the number of draws by 4 multiplies the expected value by 4 and the SE by $\sqrt{4} = 2$. The expected value for the sum of 100 draws is $4 \times 50 = 200$, and the SE is $2 \times 10 = 20$.

6. (a) is true, (b) is false: the expected value for the sum of the draws can be computed exactly, as
$$\text{number of draws} \times \text{average of box}$$
(c) is false, (d) is true: the sum will be off its expected value, and the SE tells you by about how much.

7. Yes. The chance is small, but positive. If you wait long enough, events of small probability do happen.

Set C, page 296

1. (a) Smallest, 100; largest, 400.
   (b) The average of the box is 2; the SD is 1. The sum has an expected value of $100 \times 2 = 200$; the SE for the sum is $\sqrt{100} \times 1 = 10$. The sum will be around 200, give or take 10 or so.
   (c)

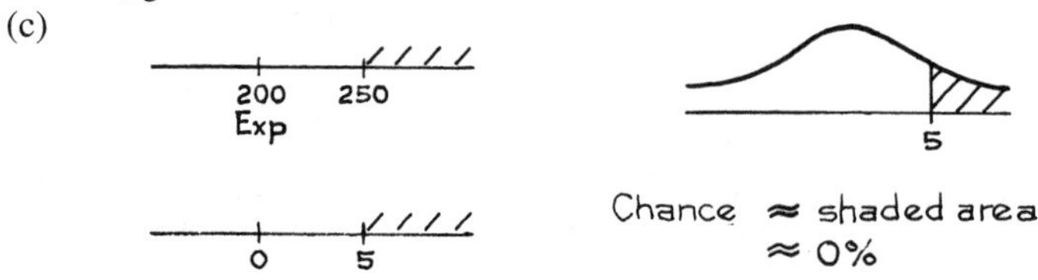

2. (a) Largest, 900; smallest, 100. (b) Chance $\approx$ 68%
3. (a) The expected value is 0, so the sum is around 0, and your best hope is chance variability in the sum—you want the sum to be far from its expected value. Chance variability goes up with the number of draws, choose 100.
   (b) Same as (a).
   (c) Now chance variability in the sum works against you, because you want the sum to be close to its expected value; choose 10.
4. (i) Expected value for sum = 500, SE for sum = 30.
   (ii) Expected value for sum = 500, SE for sum = 20.

   Both sums will be around 500, but sum (i) will be further away. In (a) and (b), chance variability helps—choose (i). In (c), chance variability hurts—choose (ii).
5. 98%.
6. Either they win \$25,000 (with chance $20/38 \approx 53\%$) or they lose \$25,000 (with chance $18/38 \approx 47\%$). The answer is 50%.

   Comment. The casino is much happier with a lot of small bets, where the profit is almost guaranteed, than with one big bet, where there is a lot of risk.
7. One number will pay off \$35,000, but the other 37 will lose, so the gambler loses \$2,000 for sure.

   Comment. The casino likes the gamblers to spread their bets.
8. Option (ii) is right; the SE doesn't go up by a full factor of 2, but only $\sqrt{2} \approx 1.4$.

Set D, page 299

1. (a) No, replace the 5 by $7 - (-2) = 9$. (b) Yes. (c) Yes.
   (d) No—the list shows 3 different numbers, so the short-cut doesn't apply.
2. The net gain is like the sum of 100 draws from the box

   | \$2 | −\$1 | −\$1 | −\$1 |

   The average of the box is $(\$2 - \$1 - \$1 - \$1)/4 = -\$0.25$. The SD is

   $$[\$2 - (-\$1)] \times \sqrt{1/4 \times 3/4} \approx \$1.30.$$

   The net gain in 100 plays will be around $100 \times (-\$0.25) = -\$25$, give or take $\sqrt{100} \times \$1.30 = \$13$ or so.
3. (a) From the point of view of the house, a dollar bet on the house special is like one draw from the box

| 5 tickets [−$6] 33 tickets [+$1] |

The average of the box is $[5 \times (-\$6) + 33 \times \$1)]/38 \approx \$.08$. So the house expects to make about 8 cents per dollar bet. As far as the house is concerned, this is a great bet.

(b) The player's net gain is like the sum of 100 draws at random with replacement from the same box with the signs reversed:

| 5 tickets [+$6] 33 tickets [−$1] |

The average of the box is −$.08; the SD is

$$[\$6 - (-\$1)] \times \sqrt{5/38 \times 33/38} \approx \$2.37.$$

The player's expected net gain in 100 plays is −$8, give or take $24 or so.

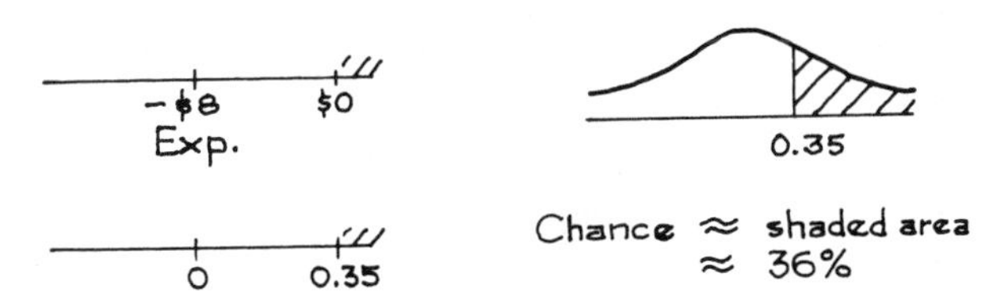

4. The expected net gain in 100 one-dollar bets on a section is −$5; the SE is $14. The expected net gain in 100 bets on red is −$5; the SE is $10. Options (i) and (ii) have the same expected net gain. But (i) has the bigger SE, that is, more variability: (a) is false, (b) and (c) are true.

Set E, page 303

1. (a) You can't add up words, so box (i) is out. With box (iii), you get 2 chances in 3 to go up each time, and it should only be 1 in 2. Box (ii) is the one.
   (b) Average of box = 0.5 and SD of box = 0.5 too. The sum of 16 draws has an expected value of $16 \times 0.5 = 8$; the SE is $\sqrt{16} \times 0.5 = 2$. The number of heads will be around 8, give or take 2 or so.

2. New box: | [0] [0] [0] [0] [1] |. It's ±3 SE, chance is about 99.7%.

3. New box: | [0] [1] |. It's 1 SE or more, chance is about 16%.

4.

Group of 100 tosses	*Observed value*	*Expected value*	*Chance error*	*Standard error*
1–100	44	50	−6	5
101–200	54	50	+4	5
201–300	48	50	−2	5
301–400	53	50	+3	5

5. Expect about 68—example 5 on p. 301; actually, you see 69.

6. (a,b) About 99.7%—it's 3 SEs.

   Comment. When the number of tosses goes up from 10,000 to 1,000,000, the percentage of heads gets closer to 50%: the 99.7%-interval shrinks from

$$50\% \pm 1.5\% \quad \text{to} \quad 50\% \pm 0.15\%.$$

7. Expected is 30, observed is 33, chance error is 3, SE is about 3.5.

8. Put in five 0's and five 1's. Tell it to draw 1,000 times.

9. It's fine. The number of aces isn't supposed to be 16.67 exactly, it's only supposed to be around 16.67.

Chapter 18. The Normal Approximation for Probability Histograms

Set A, page 312

1. Between 70 and 80 inclusive.
2. (a) Between 6.5 and 10.5.
   (b) Between 6.5 and 7.5—the left and right edges of the rectangle over 7.
3. (a) 7
   (b) 7: tallest bar in 2nd panel.
   (c) No, this is just chance variation. In fact 4 is less likely than 5, as the probability histogram in the bottom panel shows.
   (d) (iii). The top panel is an empirical histogram—it shows observed percentages, not chances.
4. (a) 3, 6
   (b) Bottom panel—the probability histogram shows chances. The values 2 and 3 are equally likely for the product.
   (c) Look at the second panel: 3 appeared more often. Chance variation again.
   (d) The value 14 is impossible for the product. Reason: there are only two ways to factor 14, as 1×14 or 2×7; no die can show 7 or 14.
   (e) The bottom panel is a probability histogram, so areas under it represent chances: 11.1% is the chance of getting a product of 6 when you roll a pair of dice.
5. A goes with (i) and B with (ii). B is lower, more spread out, and farther to the right. Box (ii) has a bigger average and a bigger SD.
6. False. The probability histogram for the sum tells you the chances for the sum. It doesn't tell you how the draws turned out. The shaded area represents the chance that the sum will be in the range from 5 to 10 inclusive. (The box had 85 tickets marked 0, 2 tickets marked 1, and 13 tickets marked 2.)

Set B, page 318

1. (i) Exactly 6 heads. (ii) 3 to 7 heads exclusive.
   (iii) 3 to 7 heads inclusive.
2. The area between 51.5 and 52.5 under the histogram gives the exact chance. The normal curve is only an approximation (but a very good one).
3. The expected number of heads is 50; the SE is 5. You want the area of the rectangle over 60 in figure 3, p. 315.

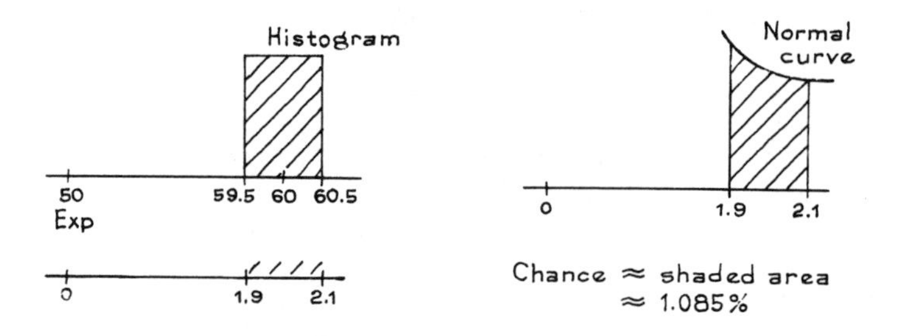

Comment. The exact chance is 1.084%.

4. From exercise 3, about one group in a hundred should have 60 heads. In fact, exactly one group in the hundred does (#6,901–7,000).

5. The expected number of heads is 5,000; the SE is 50.

   (a)

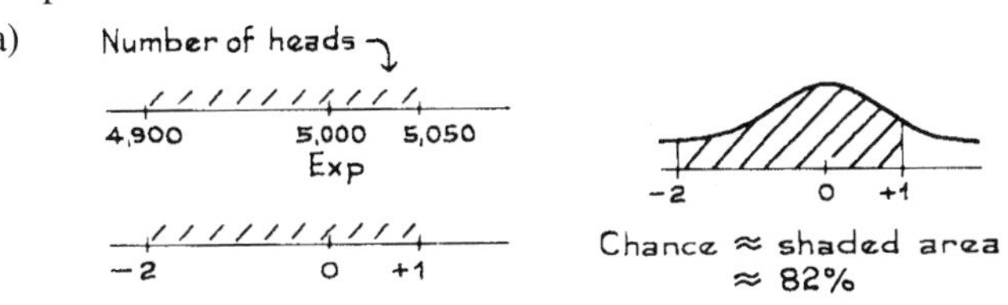

   (b) chance ≈ 2% (c) chance ≈ 16%.

6. (a) Yes. The blocks are big. (b) No. Small blocks.

   Comment on (a). Keeping track of the edges changes the estimate from 50% to 54%.

Set C, page 324

1. (a)

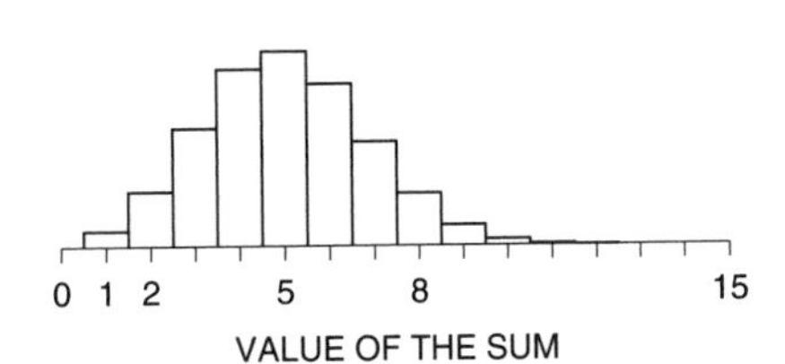

   (b) 3 is more likely than 8: the block over 3 is bigger.

2. The number of heads in 400 tosses of this biased coin is like the sum of 400 draws from the box | 9 [0]'s [1] |. The expected number of heads is 40, and the SE is 6. You want the area of the rectangle over 40, at the bottom of figure 6 on p. 320.

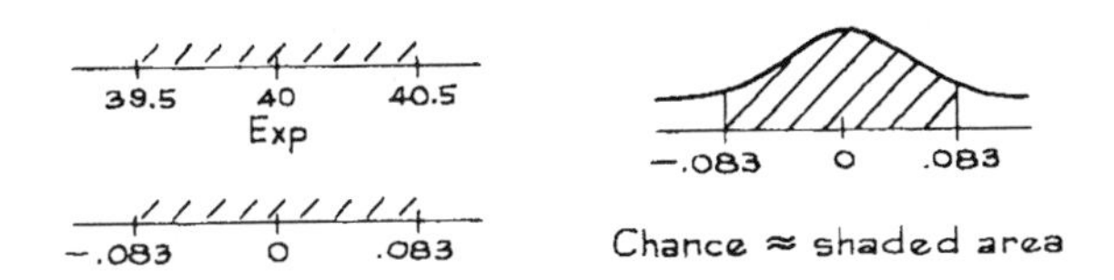

   From the table, this area is between 4% and 8%. (Actually, the area is 6.6%, and so is the chance.)

3. The normal curve is lower than the histogram around 1, so the estimate would be too low.

4. Yes. Big blocks.

5. A (ii), B (i), C (iii). The more lopsided the box, the more skewed the histogram.

   Comment. With 25 draws from the box | 24 [0]'s [1] |, you cannot expect to get many [1]'s. The leftmost rectangle in the probability histogram gives the chance that the sum will be zero—the draws are all [0]. This chance is 36%. The next rectangle gives the chance that the sum will be one—one [1] among the draws, and 24 [0]'s. This chance is 38%. And so forth. (The chances can be worked out using the binomial formula, chapter 15.)

6. (i) 100 (ii) 400 (iii) 900

   The histograms get closer to the normal curve as the number of draws goes up.

7. Choose (i).

   Comment. Chances are given by areas under probability histograms. Often, the corresponding area under the normal curve is a good approximation, but not here—the curve is much higher than the histogram, so the area under the curve is much bigger than the area under the histogram.

8. Most likely, 105; least likely, 101; expected value, 100.

   Comment. There is a trough in this histogram near the expected value. (With 100 draws the trough has disappeared.)

9. (a) Much smaller than 50%. The value 276,000 is 0.276 million, about half-way between the 0.2 and the 0.4 on the horizontal axis. The area to the right of this point is much smaller than 50%. (This histogram has a very long right-hand tail, and the expected value is a lot bigger than the median.)
   (b) $1{,}000{,}000/100 = 10{,}000$
   (c) 400,000 to 410,000 is a lot more likely, relatively speaking. The box just to the right of 400,000 is relatively much higher than the box just to the left. Products have quite irregular probability histograms.

Part VI. Sampling

Chapter 19. Sample Surveys

Set A, page 349

1. The population consists of all undergraduates registered in the current term. The parameter is the percentage of these undergraduates living at home.

2. (a) This is a probability method: it is perfectly definite, chance enters in a planned way—when you choose that random starting point between 1 and 100—and nobody has any discretion as to who gets in the sample.
   (b) The method is different from simple random sampling. For instance, two people whose names are adjacent on the list have no chance to get into the sample together. (Simple random samples are defined in section 4.)
   (c) The sample is unbiased: each person has an equal chance of getting into the sample.

3. Choose (ii). See pp. 334, 339, and 342.

4. The population and the sample are the same, namely, all men age 18 in the Netherlands in 1968; there is no room for sampling error.

5. Doing a survey by telephone could introduce bias, because telephone subscribers are probably different from non-subscribers. However, the percentage of non-subscribers is so small that this bias can usually be ignored. (If you are estimating small percentages, or are interested in the sort of people who might not have telephones, this bias can matter.) Using telephone books would introduce serious bias, since there are many unlisted numbers. See section 7.

   Comment. About 95% of households in the U.S. have telephones, according to *Statistical Abstract*, 2006, table 1117. The corresponding figure in 1980 was 93%.

6. No. You might expect the respondents interviewed by blacks to be much more critical. (And they were.)

7. No, this parish might have been quite different from the rest of the South. (It was: Plaquemines is sugar country, and sugar required more highly skilled labor than cotton.)

8. No. First, the ETS judgment about "representative" schools may have been biased. Next, the schools may not have used good methods to draw a sample of their own students.

   Comment. There are about 3,600 institutions of higher learning in the U.S., including junior colleges, community colleges, teachers' colleges. About 1,000 of them are very small, altogether enrolling only 10% of the student population. At the other end, there are about 100 schools with enrollments over 20,000—and these account for about one third of the student population.

9. Quite a bit different from. Non-respondents generally differ from respondents—early respondents probably differ from late ones. (In the study, the percentage with TB was quite a bit higher among the last 200 respondents: perhaps those people did not want to have their illness confirmed.)

10. A description of the sample design would be more reassuring than a sales pitch followed by a disclaimer.

11. With 200 replies out of 20,000 questionnaires, nonresponse bias is an overwhelming problem. With 200 responses out of 400 questionnaires, the response rate is adequate to show something important: a substantial fraction of high-school biology teachers hold creationist views.

12. False. The serious problem is non-response bias. Additional people brought into the sample to build it back up to planned size are likely to differ from non-respondents, and do not fix the problem of non-response bias.

Chapter 20. Chance Errors in Sampling

Set A, page 361

1.

population	box
population percentage	40%
sample	draws
sample size	1,000
sample number	number of 1's among the draws
sample percentage	percentage of 1's among the draws
denominator for sample percentage	1,000

2. The box model: make 400 draws from a box with 10,000 [1]'s and 15,000 [0]'s. The average of the box is 0.40, and the SD is about 0.5, so the expected value for the sum is $400 \times 0.4 = 160$ and the SE for the sum is $\sqrt{400} \times 0.5 \approx 10$.

   (a) EV for number = 160 and SE for number = 10.

   (b) EV for percent = $(160/400) \times 100\% = 40\%$, and

   $$\text{SE for percent} = (10/400) \times 100\% = 2.5\%.$$

   (c) 40%, 2.5%.

   Comments. (i) Parts (b) and (c) call for the same numbers, in part (c) you have to interpret the results. (ii) The expected value for the sample percentage is the population percentage (p. 359).

3. The SE for the number of heads is $\sqrt{10{,}000} \times 0.5 = 50$. The SE for the percent is $(50/10{,}000) \times 100\% = 0.5$ of 1%.

4. (a) and (b) are both true.

   Comment. When drawing at random from a 0–1 box, the EV for the percentage of 1's among the draws equals the percentage of 1's in the box. This is so whether the draws are made with or without replacement. The equality is exact.

5. False. They forgot to change the box. The number of 1's is like the sum of 400 draws from the box

   [0] [0] [0] [1] [0].

6. 10%+1%. The number of red marbles in the sample is 90 ± 9. If the number is 1 SE too high, it's $90 + 9$: now convert to percent out of 900. Our SE for a percentage is added to or subtracted from the expected value, not multiplied.

7. The total distance advanced equals the total number of spots thrown. This is like the sum of 200 draws (at random with replacement) from the box

   [1] [2] [3] [4] [5] [6].

   The average of this box is 3.5, and the SD is 1.7. So he can expect to advance around $200 \times 3.5 = 700$ squares, give or take $\sqrt{200} \times 1.7 \approx 24$ squares or so.

8. Sherlock Holmes is forgetting about chance error.

Set B, page 366

1. (a) The expected value for the percentage of reds in the sample equals the percentage of reds in the population. (Population = box, sample = draws.) See p. 359.

   (b) As the number of draws goes up, the SE for the number of reds in the sample goes up but the SE for the percentage of reds goes down. See p. 360.

2. The first thing to do is to set up a box model. There should be 30,000 tickets in the box, one for each registered voter; 12,000 are marked 1 (Democrat) and 18,000 are marked 0. The number of Democrats in the sample is like the sum of 1,000 draws from the box. The fraction of 1's in the box is 0.4. The expected value for the sum is $1{,}000 \times 0.4 = 400$. The SD of the box is $\sqrt{0.4 \times 0.6} \approx 0.49$. The SE for the sum is $\sqrt{1{,}000} \times 0.49 \approx 15$.

   (a) The expected value for the percent is 400 out of 1,000, or 40%. The SE for the percent is 15 out of 1,000, or 1.5%. (No surprise about the expected value: 40% of the registered voters are Democrats.)

   (b) The percentage of Democrats in the sample will be around 40.0%, give or take 1.5% or so. Parts (a) and (b) require the same calculations; in (b), you have to to interpret the results.

   (c) This is ± 0.67 SE, the chance is about 48%.

3. (a) There should be 100,000 tickets in the box, one for each person in the population, of which 60,000 are marked 1 (married) and 40,000 are marked 0. The number of married people in the sample is like the sum of 1,600 draws from the box. The expected value for the sum is $1{,}600 \times 0.6 = 960$. The SD of the box is $\sqrt{0.6 \times 0.4} \approx 0.5$. The SE for the sum is $\sqrt{1{,}600} \times 0.5 = 20$. The number of married people in the sample will be 960, give or take 20 or so. Now 960 out of 1,600 is 60%, and 20 out of 1,600 is 1.25%. So 60% of the people in the sample will be married, give or take 1.25% or so.

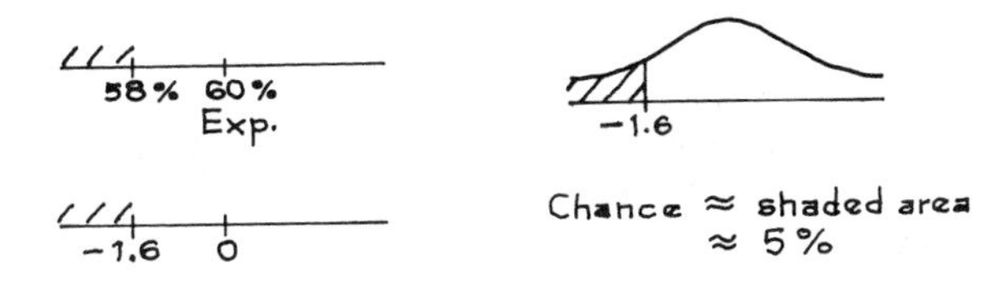

(b) There should be 100,000 tickets in the box, of which 10,000 are marked 1 (income over $75,000) and the other 90,000 are marked 0. There are 1,600 draws. The chance is about 9%.

(c) The box has 100,000 tickets, of which 20,000 are marked 1 (college degree) and the other 80,000 are marked 0. There are 1,600 draws. The chance is about 68%.

4. The shaded area represents the chance of drawing a sample in which 22% or more of the sample persons earn more than $50,000 a year.

5. (a) the chance that the sample will have 88 high earners
   (b) the chance that the sample will have 22% high earners
   (c) 88 is 22% of 400, so the same chance is described in two different ways. No coincidence at all.

Set C, page 370

1. Option (iii) is right. That is the point of the section.

2.

Number of draws	*SE for percentage of 1's among draws*
2,500	1%
25,000	0.27 of 1%
100,000	0%

Comment. After 100,000 draws, there are no more tickets in the box, and no uncertainty about the percentage of 1's among the draws.

3. The sample size should be 2,500.

4. The SE is the same for all three boxes, because all three have the same fraction of 1's, so the same SD.

5. SE with = 20%; SE without $= \sqrt{\dfrac{10-4}{10-1}} \times 20\% \approx 16\%$.

Comment. This is an artificial example where the number of draws is a large fraction of the number of tickets in the box, so the correction factor really kicks in.

Chapter 21. The Accuracy of Percentages

Set A, page 379

1. (a) observed (b,c) estimated from the data as

Comment. There is a big difference between chapter 20 and chapter 21. In chapter 20, you knew the composition of the box, and could compute the expected value and SE exactly. Here, the composition of the box has to be estimated from the data. In chapter 20, you reason forward, from the box to the draws. Here, you reason backward, from the draws to the box.

2. The first step is to set up the model. (We need the box model to compute the SE for the sum of draws.) There are 100,000 tickets in the box, some marked 1 (currently enrolled in college) and the others 0 (not enrolled). Then 500 draws are made from the box to get the sample. The number of college students in the sample is like the sum of the draws. The fraction of 1's in the box is unknown, but can be estimated by the fraction of 1's observed in the sample, which is $194/500 \approx 0.388$. So the SD of the box is estimated as $\sqrt{0.388 \times 0.612} \approx 0.49$. The SE for the sum is $\sqrt{500} \times 0.49 \approx 11$. The 11 is the likely size of the chance error in the 194. The SE for the percentage of 1's is $(11/500) \times 100\% = 2.2\%$. The percentage of persons 18–24 in the town who are college students is estimated as 38.8%. The estimate is likely to be off by 2.2% or so. The estimate is 38.8%, and the give-or-take number is 2.2%.

3. The estimate is 48%, give or take 5% or so.

4. The estimate is 2.8%, give or take 0.8 of 1% or so.

5. The estimate is 46.8%, give or take 2.5% or so.

6. No. Most people work for the few large establishments.

7. SE = 2%.

8. (a) $18.0\% \pm 1.9\%$ (b) $21.0\% \pm 2.0\%$ (c) $24.5\% \pm 2.2\%$

   Comment. The third person is off by a couple of SEs in estimating the percentage of 1's in the box; even so, the estimated standard error is only off by 0.2 of 1%. The bootstrap method is good at estimating SEs.

9.

	Known to be	*Estimated from the data as*
Observed value	30.8%	N/A
Expected value	N/A	30.8%
SE	N/A	1.5%
SD of box	N/A	0.46
Number of draws	1,000	N/A

Set B, page 383

1. (a) observed (b,c) estimated from the data as

   See exercise 1 on p. 379.

2. (a) $38.8\% \pm 4.4\%$ (b) $38.8\% \pm 6.6\%$ (c) $38.8\% \pm 3.3\%$

   Comments. As the confidence level goes up, the confidence interval gets longer. However, as the sample size goes up, the confidence interval gets shorter.

3. (a) Expect 1 red marble among the draws, give or take 1 or so.
   (b) It is impossible to draw fewer than 0 red marbles, so the chance is 0.
   (c) About 16%.
   (d) No. If the probability histogram looks like the normal curve, then the chance of drawing fewer than 0 red marbles can be read off the curve. Since $16\% \neq 0\%$—see (b) and (c)—the histogram does not look like the curve.

   Comment. The histogram is shown at the top of the next page.

4. False. The normal approximation cannot be used here. As best we can estimate from the sample, 1% of the marbles in the box are red, and 99% are blue. This is

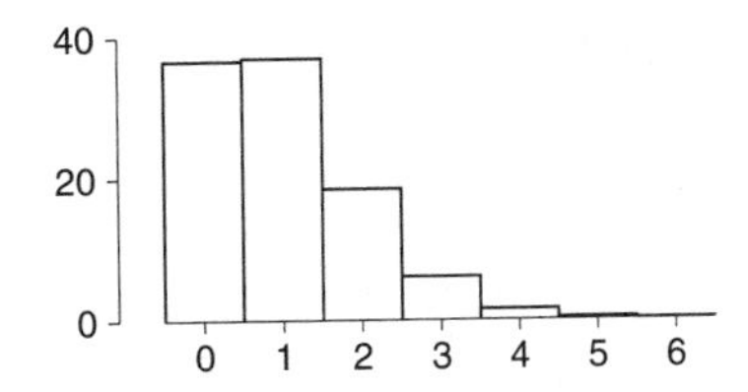

the box in exercise 3. The probability histogram for the percentage of reds among 100 marbles drawn from this box does not look like the normal curve. (With 100 draws out of 10,000, there is little difference between sampling with or without replacement.) If the sample were bigger, or the box were less lopsided, the normal curve would be fine.

Set C, page 386

1. Probabilities are used when reasoning from the <u>box</u> to the <u>draws</u>; confidence levels are used when reasoning from the <u>draws</u> to the <u>box</u>.

2. (a) The chance error is in the <u>observed</u> value.
   (b) The confidence interval is for the <u>population</u> percentage.

3. (a) $18.0\% \pm 3.8\%$, covers.
   (b) $21.0\% \pm 4.0\%$, covers.
   (c) $24.5\% \pm 4.4\%$, just misses.

4. (a) True.
   (b) False. The EV is computed exactly; the chance error is in the sample percentage of reds, not in the expected value.
   (c) True.
   (d) False. Confidence intervals are for parameters, not sample data. See pp. 385–386.
   (e) True.

   Comment on (b). The SE tells you the likely size of the chance error in the percentage of reds among the draws. The 50%, however, is a property of the box and does not depend on how the draws turn out: there is no chance error in the 50%. For instance, if you draw 100 times and get 53 reds, the sample percentage of reds is 53%, and the chance error—in the 53%—is +3%. If you get 42 reds, the percentage of reds among the draws is 42%, and the chance error in the 42% is −8%. But the expected value stays the same, no matter how the draws turn out. Also see exercise 6 on p. 294.

5. (a) True. (b) True. (c) True. (d) True.
   (e) False; the sample percentage is 53%, you don't need a confidence interval for that.

6. (a) True.
   (b) True.
   (c) False. The sample percentage is known, and in the interval.
   (d) False. If you view the interval as fixed, the chance is either 0 or 1. Moral: the chances are in the sampling procedure, not the population. That is why statisticians use the term "confidence interval."

7. False. The SE for the percentage measures the likely size of the difference between

one sample percentage and the population percentage; not the difference between two sample percentages.

Comment. The SE for the difference between two sample percentages has to be bigger, because both are subject to chance variability; by contrast, the population percentage isn't varying. See chapter 27 for more about the difference between two sample percentages.

8. True. Probabilities are used when you reason forward, from the box to the draws; confidence levels are used when reasoning backward, from the draws to the box: see pp. 385–386.

Set D, page 388

1. Theory says, watch out for this man. What population is he talking about? Why are his students like a simple random sample from the population? Until he can answer these questions, don't pay much attention to the SEs he calculates.
2. This is not a simple random sample: you are guaranteed to get 25 students from each class, a simple random sample won't do that. The procedure does not apply.

Set E, page 390

1. This isn't a simple random sample, the formulas don't apply.
2. This is fine.
3. (a) "altered voter enthusiasm"
   (b) Chance variation—the Gallup Poll is based on a random sample.
   (c) As table 2 shows, chance errors of several percentage points are quite possible. Maybe late September is not such a good guide to early November after all. (On the other hand, Bush did win.)

Chapter 22. Measuring Employment and Unemployment

Set A, page 403

1. (a) True.
   (b) False. The Bureau would divide up the sample into groups, by race, age, and so on, then weight up each group separately; section 4.
2. 151.4 million $\pm$ 0.1 million; section 5.
3. This is a simple random sample of households, and the inference is about households. The SD of the box is estimated as $\sqrt{0.80 \times 0.20} = 0.40$. The SE for the sum is $\sqrt{100} \times 0.40 = 4$. The SE for the percentage is 4%.
4. This is a simple random sample of households, but a cluster sample of people. (The household is the cluster.) The inference is about people. So, you need more information to estimate the SE—the formulas for simple random samples do not apply (section 5).

   Comment on exercises 3 and 4. In exercise 3, you have a simple random sample of households, and make an inference about households—the percent where all occupants are vaccinated. In exercise 4, you are making an inference about people from a cluster sample of people.
5. The SE for the percentage is only 0.2 of 1%, so a discrepancy of $55\% - 52\% = 3\%$

is almost impossible to explain as a chance error. People like to say they voted, even if they didn't.

6. The one for white males; it is based on a lot more people.

Chapter 23. The Accuracy of Averages

Set A, page 413

1. (a) 7,611/100 = 76.11 (b) 73.94 × 100 = 7,394
2. The SE for the average is 1. The answer to (a) is almost 100%. The answer to (b) is 68%. Don't confuse the SE for the average of the draws with the SD of the box.
3. (a) False. (b) True.
   To repeat, do not mix up the SE for the average of the draws with the SD of the box.
4. (a) The expected value for the average of the draws equals the average of the box.
   (b) As the number of draws goes up, the SE for the sum of the draws goes up but the SE for the average of the draws goes down.
5. The SE for the sum of the draws is $\sqrt{100} \times 20 = 200$. The SE for the average is $200/100 = 2$. The average of the draws will be around 50, give or take 2 or so. This is still true if the draws are made without replacement, because only a small fraction of the tickets in the box are drawn out. On the other hand, if you draw 100 tickets at random without replacement from a box of 100 tickets, the SE is 0.
6. The chance that the average of the draws is between 2.25 and 2.75.
7. The percentage of times [4] came up in the 50 draws.
8. (a) The chance that the sum will be 90.
   (b) The chance that the average will be 3.6.
   (c) 3.6 = 90/25, so the same chance is described in two different ways. No coincidence at all. See exercise 5 on p. 366.
9. (a), (c), (e) are true; (b), (d), (f) are false. You know the contents of the box; you can compute the expected value for the average without error; however, there is chance error in the average of the draws. See exercise 6 on p. 294, exercises 4–6 on pp. 386–387.
10. The average of the draws is just their sum, divided by 25 (the number of draws). So 25 changes to 1, 50 to 2, and 55 to 55/25 = 2.2.

Set B, page 420

1.

population	box
population average	average of the box
sample	draws
sample average	average of the draws
sample size	number of draws

2. (a) "SD of box" makes sense; "SE for box" does not.
   (b) "SE for average of draws" makes sense; "SE for average of box" does not.

   The term "SD" applies to a list of numbers; "SE" applies to a chance process. The tickets in the box (and their average) are fixed, but the draws are random.

3. (a,b) Estimated from the sample as. The SD of the sample is \$19,000; this is used to estimate the SD of the box. The SE is based on the estimated SD; so it too is an estimate. If you do not know what is in the box, you have to estimate the SD and the SE from the data.
   (c) observed.

4. 95% of 50 $\approx$ 48.

5. (a) Each organization takes its sample average as the center of its confidence interval. The sample averages are different, because of chance variation.
   (b) The sample SDs are different (chance variation), so the estimated SEs are different. That is why the lengths of the intervals are different.
   (c) 49.

6. The box has 30,000 tickets, one for each registered student, showing his or her age. The data are like 900 draws from the box; the sample average is like the average of the draws. The SD of the box is estimated as 4.5 years, the SE for the sum of the draws is $\sqrt{900} \times 4.5 = 135$ years, the SE for the average is $135/900 = 0.15$ years.
   (a) Estimate is 22.3 years, off by 0.15 years or so.
   (b) The interval is 22.3 ± 0.3 years.

7. (a) The interval is \$568±\$24. Even though the data don't follow the normal curve, the probability histogram for the average of the draws does.
   (b) False: \$24 is the SE for the average of the draws, not the SD of the box.

8. False. The SE for the average gives the likely size of the difference between the sample average and the population average, not the difference between two sample averages. So \$18 is the wrong margin of error. See exercise 7 on p. 387.

9. The probability histogram is about chances for the sample average; it is not about data. Here, the probability histogram is given. Part (a) asks for +1 in standard units, relative to the probability histogram. We need the center and spread of this histogram. The center is the expected value for the sample average, which equals the average of the box. This is given: it is \$61,700. The spread is the SE for the sample average. This can be worked out exactly, because the problem gives the SD of the box. This is \$50,000. So the SE for the sum of the draws is $\sqrt{625} \times \$50{,}000 = \$1{,}250{,}000$. The SE for the average of the draws is $\$1{,}250{,}000/625 = \$2{,}000$. And +1 in standard units is $\$61{,}700 + \$2{,}000 = \$63{,}700$. That is the answer to (a).

   In part (b), you are being asked to see where \$58,700 fits, on the axis of the probability histogram. It comes in below the expected value: \$58,700 is below \$61,700. So, \$58,700 is on the negative part of the axis. In fact, this value is \$3,000 below the expected value. And 1 SE is \$2,000. So \$58,700 is -1.5 in standard units. That is the answer to (b).

   Comments. (i) The key point: in this problem, the average and SD of the box are given.

   (ii) A typical sample average is around 1 SE away from the population average. Our sample average was 1.5 SE too low. We didn't get enough rich people in the sample.

   (iii) Look at figure 1 on p. 411. The histogram is about the process of drawing at random and taking the average; it is not about any particular set of draws. If you draw 25 tickets and their average happens to be 3.2, that doesn't change the histogram. This exercise illustrates the same point, in a more complicated setting.

   (iv) You would use the SD of \$50,000 to convert to standard units relative to a data histogram—for the incomes of all 25,000 families in the town. The SD of

$49,000 works relative to another data histogram—for the incomes of the 625 sample families.

Set C, page 423

1.

Number of draws	*EV for sum of draws*	*SE for sum of draws*	*EV for average of draws*	*SE for average of draws*
25	75	10	3.0	0.4
100	300	20	3.0	0.2
400	1,200	40	3.0	0.1

2. (a) True. The expected value for the average of the draws equals the average of the box (p. 410).
   (b) Can't tell; you need the SD of the box.

3. (a) Estimated from the data as; you would need the average of the box to compute the expected value exactly.
   (b) To compute the SE exactly, you need the SD of the box; even to estimate it, you would need the SD of the draws.

   Comment. The expected value applies to the process of drawing at random, rather than any particular set of draws. For example, suppose you draw 25 times at random with replacement from the box | [0] [2] [3] [4] [6] |. The expected value for the average of the draws is 3. The average of your draws could be 3.1, which is 0.1 above the expected value; or, the average of the draws could be 2.6, which is 0.4 below. There are many other possibilities. But the expected value only depends on the box, and stays the same no matter how the draws turn out.

4. (a) The SE for the sum of the draws is 7.1, and the SE for the average of the draws is 0.18.
   (b) The expected value of 100 is at the center; the next tick mark to the right is 10 boxes over, that must be 110, and so forth.

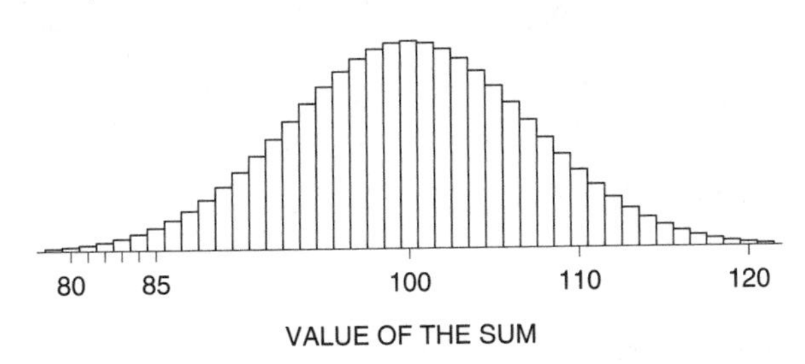

5. You can't estimate the SD of the box, so you can't get margins of error.

6. For all three boxes, the EV for the sum of 100 draws is 200. The SE for the average of the draws is

   1 from box A   1.4 from box B   2 from box C.

   (a) 203.6 is very unlikely to come from box A—it is 3.6 SEs away from the expected value for the average of 100 draws from box A. It is also quite unlikely to come from box B, because $3.6/1.4 \approx 2.6$ is too many SEs. So it comes from box C. Similarly, 198.1 comes from box B, leaving 200.4 for box A by elimination.
   (b) It could be otherwise, but that would be pushing things.

Set D, page 424

1. The 95%-confidence interval is 1.86 ± 0.06.
2. This is qualitative data, use the method of chapter 21. The interval is $60.1\% \pm 5.4\%$.
3. Can't be done with the normal curve. Suppose the sample reflects the population exactly. Then the company is drawing from a box which has 99.87% [1]'s and 0.13 of 1% [0]'s. This box is so skewed that with 750 draws, the probability histogram for the sum won't be anything like the normal curve. See exercises 3 and 4 on p. 383.
4. This is not a simple random sample of people: either you get everybody in a household, or nobody. So the SE can't be estimated by the methods of this chapter. See exercises 3 and 4 on p. 404.

   Comments. (i) This is a cluster sample of people—the household is the cluster; the half-sample method could be used to get the SE (p. 402), but more information would be needed.

   (ii) People in a household tend to be similar with respect to TV-watching, so this sample will be less informative than a simple random sample of the same size. Cluster samples are less accurate than simple random samples, but much cheaper to take.

   (iii) The usual problem with cluster samples is chance error, rather than bias; the sampling method in this exercise is unbiased.
5. (a) This is not a probability sample of any kind. It is a sample of convenience.
   (b) Same as (a).

   Comment. A cluster sample is a special kind of probability sample (pp. 340, 342).
6. The average of the box is estimated by the average of the sample: $297/100 \approx 3.0$; for the SE, you need the SD.
7. The two procedures are the same: simple random sampling means drawing at random without replacement (p. 340).

Part VII. Chance Models

Chapter 24. A Model for Measurement Error

Set A, page 444

1. Use the SE for the sum; this was figured as 60 micrograms.
2. The estimate is the average of the measurements, 82,670 pounds. This is likely to be off by the SE for the average, 100 pounds.
3. (a) 800 microns (b) 80 microns (c) 91.4402 cm $\pm$ 160 microns
4. (a) False. This range is 2 SEs, not 2 SDs, either way from the average.
   (b) False. Same reason as in (a).
   (c) True. See p. 384.
   (d) False. This is just like exercise 8, p. 421.
5. The factor is 5.

Set B, page 449

1. You would have to toss the thumbtack many times, and see whether the percentage of times it landed point down was closer to 50% or to 67%. (This will depend on the surface: in one experiment, the tack landed point down 66% of the time when tossed on linoleum, but only 50% of the time when tossed on a carpet.)
2. No. The rainy days all come close together in the rainy season. If it rains one day, it is more likely to rain the next.
3. Last digits, yes. First digits, no. For instance, in the San Francisco phone book the first digit cannot be 0. Also, many more phone numbers start with 9 than with 2.
4. No, the letters come out in alphabetical order. No box will do that.
5. Like a shot. You have a 50–50 chance to win \$5 or lose \$4.

Set C, page 452

1. In both cases, the measurement is 504 micrograms above ten grams.
2. No, as the previous exercise shows.
3. Six micrograms is the SD of the 100 measurements reported in table 1 on p. 99. This is used to estimate the SD of the error box. So, "estimated from the data as."
4. (a) Chance variation—the investigators get different sample averages.
   (b) Chance variation again—the investigators get different sample SDs.
   (c) About 95% of the 50 intervals should cover the exact weight, that is, about 48 intervals.
   (d) 48. (One of the intervals is off by quite a bit—chance variation at work.)
5. The SD of the error box is estimated as 50 micrograms.
   (a) 5 micrograms—the SE for the average.
   (b) 50 micrograms—the estimated SD of the error box.
   (c) 95%—two SEs.
6. The answer is 1.2 micrograms. See example 5, p. 451.
7. (a) 300,007 (the average); 2 (the SE for the average).
   (b) False: the average is 300,007 exactly.
   (c) True: each number on a list is off the average of the list by an SD or so.
   (d) True: the interval is "average $\pm$ 2 SEs."
   (e) False: the average of the 25 measurements is 300,007 exactly.
   (f) False: 2 is the SE, not the SD.
8. The answer is 2 inches. Here is the reason. Each measurement equals the exact length, plus a draw from the error box. The estimated distance AE is the sum of the 4 measurements, and is off the exact length AE by the sum of 4 draws from the box. The average of the error box is 0. So the sum of 4 draws will be around 0, give or take an SE or so. It is the SE for the sum which is the right give-or-take number. The SD of the box is 1 inch, so the SE for the sum is $\sqrt{4} \times 1$ inch $= 2$ inches.

   Comment. Finding the length AE involved adding the measurements, not averaging them.
9. The chance errors for different people could have different SDs. Also, if the same

person takes the test several times, the errors may be dependent. The Gauss model does not seem to apply.

Chapter 25. Chance Models in Genetics

Set A, page 461

1. Each seed has a 50% chance to get y from the y/g parent, and a 50% chance to get g. It is bound to get g from the g/g parent. So the seed has a 50% chance to be y/g, and yellow in color; it has a 50% chance to be g/g, and green in color. About 50% of the seeds should be yellow.

   The number of yellows among 1,600 seeds is like the sum of 1,600 draws from the box | [0] [1] |. The expected number of yellows is $1{,}600 \times 1/2 = 800$. The SE for the number is $\sqrt{1{,}600} \times 1/2 = 20$. Now the normal approximation can be used:

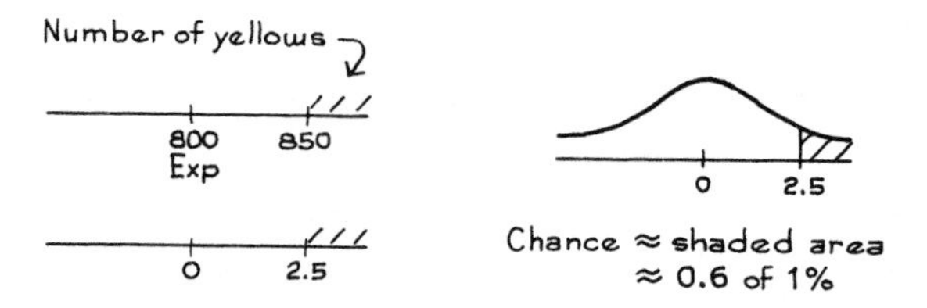

2. (a) white × red → 100% pink
   white × pink → 50% white, 50% pink
   pink × pink → 25% red, 50% pink, 25% white.
   Work for pink × pink: Each parent is r/w, so the offspring's flower color is determined by choosing a row and column at random from the table below.

	r	w
r	red	pink
w	pink	white

   (b) The expected number of pinks in 400 plants is 200, with an SE of 10. Use the normal approximation:

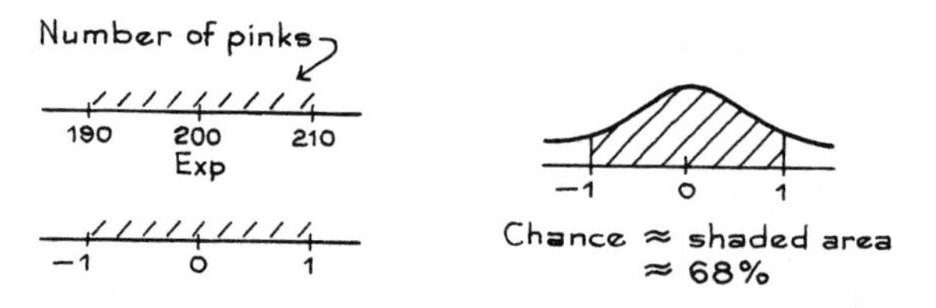

3. (a) One gene-pair controls leaf width, with variants w (wide) and n (narrow). The rules: w/w makes wide, w/n and n/w make medium, and n/n makes narrow.
   (b) narrow × narrow = $n/n \times n/n \rightarrow$ 100% n/n = narrow
   narrow × medium = $n/n \times n/w \rightarrow$

   $$50\%\ n/n = \text{narrow},\quad 50\%\ n/w = \text{medium}.$$

4. B = brown, b = blue. Husband is B/b, wife is b/b. Each child has 1 chance in 2 of having brown eyes. The three children are independent, so the chance that all three will be brown-eyed is $(1/2)^3 = 1/8$.

Part VIII. Tests of Significance

Chapter 26. Tests of Significance

Set A, page 476

1. (a) estimated from the data as
   (b) observed

2. The observed value is only 1.3 SEs below expected, Dr. Null is looking good.

3. Again, the observed value is about 1.3 SEs below expected, and Dr. Null is looking good. Moral: results depend on the observed value, the expected value, and the SE. The SE depends on the SD and the sample size.

4. If the die is fair, the total number of spots is like the sum of 100 draws from the box | [1] [2] [3] [4] [5] [6] |. The average of the box is 3.5; the SD is 1.7. So the expected value for the sum is 350, and the SE is 17. The number of spots is a little over 1 SE above its expected value, which looks like chance variation.

5. The problem can be set up like exercise 4, but this time the number of spots is over 3 SEs above its expected value. This doesn't look like chance variation.

   Comments. (i) Sample size matters; compare exercises 4 and 5.
   (ii) A more complete test for the fairness of a die will be presented in chapter 28.

Set B, page 478

1. (iii)

2. The null hypothesis says that the difference is due to chance but the alternative hypothesis says that the difference is real.

3. Choose (ii). Dr. Null and Dr. Alt both knew the data, they didn't know what was in the box. The null hypothesis is a statement about the box, and the test tells you whether this statement is plausible.

4. box. The null hypothesis is about the box.

5. The SD of the box can be estimated as 10, so the SE for the average of 100 draws is estimated as 1. If the average of the box is 20, then the average of the draws is 2.7 SEs above its expected value. This isn't plausible.

Set C, page 481

1. (a) $P = 32\%$ is best for the null.
   (b) $P = 0.1$ of 1% is best for the alternative.
   Big P is good for null; small P is bad for null.

2. (a) True. (b) False. See pp. 480–81.

3. (a) True, see pp. 480–81.
   (b) False, see pp. 480–81.

4. SE for average ≈ 1.25, so $z \approx (52.7-50)/1.25 \approx 2.16$ and P is approximately the area to the right of 2.16 under the normal curve. From the table, this is about 1.6%. The difference is hard to explain as chance variation. The alternative hypothesis is looking good.

5. box. The null hypothesis is about the box.

6. No. With 10 draws, the probability histogram for the sample average may not look like the normal curve, and the SD of the data will not be a good estimate for the SD of the box.

7. The sample is like 100 draws made at random from a box which has one ticket for each employee, showing the number of days that employee was absent. Null hypothesis: the average of the box is 6.3 days. Alternative hypothesis: the average of the box is less than 6.3 days. The SD of the box is estimated as 2.9 days, so the SE for the average is 0.29 days, and $z \approx (5.5 - 6.3)/0.29 \approx -2.8$, so $P \approx 0.3$ of 1%. This is strong evidence against the null; chance variation will not explain the drop in absenteeism.

8. Now $z \approx (5.9 - 6.3)/0.29 \approx -1.4$, so $P \approx 8\%$. The null hypothesis looks more plausible.

Set D, page 482

1. (a) False. Even if the null hypothesis is true, 1% of the time the experiment will give a result which is "highly significant."
   (b) False; pp. 480–81. (c) False; pp. 480–81.

2. (a) True. Big P is good for null.
   (b) True. Small P is bad for null.

3. (a) True; p. 482.
   (b) False; P has to be less than 1%.
   (c) True; p. 482.
   (d) True; p. 479.
   (e) True; $z = (\text{obs} - \text{exp})/\text{SE}$, and "exp" is computed on the null.

4. About 2%.

5. (a) True; $z = (\text{obs} - \text{exp})/\text{SE}$; "obs" is the average of the draws; "exp" is the average of the box, given as 50.
   (b) About 50.
   (c) About 2; and 3 of them do.
   (d) About 2%. See exercise 4.

Set E, page 486

1. Option (i) is right: "like" means, "as far as the chances are concerned." Each guess has one chance in four to be right, and each draw has one chance in four to be 1. Then the number of correct guesses is like the sum of the draws, and the square root law applies.

   Option (ii) is wrong: if there is no ESP, the chance of a correct guess is 1/4 not 1/3. Option (iii) is worse: the 2,006/7,500 is the fraction of 1's in the sample, not in the box. Option (iv) is wrong: the fraction of 1's in the sample is known, there is no argument about that. Option (v) is way off: the null corresponds to the idea that there is no ESP.

2. (a) student registered at Berkeley that term. Reason: the box corresponds to the population.
   (b) 1 = man, 0 = woman. Reason: you're counting the men. (If you want to count the women, that's fine too, but be consistent.)
   (c) There are 25,000 tickets in the box, and 100 draws. Reason: the sample is like the draws.

(d) 100 draws.
(e) 67%. Reason: You know the percentage of men in the population.

3. (a) 53 (b) 67 (c) sum
(d) $\sqrt{100} \times \sqrt{0.67 \times 0.33} \approx 4.7$
(e) $z \approx (53 - 67)/4.7 \approx -3$ and $P \approx 1/1{,}000$.

4. No. He got too many women. P is very small, so chance won't explain the difference. Taking people haphazardly isn't like a simple random sample (chapter 19).

5. (a) Computed from the null hypothesis: 100×0.67. The expected is always computed from the null hypothesis.
(b) Computed from the null hypothesis. Here, the null tells you the composition of the box. Otherwise, you might have to estimate the SD of the box from the data (p. 485).

6. (a) Null hypothesis: the number of correct guesses is like the sum of 1,000 draws from a box with one ticket marked 1 and nine 0's.
(b) $\sqrt{0.1 \times 0.9}$. The null hypothesis tells you what's in the box. Use it.
(c) $z \approx (173 - 100)/9.5 \approx 7.7$, and P is tiny.
(d) Whatever it was, it wasn't chance variation.

7. (a) Tossing the coin is like drawing at random with replacement 10,000 times from a 0–1 box, with 0 = tails and 1 = heads. The fraction of 1's in the box is unknown. Null hypothesis: this fraction equals 1/2. Alternative: the fraction is bigger than 1/2. The number of heads is like the sum of the draws.
(b) $z = 3.34$, $P \approx 4/10{,}000$.
(c) There are too many heads to explain as chance variation.

8. (a) Same as 7(a). (b) $z = 1.34$, $P \approx 9\%$.
(c) The coin looks to be fair.

9. (a) box. The null hypothesis is about the box.
(b) False; see pp. 480–81.

10. The data consist of the 25 weights. The null hypothesis says that the data are like 25 draws made at random from a box. There is one ticket in the box for each animal in the colony, showing its weight. So the average of the box is 30 grams. And its SD is 5 grams, so the SE for the average of 25 draws is 1 gram. Now $z = (33 - 30)/1 = 3$, and $P \approx 1/1{,}000$.

Comments. (i) Here, the null tells you the SD of the box, so you don't need to estimate it from the data. The SD of the data is not used in working the problem. See p. 485.

(ii) Choosing haphazardly is not like taking a simple random sample (chapter 19). When you reach into the cage to pick up an animal, probably it is the tamer ones who come to your hand, and they are a bit heavier than the others.

11. The null hypothesis says that the reduced price had no effect on sales volume. So in each pair of stores, the one with the regular price is just as likely to sell more as its partner with the reduced price. In terms of a box model, the null hypothesis says the data are like 25 draws from the box [1] [0], where 1 means the regular-price store sold more and 0 means the regular-price store sold less. The expected number of 1's is 12.5, and the SE is 2.5, so $z = (18 - 12.5)/2.5 = 2.2$ and $P \approx 1.4\%$. The evidence against the null is quite strong.

Comment. This procedure is called "the sign test." See exercise 6 on p. 258 (kangaroos) and exercise 11 on p. 262 (smokers). If the continuity correction is made, the normal approximation gives $P \approx 2.28\%$, compared to 2.16% from the binomial formula.

Set F, page 493

1. (a) 5% (b) 5% (c) 90% (d) 95%

2. From the table, the area to the right of 2.92 is 5%, and the area to the right of 6.96 is 1%. Since 4.02 is between 2.92 and 6.96, the area to the right of 4.02 is between 1% and 5%.

3. No, 3 degrees of freedom.

4. (a) degrees of freedom = 2, ave $\approx$ 72.7, $\text{SD}^+ \approx 5.7$, SE $\approx$ 3.3,
$$t \approx (72.7 - 70)/3.3 \approx 0.8,$$
P is about 25%. Inference: the calibration is fine.
(b) P is about 2.5%, recalibrate.
(c) One measurement is never enough.
(d) P is about 25%.

Comment. Two measurements are better than one; more would be even better.

5. In (a), the number 93 is an outlier, so the errors do not seem to follow the normal curve. In (c), the numbers are just switching back and forth between 69 and 71. This is not good for the Gauss model.

6. According to the Gauss model, each of the 10 new measurements equals the exact weight, plus bias, plus a draw from the error box. The null hypothesis says that the bias is zero; the alternative hypothesis says that there is some bias. The SD of the error box is estimated as $\sqrt{10/9} \times 9 \approx 9.5$. (The errors belong to the rebuilt scale, so the old SD of 7 micrograms is irrelevant.) The SE for the average $\approx$ 3 micrograms, $t \approx -2.67$. The area to the left of -2.67 under Student's curve with 9 degrees of freedom is about 1%, strong evidence against the null.

7. (a) According to the Gauss model, each of the 100 measurements equals the exact weight plus a draw from the error box. The tickets in the error box average out to 0. The unknown parameter is the exact weight. The null hypothesis says that this is still 512 micrograms above a kilogram. The alternative says that the exact weight is less.
(b) The SD of the error box can be estimated from the past data as 50 micrograms: the error box belongs to the equipment. (The new SD of 52 micrograms is irrelevant.)
(c) With 100 measurements, use z not t. The SE for the average of 100 measurements is 5 micrograms, so $z = (508 - 512)/5 = -0.8$ and $P \approx 21\%$.
(d) The drop in weight looks like a chance variation.

Chapter 27. More Tests for Averages

Set A, page 503

1. True, now the numbers are independent, so the square root law applies.

2. The expected value is $100 - 50 = 50$, and the SE is $\sqrt{2^2 + 3^2} \approx 3.6$. The square root law applies because the draws are all independent.

3. The expected value for each percent is 50%; the SEs are 2.5 and 5 percentage points. The expected value for the difference is 0, and the SE is $\sqrt{2.5^2 + 5^2} \approx$ 5.6 percentage points. The square root law applies because the two percents are independent.

4. (a,b) True.
   (c) False. The percentages are dependent: if the coin lands heads, it can't land tails. The square root law does not apply.

   Comment. The difference "number of heads – number of tails" is like the sum of 500 draws from the box | $\boxed{-1}$ $\boxed{+1}$ |, so the SE for the difference in the two numbers is about 22, and the SE for the difference in percentages is

$$(22/500) \times 100\% = 4.4\%.$$

5. True. If the draws are made with replacement, the two averages would be independent: the SE for the difference would equal $\sqrt{3^2 + 3^2}$ exactly. The box is so large that there is no practical difference between drawing with or without replacement.

6. The SD of box F can be estimated as 3, so the SE for the average of 100 draws from box F is 0.3; similarly, the SE for the average of 400 draws from box G is estimated as 0.4; the averages are independent, so the SE for the difference is $\sqrt{0.3^2 + 0.4^2} = 0.5$. If the two boxes have the same average, the observed difference $51 - 48 = 3$ is 6 SEs away from the expected value of 0. Not a likely story.

Set B, page 506

1. Two-sample z-test.

2. (a) Two-sample z-test: you're comparing two samples.
   (b) The setup is as in the text, with a 1990 box and a 2004 box. There are oodles of tickets in each box, and 1,000 draws from each. The tickets show test scores. The null hypothesis says the average of the boxes are the same. The alternative says the averages are different.
   (c) The SE for the difference is 1.37, so $z = 2/1.37 \approx 1.46$. This could easily be chance variation.

3. This difference is big, and highly significant—both practically and statistically.

4. The difference between the two sample averages is 2.1 hours, and the SE for the difference is 0.35 hours. So $z \approx 6$, and $P \approx 0$. The difference is very hard to explain away as a chance variation. Students in private universities generally come from wealthier families, and have more support from home.

5. The numerator is in percent, and the denominator is a decimal. That's a mistake. In fact, $z = (41 - 17)/4.8 = 5$. If you prefer decimals, $z = (.41 - .17)/.048 = 5$.

   Comment. Forgetting to convert the denominator to percent is a common slip. You can do the whole problem in percents or in decimals, but don't change in the middle.

6. There are two samples, so you need to make a two-sample z-test. The data consist of 1,600 0's and 1's for the men (1 = illiterate), and another 1,600 0's and 1's for the women. The model has two boxes, M and F. Box M has a ticket for every male youth in the country, marked 1 for the illiterates and 0 for the literates. Box F is similar, for the females. The data for the men are like 1,600 draws from box M, and similarly for the women. Null hypothesis: the percentage of 1's is the same in the two boxes. Alternative: the percentage of 1's is bigger in box M. The SE for the percentage of 1's in the male sample can be estimated as 0.64 of 1%; for the female sample, the SE is 0.43 of 1%. So the SE for the difference is $\sqrt{0.64^2 + 0.43^2} \approx 0.77$ of 1%. Then $z \approx (7 - 3)/0.77 \approx 5.2$ and P is almost 0. This difference is almost impossible to explain as a chance variation.

7. The SE for the difference of the two averages can be estimated as $\sqrt{0.5^2 + 0.5^2} \approx 0.7$. So $z = (26 - 25)/0.7 \approx 1.4$, and $P \approx 8\%$. The difference could well be due to chance.

8. $z = 1/0.45 \approx 2.2$ and $P \approx 1.4\%$.

   Comment. The observed significance level depends on the sample size. With large samples, even small differences will be highly statistically significant. More about this in chapter 29.

9. The treatment and control averages are dependent, because the rats came in pairs from the same litter, so if one rat has a heavy cortex, the other one in the pair is likely to also. The SE calculation does not take this pairing into account.

   Comment. See review exercise 12 in chapter 26 for a better analysis. In each pair, take the difference "treatment − control." Make the z-test on the differences.

Set C, page 511

1. (a) Two numbers. The B-number is not observed; it says what his score would have been, if he had been assigned to the control group.
   (b) Yes. The A-number says what her score would have been, if she had been assigned to the coaching group. This number is not observed, because she was in the control group. The investigators do not know what the A-number was.
   (c) Take the conservative route. The SE for the coaching average is 9.8 points. The SE for the control average is 10.3 points. The SE for the difference is $\sqrt{9.8^2 + 10.3^2} \approx 14.2$ points. The difference in average scores was 9, so $z \approx 9/14.2 \approx 0.65$, and $P \approx 26\%$. This could easily be chance variation.

   Comment. Exercise 1 is not like comparing NAEP test scores in 1990 and 2004 (section 2), because we do not have two independent samples. But it is like the vitamin C experiment (example 4). Each of the 200 students has two possible responses—one if coached and one if not coached. The investigators get to see only one of the two responses, and make their choices at random. That is why the calculation of the SE is legitimate (pp. 509–511).

2. (a) The difference is $66 - 59 = 7$ points, and the SE is 1.8 points. So $z \approx 3.9$, and $P \approx 0$. This difference is hard to explain as a chance variation. Wheaties work!
   (b) The students will know what cereal they are eating, so it is hard to blind that aspect of the study. The grading of the final could be done blind. Consent to the study should be obtained before randomization, not after, to reduce selective drop outs.

3. (a) The difference is 1 point, and the SE is 1.75 points. This looks like chance variation. The two groups are comparable—the randomization worked.
   (b) Now the difference is 9 points, with the same SE of 1.75 points. So $z \approx 5$, and $P \approx 0$. Something went wrong in the randomization.

   Comment. The difference in (b) can't be explained as the result of eating Wheaties, because the cereal feeding did not start till after the midterm. See exercise 7 on p. 22. This exercise was hypothetical; for a real study on cornflakes, see N. Vaisman et al., "Effect of breakfast timing on the cognitive functions of elementary school students," *Archives of Pediatric and Adolescent Medicine* vol. 150 (1996) pp. 1089–92; eating breakfast improves your test scores.

4. (a) The difference between the two sample averages is 0.1, and the SE is 0.13. So $z \approx 0.8$ and $P \approx 21\%$. This looks like chance variation.

(b) There is a new batch of random numbers, and other factors might be at work too—weather, new cold viruses, etc. After all, the studies involve two different groups of people, at two different times.

5. (a,b) True.
(c) False. The sample averages are dependent, so the square root law does not apply (section 1).

Set D, page 514

1. (a) $\boxed{0}\boxed{1}$
(b) Form A, prefers surgery; form B, prefers radiation.
(c) Only (ii).
(d) The number of students who got form A was $84 + 112 = 196$; of these, $112/196 \times 100\% \approx 57\%$ favored surgery. Of the students who got form B, about 83% favored surgery. The difference between the percents is 26%, and the SE is about 5.2%. So $z \approx 5$, and $P \approx 0$. The difference is hard to explain as a chance variation.

2. "Percent" means "per 100," but the rates in this problem are so small that it is more convenient to express them per 100,000. The rate in the vaccine group was 57/200,000, or 28.5 per 100,000. The SE for the number of cases is

$$\sqrt{200{,}000} \times \sqrt{\frac{57}{200{,}000} \times \left(1 - \frac{57}{200{,}000}\right)} \approx 8$$

(See section 4 of chapter 17 for the shortcut method.) So the SE for the rate is 8/200,000 or 4 per 100,000. In the placebo group, the rate was 71 per 100,000, and the SE for the rate is 6 per 100,000. The SE for the difference in rates is

$$\sqrt{4^2 + 6^2} \approx 7 \text{ per } 100{,}000$$

The difference in rates is $28.5 - 71 = -42.5$ per 100,000. On the null hypothesis, the expected difference in rates is 0. So $z \approx -42.5/7 \approx -6$. The difference in rates cannot be explained as a fluke in the randomization. The vaccine works.

3. (a) $z \approx -2.4$, $P \approx 1\%$, significant. The difference is hard to explain as chance variation. Screening prevents death from breast cancer.
(b) $z \approx -1$, $P \approx 16\%$, not significant. Breast cancer is rare: you don't see the impact of screening on the total death rate.

4. In the treatment group, 6.9% of the women experienced at least one event, compared to 7.1% in the control group. The difference is 0.2 of 1%. The SE for the difference is 0.7 of 1%. The difference is is not significant. The difference could easily be due to chance. The diet was not protective.

5. This question cannot be answered from the information given. The investigators do not have two independent samples, with one sample answering the question about Great Britain and the other the question about France. So the method of example 3 (p. 507) does not apply. The investigators have only one sample, and there are two responses for each student in the sample:

1 1	found Great Britain and France on the map
1 0	found Great Britain; could not find France
0 1	could not find Great Britain; found France
0 0	could not find either country

The investigators observe both responses when they score the test; that makes it different from the experiment in section 4, where only one of the two responses can be observed.

Comment. The question can be answered by using more advanced statistical methods, if you know the percentages in each of the 4 categories listed above.

6. (a) This is a straightforward two-sample z-test, as in section 2, because there are two independent simple random samples. The SE for the 2005 percentage is estimated as 1.6%; so is the SE for the 2000 percentage. The SE for the difference is computed from the square root law (section 1) as $\sqrt{1.6^2 + 1.6^2} \approx 2.2\%$. The observed difference is $41 - 50 = -9\%$. On the null hypothesis, the expected difference is 0%. So $z = (\text{obs} - \text{exp})/\text{SE} = -9/2.2 \approx -4.1$, $P \approx 2/100{,}000$. The difference is real. People are losing faith in the Supreme Court.
   (b) You can't tell. The method of section 2 does not apply, because you do not have two independent samples. The method of sections 3–4 does not apply, because you observe two responses for each person. See exercise 5 above.

7. (a) The difference in the two sample percents is 0.6% and the SE is about 3.6%. This looks like a chance variation. Withholding supplementation has no effect on breast feeding later.
   (b) The difference is 20.9 ml/day and the SE is 3.1 ml/day. This is almost impossible to explain as chance variation. Feeding patterns do seem to have been affected by different treatments in the nurseries.
   (c) The difference is 0.9% and the SE is 0.14%. So $z \approx 6.4$. Withholding supplementation increases weight loss: a bad side-effect.
   (d) The difference between the two sample averages is 27 grams and the SE is about 31 grams. This is chance variation: the randomization was successful.

Comments. (i) There is a tricky point in (c). Weight loss for each infant is measured in percent, relative to the birth weight. These percents are quantitative data, for which averages and SDs are computed.

(ii) The experiment shows that withholding supplementation does not promote breast feeding, and has a bad side effect—weight loss. The observational studies got it wrong. The explanation: there is an important confounding variable. Nurturing mothers are more likely to breast feed in the hospital, and their babies get less supplement. These mothers are also more likely to be breast feeding later, so there is a negative association between bottle feeding in the hospital and breast feeding later. But this association is driven by a third factor—the mother's personality.

Chapter 28. The Chi-Square Test

Set A, page 531

1. (a) 90% (b) 10% (c) 1%

2. About 10%.

   Comment. Compare this with 1(c). As the degrees of freedom go up, the curve shifts to the right and spreads out, so there is more area to the right of 15.09 with 10 degrees of freedom than with 5. See figure 1.

3. $\chi^2 = 13.2$, $d = 5$, $1\% < P < 5\%$; actually, $P \approx 2.2\%$.

   Comment. $d =$ degrees of freedom. The data do not fit the model so well.

4. $\chi^2 = 1.0$, $d = 5$, $95\% < P < 99\%$.

5. $\chi^2 = 10.0$, $d = 5$, $5\% < P < 10\%$; actually, $P \approx 7.5\%$.

   Comment. Compare exercises 4 and 5. The observed frequencies just got multiplied by 10; this doesn't change the percents. But the result of the χ^2-test depends on the sample size. With large samples, the χ^2-test will reject very reasonable models. More about this in the exercises below and in chapter 29.

6. $\chi^2 \approx 18.6$, $d = 5$, $P < 1\%$—although for most purposes, the die is as fair as could be wanted; more about this in chapter 29.

7. (a) False; the χ^2-test is preferred, see p. 524.
   (b) χ^2, see p. 525.
   (c) True.
   (d) Expected; for instance, in line 1, the expected is $0.42 \times 66 \approx 27.7$; see p. 524.
   (e) The work for the χ^2-test:

Age	*Observed*	*Expected*
21 to 40	5	27.7
41 to 50	9	15.2
51 to 60	19	10.6
61 and up	33	12.5

   $\chi^2 \approx 61$, $d = 3$, $P \approx 0$. With simple random sampling, it is almost impossible for a jury to differ this much from the county age distribution. The inference is that grand juries are not selected at random.

   Comments. (i) The expected frequencies need not be whole numbers.
   (ii) Grand juries are nominated by judges, who prefer older jurors.

8. This is not a good method. The formula for χ^2 involves frequencies—numbers not percents. Compare exercises 4 and 5 above.

9. (a) 12
   (b) You use χ^2. The χ^2 statistics: A) 15.2, B) 26.7, C) 7.5, D) 16.5. With 9 degrees of freedom, the 10% level is 14.68, the 5% level is 16.92, and the 1% level is 21.67. So A is marginal, B is way out of line, C is fine, D is marginal.
   (c) On retest, the χ^2 for set A was 14.5, and for D it was 18.8. Reject D, and maybe A as well.

10. (a) The χ^2-test will do the job.

(b) Can't be done: both boxes have the same fractions of 1's, 2's, and so forth; the test can't tell the difference.

Set B, page 534

1. Pooled $\chi^2 = 13.2 + 10 = 23.2$, $d = 5 + 5 = 10$, $P \approx 1\%$.
2. No, dependent experiments.
3. $\chi^2 \approx 0.5$, $d = 3$, $P \approx 8\%$. Inconclusive, but points to fudging.

Set C, page 539

1. It's fine. The method in the text is $(28/2{,}237) \times 1{,}170$. The method in the exercise is $(1{,}170/2{,}237) \times 28$. The result is the same, because $28 \times 1{,}170 = 1{,}170 \times 28$.

2.

Observed			*Expected*		*Difference*	
2,792	3,591	6,383	2,730.6	3,652.4	61.4	−61.4
1,486	2,131	3,617	1,547.4	2,069.6	−61.4	61.4
4,278	5,722	10,000				

$$\chi^2 \approx 6.7, \ d = 1, \ P \approx 1\%.$$

The expected frequencies are computed as on p. 538: for instance, the expected number of men who voted is $(4{,}278/10{,}000) \times 6{,}383 \approx 2{,}730.6$.

Comments. (i) 65% of the men voted, compared to 63% of the women. This is a small difference, but with a large sample it is accurately estimated. All P tells you is whether the difference can be explained by chance. More about this in chapter 29.

(ii) A 2×2 table can be handled either by the χ^2-test or by the z-test: note 3 to chapter 27.

3. Choose option (iv); the z-test is inappropriate because there are multiple categories, and the null hypothesis doesn't tell you what's in the box.

Observed			*Expected*		*Difference*	
45	30	75	35.9	39.1	9.1	−9.1
86	105	191	91.3	99.7	−5.3	5.3
12	21	33	15.8	17.2	−3.8	3.8
143	156	299				

$$\chi^2 \approx 6.8, \ d = 2, \ P \approx 3\%$$

The expected frequencies are computed as on p. 538: for instance, the expected number of never-married men is $(143/299) \times 75 \approx 35.9$. In general, women marry earlier than men; and in the age group 25–34, more women than expected are married. ("Expected" means, on the null hypothesis that men and women have the same distribution of marital status.) The extra husbands are in higher age groups, like 35–44.

4. The Current Population Survey is not a simple random sample, the formulas do not apply, the clustering would have to be taken into account.

5. You are looking at averages, so it is time for the z-test not the χ^2-test. There are two samples not just one, so option (ii) is right: $z \approx (\$50{,}000 - \$30{,}000)/\$3{,}000 \approx 6.7$, $P \approx 0$, the difference looks real. College grads make more money.

6. You are comparing a sample percent to an external standard, so option (i) is right: $z \approx (568 - 550)/15.7 \approx 1.15$, $P \approx 25\%$ (two-sided), the demographers' theory looks fine.

   You can also work this problem by method (iii): the box has 55 1's and 45 0's; there are 1,000 draws at random with replacement; make a χ^2-test.

   Comment. When there are only two kinds of tickets in the box, you can use either the z-test or the χ^2-test. The χ^2-test will give the same result as two-sided z-test because $\chi^2 = z^2$.

7. Choose option (iii). Just because the data are laid out in a 2×2 table doesn't mean you're testing independence. The χ^2-test is done below, there is only weak evidence against the null.

	Ways	*Chance*	*Expected*	*Observed*
Even, large	4, 6	2/6	200	183
Even, small	2	1/6	100	113
Odd, large	5	1/6	100	88
Odd, small	1, 3	2/6	200	216

$$\chi^2 \approx 6, \ d = 3, \ P \approx 10\%.$$

Chapter 29. A Closer Look at Tests of Significance

Set A, page 546

1. (a) True. (b) True. See p. 482.
2. (a) False. (b) False. See p. 480–481.

Set B, page 550

1. (a) About 5. (b) 8. (c) About 1.

   Comment. If you toss 100 coins, you expect to get around 50 heads. If the null hypothesis is true, the chance of getting a "significant" result is 5%; so you can expect this to happen about 5 times in 100.

2. (a) 25 (b) | [0] [0] [1] [1] |
   (c) The sum of the ranks is like the sum of 25 draws made at random with replacement from the box | [1] [2] [3] [4] |.

3. (a) About 3. On the null hypothesis, the number of hits is like the sum of 25 draws from the box | [0] [0] [0] [1] |, so the chance of a "significant" result is about 3%. (The chance is given just before the exercise, on p. 551.)
   (b) About 5.
   (c) About 5.

4. Somewhat more than. The first test by itself has about a 3% chance of getting a "significant" result; the second and third each have about a 5% chance. But the chance that at least one of the three tests finds something is bigger than 5%.

   Comment. The trouble with data snooping is that it makes significance levels close to meaningless. Those who snoop, find—even when nothing is going on.

5. Data snooping again. If 25 different hypotheses are tested, some results are likely to be significant.

6. Two-tailed.

7. One-tailed.

8. (a) Yes; $P \approx 4\%$.
   (b) No; $P \approx 96\%$.
   (c) No; $P \approx 8\%$.

9. Doctors are more likely to write a journal article if they have an unusually high fatality rate, and that is more likely with a small sample—which leaves more room for flukes. As Chalmers says, "Physicians have a tendency to report the unusual."

Set C, page 554

1. (a) False. (b) False. See pp. 552–554.

2. The question makes sense, because we are dealing with simple random samples, and it can be answered by a two-sample z-test:

$$\text{SE for men's average} \approx 1, \quad \text{SE for women's average} \approx 1$$

$$\text{SE for difference} \approx \sqrt{1^2 + 1^2} \approx 1.4, \quad z \approx 1.4, \quad P \approx 8\% \text{ (one-sided)}$$

   This could be a chance variation.

3. The SE for each average is 0.5, so the SE for the difference is 0.7, $z \approx 2.8$, and P drops to 1/4 of 1%.

   Comment. The observed significance level depends on the size of the sample. With the smaller sample, the difference was estimated as 2 ± 1.4 points; with the larger, 2 ± 0.7 points.

4. The second sentence is right. However, the null hypothesis can be rejected on the basis of a trivial difference—if the sample is large (p. 553).

5. $P = 27\%$. Big P is good for null, small P is bad.

6. (a) The test is legitimate. This is like the radiation-surgery example (section 4 of chapter 27).
   (b) If $P \approx 2\%$ (one-tailed) then $z \approx 2$. The difference is $71.5 - 25 = 46.5$ percentage points, so the SE must have been around 23 percentage points.
   (c) The difference between 71.5% and 25% is huge.
   (d) To see what the P-value adds, imagine the editors of the journal saying,

   > Look. Some of our reviewers are more critical than others. By the luck of the draw, too many critical ones were chosen to get the negative version.

   The P-value tells you that the editors cannot use the luck-of-the-draw defense with a straight face. The P-value does not help you compare 71.5% and 25%.
   (e) This study demonstrates publication bias. Reviewers are more likely to find mistakes in articles they disagree with—which is only human.

   Comment. The observed difference was 46 percentage points. The SE puts a give-or-take number of 23 percentage points on the estimate. The difference is big, but poorly estimated. (To get better accuracy, a larger sample would have been needed, and that might have been hard to arrange: there are only so many reviewers.) The P-value tells you that the difference would be hard to explain away as a chance variation.

7. The P-value does not measure the size of the difference, so there is no way of telling just from P whether the impact is weak or strong: what determines P is size relative to the SE.

8. A 99%-confidence interval is -6.0 ± 2.6 SEs, that is, -6 ± 6.5. The estimate is not very accurate. The P-value suggests that the elasticity is not exactly 0; nobody said it was. The use of tests seems questionable, and so is the model.

Set D, page 558

1. There are no probability samples here, so caution is in order. The students did rather well by comparison with the TAs.
2. Statistical significance does not make much sense here. The two inner planets do not constitute a random sample of size 2 from the population of inner planets. They *are* the inner planets. Similarly for the outer ones.
3. A test of significance is not appropriate here, unless a box model can be specified for the data.
4. The question makes sense, because we are dealing with a probability sample. However, it cannot be answered on the basis of the information given. This is a cluster sample, so the simple random sample formulas do not apply: section 4 of chapter 21 and section 5 of chapter 22.

   Comment. In this study, like many others, children's performance on intelligence tests goes up with family income.
5. The sample is so large that unimportant differences are likely to be highly significant.

   Comment. The statistical procedures in this study may be open to question too.
6. A test of significance is being done on data for a whole population—the "elites." A box model does not make much sense here.

Set E, page 561

1. In the study, absenteeism was compared to that in previous years. But this year may be different from the last (milder weather, more interesting work, etc.). It would be better to compare the amount of absenteeism among workers on flex-time and among contemporary controls. To avoid resentment by those not given flex-time, it might be a good idea to assign whole work units to treatment or control.
2. This experiment was very well designed. It is fair to conclude that the vaccine protected the children against polio. Other explanations (like the placebo effect) are ruled out by the design of the experiment.
3. No. The P-value tells you that the increase is not a fluke in the random assignment of animals to treatment or control. The P-value does not help you extrapolate from high doses in rats to low doses in humans.
4. Where's the model? Why are lower salaries evidence of discrimination? (You might have to look at experience, education, productivity, etc.) And if this expert insists on doing a test, the pairs are very dependent—for instance, there could be one highly paid man who turns up in 16 of the pairs.

Tables

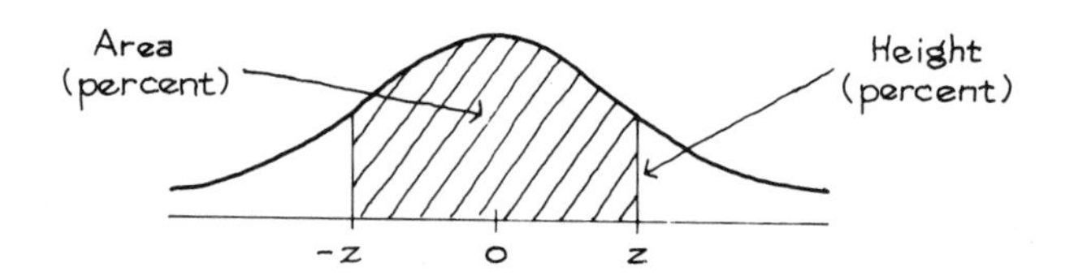

A NORMAL TABLE

z	*Height*	*Area*	*z*	*Height*	*Area*	*z*	*Height*	*Area*
0.00	39.89	0	1.50	12.95	86.64	3.00	0.443	99.730
0.05	39.84	3.99	1.55	12.00	87.89	3.05	0.381	99.771
0.10	39.69	7.97	1.60	11.09	89.04	3.10	0.327	99.806
0.15	39.45	11.92	1.65	10.23	90.11	3.15	0.279	99.837
0.20	39.10	15.85	1.70	9.40	91.09	3.20	0.238	99.863
0.25	38.67	19.74	1.75	8.63	91.99	3.25	0.203	99.885
0.30	38.14	23.58	1.80	7.90	92.81	3.30	0.172	99.903
0.35	37.52	27.37	1.85	7.21	93.57	3.35	0.146	99.919
0.40	36.83	31.08	1.90	6.56	94.26	3.40	0.123	99.933
0.45	36.05	34.73	1.95	5.96	94.88	3.45	0.104	99.944
0.50	35.21	38.29	2.00	5.40	95.45	3.50	0.087	99.953
0.55	34.29	41.77	2.05	4.88	95.96	3.55	0.073	99.961
0.60	33.32	45.15	2.10	4.40	96.43	3.60	0.061	99.968
0.65	32.30	48.43	2.15	3.96	96.84	3.65	0.051	99.974
0.70	31.23	51.61	2.20	3.55	97.22	3.70	0.042	99.978
0.75	30.11	54.67	2.25	3.17	97.56	3.75	0.035	99.982
0.80	28.97	57.63	2.30	2.83	97.86	3.80	0.029	99.986
0.85	27.80	60.47	2.35	2.52	98.12	3.85	0.024	99.988
0.90	26.61	63.19	2.40	2.24	98.36	3.90	0.020	99.990
0.95	25.41	65.79	2.45	1.98	98.57	3.95	0.016	99.992
1.00	24.20	68.27	2.50	1.75	98.76	4.00	0.013	99.9937
1.05	22.99	70.63	2.55	1.54	98.92	4.05	0.011	99.9949
1.10	21.79	72.87	2.60	1.36	99.07	4.10	0.009	99.9959
1.15	20.59	74.99	2.65	1.19	99.20	4.15	0.007	99.9967
1.20	19.42	76.99	2.70	1.04	99.31	4.20	0.006	99.9973
1.25	18.26	78.87	2.75	0.91	99.40	4.25	0.005	99.9979
1.30	17.14	80.64	2.80	0.79	99.49	4.30	0.004	99.9983
1.35	16.04	82.30	2.85	9.69	99.56	4.35	0.003	99.9986
1.40	14.97	83.85	2.90	0.60	99.63	4.40	0.002	99.9989
1.45	13.94	85.29	2.95	0.51	99.68	4.45	0.002	99.9991

A t-TABLE

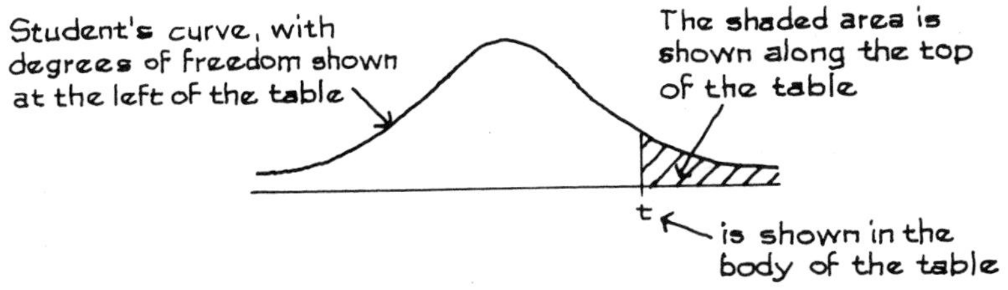

Student's curve, with degrees of freedom shown at the left of the table

The shaded area is shown along the top of the table

t is shown in the body of the table

Degrees of freedom	*25%*	*10%*	*5%*	*2.5%*	*1%*	*0.5%*
1	1.00	3.08	6.31	12.71	31.82	63.66
2	0.82	1.89	2.92	4.30	6.96	9.92
3	0.76	1.64	2.35	3.18	4.54	5.84
4	0.74	1.53	2.13	2.78	3.75	4.60
5	0.73	1.48	2.02	2.57	3.36	4.03
6	0.72	1.44	1.94	2.45	3.14	3.71
7	0.71	1.41	1.89	2.36	3.00	3.50
8	0.71	1.40	1.86	2.31	2.90	3.36
9	0.70	1.38	1.83	2.26	2.82	3.25
10	0.70	1.37	1.81	2.23	2.76	3.17
11	0.70	1.36	1.80	2.20	2.72	3.11
12	0.70	1.36	1.78	2.18	2.68	3.05
13	0.69	1.35	1.77	2.16	2.65	3.01
14	0.69	1.35	1.76	2.14	2.62	2.98
15	0.69	1.34	1.75	2.13	2.60	2.95
16	0.69	1.34	1.75	2.12	2.58	2.92
17	0.69	1.33	1.74	2.11	2.57	2.90
18	0.69	1.33	1.73	2.10	2.55	2.88
19	0.69	1.33	1.73	2.09	2.54	2.86
20	0.69	1.33	1.72	2.09	2.53	2.85
21	0.69	1.32	1.72	2.08	2.52	2.83
22	0.69	1.32	1.72	2.07	2.51	2.82
23	0.69	1.32	1.71	2.07	2.50	2.80
24	0.68	1.32	1.71	2.06	2.49	2.80
25	0.68	1.32	1.71	2.06	2.49	2.79

A CHI-SQUARE TABLE

The chi-square curve, with degrees of freedom shown along the left of the table

The shaded area is shown along the top of the table

is shown in the body of the table

Degrees of freedom	*99%*	*95%*	*90%*	*70%*	*50%*	*30%*	*10%*	*5%*	*1%*
1	0.00016	0.0039	0.016	0.15	0.46	1.07	2.71	3.84	6.64
2	0.020	0.10	0.21	0.71	1.39	2.41	4.60	5.99	9.21
3	0.12	0.35	0.58	1.42	2.37	3.67	6.25	7.82	11.34
4	0.30	0.71	1.06	2.20	3.36	4.88	7.78	9.49	13.28
5	0.55	1.14	1.61	3.00	4.35	6.06	9.24	11.07	15.09
6	0.87	1.64	2.20	3.83	5.35	7.23	10.65	12.59	16.81
7	1.24	2.17	2.83	4.67	6.35	8.38	12.02	14.07	18.48
8	1.65	2.73	3.49	5.53	7.34	9.52	13.36	15.51	20.09
9	2.09	3.33	4.17	6.39	8.34	10.66	14.68	16.92	21.67
10	2.56	3.94	4.86	7.27	9.34	11.78	15.99	18.31	23.21
11	3.05	4.58	5.58	8.15	10.34	12.90	17.28	19.68	24.73
12	3.57	5.23	6.30	9.03	11.34	14.01	18.55	21.03	26.22
13	4.11	5.89	7.04	9.93	12.34	15.12	19.81	22.36	27.69
14	4.66	6.57	7.79	10.82	13.34	16.22	21.06	23.69	29.14
15	5.23	7.26	8.55	11.72	14.34	17.32	22.31	25.00	30.58
16	5.81	7.96	9.31	12.62	15.34	18.42	23.54	26.30	32.00
17	6.41	8.67	10.09	13.53	16.34	19.51	24.77	27.59	33.41
18	7.00	9.39	10.87	14.44	17.34	20.60	25.99	28.87	34.81
19	7.63	10.12	11.65	15.35	18.34	21.69	27.20	30.14	36.19
20	8.26	10.85	12.44	16.27	19.34	22.78	28.41	31.41	37.57

Source: Adapted from p. 112 of Sir R. A. Fisher, *Statistical Methods for Research Workers* (Edinburgh: Oliver & Boyd. 1958).

Index